D0744217

# ASYMMETRIC SYNTHETIC METHODOLOGY

David J. Ager
NSC Technologies, Unit of Monsanto
Mount Prospect, Illinois

Michael B. East
FAR Research Inc.
A Division of Société Protex
Palm Bay, Florida

CRC Press
Boca Raton New York London Tokyo

**Library of Congress Cataloging-in-Publication Data**

Ager, David J.
Asymmetric synthetic methodology / David J. Ager, Michael B. East.
p. cm. -- (New directions in organic and biological chemistry)
Includes bibliographical references (p. - ) and index.
ISBN 0-8493-8942-9 (alk. paper)
1. Asymmetric synthesis. I. East, Michael B. II. Title. III. Series.
QD262.A38 1995
542.2--dc20 95-35877
CIP

This book contains information obtained from authentic and highly regarded sources. Reprinted material is quoted with permission, and sources are indicated. A wide variety of references are listed. Reasonable efforts have been made to publish reliable data and information, but the author and the publisher cannot assume responsibility for the validity of all materials or for the consequences of their use.

Neither this book nor any part may be reproduced or transmitted in any form or by any means, electronic or mechanical, including photocopying, microfilming, and recording, or by any information storage or retrieval system, without prior permission in writing from the publisher.

CRC Press, Inc.'s consent does not extend to copying for general distribution, for promotion, for creating new works, or for resale. Specific permission must be obtained in writing from CRC Press for such copying.

Direct all inquiries to CRC Press, Inc., 2000 Corporate Blvd., N.W., Boca Raton, Florida 33431.

© 1996 by CRC Press, Inc.

No claim to original U.S. Government works
International Standard Book Number 0-8493-8942-9
Library of Congress Card Number 95-35877
Printed in the United States of America 1 2 3 4 5 6 7 8 9 0
Printed on acid-free paper

## THE AUTHORS

**David J. Ager** was born in Northampton, England, in 1953. He received a B.Sc. from Imperial College, London, and a Ph.D. from the University of Cambridge, working with Dr. Ian Fleming on organosilicon chemistry. In 1977 he was awarded a Science Research Council Postdoctoral Fellowship that allowed him to collaborate with Professor Richard Cookson FRS at the University of Southampton. In 1979 he joined the faculty of the University of Liverpool as a Senior Demonstrator. This was followed by an assistant professor position at the University of Toledo in Ohio. In 1986, he joined the NutraSweet Company's Research and Development group, that has now become NSC Technologies, part of Monsanto Growth Enterprises. Dr. Ager is responsible for the chemical development of new products for the fine chemical intermediates business.

**Michael B. East** was born in Heswall, England, in 1959. He received a B.Sc. from the University of Salford, and a Ph.D. from the University of Toledo, Ohio, working with Dr. David Ager on new synthetic methodology. This was followed by a postdoctoral position with Professor Gary Posner at the Johns Hopkins University. In 1988 he joined the faculty at Florida Institute of Technology as an assistant professor. In 1993 he joined FAR Research, a division of Société Protex, in Palm Bay, Florida. Dr. East is responsible for process development and the scale up of new products in the fine chemical intermediate business.

# PREFACE

This book evolved out of two reviews, courses on "asymmetric synthetic methodology" in both academic and industrial settings, and personal needs to implement asymmetric synthesis in an industrial chemistry environment. Our experience, together with comments received from the courses, encouraged the writing of this work. In particular, there appeared to be a need for a book to survey the synthetic methodology available to perform a specific asymmetric transformation, with emphasis on its scope and limitations, especially with regard to what scale that reaction can be successfully used. To make the contents of this book useful, we have heavily referenced the material, and major topics are often augmented by material presented in tabular form. The major emphasis has been placed on synthetic methodology, and few examples of the use of a reaction in a long target molecule synthesis will be found — target molecule synthesis, usually, does not provide sufficient information about the scope and limitations of new synthetic methodologies. A side effect of this approach is a cursory coverage of historical factors in the development of asymmetric methodology, and a lack of methods that do not provide high enough enantiomeric excesses to be synthetically useful. As we have concentrated on asymmetric methodology, achiral methods have, to a large degree, been ignored. When they have been incorporated as a method to a class of starting material, for example, or as an alternative reaction pathway, a full discussion and graphical summary have been omitted. It has not always been possible to comment on the scale of a reaction in the discussion part of the chapter; it will be found in the summary section of the chapter.

We have emphasized "basic" reactions; that is, simple transformations that can be used on a variety of substrates. For some reactions, it is possible to perform one transformation, and the product can then undergo a second reaction with the conditions employed. This is, perhaps, most apparent with pericyclic reactions. Rather than discuss all of the possible permutations, reactions in "extended modes" have only been included in citations.

The inherent stereochemistry of cyclic systems is not discussed due to the large amount of material that can be found in the literature on the wide range of cyclic and polycyclic systems. However, as cyclic systems have been used to control the stereochemistry of acyclic target molecules, some discussion of these reactions is covered.

We have endeavored to make the book useful as a quick reference source: the reactions are given with extensive, although not comprehensive, literature citations; some level of selectivity is often incorporated into a reaction scheme; tables have been used for key reaction approaches, particularly when a large number of similar methodologies have given a plethora of literature citations. To consider each of these factors in turn, the literature citations should be adequate to provide enough information to start a logical literature search — whenever possible the original citations have been included rather than just a review article. However, the number of citations should also be considered to be a function of the authors' fields of interest! Thus, it is inevitable that references and information certain readers may regard as important have been omitted either through the discretion of the authors, or by a simple oversight. No examples for the introduction of fluorine have been included in this work; the special approaches that have been used to achieve this goal make them somewhat removed from general, widely applicable synthetic methodology.

Many schemes will indicate a degree of stereoselectivity — this is meant to indicate to the reader the usefulness of that particular approach. In some cases, however, no stereoselection is given for the reaction sequence. There are two main reasons for this: first, the reaction may be very dependent upon the substitution pattern of the substrate or reagent — quoting 0 to 100% stereoselection is not very useful! Second, the reaction may require very specific substrate parameters and the original paper needs to be consulted. These omissions have been done intentionally, as inclusion of any yield ranges could be very misleading. Finally, the tables are very general to give the reader some idea of the scope and limitations of that reaction; no attempt has been made to provide comprehensive coverage, such as seen in "organic reactions." Thus, the references given should be used as an initial lead into the literature. Multiple citations will be found in many cases; when closely related systems have been reported in the literature, they have been grouped together often with no specific comment. As an example, a metal hydride reduction of a carbonyl compound, where the metal carries chiral ligands from a carboxylic acid, may be illustrated graphically by an example where tartaric acid is the specific ligand, but citations to other carboxylic acid derived ligands will be found within the citation group.

The literature has been covered through 1993. However, as synthetic methods need to be applied, if not proven, the more recent citations are usually examples of established methods.

We have endeavored to provide a useful index. In addition to the cross-references within the text, index entries have been included for material covered in the text and examples provided in the schemes and figures. However, specific examples within tables have not been included as specific index entries. This approach can demand some work on the part of the reader; the entry may not be found in the text but as a reaction graphically depicted within a scheme.

The authors deal with asymmetric synthesis on a day-to-day basis, although from companies with different emphases in the chemical industry. As we are both in chemical process development, our preoccupation with scale-up of reactions is readily apparent. This seems to be one area where academic chemistry is not always completely aware of the requirements of the industrial chemist. A reaction that is relatively simple to perform in the laboratory at –78° could be a major capital expense to conduct on a multi-kilo scale. Conversely, most enzyme reactions do not require stringent environmental control, yet they are only beginning to have widespread usage in the industry. Other scale-up problems often come to the fore with this class of reactions.

Classifications of reactions are always arbitrary. In a few cases, all of the reactions of a particular class of compounds are discussed in one section, even if a few of the reactions are not pertinent to the chapter's theme. This has been done to avoid excessive fragmentation of subject matter; cross-references will lead the reader to the appropriate section. These cross-references are denoted by the appropriate section set off with brackets.

It is apparent, therefore, that we have not set out to write a textbook on asymmetric synthesis. Indeed, an understanding of the basic concepts, although discussed, is assumed — especially the nomenclature of the discipline. Some duplication has occurred because of the reference book approach, but we have assumed that few readers will read the book from start to finish in one session. To have the desired material, or plentiful cross-references at hand seemed more useful to those searching for a specific methodology. The index should also be of use, as we have tried to make this extensive.

Many individuals are worthy of our thanks. Many of my (DJA's) NSC Technologies friends have given useful criticism and feedback on various drafts of the chapters. In particular Drs. Indra Prakash, Dave Pantaleone, Dave Schaad, Scott Jenkins, and Scott Laneman (Searle) deserve acknowledgment for the hours they have spent poring over manuscripts and drafts in various states of disrepair.

We must also express thanks to our families who have put up with our many reclusive hours spent looking at a computer screen. Without their understanding this book could not have been written.

**Dave Ager, Hoffman Estates, Illinois**
**Mike East, Melbourne, Florida**
**June 1994**

# CONTENTS

# LIST OF FIGURES

# LIST OF TABLES

# NEW DIRECTIONS in ORGANIC and BIOLOGICAL CHEMISTRY

**Series Editor: C.W. Rees, CBE, FRS**
Imperial College of Science, Technology and Medicine, London, UK

## *Published and Forthcoming Titles*

**Chirality and the Biological Activity of Drugs**
Roger J. Crossley

**Enzyme-Assisted Organic Synthesis**
Manfred Schneider and Stefano Servi

***C*-Glycoside Synthesis**
Maarten Postema

**Organozinc Reagents in Organic Synthesis**
Ender Erdik

**Activated Metals in Organic Synthesis**
Pedro Cintas

**Capillary Electrophoresis: Theory and Practice**
Patrick Camilleri

**Cyclization Reactions**
C. Thebtaranonth and Y. Thebtaranonth

**Mannich Bases: Chemistry and Uses**
Maurilio Tramontini and Luigi Angiolini

**Vicarious Nucleophilic Substitution and Related Processes in Organic Synthesis**
Mieczyslaw Makosza

**Aromatic Fluorination**
James H. Clark and Tony W. Bastock

**Lewis Acids and Selectivity in Organic Synthesis**
M. Santelli and J.-M. Pons

**Dianion Chemistry in Organic Synthesis**
Charles M. Thompson

**Asymmetric Synthetic Methodology**
David J. Ager and Michael B. East

**Synthesis Using Vilsmeier Reagents**
C. M. Marson and P. R. Giles

**The Anomeric Effect**
Eusebio Juaristi

**Chiral Sulfur Reagents**
M. Mikołajczyk, J. Drabowicz, and P. Kiełbasiński

**Chemical Approaches to the Systhesis of Peptides and Proteins**
Paul Lloyd-Williams, Fernanado Albericio, and Ernest Giralt

**Concerted Organic Mechanisms**
Andrew Williams

Chapter 1

# INTRODUCTION

As many natural products exist as one stereoisomer, asymmetric synthesis is required to prepare Nature-identical material. Asymmetric synthesis is a rapidly progressing field of synthetic organic chemistry, especially in the last few years, but many methods still need to be developed or improved upon before the reactions and transformations can be considered routine. This book describes reactions that are mature and have been used to prepare a wide variety of products with predictable stereochemical outcome. On the other hand, some of the reactions described will be very new, and while they show promise, the scope and limitations have not been established, and they should be used with care. These two very different scenarios can be distinguished by comments in the text and by the number of citations to that specific method.

Although few asymmetric reactions that provided high enantiomeric excesses were known prior to 1970, the area has now seen an explosive development of methods![1,2] Industry, which now often demands the development of a product as a single enantiomer, has also fueled the fire.[3,4] The use of transition metal catalysis continues to provide a plethora of asymmetric reactions, and this trend will, no doubt, continue as organometallic chemistry sees further developments.[2,5]

Before we consider specific reactions, it is necessary to define and discuss terminology and reaction types covered in this book. A basic knowledge of stereochemical definitions and principles is assumed. This chapter, however, serves as an introduction to the methods for the achievement of a stereoselective synthesis. The remaining chapters discuss either functional group reactions or reagent types. Some chapters contain a number of related topics describing a class of transformations. Cross-references to other uses of a reaction are given in the text by section numbers.[†] The overall utility of the various reactions and approaches are discussed in the summary section at the end of each chapter. Within a chapter, sections are "self-contained." Although this may detract from the overall readability, it should be easier for a reader who just wants to find details on a specific reaction or transformation. With this philosophy in mind, the index has been made extensive. The intent is to provide a useful reference book, not a text book. General discussions are, therefore, often brief. To help with the rapid search philosophy, many reactions are summarized in tabular form. Stereoselectivity is summarized as either a ratio or excess within schemes — as long as the schemes are available from the original citation![‡]

The emphasis is placed on general methods — that is, synthetic methodology that can be applied to a number of compounds or classes of compounds. Thus, a method specific to a complex molecule is excluded. This includes the methodology specific to a particular class of compounds, such as the coupling of saccharides to produce a $\beta$-linkage.

Asymmetry can arise on a molecular level from a variety of sources, particularly in macromolecules.[6] For example, the helix of DNA is a consequence of the chiral monomer units. Also, proteins are chiral due not only to the amino acid constituents, but the ways they interact with each other — formation of $\beta$-sheets, $\alpha$-helices, etc. In addition to these types of asymmetry, spatial effects can give rise to the phenomenon. An example is the group of compounds called in-betweenanes, where the way two cyclic compounds intersect one another

† Section number cross-references are indicated by the reference in italics with brackets.

‡ In addition to the stereoselectivity not being shown, as it was not specifically given in the original citation, it has been omitted in some cases to avoid confusion or misrepresentation. For example, a reaction may be very susceptible to the structure of a nucleophile and stereoselections of 0 to 100% are given in the original paper. For specific examples, the selection may be high; the omission of general numbers, therefore, should be an indication to look at the original citation and determine the exact scope and limitations of that specific reaction.

is the origin of asymmetry.[7] However, as no functional group chemistry is directly involved in the formation of these compounds, other than in the preparation of the ring systems, they will be considered outside the scope of this book.

One other topic that is not specifically covered in this book is the introduction of asymmetry, or the consequences of chirality at an element other than carbon. Thus, although examples of the usage of chiral sulfoxides for asymmetric synthesis will be found, no specific examples of the preparation of chiral sulfoxides by, for example, the Sharpless epoxidation protocol, are cited.

As asymmetric synthesis is paramount to the achievement of organic synthesis, many reviews have been written on the topic in general, and on the use of the concept for specific classes of compounds.[8-54] Reviews of specific transformations are cited in the appropriate section.

## 1.1. NOMENCLATURE

### 1.1.1. ASYMMETRIC SYNTHESIS

The term asymmetric synthesis has been used to describe a wide variety of transformations. In its original definition, coined by Marckwald in 1904, asymmetric synthesis was described as the process for the formation of an optically active compound through reaction of an achiral substrate with a chiral reagent.[55] This definition was expanded by Morrison and Mosher in 1971 to cover a wider range of reactions. Their definition describes asymmetric synthesis as a reaction where an achiral unit in an ensemble of substrate molecules is converted by a reactant into a chiral unit in such a manner that the stereoisomeric products are formed in unequal amounts.[31] We will use the latter definition.

The new asymmetric center will often be referred to as a stereogenic center, denoting that it has the potential to give rise to stereochemistry. However, no inference should be made as to the stereochemical purity of the product nor the stereoselectivity of the reaction.[¶]

### 1.1.2. DOUBLE ASYMMETRIC INDUCTION

This term is used to describe the interactions of two chiral reactants. If the two reactants' chirality work in concert, then very high induction can be seen. In this case, the reactants are often described as a "matched pair." However, if the two reactants are not working together, they are described as a "mismatched" pair. In some cases, the stereoselection can be very low while, in other cases, it can still be high.[56,57]

A mathematical model has been proposed, with supporting experimental evidence, that high enantiomeric excesses can be obtained for a reaction where kinetic resolution is coupled with an initial asymmetric synthesis.[58] The Sharpless epoxidation is such an example [*11.1.2*].

Many examples of double asymmetric induction will be seen in the various chapters.

### 1.1.3. NOMENCLATURE CONVENTIONS

In many respects stereochemistry is a visual concept; it is often difficult to describe in words.[59] This problem is made even more difficult by a plethora of terms and naming systems.[§] For absolute and relative stereochemistry the Cahn, Ingold, Prelog (CIP) method will be used,[60-65] with its *pro-R* and *re/si* additions.[66]

¶ A stereogenic center can arise from reaction at either an enantio- or diastereotopic face.

§ The problem is compounded by the misuse of terms and slight variations in interpretation by authors. An example is provided by "homochiral" that is sometimes used as a synonym for enantiomerically pure, but has also been defined as a molecule that contains only one type of asymmetric center — for example, all of the centers have the *R*-configuration. We have defined our terminology in this section. Although more words may be needed to describe some specific stereochemical aspect, we feel that this outweighs the problems associated with the use of a term that is ambiguous or could be misinterpreted.

For diastereoisomers, a number of systems have been proposed.[67] In this book, we will use the method proposed by Carey, which is based on the CIP method.[68] This method alleviates any problems associated with some of the naming conventions where ambiguity has been introduced.

In Carey's system, the priority of the two asymmetric centers at the ends of the bond in question are assigned priorities following the CIP rules. If the relative stereochemistry of the centers is reflective, they are described as *pref* (priority reflective), and if they are not reflective they are assigned as *parf* (priority anti-reflective) (Figure 1.1).

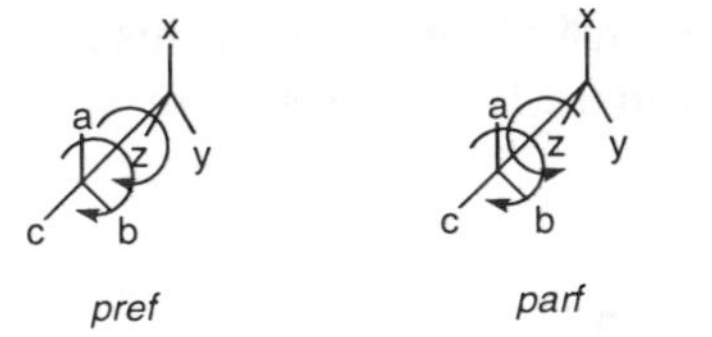

*pref* *parf*

where CIP priorities are a>b>c, and x>y>z

**FIGURE 1.1.** Diagrammatic representations of *pref* and *parf*.

This naming convention defines that when a *meso* compound is encountered, its relative configuration is *pref*. The two stereocenters do not have to be contiguous; the intervening atoms are ignored, and the centers are treated as if they were directly linked (Figure 1.2).

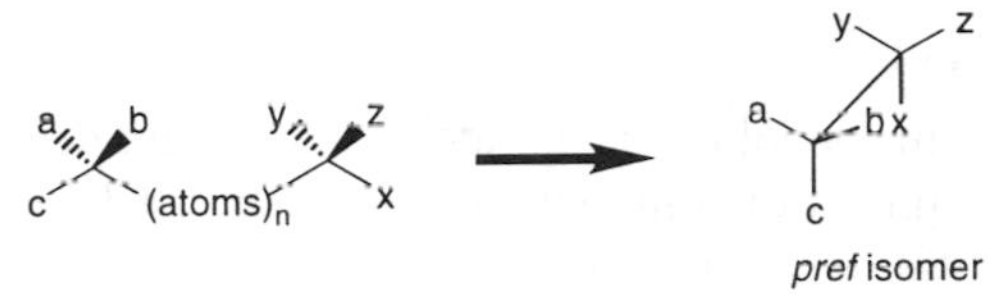

*pref* isomer

**FIGURE 1.2.** Diagrammatic representations of *pref* isomer without contiguous stereogenic centers.

To allow correlation of stereochemistry between two similar molecules, and to alleviate any ambiguity that can arise in the *erythro-threo*¶ or *syn-anti*[77] nomenclature systems, Carey also proposed the use of the informal descriptors SYNCAT and ANCAT, describing *syn* catenoid and *anti* catenoid relationships respectively (Figure 1.3).[68]

¶ Although we have avoided the ambiguities of the *erythro-threo* systems, reference to the original papers cited herein will obviously raise the issue once again. The use of *erythro* and *threo* is derived from the carbohydrates erythrose and threose which are usually drawn as Fischer projections. The system has been used, especially with regard to aldol reaction products, with reference to these carbohydrates and has been defined in these terms.[69,70] However, the terminology for aldol type products comes from an implicit definition.[71-74] A variation of the nomenclature has been proposed based on the priorities of the substituents: if all of the groups are eclipsed when the highest priority groups are paired, then the compound is *erythro*; if not, it is *threo*:[75]

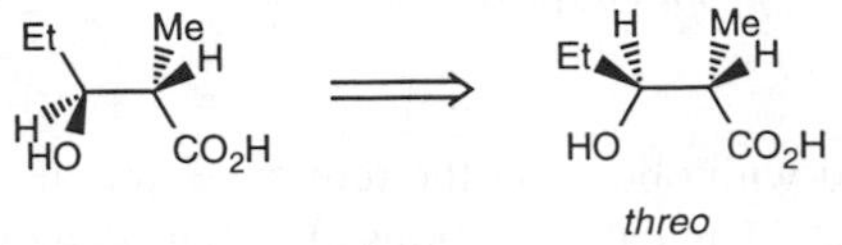

*threo*

This method is very similar to the Carey system and a further variant uses the main chain (IUPAC system) of the molecule.[76] Ambiguities usually arise when the *erythro* system is used without definition.

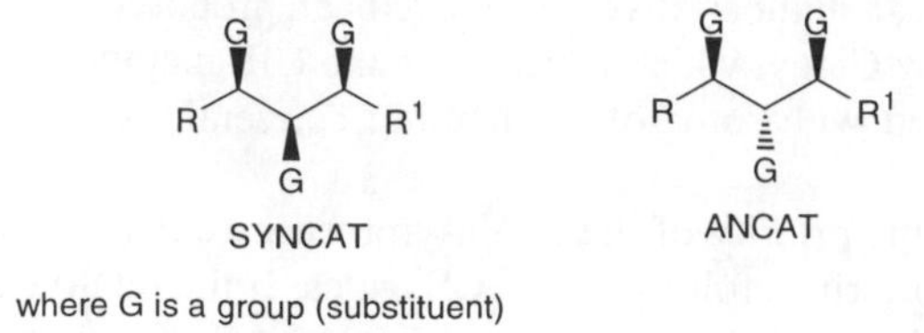

G G
R R1
G
SYNCAT

G G
R R1
G
ANCAT

where G is a group (substituent)

**FIGURE 1.3.** Diagrammatic representations of *syncat* and *ancat* isomers.

Although not all contingencies may be covered by this informal system, the chain should be written in the zig-zag form with the largest groups (by steric bulk) at the ends of the chain (Figure 1.4).

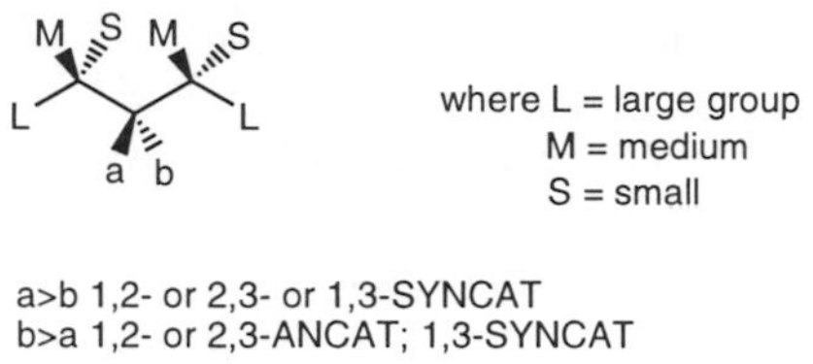

M S M S
L L
a b

where L = large group
M = medium
S = small

a>b 1,2- or 2,3- or 1,3-SYNCAT
b>a 1,2- or 2,3-ANCAT; 1,3-SYNCAT

**FIGURE 1.4.** Example of *syncat* and *ancat* isomers.

We will also use the *syncat* and *ancat* nomenclature for zig-zag structures.

### 1.1.4. OPTICAL PURITY

In addition to synthetic methodology, analytical methods have developed at a rapid rate so that antipodes of a large number of compounds can be separated and detected. The original method for the determination of enantiomeric purity was derived for polarimetry measurements — thus, enantiomeric purity and optical purity are often used interchangeably. The specific rotation of a pure, chiral compound is given by the expression:

$$[\alpha]_{\lambda}^{T} = \frac{\alpha \cdot 100}{l \cdot c}$$

where $\alpha$ = measured rotation
$l$ = path length of cell (dm)
$c$ = concentration (g/100 ml)
$T$ = temperature
$\lambda$ = wavelength of light used for measurement

To compare two samples, a reference standard must be used under identical conditions; the optical purity then being given by the equation:

$$\%\ optical\ purity = \frac{[\alpha]_{obs} \times 100}{[\alpha]_{max}}$$

The problems associated with obtaining the reference sample and the performance of the two required analyses under identical conditions are often large or even insurmountable. In addition, some of the assumptions associated with the use of the above equations are not always valid, as in the case of $\alpha$-methyl-$\alpha$-ethylsuccinic acid where the relationship deviates significantly from linearity.[78] Indeed, conversion of an optical purity to enantiomeric excess can lead to erroneous results.[16]

To determine how much one isomer is in excess over the other, analytical methods based on HPLC or GC have proven the most reliable. Chiral shift reagents for NMR are also useful, as are the optical methods that rely on more than one datum point, such as ORD and CD.[79]

A variety of methods are also available where the compound under investigation is derivatized with a chiral reagent to form diastereomeric products that can be separated or have readily detectable differences in physical attributes. However, as a mixture may result from a chemical reaction, the number of physical operations should be kept to a minimum as simple techniques, such as crystallization, sublimation, distillation, chromatography, and extraction can result in unintentional resolution. In addition, if a derivatizing agent is employed, it must be assured that the reaction with the subject molecule is quantitative, and that the derivatization reaction is run to completion; this will ensure that unintentional kinetic resolution does not occur prior to the analysis. The derivatization agent must be just one enantiomer (vide supra).

Throughout this book we will use two methods to report the asymmetric outcome of a reaction; they will refer to both enantio- and diastereomeric selectivities.[‡] The first method is to report the selectivity as a simple ratio of the two products. For example, enantioselectivity (es) could be shown as 6:1 meaning that the desired isomer was formed in an amount equivalent to 6 times the amount in which the undesired isomer was made. Diastereoselectivities (ds) will be shown in an analogous manner. Thus, "selectivity" shows the amount that one isomer is formed over the other, and is depicted as a ratio where the data can be obtained directly from an analytical method.

The second method is to report "excess," that is the excess of one isomer over the other. Note that this method does not give a linear relationship. The relationships for percentage enantiomeric excess (ee) and diastereomeric excess (de) are given by the equations:

$$\% \, ee = \left|\frac{[R] - [S]}{[R] + [S]}\right| \times 100$$

$$\% \, de = \left|\frac{[RR] - [RS]}{[RR] + [RS]}\right| \times 100$$

where [X] denotes the amount of that isomer formed.

In reactions where diastereoselectivity as well as enantioselectivity is important, ratios will be defined to avoid ambiguity.[¶]

## 1.2. WHY ASYMMETRIC SYNTHESIS?

Nature produces a wide variety of chiral compounds. The prolific functional polymers based on amino acids, carbohydrates, or nucleotides have chiral monomers, but often additional chirality is introduced in the higher molecular weight materials — an example is the

‡ Associated with the determination of enantioselectivity is the problem of the definition of "specific." It is the opinion of the authors that the problem relates to improved analytical methods and their relationship to the reporting of significant figures. A reaction that proceeds in 99.6% yield is quantitative (100%) when rounded up, but would provide an ee of 99%. The question, therefore, returns to how accurately measurements were taken in the original work and what error limits were established. We have circumvented the argument, with its associated problems, by refraining from the use of terms such as "enantiospecific," "stereospecific," etc., although we do report ratios of 100:0 if that was given in the original citation.

¶ An additional term, "relative enantioselectivity", has been proposed to allow for useful comparisons between asymmetric, catalytic reactions.[80]

**TABLE 1.1**
**Possible Benefits for Use of a Single Enantiomer for Therapeutic Uses[86,88,89]**

| Properties of Racemate | Potential Benefits of Enantiomer |
|---|---|
| One enantiomer has exclusive activity. | Reduce dose and load on metabolism. |
| Other enantiomer is toxic. | Increased latitude in dose and broader usage. |
| Enantiomers have different pharmokinetics. | Better control of kinetics and dose. |
| Enantiomers metabolized at different rates (in one person). | Wider latitude in dose setting; less variability in patient response. |
| Enantiomers metabolized at different rates (different people). | Reduction in variability of patient responses; larger confidence in dose selection. |
| One enantiomer prone to interaction with key detoxification pathways. | Reduced interactions with other (common) drugs. |
| One enantiomer is agonist, other antagonist. | Enhanced activity and reduction of dose. |
| Enantiomers vary in spectra of pharmacological action and tissue specificity. | Increased specificity and reduced side effects for one enantiomer; use of other enantiomer for different indication. |

helix of a DNA strand. Interactions with these materials, which are very important in many fields such as the development of a pharmaceutical agent, often now require the development of a chiral compound to maximize the desired effects. Public perception, and associated legislation, surrounding the development of pharmaceuticals, especially since the thalidomide (**1.1**) tragedy, is also demanding enantioselective synthesis.[81,82] In addition, one of the enantiomers may carry all, or the majority, of the activity.[83,84] Thus, preparation of the active component not only reduces the amount of the inactive enantiomer as, in essence, it is an impurity, but can also make economic sense (the specific activity can be higher). Ibuprofen (**1.2**) is an example of this type of "switch," and other compounds may also follow.[¶] The public perception of "isomeric baggage" or "isomeric ballast"[85] has been realized and the chemical business literature has discussed the arguments for and against such racemic switches.[86,87] In addition, enantioselectivity has also given a number of opportunities to pursue patents, as the development company sometimes had just protected the racemate. Some of the arguments for use of a single enantiomer over a racemate are given in Table 1.1.

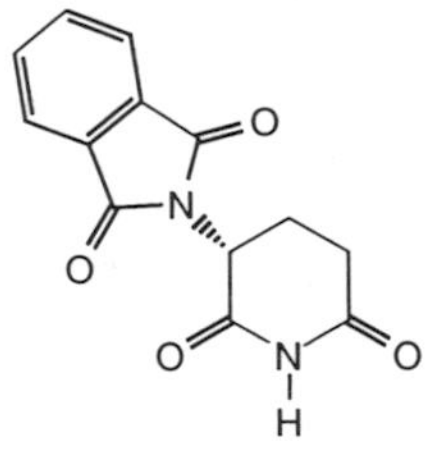

**1.1**
(*R*)-Thalidomide
Sedative,
other isomer is a teratogen

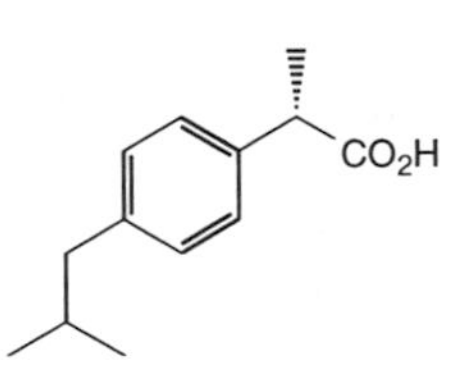

**1.2**
(*S*)-Ibuprofen

Of course, current racemic drugs were approved for use in this form. Not all of the properties listed in Table 1.1 will be available to promote a racemic switch. In some cases, the presence of the other enantiomer could be beneficial. However, as receptors are invariably chiral, the development of new therapeutic agents will increase to follow the chiral development pathway. This will become easier as chemistry and biochemistry continue to expand the methodology to perform asymmetric synthesis.

¶ This type of switch from a racemic compound to one enantiomer has been coined a "racemic switch."

Often pharmaceutical products are cited as examples for asymmetric synthesis. However, agrochemicals and food ingredients also interact with biological systems.[90] The development of chiral molecules in these areas is also becoming the norm rather than the use of a racemic mixture. In addition, cost considerations are far more crucial for these types of products; indeed, in many cases, the desired activity cannot be obtained unless a chiral compound is used.[25,28,29,91,92] Thus, the constraints on process development chemists in these industries are more severe and the number of reagents is more limited; however, the repertoire continues to expand.

In the food industry, for example, aspartame (**1.3**) is a commercial sweetener. The interactions between a sweetener and the mouth are not understood, but it is obvious that some asymmetric interaction is involved, as sweetness is only associated with the isomer derived from the L-amino acids.[93,94]

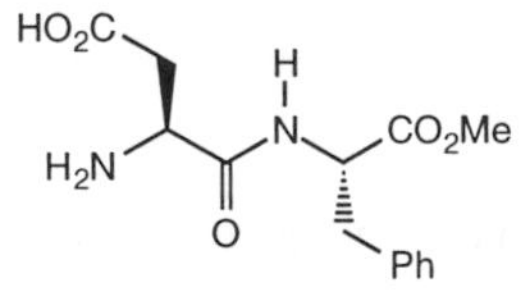

**1.3**

Other examples can also be cited, such as insect pheromones, where a compound may become inactive if only trace amounts of the antipode are present.

## 1.3. METHODS TO ACHIEVE ASYMMETRIC SYNTHESIS

To obtain an enantioselective synthesis, at least one of the agents in the system must be chiral. There are two major methods to achieve this goal: resolution or asymmetric synthesis, which includes the use of a chiral starting material, chiral auxiliaries or reagents. All have examples in the literature, but experience and our understanding of the factors involved have provided enough insight to allow the most productive methodology for a specific system to be relatively apparent. All of these methods are covered with a number of examples throughout this book. This section, therefore, will serve only as an introduction into each of the approaches. The discussion on resolution is found in the following section [*1.4*].

### 1.3.1. TRANSITION STATES

Before we embark upon a discussion on how to achieve an asymmetric synthesis, how and why does it occur? In all of the approaches a chiral substrate or agent is acted upon by another chiral entity. This interaction of two chiral agents, in turn, gives rise to diastereomeric transition states. The difference in energy between the two transition states ($\Delta\Delta G^{\ddagger}$) determines the degree of selectivity (Figure 1.5) and the lower energy pathway will provide the major antipode, even if the product is not thermodynamically favored (example B).[10,95]

Thus, asymmetric synthesis is a kinetic phenomenon. Factors that affect free energy differences can have a marked impact on the degree of induction. In many cases, the use of low temperature will increase the degree of selectivity, as the potential for the higher energy pathway to be available is reduced or removed.[96†] Small changes to the reaction parameters can greatly impact the enthalpy and entropy terms (the temperature also contributes to this term of the free energy equation); this often accounts for the reason why one isomer is preferred at higher temperatures, while selectivity looks as if it is reversed at low temperatures. Of course, the free energy argument is empirical and must be used with care.

† The theory has even been taken to a point where the prediction has been made that "the logarithm of the diastereomeric ratio is … proportional to the inverse square of the temperature." [97]

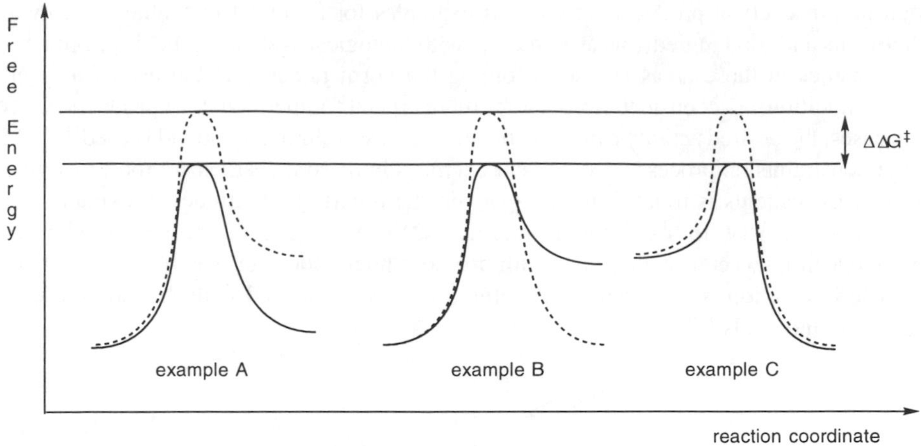


**FIGURE 1.5.** Stereoselectivity arising from diastereotopic transition states.

In addition to transition state energy differences, there can be contributions from the relative stabilities of the products. For example, a late transition state has been invoked to rationalize why additions to unhindered cyclic ketones favor the thermodynamically more stable isomer, while additions to sterically hindered ketones are under steric approach control [*4.5*].[98,99] These abstract concepts have been invoked in rationalizations for the outcome of other types of reactions. As our understanding of the factors that influence specificity increase, the models are refined and their predictive powers increase. The general principles behind some useful models are discussed in Chapter 2, while the specific arguments for a particular reaction's course will be found under that specific heading.

In many regards, progress in asymmetric synthetic methodology development is inhibited by our lack of knowledge of the factors and our understanding of how they influence transition state geometry.[100]

## 1.3.2. CHIRAL ENTITIES

Before embarking on a synthesis, careful thought must be given to how a chiral center will be introduced into a molecule. The three major options are use of a chiral reagent (chemical or biological); use of a chiral environment; and use of a chiral starting material. Chirality can also be introduced in a temporary manner through the use of a chiral auxiliary, although this is a sub-class of chiral substrates. Each approach has advantages and disadvantages. These must be considered on a case-by-case basis so that the greatest chance of success arises from the synthetic plan. In addition, a resolution can also be used to obtain chiral material anywhere in a reaction sequence.

The manipulation of functional groups and carbon–carbon bond formation are still paramount to the development of a synthetic plan, whether a retrosynthetic sequence based on a synthon, chiral template, transformation-driven (e.g., use of a certain enzyme), or a chiron approach is used.[101-104]

### 1.3.2.1. Chiral Substrates

The best scenario is to have a chiral starting material that can then control the stereoselection of the reaction itself. To achieve this, especially at the beginning of the synthetic sequence, few options are available. Nature produces chiral materials and a number of these are available in quantity (vide infra).[105,106] These compounds make up the "chiral pool." This approach is often limited to the amount of the natural product available and its price. Another consideration, sometimes overlooked, is the number of steps necessary to convert the natural product into a useful starting material for synthesis. If all of the parameters are favorable, this approach is the method of choice as it has the potential to eliminate resolutions or the necessity for an enantiospecific transformation in the synthetic design. An example of the use of a natural product for a starting material is the conversion of L-glutamic acid (**1.4**) to the chiral butyrolactone (**1.5**) (Scheme 1.1).[107]

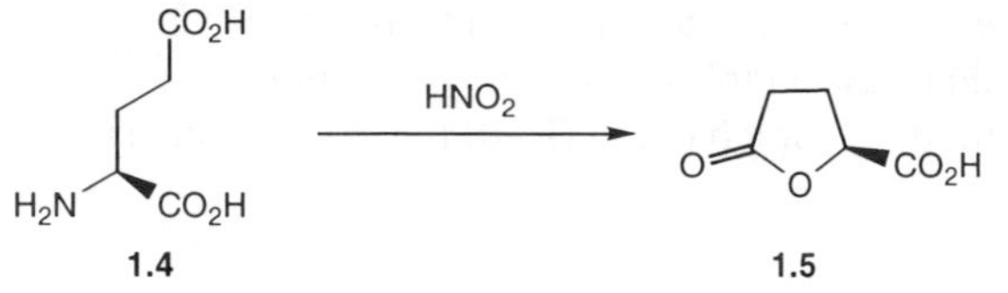

$CO_2H$ — $H_2N$ — $CO_2H$ **1.4** —($HNO_2$)→ O — O — $CO_2H$ **1.5**

**SCHEME 1.1.**

Hanessian has developed methodology, including a computer program, to look for similarities between subunits within a target structure and members of the chiral pool. These subunits are called "chirons" and are analogous to chiral synthons.[103-105]

With a chiral starting material in hand, a well-designed synthesis should then reduce to the control of relative stereochemistry — the natural product's chirality inducing the appropriate stereochemistry at the new center(s).[108] The control of relative stereochemistry is, therefore, an integral and important part of asymmetric and enantioselective synthesis. However, it must once again be stressed that the desired transformation may look trivial, but in reality must be multi-step. An example is provided by the conversion of D-glucose (**1.6**) to L-glucose (**1.7**) (Scheme 1.2).[109 §]

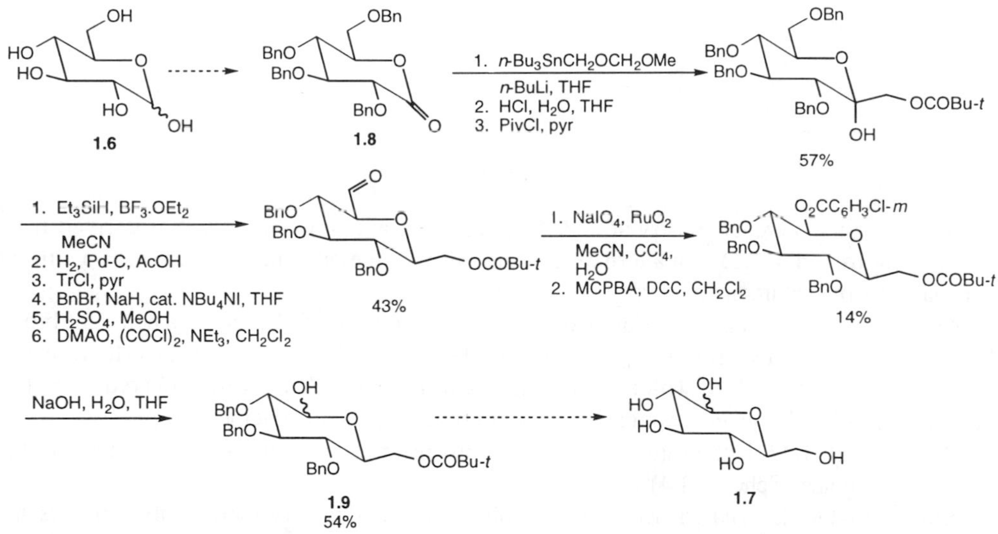

**1.6** ----→ **1.8** —(1. *n*-$Bu_3SnCH_2OCH_2OMe$, *n*-BuLi, THF; 2. HCl, $H_2O$, THF; 3. PivCl, pyr)→ 57%

—(1. $Et_3SiH$, $BF_3.OEt_2$, MeCN; 2. $H_2$, Pd-C, AcOH; 3. TrCl, pyr; 4. BnBr, NaH, cat. $NBu_4NI$, THF; 5. $H_2SO_4$, MeOH; 6. DMAO, $(COCl)_2$, $NEt_3$, $CH_2Cl_2$)→ 43% —(1. $NaIO_4$, $RuO_2$, MeCN, $CCl_4$, $H_2O$; 2. MCPBA, DCC, $CH_2Cl_2$)→ 14%

—(NaOH, $H_2O$, THF)→ **1.9** 54% ----→ **1.7**

**SCHEME 1.2.**

§ For other similar transformations, see references 110–112.

The overall strategy of this sequence is very simple; homologation is performed at C-1, while oxidative degradation at C-6 "inverts" the configuration. However, the conversion of **1.8** to **1.9** takes twelve chemical transformations, and the overall yield is 2%!

The synthetic sequence for incorporation of the carbon framework into a proposed synthetic sequence should be as short as possible.[91] There are numerous examples of long sequences for the synthesis of a target molecule from a chiral starting material. However, it should also be remembered that asymmetric methodology has made great advances in the last few years, and until recently, chiral reagents were not available. Thus, there was previously no other alternative but to use a chiral starting material as a source of a center of asymmetry.

Some syntheses that have employed a "chiron" approach are relatively short and efficient. Thus, the alkaloid, quebrachamine (**1.10**) has been synthesized from L-glutamic acid (Scheme 1.3).[113]

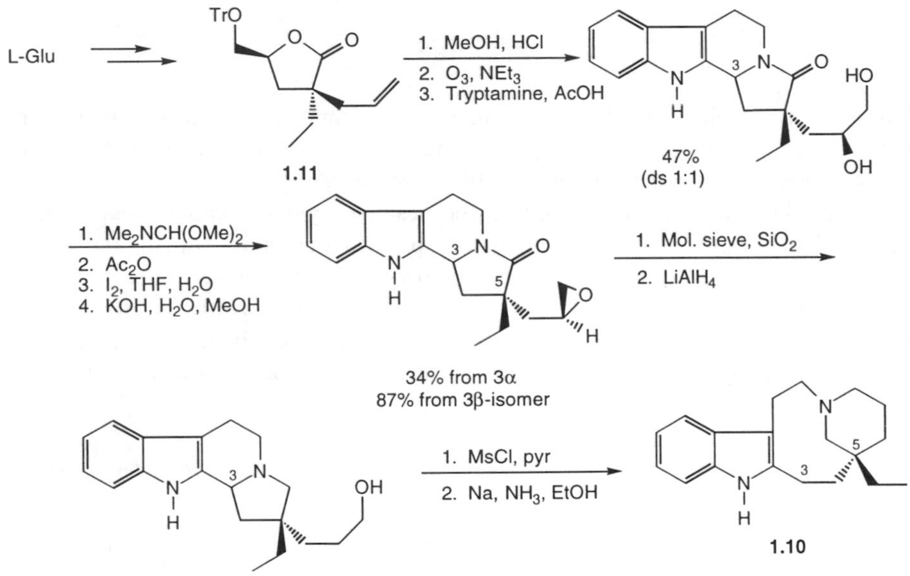

L-Glu → → TrO ... **1.11**

1. MeOH, HCl
2. $O_3$, $NEt_3$
3. Tryptamine, AcOH

47% (ds 1:1)

1. $Me_2NCH(OMe)_2$
2. $Ac_2O$
3. $I_2$, THF, $H_2O$
4. KOH, $H_2O$, MeOH

34% from 3α
87% from 3β-isomer

1. Mol. sieve, $SiO_2$
2. $LiAlH_4$

1. MsCl, pyr
2. Na, $NH_3$, EtOH

**1.10**

**SCHEME 1.3.**

Note how the stereochemistry at C-3 is not relevant and how that at C-5 is derived from the starting material, **1.11**. An alternative, rapid entry to the gibbane skeleton, uses a similar approach from L-glutamic acid.[114]

An additional example is provided by the preparation of the carbapenem antibiotic (+)-thienamycin (**1.12**) from L-aspartic acid (Scheme 1.4).[115] In this synthesis, there are a number of steps associated with the construction of the appendant five-membered ring. The synthesis can be shortened through use of a coupling reaction (Scheme 1.5)![116-118]

The product, **1.13**, is the same intermediate, just prior to cyclization, that was used in the original sequence (Scheme 1.4).

Not all materials isolated from natural sources are a single isomer. Many terpenes are obtained in "scalemic" form — one enantiomer predominates over the other. As we will see, some reactions and procedures provide a method to use just the major isomer (hydroborations based on the use of α-pinene are such examples [*8.2.2.3*]).

The chiral pool is not a stagnant pond. As enzymes and reagents are discovered and developed, they can be applied to provide large quantities of useful chiral starting materials. Examples of these reactions are given below. An example of a chiron that has entered the field,

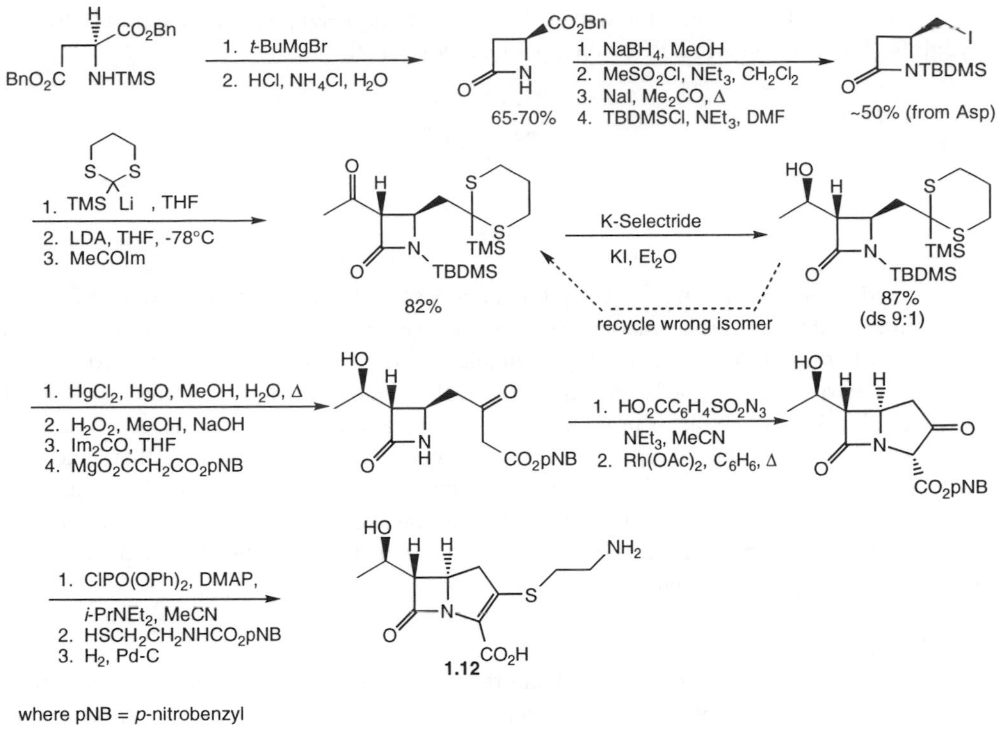
CO2Bn
BnO2C NHTMS
1. t-BuMgBr
2. HCl, NH4Cl, H2O
65-70%
1. NaBH4, MeOH
2. MeSO2Cl, NEt3, CH2Cl2
3. NaI, Me2CO, Δ
4. TBDMSCl, NEt3, DMF
NTBDMS
~50% (from Asp)
1. TMS Li , THF
2. LDA, THF, -78°C
3. MeCOIm
TBDMS
82%
K-Selectride
KI, Et2O
recycle wrong isomer
87%
(ds 9:1)
1. HgCl2, HgO, MeOH, H2O, Δ
2. H2O2, MeOH, NaOH
3. Im2CO, THF
4. MgO2CCH2CO2pNB
CO2pNB
1. HO2CC6H4SO2N3
NEt3, MeCN
2. Rh(OAc)2, C6H6, Δ
1. ClPO(OPh)2, DMAP,
i-PrNEt2, MeCN
2. HSCH2CH2NHCO2pNB
3. H2, Pd-C
NH2
CO2H
1.12
where pNB = p-nitrobenzyl

**SCHEME 1.4.**

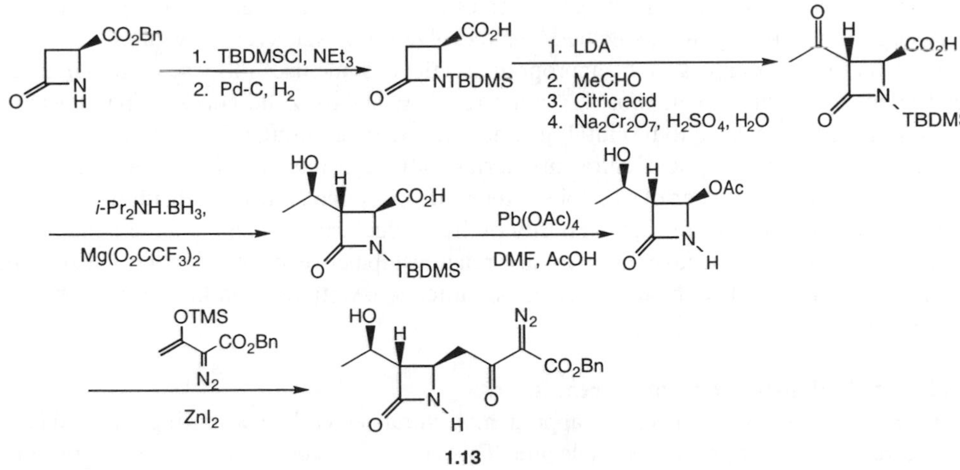
CO2Bn
1. TBDMSCl, NEt3
2. Pd-C, H2
CO2H
NTBDMS
1. LDA
2. MeCHO
3. Citric acid
4. Na2Cr2O7, H2SO4, H2O
TBDMS
i-Pr2NH.BH3,
Mg(O2CCF3)2
Pb(OAc)4
DMF, AcOH
OAc
OTMS
CO2Bn
N2
ZnI2
1.13

**SCHEME 1.5.**

and is not derived from a natural source, is the use of a Sharpless epoxidation approach to form glycidols (**1.14**) by the ARCO Chemical Company [*11.1.2*].

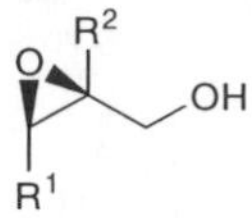

**1.14**

As noted in the section on scale [*1.5*], the amount of material required can determine how many starting materials are available.[105] For small scale reactions, with few cost constraints, thousands of chiral materials are readily available. However, not many chiral materials, even scalemic materials, are available on a multi-ton scale. A list of the most common materials is given in Table 1.2. Amino acids provide a wide variety of functionality in the side chain. The amino and carboxylic acid groups can be transformed into a wide range of other functional groups as well as aiding the formation of carbon–carbon bonds. Carbohydrates are also available in large amounts. However, unless the carbon skeleton is close to the desired target structure, the number of other reactions required to utilize these chirons often make them less desired as the starting material.

#### 1.3.2.2. Chiral Auxiliaries

As we have seen, the number of useful natural products available is not large, or the number of steps necessary to convert a cheap, readily available one to a useful intermediate in a synthesis may require many steps; some of these steps may involve expensive reagents. A number of chiral groups have been developed that can be attached to an achiral molecule. These groups then induce selectivity through a subsequent chemical reaction to afford diastereoselectivity. Removal of the "chiral auxiliary" then provides the product enriched in one enantiomer. However, this type of approach introduces two extra steps: the attachment and removal of the auxiliary. This is analogous to the problems associated with the use of a protecting group, so both of these steps must be high-yielding. However, chiral auxiliaries often act as protection for a functional group, and the two types of operation can then become consolidated and allow for a concise synthetic sequence. In addition, good chiral auxiliaries should be recovered at the end of the sequence so that they can be reused. This allows for an economical synthesis, even on only small scales, as many of these auxiliaries are expensive.

A large number of synthetic methods have evolved from relatively few chiral auxiliaries. In addition to the examples in the appropriate sections, some uses of these compounds have been reviewed, including camphor,[119,120] amino acids,[121] and oxazolidinones.[122] In many cases, an auxiliary does not have to be chiral, just not symmetrical through the absence of mirror or inversion symmetry. The use of chiral auxiliaries with a $C_2$ symmetry axis has been advocated as it can reduce the number of possible competing, diastereoisomeric transition states.[23]

It must be stressed that the optical purity of the product can depend upon the optical purity of the chiral auxiliary (or reagent).[105] A linear relationship between these two variables is often assumed, but this need not be the case, and significant deviations from linearity, for better or worse, are known.[123]

#### 1.3.2.3. Self Regeneration of Stereocenters

There is a variation on the chiron approach. A chiral center from a starting material can be transferred to another part of the molecule. This new chiral center then provides control for a stereoselective reaction, where a new center of asymmetry can be established, or the chirality at the center of the original starting material can be reestablished. Invariably, a cyclic system is involved. This approach has been used by Seebach and described as the regeneration or self-

**TABLE 1.2**
**Inexpensive Chiral Starting Materials and Resolving Agents[3,105]**

| Amino Acids | Hydroxy Acids | Carbohydrates | Terpenes | Alkaloids | Other |
|---|---|---|---|---|---|
| L-Alanine | L-Lactic acid | D-Arabinose | *l*-(–)-Borneol | Cinchonidine | L-(–)-2-Amino-1-butanol |
| L-Arginine | D-Lactic acid | L-Arabinose | *endo*-3-Bromo-*d*-camphor | Cinchonine | (4*S*,5*S*)-(+)-5-Amino-2,2-dimethyl-4-phenyl-1,3-dioxane |
| D-Asparagine | (*S*)-Malic acid | L-Ascorbic acid | (–)-Camphene | D-(+)-Ephedrine | L-(+)-2-Amino-1-phenyl-1,3-propanediol |
| L-Asparagine | (Poly)-3(*R*)-hydroxybutyrate | α-Chloralose | *d*-(+)-Camphor | *l*-Nicotine | (–)-2-Methyl-1-butanol |
| L-Aspartic acid | L-Tartaric acid | Diacetone-D-glucose | D-(+)-Camphoric acid | Quinidine | |
| L-Cysteine | D-Tartaric acid | D-Fructose | *d*-10-Camphorsulfonic acid | Quinine | |
| L-Glutamic acid | D-Threonine | D-Galactonic acid | (+)-3-Carene | D-(+)-Pseudoephedrine | |
| L-Isoleucine | L-Threonine | D-Galactonic acid, γ-lactone | *l*-(–)-Carvone | L-(–)-Pseudoephedrine | |
| L-Leucine | | D-Galactose | (+)-Citronellal | | |
| L-Lysine | | D-Glucoheptonic acid | *d*-(+)-Fenchone | | |
| L-Methionine | | α-D-Glucoheptonic acid, γ-lactone | *l*-(–)-Fenchone | | |
| L-Ornithine | | D-Gluconic acid | (+)-Isomenthol | | |
| L-Phenylalanine | | D-Gluconic acid, δ-lactone | *d*-(+)-Limonene | | |
| D-Phenylglycine | | γ-L-Gluconic acid lactone | *l*-(–)-Limonene | | |
| L-Proline | | D-Glucosamine | *l*-(–)-Menthol | | |
| L-Pyroglutamic acid | | D-Glucose | *d*-(+)-Menthol | | |
| L-Serine | | D-Glucurone | *l*-Menthone | | |
| L-Tryptophan | | D-Gluconic acid | Nopol | | |
| L-Tyrosine | | L-Glutamine | (–)-α-Phellandrene | | |
| L-Valine | | D-Isoascorbic acid | (–)-α-Pinene | | |
| | | D-Mannitol | (+)-α-Pinene | | |
| | | D-Mannose | (–)-β-Pinene | | |
| | | D-Quinic acid | (*R*)-(+)-Pulegone | | |
| | | D-Ribolactone | | | |
| | | D-Ribose | | | |
| | | D-Saccharic acid | | | |
| | | D-Sorbitol | | | |
| | | L-Sorbose | | | |
| | | D-Xylose | | | |

reproduction of stereogenic centers.[§] Examples are provided in Scheme 1.6 for a vinyl coupling reaction leading to a β-amino acid,[124-126] and Scheme 1.7 that illustrates an alkylation.[124,127-136]

Introduction of a heteroatom group, such as phenylselenyl (R=SePh), allows for an elimination to provide the chiral acetoacetate derivative (**1.15**) where the chiral center derived from the starting hydroxy acid has been destroyed. The cyclic acetal (**1.15**) can undergo a wide variety of stereoselective reactions that regenerate the original stereocenter (Figure 1.6).[127,137]

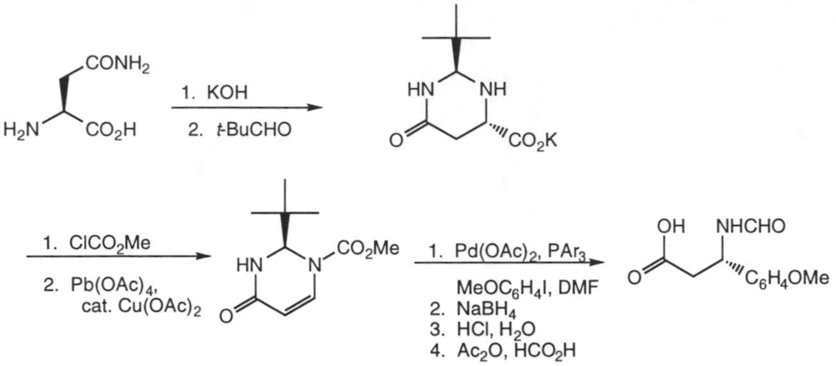

**SCHEME 1.6.**

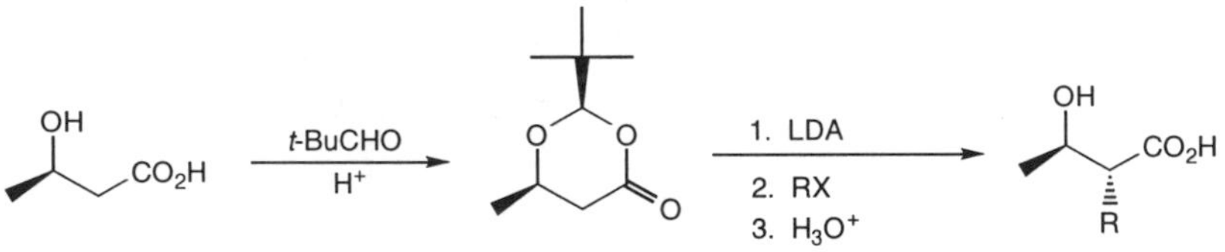

**SCHEME 1.7.**

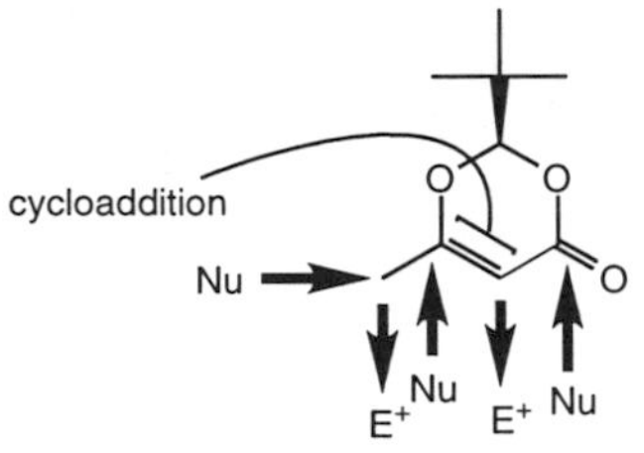

**1.15**

**FIGURE 1.6.** Possible reactions of a chiral acetal for the self-reproduction of stereogenic centers.

In this approach, 3-hydroxybutanoic acid is used as a chiral starting material. It is available as either antipode[107,138,139] and has been used as a chiron in its own right.[140] Methods to the two hydroxy acids (as their ethyl esters) are summarized in Scheme 1.8.[141-144]

### 1.3.2.4. Chiral Reagents

In many ways, this is the approach of choice as Nature utilizes this methodology through enzymes. The reagent must be selective both in terms of induction and functional group

§ All of the systems are based on cyclic systems [ *13.5.*].

specificity. The need for protection should be carefully considered as this could lead to the introduction of extra steps. Assuming that a literature precedence or analogy exists, this approach is the preferred method in many cases. The chiral reagent should allow for the expensive cost component to be recycled, if necessary, or have a very high turnover number. However, the use of a chiral catalyst that can induce the desired asymmetry in a large number of substrate molecules (just as enzymes do) has great cost implications![19] Numerous examples will be presented throughout the book.

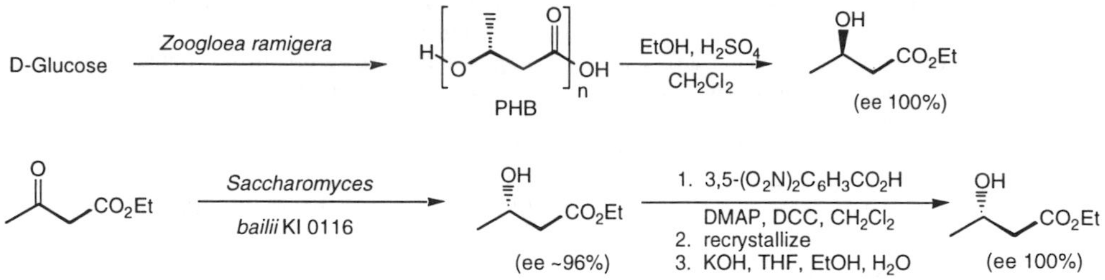

D-Glucose —*Zoogloea ramigera*→ PHB —EtOH, $H_2SO_4$, $CH_2Cl_2$→ (ee 100%)

—*Saccharomyces bailii* KI 0116→ (ee ~96%) —1. 3,5-$(O_2N)_2C_6H_3CO_2H$, DMAP, DCC, $CH_2Cl_2$; 2. recrystallize; 3. KOH, THF, EtOH, $H_2O$→ (ee 100%)

SCHEME 1.8.

### 1.3.2.5. Chiral Environments

It is possible to make the environment of a chemical reaction chiral. The majority of examples in this class utilize chiral solvents or additives. To influence the differentiation of the free energies of the diastereomeric transition states, and hence provide useful induction, these agents must be closely associated with the reaction center. In most cases, this has not been fruitful, as in the use of chiral solvents, but some reactions that use chiral ligands do provide good ee's (e.g., Scheme 1.9).[145-148]

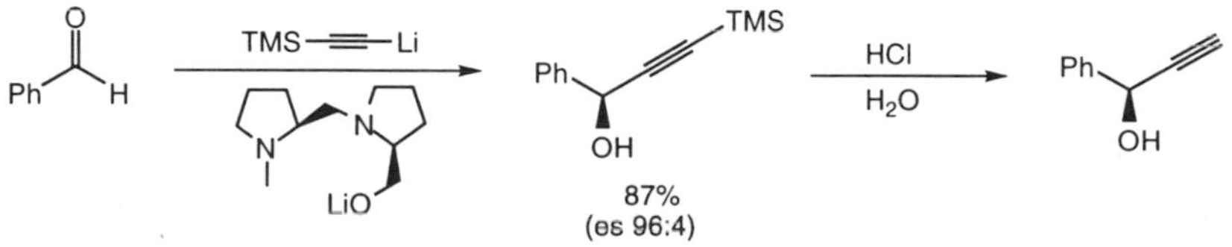

—TMS—≡—Li→ 87% (es 96:4) —HCl, $H_2O$→

SCHEME 1.9.

### 1.3.2.6. Combinations

In the synthesis of complex molecules, all of the above approaches may be combined. The marriage of two approaches also gives the potential for double asymmetric induction.

Although many methods are available for laboratory scale asymmetric synthesis, the number of methods that have been successfully employed on a large, industrial scale is still fairly limited (vide infra). The majority of chemical approaches that can be utilized on a large scale employ transition metal chemistry (although boron has provided some extremely useful methodology!), and it is surely in this arena that the significant developments will continue to be discovered.[2]

### 1.3.2.7. The Introduction of Asymmetry

Consider the addition of a nucleophile to a ketone; the nucleophile can attack from either the *Re* or *Si* face. To achieve stereoselectivity, we have to control this addition so that, for example, only the alcohol **1.16** is formed. Obviously, if no asymmetric influences are present, a racemic mixture will result (Scheme 1.10). To obtain asymmetric induction, we have a number of choices: chirality can be incorporated into $R^1$ or $R^2$; the nucleophile could be chiral; the metal counterion could be made chiral as, for example, through the use of chiral ligands; or the oxygen

**TABLE 1.3**
**Factors that Can Be Used to Achieve an Asymmetric Addition to a Carbonyl Compound**

| Chirality Present in | Advantages | Disadvantages |
|---|---|---|
| $R^1$ or $R^2$ | Anh-Felkin (or chelation controlled) addition possible | $R^1$ or $R^2$ integral parts of the substrate |
| | Predictable reaction outcome | Difficult to use chiral auxiliaries |
| | Stereoselection can be high | $R^1$ and $R^2$ must be substantially different to observe good induction |
| | Can be augmented through combination with another approach to provide double asymmetric induction | Approach is substrate specific (general methodologies may not be available) |
| $Nu^-$ | Anh-Felkin (or chelation controlled) addition possible | May be difficult to modify nucleophile structure to accommodate chirality |
| | Predictable reaction outcome | $R^1$ and $R^2$ must be substantially different to observe good induction |
| | Stereoselection can be high | |
| | Can be augmented through combination with another approach to provide double asymmetric induction | |
| | Can incorporate chiral auxiliaries | |
| | More methods are available as not substrate specific | |
| Ligands or Lewis acid | Ligands can usually be recycled | Ligand synthesis may be tedious and/or expensive |
| | Substrate can be achiral | $R^1$ and $R^2$ must be substantially different to observe good induction |
| | Allows for "general" methods | Ligand must influence diastereomeric transition states to observe good induction |
| | Allows for control of absolute stereochemistry | Lewis acid may not be compatible with reaction conditions |
| | No extra steps required to add/remove chiral auxiliaries | |

atom of the carbonyl group could have induced chirality through the use of a chiral Lewis acid. The advantages and disadvantages of each approach are summarized in Table 1.3.

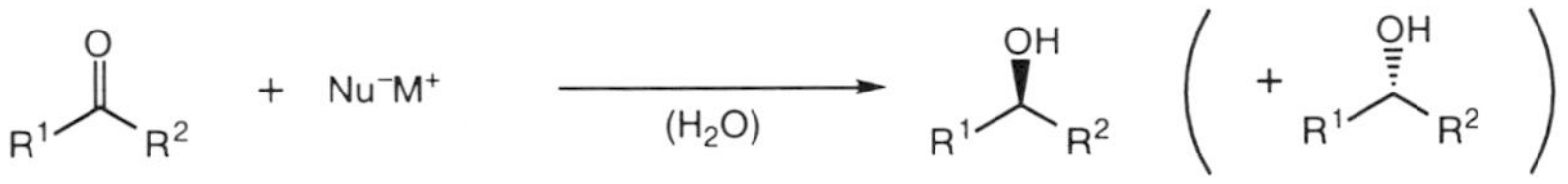

O, $R^1$, $R^2$ + $Nu^-M^+$ → $(H_2O)$ → OH, $R^1$, $R^2$ **1.16** ( + OH, $R^1$, $R^2$ )

SCHEME 1.10.

## 1.4. RESOLUTION

Although this method often is seen as the last way out of an enantioselective synthesis, and the least glamorous, there are many circumstances when it is worth looking into a resolution method. Until recently, the use of a chiral natural product as a starting material or a resolution were the only two methods available for the preparation of a chiral material.

Before embarking on a resolution through formation of diastereoisomers, especially if the reaction is not a "one-off," it is worth looking at a preferential crystallization of the enantiomers. This approach has been used successfully in a commercial synthesis of α-methylDOPA.[149,150] Unfortunately, the mixture of enantiomers has to exist as a conglomerate where only molecules of the same chirality are incorporated into a specific crystal so that a physical mixture of crystals of the two antipodes are present in a racemic mixture (*cf.* Pasteur's experiment); these are relatively rare. A racemic compound has both antipodes intimately bound in a single crystal, and thus, cannot be separated by crystallization. It is possible to determine whether a compound exists as a conglomerate through construction of a binary phase diagram; a conglomerate has a lower melting point than the enantiomers, while a racemic compound usually has a higher melting point than its constituents.[150,151]

To effect a resolution, the compound must contain a functional group that can interact or react with another chiral agent to produce diastereoisomers. These diastereoisomers can then be separated by physical means. The resolution agent then needs to be removed from the subject material. Thus, two extra chemical steps may be necessary to use the resolution agent. In addition, as only 50% of a racemic mixture can be the desired isomer, recovery is, at the very best, 50%. This detracts from the method. However, if the compound under investigation is an acid or a base, then it may be possible to form a salt that can be separated, although many factors can still affect the success of the approach (Scheme 1.11).[69,152-155] The use of a salt has many advantages, as an ionic chemical bond is replaced with high chemical yields, and simple extractive procedures often provide clean, high-yielding steps.[156] Many techniques are being developed to separate enantiomers by chromatography. In addition to analytical methods, variants are now available for larger scale work.

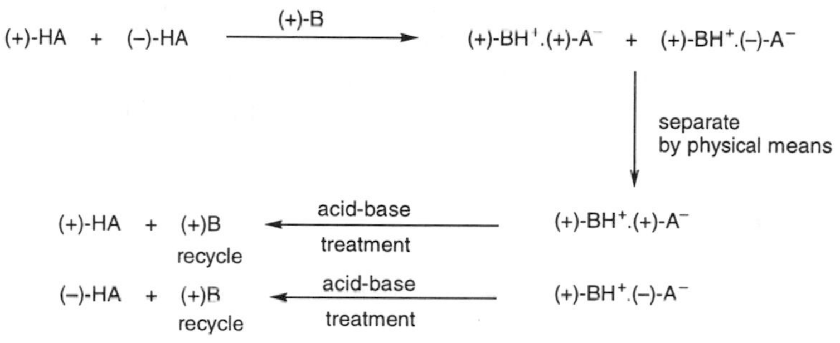


**SCHEME 1.11.**

As half of the material is discarded after a resolution step, the earlier such a process is performed in a synthetic sequence, the less will be the waste. It is not necessary to bring unwanted material through a number of steps, but if the resolution agent is expensive, a number of steps may be required to recover this material and make the method economically viable. However, there is a caveat; if the "wrong" isomer can be converted to the racemic mixture or inverted, such as through an epimerizable center, then a resolution can become a very efficient process. In addition, if the resolution agent is a cheap, readily available substance, then this may be the method of choice on the grounds of cost. For small-scale work, however, the use of recycling material can get extremely tedious and time consuming.

### 1.4.1. KINETIC RESOLUTION

In this approach a substrate is acted on by a chiral agent to produce one enantiomer or diastereoisomer of the product at a much faster rate than the other isomer. The transition states have to be of a significant energy difference for this method to be viable. In general, the enantiomeric excess of the starting material will increase as the reaction progresses, while the ee of a chiral product will decrease. As this is a resolution, only 50% of the substrate can be converted to the desired product unless it is *meso*. In this later case, all of the substrate could be converted to product due to the symmetry. This has been put to effect with enzymes (see Chapter 14). Many selective chiral reagents can be used in this type of reaction. In addition to *meso* compounds, substrates that contain a chirally labile center can also be converted to one product with high conversion. The kinetic resolution is effectively coupled to a racemization. An example is provided by the reduction of β-keto esters to β-hydroxy

**TABLE 1.4**
**Methods to Achieve Resolution**

| Chemical | Biological |
|---|---|
| Kinetic resolution by a chemical reaction | Enzymatic resolution (e.g., esterases) |
| Chromatographic separation | Catalytic antibodies |
| Selective crystallization of a single isomer | |
| Fractional crystallization | |
| Selective isomer crystallization by entrainment | |
| Selective extraction (e.g., with a chiral host compound) | |

esters [*6.4.1*].[157] As with any resolution approach, it is prudent to do this type of separation as early as possible in a synthetic sequence to avoid undue waste. The use of a recycle sequence can also increase the efficiency of this type of approach.[158,159]

### 1.4.2. METHODS AVAILABLE

There are a wide variety of methods available to achieve a resolution. Often, a number of them have to be tried to obtain a suitable method for the particular system in use. The functional groups present within the subject molecule often determine which of the methods described in Table 1.4 can be used. The other major factors are cost and scale. The factors that can influence a chemical vs. biological approach to perform a resolution are discussed in Chapter 14.

## 1.5. A QUESTION OF SCALE

Asymmetric reactions often require careful control of the reaction conditions. Most methodologies are developed on a small scale — a few grams at most. Many of the examples illustrated throughout the following chapters require the use of a low temperature to ensure selection between possible diastereomeric transition states. Use of a higher temperature would result in loss of any stereoselection. On a laboratory scale, the use of –78°C[1] is commonplace. Indeed, in many cases, –78° is used to describe a reaction flask immersed in a dry ice–acetone or isopropanol bath. The internal temperature is not measured and may be well above –78°. If the reaction is also exothermic, as invariably seen when a strong base is added to an acidic organic substrate, this temperature problem can be exacerbated.

Most organic chemists working on the discovery of new compounds for specific applications, such as the development of a new drug, are only concerned with the preparation of small amounts of their target molecule. All of the reactions described herein can be used. The problems associated with exotherms, temperature control, and mixing are all of little concern. In addition, because of the relatively small amounts of the reagents required, safety and environmental problems are usually of tolerable proportions and can be dealt with. These observations are usually true of academic research. However, commercial products usually demand the preparation of larger quantities. The problems associated with the physical parameters of the reaction are more important. Many of the reactions described below could not be scaled-up without considerable effort and cost. As both of the authors have experience in academia and industrial process scale-up, we have endeavored to include comments when a reaction has been performed on a large scale (multi-kilo) and does not require uncommon equipment (such as a low temperature reactor). As noted in the introduction, the need for asymmetric synthesis of many compounds continues to drive the development of robust asymmetric reactions. It is apparent that many gaps still need to be filled. The use of "chiral pool" materials is still the simplest entry to asymmetric synthesis on a large scale.

1 Temperatures throughout will be given on the Celsius scale.

**TABLE 1.5**
**Synthesis Methods**

| Chemical | Biological |
|---|---|
| Reductions (metal catalyzed) | Dehydrogenases |
| Rearrangements (metal catalyzed) | Yeast reductions |
| Cyclopropenation (metal catalyzed) | Transaminases |
| Oxidations (Sharpless) | Aldolases |
| Reductions (hydrides) | Oxynitrilases |
| Diels-Alder (Lewis acid catalysis) | Nitrilases |
| Hydroborations | Hydantoinases |
| Substitution reactions ($S_N2$) | Use of *meso*-substrates |
| Chiral auxiliaries | Use of prochiral substrates |
| Phase transfer chiral reagents | Fermentation |
| Pericyclic reactions | |
| Iodolactonization | |

Even when the physical parameters of a reaction can be controlled on a large scale, safety and environmental problems can still play a key role in the determination of whether a reaction is feasible. The cost of safeguards, equipment, and the need to have no environmental impact can be so overwhelming that a reaction, although chemically functional, can, in practice, be unworkable. In addition, the availability of a reagent or starting material may immediately rule out the approach. At present, this is particularly true of methods based on chiral auxiliaries, and the less common "chiral pool" materials. On a small scale, grams of a chiral pool material may be required, but for large-scale work, multi-kilos may be necessary. Only a few chiral materials are available on a very large scale (Table 1.2). However, the limits of these resources are being expanded by asymmetric synthesis, as demonstrated by the manufacture of menthol. Again, we have tried to incorporate comments when a chiral starting material, reagent, or auxiliary is readily available.[3,5,160]

As noted in the section on the choice of methods [*1.3*], there are two major approaches to achieve an asymmetric synthesis on a large scale: resolution or synthesis. The latter approach has only a limited number of methods that have been successfully employed on an industrial scale. They are summarized in Table 1.5. These methods are discussed in detail in the appropriate section of this book.

## 1.6. SUMMARY

This introductory chapter has been used to describe the scope and limitations of this book. A number of nomenclature conventions are established, in particular the use of reporting selectivities either as ratios or excesses, and the use of the stereochemical descriptors *pref* and *parf*. The general philosophies behind asymmetric synthesis are described in general terms with some emphasis on larger-scale synthetic problems. Many examples of the topics introduced in the chapter will be found throughout the book.

## 1.7. REFERENCES

1. Velluz, L.; Valls, J.; Mathieu, J. *Angew. Chem., Int. Ed. Engl.* 1967, *6*, 778.
2. Seebach, D. *Angew. Chem., Int. Ed. Engl.* 1990, *29*, 1320.
3. Crosby, J. *Tetrahedron* 1991, *47*, 4789.
4. Chan, A. S. C. *Chemtech.* 1993, 46.
5. Kagan, H. B. *Bull. Soc. Chim. Fr.* 1988, 846.

6. Buda, A. B.; Auf der Heyde, T.; Mislow, K. *Angew. Chem., Int. Ed. Engl.* 1992, *31*, 989.
7. Marshall, J. A.; Flynn, K. E. *J. Am. Chem. Soc.* 1983, *105*, 3360.
8. Ager, D. J.; East, M. B. *Tetrahedron* 1992, *48*, 2803.
9. Ager, D. J.; East, M. B. *Tetrahedron* 1993, *49*, 5683.
10. ApSimon, J. W.; Seguin, R. P. *Tetrahedron* 1979, *35*, 2797.
11. ApSimon, J. W.; Collier, T. E. *Tetrahedron* 1986, *42*, 5157.
12. Inch, T. D. *Synthesis* 1970, 466.
13. Scott, J. W.; Valentine, D. *Science* 1974, *184*, 943.
14. Eliel, E. L. *Tetrahedron* 1974, *30*, 1503.
15. Kagan, H. B.; Fiaud, J. C. *Top. Stereochem.* 1978, *10*, 175.
16. Valentine, D.; Scott, J. W. *Synthesis* 1978, 329.
17. Bartlett, P. A. *Tetrahedron* 1980, *36*, 3.
18. Mosher, H. S.; Morrison, J. D. *Science* 1983, *221*, 1013.
19. Noyori, R. *Science* 1990, *248*, 1194.
20. McGarvey, G. J.; Kimura, M.; Oh, T.; Williams, J. M. *J. Carbohydr. Chem.* 1984, *3*, 125.
21. McGarvey, G. J.; Kimura, M.; Oh, T. *J. Carbohydr. Chem.* 1984, *3*, 125.
22. Whitesell, J. K. *Acc. Chem. Res.* 1985, *18*, 280.
23. Whitesell, J. K. *Chem. Rev.* 1989, *89*, 1581.
24. Whitesell, J. K. *Chem. Rev.* 1992, *92*, 953.
25. Mori, K. In *The Total Synthesis of Natural Products* ; J. ApSimon, Ed.; John Wiley & Sons: New York, 1981; Vol. 4; pp 1.
26. Wierenga, W. In *The Total Synthesis of Natural Products* ; J. ApSimon, Ed.; John Wiley & Sons: New York, 1981; pp 263.
27. Zamojski, A.; Grynkiewicz, G. In *The Synthesis of Natural Products* ; J. ApSimon, Ed.; John Wiley & Sons: New York, 1984; Vol. 6; pp 141.
28. Henrick, C. A. *Tetrahedron* 1977, *33*, 1845.
29. Rossi, R. *Synthesis* 1978, 413.
30. Kocovsky, P.; Turecek, F.; Hajicek, J. In *Synthesis of Natural Products: Problems of Stereoselection;* CRC Press: Boca Raton, FL, 1986; Vol. 1.
31. Morrison, J. D.; Mosher, H. S. *Asymmetric Organic Reactions;* revised ed.; American Chemical Society Books: Washington D.C., 1976.
32. *Asymmetric Synthesis (Series)* ; J. D. Morrison, Ed.; Academic Press: Orlando, FL.
33. Bogdanovic, B. *Angew. Chem., Int. Ed. Engl.* 1973, *12*, 954.
34. Babievskii, K. K.; Latov, V. K. *Russ. Chem. Rev.* 1969, *38*, 456.
35. Martens, J. *Chemiker-Zeitung* 1986, *110*, 169.
36. Solladie, G. *Chimia* 1984, *38*, 233.
37. Kagan, H. B.; Rebiere, F. *Synlett* 1990, 643.
38. Davies, S. G. *Aldrichimica Acta* 1990, *23*, 31.
39. Mukaiyama, T. *Tetrahedron* 1981, *37*, 4111.
40. Davies, S. G.; Dordor-Hedgecock, I. M.; Easton, R. J. C.; Preston, S. C.; Sutton, K. H.; Walker, J. C. *Bull. Soc. Chim., Fr.* 1987, 608.
41. Davies, S. G. *Pure Appl. Chem.* 1988, *60*, 13.
42. Reissig, H.-U. *Angew. Chem., Int. Ed. Engl.* 1992, *31*, 288.
43. Porter, N. A.; Giese, B.; Curran, D. P. *Acc. Chem. Res.* 1991, *24*, 296.
44. Fuji, K. *Chem. Rev.* 1993, *93*, 2037.
45. Knochel, P.; Singer, R. D. *Chem. Rev.* 1993, *93*, 2117.
46. Ohfune, Y. *Acc. Chem. Res.* 1992, *25*, 360.
47. Brown, J. M. *Chem. Soc. Rev.* 1993, *22*, 25.
48. Walker, A. J. *Tetrahedron: Asymmetry* 1992, *3*, 961.
49. Sonawane, H. R.; Bellur, N. S.; Ahuja, J. R.; Kulkarni, D. G. *Tetrahedron: Asymmetry* 1992, *3*, 163.
50. Kim, B. H.; Curran, D. P. *Tetrahedron* 1993, *49*, 293.
51. Pommier, A.; Pons, J.-M. *Synthesis* 1993, 441.
52. Terashima, S. *Synlett* 1992, 691.
53. Duthaler, R. O. *Tetrahedron* 1994, *50*, 1539.
54. Sakai, K.; Suemune, H. *Tetrahedron: Asymmetry* 1993, *4*, 2109.
55. Marckwald, W. *Ber. Desch. Chem. Ges.* 1904, *37*, 1368.
56. Masamune, S.; Choy, W.; Petersen, J. S.; Sita, L. R. *Angew. Chem., Int. Ed. Engl.* 1985, *24*, 1.
57. Sato, T.; Abiko, A.; Masamune, S. *J. Synth. Org. Chem.* 1986, *44*, 384.
58. Schreiber, S. L.; Schreiber, T. S.; Smith, D. B. *J. Am. Chem. Soc.* 1987, *109*, 1525.
59. Maehr, H. *Tetrahedron: Asymmetry* 1992, *3*, 735.

60. Cahn, R. S.; Ingold, C. K.; Prelog, V. *Angew. Chem., Int. Ed. Engl.* 1966, *78*, 413.
61. Cahn, R. S.; Ingold, C. K.; Prelog, V. *J. Chem. Soc.* 1951, 612.
62. Cahn, R. S.; Ingold, C. K.; Prelog, V. *Experimentia* 1956, *12*, 81.
63. Dodziuk, H.; Mirowicz, M. *Tetrahedron: Asymmetry* 1990, *1*, 171.
64. Dodziuk, H. *Tetrahedron: Asymmetry* 1992, *3*, 43.
65. Mata, P.; Lobo, A. M.; Marshall, C.; Johnson, A. P. *Tetrahedron: Asymmetry* 1993, *4*, 657.
66. Hanson, K. R. *J. Am. Chem. Soc.* 1966, *88*, 2731.
67. Seebach, D.; Prelog, V. *Angew. Chem., Int. Ed. Engl.* 1982, *21*, 654.
68. Carey, F. A.; Kuehne, M. E. *J. Org. Chem.* 1982, *47*, 3811.
69. Eliel, E. L. *Stereochemistry of Carbon Compounds*; McGraw-Hill Kogakusha: Tokyo, 1962.
70. Heathcock, C. H. In *Asymmetric Synthesis*; J. D. Morrison, Ed.; Academic Press: Orlando, FL, 1984; Vol. 3; pp 111.
71. Winstein, S.; Lucas, H. J. *J. Am. Chem. Soc.* 1939, *61*, 1576.
72. Lucas, H. J.; Schlatter, M. J.; Jones, R. C. *J. Am. Chem. Soc.* 1941, *63*, 22.
73. Cram, D. J. *J. Am. Chem. Soc.* 1952, *74*, 2149.
74. House, H. O. *J. Am. Chem. Soc.* 1955, *77*, 5083.
75. Noyori, R.; Nishida, I.; Sakata, J. *J. Am. Chem. Soc.* 1981, *103*, 2106.
76. Brewster, J. H. *J. Am. Chem. Soc.* 1986, *51*, 4751.
77. Masamune, S.; Ali, S. A.; Snitman, D. L.; Garvey, D. S. *Angew. Chem., Int. Ed. Engl.* 1980, *19*, 557.
78. Horeau, A. *Tetrahedron Lett.* 1969, 3121.
79. Morrison, J. D. In *Asymmetric Synthesis*; J. D. Morrison, Ed.; Academic Press: Orlando, FL, Vol. 1; p 1.
80. Selke, R.; Facklam, C.; Foken, H.; Heller, D. *Tetrahedron: Asymmetry* 1993, *4*, 369.
81. Blashke, G.; Kraft, H. P.; Fickenscher, K.; Kohler, F. *Arzniem-Forsch/Drug Res.* 1979, *29*, 10.
82. Blashke, G.; Kraft, H. P.; Fickenscher, K.; Kohler, F. *Arzniem-Forsch/Drug Res.* 1979, *29*, 1140.
83. Powell, J. R.; Ambre, J. J.; Ruo, T. I. In *Drug Stereochemistry*; I. W. Wainer and D. E. Drayer, Eds.; Marcel Dekker: New York, 1988; pp 245.
84. Ariëns, E. J.; Soudijn, W.; Timmermas, P. B. M. W. M. *Stereochemistry and Biological Activity of Drugs*; Blackwell Scientific: Palo Alto, 1983.
85. Ariëns, E. J. *J. Clin. Pharmacol.* 1984, *26*, 663.
86. Stinson, S. C. *Chem. Eng. News* 1993, September 27, 38.
87. Stinson, S. C. *Chem. Eng. News* 1992, September 28, 46.
88. Williams, K.; Lee, E. *Drugs* 1985, *30*, 333.
89. Richards, A. *Chiral'93, USA* 1993.
90. Ramos Tombo, G. M.; Bellus, D. *Angew. Chem., Int. Ed. Engl.* 1991, *30*, 1193.
91. Mori, K. *Tetrahedron* 1989, *45*, 3233.
92. Mori, K. In *Organic Synthesis: Modern Trends*; O. Chizhov, Ed.; Blackwell Scientific: Oxford, 1987; pp 293.
93. Mazur, R. H.; Schlatter, J. M.; Goldhamp, A. H. *J. Am. Chem. Soc.* 1969, *91*, 2684.
94. *Sweeteners: Discovery, Molecular Design, and Chemoreception*; D. E. Walters, F. T. Orthoefer, G. E. Dubois, Eds.; American Chemical Society: Washington, D. C., 1991.
95. Horeau, A.; Guette, J. P. *Tetrahedron* 1979, *30*, 1923.
96. Buschmann, H.; Scharf, H.-D.; Hoffmann, N.; Esser, P. *Angew. Chem., Int. Ed. Engl.* 1991, *30*, 477.
97. Salem, L. *J. Am. Chem. Soc.* 1973, *95*, 93.
98. Dauben, W. G.; Fonken, G. H.; Noyce, D. C. *J. Am. Chem. Soc.* 1956, *78*, 2579.
99. Dauben, W. G.; Fonken, G. H.; Noyce, D. C. *J. Am. Chem. Soc.* 1956, *78*, 3752.
100. Meyers, A. I. *Pure Appl. Chem.* 1979, *51*, 1255.
101. Corey, E. J. *Chem. Soc. Rev.* 1988, *17*, 111.
102. Corey, E. J.; Xue-Min, C. *The Logic of Chemical Synthesis*; John Wiley & Sons: New York, 1989.
103. Hanessian, S. *The Total Synthesis of Natural Products: The Chiron Approach*; Pergamon Press: Oxford, 1983.
104. Hanessian, S. *Aldrichimica Acta* 1989, *22*, 3.
105. Scott, J. S. In *Asymmetric Synthesis*; J. D. Morrison, Ed.; Academic Press: Orlando, 1984; Vol. 4; pp 1.
106. Blaser, H.-U. *Chem. Rev.* 1992, *92*, 935.
107. Gringore, O. H.; Rouessac, F. P. *Org. Synth.* 1984, *63*, 121.
108. Inch, T. D. *Tetrahedron* 1984, *40*, 3161.
109. Shiozaki, M. *J. Org. Chem.* 1991, *56*, 528.
110. Valverde, S.; Garcia-Ochoa, S.; Martin-Lomas, M. *J. Chem. Soc., Chem. Commun.* 1987, 1714.
111. Szarek, W. A.; Hay, G. W.; Vyas, D. M.; Ison, E. R.; Horonowski, L. J. *Can. J. Chem.* 1984, *62*, 671.
112. Sowa, W. *Can. J. Chem.* 1969, *47*, 3931.
113. Takano, S.; Yonaga, M.; Ogasawara, K. *J. Chem. Soc., Chem. Commun.* 1981, 1153.
114. Takano, S.; Kasahara, C.; Ogasawara, K. *J. Chem. Soc., Chem. Commun.* 1981, 637.
115. Salzmann, T. N.; Ratcliffe, R. W.; Christensen, B. G.; Bouffard, F. A. *J. Am. Chem. Soc.* 1980, *102*, 6160.

116. Reider, P.; Grabowski, E. J. J. *Tetrahedron Lett.* 1982, *23*, 2293.
117. Gurjar, M.; Bhana, M. N.; Khare, V. B.; Bhandari, A.; Deshmukh, M. N.; Rao, A. V. R. *Tetrahedron* 1991, *47*, 7117.
118. Bouffard, F. A.; Salzmann, T. N. *Tetrahedron Lett.* 1985, *26*, 6285.
119. Oppolzer, W. *Tetrahedron* 1987, *43*, 1969.
120. Oppolzer, W. *Pure Appl. Chem.* 1990, *62*, 1241.
121. Martens, J. *Top. Curr. Chem.* 1984, *125*, 165.
122. Evans, D. A.; Takacs, J. M.; McGee, L. R.; Ennis, M. D.; Mathre, D. J.; Bartroli, J. *Pure Appl. Chem.* 1981, *53*, 1109.
123. Puchot, C.; Samuel, O.; Dunach, E.; Zhao, S.; Agami, C.; Kagan, H. B. *J. Am. Chem. Soc.* 1986, *108*, 2353.
124. Konopelski, J. P.; Chu, K. S.; Negrete, G. R. *J. Org. Chem.* 1991, *56*, 1355.
125. Chu, K. S.; Konopelski, J. P. *Tetrahedron* 1993, *49*, 9183.
126. Juaristi, E.; Escalante, J. *J. Org. Chem.* 1993, *58*, 2282.
127. Seebach, D.; Zimmerman, J. *Helv. Chim. Acta* 1986, *69*, 1147.
128. Polt, R.; Seebach, D. *J. Am. Chem. Soc.* 1989, *111*, 2622.
129. Herradón, B.; Seebach, D. *Helv. Chim. Acta* 1989, *72*, 690.
130. Amberg, W.; Seebach, D. *Chem. Ber.* 1990, *123*, 2413.
131. Amberg, W.; Seebach, D. *Chem. Ber.* 1990, *123*, 2429.
132. Amberg, W.; Seebach, D. *Chem. Ber.* 1990, *123*, 2439.
133. Seebach, D.; Miller, D. D.; Muller, S.; Weber, T. *Helv. Chim. Acta* 1985, *68*, 949.
134. Blank, S.; Seebach, D. *Angew. Chem., Int. Ed. Engl.* 1993, *32*, 1765.
135. Juaristi, E.; Quintana, D.; Lamatsch, B.; Seebach, D. *J. Org. Chem.* 1991, *56*, 2553.
136. Zydowsky, T. M.; de Lara, E.; Spanton, S. G. *J. Org. Chem.* 1990, *55*, 5437.
137. Harwood, L. M.; Macro, J.; Watkin, D.; Williams, C. E.; Wong, L. F. *Tetrahedron: Asymmetry* 1992, *3*, 1127.
138. Cainelli, G.; Manescalchi, F.; Martelli, G.; Panunzo, M.; Pless, L. *Tetrahedron Lett.* 1985, *26*, 3369.
139. Seebach, D.; Aebi, J.; Wasmuth, D. *Org. Synth.* 1984, *63*, 109.
140. Schnurrenberger, P.; Hungerbuhler, E.; Seebach, D. *Liebigs Ann. Chem.* 1987, 733.
141. Sugai, T.; Fujita, M.; Mori, K. *J. Chem. Soc., Jpn.* 1983, 1315.
142. Hasegawa, J.; Hamaguchi, S.; Ogura, M.; Watanabe, K. *J. Ferment. Technol.* 1981, *59*, 257.
143. Mori, K.; H., M.; Sugai, T. *Tetrahedron* 1985, *41*, 919.
144. Masamune, S.; Walsh, C. T.; Sinskey, A. J.; Peoples, O. P. *Pure Appl. Chem.* 1989, *61*, 303.
145. Mukaiyama, T.; Suzuki, K.; Soai, K.; Sato, T. *Chem. Lett.* 1979, 447.
146. Tomioka, K. *Synthesis* 1990, 541.
147. Mukaiyama, T.; Soai, K.; Sato, T.; Shimizu, H.; Suzuki, K. *J. Am. Chem. Soc.* 1979, *101*, 1455.
148. Noyori, R.; Kitamura, M. *Angew. Chem., Int. Ed. Engl.* 1991, *30*, 49.
149. Reinhold, D. F.; Firestone, R. A.; Gaines, W. A.; Chemerda, J. M.; Sletzinger, M. *J. Org. Chem.* 1968, *33*, 1209.
150. Bayley, C. R.; Vaidyin, N. A. In *Chirality in Industry*; A. N. Collins, G. N. Sheldrake and J. Crosby, Eds.; John Wiley & Sons: New York, 1992; pp 69.
151. Jacques, J.; Collet, A.; Wilen, S. H. *Enantiomers, Racemates, and Resolutions*; John Wiley & Sons: New York, 1981.
152. Secor, R. *Chem. Rev.* 1963, *63*, 297.
153. Boyle, P. H. *Quart. Rev.* 1971, *25*, 323.
154. Mazón, A.; Nájera, C.; Yus, M.; Heumann, A. *Tetrahedron: Asymmetry* 1992, *3*, 1455.
155. Leusen, F. J. J.; Noordik, J. H.; Karfunkel, H. R. *Tetrahedron* 1993, *49*, 5377.
156. Zingg, S. P.; Arnett, E. M.; McPhail, A. T.; Bothner-By, A. A.; Gilkerson, W. R. *J. Am. Chem. Soc.* 1988, *110*, 1565.
157. Kitamura, M.; Tokunaga, M.; Noyori, R. *Tetrahedron* 1993, *49*, 1853.
158. Brown, J. M. *Chem. Ind. (London)* 1988, 612.
159. Kagan, H. B.; Fiaud, J. C. *Top. Stereochem.* 1988, *18*, 249.
160. Scott, J. W. *Top. Stereochem.* 1989, *19*, 209.

Chapter 2

# GENERAL RULES

We will discuss methods to effect stereoselective change at a saturated carbon center in the following chapter. The majority of methods for asymmetric synthesis transform $sp^2$ centers to $sp^3$; a number of approaches are available to achieve this. Some general rules have been developed to rationalize the outcome of some addition reactions and allow the prediction of the product stereochemistry. The purpose of this chapter is to outline these rules; specific examples of their application will be found throughout this book. A few rules are specific to cyclic compounds; these are discussed in Chapter 13.[1] As with other empirical rules, exceptions are known and they should be applied with care.

Pericyclic reactions proceed through cyclic transition states, as do many other reactions. Some general points are discussed in this chapter, but the differences between the major reaction classes are discussed in detail in Chapter 12.

## 2.1. CARBONYL ADDITIONS

### 2.1.1. ANH-FELKIN ADDITION

Condensation of a nucleophile with a carbonyl compound, with concurrent carbon–carbon bond formation, has proven one of the most powerful reactions in organic synthesis, and many stereoselective condensations have been developed from this simple reaction.[2-4] Thus, the need to control and predict the direction of attack for the nucleophilic species has spawned many models.[5-14] The early rationalizations could be used in a predictive manner but suffered from errors due to the assumption of a perpendicular approach for the incoming nucleophile relative to the carbonyl plane. Calculations have shown that the trajectory of the nucleophile is at an angle,[15-19] and closely relates to the model of Felkin.[10,20-22] This model (Figure 2.1 and Scheme 2.1) has become known as the Anh-Felkin model.¶ [19,23,24] This model is powerful as it can also be used for α-substituted carbonyl compounds. The major shortfall centers around identification of the "large" group.[1,25-27] The interpretation of the Anh-Felkin model also gives powerful insight into ways to optimize asymmetric induction: strong electrophilic assistance (that is, complex formation between the carbonyl oxygen atom and a strong Lewis acid) aids selectivity, while use of a "hard" nucleophile will be detrimental to high asymmetric induction.[28,29]

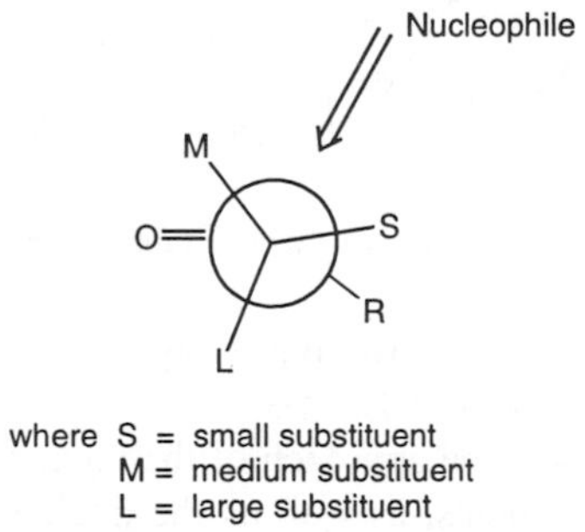

Nucleophile

M

O

S

R

L

where S = small substituent
M = medium substituent
L = large substituent

**FIGURE 2.1.** Anh-Felkin model.

¶ The direction of approach usually coincides with that determined by Cram [5] who proposed the first model. Thus, it is not uncommon to see the use of "Cram addition." The chelate model described in the following section is often referred to as "anti-Cram addition."

**SCHEME 2.1.**

As an example, consider the addition of a Grignard reagent to (*S*)-3,4,4-trimethylpentan-2-one (**2.1**). The pneumonic shown in Scheme 2.1 becomes the prediction of Scheme 2.2.

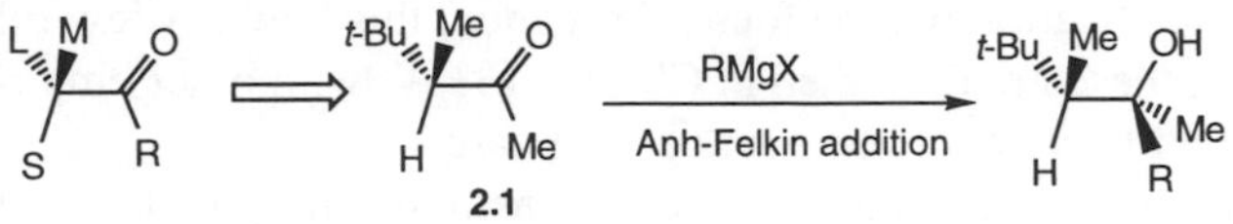

**SCHEME 2.2.**

The use of three dimensional representations clearly shows why the nucleophile approaches from this direction. Looking at the end of the molecule with all of the appendant groups, the size of the substituents is apparent. In addition, the large difference between the sizes of the *tert*-butyl and methyl groups, in terms of steric hindrance, can also be seen (Figure 2.2).[§]

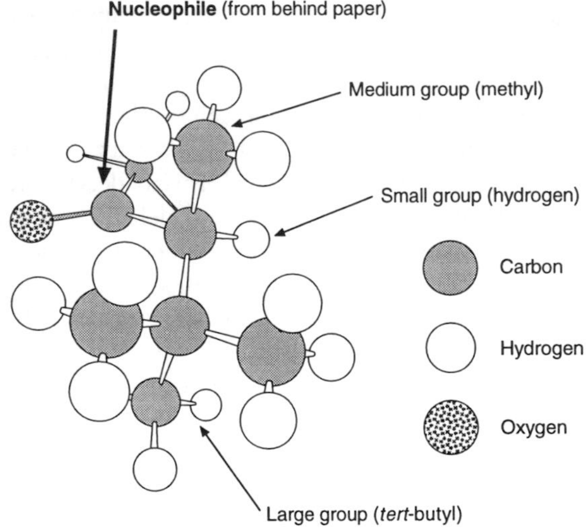


**FIGURE 2.2.** Nucleophilic approach to a carbonyl compound for Anh-Felkin addition.

From the side of the molecule, this reasoning is augmented as the sterically less demanding approach can be seen clearly (Figure 2.3).

In cyclic carbonyl compounds, the approach of the nucleophile can be influenced greatly by the nature of the ring substituents, as well as the nucleophile itself.[1,21,22,26,30-34] In many respects, therefore, when cyclic substrates are involved, each case must be considered on an individual basis.[35]

Examples of Anh-Felkin addition can be found throughout this book, especially when the additions of nucleophiles or hydride delivery agents to carbonyl compounds are discussed. Use of the rule can provide high asymmetric induction, as illustrated by the use of acyl silanes (Scheme 2.3).[36-38]

§ The structures are minima as determined with MM2 parameters.

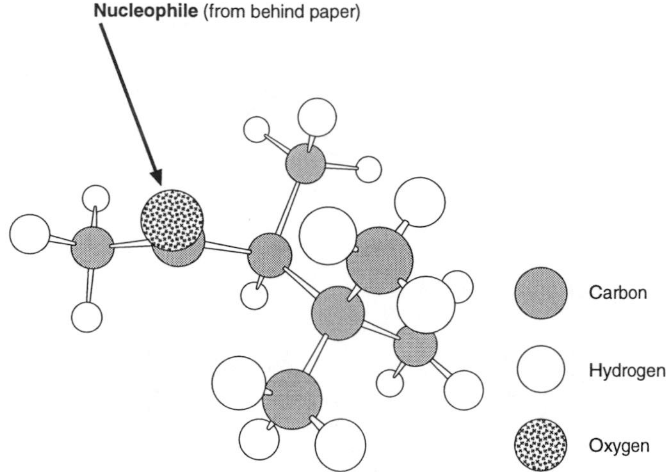


**FIGURE 2.3.** Nucleophilic approach to a carbonyl compound for Anh-Felkin addition.

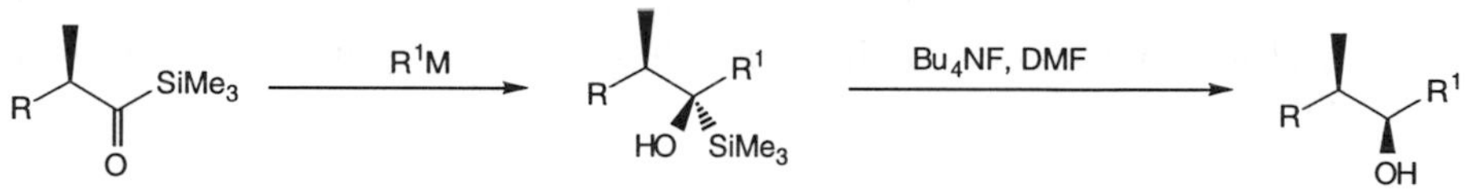

**SCHEME 2.3.**

Another example uses an oxathiane as a chiral auxiliary. The first nucleophilic addition is not stereoselective, but the second is (Scheme 2.4).[39-47] The use of ytterbium as the counterion reverses the diastereoselection of the addition.[48]

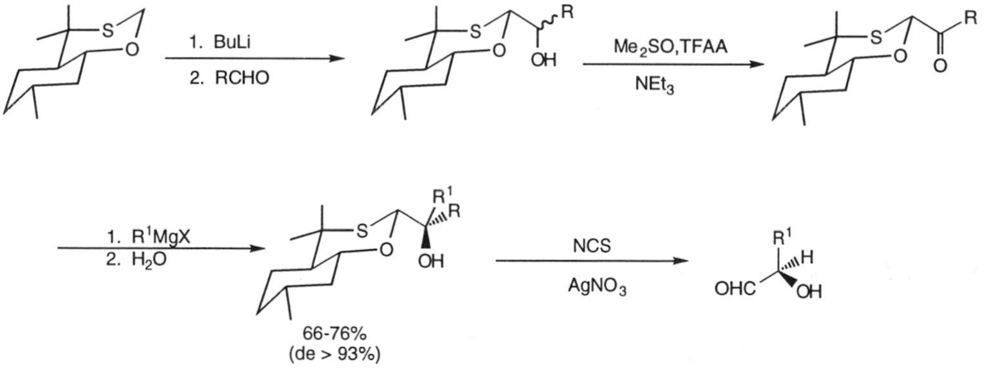

**SCHEME 2.4.**

## 2.1.2. CHELATION CONTROLLED ADDITION

Chelation controlled addition can be accomplished with a wide variety of substrates; it is simplest to consider the rules by use of the models developed for α-substituted carbonyl compounds. In this cyclic,[§] or chelation model of Cram (Figure 2.4),[¶ 7,49] formation of a chelate can reverse the stereochemical outcome of a reaction compared to when a chelate is not formed (Scheme 2.5; *cf.* Scheme 2.1)(vide infra).[50]

§ Use of this term can cause confusion when a substrate is cyclic, as opposed to acyclic; the use of chelation control or model alleviates this ambiguity.

¶ Once again the model has been modified to conform with Anh's results.

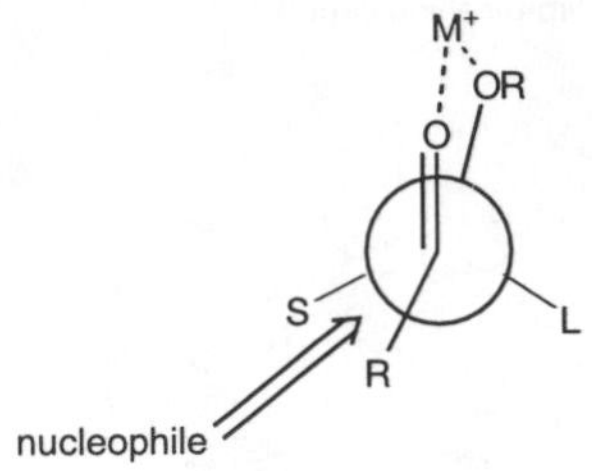

FIGURE 2.4. Preferred attack in chelation control model.

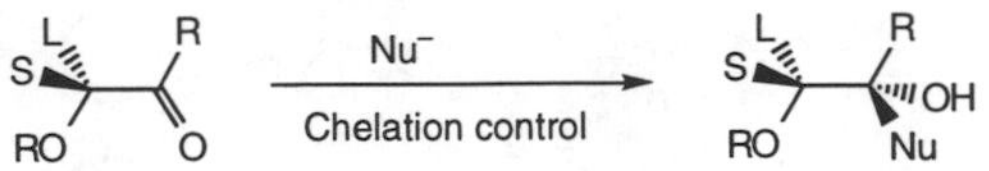

SCHEME 2.5.

Consider the example of a Grignard addition to (*R*)-3-methoxy-4,4-dimethylpentan-2-one (**2.2**) where the magnesium can form a chelate (Figure 2.5). The steric hindrance of the substituents, again, only allows for the approach of the nucleophile from one direction (Scheme 2.6).

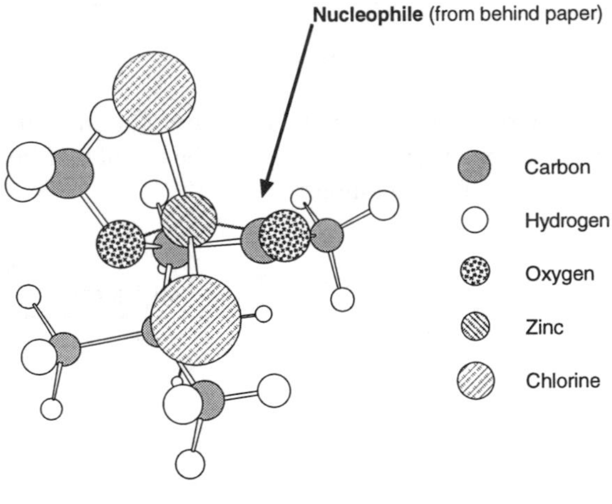

FIGURE 2.5. Chelation control model.

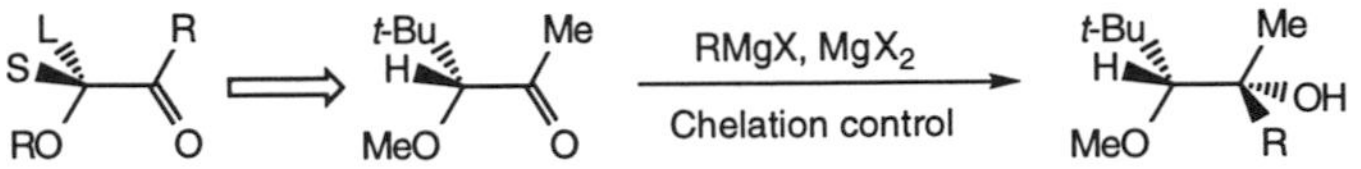

SCHEME 2.6.

In some cases, the chelate model is often invoked to explain why Anh-Felkin selectivity is not observed, but this can be in error. For example, high diastereoselectivity during an enolate reaction may not be due to a chelate-controlled mechanism, but the enolate conformation within an aggregate.[51] However, it has been demonstrated that chelates are true intermediates for the addition of an organometallic reagent to an α-alkoxy carbonyl compound.[52] In the absence of a group at the α-position of an aldehyde that can participate in complex formation, the nature of the counterion can have a marked influence on the stereochemical outcome of the reaction. For example, addition of an alkyl lithium or Grignard reagent to an aldehyde in the presence of tetra-*n*-butylammonium bromide follows Anh-Felkin addition, but the selectivity is reversed by the use of crown ethers,[53] cuprates,[54] or electron transfer processes.[55]

The mechanism of a reaction can also have important stereochemical consequences; reduction of an acyclic ketone by an electron transfer process can provide the opposite stereochemistry to the more traditional metal hydride reagents.[55]

As complex formation is often crucial for high stereoselectivity; the nature of the oxygen protecting group and solvent play an important role to determine the selectivity of the addition.[56] Obviously, this oxygen protecting group must also be easily and cleanly removed. For these reasons, the use of benzyl, benzyloxymethyl, or furfurylmethyl protection has been advocated with Grignard reagents.[57,58] The condensation of lithium enolates derived from esters and ketones with α-alkoxy aldehydes follows Anh-Felkin control, if the alkoxy substituent is assumed to be the "large" group.[59] This former mode of addition is also observed with other organometallic reagents.[60] The use of triisopropylsilyl has been advocated for the protection of an α- or β-hydroxy group when chelation controlled addition is not desired.[52,58,61,62]

The use of a Grignard reagent with a β-alkoxy aldehyde does not, however, result in high induction. This situation can be alleviated by use of a cuprate, although chemical yields may be diminished. A β-substituent, when an α-substituent is also present, can have a marked effect on selectivity.[63] An excellent illustration is provided by the cuprate addition to 2,3-*O,O*-isopropylideneglyceraldehyde (**2.3**) (Scheme 2.7).[64-67]

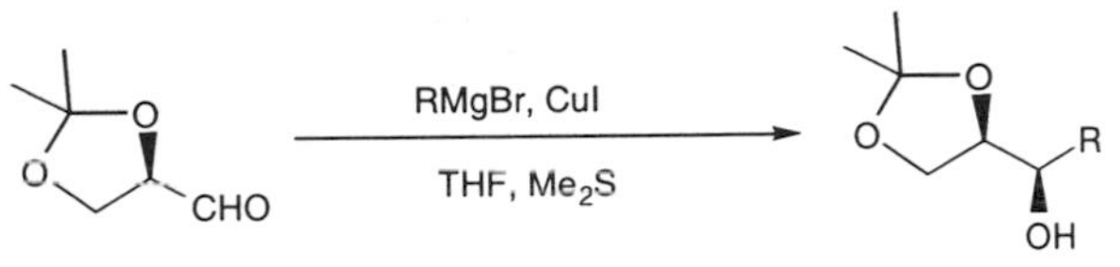

RMgBr, CuI

THF, $Me_2S$

**2.3**

(ds >10:1)

**SCHEME 2.7.**

The selective cuprate reaction with β-alkoxy aldehydes does not translate to α-alkoxy aldehydes.[63] An alternative procedure to effect chelation control with a β-alkoxy aldehyde is to use an organotitanium reagent.[68]

Another example of chelation controlled addition is given in Scheme 2.8.[69,70]

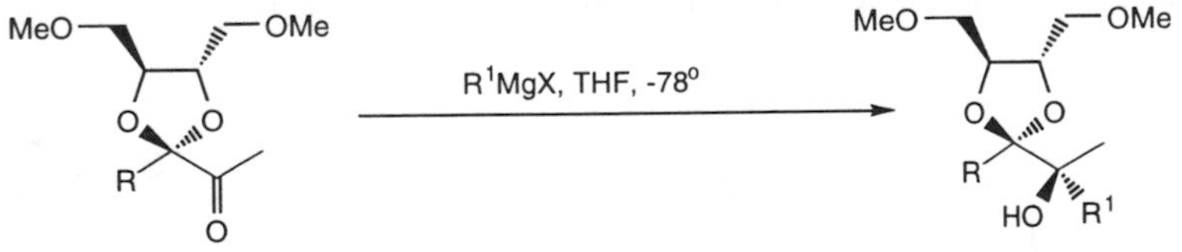

$R^1$MgX, THF, -78°

**SCHEME 2.8.**

When oxygen groups are present at the C-2 and C-3 positions, the use of protection to allow chelate formation to a specific oxygen atom can allow for good selection (Scheme 2.9). The *syncat* isomer† was the desired product, as shown by the approach model (Figure 2.6), which requires chelate formation with the oxygen at C-2; chelate formation with the C-3 oxygen atom would have given the *ancat* isomer (Figure 2.7). The C-3 oxygen was blocked with a large group, and good selection was observed (Scheme 2.9).[71]

† With respect to the oxygen functionality.

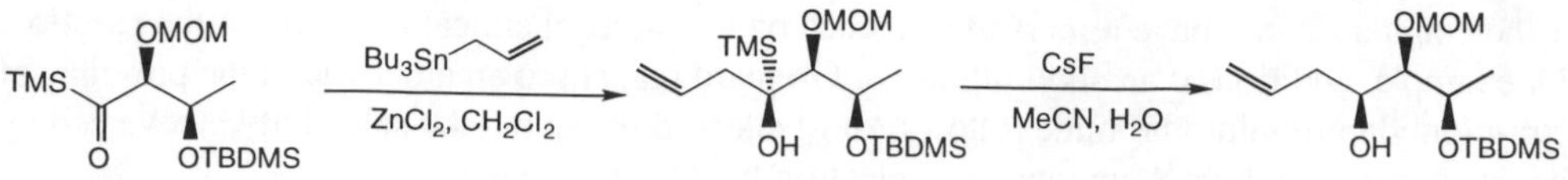

SCHEME 2.9.

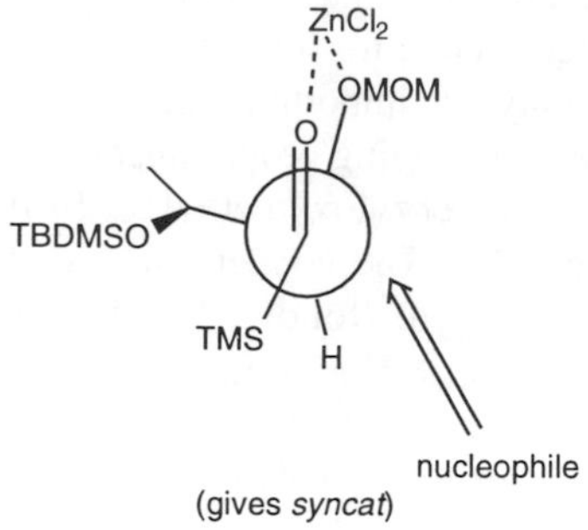

FIGURE 2.6. Approach required for formation of the *syncat* product.

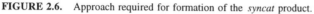

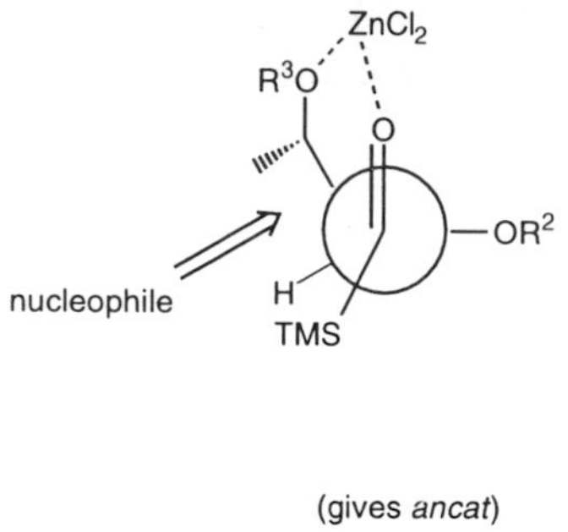

FIGURE 2.7. Approach required for formation of the *ancat* product.

The use of either Anh-Felkin or chelation controlled addition allows for the regulation of the product stereochemistry as these two modes of addition usually give rise to different diastereoisomers. The stereochemistry of the product can also be controlled by which group is nucleophilic and which electrophilic (Schemes 2.10 and 2.11).[57]

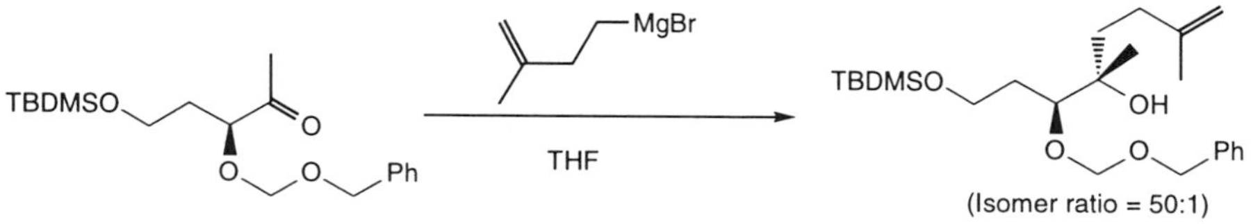

SCHEME 2.10.

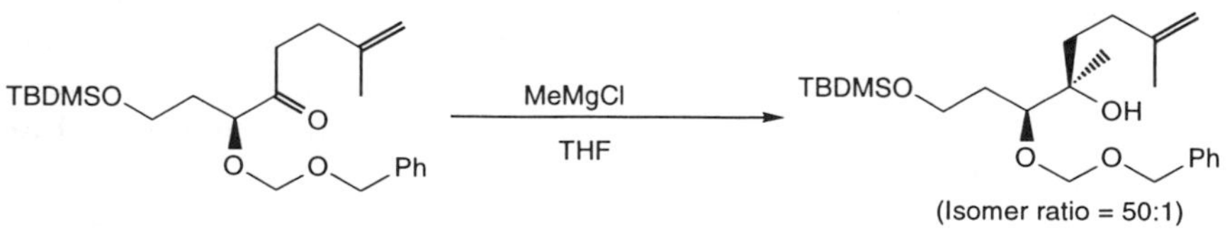

SCHEME 2.11.

The philosophy of chelation control was employed to control the aldol reactions [Chapter 7] in Still's synthesis of monensin.[72-74] Condensation of the kinetic enolate (formed from the ketone **2.4**) with the aldehyde **2.5** in the presence of magnesium bromide afforded a 5:1 mixture of the diastereomeric adducts in favor of the chelation controlled product. Subsequent transformations of the adduct, **2.6**, and removal of the minor diastereoisomer by chromatography provided the aldehyde **2.7**. This aldehyde **2.7** was subjected to a second aldol reaction, but as Anh-Felkin control was required, the branched nature of C-3 had to override the ability of the methoxy group to form a chelate. This control was achieved by the use of *cis*-2-butenyldiethylaluminum as a propanal enolate equivalent, and yielded a 3:1 diastereoisomeric mixture of the lactone **2.8** (Scheme 2.12).[73]

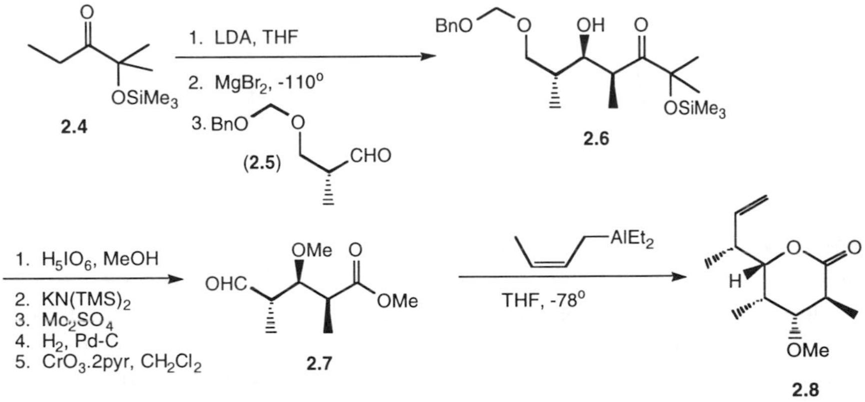

SCHEME 2.12.

Addition of a stannyl lithium reagent to an aldehyde can be made diastereoselective when a group is present that can form a complex with the metal counterion (Scheme 2.13).[64,75]

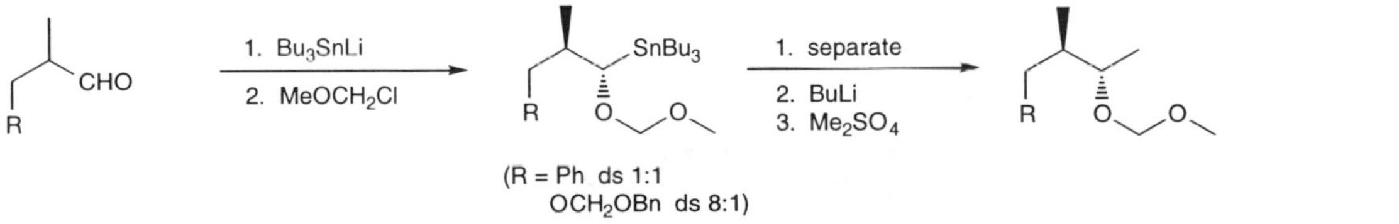

SCHEME 2.13.

The addition of a nucleophilic reagent to a carbonyl group is influenced by many variables, many of which can be determined through the reaction conditions and reagents. Although the interplay between these variables is not completely understood, the rules described above can be used in a predictive manner. For example, the size of a reducing agent for a ketone can have an effect on the facial selectivity [*4.5*].[76]

### 2.1.3. DIPOLAR ADDITION

The addition of a nucleophile to an α-haloketone, often referred to as the "dipole model,"[6] is depicted in Figure 2.8.[15] The approach has been modified from the original proposal to be consistent with the favored trajectory of the approaching nucleophile, although the overall product stereochemistry remains the same (Scheme 2.14).

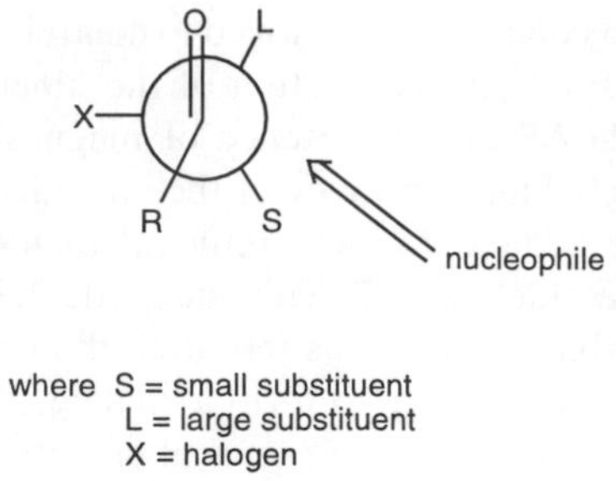

O
L
X
R
S
nucleophile
where S = small substituent
L = large substituent
X = halogen

**FIGURE 2.8.** Approach vector for dipolar addition.

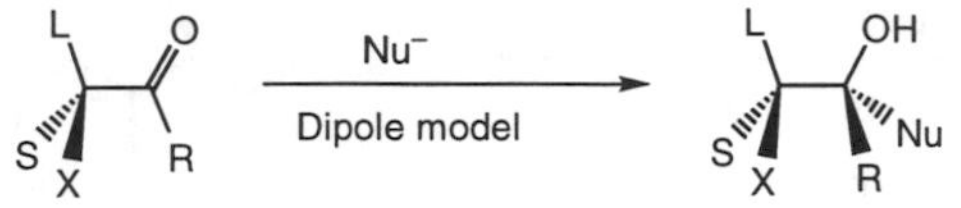

L
O
S
X
R
Nu⁻
Dipole model
OH
Nu

**SCHEME 2.14.**

## 2.1.4. PRELOG'S RULE

A variant of these rules is provided for α-keto esters by Prelog's generalization (Figure 2.9).[77-79] However, as with all of these generalizations, care must be exercised in their application as "exceptions" do exist, particularly when the models are pushed to their limits, as other stereoelectronic effects can play an important role in the determination of a reaction's outcome [*7.2.7*].[80] The chiral center in this type of reaction is remote from the reaction site, and this allows for other factors to dominate [*6.1.1 and 6.2.1*].

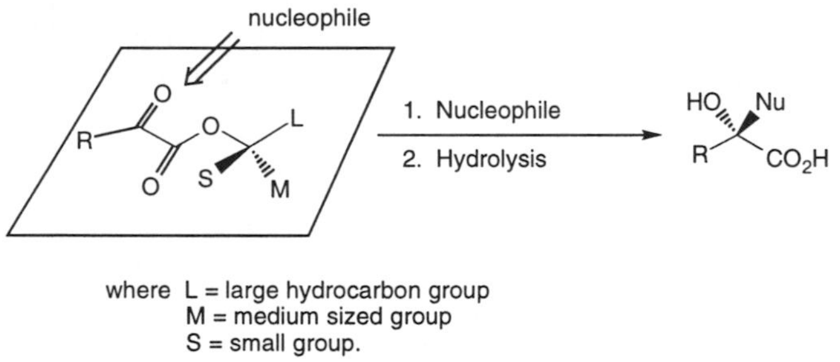

nucleophile
R
O
O
O
L
S
M
1. Nucleophile
2. Hydrolysis
HO
Nu
R
CO2H
where L = large hydrocarbon group
M = medium sized group
S = small group.

**FIGURE 2.9.** Prelog's rule for additions to α-keto esters.

An example of the addition under Prelog's rule is provided by reactions of 8-phenylmenthyl esters (Scheme 2.15).[81-83]

O
O
O
Ph
RMgX
OH
R
62-90%
(ee 90-99%)

**SCHEME 2.15.**

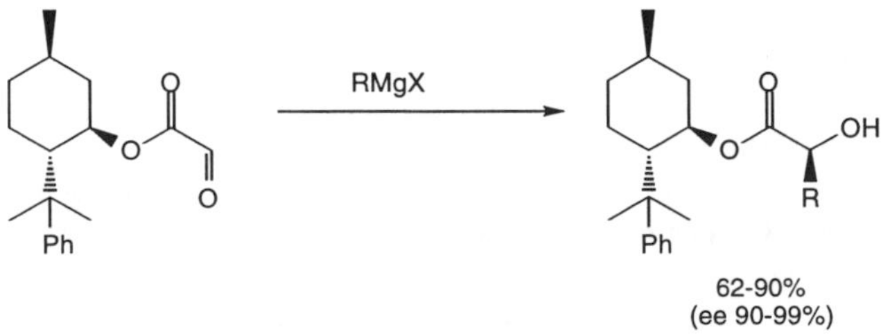

However, additions to the protected carboxylic acid compound (**2.9**) gave variable degrees of induction (Scheme 2.16).[84]

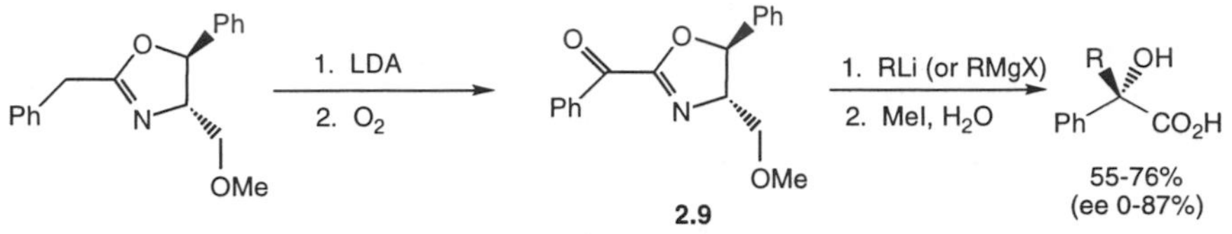

**SCHEME 2.16.**

## 2.2. HOUK'S RULE

A transformation relying on the conversion of an $sp^2$ center to $sp^3$, derives its asymmetric induction from facial selection during the reaction. This can be achieved by the reactive species being "guided" to a particular face of the unsaturated center through complex formation with an appendant functional group in the substrate. The Sharpless epoxidation is an excellent example of this type of approach [*11.1.2*]. The alternative method is to use the steric demands of groups within the substrate. Indeed, high selectivity can still be observed without prior formation of a complex between the reactant and substrate. An example of such a reaction is provided by the hydroboration of an alkene with a chiral borane [*8.2.2.3*]. The induction achieved in these latter cases has been rationalized in terms of the preferred angle of attack of the reagent on the site of unsaturation.¶ The models are summarized in Figure 2.10.

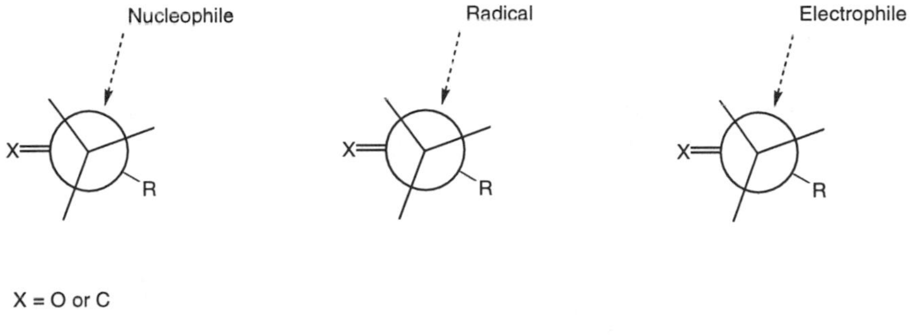

**FIGURE 2.10.** Preferred angles of attack.

These theoretical generalizations were introduced by Houk, and are now named after him.[12,19,85,86] In many respects, Houk's rule incorporates other models and rules that have been proposed for specific types of reactions. However, as an illustration, it becomes apparent that when the reagent is assumed to attack from the opposite face to a large substituent, and for X = O — that is, the substrate is a carbonyl compound — Houk's rule encompasses the Anh-Felkin model for this type of reaction (Figure 2.11).

In a similar manner, hydroboration, which uses an electrophilic boron species, changes Houk's generalization to the empirical rule formulated by Kishi [*11.4.1*].[85] Indeed the power of Houk's rule lies in the way it rationalizes the stereoselection of a wide range of reaction types.[87] Consider the oxidation of the allylsilane (**2.10**) [*9.4.1*]: the reagent is delivered from the opposite face to the large group. In this example, the silyl group is the large group(Figure 2.12). However, notice that the R group is next to the alkene hydrogen — a reversal of the small and medium groups depicted in the Anh-Felkin example of Figure 2.11. This seeming reversal is

¶ Allylic strain also provides an explanation for the observed selectivity. This is discussed in Section 2.3.

necessary to minimize the total steric interactions within the system. The differences in steric requirements between an alkene and the oxygen atom of a carbonyl group are very apparent between these two examples.

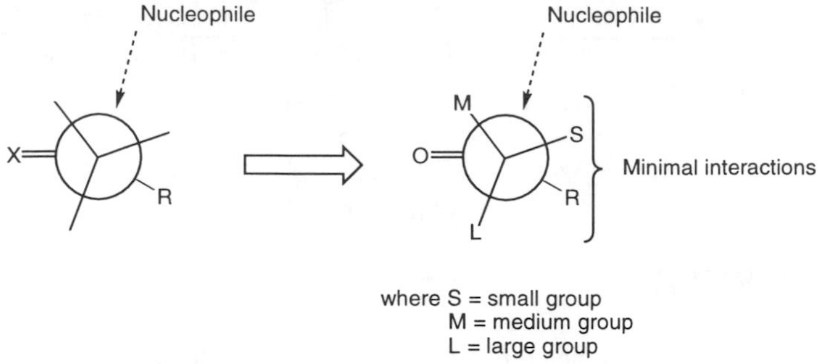

**FIGURE 2.11.** Relationship between Houk's rule and Anh-Felkin addition.

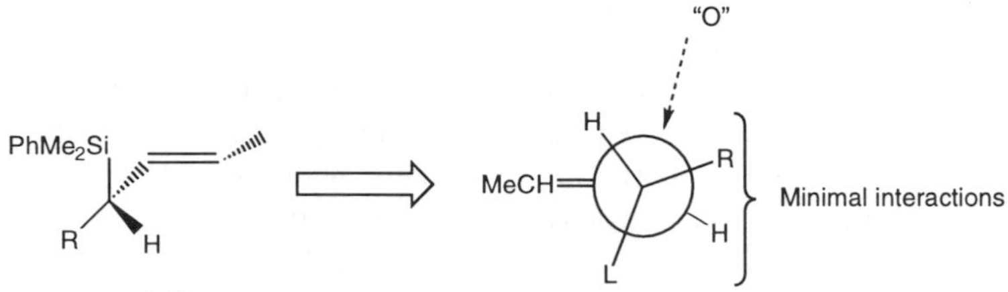

**FIGURE 2.12.** Preferred approach of oxidant to an allylsilane.

This model explains the face selectivity seen in two related reactions (Scheme 2.17).[88] The stereoselection can, however, be low for R = methyl.

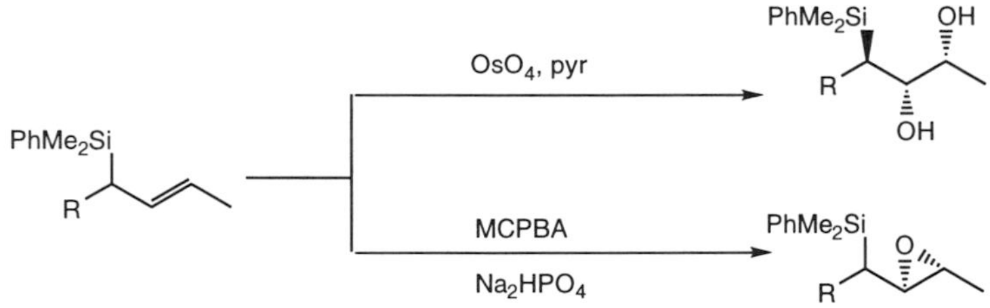

**SCHEME 2.17.**

The conjugate addition of thiophenol to the chiral α,β-unsaturated ester (**2.11**) provides a mixture of the diastereomeric β-thioesters (**2.12**). As this is a base-catalyzed addition, the intermediate is the ester enolate that has to be protonated to provide the products shown in Scheme 2.18.

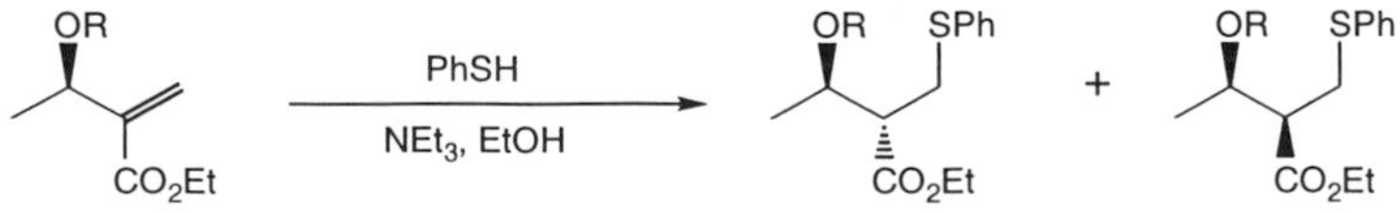

**SCHEME 2.18.**

From Houk's rule, the proton should be delivered preferentially on the face that has the least steric hindrance (Figure 2.13). Thus, as the size of the R group increases, the selectivity should reflect these changes. This is the case, as shown in Table 2.1.[89]

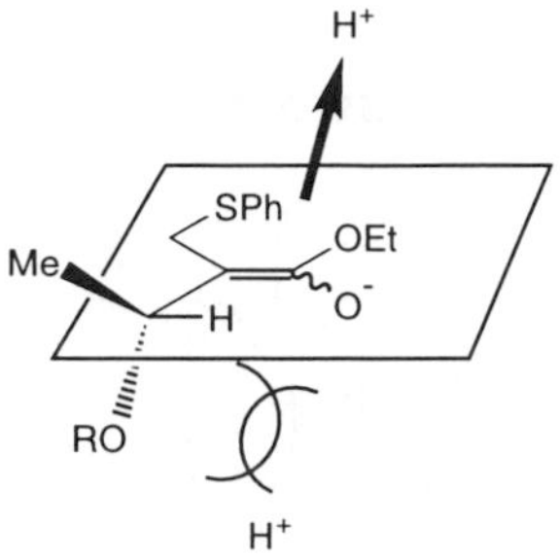

**FIGURE 2.13.** Face selectivity for ester enolate protonation.

**TABLE 2.1**
**Face Selectivity for Addition of Thiophenol to Conjugated Ester (2.11)**

| R = | % Yield of (2.12) | *ancat* Isomer | *syncat* |
|---|---|---|---|
| H | 100 | 50 | 50 |
| $SiMe_2Bu$-*t* | 99 | 79 | 21 |
| $SiMe_2Th$ | 97 | 79 | 21 |
| $SiPh_2Bu$-*t* | 96 | 89 | 11 |

This rule has been extended to ester and carbonyl enolate alkylations. Thus, the two isomeric ketones (**2.13s** and **2.13a**) are available, as the major isomers can be obtained through two complimentary approaches (Scheme 2.19).[90,91]

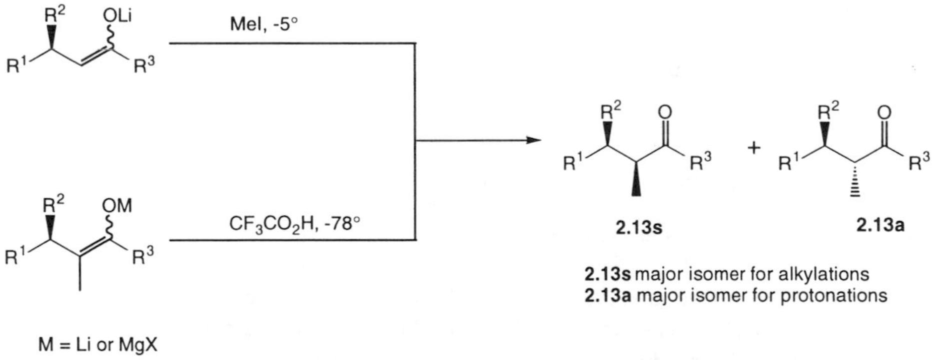

**2.13s** major isomer for alkylations
**2.13a** major isomer for protonations

M = Li or MgX

SCHEME 2.19.

In contrast, addition of the dithioacetate **2.14** to α,β-unsaturated ketones gives the *syncat* isomer as the predominant product; this would seem to be contrary to Houk's rule (Scheme 2.20). However, the intermediate enolate, **2.15**, can undergo a rapid intramolecular proton transfer (Figure 2.14). This is borne out by reaction of the enolate with methyl iodide;*S*-methylation is observed, rather than *C*-alkylation.[92]

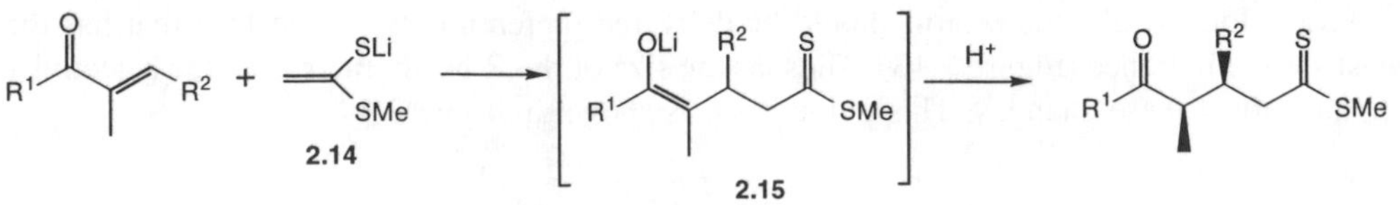

SCHEME 2.20.

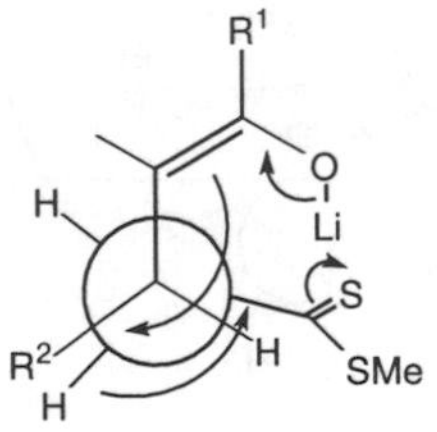

FIGURE 2.14. Intramolecular proton transfer mechanism.

As with Anh-Felkin addition, and, as we shall see for many reactions such as the aldol reaction, the size of the substituents can still be problematic in terms of the assignment of steric effects. While A-values[93] and "effective radius"[94] do provide some quantitation, anomalies can still be apparent.[90]

## 2.3. ALLYLIC STRAIN

In sterically confined systems, such as those derived from cyclic systems or containing a bond with restricted rotation, such as an alkene, substituents can interact with one another in unfavorable ways. Consider the simple example of 3-methylbut-1-ene (2.16). There are three major rotamers for the molecule, one where the terminalZ-hydrogen has a 1,3-interaction with only another hydrogen. A slight increase in energy is seen when the olefinic hydrogen interacts with a single methyl group, while a much larger increase is observed for interaction with the two methyl groups (Figure 2.15).[95]

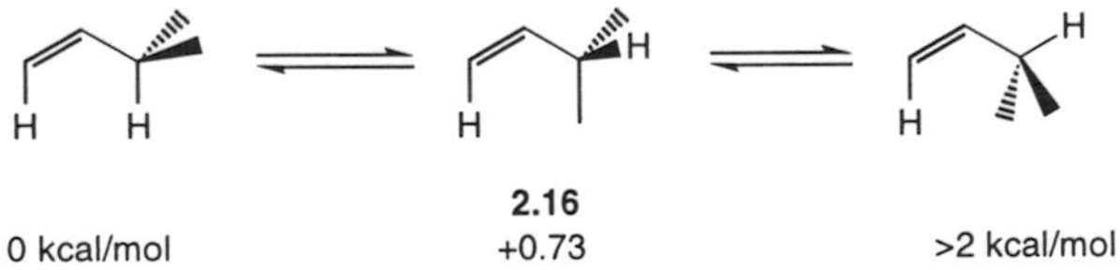

FIGURE 2.15. Relative energies of alkene rotamers.

It can be seen, therefore, that the rotamer with minimized allylic (1,3-) interactions will be energetically favored.[95] This concept has been used to rationalize a number of reactions, and it can also be used in a predictive manner.[96] For the β-substituted system (**2.17**), the allylic interactions can determine on which face a reaction takes place (Figure 2.16).[95]

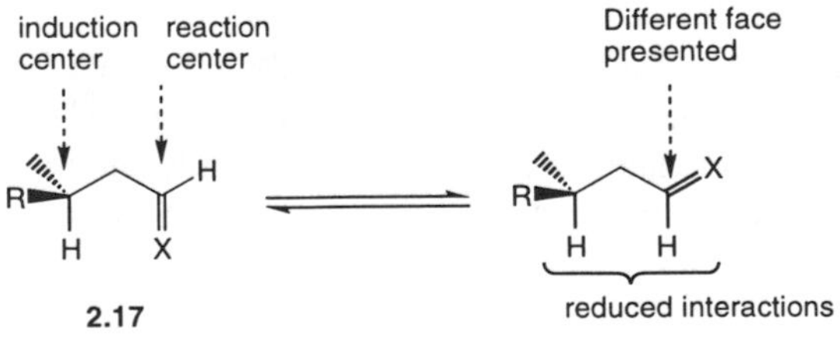

FIGURE 2.16. Face selectivity from allylic strain.

When the R group is large, attack from one face can be effectively blocked (Figure 2.17).[95]

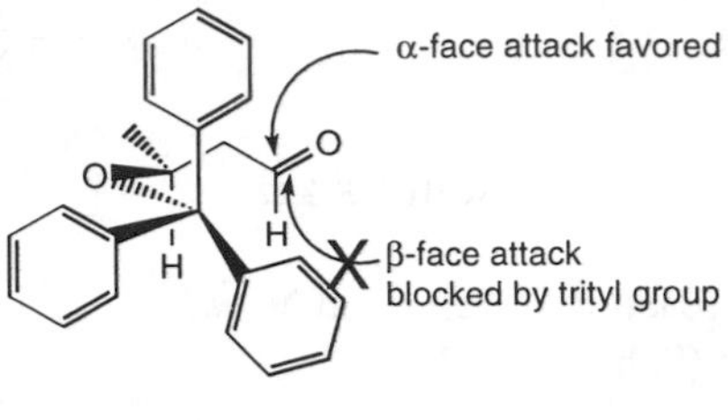

**FIGURE 2.17.** Face selectivity of a β-alkoxy aldehyde.

If β-attack is desired then chelation control, through removal of the trityl group can allow complex formation with the uncovered hydroxy group and allow attack from this face.

In a Claisen rearrangement approach with a cyclopentanoid system (Scheme 2.21), the vinyl ether reacts on the sterically more crowded face [*12.2*]. This observation can be rationalized in terms of the increased allylic strain of the system which would be necessary for a reaction to occur on the α-face.[97]

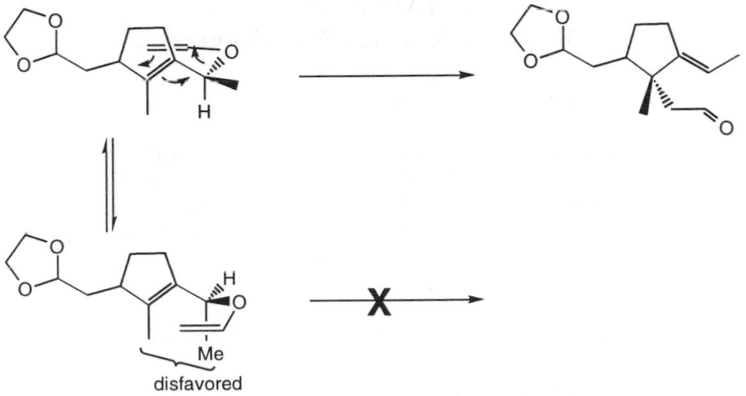

**SCHEME 2.21.**

## 2.4. BALDWIN'S RULES

Although these rules were developed to determine whether a cyclization will be favorable, they are founded on the angle of attack of a reagent at a site with defined geometry.[98-101] In many respects, therefore, they are similar to those already discussed for the addition of a nucleophile to a carbonyl center. However, due to the limitations of hybridization geometries and the ways in which certain elements react, especially if they contain d-orbitals, the rules are limited to the first row of the periodic table. Second row elements with alternative reaction pathways can seem to disobey them. The rules are also limited to the formation of medium or small rings; large rings (less than eight-membered) have enough conformational freedom to allow reactions similar to acyclic systems. In addition, exceptions are known; some of these are discussed below. Some nomenclature has to be defined to discuss these rules: the ring closure is defined as *exo* if the bond being broken during the cyclization step is outside of the resultant ring (Scheme 2.22).

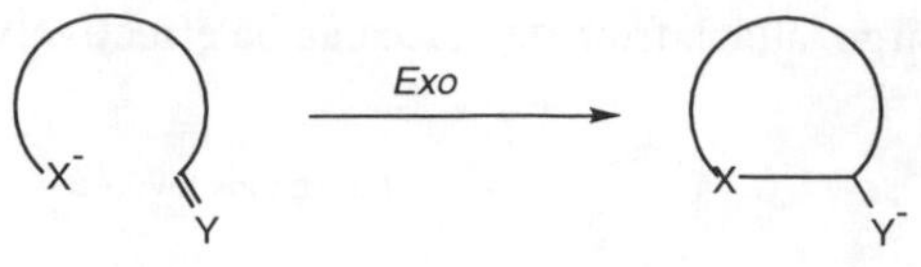

SCHEME 2.22.

In a similar manner, the reaction is described as *endo* if the bond broken during the ring formation is within the ring (Scheme 2.23).

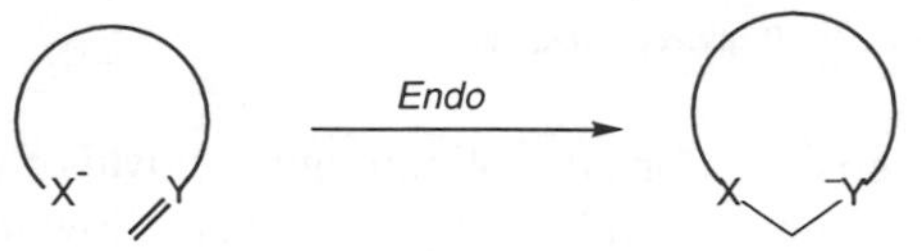

SCHEME 2.23.

The rules are summarized in Table 2.2.[99]

**TABLE 2.2**
**Baldwin's Rules for Ring Closures**

| System | Reaction Type | |
|---|---|---|
| Tetrahedral | 3- to 7-*Exo-Tet* | Favored |
| | 5- and 6-*Endo-Tet* | Disfavored |
| Trigonal | 3- to 7-*Exo-Trig* | Favored |
| | 3- to 5-*Endo-Trig* | Disfavored |
| | 6- and 7-*Endo-Trig* | Favored |
| Digonal | 3-and 4-*Exo-Dig* | Disfavored |
| | 5- to 7-*Exo-Dig* | Favored |
| | 3- to 7-*Endo-Dig* | Favored |

As an example, the cyclization of the lactone **2.18** can proceed along two possible pathways: the first, path a, is 5-*exo-trig* and favored (Scheme 2.24). This involves a direct attack of the alkoxide on the ester group. The alternative, path b, is 5-*endo-trig*, and, thus, is disfavored and the potential product **2.19** is not observed. The *exo*-methylene lactone, **2.20**, is very susceptible to Michael addition as illustrated by the addition of methoxide.[102]

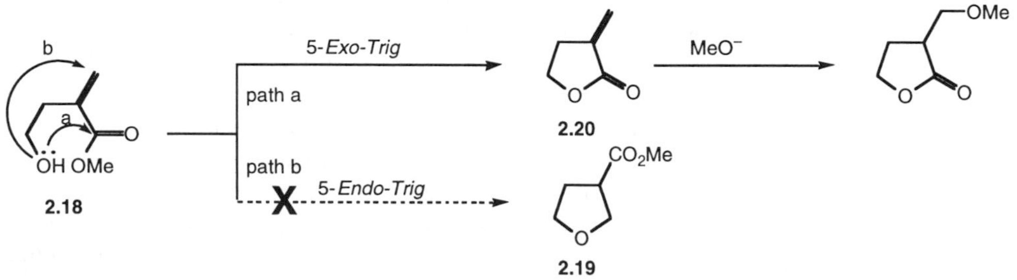

SCHEME 2.24.

To illustrate an extension of these rules, consider the addition of the amide to a stilbene derivative (Scheme 2.25).[103] The nitrogen has two possible reaction pathways; the bond that is being broken is clearly either *exo* (path a) or *endo* (path b).

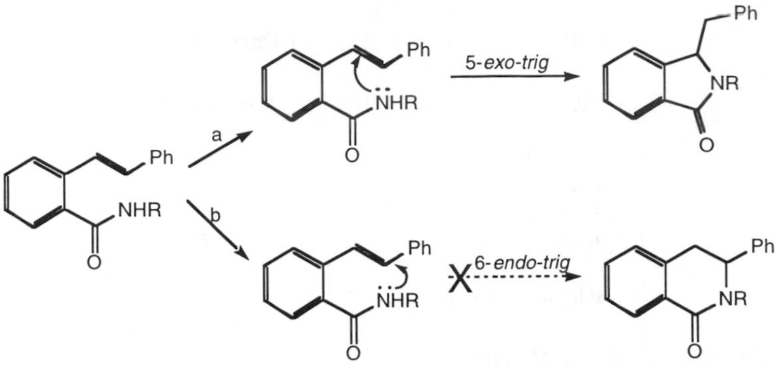

SCHEME 2.25.

An *exo*-ring closure is usually faster than a competing *endo* pathway, as the transition state for the former is easier to accommodate. This accounts for the formation of the five-membered product, although both pathways are "favored." Thus, the smaller ring is usually formed when *exo*- and *endo*- openings compete.[104] The hydroxy ester (**2.21**) can cyclize by either a 6 *exo-trig* or a 6-*endo-trig* pathway. Again, from the rules, both of these are favored. However, the *exo* product is formed preferentially because of the rate difference (Scheme 2.26)![105]

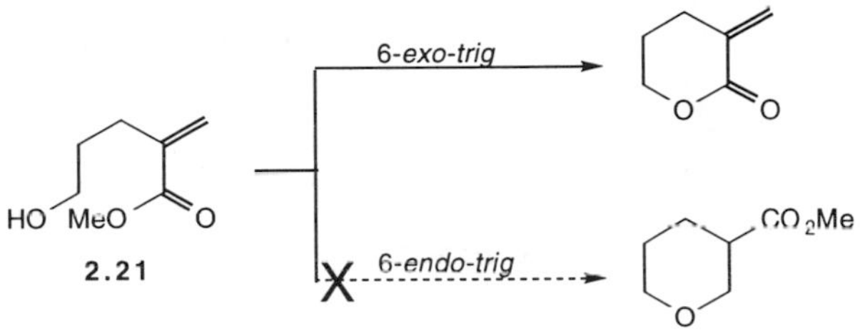

SCHEME 2.26.

Unfortunately, Baldwin's rules only provide basic geometric guidelines. Often, other factors, especially electronic ones, have to be considered![106] For the opening of the hydroxy epoxide (**2.22**) formation of a six- or seven-membered ring could result (Scheme 2.27).

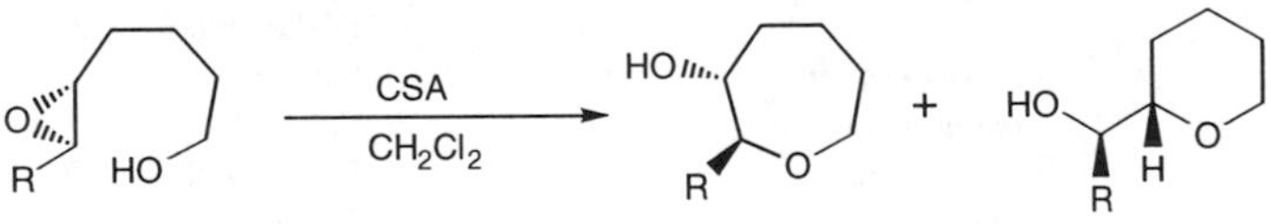

SCHEME 2.27.

The epoxide substituent, R, plays a major role to determine the reaction outcome as illustrated in Table 2.3. An electron-donating group provides the seven-membered ring as the major product.[107]

For the cyclizations of carbonyl stabilized anions (vide infra) with epoxides, the carbanion must remain perpendicular to the carbonyl group in the transition state; this geometrical constraint determines the resultant ring size![108]

**TABLE 2.3**
**Effect of R on Cyclization Pathway**

| R = | Yield (%) | Ratio 2.23:2.24 |
|---|---|---|
| $CH_2CH_2CO_2Me$ | 70 | 0:100[a] |
| $CH{=}CH_2$ | 75 | 82:18 |
| *Z*-CH=CHCl | 70 | 60:40 |
| *E*-CH=CHCl | 75 | 92:8 |

[a] Isolated as the lactone.

### 2.4.1. CARBONYL COMPOUNDS

The rules have been extended to include enols and enolates. The double bond of the enol is then defined as being either *exo* or *endo* (Scheme 2.28).

SCHEME 2.28.

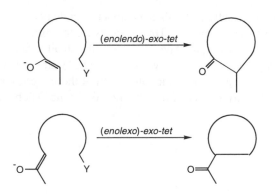

The revised rules for these systems are summarized in Table 2.4.[109,110]

**TABLE 2.4**
**Baldwin's Rules for Enol Systems**

| Reaction Type | Ring Size | |
|---|---|---|
| *(enolendo)-exo-tet* | 6–7 | Favored |
| | 3–5 | Disfavored |
| *(enolexo)-exo-tet* | 3–7 | Favored |
| *(enolendo)-exo-trig* | 6–7 | Favored |
| | 3–5 | Disfavored |
| *(enolexo)-exo-trig* | 3–7 | Favored |

The general philosophy behind these ring closures can be extended even further. For example, through the use of the bifunctional annelation reagents **2.25** (Scheme 2.29) it was experimentally determined that an (allyl *endo*)-*exo-trig* ring closure for ring sizes of 3 to 5 is not favored, but is favored for 6 and 7. The (allyl*endo*)-*exo-trig* is favored over the alternative, allowed, (allyl *exo*)-*exo-trig* pathway for three- to seven-membered rings.[109,111]

## 2.5. PERICYCLIC TRANSITION STATES

The most common transition state for a pericyclic reaction is six-membered, although five-membered transition states may also be used [Chapter 12]. In most cases, the

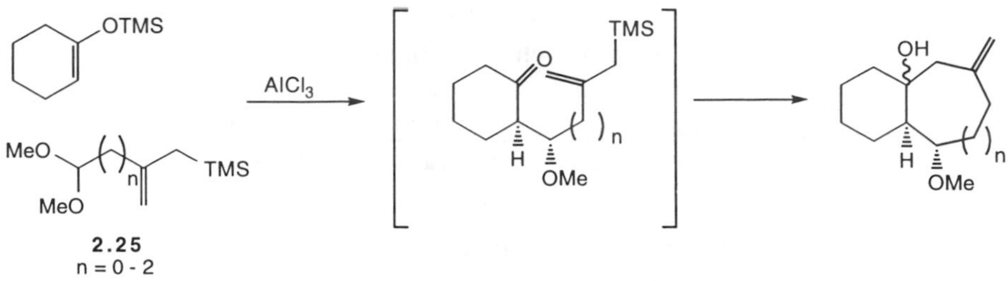

SCHEME 2.29.

six-membered transition state is in the chair form, with the preferred conformation having minimal 1,3-interactions — that is, the large substituents tend to occupy pseudo-equatorial positions.[112-114] In some cases, such as the Claisen rearrangement [*12.2*], the accommodation of appendant groups around the reaction centers can be enough of a problem to cause the preferred transition state to be based upon a boat conformation. These subtle differences are again discussed in detail in Chapter 12.

## 2.5.1. WOODWARD–HOFFMANN RULES

Pericyclic reactions are apparently concerted although not necessarily synchronous![115,116] The transition states are cyclic, and the electron flow can occur in either direction. We will discuss examples [Chapter 12] of the three major types: electrocyclic, cycloaddition, and sigmatropic rearrangement reactions. As an ordered transition state occurs for pericyclic reactions, a set of rules has been drawn up to predict the stereochemical outcome of a reaction as well as the conditions under which it will occur. These rules, known as the Woodward–Hoffmann rules, are summarized in Tables 2.5 to 2.7![117]

## 2.5.2. STEREOCHEMICAL CONSEQUENCES

For reactions proceeding through cyclic transition states, the size of the various substituents can play a pivotal role in the determination of the stereochemical outcome of the reaction. Large, bulky groups prefer to occupy pseudo-equatorial positions. The determination of the relative sizes of these substituents [*2.2*] is the problematic step to predict a reaction's outcome. This aspect is also discussed in Chapter 12. As this type of six-membered transition state is quite common, examples will also be found for other reactions, such as the addition of an allylborane to a carbonyl compound [*4.1.5.1*].

In addition to the Woodward–Hoffmann rules that describe first-order orbital interactions that are the primary source of stereoselection, secondary interactions can have a marked influence on stereochemistry as illustrated by the Alder rule that predicts that the*endo*-isomer is the major product through symmetry allowed mixing of orbitals at secondary reaction centers (Scheme 2.30).[117]

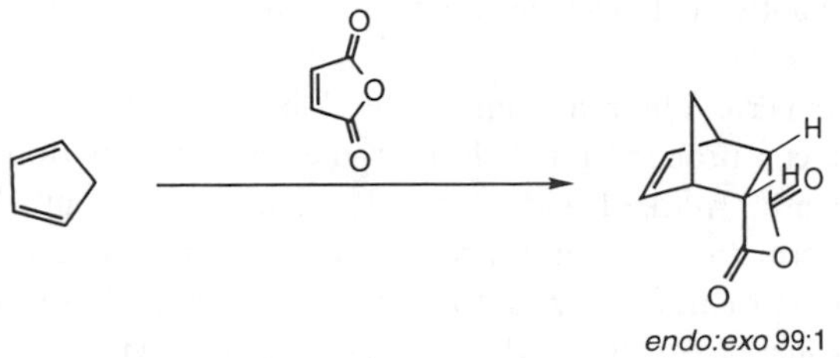

SCHEME 2.30.

**TABLE 2.5**
**Electrocyclic Reactions**

| Number of Electrons in Cyclic Transition State | Allowed, or Preferred Mode | |
|---|---|---|
| | Thermal | Photochemical |
| 4n | Conrotatory | Disrotatory |
| 4n + 2 | Disrotatory | Conrotatory |

**TABLE 2.6**
**Cycloaddition Reactions**

| Number of Electrons in Cyclic Transition State | Allowed, or Preferred Mode | |
|---|---|---|
| | Thermal | Photochemical |
| 4n | Antarafacial-suprafacial | Suprafacial-suprafacial |
| 4n+2 | Suprafacial-suprafacial | Antarafacial-suprafacial |

**TABLE 2.7**
**Sigmatropic Reactions**

| | Allowed, or Preferred Mode Hydrogen Migration with | | | |
|---|---|---|---|---|
| | Retention | | Inversion | |
| Number of Electrons | Thermal | Photochemical | Thermal | Photochemical |
| 4n | Antarafacial | Suprafacial | Suprafacial | Antarafacial |
| 4n+2 | Suprafacial | Antarafacial | Antarafacial | Suprafacial |

Electronic effects can also have significance as observed for the regiochemical influences of electron rich substituents on dienes in the Diels–Alder reaction.[118,119] As first- and second-order interactions still do not account for all examples, more subtle effects, such as polar group influences, have been proposed to rationalize some cycloaddition reactions [*12.1*].[120-122]

## 2.6. SUMMARY

This chapter describes some general rules and principles that, with care, can be used to predict the stereochemical outcome of a wide variety of reactions, but especially additions to $sp^2$ centers. Examples and extensions of Houk's rule can be seen in the chapters covering the reactions of carbonyl compounds and alkenes, as this rule rationalizes a number of more specific models.

The concept of allylic strain describes much of cyclic chemistry and ring conformational analysis, but the resultant effects can be more subtle in acyclic cases due to the removal of rotational restrictions.

Many models, from pericyclic reactions to the aldol reaction to those of allyl anions with carbonyl compounds, can proceed through a six-membered transition state. In these cases, substitution effects can be marked with sterically demanding groups having a tendency to prefer equatorial conformations to minimize allylic (1,3-) interactions.

As our understanding of the many factors that influence and control the stereochemical outcome of reactions increases, the models described above will continue to be refined. As conflicts can occur within some substrates, as for example, the addition to a carbonyl group within a rigid polycyclic system, the effects of Anh-Felkin or chelation controlled addition can be overwhelmed by the steric demands of the parent ring systems. With the potential for "exceptions," all of these rules should be applied with caution until experimental verification is obtained.

## 2.7. REFERENCES


1. Wipke, W. T.; Gund, P. *J. Am. Chem. Soc.* 1976, *98*, 8107.
2. Eliel, E. L. In *Asymmetric Synthesis*; J. D. Morrison, Ed.; Academic Press: Orlando, 1983; Vol. 2; pp 125.
3. Solladie, G. In *Asymmetric Synthesis*; J. D. Morrison, Ed.; Academic Press: Orlando, 1983; Vol. 2; pp 157.
4. Martin, S. F. *Synthesis* 1979, 633.
5. Cram, D. J.; Elhafez, F. A. A. *J. Am. Chem. Soc.* 1952, *74*, 5828.
6. Cornforth, J. W.; Cornforth, R. H.; Mathew, K. K. *J. Chem. Soc.* 1959, 112.
7. Leitereg, T. J.; Cram, D. J. *J. Am. Chem. Soc.* 1968, *90*, 4019.
8. Karabatsos, G. J. *J. Am. Chem. Soc.* 1967, *89*, 1367.
9. Karabatsos, G. J.; Zioudrou, C.; Moustakali, I. *Tetrahedron Lett.* 1972, 5289.
10. Chérest, M.; Felkin, H.; Prudent, N. *Tetrahedron Lett.* 1968, 2199.
11. Kruger, D.; Sopchik, A. E.; Kingsbury, C. A. *J. Org. Chem.* 1984, *49*, 778.
12. Houk, K. N.; Padden-Row, M. N.; Rondan, N. G.; Wu, Y.-D.; Brown, F. K.; Spellmeyer, D. C.; Metz, J. T.; Li, Y.; Loncharich, R. J. *Science* 1986, *231*, 1108.
13. Frenking, G.; Köhler, K. F.; Reetz, M. T. *Tetrahedron* 1993, *49*, 3983.
14. Frenking, G.; Köhler, K. F.; Reetz, M. T. *Tetrahedron* 1993, *49*, 3971.
15. Anh, N. T.; Eisenstein, O. *Nouv. J. Chim.* 1977, *1*, 61.
16. Bürgi, H. B.; Dunitz, J. D.; Lehn, J. M.; Wipff, G. *Tetrahedron* 1974, *30*, 1563.
17. Bürgi, H. B.; Lehn, J. M.; Wipff, G. *J. Am. Chem. Soc.* 1974, *96*, 1956.
18. Baldwin, J. E. *J. Chem. Soc., Chem. Commun.* 1976, 738.
19. Paddon-Row, M. N.; Rondan, N. G.; Houk, K. N. *J. Am. Chem. Soc.* 1982, *104*, 7162.
20. Anh, N. T.; Eisenstein, O. *Tetrahedron Lett.* 1976, 155.
21. Chérest, M.; Felkin, H. *Tetrahedron Lett.* 1968, 2205.
22. Huet, J.; Maroni-Barnaud, Y.; Anh, N. T.; Seyden-Penne, J. *Tetrahedron Lett.* 1976, 159.
23. Bartlett, P. A. *Tetrahedron* 1980, *36*, 3.
24. Anh, N. T. *Top. Curr. Chem.* 1980, *88*, 145.
25. Komatsukai, T.; Sakakibara, K.; Hirota, M. *Tetrahedron Lett.* 1989, *30*, 3309.
26. Wipke, W. T.; Gund, P. *J. Am. Chem. Soc.* 1974, *96*, 299.
27. Wong, S. S.; Paddon-Row, M. N. *J. Chem. Soc., Chem. Commun.* 1990, 456.
28. Ho, T.-L. *Tetrahedron* 1985, *41*, 3.
29. Ashby, E. C.; Boone, J. R. *J. Am. Chem. Soc.* 1976, *98*, 5524.
30. Noyce, D. S.; Denny, D. B. *J. Am. Chem. Soc.* 1950, *72*, 5743.
31. Ashby, E. C.; Laemmle, J. T. *Chem. Rev.* 1975, *75*, 521.
32. Chérest, M.; Felkin, H.; Tacheau, P.; Jacques, J.; Varech, D. *J. Chem. Soc., Chem. Commun.* 1977, 372.
33. Boone, J. R.; Ashby, E. C. *Top. Stereochem.* 1979, *11*, 53.
34. Wigfield, D. C. *Tetrahedron* 1979, *35*, 449.
35. Mukherjee, D.; Wu, Y.-D.; Fronczek, F. R.; Houk, K. N. *J. Am. Chem. Soc.* 1988, *110*, 3328.
36. Nakada, M.; Urano, Y.; Kobayashi, S.; Ohno, M. *J. Am. Chem. Soc.* 1988, *110*, 4826.
37. Hwu, J. R.; Wang, N. *Chem. Rev.* 1989, *89*, 1599.
38. Buynak, J. D.; Strickland, J. B.; Hurd, T.; Phan, A. *J. Chem. Soc., Chem. Commun.* 1989, 89.
39. Eliel, E. L.; Morris-Natschke, S. *J. Am. Chem. Soc.* 1984, *106*, 2937.
40. Eliel, E. L.; Soai, K. *Tetrahedron Lett.* 1981, *22*, 2859.
41. Eliel, E. L.; Lynch, J. E. *Tetrahedron Lett.* 1981, *22*, 2855.
42. Eliel, E. L.; Koskimies, J. K.; Lohri, B. *J. Am. Chem. Soc.* 1978, *100*, 1614.
43. Lynch, J. E.; Eliel, E. L. *J. Am. Chem. Soc.* 1984, *106*, 2943.
44. Kogure, T.; Eliel, E. L. *J. Org. Chem.* 1984, *49*, 576.
45. Kaulen, J. *Angew. Chem., Int. Ed. Engl.* 1989, *28*, 462.
46. Eliel, E. L.; Bai, X.; Abdel-Magid, A. F.; Hutchins, R. O. *J. Org. Chem.* 1990, *55*, 4951.
47. Wei, J.; Hutchins, R. O.; Prol, J. *J. Org. Chem.* 1993, *58*, 2920.
48. Utimoto, K.; Nakamura, A.; Matsubara, S. *J. Am. Chem. Soc.* 1990, *112*, 8189.
49. Cram, D. J.; Kopecky, K. R. *J. Am. Chem. Soc.* 1959, *81*, 2748.

50. Cram, D. J.; Wilson, D. R. *J. Am. Chem. Soc.* 1963, *85*, 1245.
51. Mulzer, J.; Steffen, U.; Zorn, L.; Schneider, C.; Weinhold, E.; Munch, W.; Rudert, R.; Luger, P.; Harth, H. *J. Am. Chem. Soc.* 1988, *110*, 4640.
52. Chen, X.; Hortelano, E. R.; Eliel, E. L.; Frye, S. V. *J. Am. Chem. Soc.* 1990, *112*, 6130.
53. Yamamoto, Y.; Maryama, K. *J. Am. Chem. Soc.* 1985, *107*, 6411.
54. Yamamoto, Y.; Matsuoka, K. *J. Chem. Soc., Chem. Commun.* 1987, 923.
55. Yamamoto, Y.; Matsuoka, K.; Nemoto, H. *J. Am. Chem. Soc.* 1988, *110*, 4475.
56. Reetz, M. T.; Raguse, B.; Seltz, T. *Tetrahedron* 1993, *49*, 8561.
57. Still, W. C.; McDonald, J. H. *Tetrahedron Lett.* 1980, *21*, 1031.
58. Keck, G. E.; Castellino, S. *Tetrahedron Lett.* 1987, *28*, 281.
59. Heathcock, C. H.; Young, S. D.; Hagen, J. P.; Pirrung, M. C.; White, C. T.; Van Derveer, D. *J. Org. Chem.* 1980, *45*, 3846.
60. Mulzer, J.; Angermann, A. *Tetrahedron Lett.* 1983, *24*, 2843.
61. Frye, S. V.; Eliel, E. L. *Tetrahedron Lett.* 1986, *27*, 3223.
62. Kahn, S. D.; Keck, G. E.; Hehre, W. J. *Tetrahedron Lett.* 1987, *28*, 279.
63. Still, W. C.; Schneider, J. A. *Tetrahedron Lett.* 1980, *21*, 1035.
64. Jurczak, J.; Pikul, S.; Bauer, T. *Tetrahedron* 1986, *42*, 447.
65. Sato, F.; Kobayashi, Y.; Takahashi, O.; Chiba, T.; Takeda, Y.; Kusakabe, M. *J. Chem. Soc., Chem. Commun.* 1985, 1636.
66. Metz, P.; Schoop, A. *Tetrahedron* 1993, *49*, 10597.
67. López-Herrera, F. J.; Sarabia-García, F. *Tetrahedron Lett.* 1993, *34*, 3467.
68. Reetz, M. T.; Jung, A. *J. Am. Chem. Soc.* 1983, *105*, 4833.
69. Tamura, Y.; Ko, T.; Kondo, H.; Annoura, H.; Fuji, M.; Takeuchi, R.; Fujioka, H. *Tetrahedron Lett.* 1986, *27*, 2117.
70. Heitz, M. P.; Gellibert, F.; Mioskowski, C. *Tetrahedron Lett.* 1986, *27*, 3859.
71. Cirillo, P. F.; Panek, J. S. *J. Org. Chem.* 1990, *55*, 6071.
72. Collum, D. C.; McDonald, J. H.; Still, W. C. *J. Am. Chem. Soc.* 1980, *102*, 2117.
73. Collum, D. C.; McDonald, J. H.; Still, W. C. *J. Am. Chem. Soc.* 1980, *102*, 2118.
74. Collum, D. C.; McDonald, J. H.; Still, W. C. *J. Am. Chem. Soc.* 1980, *102*, 2120.
75. Still, W. C.; Sreekumar, C. *J. Am. Chem. Soc.* 1980, *102*, 1201.
76. Midland, M. M. In *Asymmetric Synthesis*; J. D. Morrison, Ed.; Academic Press: Orlando, 1983; Vol. 2; pp 45.
77. Prelog, V. *Helv. Chim. Acta* 1953, *36*, 308.
78. Prelog, V. *Bull. Soc. Chim., Fr.* 1956, 987.
79. Weill-Raynal, J.; Mathieu, J. *Bull. Soc. Chim., Fr.* 1969, 115.
80. Brandange, S.; Josephson, S.; Vallen, S. *Acta Chem. Scand.* 1977, *31B*, 179.
81. Whitesell, J. K.; Deyo, D.; Bhattacharya, A. *J. Chem. Soc., Chem. Commun.* 1983, 802.
82. Yamamoto, Y.; Maeda, N.; Maruyama, K. *J. Chem. Soc., Chem. Commun.* 1983, 774.
83. Ojima, I.; Miyazawa, Y.; Kumagai, M. *J. Chem. Soc., Chem. Commun.* 1976, 927.
84. Meyers, A. I.; Slade, J. *J. Org. Chem.* 1980, *45*, 2785.
85. Houk, K. N.; Rondan, N. G.; Wu, Y.-D.; Metz, J. T.; Padden-Row, M. N. *Tetrahedron* 1984, *40*, 2257.
86. Houk, K. N.; R., M. S.; Wu, Y.-D.; Rondan, N. G.; Jager, V.; Schohe, R.; Fronczek, F. R. *J. Am. Chem. Soc.* 1984, *106*, 3880.
87. Fujita, M.; Ishida, M.; Manako, K.; Sato, K.; Ogura, K. *Tetrahedron Lett.* 1993, *34*, 645.
88. Fleming, I.; Sarkar, A. K.; Thomas, A. P. *J. Chem. Soc., Chem. Commun.* 1987, 157.
89. Kita, Y.; Shibata, N.; Miki, T.; Takemura, Y.; Tamura, O. *J. Chem. Soc., Chem. Commun.* 1990, 727.
90. Fleming, I.; Lewis, J. J. *J. Chem. Soc., Chem. Commun.* 1985, 149.
91. McGarvey, G. J.; Williams, J. M. *J. Am. Chem. Soc.* 1985, *107*, 1435.
92. Berrada, S.; Metzner, P. *Tetrahedron Lett.* 1987, *28*, 409.
93. Eliel, E. L.; Allinger, N. L.; Angyal, S. J.; Morrison, G. A. *Conformational Analysis*; John Wiley & Sons: New York, 1965, pp 44.
94. Bott, G.; Field, L. D.; Sternhill, S. *J. Am. Chem. Soc.* 1980, *102*, 5618.
95. Hoffmann, R. W. *Chem. Rev.* 1989, *89*, 1841.
96. Eg. Gung, B. W.; Wolf, M. A.; Ohm, K.; Peat, A. J. *Tetrahedron Lett.* 1993, *34*, 1417.
97. Takahashi, T.; Yamada, H.; Tsuji, J. *Tetrahedron Lett.* 1982, *23*, 233.
98. Baldwin, J. E.; Thomas, R. C.; Kruse, L. I.; Silberman, L. *J. Org. Chem.* 1977, *42*, 3846.
99. Baldwin, J. E. *J. Chem. Soc., Chem. Commun.* 1976, 734.
100. Johnson, C. D. *Acc. Chem. Res.* 1993, *26*, 476.
101. Ogura, K.; Kayano, A.; Fujino, T.; Sumitani, N.; Fujita, M. *Tetrahedron Lett.* 1993, *34*, 8313.
102. Baldwin, J. E.; Cutting, J.; Dupont, W.; Kruse, L.; Silberman, L.; Thomas, R. C. *J. Chem. Soc., Chem. Commun.* 1976, 736.
103. Napolitano, E.; Fiaschi, R.; Marsili, A. *Tetrahedron Lett.* 1983, *24*, 1319.

104. Martin, V. S.; Nunez, M. T.; Ramirez, M. A.; Soler, M. A. *Tetrahedron Lett.* 1990, *31*, 763.
105. Baldwin, J. E.; Reiss, J. A. *J. Chem. Soc., Chem. Commun.* 1977, 77.
106. Nicolaou, K. C.; Prasad, C. V. C.; Somers, P. K.; Hwang, C.-K. *J. Am. Chem. Soc.* 1989, *111*, 5330.
107. Nicolaou, K. C.; Prasad, C. V. C.; Somers, P. K.; Hwang, C.-K. *J. Am. Chem. Soc.* 1989, *111*, 5335.
108. Stork, G.; Kobayashi, Y.; Suzuki, T.; Zhao, K. *J. Am. Chem. Soc.* 1990, *112*, 1661.
109. Baldwin, J. E.; Lusch, M. J. *Tetrahedron* 1982, *38*, 2939.
110. Baldwin, J. E.; Kruse, L. I. *J. Chem. Soc., Chem. Commun.* 1977, 233.
111. Lee, T. V.; Roden, F. S.; Yeoh, H. T.-L. *Tetrahedron Lett.* 1990, *31*, 2063.
112. Johnson, F. *Chem. Rev.* 1968, *68*, 375.
113. Houk, K. N.; Li, Y.; Evanseck, J. D. *Angew. Chem., Int. Ed. Engl.* 1992, *31*, 682.
114. Berson, J. A. *Acc. Chem. Res.* 1991, *24*, 215.
115. Dewar, M. J. S. *J. Am. Chem. Soc.* 1984, *106*, 209.
116. Tolbert, L. M.; Ali, M. B. *J. Am. Chem. Soc.* 1981, *103*, 2104.
117. Woodward, R. B.; Hoffmann, R. *The Conservation of Orbital Symmetry*; Verlag Chemie: Weinhelm, 1970.
118. Houk, K. N. *Acc. Chem. Res.* 1975, *8*, 361.
119. Houk, K. N. *Chem. Rev.* 1976, *76*, 1.
120. Gleiter, R.; Paquette, L. A. *Acc. Chem. Res.* 1983, *16*, 328.
121. Kalo, J.; Vogel, E.; Ginsburg, D. *Tetrahedron* 1977, *33*, 1177.
122. Gleiter, R.; Ginsburg, D. *Pure Appl. Chem.* 1979, *51*, 1301.

Chapter 3

# ONE-CARBON TRANSFORMATIONS

Although the emphasis of this treatise is focused on the synthesis of contiguous chiral centers, some transformations at a single asymmetric center are relevant as they extend the use of other methodologies. Most of the reactions that have been developed for asymmetric synthesis involve the transformation of one or more $sp^2$ centers to $sp^3$ through face-selective reactions. A few reactions are available to interchange saturated, $sp^3$ atoms. Obviously, the reaction center in the starting material must be asymmetric or induction will not be observed. The use of one of the substitution reactions does, however, allow for a "correction" in stereochemistry when it is necessary. These stereospecific substitutions are also important for epoxide and related chemistry, but these reactions will be described in the appropriate subsequent chapters.

This chapter is concerned only with reactions of $sp^3$ centers, where the stereochemistry has already been established. The conversion of an $sp^2$ center to $sp^3$ can also provide, in principle, a one-carbon transformation. As an example, reduction of a ketone may generate a stereogenic center. Electrophilic additions to alkenes can provide 1,2-functionality. In other cases, such as reaction of an alkene with water, only one functional group is introduced. Two $sp^3$ centers always have to be involved in the reactions of carbonyl compounds and alkenes; the reactions of these compounds are discussed in Chapters 4 and 8.

Nucleophilic substitution at a stereogenic center allows for inversion of configuration. This approach can remedy a condensation reaction that provides the "wrong" isomer as the major product. In addition to nucleophilic reactions, as with the classical $S_N2$ reaction, a chiral one-carbon unit can also be a nucleophile. The integrity of the stereocenter must be maintained throughout the reaction sequence. Although examples are not widespread, the ones that have been developed can be useful to extend other methodologies that employ a nucleophile (vide infra). In other cases, the reaction is required to proceed with retention of configuration at the reaction site. An example is provided by the conversion of an α-amino acid to an α-hydroxy acid (*cf.* Scheme 1.1).[1,2]

## 3.1. INVERSIONS AT A SINGLE CARBON CENTER

The conversion of a pre-existing stereocenter to its antipode requires a reaction sequence that has high stereoselectivity. The mechanistic variables that differentiate an $S_N1$ reaction from $S_N2$ demand that the latter be the pathway of choice, with its concerted displacement, rather than the former's carbocation intermediate that loses its stereochemical integrity. In general terms, therefore, the examples and methods described below involve an $S_N2$ reaction — the classical Walden inversion.[3] It should also be remembered that elimination reactions can effectively compete with substitution [Chapter 8].[4]

The converse is also important; when the integrity of a stereocenter must be retained, as for example, with secondary alcohols, reactions and reagents must be employed that have been shown to react exclusively at oxygen, such as deprotonation, tosylation, and esterification.[4,5]

The classic $S_N2$ reaction is a useful, ubiquitous tool for the stereospecific introduction of a variety of functional groups at a specific carbon atom.[6,7] As always, care must be exercised with the choice of reagent, solvent, and other reaction parameters.[4,8] With certain reaction conditions and substrates, the mechanism of a substitution reaction may change from the linear, concerted $S_N2$ transition state to one involving single electron transfers.[9,10]

Some control over stereochemistry, especially the conversion of an alcohol to an alkyl halide, can be exerted through application of the $S_Ni$ mechanism, where the consequences of the cyclic transition state can ensure high selectivity. The reaction proceeds with the overall retention of configuration. The best known example of this reaction is the formation of alkyl chlorides by reaction of alcohols with thionyl chloride, but otherwise is relatively rare.[4] In addition, the reaction conditions must be controlled as use of a base, such as pyridine, can result in the liberation of the chloride ion that can act as a nucleophile and, through an $S_N2$ reaction, could provide the alkyl chloride with an inversion of configuration.

The use of cyclic systems coupled with an $S_N2$ reaction can ensure that reactions proceed with high stereoselection.[11,12] Of particular note are the reactions of epoxides [*10.2*] and cyclic sulfates [*10.4*].

When a cyclic intermediate or transition state is not available, the Mitsunobu reaction has proven the most reliable method to perform an inversion (vide infra).

Representative examples of the $S_N2$ methodology are given in Schemes 3.1[13-15] and 3.2.[16-23]

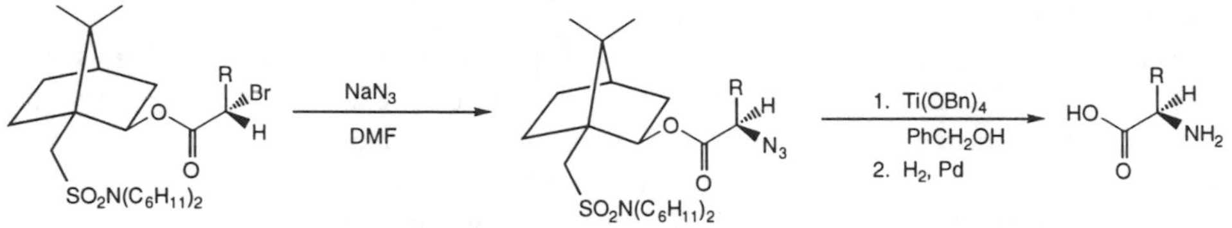

**SCHEME 3.1.**

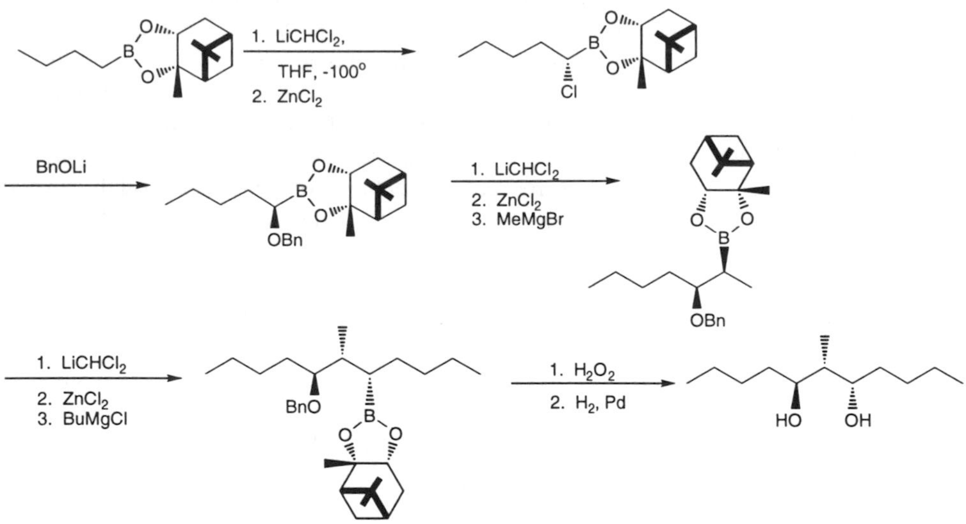

**SCHEME 3.2.**

The former approach is particularly useful for the introduction of a nitrogen moiety adjacent to a carbonyl group.[24-26] The latter method has been extended to prepare 1,2-diols (Scheme 3.3)[27] and 1,2,3-triols.[21]

In Scheme 3.3, the group on boron is now derived from diisopropylethanediol (DIPED), and also provides alcohols with high enantio- and diastereoselectivities (Scheme 3.4). The stereochemistry of the chlorine is important, as an alternative rearrangement can occur (Scheme 3.5).[19] The use of DIPED circumvents some of the problems associated with the recovery of pinanediol (Scheme 3.2).[28]

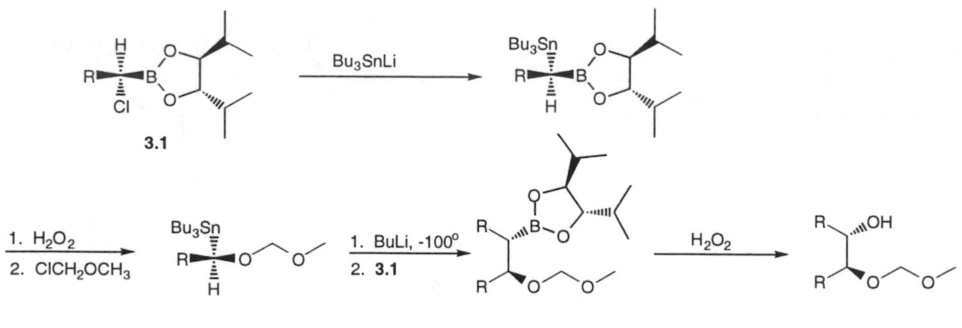

SCHEME 3.3.

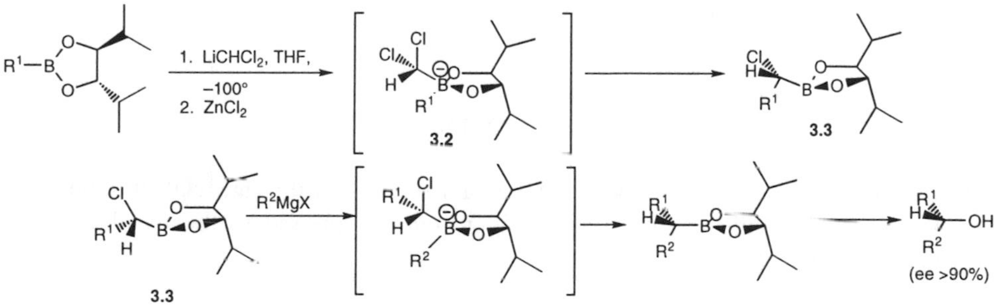

SCHEME 3.4.

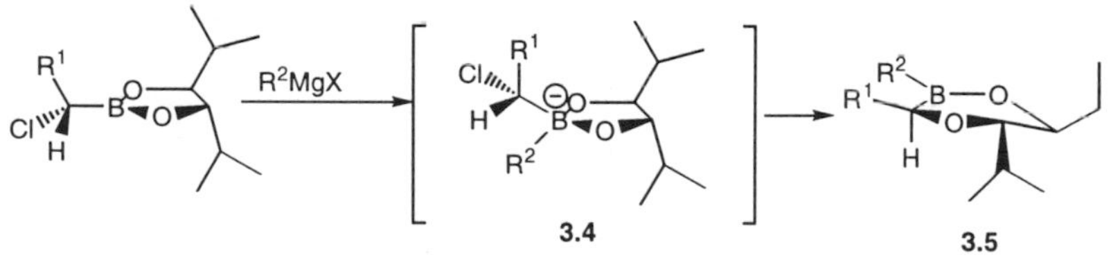

SCHEME 3.5.

The rearrangement has been rationalized by the choice of pathway arising from an exothermic reaction with a very early transition state,[29] so that bond energy differences have little effect and steric factors take precedence. When displacement of the first chlorine atom (from **3.2**) occurs, the carbon is inverted so that the remaining chlorine is now set up for displacement. However, if the diastereoisomer is formed, the favored conformation,**3.4**, now does not have the chlorine and $R^2$ in an *anti* situation, and ring expansion to **3.5** occurs.[19]

The methodology has also been used to prepare functionalized boranes that can then be reacted with aldehydes [*4.1.5.1*].[17,18,30,31]

On occasion, inversion can occur with high specificity, but is not expected. An example is provided by the reaction of the acid chloride, **3.6**, prepared from ethyl lactate via **3.7**, with *m*-difluorobenzene in the presence of aluminum trichloride (Scheme 3.6).[32]

Retention of configuration at a specific center is often best achieved by the use of two inversions.[33] This is adequately illustrated by a method based on selenium chemistry for the conversion of an alcohol to an alkyl bromide with an alkyl selenide as the intermediate (Scheme 3.7).[33]

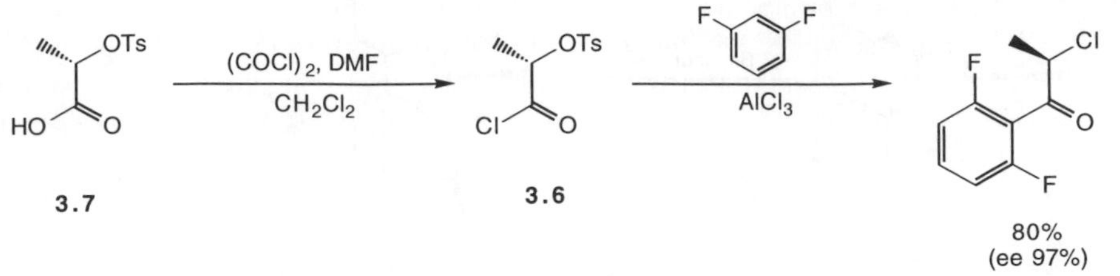

SCHEME 3.6.

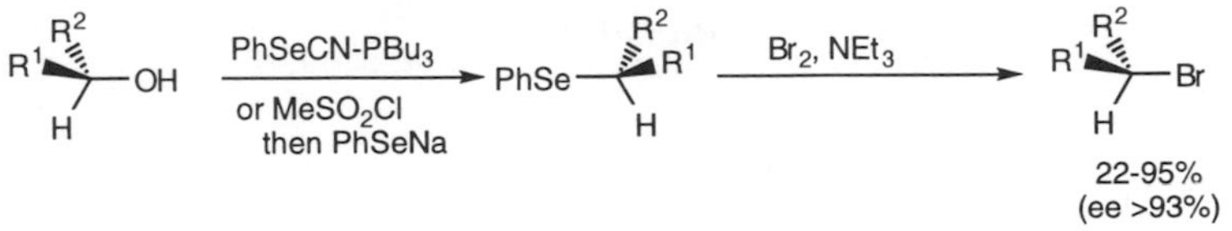

SCHEME 3.7.

The use of a copper catalyst for the reaction of a metal azide with a mesylate has, however, been observed to provide retention of configuration as opposed to the usual inversion (Scheme 3.8) [*9.6*].[34,35]

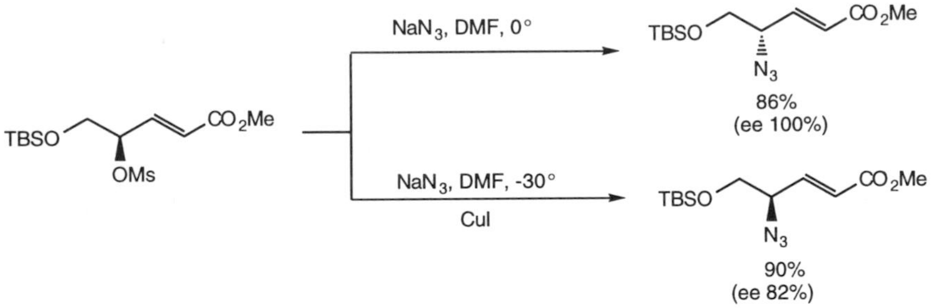

SCHEME 3.8.

Not all substitution reactions can be accomplished with a high degree of asymmetric induction. However, results continue to accumulate, as for the coupling of an organometallic reagent with an alkyl halide, and hold promise for future application.[36] The reaction of the lithium enolate derived from the chiral iron acetyl complex **3.8** shows chiral discrimination in the $S_N2$ reaction with *tert*-butyl 2-propionate (Scheme 3.9).[37]

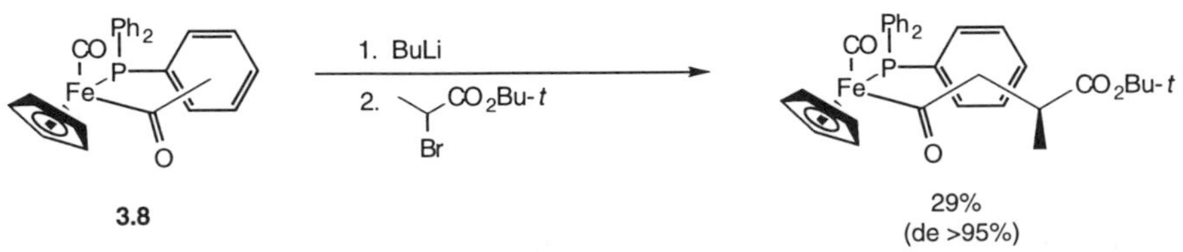

SCHEME 3.9.

A wide variety of methods exist for the homologation of carbon chains through nucleophilic displacement of a leaving group at a primary center.[38-40] Under many circumstances, the use of a substitution reaction at a secondary center is not a simple extension of methodology developed for primary carbon atoms. The stereochemical integrity of the reaction is important while competing reactions, such as elimination, need to be minimized, if not eradicated.

For the formation of a carbon–carbon bond, the use of an organocuprate, particularly with a tosylate, has been advocated as inversion at a secondary center is clean, particularly when a heteroatom group is on an adjacent carbon atom.[41-46]

### 3.1.1. INVERSION OF A HYDROXY GROUP

The inversion of a hydroxy group is not a simple procedure. The most reliable methods are based on Mitsunobu methodology (vide infra). Inversion can be achieved by conversion of the alcohol to a mesylate or tosylate, then by nucleophilic displacement with potassium superoxide or nitrite in dimethyl sulphoxide.[47-53] The nucleophile can also be introduced as a quaternary ammonium salt (Scheme 3.10).[54]

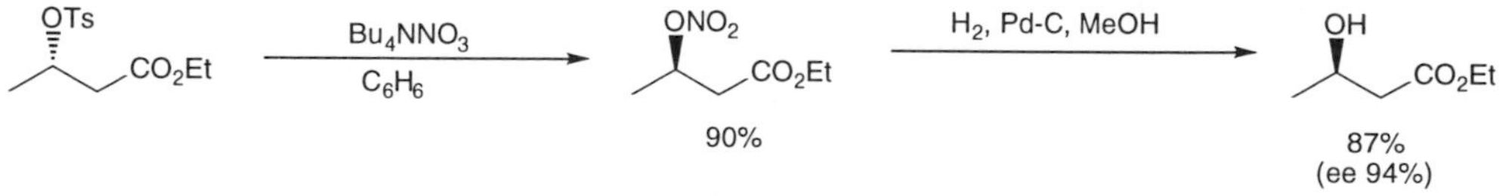

OTs, $CO_2Et$ → ($Bu_4NNO_3$, $C_6H_6$) → $ONO_2$, $CO_2Et$ (90%) → ($H_2$, Pd-C, MeOH) → OH, $CO_2Et$ (87%, (ee 94%))

SCHEME 3.10.

Cesium or potassium carboxylates in *N,N*-dimethylformamide, or in toluene with a crown ether, also give clean $S_N2$ reactions with mesylates or tosylates.[55-57] The use of delocalized nucleophiles minimizes side reactions that could occur through the nucleophile also acting as a strong base. Potassium acetate has also been used to effect an $S_N2$ reaction of mesylates derived from 1,3- or 1,4-diols.[58]

In some cyclic examples, the thermodynamics of the ring system can be utilized to bring about overall inversion of an alcohol through an oxidation-reduction equilibrium; a nickel-based system has been used effectively.[59]

### 3.1.2. THE MITSUNOBU REACTION

The Mitsunobu reaction allows substitution of a hydroxy group by a wide variety of nucleophiles with inversion of configuration.[60-62] This Mitsunobu protocol often provides very high yields for unhindered alcohols.[60-67] A useful version for the inversion of configuration of a secondary alcohol involves the use of zinc tosylate, diethyl azodicarboxylate, and triphenylphosphine,[68] although other variants can be just as effective.[69] The approach can result in mixtures if the alcohol is hindered.[70,71]

A wide variety of nucleophiles can be employed in the reaction, including azides, phthalimides, and sulfonamides for the preparation of amines,[62,72-76] cyanides, thiocyanates, carboxylic acid salts, and thioalkoxides.[60,61] The reaction conditions favor inversion through the use of the $S_N2$ reaction mechanism.[61,67,77]

The Mitsunobu approach can complement other methods, as illustrated by the inversion of the epoxy-alcohol prepared by a kinetic resolution with Sharpless epoxidation methodology [*11.1.2*] (Scheme 3.11).[78,79]

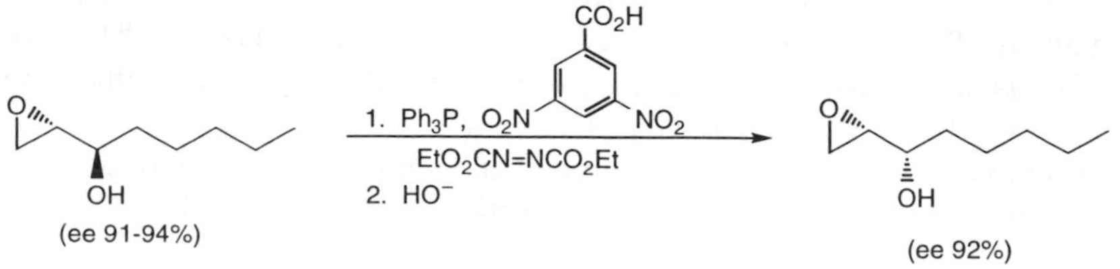

OH (ee 91-94%) → (1. $Ph_3P$, $O_2N$–C6H3($CO_2H$)–$NO_2$, $EtO_2CN{=}NCO_2Et$; 2. $HO^-$) → OH (ee 92%)

SCHEME 3.11.

Amino alcohols provide for nucleophilic substitution under Mitsunobu conditions with stereochemistry being controlled by the nitrogen protection. Use of an amide provides retention of configuration through formation of an intermediate oxazoline; this results in two substitutions with inversion to give overall retention. In contrast, use of a carbamate to protect the nitrogen provides overall inversion at the reaction center, as no intramolecular reaction is involved (Scheme 3.12).[80]

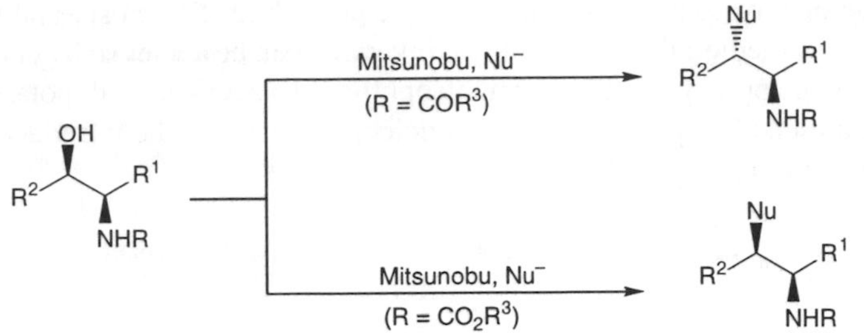

Mitsunobu, $Nu^-$ ($R = COR^3$)

Mitsunobu, $Nu^-$ ($R = CO_2R^3$)

OH, $R^2$, $R^1$, NHR, Nu

**SCHEME 3.12.**

## 3.2. CHIRAL ORGANOMETALLIC REAGENTS

It is possible to use a nucleophile where the reaction center bears the chirality. The organometallic species must have a stable configuration, otherwise racemization will be observed. One very successful approach is based on the use of a chiral sulfoxide to stabilize an adjacent carbanion center through complex formation.[81] The majority of examples of this class of organometallic reagent have an additional functional group present to allow for a wealth of subsequent transformations.[§][82-86] For a "simple" α-sulfinyl carbanion, the configuration of the anionic center is strongly pyramidalized.[87-94] The reactions of these carbanions also provide some insight into the structure of the anion; electrophiles containing oxygen react, in THF, with retention of configuration, while methyl iodide gives rise to inversion (Figure 3.1).[83,87,95-101]

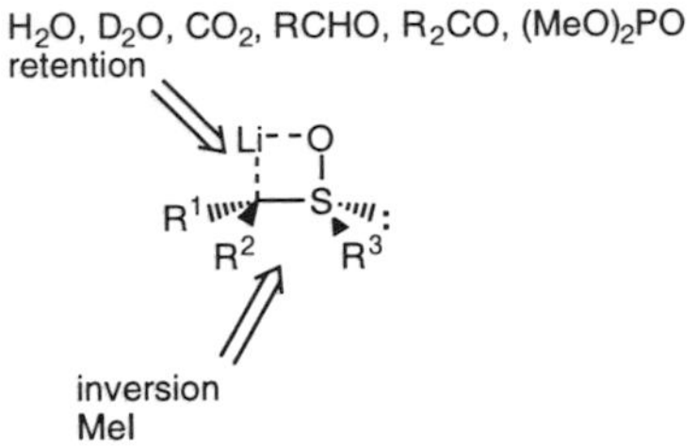

$H_2O$, $D_2O$, $CO_2$, RCHO, $R_2CO$, $(MeO)_2PO$
retention
Li, O, S, $R^1$, $R^2$, $R^3$
inversion
MeI

**FIGURE 3.1.** Reactions of sulfoxide stabilized carbanions.

However, a crystal structure where TMEDA was also present showed that the lithium was bound to oxygen, but the carbon of the "carbanion" was not planar.[92,102] The exact structures of these synthetically useful carbanions will continue to be investigated,[93,103,104] but sufficient understanding is already available to allow for their reliable usage in a synthetic sequence; many examples will be found in other chapters.

Other systems also allow for the preparation of alkyl lithiums with stereochemical control, although the direct method from a reaction of a metal with an alkyl halide is not available, as

§ As the chirality is not derived from a carbon atom, these reactions are discussed in a number of places based on the carbon framework of the rest of the molecule.

extensive racemization occurs.[105-108] The resultant carbanion must be configurationally stable, as for vinyl,[109] α-alkoxyalkyl,[110-113] or cyclopropyl systems,[114] or show strong preferences for one stereoisomer through chelation[115-124] or stereoelectronic reasons.[124-126]

The use of these chiral organometallic reagents allows for a wide range of synthetic methods, including the preparation of alcohols by silicon chemistry (Scheme 3.13),[116,127] and α-alkoxy carboxylic acids (Scheme 3.14).[128] In the latter example, the asymmetry is introduced by a chiral reduction of an acyl stannane,[129] while the chelate formed with the oxygen atoms allows for stereochemical control during reactions with other electrophiles [2.1.2].[111,130-134] The selectivity may be augmented by the use of a chiral ligand, such as sparteine.[135]

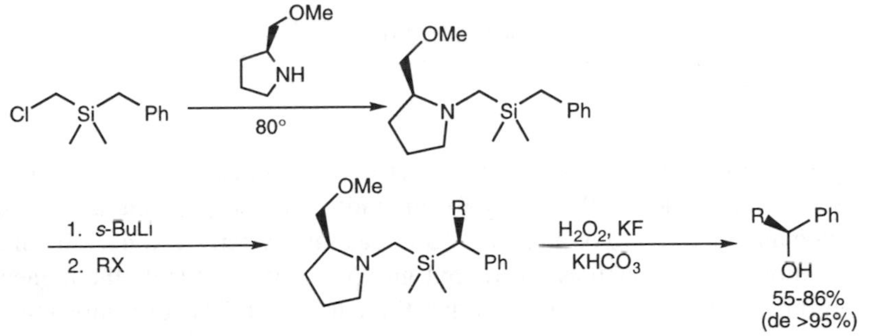
OMe
NH
Cl
Si
Ph
80°
OMe
N
Si
Ph
1. s-BuLi
2. RX
OMe
R
N
Si
Ph
H2O2, KF
KHCO3
R
Ph
OH
55-86%
(de >95%)

**SCHEME 3.13.**

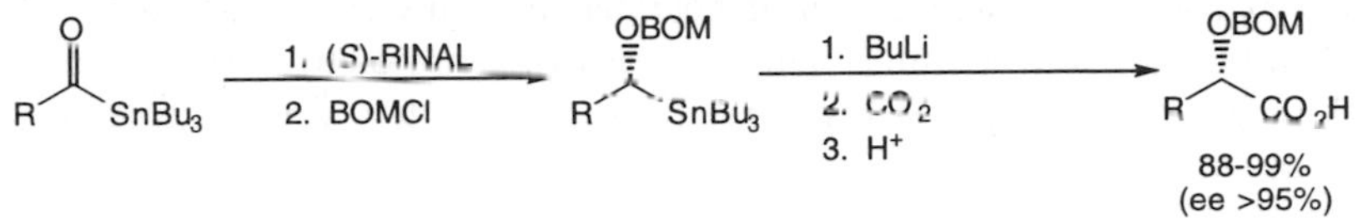
O
R
SnBu3
1. (S)-BINAL
2. BOMCl
OBOM
R
SnBu3
1. BuLi
2. CO2
3. H+
OBOM
R
CO2H
88-99%
(ee >95%)

**SCHEME 3.14.**

In addition to oxygen and sulfur, other heteroatoms can be incorporated into the organometallic reagent (vide supra), including nitrogen,[136,137] selenium,[138] silicon,[139] and halogens (Scheme 3.15).[140]

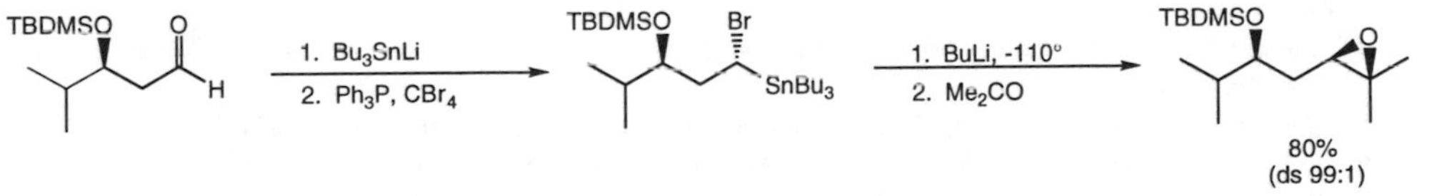
TBDMSO
O
H
1. Bu3SnLi
2. Ph3P, CBr4
TBDMSO
Br
SnBu3
1. BuLi, -110°
2. Me2CO
TBDMSO
O
80%
(ds 99:1)

**SCHEME 3.15.**

## 3.3. OTHER REACTIONS

In many instances, the geometric integrity of an $sp^2$ carbon atom must be kept during a reaction. There are many examples of nucleophilic substitution reactions of vinyl anions; these are discussed in detail in Chapter 9 and will not be duplicated here.

For the nucleophilic substitutions described above, we have, by necessity, been concerned with displacement of a heteroatom. A wide variety of reagents are available that react with the oxygen of an alcohol, and thus, leave the stereochemical integrity of that alcoholintact. An example is provided by the kinetic resolution of phenylethanol with the chiral carboxylic acid **3.9** (Scheme 3.16).[5]

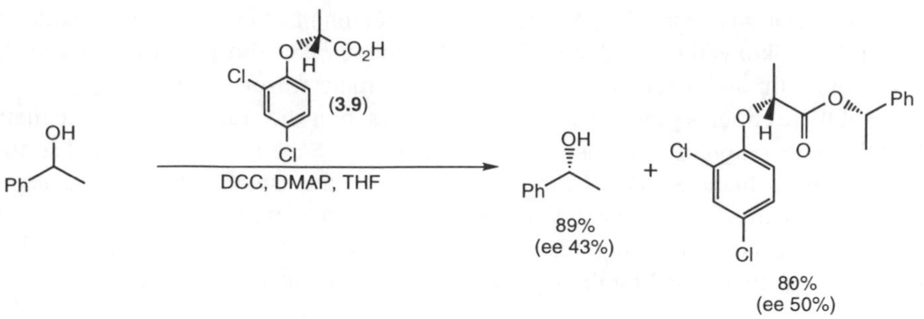

(3.9)

DCC, DMAP, THF

89% (ee 43%)

80% (ee 50%)

SCHEME 3.16.

## 3.4. SUMMARY

Substitution at an $sp^3$ center is a reaction that has to be undertaken with care. Although the Mitsunobu reaction provides methodology for unhindered centers, problems can still occur when steric demands are high. For a classical $S_N2$ reaction, systems that contain an adjacent activating functional group, such as a carbonyl group or boron, provide the highest stereochemical control. Other than cost considerations, these nucleophilic substitution reactions can be scaled up.

As the stereochemical integrity of an organolithium can be assured by complex formation, or, in a few cases by the structure of substrate, it is possible to generate chiral organolithium agents, thus extending the stereochemical potential of many reactions that employ these reagents. However, the stereochemical integrity has to be ensured by the use of low temperature, which means that this approach will not be a trivial undertaking on more than a small laboratory scale.

## 3.5. REFERENCES


1. Zhang, W.; Loebach, J. L.; Wilson, S. R.; Jacobsen, E. N. *J. Am. Chem. Soc.* 1990, *112*, 2801.
2. Gringore, O. H.; Rouessac, F. P. *Org. Synth.* 1984, *63*, 121.
3. Beak, P. *Acc. Chem. Res.* 1992, *25*, 215.
4. March, J. *Advanced Organic Chemistry: Reactions Mechanisms and Structure;* 3rd ed.; John Wiley & Sons: New York, 1985.
5. Chinchilla, R.; Najera, C.; Yus, M.; Heumann, A. *Tetrahedron: Asymmetry* 1990, *1*, 851.
6. Wolfe, S.; Mitchell, D. J.; Schlegel, H. B. *J. Am. Chem. Soc.* 1981, *103*, 7692.
7. Wolfe, S.; Mitchell, D. J.; Schlegel, H. B. *J. Am. Chem. Soc.* 1981, *103*, 7694.
8. Deslongchamps, P. *Stereoelectronic Effects in Organic Chemistry*; Pergamon Press: Oxford, 1983.
9. Lewis, E. S. *J. Am. Chem. Soc.* 1989, *111*, 7576.
10. Bordwell, F. G.; Harrelson, J. A. *J. Org. Chem.* 1989, *54*, 4893.
11. Houge-Frydrych, C. S. V.; Pinto, I. L. *Tetrahedron Lett.* 1989, *30*, 3349.
12. Sunay, U.; Mootoo, D.; Molino, B.; Fraser-Reid, B. *Tetrahedron Lett.* 1986, *27*, 4697.
13. Oppolzer, W.; Pedrosa, R.; Moretti, R. *Tetrahedron Lett.* 1986, *27*, 831.
14. Koh, K.; Ben, R. N.; Durst, T. *Tetrahedron Lett.* 1993, *34*, 4473.
15. Kitagawa, O.; Hanano, T.; Kikuchi, N.; Taguchi, T. *Tetrahedron Lett.* 1993, *34*, 2165.
16. Matteson, D. S.; Sadhu, K. M.; Peterson, M. L. *J. Am. Chem. Soc.* 1986, *108*, 810.
17. Matteson, D. S.; Majumdar, D. *J. Am. Chem. Soc.* 1980, *102*, 7588.
18. Matteson, D. S. *Chem. Rev.* 1989, *89*, 1535.
19. Tripathy, P. B.; Matteson, D. S. *Synthesis* 1990, 200.
20. Matteson, D. S.; Sadhu, K. M.; Ray, R.; Peterson, M. L.; Majumdar, D.; Hurst, G. D.; Jesthi, P. K.; Tsai, D. J. S.; Erdik, E. *Pure Appl. Chem.* 1985, *57*, 1741.
21. Matteson, D. S.; Kandil, A. A.; Soundararajan, R. *J. Am. Chem. Soc.* 1990, *112*, 3964.

22. Brown, H. C.; Rangaishenvi, M. V. *Tetrahedron Lett.* 1990, *31*, 7113.
23. Brown, H. C.; Rangaishenvi, M. V. *Tetrahedron Lett.* 1990, *31*, 7115.
24. Hoffman, R. V.; Kim, H.-O. *Tetrahedron Lett.* 1990, *31*, 2953.
25. Dharanipragada, R.; Nicolas, E.; Toth, G.; Hruby, V. J. *Tetrahedron Lett.* 1989, *30*, 6841.
26. Nicolas, E.; Dharanipragada, R.; Toth, G.; Hruby, V. J. *Tetrahedron Lett.* 1989, *30*, 6845.
27. Matteson, D. S.; Tripathy, P. B.; Sarkar, A.; Sadhu, K. M. *J. Am. Chem. Soc.* 1989, *111*, 4399.
28. Brown, H. C.; Rangaishenvi, M. V. *J. Organometal. Chem.* 1988, *358*, 15.
29. Matteson, D. S.; Mak, R. W. H. *J. Org. Chem.* 1963, *28*, 2171.
30. Sturmer, R.; Hoffmann, R. W. *Synlett* 1990, 759.
31. Rangaishenvi, M. V.; Singaram, B.; Brown, H. C. *J. Org. Chem.* 1991, *56*, 3286.
32. Konosu, T.; Tajima, Y.; Miyaoka, T.; Oida, S. *Tetrahedron Lett.* 1991, *32*, 7545.
33. Sevrin, M.; Krief, A. *J. Chem. Soc., Chem. Commun.* 1980, 656.
34. Yamamoto, Y.; Asao, N. *J. Org. Chem.* 1990, *55*, 5303.
35. Scriven, E. F. V.; Turnbull, K. *Chem. Rev.* 1988, *88*, 297.
36. Hayashi, T.; Konishi, M.; Fukushima, M.; Mise, T.; Kagotani, M.; Tajika, M.; Kumada, M. *J. Am. Chem. Soc.* 1982, *104*, 180.
37. Collingwood, S. P.; Davies, S. G.; Preston, S. C. *Tetrahedron Lett.* 1990, *31*, 4067.
38. Kurth, M.; Soares, C. J. *Tetrahedron Lett.* 1987, *28*, 1031.
39. Kotsuki, H.; Kadota, I.; Ochi, M. *Tetrahedron Lett.* 1989, *30*, 1281.
40. Gingras, M.; Chan, T. H. *Tetrahedron Lett.* 1989, *30*, 279.
41. Posner, G. H. *An Introduction to Synthesis Using Organocopper Reagents* ; John Wiley & Sons: New York, 1980.
42. Posner, G. H. *Org. React.* 1975, *22*, 253.
43. Lipshutz, B. H. *Synthesis* 1987, 325.
44. Lipshutz, B. H.; Kozlowski, J.-A. *Tetrahedron* 1984, *40*, 5005.
45. Hanessian, S.; Thavonekham, B.; DeHoff, B. *J. Org. Chem.* 1989, *54*, 5831.
46. Petit, Y.; Sanner, C.; Larcheveque, M. *Tetrahedron Lett.* 1990, *31*, 2149.
47. Filippo, J. S.; Chern, C.-I.; Valentine, J. S. *J. Org. Chem.* 1975, *40*, 1680.
48. Corey, E. J.; Nicolaou, K. C.; Shibasaki, M.; Machida, Y.; Shiner, C. S. *Tetrahedron Lett.* 1975, 3183.
49. Sawyer, D. T.; Gibian, M. J. *Tetrahedron* 1979, *35*, 1471.
50. Fukuyama, T.; Wang, C.-L. J.; Kishi, Y. *J. Am. Chem. Soc.* 1979, *101*, 260.
51. Hatakeyama, S.; Sakurai, K.; Takano, S. *J. Chem. Soc., Chem. Commun.* 1985, 1759.
52. Raduchel, B. *Synthesis* 1980, 292.
53. Moriarty, R. M.; Zhuang, H.; Penmasta, R.; Liu, K.; Awasthi, A. K.; Tuladar, S. M.; Rao, M. S. C.; Singh, V. K. *Tetrahedron Lett.* 1993, *34*, 8029.
54. Cainelli, G.; Manescalchi, F.; Martelli, G.; Panunzo, M.; Pless, L. *Tetrahedron Lett.* 1985, *26*, 3369.
55. Kruizinga, W. H.; Strijtveen, B.; Kellogg, R. M. *J. Org. Chem.* 1981, *46*, 4321.
56. Willis, C. L. *Tetrahedron Lett.* 1987, *28*, 6705.
57. Zak, H.; Schmidt, U. *Angew. Chem., Int. Ed. Engl.* 1975, *14*, 432.
58. Takano, S.; Hirama, M.; Seya, K.; Agasawa, K. *Tetrahedron Lett.* 1983, *24*, 4233.
59. Feghouli, G.; Vanderesse, R.; Fort, Y.; Caubere, P. *Tetrahedron Lett.* 1988, *29*, 1383.
60. Mitsunobu, O. *Synthesis* 1981, 1.
61. Loibner, H.; Zbiral, E. *Helv. Chim. Acta* 1976, *59*, 2100.
62. Viaud, M. C.; Rollin, P. *Synthesis* 1990, 130.
63. Loibner, H.; Zbiral *Helv. Chim. Acta* 1977, *60*, 417.
64. Bose, A. K.; Lal, B.; Hoffman, W. A.; Manhas, M. S. *Tetrahedron Lett.* 1973, 1619.
65. Isobe, M.; Ichikawa, Y.; Funabashi, Y.; Mio, S.; Goto, T. *Tetrahedron* 1986, *42*, 2863.
66. Hauser, F. M.; Ellenberger, S. R.; Ellenberger, W. P. *Tetrahedron Lett.* 1988, *29*, 4939.
67. Sliwka, H.-R.; Liaaen-Jensen, S. *Acta Chem. Scand.* 1987, *41B*, 518.
68. Galynker, I.; Still, W. C. *Tetrahedron Lett.* 1982, *23*, 4461.
69. De Witt, P.; Misiti, D.; Zappia, G. *Tetrahedron Lett.* 1989, *30*, 5505.
70. Farina, V. *Tetrahedron Lett.* 1989, *30*, 6645.
71. Subramanian, R. S.; Balasubramanian, K. K. *Tetrahedron Lett.* 1990, *31*, 2201.
72. Golding, B. T.; Howes, C. *J. Chem. Res. (S)* 1984, 1.
73. Henry, J. R.; Marcin, L. R.; McIntosh, M. C.; Scola, P. M.; Harris, G. D.; Weinreb, S. M. *Tetrahedron Lett.* 1989, *30*, 5709.
74. Edwards, M. L.; Stemerick, D. M.; McCarthy, J. R. *Tetrahedron Lett.* 1990, *31*, 3417.
75. Brown, R. F. C.; Jackson, W. R.; McCarthy, T. D. *Tetrahedron Lett.* 1993, *34*, 1195.
76. Campbell, J. A.; Hart, D. J. *J. Org. Chem.* 1993, *58*, 2900.
77. Crich, D.; Dyker, H.; Harris, R. J. *J. Org. Chem.* 1989, *54*, 257.
78. Page, P. C. B.; Carefull, J. F.; Powell, L. H.; Sutherland, I. O. *J. Chem. Soc., Chem. Commun.* 1985, 822.
79. Mori, K.; Otsuka, T. *Tetrahedron* 1985, *41*, 553.
80. Lipshutz, B. H.; Miller, T. A. *Tetrahedron Lett.* 1990, *31*, 5253.

81. Nakamura, K.; Higaki, M.; Adachi, S.; Oka, S.; Ohro, A. *J. Org. Chem.* 1987, *52*, 1414.
82. Solladié, G. *Chimia* 1984, *38*, 233.
83. Solladié, G. *Synthesis* 1981, 185.
84. Kagan, H. B.; Rebiere, F. *Synlett* 1990, 643.
85. Williams, D. R.; Phillips, J. G.; Huffman, J. C. *J. Org. Chem.* 1981, *46*, 4101.
86. Solladié-Cavallo, A.; Adib, A.; Schmitt, M.; Fischer, J.; DeCian, A. *Tetrahedron: Asymmetry* 1992, *3*, 1597.
87. Boche, G. *Angew. Chem., Int. Ed. Engl.* 1989, *28*, 277.
88. Wolfe, S.; Stolow, A.; La John, L. A. *Can. J. Chem.* 1984, *62*, 1470.
89. Wolfe, S. *Acc. Chem. Res.* 1972, *5*, 102.
90. Wolfe, S.; Stolow, A.; La John, L. A. *Tetrahedron Lett.* 1983, *24*, 4071.
91. Wolfe, S.; La John, L. A.; Weaver, D. F. *Tetrahedron Lett.* 1984, *25*, 2863.
92. Marsch, M.; Massa, W.; Harms, K.; Baum, G.; Boche, G. *Angew. Chem., Int. Ed. Engl.* 1986, *25*, 1011.
93. Gais, H. J.; Hellmann, G.; Lindner, H. J. *Angew. Chem., Int. Ed. Engl.* 1990, *29*, 100.
94. Walker, A. J. *Tetrahedron: Asymmetry* 1992, *3*, 961.
95. Chassaing, G.; Marquet, A.; Corset, J.; Froment, F. *J. Organometal. Chem.* 1982, *232*, 293.
96. Nishihata, K.; Nishio, M. *Tetrahedron Lett.* 1976, 1695.
97. Biellmann, J. F.; Viciens, J. J. *Tetrahedron Lett.* 1974, 2915.
98. Biellmann, J. F.; Viciens, J. J. *Tetrahedron Lett.* 1978, 467.
99. Durst, T.; Molin, M. *Tetrahedron Lett.* 1975, 63.
100. Chassaing, G.; Lett, R.; Marquet, A. *Tetrahedron Lett.* 1978, 471.
101. Hutchinson, B. J.; Andersen, K. K.; Katritzky, A. R. *J. Am. Chem. Soc.* 1969, *91*, 3839.
102. Najera, C.; Yus, M.; Hässig, R.; Seebach, D. *Helv. Chim. Acta* 1984, *67*, 1100.
103. Gais, H.-J.; Hellmann, G.; Günther, H.; Lopez, F.; Lindner, H. J.; Braun, S. *Angew. Chem., Int. Ed. Engl.* 1989, *28*, 1025.
104. Gais, H.-J.; Hellman, G. *J. Org. Chem.* 1992, *57*, 4439.
105. Rachon, J.; Walborsky, H. M. *Tetrahedron Lett.* 1989, *30*, 7345.
106. Dewar, M. J. S.; Harris, J. M. *J. Am. Chem. Soc.* 1969, *91*, 3652.
107. Grovenstein, E.; Cheng, Y.-M. *J. Chem. Soc., Chem. Commun.* 1970, 101.
108. Letsinger, R. L. *Angew. Chem.* 1958, *70*, 151.
109. Seyferth, D.; Vaughan, L. G. *J. Am. Chem. Soc.* 1964, *86*, 883.
110. Cohen, T.; Bhupathy, M. *Acc. Chem. Res.* 1989, *22*, 152.
111. Still, W. C.; Sreekumar, C. *J. Am. Chem. Soc.* 1980, *102*, 1201.
112. McGarvey, G. J.; Kimura, M. *J. Org. Chem.* 1985, *50*, 4655.
113. Rychnovsky, S. D.; Mickus, D. E. *Tetrahedron Lett.* 1989, *30*, 3011.
114. Applequist, D. E.; Peterson, A. H. *J. Am. Chem. Soc.* 1961, *83*, 862.
115. Hoppe, D.; Hintze, F.; Tebben, P. *Angew. Chem., Int. Ed. Engl.* 1990, *29*, 1422.
116. Chan, T. H.; Pellon, P. *J. Am. Chem. Soc.* 1989, *111*, 8737.
117. Hoffmann, R. W.; Bewersdorf, M. *Tetrahedron Lett.* 1990, *31*, 67.
118. Hoffmann, R. W.; Bewersdorf, M.; Ditrich, K.; Krüger, M.; Stümer, R. *Angew. Chem., Int. Ed. Engl.* 1988, *27*, 1176.
119. McDougal, P. G.; Condon, B. D.; Raffosse, M. D.; Lauro, A. M.; Van Derveer, D. *Tetrahedron Lett.* 1988, *29*, 2547.
120. McDougal, P. G.; Condon, B. D. *Tetrahedron Lett.* 1989, *30*, 789.
121. Ritter, R. H.; Cohen, T. *J. Am. Chem. Soc.* 1986, *108*, 3718.
122. Meyers, A. I.; Miller, D. B.; White, F. H. *J. Am. Chem. Soc.* 1988, *110*, 4778.
123. Krief, A.; Dumont, W.; Clarenbeau, M.; Bernard, G.; Badaoui, E. *Tetrahedron Lett.* 1989, *30*, 2005.
124. Reich, H. J.; Bowe, M. D. *J. Am. Chem. Soc.* 1990, *112*, 8994.
125. Eliel, E. L.; Hartmann, A. A.; Abatjoglou, A. G. *J. Am. Chem. Soc.* 1974, *96*, 1807.
126. Eliel, E. L. *Tetrahedron* 1974, *30*, 1503.
127. Chan, T. H.; Nwe, K. T. *J. Org. Chem.* 1992, *57*, 6107.
128. Chan, P. C.-M.; Chong, J. M. *Tetrahedron Lett.* 1990, *31*, 1985.
129. Chan, P. C.-M.; Chong, J. M. *J. Org. Chem.* 1988, *53*, 5584.
130. Yamada, J.-I.; Abe, H.; Yamamoto, Y. *J. Am. Chem. Soc.* 1990, *112*, 6118.
131. Schwerdtfeger, J.; Hoppe, D. *Angew. Chem., Int. Ed. Engl.* 1992, *31*, 1505.
132. Knochel, P. *Angew. Chem., Int. Ed. Engl.* 1992, *31*, 1459.
133. Burchat, A. F.; Chong, J. M.; Park, S. B. *Tetrahedron Lett.* 1993, *34*, 51.
134. Sommerfeld, P.; Hoppe, D. *Synlett* 1992, 764.
135. Beak, P.; Du, H. *J. Am. Chem. Soc.* 1993, *115*, 2516.
136. Pearson, W. H.; Lindbeck, A. C. *J. Org. Chem.* 1989, *54*, 5651.
137. Pearson, W. H.; Lindbeck, A. C.; Kampf, J. W. *J. Am. Chem. Soc.* 1993, *115*, 2622.
138. Ruhland, T.; Dress, R.; Hoffmann, R. W. *Angew. Chem., Int. Ed. Engl.* 1993, *32*, 1467.
139. Reich, H. J.; Dykstra, R. R. *Angew. Chem., Int. Ed. Engl.* 1993, *32*, 1469.
140. Hoffmann, R. W.; Ruhland, T.; Bewersdorf, M. *J. Chem. Soc., Chem. Commun.* 1991, 195.

Chapter 4

# REACTIONS WITH CARBONYL COMPOUNDS

The carbonyl group is extremely versatile for the introduction of functionality, as reaction can occur at either the carbonyl carbon atom — the carbonyl group is an electrophile — or, through enol(ate) formation, at the adjacent carbon atom, the carbonyl group being an integral part of the nucleophile. The carbonyl group can also be part of a complex molecule where other functional groups can influence its chemical reactivity. For an alkene conjugated with a carbonyl system, and where reaction occurs at the alkene moiety, the reactions are discussed in Chapter 9, while reactions of functionalized carbonyl compounds are to be found in Chapter 6. This chapter and the following one, are concerned with the reactions of "simple" carbonyl compounds, which we have taken as any functional group that contains the "C=O" moiety.[†] Reactions where the carbonyl group acts as an electrophile are to be found in this chapter, which includes the addition of a nucleophile to a carbonyl carbon atom and reductions. For the nucleophilic additions, complex nucleophiles have been included here, as well as extensions of Anh-Felkin and chelation-controlled additions [*2.1*]. The nucleophilic reactions of carbonyl compounds are to be found in Chapter 5.

Reactions of closely related systems to the carbonyl group, such as imines and enamines, are also to be found in the corresponding chapter to the analogous carbonyl chemistry.

## 4.1. REACTIONS OF CARBONYL COMPOUNDS WITH NUCLEOPHILES

A wide variety of nucleophiles has been reacted with carbonyl compounds to provide an alcohol. As we are concerned with asymmetric methodology, the pivotal rules are those of Anh-Felkin addition [*2.1.1*] and chelation-control [*2.1.2*] that can predict facial selectivity. Of course, some nucleophiles that contain functionality, such as an enolate [*7.1*], provide for very powerful methodology; only simple reactions are discussed below with the nature of the nucleophile being the major variable.

### 4.1.1. ADDITIONS OF ORGANOMETALLIC SPECIES TO CARBONYL COMPOUNDS

To circumvent some of the problems associated with the tendencies of organolithium or Grignard reagents[‡] to act as bases with carbonyl compounds, the use of organocerium and organotitanium reagents has been advocated.[2-7] Organoytterbium reagents have also been promoted as they enhance diastereoselectivity.[8,9]

The factors that control the addition of a nucleophile to a carbonyl group with an adjacent asymmetric center have already been discussed [*2.1*].[10] Indeed, many reagents have been investigated to improve Anh-Felkin selectivity in simple systems.[11,12] A similar scenario can be visualized if an asymmetric center is present in the nucleophilic moiety adjacent to the reaction site; examples of this type of addition are, however, not as common.[13-19] An example is provided by the use of a chiral sulfoxide where the best selectivity is seen when an aryl ketone is employed as the electrophile and no branching is present in the sidechain (Scheme 4.1).[17]

† However, examples of carboxylic acid derivatives will not be common in this chapter as few reactions result in the transformation of an acid derivative into a new asymmetric center.

‡ In addition, the ligand associated with the magnesium can affect the reaction's outcome.[1]

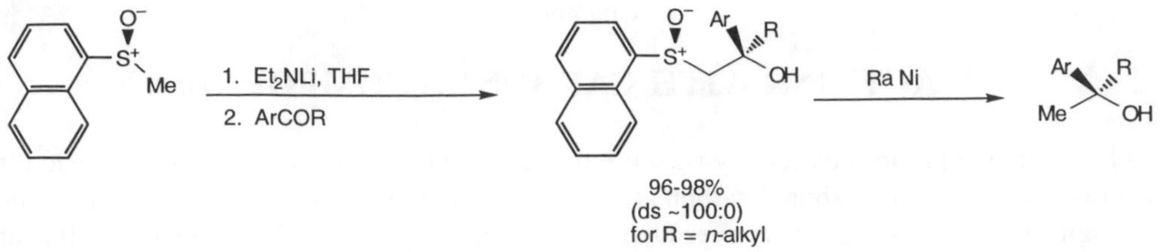

1. $Et_2NLi$, THF
2. ArCOR

Ra Ni

96-98%
(ds ~100:0)
for R = *n*-alkyl

**SCHEME 4.1.**

To circumvent some of the problems associated with the addition of functionalized organometallic reagents to complex carbonyl compounds, nucleophilic reagents are available that contain masked functionality (vide infra). Once the condensation has been performed, this functionality can be used for further elaboration (Scheme 4.2) [*10.1.4*].[20-27]

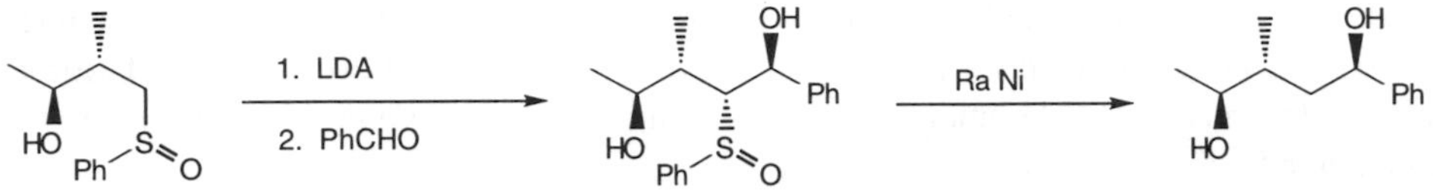

1. LDA
2. PhCHO

Ra Ni

**SCHEME 4.2.**

## 4.1.2. ALKYLATING AGENTS

For asymmetric induction, the nucleophile or carbonyl group must be under a chiral influence [*1.3*]. Face selectivity under Anh-Felkin or chelation control can influence the diastereoselectivity of the reaction outcome [*2.1*]. Some additional examples of these modes of addition are given here.[#] Enhanced Anh-Felkin selectivity may be observed for the reaction of an organocuprate with a carbonyl compound in the presence of trimethylsilyl chloride; this reagent also accelerates 1,2-addition [*9.5*].[28]

Chiral acyl silanes show high diastereofacial selectivity, which can be used to prepare alcohols in a highly stereoselective manner (Scheme 2.3).[29-32] The silyl group acts as a large, sterically-demanding substituent [*9.4.1*].

α-Alkoxy esters can react with nucleophiles in the presence of a reducing agent to provide either isomer of the protected diol (Scheme 4.3).[33]

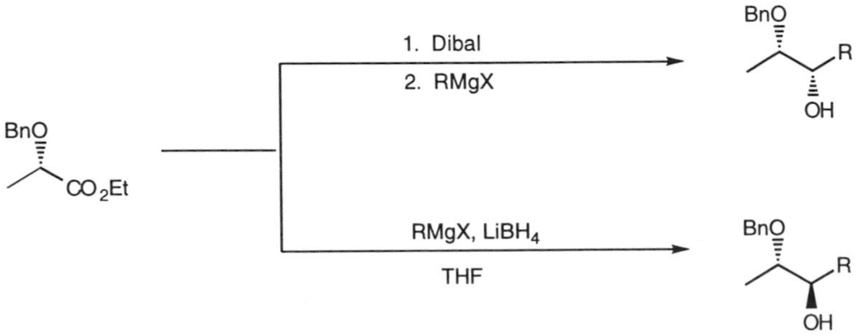

1. Dibal
2. RMgX

$RMgX, LiBH_4$
THF

**SCHEME 4.3.**

A 1,3-oxathiane chiral auxiliary allows excellent diastereoselective addition of Grignard reagents (Scheme 2.4).[34-41] The use of this reaction has been extended further by the addition

[#] Some examples in this chapter include α-alkoxy carbonyl compounds as a reactant. The chemistry of these systems are discussed in detail in Chapter 6.

of ytterbium as this reverses the diastereoselectivity.[42] A further variation is provided by α-keto-1,3-dithiane-1-oxides.¶ [43,44]

Complex formation of chiral oxazolines with Grignard reagents does allow some enantioselectivity in the subsequent reaction with carbonyl compounds, but the ee's are low.[45] Good Anh-Felkin diastereoselectivity is observed with acyl iron compounds.[46]

The nucleophile can be the source of chirality at the reaction center itself [*3.2*] (e.g., Scheme 4.4, *cf.* Scheme 2.13).[47,48]

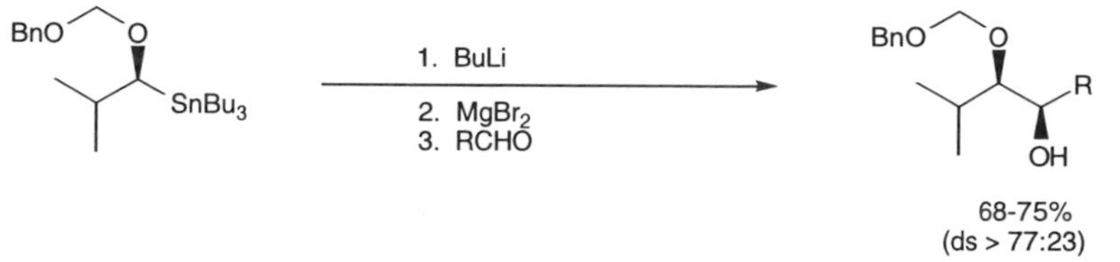

BnO O SnBu$_3$ — 1. BuLi; 2. MgBr$_2$; 3. RCHO → BnO O R OH

68-75%
(ds > 77:23)

**SCHEME 4.4.**

Chiral ligands can be used to complex with a metal counterion.[10,49-52] Thus, addition of *n*-butyllithium to benzaldehyde in the presence of a chiral ligand derived from two proline molecules gives rise to moderate optical yields. Functionalized nucleophiles could also be used with this class of chiral ligand (Scheme 1.9).[50,53-58] Similar findings are observed with chiral auxiliaries derived from tartaric acid,[59] carbohydrates,[60] and other sources.[45,61-64] However, a closely related reaction between diethylzinc and benzaldehyde proceeds with high asymmetric induction,[10,65-105] and closely related carbonyl compounds.[82,84,85,87,106-112] Methyltitanium reagents also show high selectivity.[78,113] With all of these chiral ligands, high optical yields are observed with only specific aldehydes;[114] this approach, as yet, does not seem to be general.[56,63,77,115] The use of organozinc reagents allows for a catalytic cycle as the organometallic reagent will not react with the carbonyl compound unless "activated," usually through complex formation with a nitrogen ligand.[10] This is in contrast to other metals, such as lithium or magnesium, where the nucleophilic species will react very rapidly with a carbonyl compound whether complexed to a chiral ligand or not; electron transfer reactions may also be involved with these reagents.[116]

The TADDOL-based approach with titanium-based reagents (vide infra) has allowed for the problems associated with the zinc approach to be circumvented, particularly as less discrimination between the small and large groups of the carbonyl compound seems to be required for a good degree of induction to be observed. The use of titanium has also proven useful,[117] especially when a number of oxygen coordination sites are available within the substrate molecule [*7.2.6*].[118]

The TADDOL system provides the most promise for development as it has extended the asymmetric addition of a nucleophile to carbonyl compounds other than aryl aldehydes (Scheme 4.5).[88]

The chiral titanium complex is used in catalytic amounts and the mechanism proceeds through exchange with the less hindered isopropoxide to give Ti(TADDOL)(OPr-$i$)$_2$. This is the active catalyst, and is much more reactive than the less hindered tetraisopropoxide. This latter compound is then available to regenerate the active catalyst by a disproportionation mechanism, and thus, acts as a sink for the product. The reaction can also be performed with Grignard reagents (Scheme 4.6).[119,120]

¶ As the sulfur unit is also a masked carbonyl group, other reactions of this type can be found in Chapter 6.

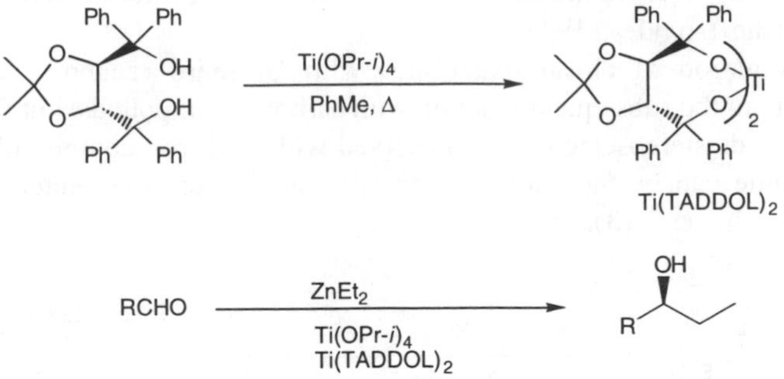

$Ti(OPr\text{-}i)_4$, PhMe, Δ → $Ti(TADDOL)_2$

RCHO → ($ZnEt_2$, $Ti(OPr\text{-}i)_4$, $Ti(TADDOL)_2$)

SCHEME 4.5.

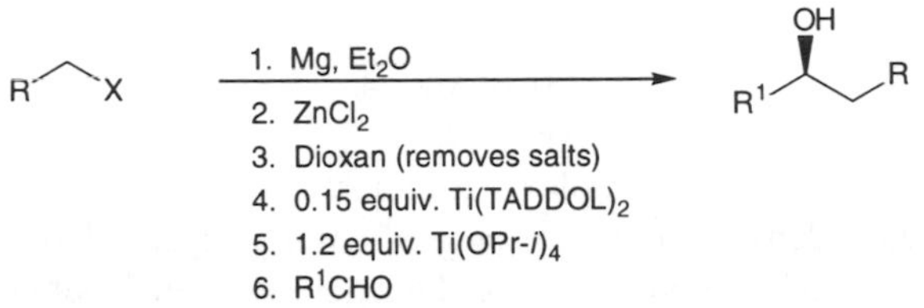

1. Mg, $Et_2O$
2. $ZnCl_2$
3. Dioxan (removes salts)
4. 0.15 equiv. $Ti(TADDOL)_2$
5. 1.2 equiv. $Ti(OPr\text{-}i)_4$
6. $R^1CHO$

SCHEME 4.6.

With diethylzinc, the TADDOL approach gives ee's >70% even with alkyl ketones, although some differentiation between the size of the two alkyl groups is still required. In addition, the TADDOLs can act as cathrates and increase the observed ee during the isolation procedure. Certainly, these TADDOL reactions are as good as, and in some cases better, than comparable transformations with enzymatic systems.

BINOL has been used in a similar series of reactions in place of TADDOL.[107,117,121]

Many examples of high diastereoselection are a result of the use of a chiron, such as 2,3-*O,O*-isopropylideneglyceraldehyde (e.g., Scheme 2.7) [Chapter 6].[122]

### 4.1.3. HOMOLOGATION PROCEDURES

Homologation of an aldehyde can produce a new chiral center at the original carbonyl site. Many procedures accomplish this transformation and involve *umpolung* reagents.[123-131] Methodology has been developed which employs a metal catalyzed addition of carbon monoxide.[132,133]

Stereochemical control for the addition of an umpolung formyl reagent can be achieved through chelate formation (Scheme 4.7),[134] or a similar strategy.[135-137]

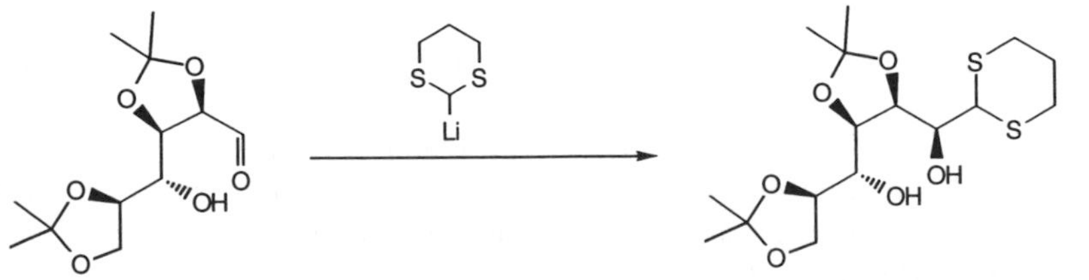

SCHEME 4.7.

The use of 2-trimethylsilylthiazole (ThTMS) (**4.1**) has been advocated as a useful formyl anion reagent, as additions to asymmetric aldehydes proceed with high diastereoselectivity[124,138-140] to provide the *ancat*-isomer.[141] The use of this iterative process is illustrated by the synthesis of the *meso*-octitol derivative **4.2** (Scheme 4.8).[139,142,143]

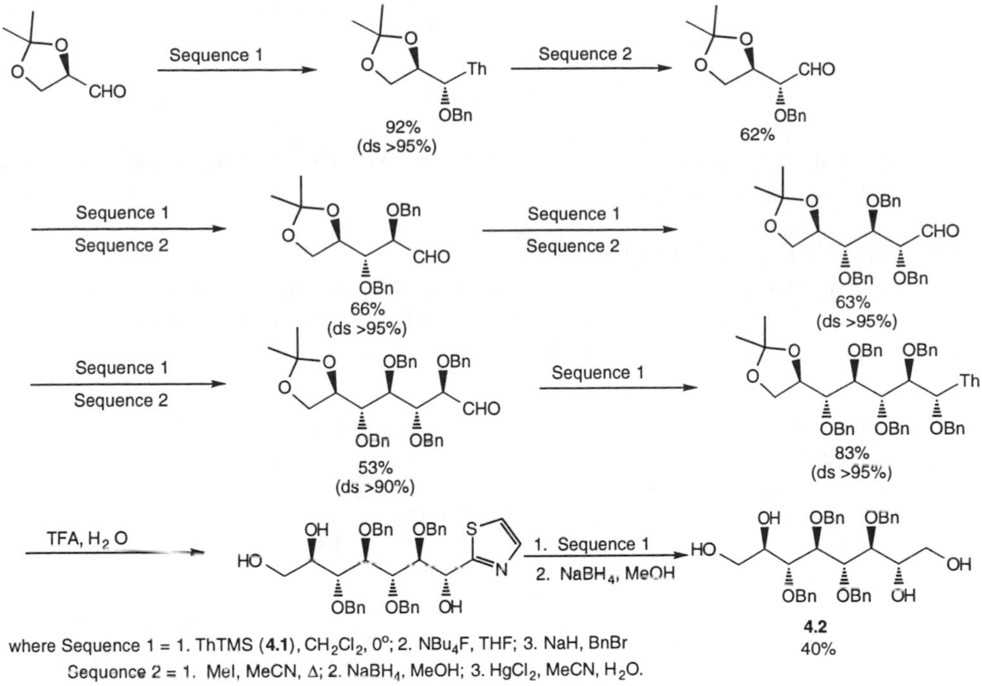
Sequence 1
Th
92%
(ds >95%)
OBn
Sequence 2
CHO
OBn
62%
Sequence 1
Sequence 2
OBn
CHO
OBn
66%
(ds >95%)
Sequence 1
Sequence 2
OBn
CHO
OBn OBn
63%
(ds >95%)
Sequence 1
Sequence 2
OBn OBn
CHO
OBn OBn
53%
(ds >90%)
Sequence 1
OBn OBn
Th
OBn OBn OBn
83%
(ds >95%)
TFA, H2O
HO
OH OBn OBn
OBn OBn OH
1. Sequence 1
2. NaBH4, MeOH
HO
OH OBn OBn
OH
OBn OBn OH
4.2
40%
where Sequence 1 = 1. ThTMS (4.1), CH2Cl2, 0°; 2. NBu4F, THF; 3. NaH, BnBr
Sequence 2 = 1. MeI, MeCN, Δ; 2. NaBH4, MeOH; 3. HgCl2, MeCN, H2O.

**SCHEME 4.8.**

The limitations of this stereoselective addition can be overcome by an oxidation–reduction procedure (Scheme 4.9).[144,145]

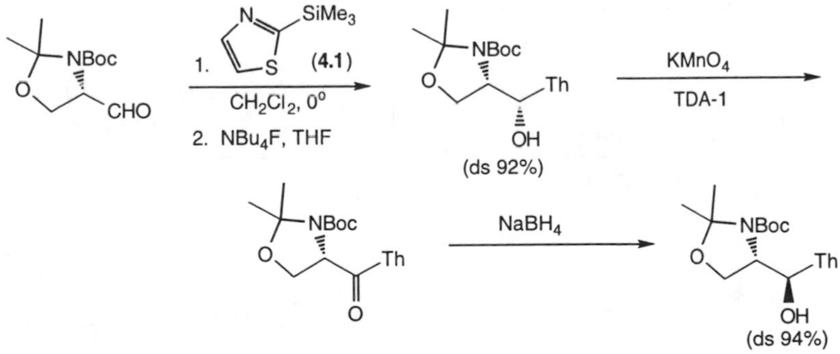
NBoc
CHO
SiMe3
1. (4.1)
CH2Cl2, 0°
2. NBu4F, THF
NBoc
Th
OH
(ds 92%)
KMnO4
TDA-1
NBoc
Th
NaBH4
NBoc
Th
OH
(ds 94%)

**SCHEME 4.9.**

In many respects, the oxathiane is an asymmetric homologation reagent, although the original condensation tends not to be stereospecific; further manipulations allow a wide range of target compounds to be reached (Scheme 2.4).[34-41] The use of the chiral acyl anion equivalent, *trans*-1,3-dithiane-*S,S*-dioxide does allow for some selection by use of kinetic or thermodynamic conditions, but separation of the carbonyl addition products is still necessary to obtain just one isomer.[146]

The chiral sulfoxide **4.3** has been used for a chiral homologation (≥70% ee) (Scheme 4.10).[23]

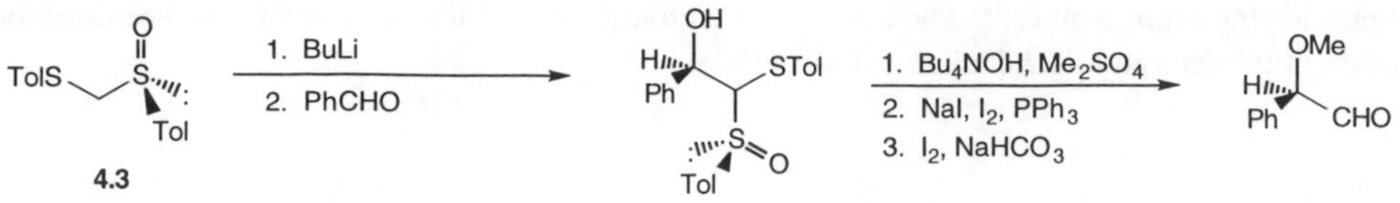

SCHEME 4.10.

The Strecker type synthesis of α-amino acids does provide some asymmetric induction when a chiral amine is employed.[147-149]

### 4.1.4. REACTIONS OF VINYL ANIONS

Vinyl anions have also been used in stereoselective additions to carbonyl compounds.[150-157] Reaction of the vinyllithium **4.4**, obtained from ethyl (*S*)-lactate, with benzaldehyde resulted in formation of the enone **4.5**, after hydrolysis (Scheme 4.11).[158]

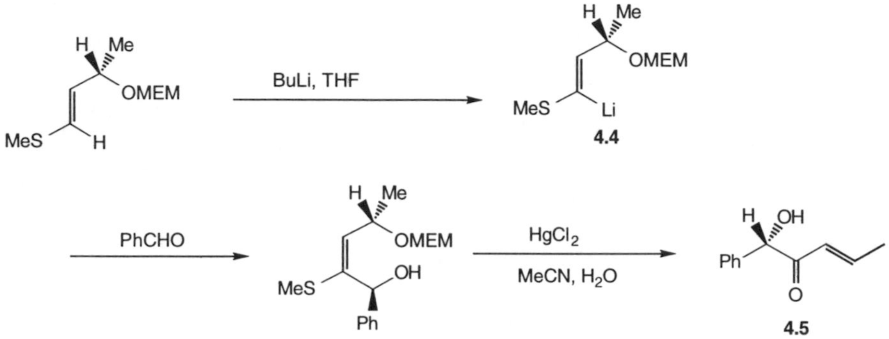

SCHEME 4.11.

A related reaction allows for the preparation of 1,2-diols (Scheme 4.12),[159,160] 1,3-diols,[161] triols,[162] and carbohydrate derivatives.[163,164]

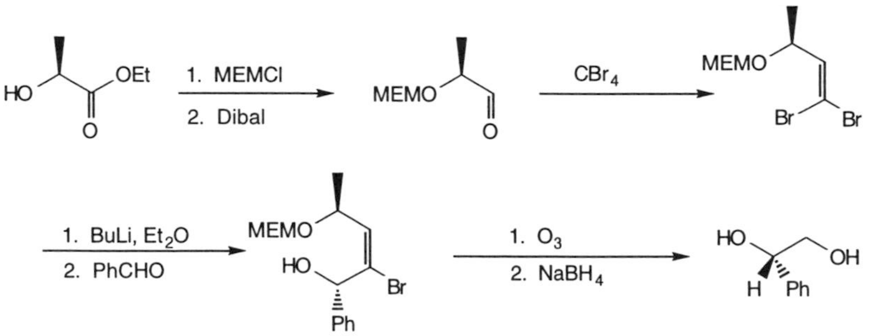

SCHEME 4.12.

Reasonable selectivity can be realized for the addition of a vinyl anion to an aldehyde, especially if a heteroatom is also present in the electrophilic moiety to allow for chelation control.[161,165,166]

Functionalized vinyl anions also provide a wide variety of methods for the homologation of carbonyl compounds, but few provide a high degree of asymmetric induction.[124] One example of the incorporation of a chiral moiety is the use of a chiral sulfoxide that can provide some diastereoselectivity (Scheme 4.13).[150,151,167,168]

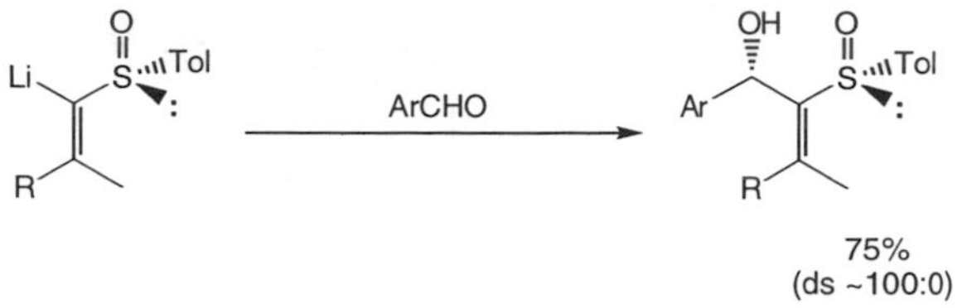

ArCHO

75%
(ds ~100:0)

**SCHEME 4.13.**

An interesting example of a diastereoselective addition to an aldehyde involves, not a vinyl anion, but an ylide (Scheme 4.14.).[169,170]

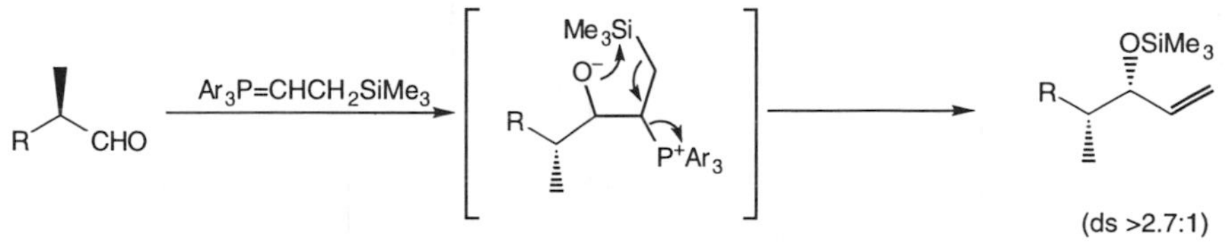

$Ar_3P{=}CHCH_2SiMe_3$

(ds >2.7:1)

**SCHEME 4.14.**

## 4.1.5. USE OF "EXTENDED" ORGANOMETALLIC REAGENTS

The use of an organometallic reagent that contains functionality allows further modification of the adducts derived from addition to carbonyl compounds.[171-177] This approach has provided alternatives to the aldol [*7.1*], and the homoaldol reactions.[138] In this regard, many heteroatoms have been incorporated into the organometallic unit along with unsaturation to afford a substituted allyl anion.[171]

The flexibility of such an approach does rely somewhat upon the configurational stability of the ambident allyl anion and its site of reaction.[178-181] Examples of metals that fulfill these regiochemical criteria are included in Table 4.1. A wide variety of electrophiles have been condensed with allyl anions.[182-184]

The addition of allyl organometallic reagents to a carbonyl compound results in the formation of a homoallyl alcohol;[274,289,296,341-344] epoxides can also be used as the electrophile to provide alk-4-en-1-ols.[345-347] If a chiral ligand is present on the metal, particularly with boranes,[279,348-351] asymmetric induction may be observed,[352-354] as illustrated in Schemes 4.15,[279] and 4.16.[300,302]

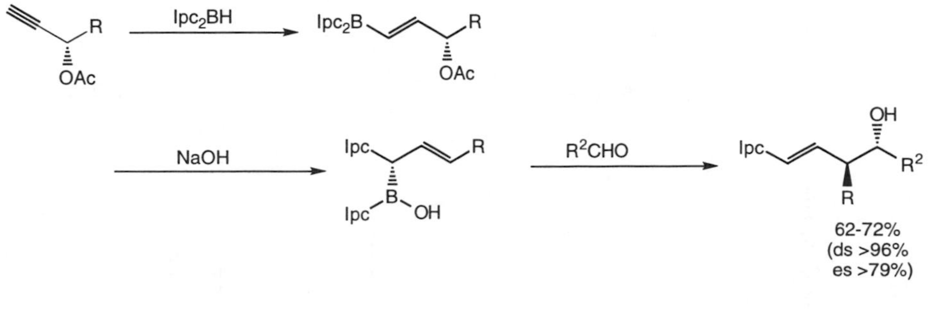

$Ipc_2BH$

NaOH

$R^2CHO$

62-72%
(ds >96%
es >79%)

**SCHEME 4.15.**

# TABLE 4.1
## Examples of Allylic Moieties that Condense with Carbonyl Compounds

| Allylic Compound | Base or Catalyst | Major Isomer[a] | Ref. |
|---|---|---|---|
| R–CH=CH–CH$_2$–X$_b$ | Cr(II) | A | 185–189 |
| | LDMAN-$CeCl_3$[c] | —[d] | 191,192 |
| | Li | S | 197 |
| | Li or Mg, $Cp_2ZrCl_2$ | A | 186,201 |
| | Mg | — | 186 |
| | Mo | A | 205 |
| | Sb | S | 207 |
| | $SbCl_3$-Fe | — | 208 |
| | $SbCl_3$-Al | — | 208 |
| | Sn-Al | A | 211 |
| | $SnX_2$ | — | 189,213 |
| | Ti | A | 186,188, 215–218 |
| | Zn | S | 219 |
| | Zn-$NH_4Cl$ | —[e] | 221 |
| R–CH=CH–CH(R$^1$)–OH | $PdCl_2$-$(PhCN)_2$, $SnCl_2$ | S[f] | 222,223 |
| | | A[g] | 223 |
| R–CH=CH–CH$_2$–OR$^1$ | Al | A | 206,226 |
| | Base[h] | — | 228 |
| | Li | —[i] | 228 |
| | Li, $BF(OMe)_2$ | S | 230 |
| | Ti | A | 231 |
| | Zr | S | 206 |
| CH$_2$=CH–CH$_2$–OImd$_j$ | Cd | A | 172,190 |
| R$^1$–CH=C(R$^2$)–CH(R$^3$)–OCONR$^4{}_2$ | Al | S | 193–196 |
| | Ti | A | 194,198–200 |
| R–CH=CH–CH$_2$–SR$^1$ | Base | γ[k] | 182,202–204 |
| | Li | A | 203 |
| | Li, $Et_3Al$ | S | 206 |
| | Sn, $BF_3$ | A | 206 |
| | Ti | A | 206,209,210 |
| | Zr | S | 206 |
| CH$_2$=CH–CH$_2$–SOR$^1$ | LDA, HMPA | S | 212 |
| | | A[l] | 214 |
| | LDA, $Me_3P$ | —[m] | 212 |
| CH$_2$=CH–CH$_2$–SeR$^1$ | Al | — | 220 |
| | Li | γ | 220 |
| CH$_2$=CH–CH$_2$–NHCO$_2$Bu-*t* | Li | — | 224 |
| CH$_2$=CH–CH$_2$–NRCOR$^1$ | Li | γ | 225 |
| 2-(methoxymethyl)pyrrolidin-1-yl–CH=CH–CH$_2$–Ph (OMe, N, Ph) | Li | — | 227 |

| Allylic Compound | Base or Catalyst | Major Isomer[a] | Ref. |
|---|---|---|---|
| 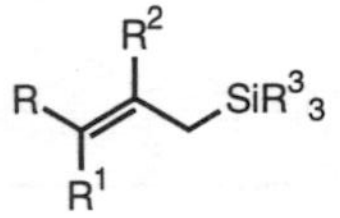 R, $R^2$, $SiR^3_3$, $R^1$ | — | — | 232 |
| | — | A | 234 |
| | — | S[n] | 238 |
| | Al | — | 220 |
| | $Bu_3SnCl\text{-}BF_3$ | A | 242 |
| | $EtAlCl_2$ | S | 242 |
| | $F^-$ | A | 252–256 |
| | Li | γ | 220 |
| | $MgBr_2$ | — | 263 |
| | ROH, $NEt_3$[o] | A | 265 |
| | $SnCl_4$ | — | 267–269 |
| | Ti | A | 231,270 |
| | $TiCl_4$ | γ | 254,272–275 |
| | | S[p] | 263 |
| 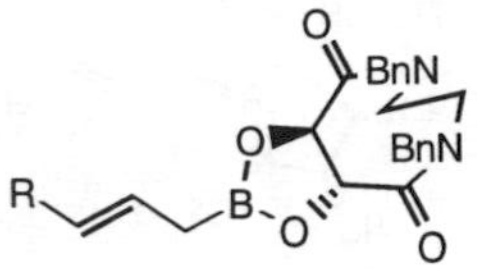 R, O, B, O, BnN, BnN, O, O | — | — | 280 |
| 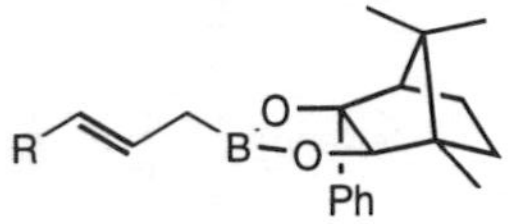 R, B, O, O, Ph | — | A | 282–285 |
| 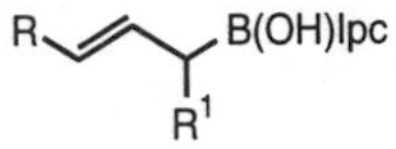 R, B(OH)Ipc, $R^1$ | — | A | 279 |

| Allylic Compound | Base or Catalyst | Major Isomer | Ref. |
|---|---|---|---|
|  $R^*N$, $R^1$ | Li | γ | 229 |
| 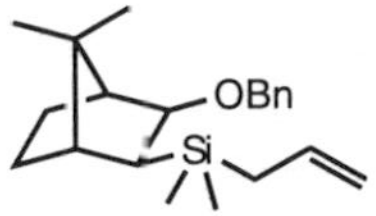 OBn, Si | Ti | — | 233 |
| 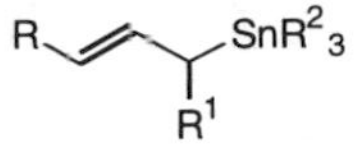 R, $SnR^2_3$, $R^1$ | — | A[q] | 235–237 |
| | — | S | 234,239, 240 |
| | $AlCl_3$-*i*PrOH | — | 241 |
| | $BF_3{\cdot}OEt_2$ | S | 235,243–250 |
| | $MgBr_2$ | S | 245,251 |
| 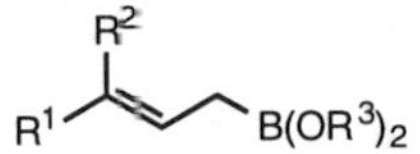 $R^2$, $R^1$, $B(OR^3)_2$ | — | S[n] | 257,258 |
| 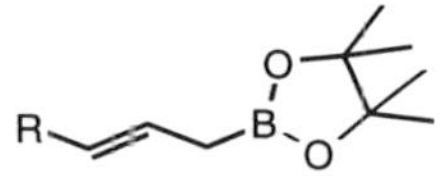 R, O, B, O | — | A | 174,259–262 |
| | | S[q] | 174,259–262,264 |
| 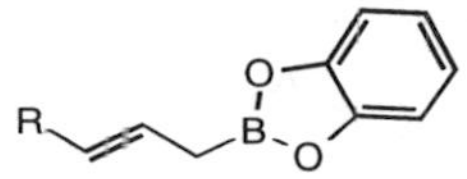 R, O, B, O | — | A | 266 |
| | $Pd(PPh_3)_4$ | S[r] | 266 |
| 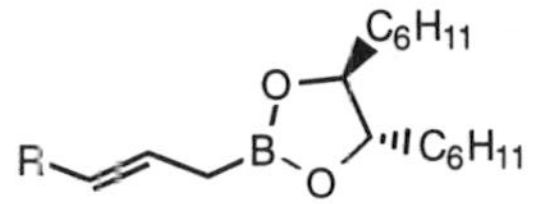 R, O, B, O, $C_6H_{11}$, $C_6H_{11}$ | — | S[q] | 259,271 |

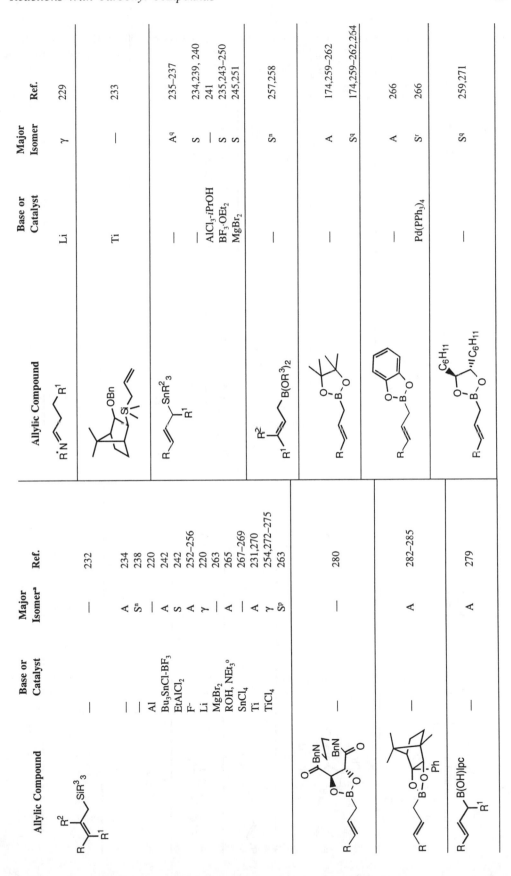

**TABLE 4.1 (continued)**

| Allylic Compound | Base or Catalyst | Major Isomer[a] | Ref. |
|---|---|---|---|
|  R, B(NMe$_2$)$_2$ | — | A<br>S[q] | 282<br>282 |
| 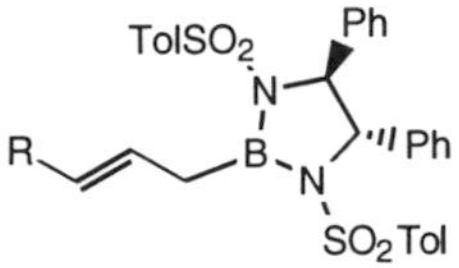 R, TolSO$_2$, N, Ph, B, N, Ph, SO$_2$Tol | — | — | 291,292 |
| 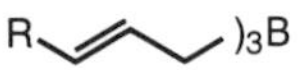 R, )$_3$B [s] | — | A | 293 |
|  R, 9-BBN | — | A | 246,295,296 |
| 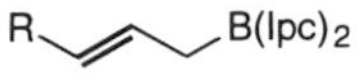 R, B(Ipc)$_2$ | — | —<br>A<br>S[q] | 293,298–301<br>302<br>302 |
| 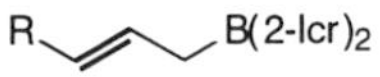 R, B(2-Icr)$_2$ | —<br>— | —<br>A<br>S[q] | 305<br>305<br>305 |
| 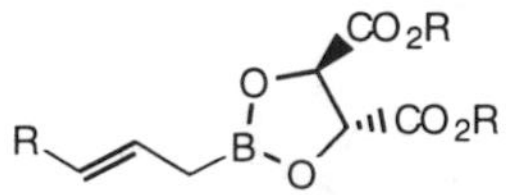 R, B, O, O, CO$_2$R, CO$_2$R | — | A<br>S[q] | 276–279<br>278 |
| 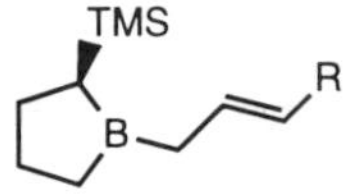 TMS, B, R | — | — | 281 |
| 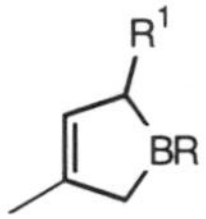 R[1], BR | — | S[t] | 286 |
|  Cl, SiMe$_3$ | AlCl$_3$ | — | 287 |
| 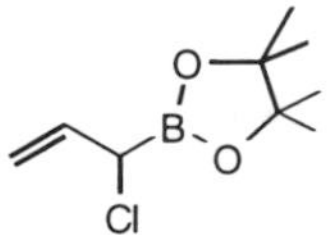 B, O, O, Cl | — | A | 288,289 |
| 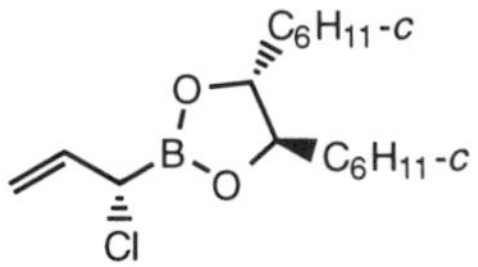 B, O, O, C$_6$H$_{11}$-*c*, C$_6$H$_{11}$-*c*, Cl | — | — | 290 |

| Allylic Compound | Base or Catalyst | Major Isomer[a] | Ref. |
|---|---|---|---|
| 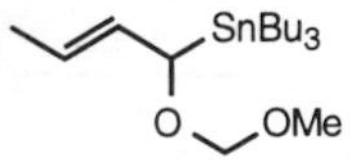 SnBu3, O, OMe | — | A | 306,307 |
| | Δ | A | 307,309–311 |
| 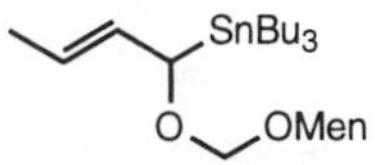 SnBu3, O, OMen | — | A | 306,310 |
| | Δ | A | 306 |
| 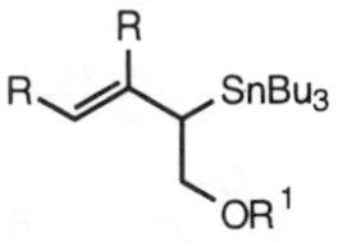 R, R, SnBu3, OR 1 | — | S | 316 |
| 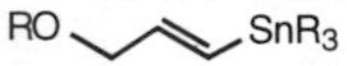 RO, SnR3 | — | S | 318 |
| 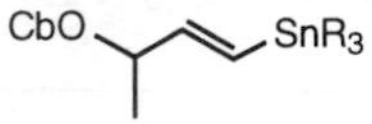 CbO, SnR3 | $TiCl_4$ | A | 321,322 |
| 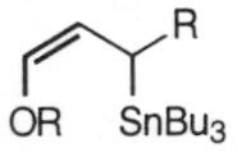 R, OR, SnBu3 | — | S | 324 |

| Allylic Compound | Base or Catalyst | Major Isomer | Ref. |
|---|---|---|---|
| 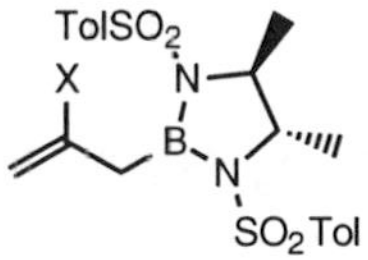 TolSO2, X, N, B, N, SO2Tol; X = Cl or Br | — | — | 291 |
| 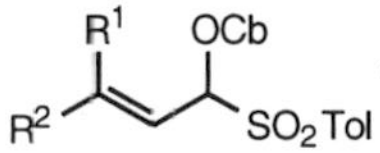 R1, OCb, R2, SO2Tol [u] | Li | S | 294 |
| | Ti | A | 297 |
| | | S[p] | 294 |
| 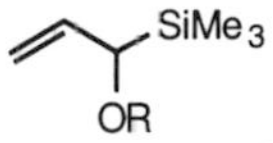 SiMe3, OR | Li | — | 303 |
| 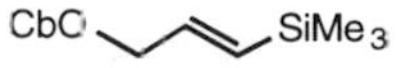 CbO, SiMe3 | Al | A[q] | 304 |
| | Li | A[v] | 304 |
| | Ti | A[q] | 304 |
| 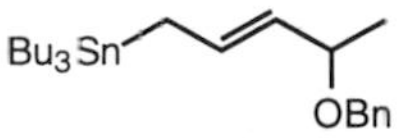 Bu3Sn, OBn | $SnCl_4$ | — | 308 |
| 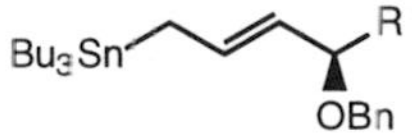 Bu3Sn, R, OBn | $SnCl_2$ | — | 312 |
| 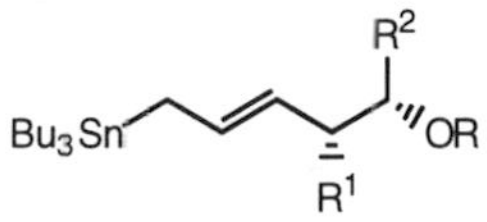 Bu3Sn, R2, OR, R1 | $SnBr_4$ | — | 313,314 |

**TABLE 4.1 (continued)**

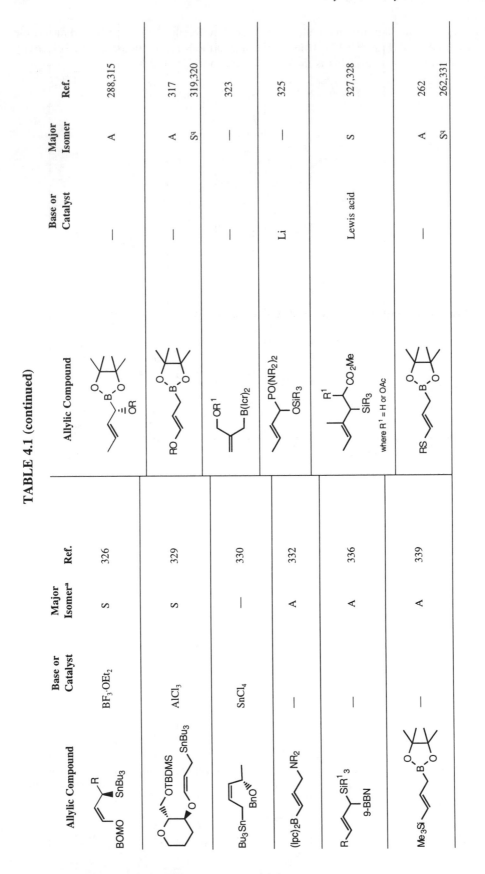

| Allylic Compound | Base or Catalyst | Major Isomer[a] | Ref. |
|---|---|---|---|
| BOMO, R, $SnBu_3$ | $BF_3 \cdot OEt_2$ | S | 326 |
| O, OTBDMS, O, $SnBu_3$ | $AlCl_3$ | S | 329 |
| $Bu_3Sn$, BnO | $SnCl_4$ | — | 330 |
| $(Ipc)_2B$, $NR_2$ | — | A | 332 |
| R, $SiR^1_3$, 9-BBN | — | A | 336 |
| $Me_3Si$, B, O, O | — | A | 339 |

| Allylic Compound | Base or Catalyst | Major Isomer | Ref. |
|---|---|---|---|
| 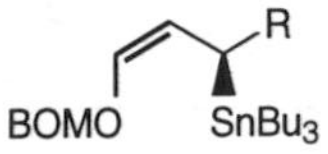 B, O, O, OR | 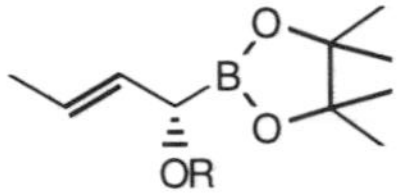 — | A | 288,315 |
| 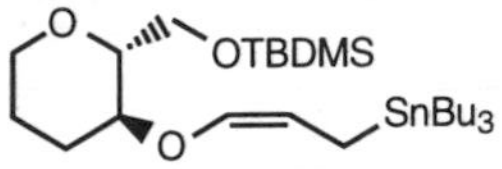 RO, B, O, O | 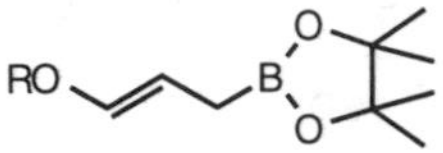 — | A<br>S[q] | 317<br>319,320 |
| $OR^1$, $B(Icr)_2$ | — | — | 323 |
| $PO(NR_2)_2$, $OSiR_3$ | Li | — | 325 |
| 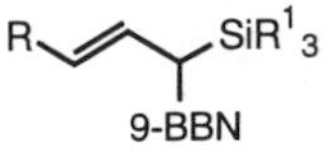 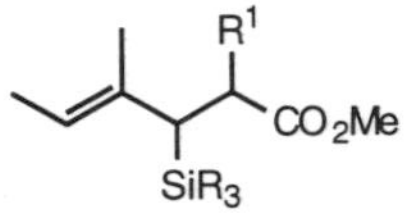 $R^1$, $CO_2Me$, $SiR_3$ where $R^1$ = H or OAc | Lewis acid | S | 327,328 |
| 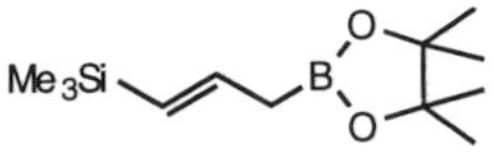 RS, B, O, O | 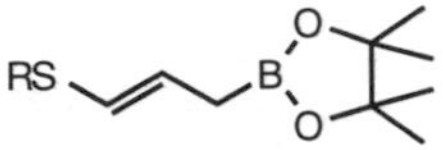 — | A<br>S[q] | 262<br>262,331 |

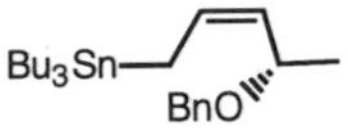

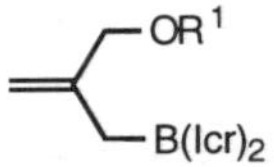

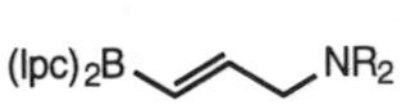

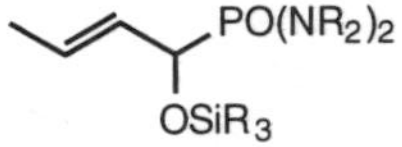

| Allylic Compound | Base or Catalyst | Major Isomer[a] | Ref. | Allylic Compound | Base or Catalyst | Major Isomer | Ref. |
|---|---|---|---|---|---|---|---|
| $Me_3Si$, R, 9-BBN | — | —[w] | 337,338 | $C_6H_{11}OMe_2Si$, B, O, O, $CO_2Pr$-*i*, $CO_2Pr$-*i* | — | A | 333–335 |

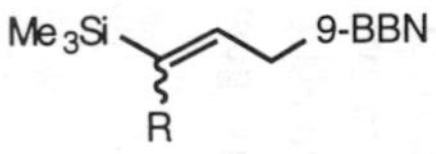

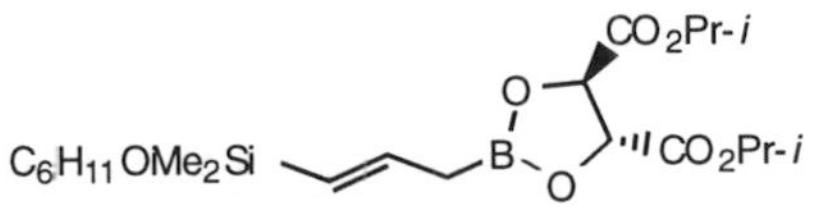

[a] The isomer can vary depending on the substrate, refer to the original citation. The abbreviations S = *syncat*, and A = *ancat* have been used. [b] Where X = a halogen. [c] For X = SPh. [d] Signifies not applicable or not given. [e] Little selectivity is observed. [f] In the presence of a small amount of water. [g] In the presence of a large amount of water. [h] No specific base is quoted. [i] A Brook rearrangement is also available when $R^1 = SiR_3$.[340] [j] Where Imd = (N, N, R) [k] Denotes g-addition of the allyl anion is the major reaction pathway. [l] Where $R^1$ = 2-pyridyl. [m] An Evans rearrangement also occurs under the reaction conditions. [n] With respect to $R^2$. [o] For $R^1$ = F. [p] With a-alkoxy aldehydes. [q] From the *cis*-alkene, otherwise the *ancat* isomer is formed. [r] With respect to the alkyl group. [s] From allyllithium with $BR^1_3$. [t] With respect to the sidechain delivered from the borolene and the hydroxy group. [u] Where Cb = CON(Pr-*i*)$_2$. [v] The geometry of the resultant alkene is determined by the use of *i*-$Bu_2AlOSO_2Me$ (gives *E*-isomer) or $(Me_2N)_3TiCl$ (gives *Z*-isomer). [w] Elimination occurs during work-up to give the diene.

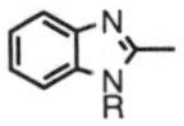

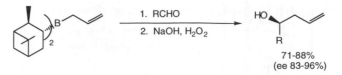

**SCHEME 4.16.**

The additions of allyl organometallic reagents to carbonyl compounds can proceed through four possible cyclic transition states (Scheme 4.17).[171] In these cases, the ligand, L, can determine the stereochemical outcome of the reaction. Diastereoselectivity decreases as the length of the metal–oxygen bond increases.[355] The selectivity is dependent upon the minimization of steric interactions within the chair transition state.[174,356] However, it is not certain that a cyclic transition state is involved in every case; an $S_E2'$ reaction pathway could be preferred, particularly when the heteroatom is silicon or tin.[234,357-362] The use of organosilanes and stannanes in the presence of a Lewis acid provides homoallylic alcohols.[269]



**SCHEME 4.17.**

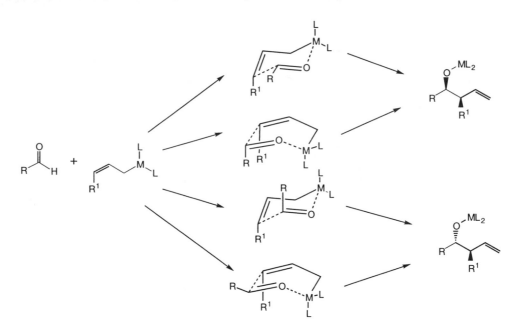

Chiral carbonyl moieties provide the opportunity for either Anh-Felkin or chelation control addition,[317] but the best selectivity is often seen with an alkoxy aldehyde (Chapter 6).[243,245,360,363,364] Of the many allylic nucleophiles, allyltrimethylsilanes provide the best selectivity (vide infra); this outcome has been rationalized in terms of a strong chelation effect.[254]

#### 4.1.5.1. Allylboranes

The use of chiral allylboranes allows transfer of the allyl group in an asymmetric manner.[177,259-261,276,279,280,300,305,306,365,366] Indeed, the use of substituted allylboron systems, such as crotyl, with α-substituted aldehydes has been the subject of considerable investigation. The diastereoselectivity does not depend upon the size of the substituent in the reagent, but only on the alkene geometry (vide infra).

The presence of a chiral center within the electrophilic moiety provides the possibility for double asymmetric induction (Scheme 4.18).[335,356,367-374] The selectivity observed with the mismatched pair (**4.6** with **4.7**) is lower than that for the corresponding matched pair (**4.6** and **4.8**).

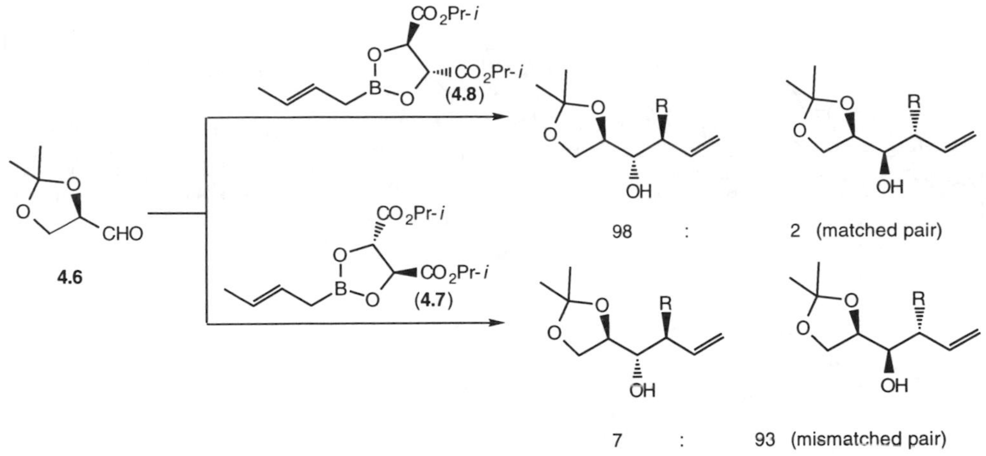

**SCHEME 4.18.**

These observations have been interpreted in terms of the minimization of strain energies for a chair transition state.[375-377] The high selectivity could arise from n/n electronic repulsive interactions between the aldehyde oxygen atom and an ester carbonyl group. These interactions have been exploited by the use of the boron compound **4.9** (Scheme 4.19).[278]

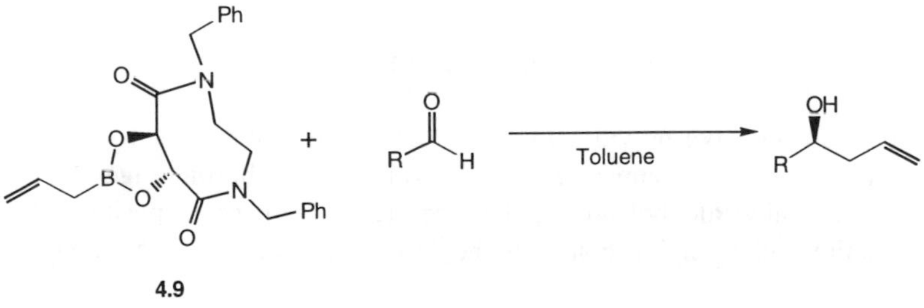

**SCHEME 4.19.**

The tartrate esters, although they do not provide the highest diastereo- and enantioselectivity, are very reactive and provide very short reaction times, certainly when compared to compound **4.9**.[348,367] A hypothesis which accounts for the effect of reaction variables (temperature, solvent, moisture), and based on lone pair interactions has been proposed (Scheme 4.20).[276,277,280,367]

The enantioselectivity of the allylborations of aromatic and propargylic aldehydes is enhanced by the use of metal carbonyl complexes of the aldehyde substrates.[378]

The use of the "stien" control group allows recovery of the chiral reagent in an expeditious manner (Scheme 4.21). Substituted allyl anions can also be employed in this approach.[291]

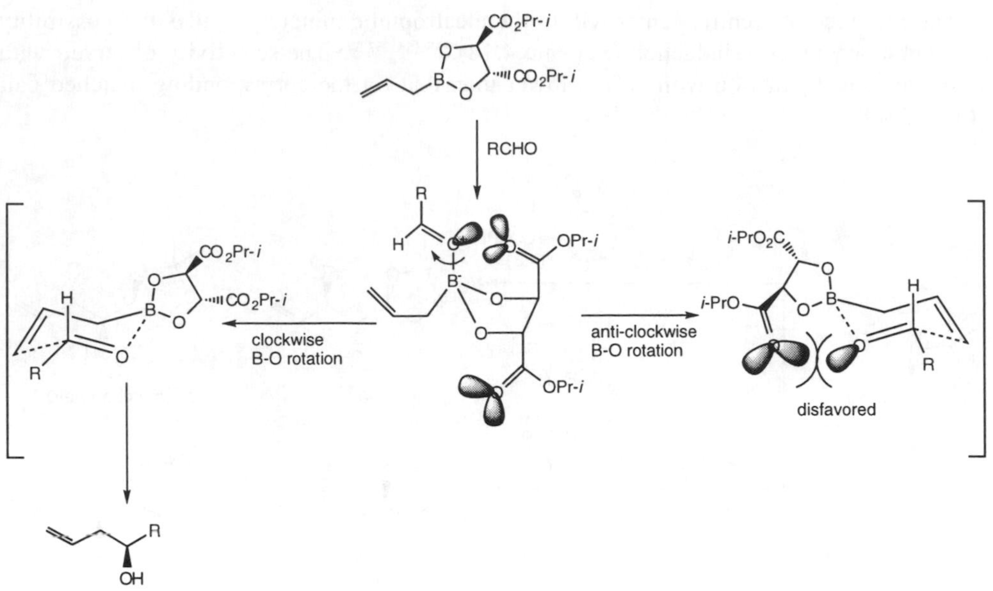

SCHEME 4.20.

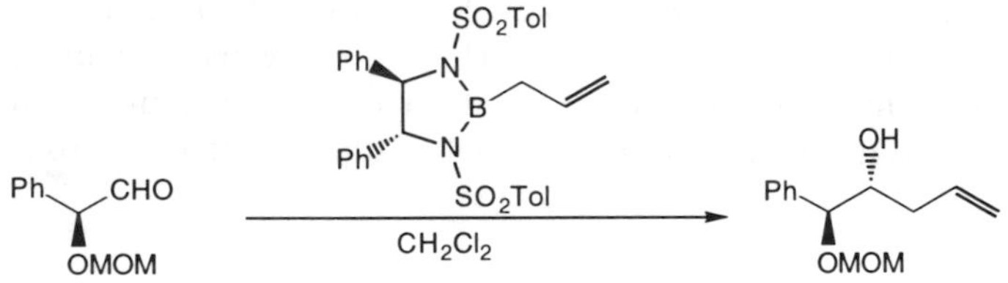

SCHEME 4.21.

Isocaranyl boranes have provided excellent selectivity for the control of a single chiral center (Scheme 4.22), and compare extremely well to other boron reagents (Table 4.1). In addition, the homoallyl alcohol product has the stereochemistry opposite of that from the analogous reaction with *B*-allyldiisopinocampheylborane or *B*-allylbis(4-isocaranyl)borane.[301,305]

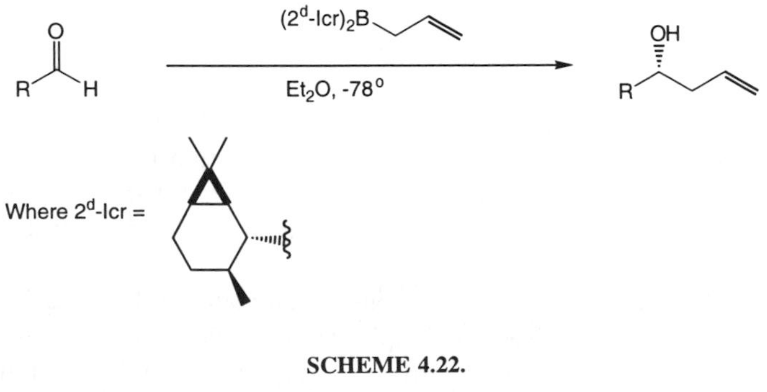

SCHEME 4.22.

When a substituted allyl system is employed with 2-isocaranylboranes, a high degree of selectivity is observed (Schemes 4.23 and 4.24). Again, the resultant diastereoselectivities compare favorably to those obtained with other chiral auxiliaries on boron.[305,351]

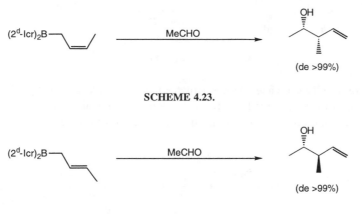

**SCHEME 4.23.**

**SCHEME 4.24.**

The use of the borolane derivative **4.10** also provides a means for allylboration of aldehydes, as a resolution process can be incorporated into the reaction sequence (Scheme 4.25).[281]

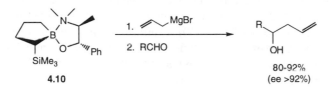

**SCHEME 4.25.**

### 4.1.5.2. Other "Neutral" Allylic Nucleophiles

Some of the examples in Table 4.1 include allylic species that do not require formation of an allylic anion; allylboranes are such an example. The heteroatom substituent provides the driving force for the reaction, often through stabilization of an intermediate carbocation. However, many of the models used to rationalize the reaction accommodate this difference (vide infra); a Lewis acid can also play a variety of roles.[379,380] Examples are provided by allylstannanes with methodology to a number of carbohydrate derivatives as illustrated by the expeditious synthesis of the Prelog-Djerassi lactone (**4.11**) (Scheme 4.26).[248,251,381]

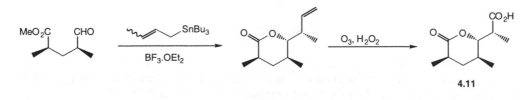

**SCHEME 4.26.**

Reaction of an allylstannane with an aldehyde results in the formation of the *ancat* product; the presence of a stereogenic center within the nucleophilic substrate can be used to control the resultant stereochemistry (Scheme 4.27),[306,310,382] or a chiral catalyst can be employed.[383-385]

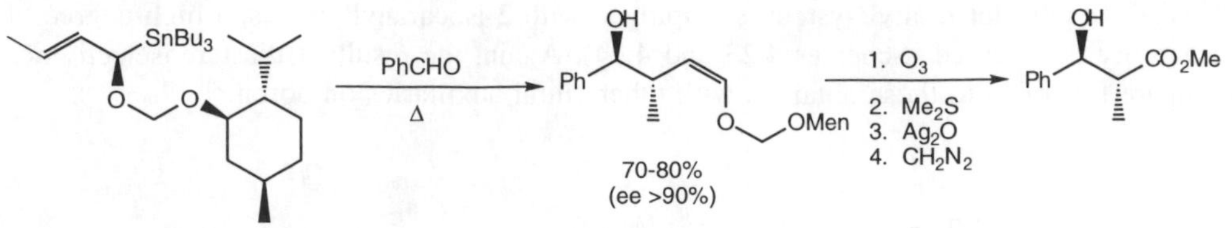

SCHEME 4.27.

The reaction of allylstannanes with aldehydes has been investigated by NMR.[359,363,364]

L-Hexose derivatives are available from the condensation of allyltrimethylsilane with a chiral aldehyde in the presence of magnesium bromide (Scheme 4.28).[254]

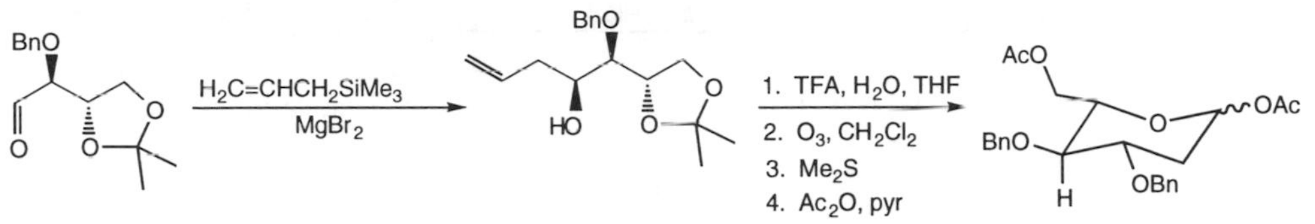

SCHEME 4.28.

The alkene unit of the condensation product can be used for further modifications. In the first example (Scheme 4.29), an aldol-type product is obtained without the use of a strong base.[250] For the latter example (Scheme 4.30), a chiral silane is used which ultimately provides a β-hydroxy ester with good enantioselectivity.[272] A chiral allylsilane based on an asymmetric silicon also affords enantioselectivity when reacted with an aldehyde.[269,362,386] Chiral catalysts can also be used as the source of asymmetry.[387]

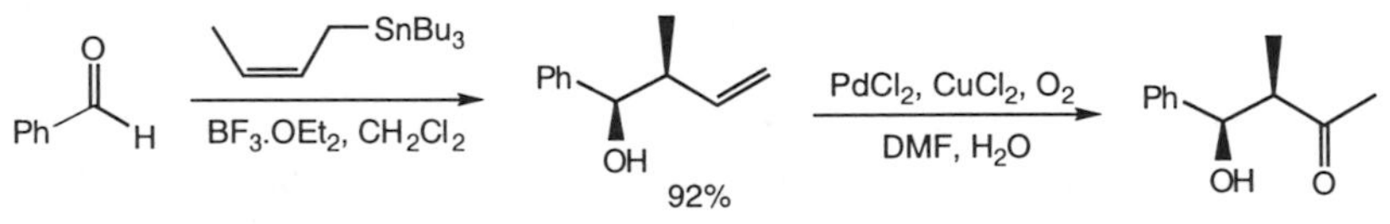

SCHEME 4.29.

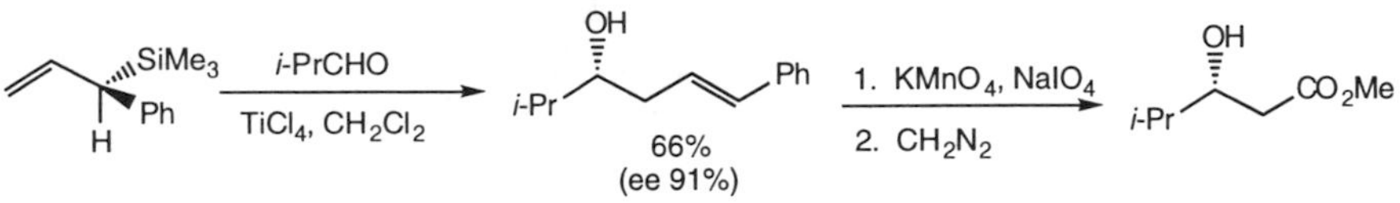

SCHEME 4.30.

#### 4.1.5.3. Allyl Anions

Other metals can provide high diastereoselectivity, and hence, enantioselectivity, such as molybdenum (Scheme 4.31),[205,388,389] and titanium (Scheme 4.32).[215-218,390-392]

In addition, other functional groups can be tolerated in the nucleophilic species;[170,175,286,393-403] this then allows a rapid entry to highly functionalized compounds (Table 4.1). Examples are provided by the γ-alkoxyallylaluminum series, which show high diastereoselectivity through formation of a chair-like transition state (**4.12**) where the large group adopts a quasi-equatorial orientation (Scheme 4.33).[226]

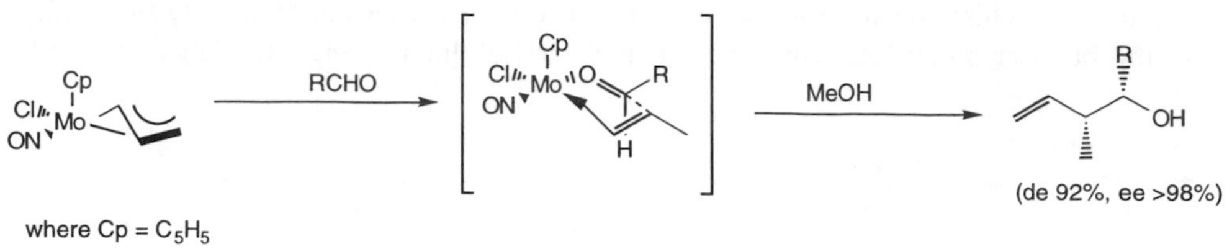

where Cp = $C_5H_5$

SCHEME 4.31.

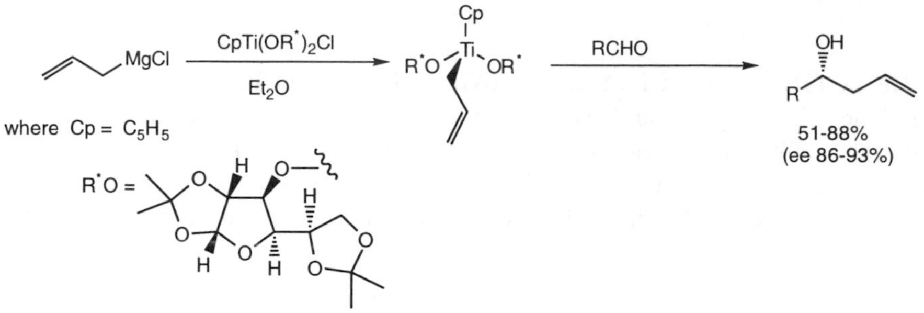

where Cp = $C_5H_5$

SCHEME 4.32.

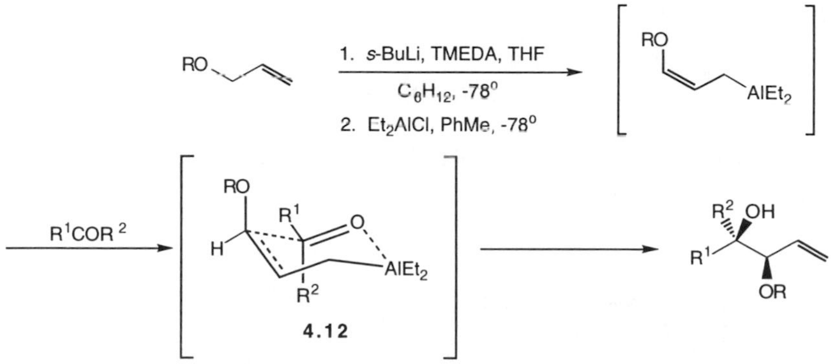

SCHEME 4.33.

Of course, the carbonyl moiety can contain additional functionality and many useful chirons have been used in this context, such as the acetonide of L-glyceraldehyde (**4.6**), as illustrated by a synthesis of D-ribose (**4.13**) (Scheme 4.34).[10,213,404-408]

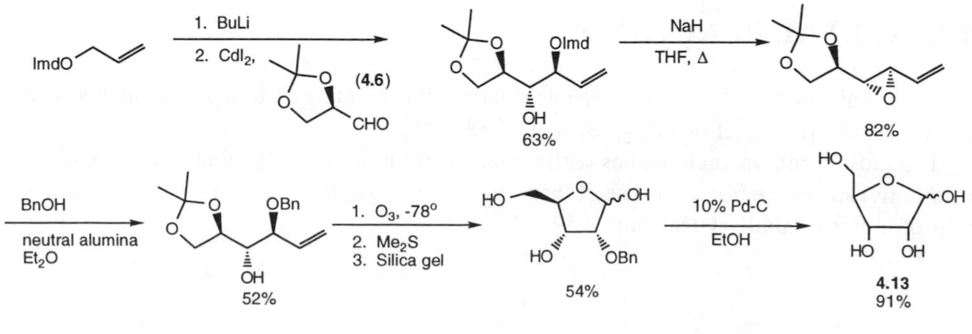

SCHEME 4.34.

The use of a chiral ligand, such as sparteine, can enhance enantioselectivity of an allyl anion; this has been utilized to prepare the functionalized allylic stannane **4.14** (Scheme 4.35).[321]

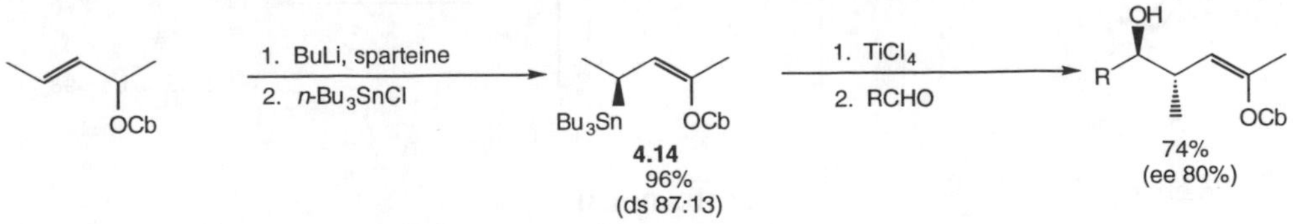

1. BuLi, sparteine
2. $n$-$Bu_3SnCl$
OCb
$Bu_3Sn$ OCb
**4.14**
96%
(ds 87:13)
1. $TiCl_4$
2. RCHO
OH
R
OCb
74%
(ee 80%)

SCHEME 4.35.

### 4.1.6. REACTIONS OF ACETYLENE AND ALLENIC ANIONS

Acetylene anions rely on Anh-Felkin control, as their steric demands are relatively small.[409] The addition of alkynyl zinc reagents to aldehydes can be brought about by a chiral catalyst, but the enantioselectivity is usually moderate at best (*cf.* Scheme 1.9).[55,75,410] Much higher selectivity is observed with chiral aldehydes as the electrophilic unit (Scheme 4.36).[411,412]

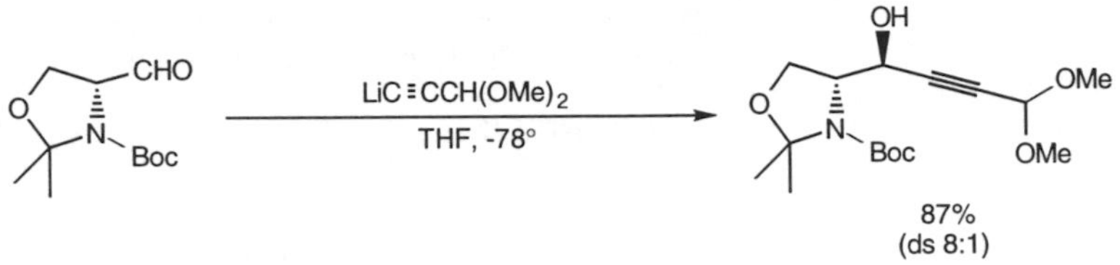

CHO
O
N
Boc
LiC≡CCH(OMe)$_2$
THF, -78°
OH
OMe
OMe
Boc
87%
(ds 8:1)

SCHEME 4.36.

Allenes, in contrast, are sterically more demanding and useful selectivity has been observed with allenic boranes (Scheme 4.37).[154,413]

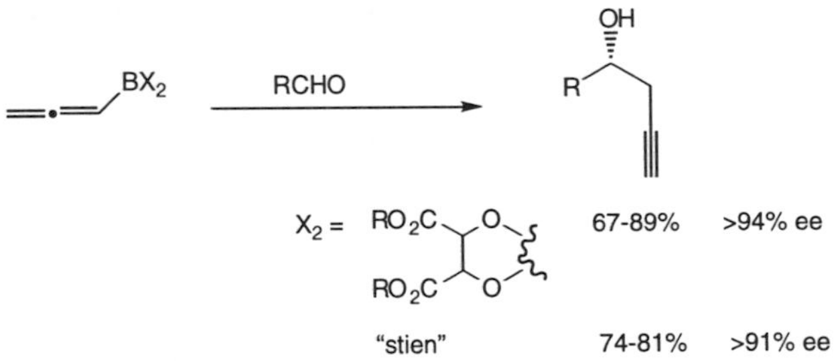

$BX_2$
RCHO
OH
R
$X_2$ = $RO_2C$ O, $RO_2C$ O — 67-89% >94% ee
"stien" — 74-81% >91% ee

SCHEME 4.37.

## 4.2. ACETAL REACTIONS

An acetal functionalized within the auxiliary unit can act as carbonyl equivalents to allow for stereoselective reactions (e.g., Scheme 4.38).[414-420]

The most common nucleophiles with acetal electrophiles are allylsilanes; the acetal acts as a carbonyl surrogate.[421,422] The use of these masked electrophiles also allows for the introduction of a chiral auxiliary (Scheme 4.39).[153,414,415,417,422-428]

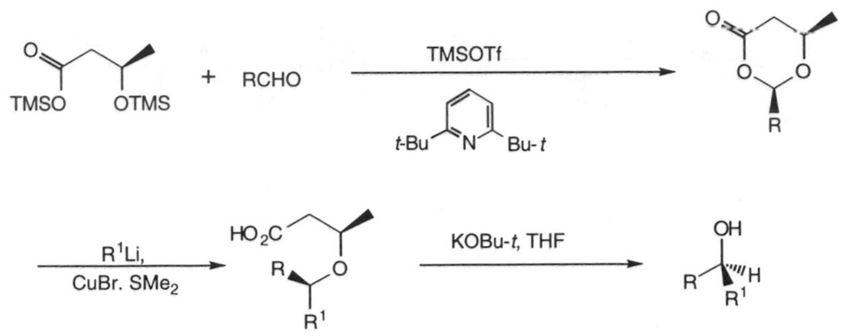

SCHEME 4.38.

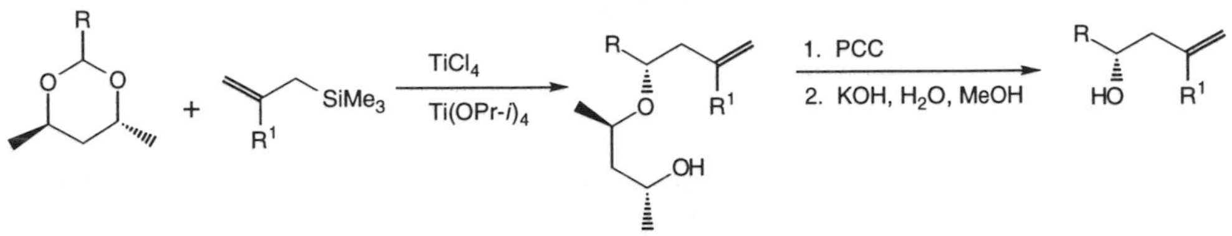

SCHEME 4.39.

A variant is the use of 2-methoxyoxazolidines where the Lewis acid employed can control the stereochemical outcome of the reaction (Scheme 4.40). In this case, the addition could be selective, while the Lewis acid promotes an equilibration; treatment of **4.15** with titanium tetrachloride provided exclusively **4.16**.[429,430]

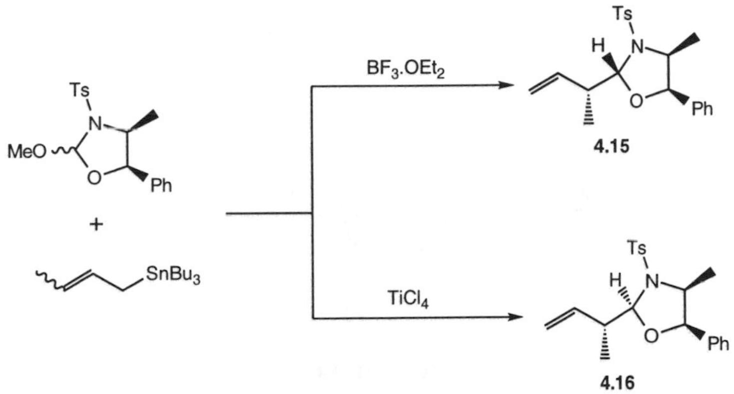

SCHEME 4.40.

In addition, if the α-group contains a heteroatom, this can influence the stereochemical outcome of the reaction through chelate formation [*2.1.2*].[256,431-438]

The use of an aryl-metal complex can provide facial selectivity, as illustrated by the synthesis of the chiral amide **4.17** by Ritter methodology (Scheme 4.41).[439-441]

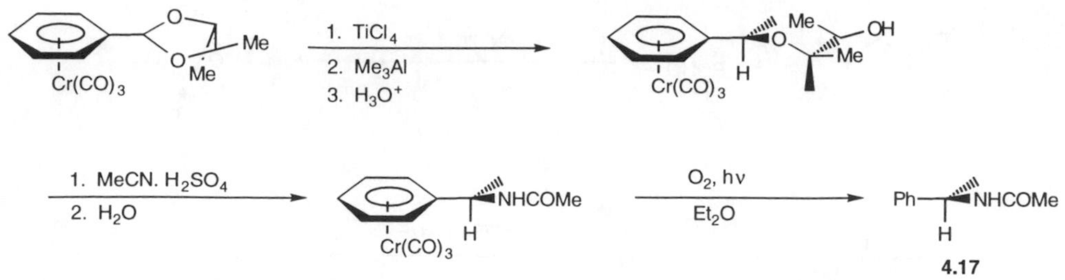

SCHEME 4.41.

An oxazolidine can be used in place of an acetal (Scheme 4.42); the addition is thought to occur through ring opening and subsequent reaction of the resultant imine (vide infra).[442,443]

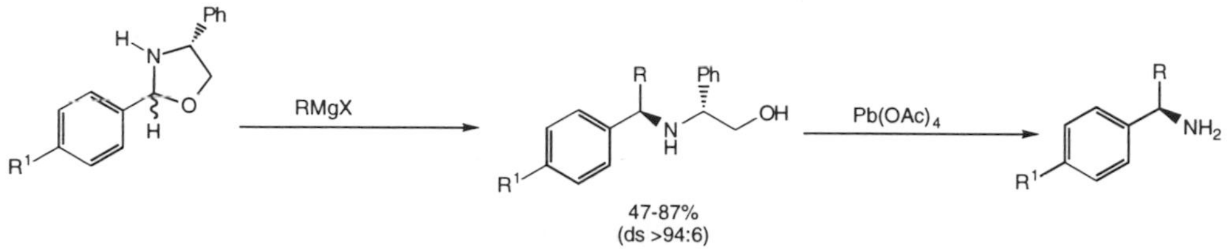

SCHEME 4.42.

An alternative methodology is described in Scheme 4.43.[444]

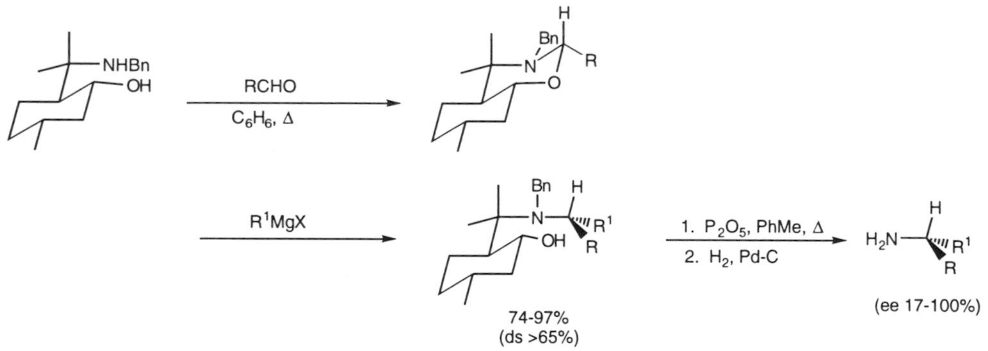

SCHEME 4.43.

## 4.3. OTHER ELECTROPHILES

There are many variants that allow for the asymmetric addition of a nucleophilic species to a carbonyl group equivalent,[49,444] as illustrated by an approach to α-amino acids (Scheme 4.44).[445-452] The use of a nitrogen derivative of a carbonyl compound allows for the incorporation of asymmetry into the electrophilic moiety.[90,443,453-455]

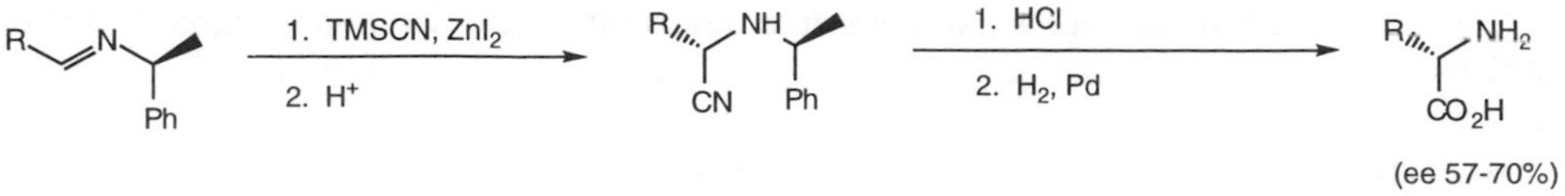

1. TMSCN, $ZnI_2$
2. $H^+$

1. HCl
2. $H_2$, Pd

(ee 57-70%)

**SCHEME 4.44.**

Conversely, the nucleophile can contain the asymmetry (Scheme 4.45),[47,456-465] and it should be possible to realize the potential of double asymmetric induction. Chiral ligands can also be used to provide some induction for the addition of organolithium reagents to imines.[451,466]

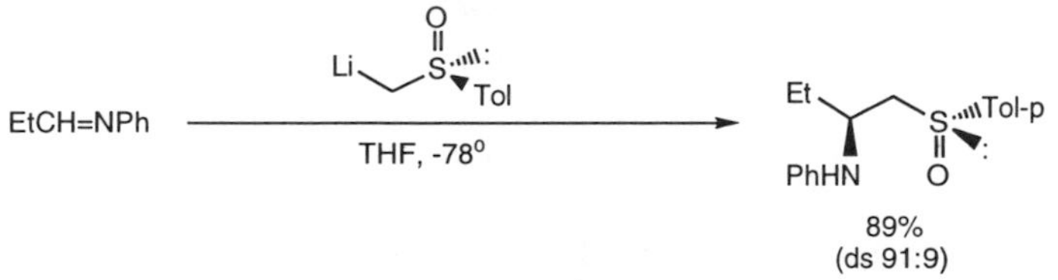

EtCH=NPh

THF, -78°

89%
(ds 91:9)

**SCHEME 4.45.**

The use of organocopper, lead, or zinc reagents provides for Anh-Felkin selectivity,[467,468] without competition from the organometallic reagent acting as a base.[469-472]

The addition of allyl organometallic reagents to imines affords homoallyl amines.[442,473-476] The allyl unit can be bifunctional; the imine nitrogen also allows for the introduction of a chiral auxiliary (Scheme 4.46).[460,477,478]

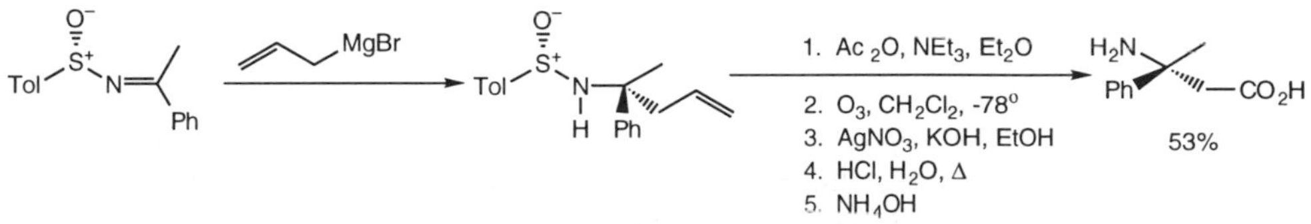

1. $Ac_2O$, $NEt_3$, $Et_2O$
2. $O_3$, $CH_2Cl_2$, -78°
3. $AgNO_3$, KOH, EtOH
4. HCl, $H_2O$, Δ
5. $NH_4OH$

53%

**SCHEME 4.46.**

The addition of an oxazoline derivative to an imine provides a route to a β-amino acid derivative (Scheme 4.47).[448]

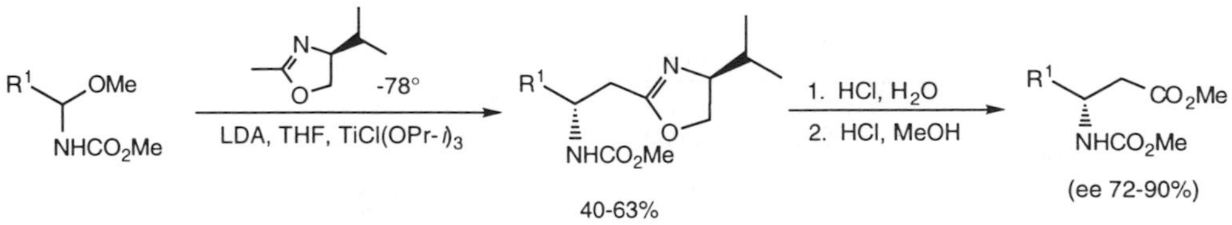

-78°
LDA, THF, $TiCl(OPr\text{-}i)_3$

40-63%

1. HCl, $H_2O$
2. HCl, MeOH

(ee 72-90%)

**SCHEME 4.47.**

A carbohydrate template allows for high stereoselectivity in the addition reaction (Scheme 4.48).[479]

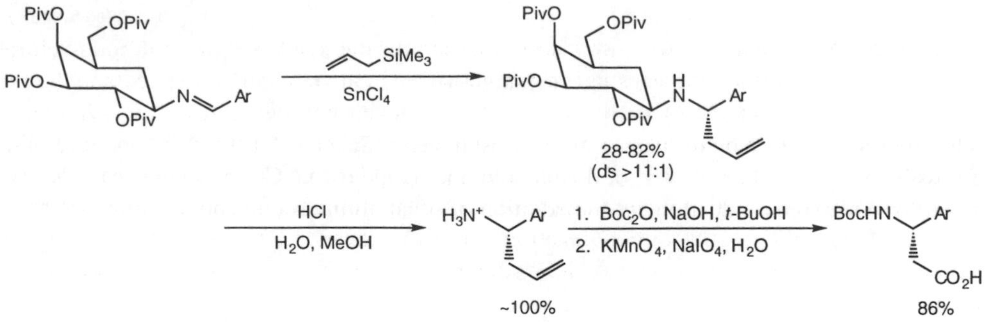

SCHEME 4.48.

Hydrazone derivatives also allow for selective additions (Scheme 4.49).[480,481]

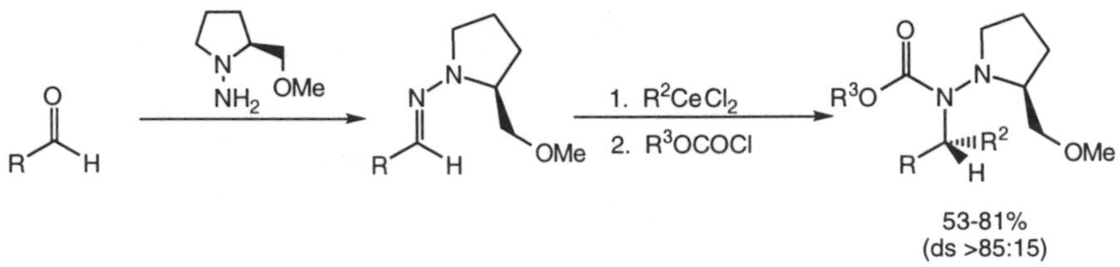

SCHEME 4.49.

## 4.4. OTHER REACTIONS

In addition to alkyl nucleophiles, the use of a functionalized unit allows for a wide variety of products. For example, the addition of cyanide to an aldehyde provides a cyanohydrin that can be hydrolyzed to an α-hydroxy acid (*cf.* Scheme 4.44). In addition to biochemical methods [Chapter 14], chiral ligands[482,483] and oxazolidinium salts,[420] as acetal equivalents, can be used to provide stereoselection.

Pinacol-type coupling of two aldehydes can provide high selection for the *syncat* diol. Such an approach has been used to prepare *N*-benzyl-D-3-*epi*-daunosamine (**4.18**) (Scheme 4.50).[484]

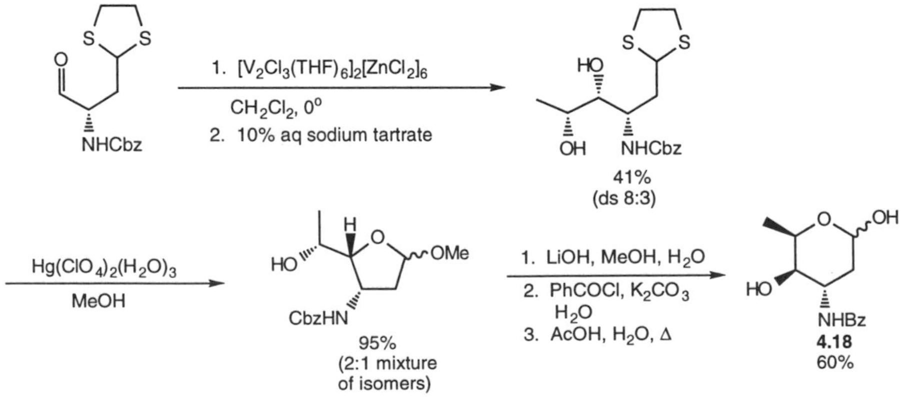

SCHEME 4.50.

## 4.5. REDUCTIONS

A wide variety of reducing agents have been reacted with carbonyl compounds to provide an alcohol. Many classical reductions allow for equilibration and cannot be employed for asymmetric reductions. The reduction of functionalized carbonyl compounds, which have realized more general methods and applications than for those of simple carbonyl compounds, can be accessed in Chapter 6.

### 4.5.1. ADDITION TO CARBONYL COMPOUNDS

Most enantioselective approaches for the reduction of simple carbonyl compounds are based on the use of chiral ligands on the reducing agent.[485] The asymmetric reduction of carbonyl compounds relies on the ability of the reductant to differentiate between the two groups adjacent to the carbonyl, usually based on steric effects. It should be noted that in many examples given below, dialkyl ketones have not been mentioned in the original citations, or show much lower enantioselectivity than aryl ketones. With diastereoselective reductions, the presence of a chiral center within the substrate can have a significant effect on the asymmetric induction. This is especially true if the chiral center is α to the carbonyl group. The use of unsaturation, as in a conjugated enone, has also been used as a means to allow steric differentiation between the two carbonyl substituents.¶

#### 4.5.1.1. Addition to Acyclic Carbonyl Compounds

As with other nucleophiles, the pivotal rules for the delivery of hydride are those of Anh-Felkin addition [*2.1.1*] and chelation control [*2.1.2*] that predict facial selectivity. Much more theoretical work has been conducted on cyclic rather than acyclic ketones and is discussed below.

The effect of Anh-Felkin addition is readily apparent for the reduction of the ketones **4.19** with lithium aluminum hydride (Scheme 4.51). As the size of the alkyl group increases, so does the *ancat* selectivity.[486-488] The size and electronic nature of the metal hydride and the counterion can also influence the reaction stereoselectivity (vide infra).[487-493]

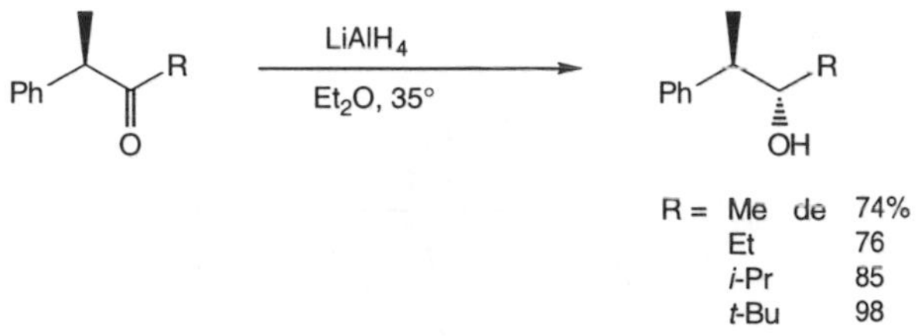

SCHEME 4.51.

The mechanism of a reaction can have important stereochemical consequences; for example, reduction of an acyclic ketone by an electron transfer process can provide the opposite stereochemistry to the more traditional metal hydride reagents.[494,495]

#### 4.5.1.2. Addition to Cyclic Carbonyl Compounds

For cyclic ketones, factors other than the steric requirements around the carbonyl group have been called upon to play a major role, such as the size of the nucleophile and whether the transition state is early or late. The stereochemical outcome of a reaction may be a result, for example, of whether an intermediate complex is formed with the carbonyl oxygen

¶ The reductions of enones are discussed elsewhere [*9.5.1.2*]. Some examples have also been included in this section to illustrate general principles.

compound prior to reaction so that a pseudo-equatorial conformation is favored with axial delivery of the nucleophilic species, or, if the transition state is late and the nucleophile is large, then an equatorial disposition may be preferred. These arguments imply that substitution on the cyclic system can have a profound effect on the reaction outcome, especially if the conformation mobility of the system is extremely limited.

The prediction of axial vs. equatorial attack has resulted in the proposal of a number of models. One was inferred above where steric approach control implies a reactant-like transition state and formation of the equatorial product (path a), or product development control where the equatorial alcohol is formed (path b) (Scheme 4.52).[496] It has also been proposed that if the torsional strain of the ring systems is taken into account, then the acyclic models can be used to predict the reaction outcome of cyclic substrates.[497-499] This model has been "refined" further by the postulate that distortion of the carbonyl $\pi$ and $\pi^*$ orbitals enhances axial selectivity.[500-506] One model that correlates well with observed product ratios for sterically hindered, nonconjugated ketones, and usually for unhindered ketones if a torsional correction term is added, is shown by the cone of preferred approach (Figure 4.1); equations describing the accessibility were derived from the area of a sphere cut by this cone.[507-509]

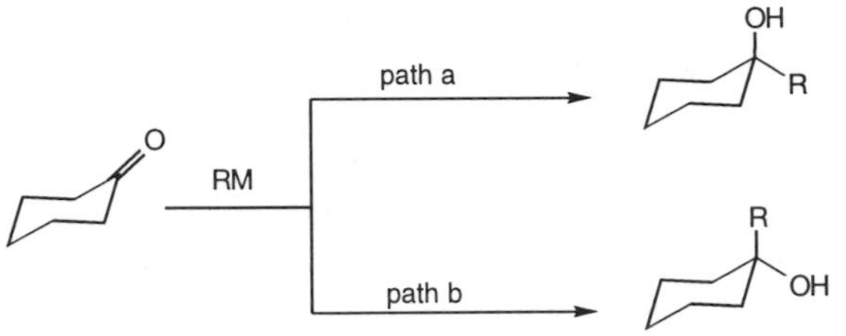

**SCHEME 4.52.**

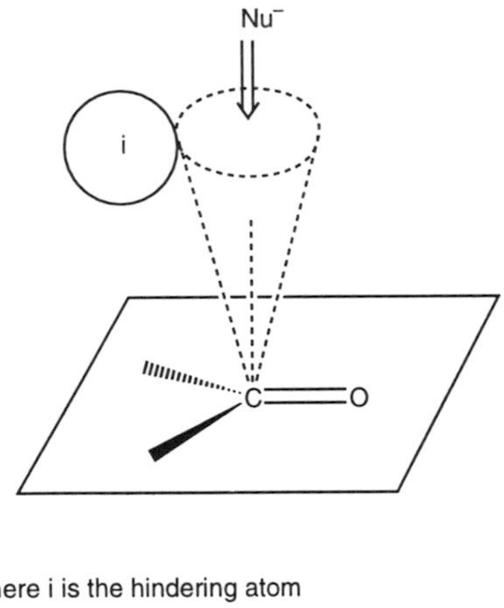

**FIGURE 4.1.** Cone of preferred approach for a nucleophile to a hindered carbonyl group.

Another hypothesis is based on the premise that the nucleophile attacks *anti* to the best electron donating bond.[510]

Calculations have been performed that suggest that the torsional strain model (Felkin), that also correlates to acyclic systems, provides the most reliable predictions.[511,512] This implies that cyclic compounds are not special cases for nucleophilic additions to carbonyl compounds, as acyclic stereochemical predictions also have to take into account whether an intermediate chelate is formed [*2.1.2*].[513-517]

### 4.5.2. ADDITION OF HYDROGEN

Carbonyl compounds can be reduced by hydrogenation, although it should be noted that it is not chemoselective and other functional groups can be reduced. As with alkenes, asymmetric induction is usually not high [*8.2.2.1*].[518-524] Again, as with alkene reduction, the presence of additional heteroatoms in the substrate greatly increases asymmetric induction [*6.2.1*].

### 4.5.3. HYDROSILYLATION

In contrast to hydrogenation, hydrosilylation proceeds in good optical yield and affords silyl ethers, that are readily hydrolyzed to the corresponding alcohols. A number of chiral auxiliaries have been used with thiazolidine systems giving the highest enantiomeric excesses (Scheme 4.53).[521,525-536]

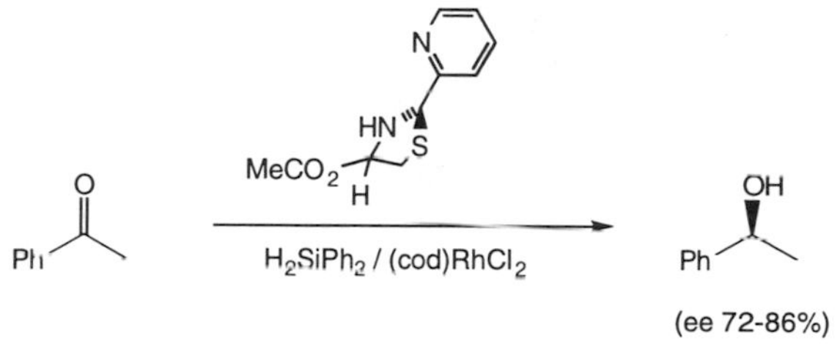

Ph—CO—CH₃ → ($H_2SiPh_2$ / (cod)$RhCl_2$, with MeCO₂-thiazolidine–pyridyl ligand) → Ph—CH(OH)—CH₃ (ee 72-86%)

**SCHEME 4.53.**

### 4.5.4. METAL HYDRIDES

Metal hydrides, and especially modified hydrides, are common reducing agents for the carbonyl compound, and have proved useful in selective reductions.[537-541] Reduction by metal hydrides can either take place by delivery of the "free" hydride ion, or through electron migration (Figure 4.2). Many ligands have been used for metal hydrides ranging from amino acids and sugars to clay.[55,125,542-554] Unfortunately, although good ee's can be realized with specific examples, the methodology is not generally applicable to a wide range of carbonyl compounds, which detracts from its synthetic utility.

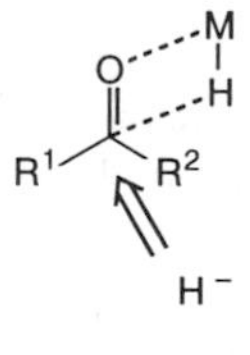

**FIGURE 4.2.** Methods for hydride delivery to a carbonyl group.

#### 4.5.4.1. Borohydrides

A large number of modified borohydrides have been investigated. Many of these have been based on boranes that gave high asymmetric induction in other systems such as olefins [*8.2.2.3*]. Chiral alcohols, acids, amines, sugars, terpenes and phase transfer catalysts have been utilized in the synthesis of potential chiral reductants.[542,549,553,555-562] Almost without exception, the enantioselectivity observed for these chiral reductants has been low with simple carbonyl compounds.[549,550,555,558,563-571] Some success has been realized for the borohydride reductions of functionalized carbonyl compounds [Chapter 6].[572,573]

### 4.5.4.2. Lithium Aluminum Hydrides

As with the borohydrides, a large number of modified lithium aluminum hydrides have been investigated.[518,554,574] Although some success has been achieved with carbonyl compounds where the two groups are significantly different in size, a general method is still elusive. As an illustration, the reduction of aryl alkyl ketones with lithium aluminum hydride in the presence of an amino alcohol modifier can provide acceptable enantioselectivity that is continued with propargylic carbonyl compounds.[† 576-586] Acyclic dialkyl ketones gave low selectivity. Additional examples are provided by the use of a proline derivative (Scheme 4.54),[587-589] and binaphthol (Scheme 4.55)[590-593] as ligands that allow for high asymmetric induction, especially with aryl ketones.

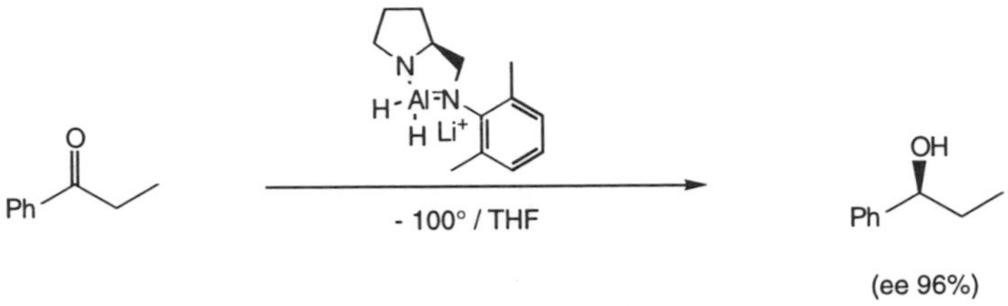

- 100° / THF

(ee 96%)

SCHEME 4.54.

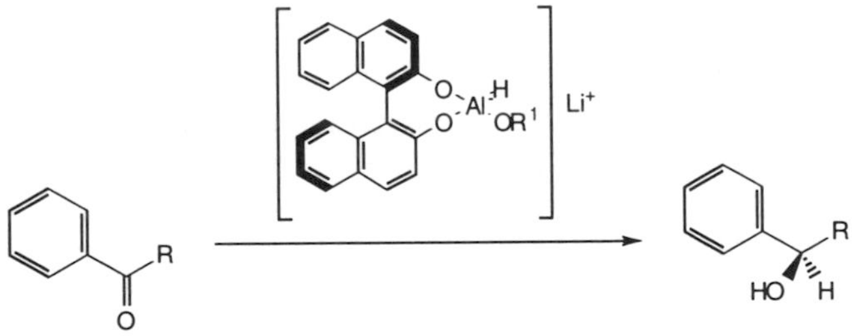

SCHEME 4.55.

## 4.5.5. BORON REAGENTS

In sharp contrast to borohydrides, boranes have found much more success in asymmetric reductions of carbonyl compounds, and provide efficient access to alcohols.[542,594-598] Chiral boranes, such as the isocampheyl derivatives, $(Ipc)_2BH$ and $IpcBH_2$, can provide asymmetric induction, but the levels are moderate.[599-603] The use of trialkylboranes does allow for the stereoselective reductions of aldehydes,[604-606] although the reduction of ketones is not as selective.[542,607,608] The use of elevated pressures has shown encouraging results with *B*-3-pinanyl-9-borabicyclo[3.3.1]nonane (Alpine-Borane) even with ketones that are slow to react with the milder conditions more commonly employed.[485,598,609-611] Diisopinocampheylchloroborane can provide high induction in the reduction of aryl ketones.[612-617] To date, the system that can provide high selectivity, even in alkyl cases, is based on oxazaborolidines **4.20** (Scheme 4.56).[485,618-628]

† It should also be noted that the selectivity can depend upon the *exact* experimental conditions used for the preparation of the reagent.[575]

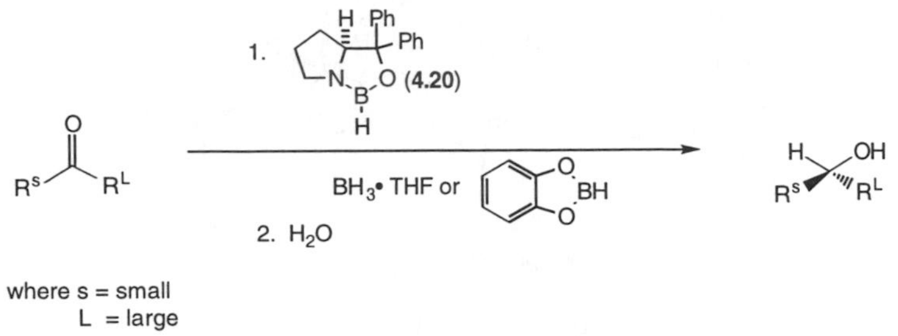

1. (4.20)

$BH_3$• THF or

2. $H_2O$

where s = small
L = large

**SCHEME 4.56.**

The proposed mechanism for the reduction is summarized in Scheme 4.57.[619,620,629-635]

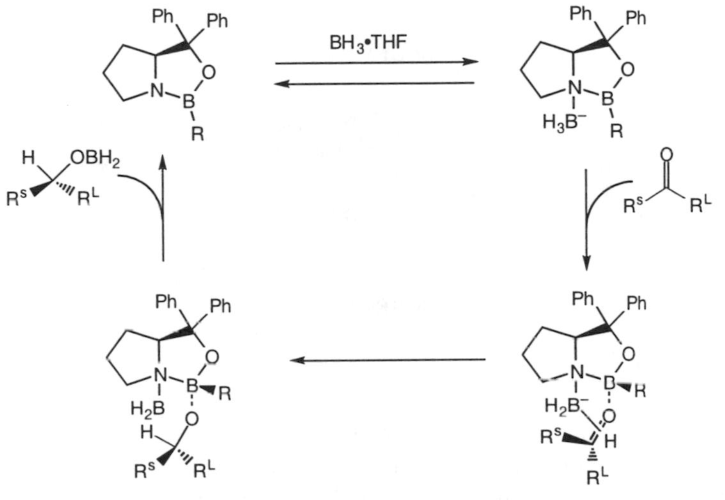

$BH_3$•THF

**SCHEME 4.57.**

Other oxazaborolidines, such as **4.21** to **4.23**, are also excellent catalysts for the asymmetric reduction of ketones to secondary alcohols.[485,602,618-620,624,625,636-645] Borolanes **4.24** also give a high degree of asymmetric reduction with dialkyl ketones,[646-648] as do oxazaborole-boranes, which are stable and amenable to scale-up,[643] and terpene-based oxazaborolidines, although enantioselectivity is not as high with the latter.[649] The addition of triethylamine to oxazaborolidine catalyzed reactions has been shown to increase enantioselectivity, especially in dialkyl ketones.[650] Boronate reductions, especially those involving chiral boronate esters, have shown high diastereoselectivity (Scheme 4.58).[651,652]

## 4.5.6. OTHER REDUCING AGENTS

Grignard reductions have shown little asymmetric induction.[653-662] The Meerwein-Ponndorf-Verley reaction, when used in conjunction with a chiral catalyst, can be highly enantioselective (Scheme 4.59).[491,663] The use of $C_2$-symmetric diamines as ligands on a rhodium catalyst has shown encouraging results.[664] The enantioselective reduction of ketones using chiral lanthanoid catalysts has also shown promise.[665] In addition, baker's yeast and enzymes have proved useful in the enantioselective reduction of ketones [*14.2*]

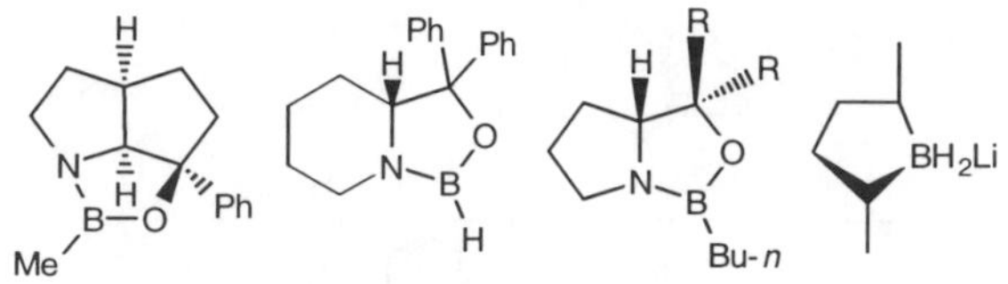

4.21 4.22 4.23 4.24

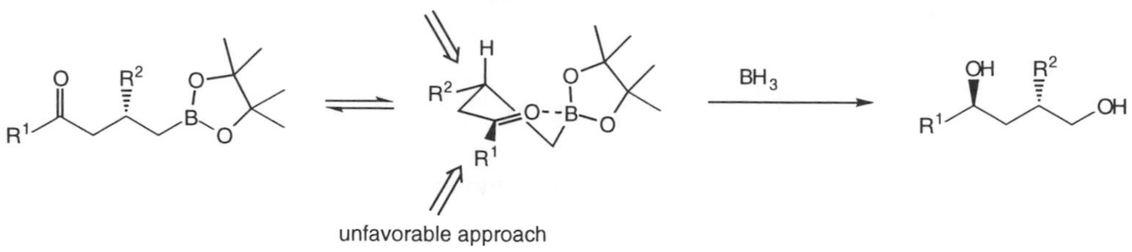

favorable approach

unfavorable approach

SCHEME 4.58.

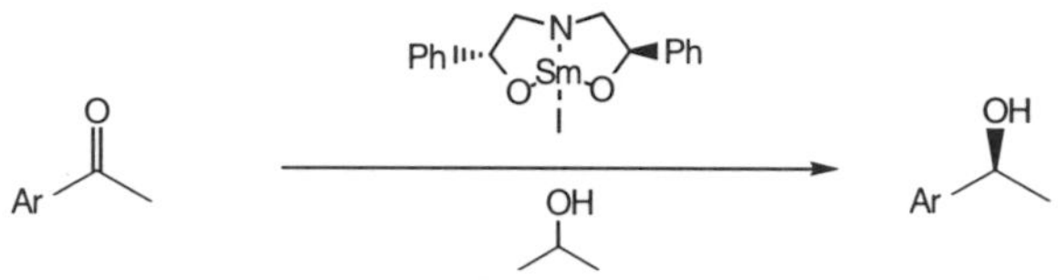

SCHEME 4.59.

## 4.6. SUMMARY

Facial selectivity is relatively simple to predict and practice for the reaction of a nucleophile with a carbonyl compound. A wide variety of nucleophiles is available to provide enantioselective additions, especially those based on chiral sulfur reagents. A wide variety of ligands has been developed to influence the reaction of a simple carbon nucleophile with a carbonyl electrophile. Although developments are still occurring in this area, methodology already exists to allow for reasonable asymmetric induction with a good confidence level for a variety of reactants. Use of an allylic nucleophile increases the comfort level and induction can be high in these cases, especially with allylboranes. However, other metals and chiral ligands can also be used to provide high ee's. However, low temperatures and exothermic reactions are usually involved in this class of reactions; this can make scale-up problematic. On a laboratory scale, these reactions should be considered when designing a synthesis.

The reduction of simple carbonyl compounds has produced a large number of reagents, in particular modified metal hydrides, but substrate specificity still remains narrow if high facial selectivity is desired. Boranes have proven more successful, but these are being succeeded by oxazaborolidines. With these latter reagents, it is now practical to reduce a ketone to an alcohol with excellent enantioselectivity, although steric differentiation between the two carbonyl substituents is still necessary.

Most reducing agents employ reactive hydrides, and care has to be exercised in scale-up as safety becomes a significant issue.

## 4.7. REFERENCES


1. Reetz, M.; Harmat, N.; Mahrwald, R. *Angew. Chem., Int. Ed. Engl.* 1992, *31*, 342.
2. Imamoto, T. *Pure Appl. Chem.* 1990, *62*, 747.
3. Imamoto, T.; Takiyama, N.; Nakamura, K.; Hatajima, T.; Kamiya, Y. *J. Am. Chem. Soc.* 1989, *111*, 4392.
4. Imamoto, T.; Kusumoto, T.; Tawarayama, Y.; Sugiura, Y.; Mita, T.; Hatanaka, Y.; Yokoyama, M. *J. Org. Chem.* 1984, *49*, 3904.
5. Imamoto, T.; Kusumoto, T.; Yokoyama, M. *J. Chem. Soc., Chem. Commun.* 1982, 1042.
6. Reetz, M. T.; Jung, A. *J. Am. Chem. Soc.* 1983, *105*, 4833.
7. Reetz, M. T.; Westermann, J.; Steinbach, R.; Wenderoth, B.; Peter, R.; Ostarek, R.; Maus, S. *Chem. Ber.* 1985, *115*, 1421.
8. Molander, G. A.; Burkhardt, E. R.; Weinig, P. *J. Org. Chem.* 1990, *55*, 4990.
9. Collins, S.; Hong, Y.; Hoover, G. J.; Veit, J. R. *J. Org. Chem.* 1990, *55*, 3565.
10. Noyori, R.; Kitamura, M. *Angew. Chem., Int. Ed. Engl.* 1991, *30*, 49.
11. Abenhaim, D.; Boireau, G.; Bernardon, C.; Deberly, A.; Germain, C. *Tetrahedron Lett.* 1976, 993.
12. Boireau, G.; Abenhaim, D.; Bourdais, J.; Henry-Basch, E. *Tetrahedron Lett.* 1976, 4781.
13. Hoppe, D.; Carstens, A.; Kramer, T. *Angew. Chem., Int. Ed. Engl.* 1990, *29*, 1424.
14. Hoffmann, R. W.; Ruhland, T.; Bewersdorf, M. *J. Chem. Soc., Chem. Commun.* 1991, 195.
15. Hoffmann, R. W.; Julius, M.; Oltmann, K. *Tetrahedron Lett.* 1990, *31*, 7419.
16. Hoppe, D.; Hintze, F.; Tebben, P. *Angew. Chem., Int. Ed. Engl.* 1990, *29*, 1422.
17. Sakuraba, H.; Ushiki, S. *Tetrahedron Lett.* 1990, *31*, 5349.
18. Chan, P. C.-M.; Chong, J. M. *Tetrahedron Lett.* 1990, *31*, 1985.
19. Kingsbury, C. A. *J. Org. Chem.* 1972, *37*, 102.
20. Williams, D. R.; Phillips, J. G.; White, F. H.; Huffman, J. C. *Tetrahedron* 1986, *42*, 3003.
21. Williams, D. R.; Phillips, J. G.; Huffman, J. C. *J. Org. Chem.* 1981, *46*, 4101.
22. Satoh, T.; Oohara, T.; Yamakawa, K. *Tetrahedron Lett.* 1988, *29*, 2851.
23. Colombo, L.; Gennari, C.; Scolastico, C.; Guanti, G.; Narisano, E. *J. Chem. Soc., Chem. Commun.* 1979, 591.
24. Satoh, T.; Kaneko, Y.; Yamakawa, K. *Tetrahedron Lett.* 1986, *27*, 2379.
25. Satoh, T.; Oohara, T.; Ueda, Y.; Yamakawa, K. *J. Org. Chem.* 1989, *54*, 3130.
26. Bravo, P.; Frigerio, M.; Resnati, G. *J. Org. Chem.* 1990, *55*, 4216.
27. Satoh, T.; Motohashi, S.; Kimura, S.; Tokutake, N.; Yamakawa, K. *Tetrahedron Lett.* 1993, *34*, 4823.
28. Matsuzawa, S.; Isaka, M.; Nakamura, E.; Kuwajima, I. *Tetrahedron Lett.* 1989, *30*, 1975.
29. Nakada, M.; Urano, Y.; Kobayashi, S.; Ohno, M. *J. Am. Chem. Soc.* 1988, *110*, 4826.
30. Hwu, J. R.; Wang, N. *Chem. Rev.* 1989, *89*, 1599.
31. Buynak, J. D.; Strickland, J. B.; Hurd, T.; Phan, A. *J. Chem. Soc., Chem. Commun.* 1989, 89.
32. Page, P. C. B.; Klair, S. S.; Rosenthal, S. *Chem. Soc. Rev.* 1990, *19*, 147.
33. Burke, S. D.; Deaton, D. N.; Olsen, R. J.; Armistead, D. M.; Blough, B. E. *Tetrahedron Lett.* 1987, *28*, 3905.
34. Lynch, J. E.; Eliel, E. L. *J. Am. Chem. Soc.* 1984, *106*, 2943.
35. Eliel, E. L.; Morris-Natschke, S. *J. Am. Chem. Soc.* 1984, *106*, 2937.
36. Eliel, E. L.; Soai, K. *Tetrahedron Lett.* 1981, *22*, 2859.
37. Eliel, E. L.; Lynch, J. E. *Tetrahedron Lett.* 1981, *22*, 2855.
38. Eliel, E. L.; Koskimies, J. K.; Lohri, B. *J. Am. Chem. Soc.* 1978, *100*, 1614.
39. Kogure, T.; Eliel, E. L. *J. Org. Chem.* 1984, *49*, 576.
40. Kaulen, J. *Angew. Chem., Int. Ed. Engl.* 1989, *28*, 462.
41. Eliel, E. L.; Bai, X.; Abdel-Magid, A. F.; Hutchins, R. O. *J. Org. Chem.* 1990, *55*, 4951.
42. Utimoto, K.; Nakamura, A.; Matsubara, S. *J. Am. Chem. Soc.* 1990, *112*, 8189.
43. Page, P. C. B.; Westwood, D.; Slawin, A. M. Z.; Williams, D. J. *J. Chem. Soc., Perkin Trans. I* 1989, 1158.
44. Page, P. C. B.; Namwindwa, E.; Klair, S. S.; Westwood, D. *Synlett* 1990, 457.
45. Meyers, A. I.; Ford, M. E. *Tetrahedron Lett.* 1974, 1341.
46. Burton, C. A.; Crawford, W.; Watts, W. E. *Tetrahedron Lett.* 1977, 3755.
47. Still, W. C.; *J. Am. Chem. Soc.* 1978, *100*, 1481.
48. McGarvey, G. J.; Kimura, M. *J. Org. Chem.* 1982, *47*, 5720.
49. Takahashi, H.; Inagaki, H. *Chem. Pharm. Bull.* 1982, *30*, 922.
50. Mukaiyama, T.; Soai, K.; Sato, T.; Shimizu, H.; Suzuki, K. *J. Am. Chem. Soc.* 1979, *101*, 1455.
51. Whitesell, J. K.; Jaw, B.-R. *J. Org. Chem.* 1981, *46*, 2798.
52. Fujisawa, T.; Takemura, I.; Ukaji, Y. *Tetrahedron Lett.* 1990, *31*, 5479.
53. Colombo, L.; Gennari, C.; Poli, G.; Scolastico, C.; Annunziata, R.; Cinquini, M.; Cozzi, F. *J. Chem. Soc., Chem. Commun.* 1983, 403.
54. Mukaiyama, T.; Soai, K.; Kobayashi, S. *Chemistry Lett.* 1978, 219.

55. Mukaiyama, T. *Tetrahedron* 1981, *37*, 4111.
56. Tomioka, K. *Synthesis* 1990, 541.
57. Soai, F.; Mukaiyama, T. *Bull. Chem. Soc., Jpn.* 1979, *52*, 3371.
58. Colombi, L.; Gennari, C.; Poli, G.; Scolastico, C. *Tetrahedron* 1982, *38*, 2725.
59. Seebach, D.; Kalinowski, H.-O.; Bastani, B.; Crass, G.; Daum, H.; Dorr, H.; DuPreez, N. P.; Ehrig, V.; Langer, W.; Nussler, C.; Oei, H.-A.; Schmidt, M. *Helv. Chim. Acta* 1977, *60*, 301.
60. Inch, T. D.; Lewis, G. J.; Sainsbury, G. L.; Sellers, D. J. *Tetrahedron Lett.* 1969, 3657.
61. Seebach, D.; Dorr, H.; Bastini, B.; Ehrig, V. *Angew. Chem., Int. Ed. Engl.* 1969, *8*, 982.
62. Soai, K.; Niwa, S.; Hori, H. *J. Chem. Soc., Chem. Commun.* 1990, 982.
63. Mazaleyrat, J.-P.; Cram, D. J. *J. Am. Chem. Soc.* 1981, *103*, 4585.
64. Nozaki, H.; Aratani, T.; Toraya, T.; Noyori, R. *Tetrahedron* 1971, *27*, 905.
65. Soai, K.; Niwa, S. *Chem. Rev.* 1992, *92*, 833.
66. Oppolzer, W.; Radinov, R. N. *Tetrahedron Lett.* 1988, *29*, 5645.
67. Soai, K.; Niwa, S.; Yamada, Y.; Inoue, H. *Tetrahedron Lett.* 1987, *28*, 4841.
68. Soai, K.; Ookawa, A.; Kaba, T.; Ogawa, K. *J. Am. Chem. Soc.* 1987, *109*, 7111.
69. Oguni, N.; Matsuda, Y.; Kaneko, T. *J. Am. Chem. Soc.* 1988, *110*, 7877.
70. Corey, E. J.; Hannon, F. J. *Tetrahedron Lett.* 1987, *28*, 5233.
71. Muchow, G.; Vannoorenberghe, Y.; Buono, G. *Tetrahedron Lett.* 1987, *28*, 6163.
72. Chaloner, P. A.; Perera, S. A. R. *Tetrahedron Lett.* 1987, *28*, 3013.
73. Soai, K.; Watanabe, M. *J. Chem. Soc., Chem. Commun.* 1990, 43.
74. Bolm, C.; Zehnder, M.; Bur, D. *Angew. Chem., Int. Ed. Engl.* 1990, *29*, 205.
75. Niwa, S.; Soai, K. *J. Chem. Soc., Perkin Trans. I* 1990, 937.
76. Tanaka, K.; Ushino, H.; Suzuki, H. *J. Chem. Soc., Chem. Commun.* 1989, 1700.
77. Soai, K.; Niwa, S.; Hatanaka, T. *J. Chem. Soc., Chem. Commun.* 1990, 709.
78. Yoshioka, M.; Kawakita, T.; Ohno, M. *Tetrahedron Lett.* 1989, *30*, 1657.
79. Soai, K.; Watanabe, M.; Koyana, M. *J. Chem. Soc., Chem. Commun.* 1989, 534.
80. Chelucci, G.; Falorni, M.; Giacomelli, G. *Tetrahedron: Asymmetry* 1990, *1*, 843.
81. Hayashi, M.; Kaneko, T.; Oguni, N. *J. Chem. Soc., Perkin Trans. I* 1991, 25.
82. Soai, K.; Kawase, Y. *J. Chem. Soc., Perkin Trans. I* 1990, 3214.
83. Chaloner, P. A.; Langadianou, E. *Tetrahedron Lett.* 1990, *31*, 5185.
84. Rosini, C.; Franzini, L.; Pini, D.; Salvadori, P. *Tetrahedron: Asymmetry* 1990, *1*, 587.
85. Soai, K.; Hori, H.; Kawahara, M. *Tetrahedron: Asymmetry* 1990, *1*, 769.
86. Takahashi, H.; Kawakita, T.; Yoshioka, M.; Kobayashi, S.; Ohno, M. *Tetrahedron Lett.* 1990, *31*, 7095.
87. van Oeveren, A.; Menge, W.; Feringa, B. L. *Tetrahedron Lett.* 1989, *30*, 6427.
88. Schmidt, B.; Seebach, D. *Angew. Chem., Int. Ed. Engl.* 1991, *30*, 99.
89. Joshi, N. N.; Srebnik, M.; Brown, H. C. *Tetrahedron Lett.* 1989, *30*, 5551.
90. Itsuno, S.; Sakurai, Y.; Ito, K.; Maryama, T.; Nakahama, S.; Frechet, J. M. J. *J. Org. Chem.* 1990, *55*, 304.
91. Soai, K.; Watanabe, M.; Yamamoto, A. *J. Org. Chem.* 1990, *55*, 4832.
92. Kitamura, M.; Okada, S.; Suga, S.; Noyori, R. *J. Am. Chem. Soc.* 1989, *111*, 4028.
93. Soai, K.; Watanabe, M. *Tetrahedron: Asymmetry* 1991, 2, 97.
94. Watanabe, M.; Araki, S.; Butsugan, Y.; Uemura, M. *J. Org. Chem.* 1991, *56*, 2218.
95. Corey, E. J.; Yuen, P.-W.; Hannon, F. J.; Wierda, D. A. *J. Org. Chem.* 1990, *55*, 784.
96. Itsuno, S.; Frechet, J. M. J. *J. Org. Chem.* 1987, *52*, 4140.
97. Soai, K.; Yokoyama, S.; Ebihara, K.; Hayashi, T. *J. Chem. Soc., Chem. Commun.* 1987, 1690.
98. Takahashi, H.; Kawakita, T.; Yoshioka, M.; Kobayashi, S.; Ohno, M. *Tetrahedron Lett.* 1989, *30*, 7095.
99. Briedene, W.; Ostwald, R.; Knochel, P. *Angew. Chem., Int. Ed. Engl.* 1993, *32*, 582.
100. ShengJian, L.; Yaozhong, J.; Aiqiao, M. *Tetrahedron: Asymmetry* 1992, *3*, 1467.
101. Ito, K.; Kimura, Y.; Okamura, H.; Katsuki, T. *Synlett* 1992, 573.
102. Nakajima, M.; Tomioka, K.; Koga, K. *Tetrahedron* 1993, *49*, 9751.
103. Carreño, M. C.; Ruano, J. L. G.; Maestro, M. C.; Cabrejas, L. M. M. *Tetrahedron: Asymmetry* 1993, *4*, 727.
104. de Vries, E. F. J.; Brusse, J.; Kruse, C. G.; van der Gen, A. *Tetrahedron: Asymmetry* 1993, *4*, 1987.
105. Falorni, M.; Satta, M.; Conti, S.; Giacomelli, G. *Tetrahedron: Asymmetry* 1993, *4*, 2389.
106. Schmidt, B.; Seebach, D. *Angew. Chem., Int. Ed. Engl.* 1991, *30*, 1321.
107. Pini, D.; Mastantuono, A.; Uccello-Barretta, G.; Iuliano, A.; Salvadori, P. *Tetrahedron* 1993, *49*, 9613.
108. Wallbaum, S.; Martens, J. *Tetrahedron: Asymmetry* 1993, *4*, 637.
109. Allen, J. V.; Frost, C. G.; Williams, J. M. J. *Tetrahedron: Asymmetry* 1993, *4*, 649.
110. Soai, K.; Hirose, Y.; Ohno, Y. *Tetrahedron: Asymmetry* 1993, *4*, 1473.
111. Behnen, W.; Mehler, T.; Martens, J. *Tetrahedron: Asymmetry* 1993, *4*, 1413.
112. Matsumoto, Y.; Ohno, A.; Lu, S.-J.; Hayashi, T. *Tetrahedron: Asymmetry* 1993, *4*, 1763.
113. Reetz, M. T.; Kukenhohner, T.; Weinig, P. *Tetrahedron Lett.* 1986, *27*, 5711.

114. Barr, D.; Berrisford, D. J.; Jones, R. V. H.; Slawin, A. M. Z.; Snaith, R.; Stoddart, J. F.; Williams, D. J. *Angew. Chem., Int. Ed. Engl.* 1989, *28*, 1044.
115. Tombo, G. M. R.; Didier, E.; Loubinoux, B. *Synlett* 1990, 547.
116. Yamataka, H.; Kawafuji, Y.; Nagareda, K.; Miyano, N.; Hanafusa, T. *J. Org. Chem.* 1989, *54*, 4706.
117. Wang, J.-T.; Fan, X.; Feng, X.; Qian, Y.-M. *Synthesis* 1989, 291.
118. Evans, D. A.; Urpi, F.; Somers, T. C.; Clark, J. S.; Bilodeau, M. T. *J. Am. Chem. Soc.* 1990, *112*, 8215.
119. Weber, B.; Seebach, D. *Angew. Chem., Int. Ed. Engl.* 1992, *31*, 84.
120. Seebach, D.; Behrendt, L.; Felix, D. *Angew. Chem., Int. Ed. Engl.* 1991, *30*, 1008.
121. Chibale, K.; Greeves, N.; Lyford, L.; Pease, J. E. *Tetrahedron: Asymmetry* 1993, *4*, 2407.
122. Jurczak, J.; Pikul, S.; Bauer, T. *Tetrahedron* 1986, *42*, 447.
123. Martin, S. F. *Synthesis* 1979, 633.
124. Ager, D. J. In *Umpoled Synthons: A Survey of Sources and Uses in Synthesis*; T. A. Hase, Ed.; Wiley-Interscience: New York, 1987; pp 19.
125. Seebach, D.; Daum, H. *Chem. Ber.* 1974, *107*, 1748.
126. Adamozyk, M.; Dolence, E. K.; Watt, D. S.; Christy, M. R.; Reibenspies, J. H.; Andersen, O. P. *J. Org. Chem.* 1984, *49*, 1378.
127. Vatele, J.-M. *Tetrahedron Lett.* 1984, *25*, 5997.
128. Posner, G. H.; Tang, P.-W. *J. Org. Chem.* 1978, *43*, 4131.
129. Iriuchijima, S.; Maniwa, K.; Tsuchihashi, G.-i. *J. Am. Chem. Soc.* 1974, *96*, 4280.
130. Field, L. *Synthesis* 1978, 713.
131. Brown, H. C.; Imai, T. *J. Am. Chem. Soc.* 1983, *105*, 6285.
132. Murai, S.; Kato, T.; Sonoda, N.; Seki, Y.; Kawamoto, K. *Angew. Chem., Int. Ed. Engl.* 1979, *18*, 393.
133. Murai, S.; Sonoda, N. *Angew. Chem., Int. Ed. Engl.* 1979, *18*, 837.
134. Redlich, H.; Thormahlen, S. *Tetrahedron Lett.* 1985, *26*, 3685.
135. Mori, M.; Chuman, T.; Kato, K.; Mori, K. *Tetrahedron Lett.* 1982, *23*, 4593.
136. Walba, D. M.; Ward, D. M. *Tetrahedron Lett.* 1982, *23*, 4995.
137. Paulsen, H.; Schuller, M.; Nashed, M. A.; Heitman, A.; Redlich, H. *Tetrahedron Lett.* 1985, *26*, 3689.
138. Ager, D. J.; East, M. B. *Tetrahedron* 1993, *49*, 5683.
139. Dondoni, A.; Fantin, G.; Fogagnolo, M.; Medici, A.; Pedrini, P. *J. Org. Chem.* 1989, *54*, 693.
140. Wagner, A.; Mollath, M. *Tetrahedron Lett.* 1993, *34*, 619.
141. Dondoni, A.; Fantin, G.; Fogagnolo, M.; Medici, A. *J. Chem. Soc., Chem. Commun.* 1988, 10.
142. Dondoni, A. *Pure Appl. Chem.* 1990, *62*, 643.
143. Dondoni, A.; Merino, P. *Synthesis* 1993, 903.
144. Dondoni, A.; Fantin, G.; Fogagnolo, M.; Medici, A.; Pedrini, P. *J. Org. Chem.* 1989, *54*, 702.
145. Dondoni, A.; Perrone, D. *Synthesis* 1993, 1162.
146. Aggarwal, V. K.; Davies, I. W.; Maddock, J.; Mahon, M. F.; Molloy, K. C. *Tetrahedron Lett.* 1990, *31*, 135.
147. Harada, K.; Okawara, T.; Matsumoto, K. *Bull. Chem. Soc., Jpn.* 1973, *46*, 1865.
148. Tsuloyama, S. *Bull. Chem. Soc., Jpn.* 1962, *35*, 1004.
149. Babievskii, K. K.; Latov, V. K. *Russ. Chem. Rev.* 1969, *38*, 456.
150. Solladie, G.; Moine, G. *J. Am. Chem. Soc.* 1984, *106*, 6097.
151. Takeda, T.; Furukawa, H.; Fujimori, M.; Suzuki, K.; Fujiwara, T. *Bull. Chem. Soc., Jpn.* 1984, *57*, 1863.
152. Yamakado, Y.; Ishiguro, M.; Ikeda, N.; Yamamoto, H. *J. Am. Chem. Soc.* 1981, *103*, 5568.
153. Johnson, W. S.; Elliott, R.; Elliott, J. D. *J. Am. Chem. Soc.* 1983, *105*, 2904.
154. Ikeda, N.; Arai, I.; Yamamoto, H. *J. Am. Chem. Soc.* 1986, *108*, 483.
155. Ikeda, N.; Omori, K.; Yamamoto, H. *Tetrahedron Lett.* 1986, *27*, 1175.
156. Braun, M.; Mahler, H. *Angew. Chem., Int. Ed. Engl.* 1989, *28*, 896.
157. Kusuda, S.; Kawamura, K.; Ueno, Y.; Toru, T. *Tetrahedron Lett.* 1993, *34*, 6587.
158. Braun, M.; Hild, W. *Angew. Chem., Int. Ed. Engl.* 1984, *23*, 723.
159. Mahler, H.; Braun, M. *Tetrahedron Lett.* 1987, *28*, 5145.
160. Braun, M.; Opdenbusch, K. *Angew. Chem., Int. Ed. Engl.* 1993, *32*, 578.
161. Braun, M.; Mahler, H. *Synlett* 1990, 587.
162. Braun, M.; Moritz, J. *Chem. Ber.* 1989, *122*, 1215.
163. Esswein, A.; Betz, R.; Schmidt, R. R. *Helv. Chim. Acta* 1989, *72*, 213.
164. Datta, A.; Schmidt, R. R. *Tetrahedron Lett.* 1993, *34*, 4161.
165. Boeckman, R. K.; O'Connor, K. J. *Tetrahedron Lett.* 1989, *30*, 3271.
166. Thompson, W. J.; Tucker, T. J.; Schwering, J. E.; Barnes, J. L. *Tetrahedron Lett.* 1990, *31*, 6819.
167. Cheng, H. C.; Yan, T.-H. *Tetrahedron Lett.* 1990, *31*, 673.
168. Anderson, M. B.; Fuchs, P. L. *J. Org. Chem.* 1990, *55*, 337.
169. Tsukamoto, M.; Iio, H.; Tokoroyama, T. *Tetrahedron Lett.* 1985, *26*, 4471.
170. Iio, H.; Mizobuchi, T.; Tsukamoto, M.; Tokoroyama, T. *Tetrahedron Lett.* 1986, *27*, 6373.

171. Hoffmann, R. W. *Angew. Chem., Int. Ed. Engl.* 1982, *21*, 555.
172. Yamamoto, Y.; Maruyama, K. *Heterocycl.* 1982, *18*, 357.
173. Consiglio, G.; Waymouth, R. M. *Chem. Rev.* 1989, *89*, 257.
174. Hoffmann, R. W.; Niel, G.; Schlapbach, A. *Pure Appl. Chem.* 1990, *62*, 1993.
175. Ager, D. J.; East, M. B. *Tetrahedron* 1992, *48*, 2803.
176. Yamamoto, Y. *Acc. Chem. Res.* 1987, *20*, 243.
177. Hoffmann, R. W. *Pure Appl. Chem.* 1988, *60*, 123.
178. Schlosser, M.; Hartmann, J.; David, V. *Helv. Chim. Acta* 1974, *57*, 1567.
179. Schlosser, M.; Hartmann, J. *J. Am. Chem. Soc.* 1976, *98*, 4674.
180. Thompson, T. B.; Ford, W. T. *J. Am. Chem. Soc.* 1979, *101*, 5459.
181. Schlosser, M.; Desponds, O.; Lehmann, R.; Moret, E.; Rauchschwalbe, G. *Tetrahedron* 1993, *49*, 10175.
182. Atlani, P. M.; Biellmann, J. F.; Dude, S.; Vicens, J. J. *Tetrahedron Lett.* 1974, 2665.
183. Hirama, M. *Tetrahedron Lett.* 1981, *22*, 1905.
184. Hoppe, D.; Hanko, R.; Bronneke, A. *Angew. Chem., Int. Ed. Engl.* 1980, *19*, 625.
185. Hiyama, T.; Kimura, K.; Nozaki, H. *Tetrahedron Lett.* 1981, *22*, 1037.
186. Martin, S. F.; Li, W. *J. Org. Chem.* 1989, *54*, 6129.
187. Hiyama, T.; Okude, Y.; Kimura, K.; Nozaki, H. *Bull. Chem. Soc., Jpn.* 1982, *55*, 561.
188. Takai, K.; Kataoka, Y.; Utimoto, K. *Tetrahedron Lett.* 1989, *30*, 4389.
189. Nagaoka, H.; Kishi, Y. *Tetrahedron* 1981, *37*, 3873.
190. Yamaguchi, M.; Mukaiyama, T. *Chem. Lett.* 1979, 1279.
191. Guo, B.-S.; Doubleday, W.; Cohen, T. *J. Am. Chem. Soc.* 1987, *109*, 4710.
192. Cohen, T.; Bhupathy, M. *Acc. Chem. Res.* 1989, *22*, 152.
193. Hoppe, D.; Lichtenberg, F. *Angew. Chem., Int. Ed. Engl.* 1982, *21*, 372.
194. Hoppe, D.; Hanko, R.; Bronneke, A.; Lichtenberg, F. *Angew. Chem., Int. Ed. Engl.* 1981, *20*, 1024.
195. Hanko, R.; Hoppe, D. *Angew. Chem., Int. Ed. Engl.* 1982, *21*, 372.
196. Hoppe, D.; Lichtenberg, F. *Angew. Chem., Int. Ed. Engl.* 1984, *23*, 239.
197. Mallaiah, K.; Satanarayana, J.; Ila, H.; Junjappa, H. *Tetrahedron Lett.* 1993, *34*, 3145.
198. Kramer, T.; Hoppe, D. *Tetrahedron Lett.* 1987, *28*, 5149.
199. Hoppe, D.; Bronneke *Tetrahedron Lett.* 1983, *24*, 1687.
200. Krämer, T.; Hoppe, D. *Tetrahedron Lett.* 1989, *30*, 7037.
201. Yamamoto, Y.; Maryama, K. *Tetrahedron Lett.* 1981, *22*, 2895.
202. Ridley, D. D.; Smal, M. A. *Aust. J. Chem.* 1980, *33*, 1345.
203. Hayashi, T.; Fujitaka, N.; Oishi, T.; Takeshima, T. *Tetrahedron Lett.* 1980, *21*, 303.
204. Biellmann, J. F.; Ducep, J. B. *Tetrahedron Lett.* 1968, 5629.
205. Faller, J. W.; John, J. A.; Mazzieri, M. R. *Tetrahedron Lett.* 1989, *30*, 1769.
206. Yamamoto, Y.; Saito, Y.; Maruyama, K. *Tetrahedron Lett.* 1982, *23*, 4959.
207. Butsugan, Y.; Ito, H.; Araki, S. *Tetrahedron Lett.* 1987, *28*, 3707.
208. Wang, W.-B.; Shi, L.-L.; Huang, Y.-Z. *Tetrahedron* 1990, *46*, 3315.
209. Ikeda, Y.; Furuta, K.; Meguriya, N.; Ikeda, N.; Yamamoto, H. *J. Am. Chem. Soc.* 1982, *104*, 7663.
210. Furuta, K.; Ikeda, Y.; Meguriya, N.; Ikeda, N.; Yamamoto, H. *Bull. Chem. Soc., Jpn.* 1984, *57*, 2781.
211. Coxon, J. M.; Juan Eyk, S.; Steel, P. *Tetrahedron* 1989, *45*, 1029.
212. Annunziata, R.; Cinquini, M.; Cozzi, F.; Raimondi, L. *J. Chem. Soc., Chem. Commun.* 1986, 366.
213. Mukaiyama, T. In *Trends in Synthetic Carbohydrate Chemistry*; D. Horton and L. D. Hawkins, Eds.; American Chemical Society: Washington, D. C., 1989; pp 278.
214. Annunziata, R.; Cinquini, M.; Cozzi, F.; Raimondi, L.; Stefanelli, S. *Tetrahedron* 1986, *42*, 5451.
215. Sato, F.; Iijima, S.; Sato, M. *Tetrahedron Lett.* 1981, *22*, 243.
216. Kobayashi, Y.; Umeyama, K.; Sato, F. *J. Chem. Soc., Chem. Commun.* 1984, 621.
217. Riediker, M.; Duthaler, R. O. *Angew. Chem., Int. Ed. Engl.* 1989, *28*, 494.
218. Duthaler, R. O.; Hafner, A. *Chem. Rev.* 1992, *92*, 807.
219. Nagaoka, H.; Rutsch, W.; Schmid, G.; Iio, H.; Johnson, M. R.; Kishi, Y. *J. Am. Chem. Soc.* 1980, *102*, 7962.
220. Yamamoto, Y.; Saito, Y.; Maryama, K. *Tetrahedron Lett.* 1982, *23*, 4597.
221. Wilson, S. R.; Guazzaroni, M. E. *J. Org. Chem.* 1989, *54*, 3087.
222. Masuyama, Y.; Takahara, J. P.; Kurusu, Y. *J. Am. Chem. Soc.* 1988, *110*, 4473.
223. Masuyama, Y.; Takahara, J. P.; Kurusu, Y. *Tetrahedron Lett.* 1989, *30*, 3437.
224. Resek, J. E.; Beak, P. *Tetrahedron Lett.* 1993, *34*, 3043.
225. Beak, P.; Lee, B. *J. Org. Chem.* 1989, *54*, 458.
226. Koreeda, M.; Tanaka, Y. *J. Chem. Soc., Chem. Commun.* 1982, 845.
227. Ahlbrecht, H.; Enders, D.; Santowski, L.; Zimmermann, G. *Chem. Ber.* 1989, *122*, 1995.
228. Evans, D. A.; Andrews, G. C.; Buckwalter, B. *J. Am. Chem. Soc.* 1974, *96*, 5560.
229. Meyers, A. I.; Brich, Z.; Erickson, G. W.; Traynor, S. G. *J. Chem. Soc., Chem. Commun.* 1979, 566.
230. Roush, W. R.; Michaelides, M. R.; Tai, D. F.; Chong, W. K. M. *J. Am. Chem. Soc.* 1987, *109*, 7575.

231. Sato, F.; Uchiyama, H.; Iida, K.; Kobayashi, Y.; Sato, M. *J. Chem. Soc., Chem. Commun.* 1983, 921.
232. Hosomi, A.; Kohraga, S.; Tominaga, Y. *J. Chem. Soc., Chem. Commun.* 1987, 1517.
233. Nativi, C.; Ravida, N.; Ricci, A.; Seconi, G.; Taddei, M. *J. Org. Chem.* 1991, *56*, 1951.
234. Mikami, K.; Kawamoto, K.; Loh, T.-P.; Nakai, T. *J. Chem. Soc., Chem. Commun.* 1990, 1161.
235. Yamamoto, Y.; Maruyama, K.; Matsumoto, K. *J. Chem. Soc., Chem. Commun.* 1983, 489.
236. Hull, C.; Mortlock, S. V.; Thomas, E. J. *Tetrahedron Lett.* 1987, *28*, 5343.
237. Hull, C.; Mortlock, S. V.; Thomas, E. J. *Tetrahedron* 1989, *45*, 1007.
238. Kobayashi, S.; Nishio, K. *Tetrahedron Lett.* 1993, *34*, 3453.
239. Marshall, J. A.; Tang, Y. *Synlett* 1992, 653.
240. Yamamoto, Y.; Yatagai, H.; Ishihara, Y.; Maeda, N.; Maruyama, K. *Tetrahedron* 1984, *40*, 2239.
241. Yamamoto, Y.; Maeda, N.; Maruyama, K. *J. Chem. Soc., Chem. Commun.* 1983, 742.
242. Yamamoto, Y.; Saito, Y.; Maruyama, K. *J. Chem. Soc., Chem. Commun.* 1982, 1326.
243. Keck, G. E.; Abbott, D. E. *Tetrahedron Lett.* 1984, *25*, 1883.
244. Keck, G. E.; Boden, E. P. *Tetrahedron Lett.* 1984, *25*, 265.
245. Keck, G. E.; Boden, E. P. *Tetrahedron Lett.* 1984, *25*, 1879.
246. Yamamoto, Y.; Komatsu, T.; Maruyama, K. *J. Chem. Soc., Chem. Commun.* 1983, 191.
247. Keck, G. E.; Abbott, D. E.; Boden, E. P.; Enholm, E. J. *Tetrahedron Lett.* 1984, *25*, 3927.
248. Maruyama, K.; Ishihara, Y.; Yamamoto, Y. *Tetrahedron Lett.* 1981, *22*, 4235.
249. Yamamoto, Y.; Yatagai, H.; Naruta, Y.; Maruyama, K. *J. Am. Chem. Soc.* 1980, *102*, 7107.
250. Yatagai, H.; Yamamoto, Y.; Maruyama, K. *J. Am. Chem. Soc.* 1980, *102*, 4548.
251. Schlessinger, R. H.; Graves, D. D. *Tetrahedron Lett.* 1987, *28*, 4381.
252. Kira, M.; Kobayashi, M.; Sakurai, H. *Tetrahedron Lett.* 1987, *28*, 4081.
253. Hosomi, A.; Shirahata, A.; Sakurai, H. *Tetrahedron Lett.* 1978, 3043.
254. Hosomi, A. *Acc. Chem. Res.* 1988, *21*, 200.
255. Hosomi, A.; Kohra, S.; Ogata, K.; Yanagi, T.; Tominaga, Y. *J. Org. Chem.* 1990, *55*, 2415.
256. Sato, K.; Kira, M.; Sakurai, H. *J. Am. Chem. Soc.* 1989, *111*, 6429.
257. Sato, M.; Yamamoto, Y.; Hara, S.; Suzuki, A. *Tetrahedron Lett.* 1993, *34*, 7071.
258. Moret, E.; Schlosser, M. *Tetrahedron Lett.* 1984, *25*, 4491.
259. Hoffmann, R. W.; Ditrich, K.; Strumer, R. *Chem. Ber.* 1989, *122*, 1783.
260. Andersen, M. W.; Hildebrandt, B.; Koster, G.; Hoffmann, R. W. *Chem. Ber.* 1989, *122*, 1777.
261. Hoffmann, R. W.; Zeiss, H.-J. *J. Org. Chem.* 1981, *46*, 1296.
262. Hoffmann, R. W.; Kemper, B. *Tetrahedron Lett.* 1980, *21*, 4883.
263. Williams, D. R.; Klingler, F. D. *Tetrahedron Lett.* 1987, *28*, 869.
264. Hoffmann, R. W.; Zeiss, H.-J. *Angew. Chem., Int. Ed. Engl.* 1979, *18*, 306.
265. Kira, M.; Sato, K.; Sakurai, H. *J. Am. Chem. Soc.* 1990, *112*, 257.
266. Satoh, M.; Nomoto, Y.; Miyaura, N.; Suzuki, A. *Tetrahedron Lett.* 1989, *30*, 3789.
267. Kiyooka, S.-i.; Heathcock, C. H. *Tetrahedron Lett.* 1983, *24*, 4765.
268. Reetz, M. T.; Kesseler, K.; Jung, A. *Tetrahedron Lett.* 1984, *25*, 729.
269. Coppi, L.; Mordini, A.; Taddei, M. *Tetrahedron Lett.* 1987, *28*, 969.
270. Sato, F.; Suzuki, Y.; Sato, M. *Tetrahedron Lett.* 1982, *23*, 4589.
271. Sturmer, R. *Angew. Chem., Int. Ed. Engl.* 1990, *29*, 59.
272. Hayashi, T.; Konishi, M.; Kumada, M. *J. Org. Chem.* 1983, *48*, 281.
273. Hayashi, T.; Kabeta, K.; Hamachi, I.; Kumada, M. *Tetrahedron Lett.* 1983, *28*, 2865.
274. Hosomi, A.; Sakurai, H. *Tetrahedron Lett.* 1976, 1295.
275. Tietze, L. F.; Dölle, A.; Schiemann, K. *Angew. Chem., Int. Ed. Engl.* 1992, *31*, 1372.
276. Roush, W. R.; Walts, A. E.; Hoong, L. K. *J. Am. Chem. Soc.* 1985, *107*, 8186.
277. Roush, W. R.; Halterman, R. L. *J. Am. Chem. Soc.* 1986, *108*, 294.
278. Roush, W. R.; Ando, K.; Powers, D. B.; Halterman, R. L.; Palkowitz, A. D. *Tetrahedron Lett.* 1988, *29*, 5579.
279. Midland, M. M.; Preston, S. B. *J. Am. Chem. Soc.* 1982, *104*, 2330.
280. Roush, W. R.; Banfi, L. *J. Am. Chem. Soc.* 1988, *110*, 3979.
281. Short, R. P.; Masamune, S. *J. Am. Chem. Soc.* 1989, *111*, 1892.
282. Hoffmann, R. W.; Zeiss, H.-J. *Angew. Chem., Int. Ed. Engl.* 1980, *19*, 218.
283. Hoffmann, R. W.; Zeiss, H.-J.; Ladner, W.; Tabche, S. *Chem. Ber.* 1982, *115*, 2357.
284. Hoffmann, R. W.; Endesfelder, A.; Zeiss, H.-J. *Carbohydr. Res.* 1983, *123*, 320.
285. Herold, T.; Hoffmann, R. W. *Angew. Chem., Int. Ed. Engl.* 1978, *17*, 768.
286. Zweifel, G.; Shoup, T. M. *J. Am. Chem. Soc.* 1988, *110*, 5578.
287. Ochiai, M.; Fujita, E. *J. Chem. Soc., Chem. Commun.* 1980, 1118.
288. Hoffmann, R. W.; Dresely, S. *Tetrahedron Lett.* 1987, *28*, 5303.
289. Hoffmann, R. W.; Landmann, B. *Tetrahedron Lett.* 1983, *24*, 3209.
290. Sturmer, R.; Hoffmann, R. W. *Synlett* 1990, 759.
291. Corey, E. J.; Yu, C.-M.; Kim, S. S. *J. Am. Chem. Soc.* 1989, *111*, 5495.

292. Dumartin, G.; Pereyre, M.; Quintard, J.-P. *Tetrahedron Lett.* 1987, *28*, 3935.
293. Yamamoto, Y.; Yatagai, H.; Maruyama, K. *J. Chem. Soc., Chem. Commun.* 1980, 1072.
294. Tebben, P.; Reggelin, M.; Hoppe, D. *Tetrahedron Lett.* 1989, *30*, 2919.
295. Kramer, G. W.; Brown, H. C. *J. Org. Chem.* 1977, *42*, 2292.
296. Yamamoto, Y.; Yatagai, H.; Maruyama, K. *J. Am. Chem. Soc.* 1981, *103*, 1969.
297. Reggelin, M.; Tebben, P.; Hoppe, D. *Tetrahedron Lett.* 1989, *30*, 2915.
298. Brown, H. C.; Bhat, K. S. *J. Am. Chem. Soc.* 1986, *108*, 293.
299. Brown, H. C.; Jadhav, P. K.; Bhat, K. S. *J. Am. Chem. Soc.* 1988, *110*, 1535.
300. Brown, H. C.; Jadhav, P. K. *J. Am. Chem. Soc.* 1983, *105*, 2092.
301. Jadhav, P. K. *Tetrahedron Lett.* 1989, *30*, 4763.
302. Brown, H. C.; Bhat, K. S.; Randad, R. S. *J. Org. Chem.* 1989, *54*, 1570.
303. Murai, A.; Abiko, A.; Shimada, N.; Masamune, T. *Tetrahedron Lett.* 1984, *25*, 4951.
304. Hulsen, E. v.; Hoppe, D. *Tetrahedron Lett.* 1985, *26*, 411.
305. Brown, H. C.; Randad, R. S.; Bhat, K. S.; Zaidlewicz, M.; Racherla, U. S. *J. Am. Chem. Soc.* 1990, *112*, 2389.
306. Jephcote, V. J.; Pratt, A. J.; Thomas, E. J. *J. Chem. Soc., Perkin Trans. 1* 1989, 1529.
307. Pratt, A. J.; Thomas, E. J. *J. Chem. Soc., Perkin Trans. 1* 1989, 1521.
308. McNeill, A. H.; Thomas, E. J. *Tetrahedron Lett.* 1992, *33*, 1369.
309. Pratt, A. J.; Thomas, E. J. *J. Chem. Soc., Chem. Commun.* 1982, 1115.
310. Jephcote, V. J.; Pratt, A. J.; Thomas, E. J. *J. Chem. Soc., Chem. Commun.* 1984, 800.
311. Marshall, J. A.; DeHoff, B. S.; Crooks, S. L. *Tetrahedron Lett.* 1987, *28*, 527.
312. McNeill, A. H.; Thomas, E. J. *Tetrahedron Lett.* 1993, *34*, 1669.
313. Carey, J. S.; Coulter, T. S.; Thomas, E. J. *Tetrahedron Lett.* 1993, *34*, 3933.
314. Carey, J. S.; Thomas, E. J. *Tetrahedron Lett.* 1993, *34*, 3935.
315. Hoffmann, R. W.; Dresely, S. *Chem. Ber.* 1989, *122*, 903.
316. Gung, B. W.; Smith, D. T.; Wolf, M. A. *Tetrahedron Lett.* 1991, *32*, 13.
317. Roush, W. R.; Adam, M. A.; Walts, A. E.; Harris, D. J. *J. Am. Chem. Soc.* 1986, *108*, 3422.
318. Koreeda, M.; Tanaka, Y. *Tetrahedron Lett.* 1987, *28*, 143.
319. Hoffmann, R. W.; Kemper, B. *Tetrahedron Lett.* 1982, *23*, 845.
320. Roush, W. R.; Michaelides, M. R. *Tetrahedron Lett.* 1986, *27*, 3353.
321. Zschage, O.; Schwark, J.-R.; Hoppe, D. *Angew. Chem., Int. Ed. Engl.* 1990, *29*, 296.
322. Kramer, T.; Schwark, J.-R.; Hoppe, D. *Tetrahedron Lett.* 1990, *31*, 7037.
323. van der Heide, T. A. J.; van der Baan, J. L.; Bijpost, E. A. *Tetrahedron Lett.* 1993, *34*, 4655.
324. Marshall, J. A.; Luke, G. P. *Synlett* 1992, 1007.
325. Evans, D. A.; Takacs, J. M.; Hurst, K. M. *J. Am. Chem. Soc.* 1979, *101*, 371.
326. Marshall, J. A.; Gung, W. Y. *Tetrahedron Lett.* 1989, *30*, 2183.
327. Panek, J. S.; Yang, M.; Solomono, J. S. *J. Org. Chem.* 1993, *58*, 1003.
328. Panek, J. S.; Cirillo, P. F. *J. Org. Chem.* 1993, *58*, 999.
329. Yamamoto, Y.; Kobayashi, K.; Okano, H.; Kadota, I. *J. Org. Chem.* 1992, *57*, 7003.
330. Carey, J. S.; Thomas, E. J. *Synlett* 1992, 585.
331. Hoffmann, R. W.; Kemper, B. *Tetrahedron* 1984, *40*, 2219.
332. Barrett, A. G. M.; Seefeld, M. A. *Tetrahedron* 1993, *49*, 7857.
333. Roush, W. R.; Grover, P. T.; Lin, X. *Tetrahedron Lett.* 1990, *31*, 7563.
334. Roush, W. R.; Grover, P. T. *Tetrahedron Lett.* 1990, *31*, 7567.
335. Roush, W. R.; Ando, K.; Powers, D. B.; Palkowitz, A. D.; Halterman, R. L. *J. Am. Chem. Soc.* 1990, *112*, 6339.
336. Yamamoto, Y.; Yatagai, H.; Maruyama, K. *J. Am. Chem. Soc.* 1981, *103*, 3229.
337. Wang, K. K.; Gu, Y. G.; Lui, C. *J. Am. Chem. Soc.* 1990, *112*, 4424.
338. Wang, K. K.; Lui, C.; Gu, Y. G.; Burnett, F. N.; Satteangi, P. D. *J. Org. Chem.* 1991, *56*, 1914.
339. Tsai, D. J. S.; Matteson, D. S. *Tetrahedron Lett.* 1981, *22*, 2751.
340. Hosomi, A.; Hashimoto, H.; Sakurai, H. *J. Org. Chem.* 1978, *43*, 2551.
341. Mukaiyama, T.; Harada, T.; Shoda, S.-I. *Chemistry Lett.* 1980, 1507.
342. Bartlett, P. A.; Johnson, W. S.; Elliott, J. D. *J. Am. Chem. Soc.* 1983, *105*, 2088.
343. Marshall, J. A.; Wang, X.-J. *J. Org. Chem.* 1990, *55*, 6246.
344. Hiroi, K.; Abe, J. *Tetrahedron Lett.* 1990, *31*, 3623.
345. Schaumann, E.; Kirschning, A. *Tetrahedron Lett.* 1988, *29*, 4281.
346. Schaumann, E.; Kirschning, A. *J. Chem. Soc., Chem. Commun.* 1990, 419.
347. Fujimura, O.; Takai, K.; Utimoto, K. *J. Org. Chem.* 1990, *55*, 1705.
348. Roush, W. R.; Hoong, L. K.; Palmer, M. A. J.; Park, J. C. *J. Am. Chem. Soc.* 1990, *112*, 4109.
349. Haruta, R.; Ishiguro, M.; Ikeda, N.; Yamamoto, H. *J. Am. Chem. Soc.* 1982, *104*, 7667.
350. Brown, H. C.; Randad, R. S. *Tetrahedron* 1990, *46*, 4463.
351. Roush, W. R.; Banfi, L.; Park, J. C.; Hoong, L. K. *Tetrahedron Lett.* 1989, *30*, 6457.

352. Hoffmann, R. W. *Chem. Rev.* 1989, *89*, 1841.
353. Wuts, P. G. M.; Bigelow, S. S. *J. Chem. Soc., Chem. Commun.* 1984, 736.
354. Andersen, M. W.; Hildebrandt, B.; Hoffmann, R. W. *Angew. Chem., Int. Ed. Engl.* 1991, *30*, 97.
355. Evans, D. A.; Nelson, J. V.; Taber, T. R. *Top. Stereochem.* 1982, *13*, 1.
356. Vulpetti, A.; Gardner, M.; Gennari, C.; Bernardi, A.; Goodman, J. M.; Paterson, I. *J. Org. Chem.* 1993, *58*, 1711.
357. Yamamoto, Y.; Maruyama, K. *Tetrahedron Lett.* 1980, *21*, 4607.
358. Schinzer, D. *Synthesis* 1988, 263.
359. Denmark, S. E.; Weber, E. J.; Wilson, T. M.; Willson, T. M. *Tetrahedron* 1989, *45*, 1053.
360. Marshall, J. M.; Gung, W. Y. *Tetrahedron* 1989, *45*, 1043.
361. Yamamoto, Y.; Saito, K.-I. *J. Chem. Soc., Chem. Commun.* 1989, 1676.
362. Chan, T. H.; Wang, D. *Tetrahedron Lett.* 1989, *30*, 3041.
363. Denmark, S. E.; Wilson, T.; Willson, T. M. *J. Am. Chem. Soc.* 1988, *110*, 984.
364. Keck, G. E.; Andrus, M. B.; Castellino, S. *J. Am. Chem. Soc.* 1989, *111*, 8136.
365. Hoffmann, R. W.; Sander, T. *Chem. Ber.* 1990, *123*, 145.
366. Roush, W. R. In *Trends in Synthetic Carbohydrate Chemistry*; D. Horton and L. D. Hawkins, Eds.; American Chemical Society: Washington D. C., 1989; pp 242.
367. Roush, W. R.; Hoong, L. K.; Palmer, M. A. J.; Straub, J. A.; Palkowitz, A. D. *J. Am. Chem. Soc.* 1990, *112*, 4117.
368. Trost, B. M.; Kondo, Y. *Tetrahedron Lett.* 1991, *32*, 1613.
369. Roush, W. R.; Palkowitz, A. D.; Ando, K. *J. Am. Chem. Soc.* 1990, *112*, 6348.
370. White, J. D.; Johnson, A. T. *J. Org. Chem.* 1990, *55*, 5938.
371. Akita, H.; Yamada, H.; Matsukura, H.; Nakata, T.; Oishi, T. *Tetrahedron Lett.* 1990, *31*, 1735.
372. Mulzer, J.; Kirstein, H. M.; Buschmann, J.; Lehmann, C.; Luger, P. *J. Am. Chem. Soc.* 1991, *113*, 910.
373. Roush, W. R.; Straub, J. A.; VanNieuwenhze, M. S. *J. Org. Chem.* 1991, *56*, 1636.
374. Roush, W. R.; Lin, X.; Straub, J. A. *J. Org. Chem.* 1991, *56*, 1649.
375. Hoffmann, R. W.; Brinkmann, H.; Frenking, G. *Chem. Ber.* 1990, *123*, 2387.
376. Brinkmann, H.; Hoffmann, R. W. *Chem. Ber.* 1990, *123*, 2395.
377. Li, Y.; Houk, K. N. *J. Am. Chem. Soc.* 1989, *111*, 1236.
378. Roush, W. R.; Park, J. C. *J. Org. Chem.* 1990, *55*, 1143.
379. Nishigaichi, Y.; Takuwa, A.; Naruta, Y.; Maruyama, K. *Tetrahedron* 1993, *49*, 7395.
380. Denmark, S. E.; Almstead, N. G. *J. Am. Chem. Soc.* 1993, *115*, 3133.
381. Marshall, J. A.; Luke, G. P. *J. Org. Chem.* 1991, *56*, 483.
382. Guanti, G.; Banfi, L.; Zannetti, M. T. *Tetrahedron Lett.* 1993, *34*, 5487.
383. Keck, G. E.; Geraci, L. S. *Tetrahedron Lett.* 1993, *34*, 7827.
384. Costa, A. L.; Piazza, M. G.; Tagliavini, E.; Trombini, C.; Umani-Ronchi, A. *J. Am. Chem. Soc.* 1993, *115*, 7001.
385. Keck, G. E.; Krishnamurthy, D.; Grier, M. C. *J. Org. Chem.* 1993, *58*, 6543.
386. Wei, Z. Y.; Wang, D.; Li, J. S.; Chan, T. H. *J. Org. Chem.* 1989, *54*, 5768.
387. Ishihara, K.; Mouri, M.; Gao, Q.; Maruyama, T.; Furuta, K.; Yamamoto, H. *J. Am. Chem. Soc.* 1993, *115*, 11490.
388. Faller, J. W.; Linebarrier, D. L. *J. Am. Chem. Soc.* 1989, *111*, 1937.
389. Faller, J. W.; DiVerdi, M. J.; John, J. A. *Tetrahedron Lett.* 1991, *32*, 1271.
390. Riediker, M.; Hafner, A.; Piantini, U.; Rihs, G.; Togni, A. *Angew. Chem., Int. Ed. Engl.* 1989, *28*, 499.
391. Duthaler, R. O.; Hafner, A.; Riediker, M. *Pure Appl. Chem.* 1990, *62*, 631.
392. Collins, S.; Kuntz, B. A.; Hong, Y. *J. Org. Chem.* 1989, *54*, 4154.
393. Sato, F.; Takahashi, O.; Kato, T.; Kobayashi, Y. *J. Chem. Soc., Chem. Commun.* 1985, 1638.
394. Soai, K.; Ishizaki, M. *J. Chem. Soc., Chem. Commun.* 1984, 1016.
395. Trost, B. M.; Merlic, C. A. *J. Am. Chem. Soc.* 1988, *110*, 5216.
396. Tanaka, K.; Yoda, H.; Kaji, A. *Tetrahedron Lett.* 1985, *26*, 4751.
397. Nakamura, E.; Sekiya, K.; Kuwajima, I. *Tetrahedron Lett.* 1987, *28*, 337.
398. Hoppe, D.; Zschage, O. *Angew. Chem., Int. Ed. Engl.* 1989, *28*, 69.
399. Verlhac, J.-B.; Pereyre, M.; Quintard, J.-P. *Tetrahedron* 1990, *46*, 6399.
400. Hoppe, D.; Kramer, T.; Schwark, J.-R.; Zschage, O. *Pure Appl. Chem.* 1990, *62*, 1999.
401. McNeill, A. H.; Thomas, E. J. *Tetrahedron Lett.* 1990, *31*, 6239.
402. Hoppe, D.; Kramer, T.; Erdbrugger, C. F.; Egert, E. *Tetrahedron Lett.* 1989, *30*, 1233.
403. Marsch, M.; Harms, K.; Zschage, O.; Hoppe, D.; Boche, G. *Angew. Chem., Int. Ed. Engl.* 1991, *30*, 321.
404. Yamaguchi, M.; Mukaiyama, T. *Chemistry Lett.* 1981, 1005.
405. Mukaiyama, T.; Yamada, T.; Suzuki, K. *Chemistry Lett.* 1983, 5.
406. Kang, S. H.; Kim, W. J. *Tetrahedron Lett.* 1989, *30*, 5915.
407. Suzuki, K.; Yuki, Y.; Mukaiyama, T. *Chemistry Lett.* 1981, 1529.

408. McGarvey, G. J.; Kimura, M.; Oh, T. *J. Carbohydr. Chem.* 1984, *3*, 125.
409. Yamamoto, Y.; Nishii, S.; Maruyama, K. *J. Chem. Soc., Chem. Commun.* 1986, 102.
410. Mukaiyama, T.; Suzuki, K.; Soai, K.; Sato, T. *Chemistry Lett.* 1979, 447.
411. Garner, P.; Park, J. M. *J. Org. Chem.* 1990, *55*, 3772.
412. Koga, K.; Yamada, S.-I.; Yoh, M.; Mizoguchi, T. *Carbohydr. Res.* 1974, *36*, C9.
413. Corey, E. J.; Yu, C.-M.; Lee, D.-H. *J. Am. Chem. Soc.* 1990, *112*, 878.
414. Choi, V. M. F.; Elliott, J. D.; Johnson, W. S. *Tetrahedron Lett.* 1984, *25*, 591.
415. Lindell, S. D.; Elliott, J. D.; Johnson, W. S. *Tetrahedron Lett.* 1984, *25*, 3947.
416. Schreiber, S. L.; Reagan, J. *Tetrahedron Lett.* 1986, *27*, 2945.
417. Alexakis, A.; Mangeney, P.; Ghribi, A.; Marek, I.; Sedrani, R.; Guir, C.; Normant, J. *Pure Appl. Chem.* 1988, *60*, 49.
418. Corcoran, R. C. *Tetrahedron Lett.* 1990, *31*, 2101.
419. Alexakis, A.; Mangeney, P. *Tetrahedron: Asymmetry* 1990, *1*, 477.
420. Andrés, C.; Delgado, M.; Pedrosa, R.; Rodríguez, R. *Tetrahedron Lett.* 1993, *34*, 8325.
421. Kiyooka, S.-I.; Sasaoka, H.; Fujiyama, R.; Heathcock, C. H. *Tetrahedron Lett.* 1984, *25*, 5331.
422. Denmark, S. E.; Willson, T. M. *J. Am. Chem. Soc.* 1989, *111*, 3475.
423. Johnson, W. J.; Crackett, P. H.; Elliott, J. D.; Jagodzinski, J. J.; Lindell, S. D.; Natarajan, S. *Tetrahedron Lett.* 1984, *25*, 3951.
424. Yamamoto, Y.; Nishii, S.; Yamada, J.-I. *J. Am. Chem. Soc.* 1986, *108*, 7116.
425. Denmark, S. E.; Willson, T. M.; Almstead, N. G. *J. Am. Chem. Soc.* 1989, *111*, 9258.
426. Chan, T.; Wang, D. *Chem. Rev.* 1992, *92*, 995.
427. Hosomi, A.; Sakurai, H. *Tetrahedron Lett.* 1978, 2589.
428. Tsunoda, T.; Suzuki, M.; Noyori, R. *Tetrahedron Lett.* 1980, *21*, 71.
429. Pasquarello, A.; Poli, G.; Potenza, D.; Scolastico, C. *Tetrahedron: Asymmetry* 1990, *1*, 429.
430. Conde-Frieboes, K.; Hoppe, D. *Synlett* 1990, 99.
431. Cirillo, P. F.; Panek, J. S. *J. Org. Chem.* 1990, *55*, 6071.
432. Sato, T.; Otera, J.; Nozaki, H. *J. Org. Chem.* 1990, *55*, 6116.
433. Kiyooka, S.-I.; Nakano, M.; Shiota, F.; Fujiyama, R. *J. Org. Chem.* 1989, *54*, 5409.
434. DeCamp, A. E.; Kawaguchi, A. T.; Volante, R. P.; Shinkai, I. *Tetrahedron Lett.* 1991, *32*, 1867.
435. Roush, W. R.; Palkowitz, A. D. *J. Org. Chem.* 1989, *54*, 3009.
436. Mulzer, J.; Kattner, L. *Angew. Chem., Int. Ed. Engl.* 1990, *29*, 679.
437. Prasad, J. V. N.; Rich, D. H. *Tetrahedron Lett.* 1990, *31*, 1803.
438. Roush, W. R.; Michaelides, M. R.; Tai, D. F.; Lesur, B. M.; Chong, W. K. M.; Harris, D. J. *J. Am. Chem. Soc.* 1989, *111*, 2984.
439. Davies, S. G.; Newton, R. F.; Williams, J. M. J. *Tetrahedron Lett.* 1989, *30*, 2967.
440. Davies, S. G.; Goodfellow, C. L. *J. Chem. Soc., Perkin Trans. 1* 1990, 393.
441. Davies, S. G.; Goodfellow, C. L. *Synlett* 1989, 59.
442. Wu, M.-J.; Pridgen, L. N. *Synlett* 1990, 636.
443. Wu, M.-J.; Pridgen, L. N. *J. Org. Chem.* 1991, *56*, 1340.
444. Alberola, A.; Andres, C.; Pedrosa, R. *Synlett* 1990, 763.
445. Ojima, I.; Inaba, S.-I.; Nagai, Y. *Chem. Lett.* 1975, 737.
446. Fiaud, J.-C.; Kagan, H. B. *Tetrahedron Lett.* 1971, 1019.
447. Fiaud, J.-C.; Kagan, H. B. *Tetrahedron Lett.* 1970, 1813.
448. Shono, T.; Kise, N.; Sanda, F.; Ohi, S.; Tsubato, K. *Tetrahedron Lett.* 1988, *29*, 231.
449. Harding, K. E.; Davis, C. S. *Tetrahedron Lett.* 1988, *29*, 1891.
450. Bringmann, G.; Geisler, J.-P. *Synthesis* 1989, 608.
451. Tomioka, K.; Inoue, I.; Shindo, M.; Koga, K. *Tetrahedron Lett.* 1990, *31*, 6681.
452. Thiam, M.; Chastrette, F. *Tetrahedron Lett.* 1990, *31*, 1429.
453. Saito, K.; Harada, K. *Tetrahedron Lett.* 1989, *30*, 4535.
454. Chang, Z.-Y.; Coates, R. M. *J. Org. Chem.* 1990, *55*, 3475.
455. Chang, Z.-Y.; Coates, R. M. *J. Org. Chem.* 1990, *55*, 3464.
456. Still, W. C.; McDonald, J. H. *Tetrahedron Lett.* 1980, *21*, 1031.
457. Abenhaim, D.; Boireau, G.; Sabourault, B. *Tetrahedron Lett.* 1980, *21*, 3043.
458. Pyne, S. G.; Dikic, B. *J. Chem. Soc., Chem. Commun.* 1989, 826.
459. McGarvey, G. J.; Kimura, M. *J. Org. Chem.* 1982, *47*, 5422.
460. Pyne, S. G.; Dikic, B. *Tetrahedron Lett.* 1990, *31*, 5231.
461. Mash, E. A.; Fryling, J. A. *J. Org. Chem.* 1991, *56*, 1094.
462. Yamada, J.-I.; Abe, H.; Yamamoto, Y. *J. Am. Chem. Soc.* 1990, *112*, 6118.
463. Polniaszek, R. P.; Belmont, S. E.; Alvarez, R. *J. Org. Chem.* 1990, *55*, 215.
464. Tsuchihashi, G.-I.; Iruchijima, S.; Maniwa, K. *Tetrahedron Lett.* 1973, 3389.
465. Polniaszek, R. P.; McKee, J. A. *Tetrahedron Lett.* 1987, *28*, 4511.

466. Inoue, I.; Shindo, M.; Koga, K.; Tomioka, K. *Tetrahedron: Asymmetry* 1993, *4*, 1603.
467. Yamada, J.-I.; Sato, H.; Yamamoto, Y. *Tetrahedron Lett.* 1989, *30*, 5611.
468. Wada, M.; Sakurai, Y.; Akiba, K. *Tetrahedron Lett.* 1984, *25*, 1079.
469. Stork, G.; Dowd, S. *J. Am. Chem. Soc.* 1963, *85*, 2178.
470. Layer, R. *Chem. Rev.* 1963, *63*, 489.
471. Hosomi, A.; Araki, Y.; Sakurai, H. *J. Am. Chem. Soc.* 1982, *104*, 2081.
472. Boga, C.; Savoia, D.; Umani-Ronchi, A. *Tetrahedron: Asymmetry* 1990, *1*, 291.
473. Tanaka, H.; Inoue, K.; Pokorski, U.; Taniguchi, M.; Torii, S. *Tetrahedron Lett.* 1990, *31*, 3023.
474. Ciufolini, M. A.; Spencer, G. O. *J. Org. Chem.* 1989, *54*, 4739.
475. Bocoum, A.; Boga, C.; Savoia, D.; Umani-Ronchi, A. *Tetrahedron Lett.* 1991, *32*, 1367.
476. Wuts, P. G. M.; Jung, Y.-W. *J. Org. Chem.* 1991, *56*, 365.
477. Hua, D. H.; Miao, S. W.; Chen, J. S.; Iguchi, S. *J. Org. Chem.* 1991, *56*, 4.
478. Demélé, Y. A.; Belaud, C.; Villiéras, J. *Tetrahedron: Asymmetry* 1992, *3*, 511.
479. Laschat, S.; Kunz, H. *Synlett* 1990, 51.
480. Weber, T.; Edwards, J. P.; Denmark, S. E. *Synlett* 1989, 20.
481. Denmark, S. E.; Edwards, J. P.; Nicaise, O. *J. Org. Chem.* 1993, *58*, 569.
482. Callant, D.; Stanssens, D.; de Vries, J. G. *Tetrahedron: Asymmetry* 1993, *4*, 185.
483. Hayashi, M.; Miyamoto, Y.; Inoue, T.; Oguni, N. *J. Org. Chem.* 1993, *58*, 1515.
484. Konradi, A. W.; Pedersen, S. F. *J. Org. Chem.* 1990, *55*, 4506.
485. Singh, V. K. *Synthesis* 1992, 605.
486. Chérest, M.; Felkin, H.; Prudent, N. *Tetrahedron Lett.* 1968, 2199.
487. Cram, D. J.; Abd Elhafez, F. A.; Weingartner, H. *J. Am. Chem. Soc.* 1953, *75*, 2293.
488. Cram, D. J.; Allinger, J. *J. Am. Chem. Soc.* 1954, *76*, 4516.
489. Kruger, D.; Sopchik, A. E.; Kingsbury, C. A. *J. Org. Chem.* 1984, *49*, 778.
490. Suzuki, K.; Katayama, E.; Tsuchihashi, G.-I. *Tetrahedron Lett.* 1984, *25*, 2479.
491. Gault, Y.; Flekin, H. *Bull. Soc. Chim. Fr.* 1960, 1342.
492. Midland, M. M.; Kwon, Y. C. *J. Am. Chem. Soc.* 1983, *105*, 3725.
493. Guyon, R.; Villa, P. *Bull. Soc. Chim. Fr.* 1977, 152.
494. Yamamoto, Y.; Matsuoka, K.; Nemoto, H. *J. Am. Chem. Soc.* 1988, *110*, 4475.
495. Wu, Y.; Houk, K. N. *J. Am. Chem. Soc.* 1992, *114*, 1656.
496. Dauben, W. G.; Fonken, G. H.; Noyce, D. C. *J. Am. Chem. Soc.* 1956, *78*, 3752.
497. Chérest, M.; Felkin, H. *Tetrahedron Lett.* 1968, 2205.
498. Huet, J.; Maroni-Barnaud, Y.; Anh, N. T.; Seyden-Penne, J. *Tetrahedron Lett.* 1976, 159.
499. Johnson, F. *Chem. Rev.* 1968, *68*, 375.
500. Liotta, C. L. *Tetrahedron Lett.* 1975, 519.
501. Klein, J. *Tetrahedron Lett.* 1973, 4307.
502. Anh, N. T.; Eisenstein, O.; Lefour, J. M.; Dâu, M. E. T. H. *J. Am. Chem. Soc.* 1973, *95*, 6146.
503. Liotta, C. L. *Tetrahedron Lett.* 1975, 525.
504. Liotta, C. L.; Burgess, E. M.; Eberhardt, W. H. *J. Am. Chem. Soc.* 1984, *106*, 4849.
505. Ashby, E. C.; Boone, J. R. *J. Org. Chem.* 1976, *41*, 2890.
506. Gidding, M. R.; Hudec, J. *Can. J. Chem.* 1981, *59*, 459.
507. Wipke, W. T.; Gund, P. *J. Am. Chem. Soc.* 1974, *96*, 299.
508. Wipke, W. T.; Gund, P. *J. Am. Chem. Soc.* 1976, *98*, 8107.
509. Perlburger, J.-C.; Müller, P. *J. Am. Chem. Soc.* 1977, *99*, 6316.
510. Cieplak, A. S. *J. Am. Chem. Soc.* 1981, *103*, 4540.
511. Mukherjee, D.; Wu, Y.-D.; Fronczek, F. R.; Houk, K. N. *J. Am. Chem. Soc.* 1988, *110*, 3328.
512. Wu, Y.-D.; Houk, K. N. *J. Am. Chem. Soc.* 1987, *109*, 908.
513. Wigfield, D. C. *Tetrahedron* 1979, *35*, 449.
514. Loupy, A.; Seyden-Penne, J. *Tetrahedron* 1980, *36*, 1937.
515. Carreño, M. C.; Ruano, J. L. G.; Garrido, M.; Ruiz, M. P.; Solladié, G. *Tetrahedron Lett.* 1990, *31*, 6653.
516. Marouka, K.; Itoh, T.; Yamamoto, H. *J. Am. Chem. Soc.* 1985, *107*, 4573.
517. D'Incan, E.; Loupy, A.; Maïa, A. *Tetrahedron Lett.* 1981, *22*, 941.
518. ApSimon, J. W.; Collier, T. E. *Tetrahedron* 1986, *42*, 5157.
519. Soladar, J. *Chemtech.* 1975, 421.
520. Levi, A.; Modena, G.; Scorrano, G. *J. Chem. Soc., Chem. Commun.* 1975, 6.
521. Hayashi, T.; Yamamoto, K.; Kumada, M. *J. Organometal. Chem.* 1976, *112*, 253.
522. Bonvicini, P.; Levi, A.; Modena, G.; Scorrano, G. *J. Chem. Soc., Chem. Commun.* 1972, 1188.
523. Hayashi, T.; Mise, T.; Kumada, M. *Tetrahedron Lett.* 1976, 4351.
524. Joó, F.; Trócsányi, E. *J. Organometal. Chem.* 1982, *231*, 63.
525. Valentine, D.; Scott, J. W. *Synthesis* 1978, 329.
526. Corriu, R. J. P.; Moreau, J. J. E. *J. Organometal. Chem.* 1975, *91*, C27.

527. Corriu, R. J. P.; Moreau, J. J. E. *J. Organometal. Chem.* 1975, *85*, 19.
528. Ojima, I.; Kogure, T.; Kumagai, M. *J. Org. Chem.* 1977, *42*, 1671.
529. Hayashi, T.; Yamamoto, K.; Kumada, M. *Tetrahedron Lett.* 1974, 331.
530. Ojima, I.; Kogure, T.; Nagai, Y. *Tetrahedron Lett.* 1974, 1889.
531. Brunner, H.; Riepl, G. *Angew. Chem., Int. Ed. Engl.* 1982, *21*, 377.
532. Hayashi, T.; Yamamoto, K.; Kasuga, K.; Omizu, H.; Kumada, M. *J. Organometal. Chem.* 1976, *113*, 127.
533. Peyronel, J.-F.; Fiaud, J.-C.; Kagan, H. B. *J. Chem. Research (S)* 1980, 320.
534. Karim, A.; Mortreux, A.; Petit, F. *Tetrahedron Lett.* 1986, *27*, 345.
535. Brunner, H.; Riepl, G.; Weitzner, H. *Angew. Chem., Int. Ed. Engl.* 1983, *22*, 331.
536. Balavoine, G.; Clinet, J. C.; Lellouche, I. *Tetrahedron Lett.* 1989, *30*, 5141.
537. Brown, H. C.; Krishnamurthy, S. *J. Am. Chem. Soc.* 1972, *94*, 7159.
538. Malék, J. *Org. React.* 1985, *34*, 1.
539. Malék, J. *Org. React.* 1988, *36*, 249.
540. Lane, C. F. *Synthesis* 1975, 135.
541. Krishnamurthy, S.; Brown, H. C. *J. Am. Chem. Soc.* 1976, *98*, 3383.
542. Midland, M. M. In *Asymmetric Synthesis*; J. D. Morrison, Ed.; Academic Press: Orlando, FL, 1983; Vol. 2; pp 45.
543. Haller, R.; Schneider, H. J. *Chem. Ber.* 1973, *106*, 1312.
544. Yamaguchi, S.; Mosher, H. S. *J. Org. Chem.* 1973, *38*, 1370.
545. Giacomelli, G.; Menicagli, R.; Caporusso, A. M.; Lardicci, L. *J. Org. Chem.* 1978, *43*, 1790.
546. Evans, R. J. D.; Landor, S. R.; Regan, J. P. *J. Chem. Soc., Perkin Trans. 1* 1974, 552.
547. Landor, S. R.; Sonola, O. O.; Tatchell, A. R. *J. Chem. Soc., Perkin Trans. 1* 1974, 1902.
548. Nasipuri, D.; Bhattacharya, P. K. *J. Chem. Soc., Perkin Trans. 1* 1977, 576.
549. Colonna, S.; Fornasier, R. *J. Chem. Soc., Perkin Trans. 1* 1978, 371.
550. Balcells, J.; Colonna, S.; Fornasier, R. *Synthesis* 1976, 266.
551. Meyers, A. I.; Kendall, P. M. *Tetrahedron Lett.* 1974, 1337.
552. Bothner-By, A. A. *J. Am. Chem. Soc.* 1951, *73*, 846.
553. Krishnamurthy, S.; Vogel, F.; Brown, H. C. *J. Org. Chem.* 1977, *42*, 2534.
554. Grandbois, E. R.; Howard, S. I.; Morrison, J. D. In *Asymmetric Synthesis*; J. D. Morrison, Ed.; Academic Press: Orlando, FL, 1983; Vol. 2; pp 71.
555. Corey, E. J.; Becker, K. B.; Varma, R. K. *J. Am. Chem. Soc.* 1972, *94*, 8616.
556. Massé, J. P.; Parayre, E. R. *J. Chem. Soc., Chem. Commun.* 1976, 438.
557. Sugimoto, T.; Matsumara, Y.; Tanimoto, S.; Okane, M. *J. Chem. Soc., Chem. Commun.* 1978, 926.
558. Morrison, J. D.; Grandbois, E. R.; Howard, S. I. *J. Org. Chem.* 1980, *45*, 4229.
559. Brown, H. C.; Krishnamurthy, S.; Hubbard, J. L. *J. Organometal. Chem.* 1979, *166*, 271.
560. Brown, H. C.; Kramer, G. W.; Hubbard, J. L.; Krishnamurthy, S. *J. Organometal. Chem.* 1980, *188*, 1.
561. Brown, H. C.; Singram, B.; Mathew, P. C. *J. Org. Chem.* 1981, *46*, 2712.
562. Midland, M. M.; Kazubski, A. *J. Org. Chem.* 1982, *47*, 2814.
563. Hirao, A.; Mochizuki, H.; Nakahama, S.; Yamazaki, N. *J. Org. Chem.* 1979, *44*, 1720.
564. Hirao, A.; Nakahama, S.; Mochizuki, D.; Itsuno, S.; Ohowa, M.; Yamazaki, N. *J. Chem. Soc., Chem. Commun.* 1979, 807.
565. Hirao, A.; Nakahama, S.; Mochizucki, H.; Itsuno, S.; Yamazaki, N. *J. Org. Chem.* 1980, *45*, 4231.
566. Hirao, A.; Itsuno, S.; Owa, M.; Nagami, S.; Mochizuki, H.; Zoorov, H. H. A.; Niakahama, S.; Yamazaki, N. *J. Chem. Soc., Perkin Trans. 1* 1981, 900.
567. McMahon, R. J.; Wiegers, K. E.; Smith, S. G. *J. Org. Chem.* 1981, *46*, 99.
568. Soai, K.; Oyamada, H.; Yamanoi, T. *Chem. Lett.* 1984, 251.
569. Yamada, K.; Takeda, M.; Iwakuma, T. *Tetrahedron Lett.* 1981, *21*, 3869.
570. Massé, J. P.; Paryre, E. R. *J. Chem. Soc., Chem. Commun.* 1976, 438.
571. Horner, L.; Brich, W. *Liebigs Ann. Chem.* 1978, 710.
572. Evans, D. A.; Chapman, K. T.; Carreira, E. M. *J. Am. Chem. Soc.* 1988, *110*, 3560.
573. Evans, D. A.; Chapman, K. T. *Tetrahedron Lett.* 1986, *27*, 5939.
574. ApSimon, J. W.; Seguin, R. P. *Tetrahedron* 1979, *35*, 2797.
575. Yamaguchi, S.; Mosher, H. S. *J. Org. Chem.* 1973, *38*, 1870.
576. Jacquet, I.; Vigneron, J.-P. *Tetrahedron Lett.* 1974, 2065.
577. Vigneron, J.-P.; Jacquet, I. *Tetrahedron* 1976, *32*, 939.
578. Vigneron, J.-P.; Bloy, V. *Tetrahedron Lett.* 1979, 2683.
579. Yamada, S.; Kitamoto, M.; Terashima, S. *Tetrahedron Lett.* 1976, 3165.
580. Kitamoto, M.; Kameo, K.; Terasjo, S.; Yamada, S. *Chem. Pharm. Bull.* 1977, *25*,
581. Terashima, S.; Tanno, N.; Koga, K. *Chem. Lett.* 1980, 981.
582. Kawasaki, M.; Suzuki, Y.; Terashima, S. *Chem. Lett.* 1984, 239.
583. Pohland, A.; Sullivan, H. R. *J. Am. Chem. Soc.* 1953, *75*, 4453.

584. Yamaguchi, S.; Mosher, H. S.; Pohland, A. *J. Am. Chem. Soc.* 1972, *94*, 9254.
585. Brinkmeyer, R. S.; Kapoor, V. M. *J. Am. Chem. Soc.* 1977, *99*, 8341.
586. Cohen, N.; Lopresti, R. J.; Neukon, C.; Saucy, G. *J. Org. Chem.* 1980, *45*, 582.
587. Asami, M.; Mukaiyama, T. *Heterocycles* 1979, *12*, 499.
588. Asami, M.; Ohno, H.; Kobayashi, S.; Mukaiyama, T. *Bull. Chem. Soc., Jpn.* 1978, *51*, 1869.
589. Mukaiyama, T.; Asami, M.; Hanna, M.; Kobayashi, S. *Chem. Lett.* 1977, 783.
590. Noyori, R. *Pure Appl. Chem.* 1981, *53*, 2315.
591. Noyori, R.; Tomino, I.; Tanimoto, Y. *J. Am. Chem. Soc.* 1979, *101*, 3129.
592. Noyori, N.; Tomino, I.; Yamada, M.; Nishizawa, M. *J. Am. Chem. Soc.* 1984, *106*, 6709.
593. Noyori, R.; Tmino, I.; Yamada, M.; Nishizawa, M. *J. Am. Chem. Soc.* 1984, *106*, 6717.
594. Brown, H. C.; Jadhaw, P. K.; Mandal, A. K. *Tetrahedron* 1981, *37*, 3547.
595. Midland, M. M. *Chem. Rev.* 1989, *89*, 1553.
596. Brown, H. C.; Bigley, D. B. *J. Am. Chem. Soc.* 1961, *83*, 486.
597. Eleveld, M. B.; Hogeveen, H. *Tetrahedron Lett.* 1986, *27*, 631.
598. Midland, M. M.; McLoughlin, J. I.; Gabriel, J. *J. Org. Chem.* 1989, *54*, 159.
599. Brown, H. C.; Mandal, A. K. *J. Org. Chem.* 1977, *42*, 2996.
600. Wolfe, S.; Rank, A. *Can. J. Chem.* 1966, *44*, 2591.
601. Varma, K. R.; Caspi, E. *J. Org. Chem.* 1969, *34*, 2489.
602. Youn, I. K.; Lee, S. W.; Pak, C. S. *Tetrahedron Lett.* 1988, *29*, 4453.
603. Brown, H. C.; Jadhav, P. K.; Mandal, A. K. *Tetrahedron Lett.* 1981, *21*, 554.
604. Midland, M. M.; Tramontano, A.; Zderic, S. A. *J. Organometal. Chem.* 1978, *156*, 203.
605. Midland, M. M.; Zderic, S. A. *J. Am. Chem. Soc.* 1982, *104*, 525.
606. Midland, M. M.; Petre, J. E.; Tramontano, A.; Zderic, S. A. *J. Am. Chem. Soc.* 1979, *101*, 2352.
607. Brown, M. C.; Pai, G. G. *J. Org. Chem.* 1982, *47*, 1606.
608. Yamamoto, K.; Kimura, T.; Tomo, Y. *Tetrahedron Lett.* 1985, *26*, 4505.
609. Midland, M. M.; McLoughlun, J. I. *J. Org. Chem.* 1984, *49*, 1316.
610. Midland, M. M.; Greer, S. *Synthesis* 1978, 845.
611. Brown, H. C.; Pai, G. G. *J. Org. Chem.* 1985, *50*, 1384.
612. Brown, H. C.; Chandrasekharan, J.; Ramachandran, P. V. *J. Org. Chem.* 1986, *51*, 3394.
613. Chandrasekharan, J.; Ramachandran, J. V.; Brwon, H. C. *J. Org. Chem.* 1985, *50*, 5446.
614. Srebnik, M.; Ramachandran, P. V. *Aldrichimica Acta* 1987, *20*, 9.
615. Brown, H. C.; Chandrasekharan, J.; Ramachandran, P. V. *J. Am. Chem. Soc.* 1988, *110*, 1539.
616. Brown, H. C.; Srebnik, M.; Ramachandran, P. V. *J. Org. Chem.* 1989, *54*, 1577.
617. King, A. O.; Corely, E. G.; Anderson, R. K.; Larsen, R. D.; Verhoeven, T. R.; Reider, P. J.; Xiang, Y. B.; Belley, M.; Leblanc, Y.; Labelle, M.; Prasit, P.; Zamboni, R. J. *J. Org. Chem.* 1993, *58*, 3731.
618. Corey, E. J.; Bakshi, R. K. *Tetrahedron Lett.* 1990, *31*, 611.
619. Corey, E. J.; Bakshi, R. K.; Shibata, S. *J. Am. Chem. Soc.* 1987, *109*, 5551.
620. Corey, E. J.; Bakshi, R. K.; Shibata, S.; Chen, C.-P.; Singh, V. K. *J. Am. Chem. Soc.* 1987, *109*, 7925.
621. DeNinno, M. P.; Perneer, R. J.; Lijewski, L. *Tetrahedron Lett.* 1990, *31*, 7415.
622. Corey, E. J. *Pure Appl. Chem.* 1990, *62*, 1209.
623. Nevalainen, V. *Tetrahedron: Asymmetry* 1991, *2*, 63.
624. Mathre, D. J.; Jones, T. K.; Xavier, L. C.; Blacklock, T. J.; Reamer, R. A.; Mohan, J. J.; Jones, E. T. T.; Hoogsteen, K.; Baum, M. W.; Grabowski, E. J. J. *J. Org. Chem.* 1991, *56*, 751.
625. Corey, E. J.; Chen, C.-P.; Reichard, G. A. *Tetrahedron Lett.* 1989, *30*, 5547.
626. Corey, E. J.; Shibata, S.; Bakshi, R. K. *J. Org. Chem.* 1988, *53*, 2861.
627. Corey, E. J.; Link, J. O. *Tetrahedron Lett.* 1989, *30*, 6275.
628. Itsuno, S.; Ito, K.; Hirao, A.; Nakahama, S. *J. Org. Chem.* 1984, *49*, 555.
629. Corey, E. J.; Gavai, A. V. *Tetrahedron Lett.* 1988, *29*, 3201.
630. Corey, E. J.; Su, W.-G. *Tetrahedron Lett.* 1988, *29*, 3423.
631. Nevalainen, V. *Tetrahedron: Asymmetry* 1992, *3*, 921.
632. Nevalainen, V. *Tetrahedron: Asymmetry* 1992, *3*, 1563.
633. Jones, D. K.; Liotta, D. C.; Shinkai, I.; Mathre, D. J. *J. Org. Chem.* 1993, *58*, 799.
634. Nevalainen, V. *Tetrahedron: Asymmetry* 1992, *3*, 1441.
635. Nevalainen, V. *Tetrahedron: Asymmetry* 1992, *3*, 933.
636. Rao, A. V. R.; Gurjar, M. K.; Sharma, P. A.; Kaiwar, V. *Tetrahedron Lett.* 1990, *31*, 2341.
637. Corey, E. J.; Link, J. O. *Tetrahedron Lett.* 1990, *31*, 601.
638. Itsuno, S.; Sakarai, Y.; Ito, K.; Hirao, A.; Nakahama, S. *Bull. Chem. Soc., Jpn.* 1987, *60*, 395.
639. Corey, E. J.; Bakshi, R. K.; Shibata, S. *J. Org. Chem.* 1988, *53*, 2861.
640. Corey, E. J.; Link, J. O. *Tetrahedron Lett.* 1988, *29*, 6275.
641. Corey, E. J.; Link, J. O. *J. Org. Chem.* 1991, *56*, 442.
642. Corey, E. J.; Crimprich, K. A. *Tetrahedron Lett.* 1992, *33*, 4099.

643. Mathre, D. J.; Thompson, A. S.; Douglas, A. W.; Hoogsteen, K.; Carrol, J. D.; Corley, E. G.; Grabowski, E. J. J. *J. Org. Chem.* 1993, *58*, 2880.
644. Kim, Y. H.; Park, D. H.; Byum, I. S. *J. Org. Chem.* 1993, *58*, 4511.
645. Youn, I. K.; Lee, S. W.; Pak, C. S. *Tetrahedron Lett.* 1990, *31*, 4453.
646. Imai, T.; Tamura, T.; Yamamuro, A.; Sato, T.; Wollman, T. A.; Kennedy, R. M.; Masamune, S. *J. Am. Chem. Soc.* 1986, *108*, 7402.
647. Masamune, S.; Kennedy, R. M.; Petersen, J. S.; Houk, K. N.; Wu, Y.-D. *J. Am. Chem. Soc.* 1986, *108*, 7404.
648. Ramachandran, P. V.; Brown, H. C.; Swaminathan, S. *Tetrahedron: Asymmetry* 1990, *1*, 433.
649. Midland, M. M.; Kazubski, A. *J. Org. Chem.* 1992, *57*, 2953.
650. Cai, D.; Tschaen, D.; Shi, Y.; Verhoeven, T. R.; Reamer, R. A.; Douglas, A. W. *Tetrahedron Lett.* 1993, *34*, 3243.
651. Mears, R. J.; Whiting, A. *Tetrahedron Lett.* 1993, *34*, 8155.
652. Molander, G. A.; Bobbitt, K. L.; Murray, C. K. *J. Am. Chem. Soc.* 1992, *114*, 2759.
653. Inch, T. D. *Synthesis* 1970, 466.
654. Denise, B.; Ducom, J.; Fauvarque, J.-F. *Bull. Soc. Chim., Fr.* 1972, 990.
655. Lardicci, L.; Giacomelli, G. *J. Chem. Soc., Perkin Trans. 1* 1974, 337.
656. Morrison, J. D.; Tomaszewski, J. E.; Mosher, H. S.; Dale, J.; Miller, D.; Elsenbaumer, R. L. *J. Am. Chem. Soc.* 1977, *99*, 3167.
657. Macleod, R.; Welsh, F. J.; Mosher, H. S. *J. Am. Chem. Soc.* 1960, *82*, 876.
658. Tatibouet, M. F. *Bull. Soc. Chim. Fr.* 1951, 868.
659. Foley, W. M.; Welsh, F. J.; LaCombe, E. M.; Mosher, H. S. *J. Am. Chem. Soc.* 1959, *81*, 2779.
660. Giacomelli, G.; Lardicci, L.; Caporusso, A. M. *J. Chem. Soc., Perkin Trans. 1* 1975, 1795.
661. Capillon, J.; Guetté, J. P. *Tetrahedron* 1979, *35*, 1807.
662. Capillon, J.; Guetté, J. P. *Tetrahedron* 1979, *35*, 1817.
663. Evans, D. A.; Nelson, S. G.; Gagné, M. R.; Muci, A. R. *J. Am. Chem. Soc.* 1993, *115*, 9800.
664. Gamel, P.; Fache, F.; Maneney, P.; Lemaire, M. *Tetrahedron Lett.* 1993, *34*, 6897.
665. Okawa, H.; Katsuki, T.; Nakamura, M.; Kumagai, N.; Shuin, Y.; Shinmyozu, T.; Kida, S. *J. Chem. Soc., Chem. Commun.* 1989, 139.

Chapter 5

# ENOLATE REACTIONS OF CARBONYL COMPOUNDS

The carbonyl group is extremely versatile for the introduction of functionality. The powerful transformation of nucleophilic addition to afford an alcohol derivative was discussed in Chapter 4. The reactions of functionalized carbonyl compounds are discussed in Chapter 6, although some examples have been included in this chapter to avoid unnecessary fragmentation.

The emphasis of this chapter is placed on the introduction of an additional substituent, such as a heteroatom or alkyl group, at the carbon atom next to the carbonyl group of a simple compound. The aldol reaction and its analogs are discussed in Chapter 7. This demarcation has resulted in some duplication as a chiral auxiliary that allows both high selectivity for the reaction of a ketone equivalent with an alkyl halide and aldehyde will appear in this chapter and the one on the aldol reaction. Some fragmentation of topics has also occurred: as an example, the synthesis of α-amino acids by reaction of a carboxylic enolate derivative with a nitrogen electrophile will be found in this chapter, while the alkylation reactions of glycine derivatives are discussed in Chapter 6.

The incorporation of nitrogen, through the utilization of imines, enamines, and hydrazones, has allowed for the introduction of a large number of chiral auxiliaries to be used in alkylation-type chemistry. The appropriate chemistry, in keeping with the limitations outlined above, is also discussed in this chapter.

The chemistry of chiral bases, especially with regard to enolate formation, continues to expand. A section in this chapter discusses the general usage of this approach and its closely related asymmetric protonation methodology.

## 5.1. ENOL ETHERS AND ENOLATE FORMATION

The chemistry of enol ethers and enolate formation are closely related as the preparation of the former often proceeds by way of the latter. As we are concerned with asymmetric methodology, there will be little discussion of the chemistry of enols, as their formation and reactions are often under thermodynamic control — although this can be useful in cyclic systems.[1]

The specific formation of an enolate from a carbonyl compound is not a trivial task; regiochemical control has to be exercised, as well as stereochemical. This usage of a masked or control group often requires additional transformations that detract from synthetic eloquence or result in scale-up problems. Sometimes, these factors are simpler to control in cyclic systems, and this has been put to elegant usage in Woodward's synthesis of erythromycin where cyclic sulfides were used as masked alkyl substituents on an acyclic chain so that the chemistry of the cyclic system could be exploited to control stereochemistry [*13.4*].[2-5]

Deprotonation of the carbonyl substrate with a hindered base, such as LDA, usually ensures formation of the kinetic enolate.[6-20] The use of low temperature eliminates problems due to self condensation of the carbonyl compound.[21] Reaction of (*S*)-3-methyl-2-pentanone with LDA proceeded with deprotonation at the less hindered methyl group; less than 10% racemization occurred.[22] The kinetic enolate can be trapped *in situ* by generation in the presence of trimethylsilyl chloride.[23]

The thermodynamic enolate is formed when sodium hydride is employed as a base with the carbonyl compound;[24,25] the silyl enol ether is available through treatment with chlorotrimethylsilane and triethylamine in DMF.[18,21,26]

Silyl enol ethers can be converted to the corresponding lithium enolate without loss of stereochemical integrity by reaction with methyllithium or the fluoride ion.[24,27-31]

Regioselective generation of enolates can be accomplished from conjugated enones by the Michael addition of an organometallic reagent that delivers an alkyl group [*9.5.1*],[32] or a 1,4-reduction [*9.5.1.2*].[33,34] In addition to lithium enolates, other metals such as boron can also be incorporated into this approach.[35,36] An alternative addition reaction approach to an enolate relies on a 1,2-addition of an organometallic reagent to a ketene,[37-39] while reaction of a symmetrical α,α′-dibromoketone with a cuprate leads to introduction of an alkyl group from a cuprate to displace on bromine, and enolate generation by elimination of the other halogen atom.[40] The reaction proceeds through an intermediate cyclopropanone. This latter procedure is a more advanced version of the conversion of an α-halocarbonyl compound to enolate by reaction with a metal;[41] Reformatsky reagents are well known examples [*7.2.7*].[42-44]

In cyclic examples, the subsequent reactions of an enolate can be controlled by face selectivity. For the deprotonation step itself, an asymmetric base can ensure face selectivity [*5.4.1*]. Indeed, this philosophy of face selectivity is behind many of the successful systems for chiral auxiliaries or methods that use the self-regeneration of stereogenic centers [*1.3.2.3*]. For enolate formation, even in acyclic cases, facial selectivity during the deprotonation step may determine the resultant enolate geometry;[10,45-49] however, this assumes that no proton source is available for equilibration.[22] Recent studies suggest that it is the enolate complex that determines facial selectivity during subsequent reaction, as long as the geometry of the enolate is controlled during its formation.[50]

One aspect now beginning to come to fruition pertains to the structure determination of lithium enolates,[8-12,18,51-55] and the nitrogen bases.[56-59] The insight gained in this area should enable asymmetric induction to be increased.[60] Lithium enolates are the most commonly used species to perform functionalization of a carbon atom juxtaposed to a carbonyl moiety (vide supra).

Silyl enol ethers have also found extensive application due to their ready availability.[43,49,61-64] As isomeric enol ethers can be separated, this class of compounds provides a powerful method to control the regiochemical outcome of many carbonyl reactions. Many methods have now been proposed for the generation of regioselective enol ethers for a variety of carbonyl compounds containing functionality.[37,38,43,44,46,65-72] However, the geometry of the enolate may not be controlled during these preparations.[47] Vinyl anions can be oxidized to provide silyl enol ethers with retention of double bond geometry.[73,74] Alternative approaches rely on the thermal rearrangements of α- and β-keto silanes,[64,66,67,75-77] β-keto esters,[78] conjugate additions of organometallic reagents to enones in the presence of trimethylsilyl chloride (vide supra), and stereoselective eliminations of β-hydroxy esters.[70,79] α,β-Epoxy silanes rearrange to silyl enol ethers under Lewis acid,[80,81] base catalysis,[68,82] or through thermolytic conditions.[83] Brook rearrangement methodology can also be employed.[66,75,78,84]

In addition to physical separations,[24] isomeric silyl enol ethers can be separated by a kinetic resolution procedure that employs reaction with nitrosostyrene.[85] Studies on the isomerization of propionic ester silyl ketene acetals with trialkylammonium perchlorates have shown that the *Z*-isomer is thermodynamically more stable.[86] Chiral silicon moieties have been incorporated into silyl enol ethers;[87-90] however, the chemistry of these chiral compounds has yet to be exploited.[91]

Studies on the Claisen ester enolate rearrangement have shown that the ester *E*-(*trans*)-enolate is formed in THF, while in HMPA-THF the *Z*-enolate predominates (Scheme 5.1).[48,92-95] *N,N′*-Dimethyl-*N,N′*-propyleneurea (DMPU) in THF also affords the *Z*-enolate with high diastereoselection, and provides an alternative to the use of HMPA.[96-98] As a comparison, ethyl ketones react with LDA to form the Z-enolate.[99]

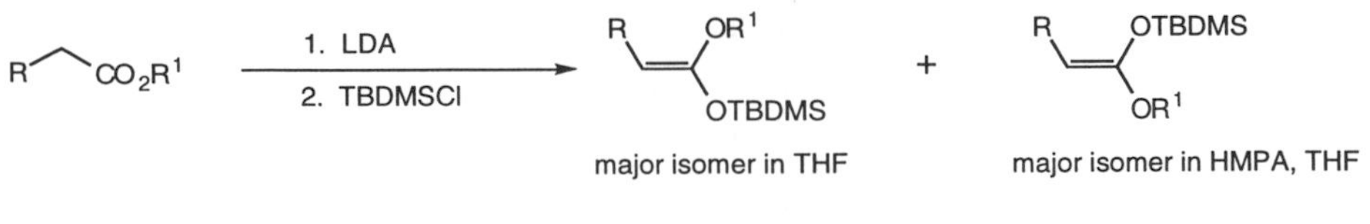

R CO$_2$R$^1$ → 1. LDA 2. TBDMSCl → R, OR$^1$, OTBDMS (major isomer in THF) + R, OTBDMS, OR$^1$ (major isomer in HMPA, THF)

**SCHEME 5.1.**

The observation that HMPA can provide access to the alternative enolate geometry[8] has been applied to other systems including ketones,[100,101] hydrazones,[102] and oxazolines.[103,104] The formation of the *E*-enolate in THF has been the subject of various interpretations.[10,47] The control of enolate geometry can be used to form silyl enol ethers with a high degree of stereoselection.[23,105] Some variations in formation of the enolate are observed when sodium hexamethyldisilazide is used as the base.[106] Silyl ketene acetals can undergo thermodynamic equilibration in the presence of ammonium salts.[86]

Ketones can be converted to either the corresponding (*E*)- or (*Z*)-boron enolate (**5.1** and **5.2**) by use of either dicyclohexylchloroborane ($Chx_2BCl$) or *B*-chloro-9-borabicyclo[3.3.1]nonane (*B*-Cl-9-BBN) (Scheme 5.2).[107-109] This stereoselectivity is particularly useful for application with the aldol reaction [*7.2.1*].[110] Z-Boron enolates are also available by the 1,4-hydroboration of conjugated enones.[111-113]

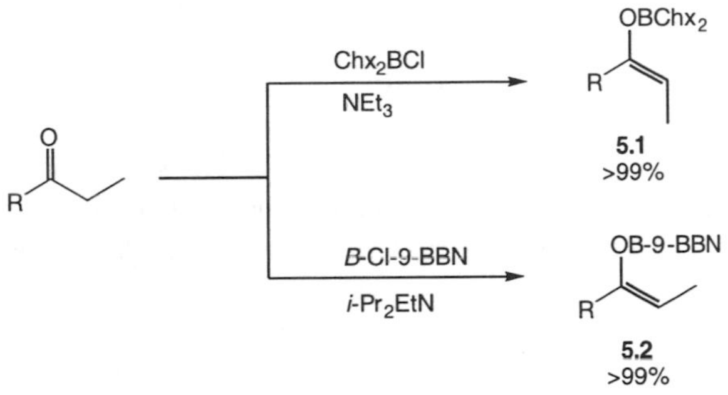

**SCHEME 5.2.**

In addition to the control of simple ester enolates as outlined in Scheme 5.1, many of the successful chiral auxiliaries (*cf.* Table 5.1) also provide a high degree of stereoselection in the enolate (or equivalent) formation step. Amides provide a high (e.g., 97:3) degree of selection for the formation of *Z*-enolates that can be increased through intramolecular complex formation.[114]

Enolates undergo numerous reactions, including alkylations (vide supra).[18] However, side reactions such as alkene formation from the alkyl halide can be minimized by use of an enol ether, and the opportunity is available to obtain the enol ether as a single stereoisomer. An additional advantage is that alkylation of a silyl enol ether is usually conducted in the presence of a Lewis acid, which allows for the use of functional groups not compatible with enolate chemistry, such as tertiary alkyl halides.[7,43,62,115-120]

Although the majority of examples in this section have centered around silyl enol ethers, other elements, such as boron, can be used in a similar context.[121,122]

## 5.2. REACTIONS OF ENOL ETHERS AND ENOLATES

### 5.2.1. ALKYLATION REACTIONS

Carbonyl compounds are versatile in synthesis due to their inherent ability to undergo nucleophilic addition and also to act as nucleophiles. The aldol reaction involves the combination of a nucleophilic carbonyl species with an electrophilic carbonyl compound and results in a 1,3-difunctional product. Similar methodology can be applied to other electrophiles, including alkyl halides, and again results in the introduction of 1,2-functionality.[123-125] In addition to anions, enolate radicals can be used for alkylations and similar reactions; the *ancat* isomer usually results.[126,127]

α-Alkylation of a carbonyl compound can be achieved by formation of an enolate,[16,42,61,128] followed by condensation with an alkyl halide.[17,25,33,129-131] This methodology not only requires regioselective formation of the enolate (vide supra),[6,18,106] but also minimization of competing elimination reactions in the alkyl halide fragment.[132] This latter problem can be circumvented, to a certain degree, by use of an enol ether and a Lewis acid catalyst.[7] In addition to these stereo- and regiochemical concerns, polyalkylation can still be a problem.[133,134] This is diminished, however, by the use of carbonyl modifiers and enol ethers (vide infra).[17,31,43,130] The use of counterions such as lithium and boron that form covalent bonds with the enolate oxygen atom also minimize polyalkylation.[129,135,136] The presence of diethylzinc in the reaction has been shown to suppress α-proton exchange and enhance alkylation.[137] Sometimes, a relatively unique approach is available to a specific metal enolate, as illustrated by the use of the Hooz reaction for the formation of boron enolates (Scheme 5.3);[138] note the isomerization methodology.

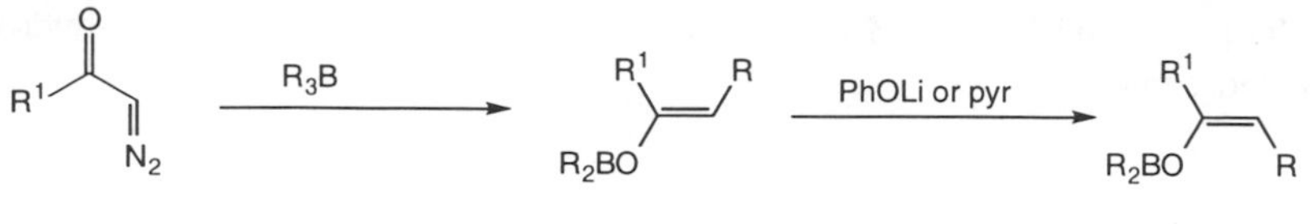

$R^1COCHN_2$ —($R_3B$)→ $R_2BO(R^1)C{=}CHR$ —(PhOLi or pyr)→ $R_2BO(R^1)C{=}CHR$

SCHEME 5.3.

An elegant solution to the regiochemical problems, and those arising from an enolate acting as a base with an alkyl halide, is to use a silyl enol ether with an α-chloro alkyl sulfide. This allows a wide variety of alkyl side chains to be introduced (Scheme 5.4.).[7,62,139-150]

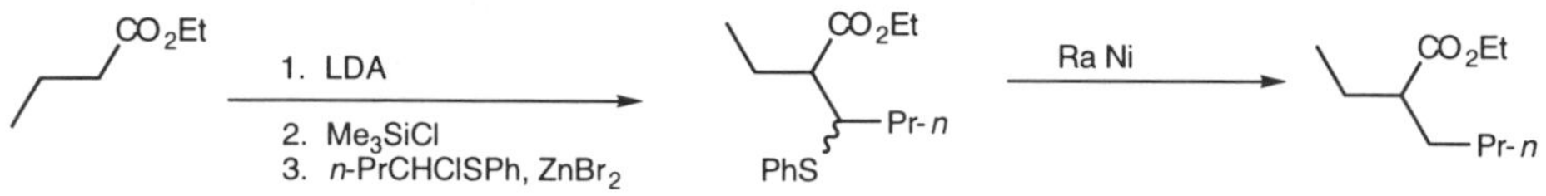

Ethyl butanoate —(1. LDA; 2. $Me_3SiCl$; 3. n-PrCHClSPh, $ZnBr_2$)→ PhS / Pr-n / $CO_2Et$ product —(Ra Ni)→ $CO_2Et$ / Pr-n product

SCHEME 5.4.

If an enolate still has to be employed, additives that change the structure of the enolate complex, such as hexamethylphosphoric triamide, can increase reactivity and allow alkylation with alkyl halides, especially iodides.[151]

The regiochemical limitations are lifted for carboxylic acid derivatives and only the reaction with the alkylating agent, while avoiding self-condensation, is of concern.[13,19,152] Indeed, most of the asymmetric methods for alkylations of enolates are designed for carboxylic acid derivatives (Table 5.1). As silicon has a high affinity for oxygen, silylation of an ester enolate (or ketone enolate) usually occurs at oxygen, but this can be overcome by the presence of a large group on the α-carbon atom.[†] [19]

One aspect now beginning to come to fruition pertains to the structure determination of lithium enolates (vide supra). For more complex cases, the substrate enolate can enforce stereoselection through intramolecular complex formation,[153] or face selectivity through the conformation of a ring or a substituent.[154]

In simple cases, Houk's rule can determine face selectivity [2.2] (e.g., Scheme 2.19).[155-158] The effect of a large group is readily apparent (e.g., Scheme 5.5).[159,160]

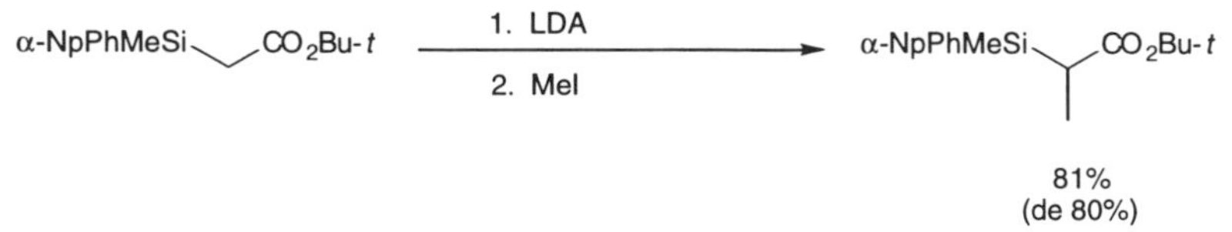

α-NpPhMeSi–$CH_2CH_2$–$CO_2Bu$-t —(1. LDA; 2. MeI)→ α-NpPhMeSi–CH($CH_3$)–$CH_2$–$CO_2Bu$-t, 81% (de 80%)

SCHEME 5.5.

† With respect to the former carbonyl carbon.

**TABLE 5.1**
**Chiral Auxiliaries that Have Been Employed in Collaboration with the α-Alkylation of Carbonyl Compounds**

| Auxiliary System | Equivalence | Reference | Auxiliary System | Equivalence | Reference |
|---|---|---|---|---|---|
| 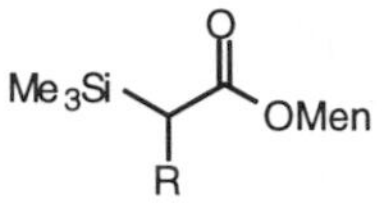 | $RCH^{-}CO_2R^1$ | 187 | 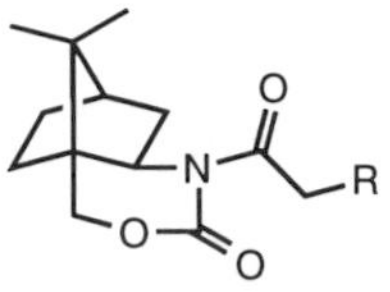 | $RCH^{-}CO_2H$ | 188 |
| 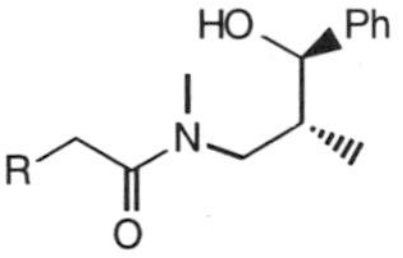 | $RCH^{-}CO_2H$ | 189 | 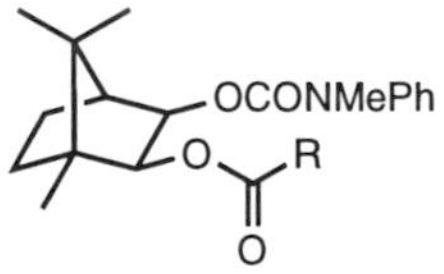 | $RCH^{-}CO_2R^1$ | 190 |
| 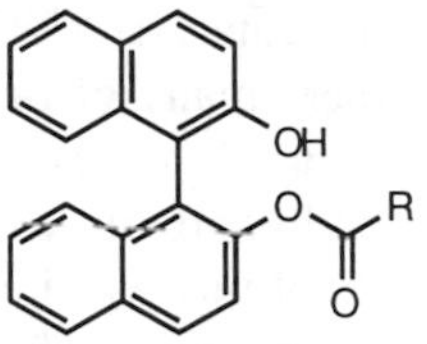 | $RCH^{-}CO_2R^1$ | 191,192 | 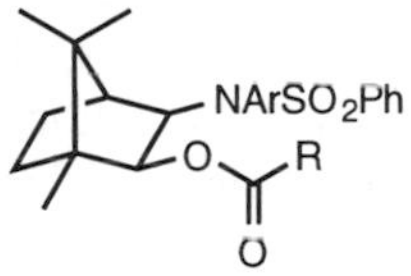 | $RCH^{-}CO_2R^1$ | 190 |
| 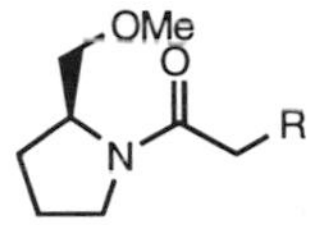 | $RCH^{-}CO_2H$ | 193 | 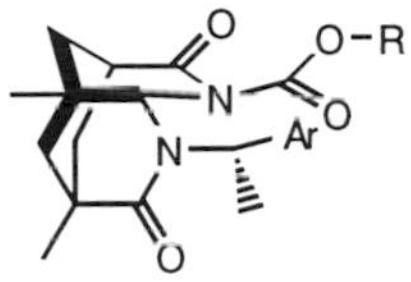 | $RCH^{-}CO_2R^1$ | 194 |
| 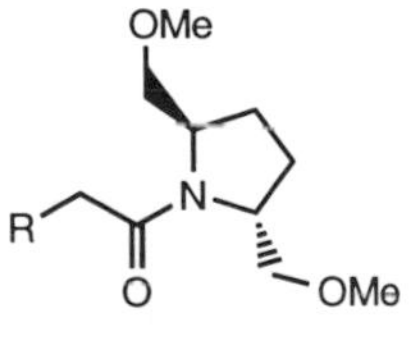 | $RCH^{-}CO_2H$ | 195 | 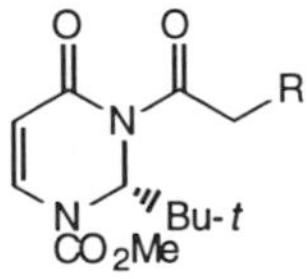 | $RCH^{-}CO_2H$ | 196 |
| 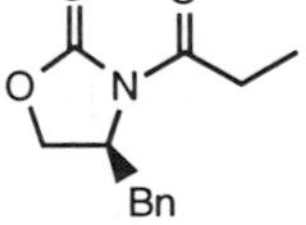 | $RCH^{-}CO_2H$ | 183,197 | 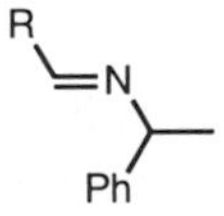 | $RCH^{-}CHO$ | 198 |
| 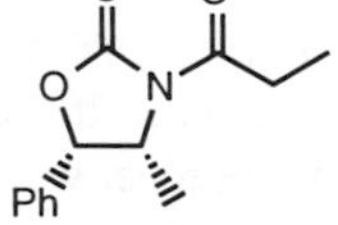 | $RCH^{-}CO_2H$ | 183 | 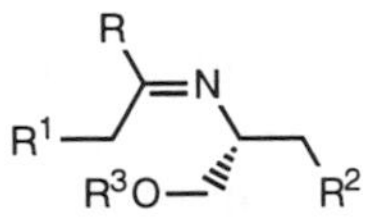 | 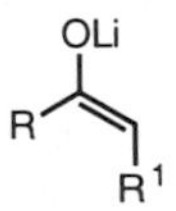 | 199–202 |
| 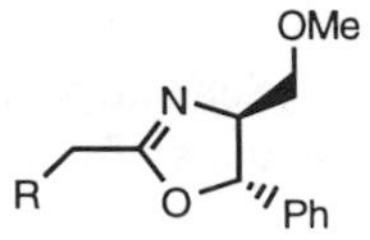 | $RCH^{-}CO_2R^1$ | 199, 203–205 | 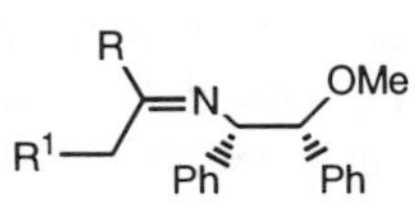 | 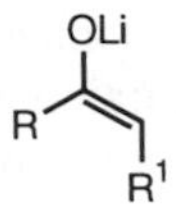 | 206 |

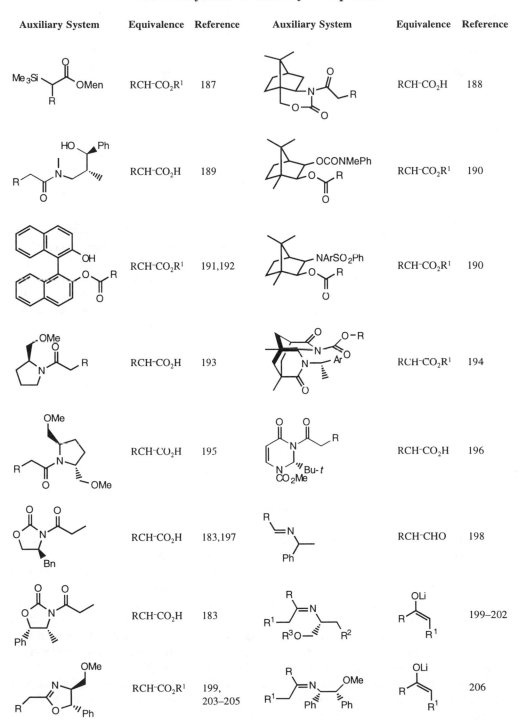

**TABLE 5.1 (continued)**

| Auxiliary System | Equivalence | Reference | Auxiliary System | Equivalence | Reference |
|---|---|---|---|---|---|
| 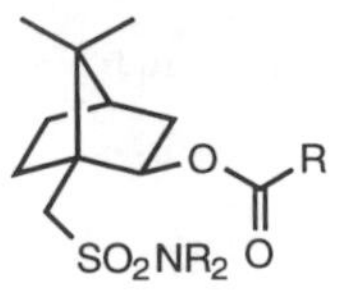 | $RCH^{-}CO_2R^1$ | 190 | 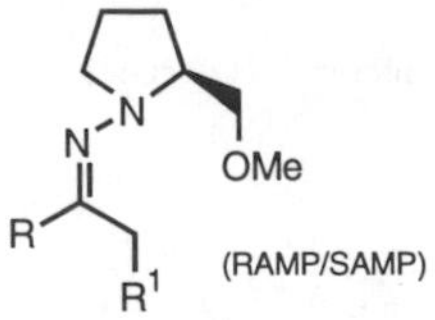 (RAMP/SAMP) | 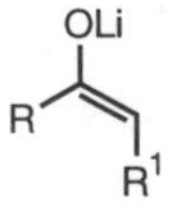 | 207,208 |
| 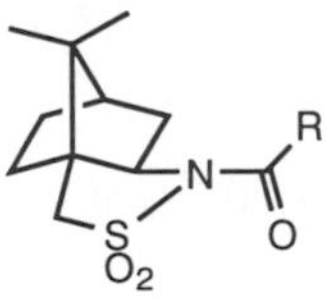 | $RCH^{-}CO_2R^1$ | 190,209 | 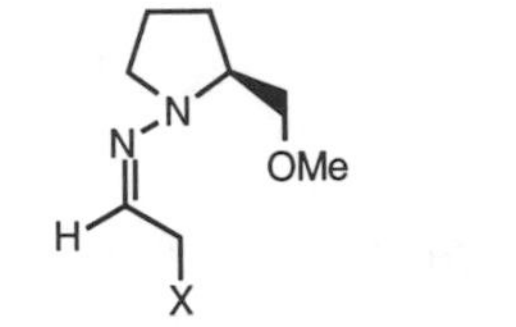 where X= SPh or $t$-BuMe$_2$Si | $XCH^{-}CHO$ | 210 |

Another area currently being exploited involves use of chiral phase transfer catalysts.[161-163] This has been exploited for the synthesis of α-amino acid derivatives, where many of the methods rely on the use of masked functionality.[164-179] The approach can be used, however, for simple cyclic ketones and, because of the absence of a strong base and exotherms, is amenable to low cost scale-up.[161,162,180-182] One of the better known examples of this reaction is the Merck synthesis of (+)-indacrinone that relies on the alkylation of the aryl ketone **5.3** (Scheme 5.6).[180]

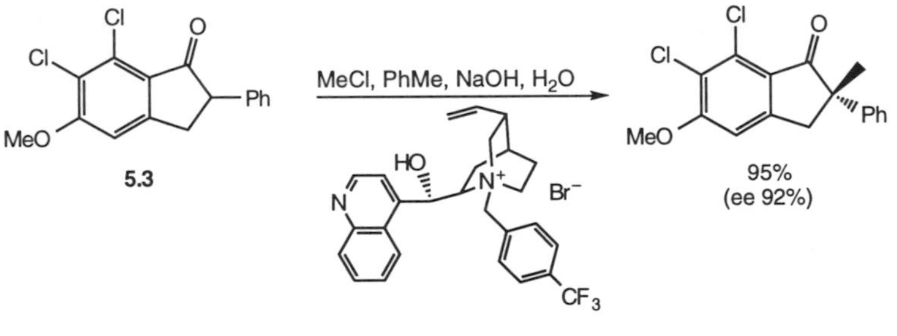

MeCl, PhMe, NaOH, $H_2O$

**5.3** → 95% (ee 92%)

**SCHEME 5.6.**

Unfortunately, this methodology has limited application, as face selectivity depends on strong interactions between the enolate and phase transfer catalyst.

This problem of facial selectivity with a planar enolate has led to the incorporation of various chiral auxiliaries into the carbonyl precursor.[§] The majority of these auxiliaries are nitrogen based [*5.3*], but there is still a wide variety available based on alternative systems. Representative examples are given in Table 5.1. There is considerable overlap between these alkylation reactions and other chemistry, as these chiral enolate equivalents usually react with a wide variety of electrophiles, including carbonyl compounds [Chapter 7]. Other systems can provide complementary selectivity as illustrated in Scheme 5.7.[183,184]

Sulfur-containing masking groups also allow for high diastereoselectivity (Scheme 5.8). The induction has been rationalized in terms of a chelate between the enolate and sulfoxide oxygen atoms leading to a chair transition state.[185,186]

§ Face selectivity can occur in cyclic substrates due to the inherent characteristics of the ring system, see Chapter 13.

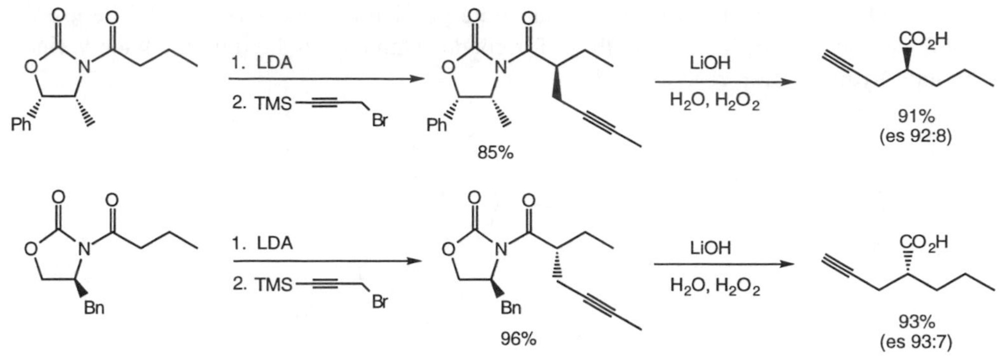

1. LDA
2. TMS—≡—Br
85%
LiOH
$H_2O$, $H_2O_2$
91%
(es 92:8)

1. LDA
2. TMS—≡—Br
96%
LiOH
$H_2O$, $H_2O_2$
93%
(es 93:7)

**SCHEME 5.7.**

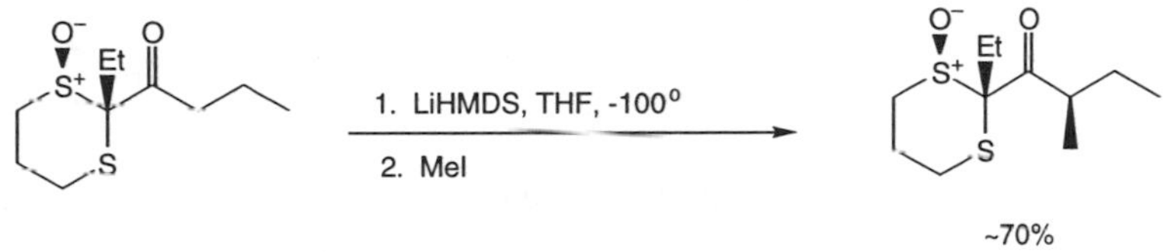

1. LiHMDS, THF, -100°
2. MeI
~70%

**SCHEME 5.8.**

Alkylations of organotransition metal complexes can often be highly stereoselective (Scheme 5.9).[211-216] Organometallic reagents can also provide useful electrophiles as the functionality can be masked, or one face can be made extremely bulky. This is illustrated by the use of a manganese arene complex in the preparation of 2-arylpropionic acids (Scheme 5.10).[217]

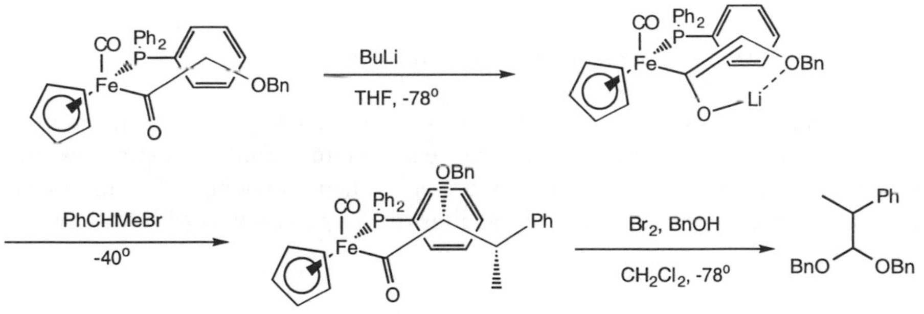

BuLi
THF, -78°
PhCHMeBr
-40°
$Br_2$, BnOH
$CH_2Cl_2$, -78°

**SCHEME 5.9.**

Asymmetric protonations can perform the operational equivalent of an asymmetric alkylation [*5.4.2*].[218,219] The substitution of functionality adjacent to a carbonyl group [*3.1*] can also be the equivalent of an alkylation,[220] as can any method that masks the carbonyl group as in an *umpolung* reagent.[221,222] The alkylation agent can also contain additional functionality — the aldol reaction can be considered such an example that allows for further elaboration once the alkylation has occurred.[223-226]

## 5.2.2. α-HYDROXYLATION OF CARBONYL COMPOUNDS

Various methodologies have been developed to accomplish hydroxylation adjacent to a carbonyl group, although there has been limited development of asymmetric methods except when the inherent properties of a cyclic substrate provide for selectivity.[227-229] The use of chiral *N*-sulfonyloxaziridines now allows for good asymmetric induction in a wide variety of enolate hydroxylation reactions.[230]

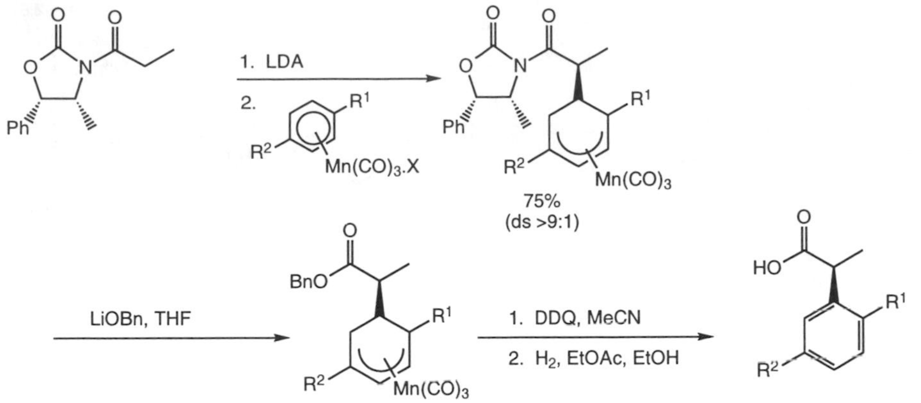

**SCHEME 5.10.**

The introduction of an oxygen functional group juxtaposed to a carbonyl moiety can be accomplished though reaction of the derived enol acetate, or other ester, with a peroxy acid or dioxiranes.[231-240] The enolate derived from the corresponding carbonyl compound can also be oxidized with an oxygen donor, such as a molybdenum peroxide reagent or dimethyldioxirane.[241-246] These silyl enol ethers of aldehydes and ketones can be oxidized by a peroxy acid to α-hydroxy carbonyl compounds.[231,247-253] Lead(IV) and manganese(III) reagents have been used in place of the peroxy acid.[254-256]

An alternative methodology is provided by hypervalent iodine compounds.[257-260] Some of these compounds, such as *o*-iodosylbenzoic acid,[261] react directly with ketones and alleviate the need to prepare the silyl enol ether.[259] Esters and lactones can also be used as substrates.[262,263] Variants of this methodology allow for the oxygen atom to be functionalized during its introduction.[264-270]

However, none of these methods allows for stereoselection unless it is inherent within the substrate.[234] The silyl group of the enol ether can contain chirality, but this does not seem to provide a means for high asymmetric induction.[91] The introduction of a chiral auxiliary can provide excellent diastereoselectivity as illustrated by a reaction with lead(IV) acetate (Scheme 5.11).[271,272]

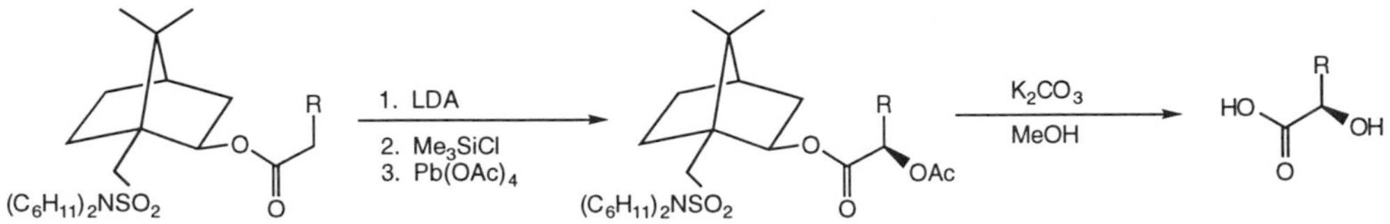

**SCHEME 5.11.**

Osmium tetroxide in the presence of *N*-methylmorpholine-*N*-oxide has also been used to effect this transformation,[257] while the use of a chiral diamine, derived from tartaric acid, as the ligand can lead to some chiral induction.[273]

Treatment of a silyl enol ether with ozone usually results in oxidative cleavage of the unsaturation, although in a few cases, an α-hydroxy carbonyl compound can be formed.[274,275] This oxidative cleavage problem can be overcome by use of a hydroboration-oxidation protocol, a dioxirane (vide supra),[232] or an ene reaction with singlet oxygen [*12.3*].[276] As with alkylations, chiral phase transfer catalysts provide for the asymmetric synthesis of α-hydroxy ketones, the oxygen being derived from molecular oxygen (Scheme 5.12).[277,278]

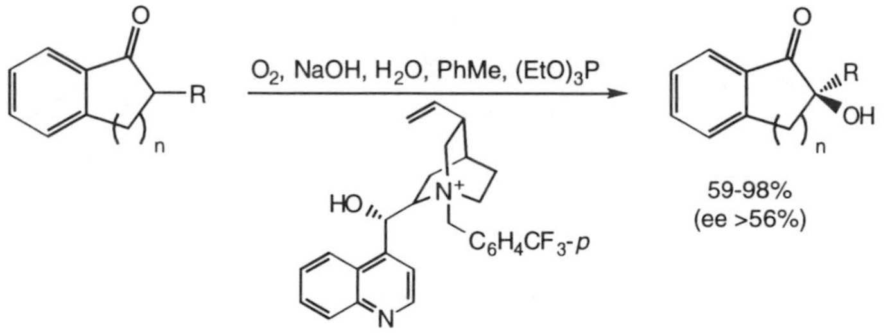

O
R
n
O2, NaOH, H2O, PhMe, (EtO)3P
HO
N+
C6H4CF3-p
N
O
R
OH
n
59-98%
(ee >56%)

**SCHEME 5.12.**

2-Phenylsulfonyl-3-phenyloxaziridine (**5.4**) reacts with ketone enolates to provide α-hydroxy carbonyl compounds.[227,230,279-282] This latter methodology has been adapted, through the use of a chiral oxaziridine (**5.5**), to introduce the hydroxy functionality in a synthesis of kjellmanianone (**5.6**) with an enantiomeric excess of 33 to 37%.[283] The asymmetric induction has been improved through use of the oxaziridine **5.7** (Scheme 5.13) or chiral auxiliaries with **5.4** (Scheme 5.14).[284-292]

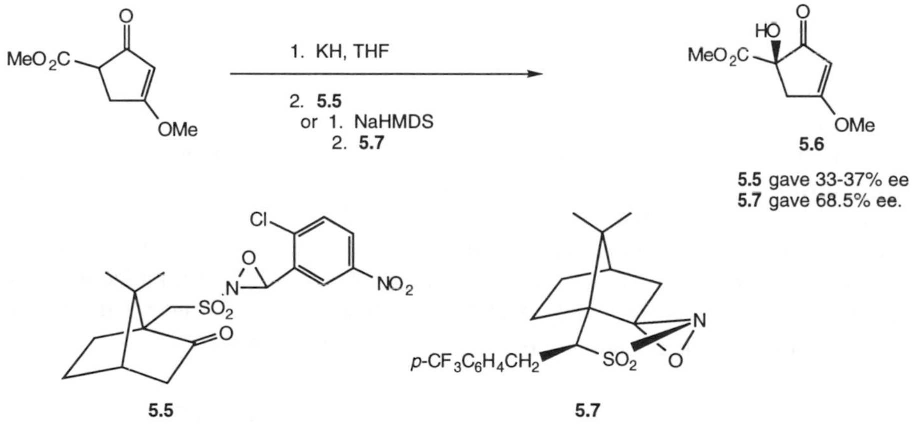

O
MeO2C
OMe
1. KH, THF
2. 5.5
or 1. NaHMDS
2. 5.7
HO
O
MeO2C
OMe
5.6
5.5 gave 33-37% ee
5.7 gave 68.5% ee.
Cl
O
N
NO2
SO2
O
5.5
N
p-CF3C6H4CH2
SO2
O
5.7

**SCHEME 5.13.**

The stereoselectivity is not only dependent upon the structure of the oxaziridine, but on the enolate substitution pattern and solution structure of the enolate, as well as, but to a lesser degree, on the geometry of the enolate.[230,290,293,294] Kinetic resolutions can also be accomplished.[295]

High enantiomeric excesses are also possible when a hydrazone chiral auxiliary is employed to provide α-substituted ketones (vide infra).[296] This approach has been used to introduce a silyl group with regio- and enantioselective control. The resultant α-silyl ketone was then transformed to the alternative regioisomeric α-hydroxy ketone (Scheme 5.15).[297-299]

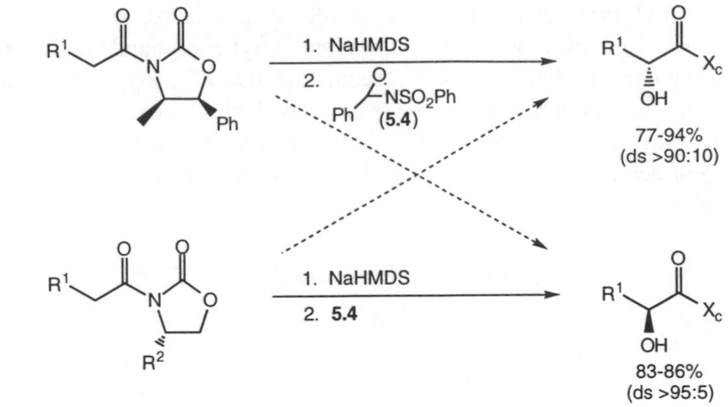

1. NaHMDS
2. (5.4)

77-94%
(ds >90:10)

1. NaHMDS
2. **5.4**

83-86%
(ds >95:5)

**SCHEME 5.14.**

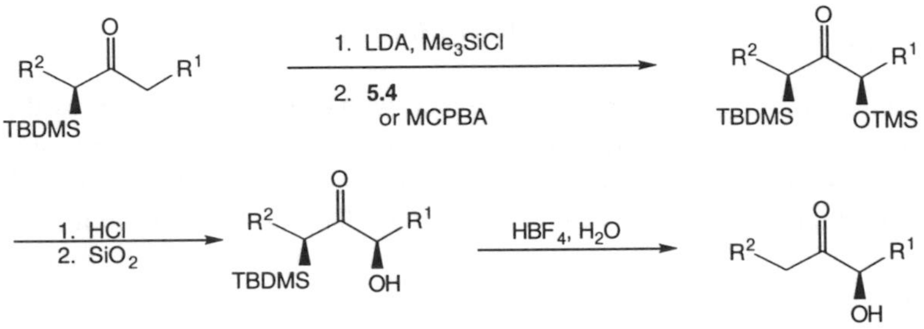

1. LDA, $Me_3SiCl$
2. **5.4** or MCPBA

1. HCl
2. $SiO_2$

$HBF_4$, $H_2O$

**SCHEME 5.15.**

The α-hydroxylation of lactones and esters has been achieved by reaction of the corresponding enolate with an oxaziridine,[227,279,300] and most of the oxidation reactions described for silyl enol ethers can be used with ketene acetals.[234] Use of a chiral imide enolate with this protocol results in good diastereoselectivity.[291] Chiral esters also provide some diastereoselectivity when oxidized with DDQ in acetic acid.[301]

α-Hydroxylation of an ester or carboxylic acid may also be achieved by deprotonation followed by reaction with oxygen or an oxygen donor.[241,244,277,302-309] With a lactone, a study of a variety of oxidants showed that the MoOPH system gave the highest diastereoselectivity; this was also improved as the amount of base increased (Scheme 5.16).[300]

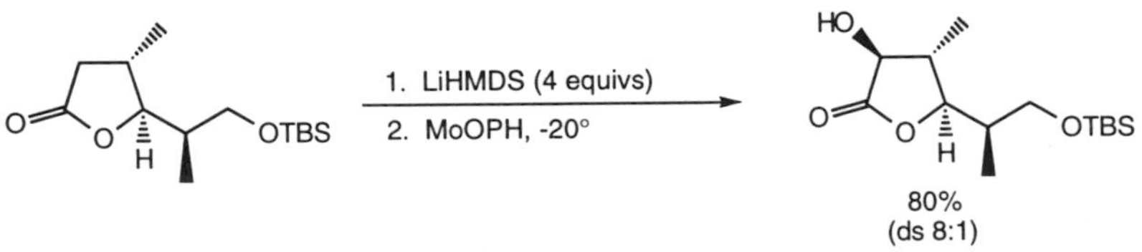

1. LiHMDS (4 equivs)
2. MoOPH, -20°

80%
(ds 8:1)

**SCHEME 5.16.**

An alternative strategy, which does allow asymmetric induction, is the reaction of an organolithium or Grignard reagent with the ketooxazoline **5.8** (Scheme 5.17) [*6.1.1*].[310]

The use of a chiral amide allows for the hydroxylation procedure by an adaptation of the methodology used to prepare α,β-epoxy carbonyl compounds [*11.3*] (Scheme 5.18).[311,312]

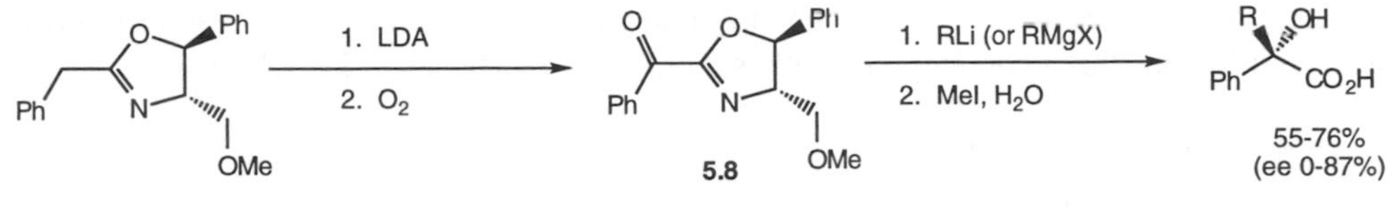

1. LDA
2. $O_2$

**5.8**

1. RLi (or RMgX)
2. MeI, $H_2O$

55-76%
(ee 0-87%)

SCHEME 5.17.

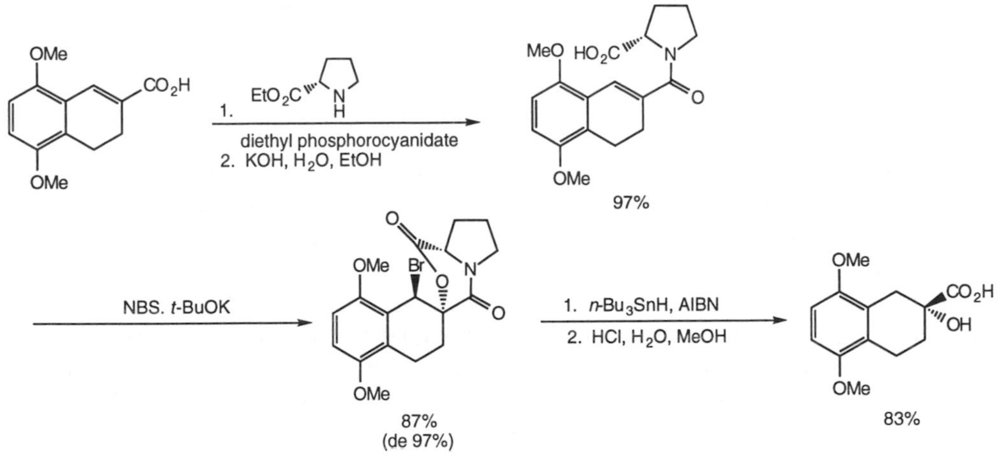

1. $EtO_2C$ (prolinate)
diethyl phosphorocyanidate
2. KOH, $H_2O$, EtOH

97%

NBS. *t*-BuOK

87%
(de 97%)

1. *n*-$Bu_3SnH$, AIBN
2. HCl, $H_2O$, MeOH

83%

SCHEME 5.18.

### 5.2.2.1. Other Methods to α-Hydroxy Carbonyl Compounds

A conjugate addition of an oxygen nucleophile to an α,β-unsaturated nitro compound, followed by conversion of the nitro group to a carbonyl moiety through the Nef reaction also provides α-hydroxycarbonyl compounds through *umpolung* methodology.[221,222] Other umpolung methodologies, such as the use of α-amino nitriles and vinyl sulfoxides, can also lead to α-hydroxy carbonyl compounds.[222,313,314]

## 5.2.3. REACTIONS OF ENOL ETHERS WITH OTHER ELECTROPHILES

In addition to carbon and oxygen nucleophiles, other useful groups and functionality can be introduced α to a carbonyl group through enolate or enol ether chemistry.

Sulfur electrophiles allow for a later elimination of a sulfoxide to provide the enone [*8.1.3*]. In an analogous manner, selenium can be used in place of sulfur.[315,316] However, the rich chemistry of these heteroatoms can also be exploited.[317-320]

Reactions of carbonyl compounds with halogens are very well known.[1] Use of a chiral auxiliary does allow for face selectivity (e.g., Scheme 5.19),[121,321-323] with the resultant α-halo carbonyl product being a useful precursor for α-amino acids and epoxides.[190,321,324] In the example illustrated, some increase in "apparent selectivity" was seen during crystallization of the product.

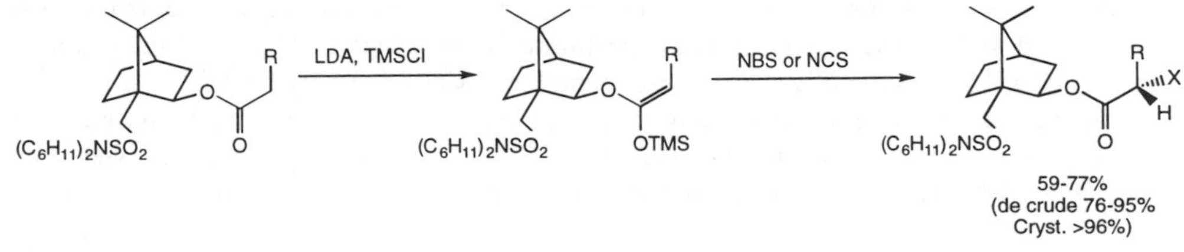

LDA, TMSCl

NBS or NCS

$(C_6H_{11})_2NSO_2$

59-77%
(de crude 76-95%
Cryst. >96%)

SCHEME 5.19.

Acetals can also be used in a halogenation reaction and allow for the inclusion of asymmetry as illustrated in a commercial synthesis of naproxen (Scheme 5.20).[325,326]

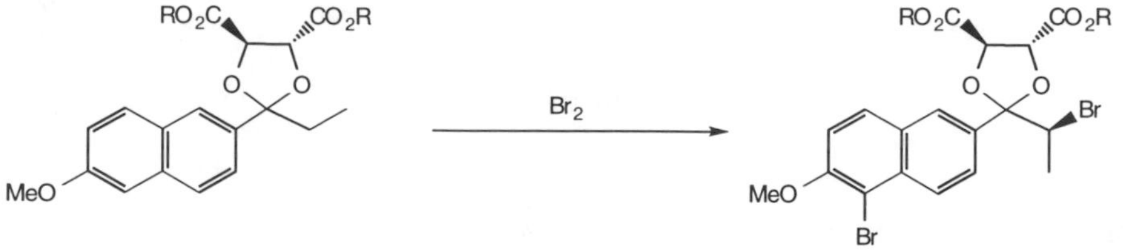

SCHEME 5.20.

Nitrogen electrophiles present a problem, just as oxygen does, because the element tends to be nucleophilic rather than electrophilic. However, some nitrogen electrophiles are available, and, when used in conjunction with a chiral auxiliary, provide rapid entry to α-amino acid derivatives. These nitrogen electrophiles include 1-chloro-1-nitrosocyclohexane,[327] ethyl azidoformate,[328] azodicarboxylate esters,[329-331] and 2,4,6-triisopropylbenzenesulfonyl azide.[323] The lack of cheap, readily available nitrogen electrophiles[332-334] is often circumvented by use of a halogen as an electrophile followed by inversion with a nitrogen nucleophile, such as azide [*3.1*].[190] An ene-type process using tosyldiazoselenide or tosyl isocyanate and selenium dioxide has been used to functionalize hindered, cyclic silyl enol ethers.[335]

## 5.3. USE OF NITROGEN DERIVATIVES

The formation of an imine or other nitrogen analogs from a ketone allows incorporation of a chiral auxiliary which can be removed by hydrolysis after the alkylation.[336-339] A wide variety of groups has been used in the amine moiety, including ones that can form a chelate with the metal counterion during the alkylation procedure.[191,192,195,198,200,206,340-345] However, many of the methods have been developed with cyclic ketones as substrates and have not been extended to acyclic cases.[322,346-349] Acyclic ketones can be alkylated by use of a hydrazone, although optical yields can be variable (Scheme 5.21).[102,208,339,341,345,350-356]

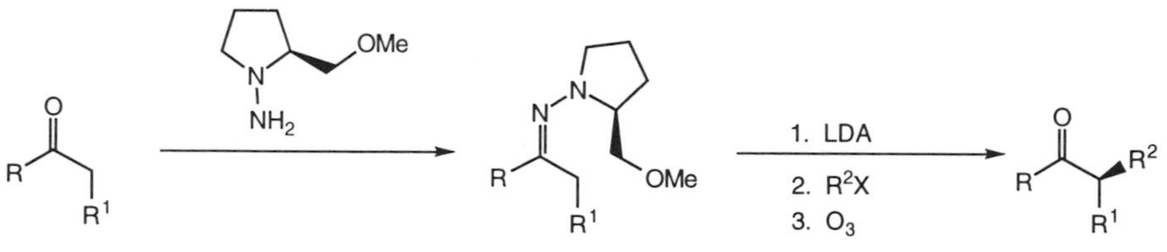

SCHEME 5.21.

With these hydrazones, the regiochemical problem is overcome through complex formation between the lithium, at the center where reaction is desired, and the methoxy group of the auxiliary.[352] If an α-proton is present in R, however, the correct isomer of the hydrazone must be employed in the alkylation reaction or substitution at the "incorrect" center may occur.[102] With cyclic substrates there is a strong preference for axial selectivity.[345,351,357]

Enamines also provide an opportunity to incorporate a chiral auxiliary while masking the carbonyl group.[201,202,358,359] In general, induction is not as high as for that observed with hydrazones.[348,360] Enamine alkylations, however, do solve the polyalkylation problem associated with the parent carbonyl compounds (vide supra),[361-364] and do not require hazardous compound handling as a simple acid hydrolysis unmasks the carbonyl group. The use of β-lithio enamines does overcome some of the problems associated with potential problems arising from the ambident properties of anions derived from enamines or imines.[365,366]

Imines have been used as a vehicle to transfer chirality from an α-amino acid, or similar precursor, to an α-keto acid.[367-372] Some asymmetric induction has been observed during the hydrolysis of an enamine in the presence of a chiral acid;[373] an alternative is to use a chiral amine or other auxiliary for the formation of the imine.[177,346,374,375] For cyclic ketones, imine formation of the ketone substrate with the ester of an α-amino acid can provide high face selection in an alkylation reaction.[376,377]

Chiral enolates derived from imines, imides, amides, and sultams provide entry into α-substituted carboxylic acids (Schemes 5.22 and 5.23),[110,183,193,194,209,217,223,378-380] including α-amino acids,[121,327,381-383] aldehydes,[103,198,384,385] and ketones (Scheme 5.24).[201,358] Examples for nitrogen derivatives are included in Table 5.1.

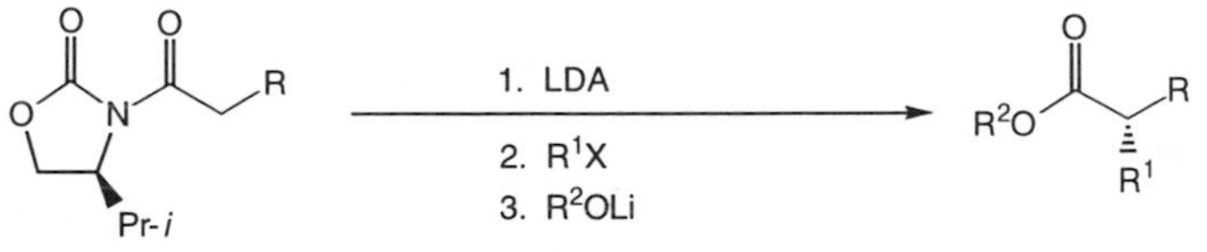

**SCHEME 5.22.**

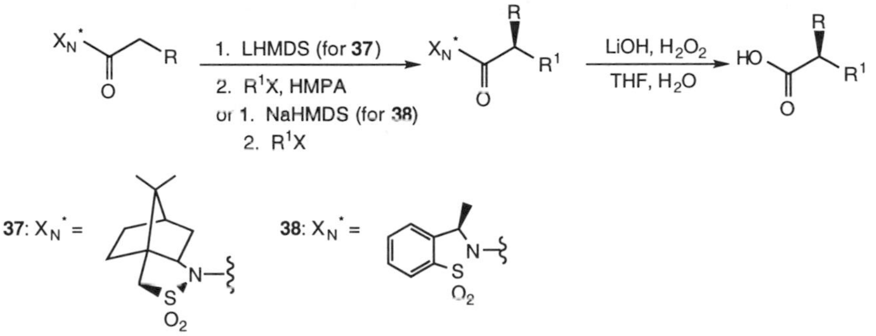

**SCHEME 5.23.**

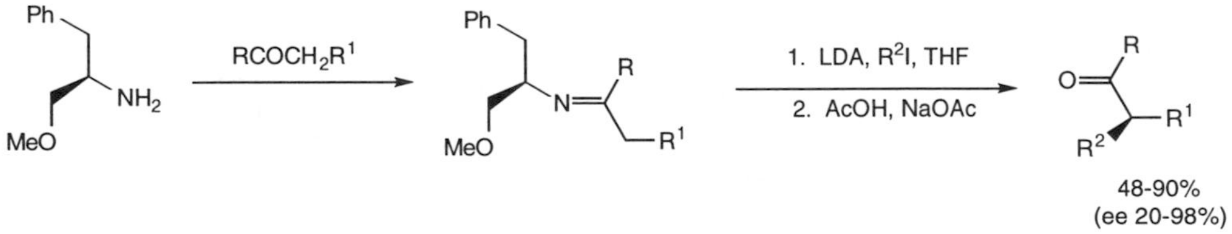

**SCHEME 5.24.**

Oxazolines are useful intermediates for the asymmetric synthesis of many classes of compounds. They provide useful methods for the α-alkylation of a masked carbonyl derivative (Scheme 5.25).[104,203-205,386-389] Indeed, this methodology can be used for kinetic resolution.[204,390]

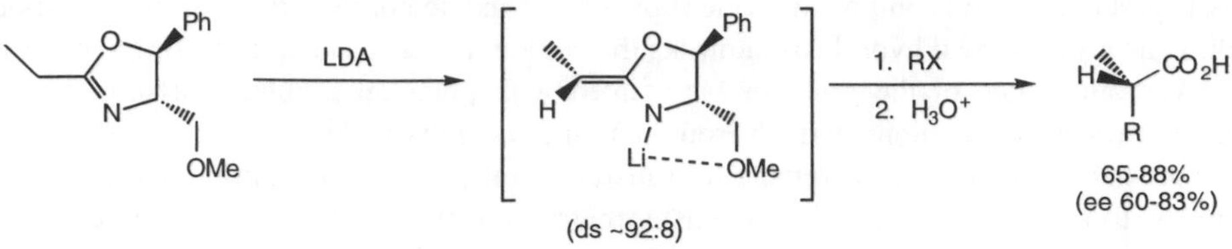

SCHEME 5.25.

There is also control of the stereochemistry of the α-center by the order in which the electrophilic units are added, as the face selectivity is ensured by the auxiliary rather than the substitution pattern.[205] NMR studies showed that for the reaction shown in Scheme 5.25, two lithiated oxazolines are present in a ratio of about 92:8, and that equilibration does not occur over the usual reaction temperature range while the reactivity of the two species to alkyl halides is similar.[388,389] The nature of the base can also affect the ratio of the *E*- and *Z*-enolate equivalents formed.[104]

As with carbonyl compounds, various metal counterions are also available for the nitrogen analogs,[197] as are chiral ligands.[391]

Nitrogen based carbonyl derivatives can also be incorporated into cyclic systems — that is, the carbonyl group of interest becomes part of the ring system itself; an example is provided by the self-regeneration of the stereogenic center approach [*1.3.2.3*].

## 5.4. ASYMMETRIC DEPROTONATIONS AND PROTONATIONS

### 5.4.1. ASYMMETRIC DEPROTONATIONS

A *meso*-ketone can be enantioselectively deprotonated by a chiral base;[392-397] reaction with anything other than a *meso*-substrate gives rise to diastereoisomer problems (vide infra). The enolate can then react with an electrophile such as an alkyl halide or carbon dioxide, or it can serve as a precursor to an enol ether (Scheme 5.26).[393,396,398-415]

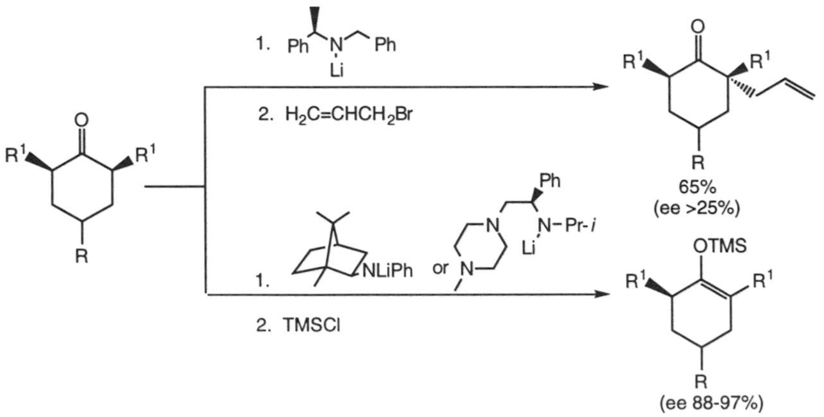

SCHEME 5.26.

The use of 4-siloxythiane oxides as cyclic sulfoxides allows for asymmetric deprotonations with chiral bases.[416]

The use of a carbonyl compound with just one acidic center does circumvent some of the diastereoisomer problems (e.g., Scheme 5.27),[402,417-419] and kinetic resolutions are possible.[420]

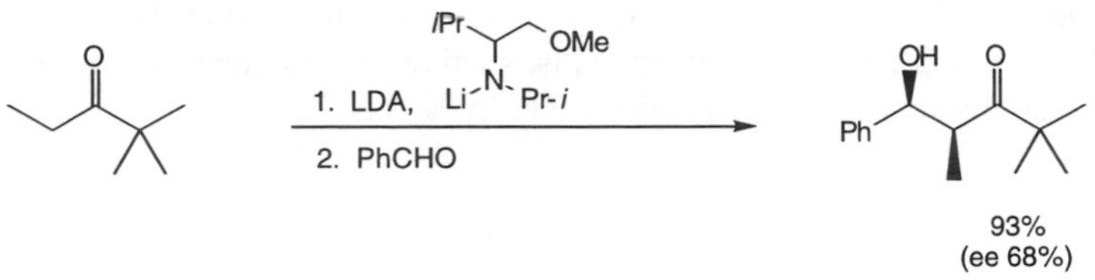

1. LDA, Li–N(Pr-*i*)–CH(*i*Pr)CH$_2$OMe
2. PhCHO

93%
(ee 68%)

**SCHEME 5.27.**

The chiral influence can be derived from a ligand that forms a complex with the anion; although an asymmetric deprotonation may be involved, the ligand complex to the anion can force face selectivity (e.g., Scheme 5.28).[421,422] Thus, an asymmetric protonation, or analogous reaction may be involved. Indeed, a preformed enolate, in the presence of lithium bromide, can still undergo enantioselective alkylations after the addition of a chiral amine base.[403]

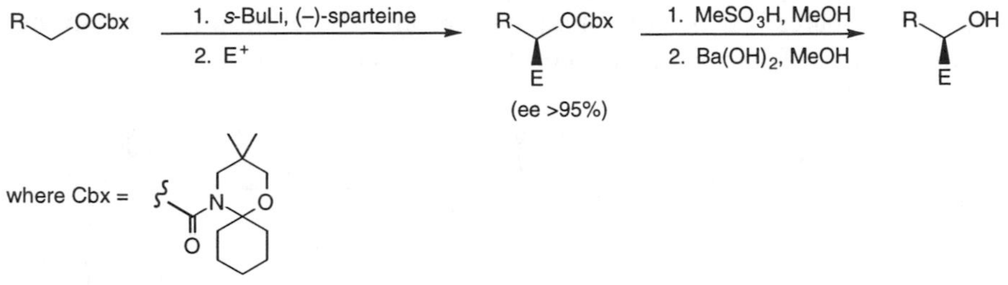

R–OCbx → 1. *s*-BuLi, (–)-sparteine; 2. E$^+$ → R–CH(E)–OCbx (ee >95%) → 1. $MeSO_3H$, MeOH; 2. $Ba(OH)_2$, MeOH → R–CH(E)–OH

where Cbx =

**SCHEME 5.28.**

Chiral bases have also been used to promote eliminations,[417] and to open epoxides.[423-425] Many of these reactions provide high regioselectivity.[405-407] To date, the use of chiral bases with asymmetric substrates has met with little general success for asymmetric transformations, although it holds much potential.[426,427]

## 5.4.2. ASYMMETRIC PROTONATIONS

As a complement to asymmetric deprotonations, asymmetric protonations can be used to achieve asymmetric induction. In the case of protonations, a *meso*-compound is not a prerequisite, as face selectivity is the key requirement; this was used as an illustration of Houk's rule [2.2] (e.g., Scheme 2.18). An additional example is given by the reaction of Scheme 5.29; partial resolution was observed during imine formation, while the deprotonation caused racemization.[428,429]

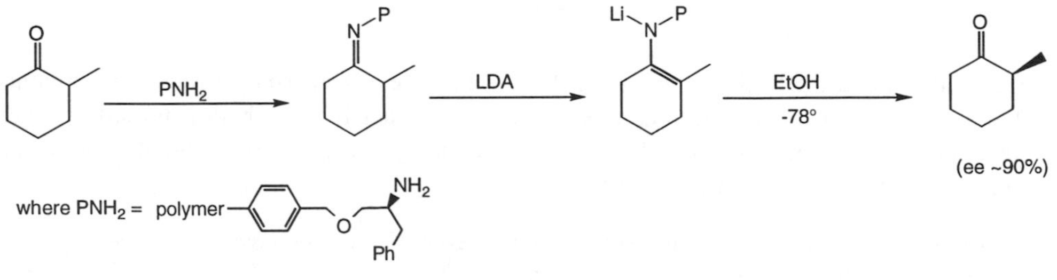

PNH$_2$ → LDA → EtOH, -78°

(ee ~90%)

where PNH$_2$ = polymer–C$_6$H$_4$–CH$_2$–O–CH$_2$–CH(NH$_2$)–CH$_2$Ph

**SCHEME 5.29.**

A study of the protonation of the lactone **5.9** enolate, generated by LDA, has shown that face selectivity is dependent upon the acid source; oxygen-based acids give little selectivity while ratios up to 13:1 (β-face reaction) are possible with carbon-based acids, such as malonates.† The low selectivity of oxygen acids, such as tartaric acid, is also seen during the hydrolysis of enamines and related reactions.[374,430-432]

O
O
ZHN
Pr-i


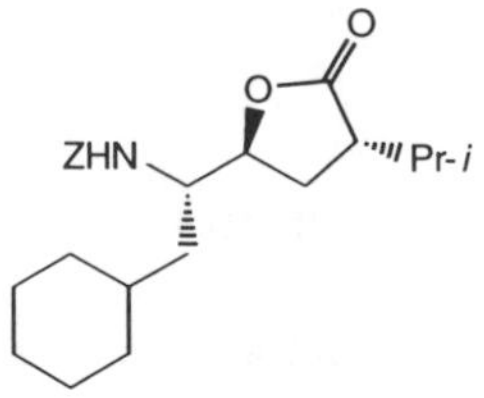

**5.9**

The asymmetric protonations of enolates with chiral amines as the proton source have also given reasonable ee's.[433,434]

This face-selective protonation is, in many regards, similar to asymmetric alkylations, when the proton source is derived from a protic solvent. However, in aprotic solvents, the reaction itself can create a chiral proton donor (e.g., Scheme 5.30).[435-437]

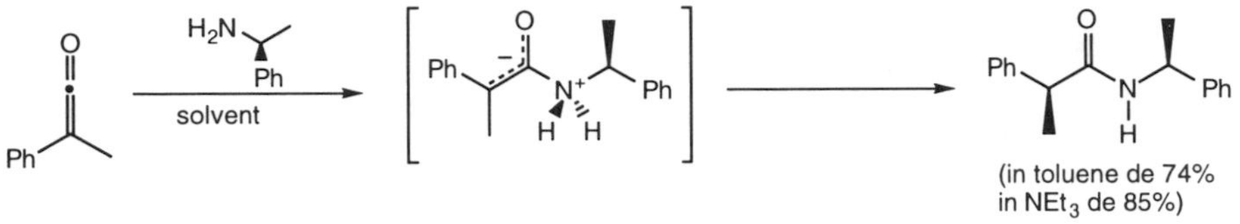

O
Ph
H2N
Ph
solvent
O
Ph
N+
H H
Ph
O
Ph
N
H
Ph
(in toluene de 74%
in NEt3 de 85%)


SCHEME 5.30.

The addition of a chiral alcohol to a ketene has been improved by optimization of the alcohol.[438-440] The use of an α-hydroxy ester provides for high diastereoselection in the preparation of 2-arylpropionic acids (Scheme 5.31).[441]

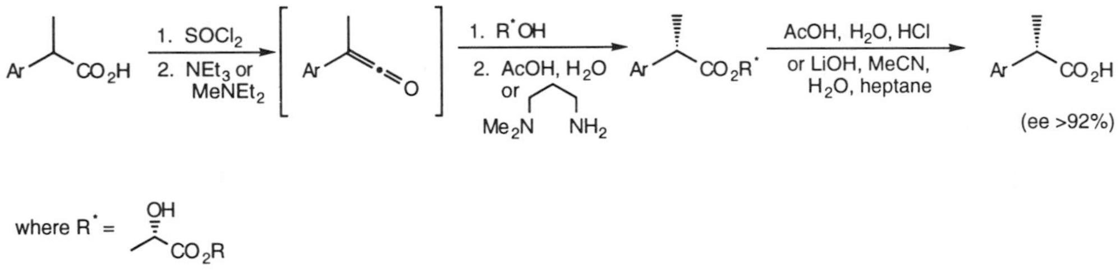

Ar
CO2H
1. SOCl2
2. NEt3 or
MeNEt2
Ar
O
1. R*OH
2. AcOH, H2O
or
Me2N NH2
Ar
CO2R*
AcOH, H2O, HCl
or LiOH, MeCN,
H2O, heptane
Ar
CO2H
(ee >92%)
where R* =
OH
CO2R


SCHEME 5.31.

Of course, the face selectivity during a protonation can be influenced by adjacent substituents (Scheme 5.32) — the *syncat* isomer was the major product if the β-hydro nitroalkene was subjected to an analogous reaction[442,443] — and the thermodynamically less favorable isomer can become available (e.g., Scheme 5.33).[234,444-446]

† In this specific example, asymmetric protonation was the method of choice as hydroboration of the alkene **5.10** gave a diastereoselectivity of 2.6:1 at best.

OR
Pr-i
ZHN


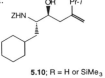

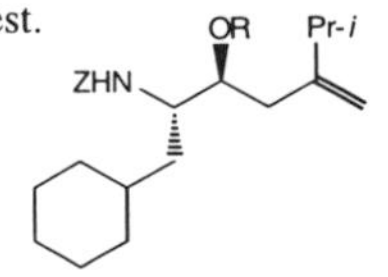

**5.10**; R = H or SiMe$_3$

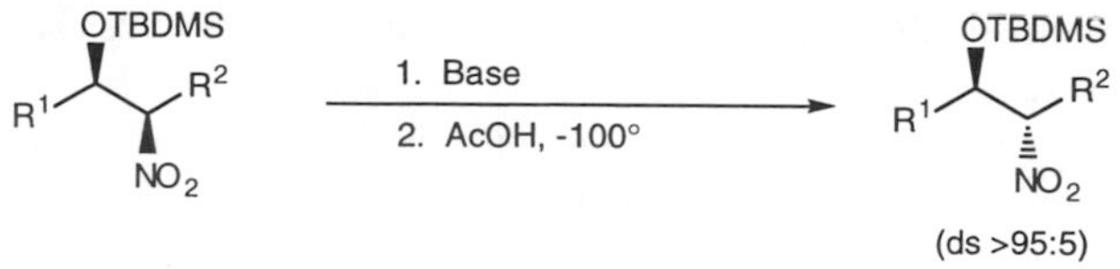

SCHEME 5.32.

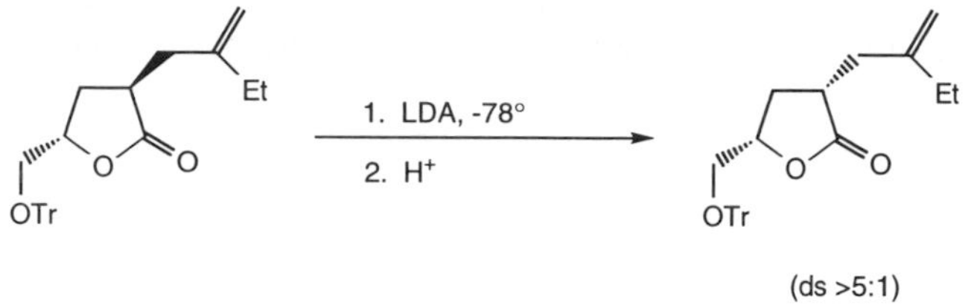

SCHEME 5.33.

Cyclic systems offer the opportunity for increased facial selectivity, but as the reaction is rapid, intermolecular examples often do not provide higher ee's than acyclic cases.[168] An example of an asymmetric protonation is given by the symmetrical enediol dianion **5.11** with (2*R*,3*R*)-*O*,*O*-dipivaloyltartaric acid (DPTA) at low temperature which afforded the optically active α-hydroxy ketone (Scheme 5.34);[447] a derivative of Kemp's triacid has also been used with success in an analogous reaction.[219]

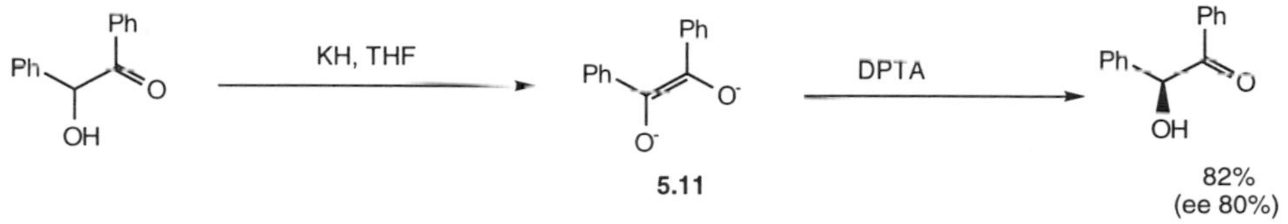

SCHEME 5.34.

Unfortunately, asymmetric protonation by a chiral acid is far from a general method; ee's are generally low.[218,409]

## 5.5. SUMMARY

The formation of a specific enolate isomer can be overcome through control of the reaction conditions. As temperature can play an important role, some of the reactions present problems during scale-up. In some cases, phase transfer methodology may be available, and this can be performed on a large scale.

Reactions of enolates with alkyl halides can be thwarted by a plethora of alternative reaction pathways. The use of enol ethers can circumvent many of these problems, and, as the reactions are not usually as exothermic, scale-up can be simpler. The introduction of a chiral auxiliary allows for asymmetric induction that can be predictable. Enolate reactions, for the most part, are useful at a laboratory scale and for higher-priced products.

The introduction of heteroatoms adjacent to a carbonyl group has to overcome the electronic demands of the element itself for oxygen and nitrogen electrophiles. Many methods have been proposed for oxygen, and asymmetric methods are now available. Again, scale-up, especially with regard to cost, is not trivial unless a phase transfer approach is adopted.

Nitrogen-based analogs of carbonyl compounds provide a relatively simple methodology to introduce a chiral moiety that is close to the reaction center. The scale-up problems are not alleviated, and some methods involve intermediates whose safety must be carefully considered for larger scale reactions.

The usage of asymmetric deprotonations and protonations is relatively new. In the former case, scale-up problems still abound. Asymmetric protonations, however, provide some exciting potential. As our understanding of these reactions increases, they will surely find widespread usage; recovery of the chiral moiety would be relatively trivial.

For laboratory usage, enolate chemistry provides a plethora of methodologies that are predictable in outcome, and are relatively simple to implement. Phase transfer methods can be performed on a large scale, but other methods require use of low temperatures to control stereochemistry.

## 5.6. REFERENCES


1. March, J. *Advanced Organic Chemistry: Reactions Mechanisms and Structure;* 3rd ed.; John Wiley & Sons: New York, 1985.
2. Woodward, R. B.; Logusch, E.; Nambiar, K. P.; Sakan, K.; Ward, D. E.; Au-Yeung, B.-W.; Balaram, P.; Browne, L. J.; Card, P. J.; Chen, C. H.; Chenevert, R. B.; Fliri, A.; Frobel, K.; Gais, H.-J.; Garratt, D. G.; Hayakawa, K.; Heggie, W.; Hesson, D. P.; Hoppe, D.; Hoppe, I. et al. *J. Am. Chem. Soc.* 1981, *103*, 3210.
3. Woodward, R. B.; Logusch, E.; Nambiar, K. P.; Sakan, K.; Ward, D. E.; Au-Yeung, B.-W.; Balaram, P.; Browne, L. J.; Card, P. J.; Chen, C. H.; Chenevert, R. B.; Fliri, A.; Frobel, K.; Gais, H.-J.; Garratt, D. G.; Hayakawa, K.; Heggie, W.; Hesson, D. P.; Hoppe, D.; Hoppe, I. et al. *J. Am. Chem. Soc.* 1981, *103*, 3213.
4. Woodward, R. B.; Logusch, E.; Nambiar, K. P.; Sakan, K.; Ward, D. E.; Au-Yeung, B.-W.; Balaram, P.; Browne, L. J.; Card, P. J.; Chen, C. H.; Chenevert, R. B.; Fliri, A.; Frobel, K.; Gais, H.-J.; Garratt, D. G.; Hayakawa, K.; Heggie, W.; Hesson, D. P.; Hoppe, D.; Hoppe, I. et al. *J. Am. Chem. Soc.* 1981, *103*, 3215.
5. Vedejs, E.; Krafft, G. A. *Tetrahedron* 1982, *38*, 2857.
6. Stork, G. *Pure Appl. Chem.* 1975, *43*, 553.
7. Reetz, M. T. *Angew. Chem., Int. Ed. Engl.* 1982, *21*, 96.
8. Seebach, D. *Angew. Chem., Int. Ed. Engl.* 1988, *27*, 1624.
9. Seebach, D.; Amstutz, R.; Dunitz, J. D. *Helv. Chim. Acta* 1981, *64*, 2622.
10. Narula, A. S. *Tetrahedron Lett.* 1981, *22*, 4119.
11. Amstutz, R.; Schweizer, W. B.; Seebach, D.; Dunitz, J. D. *Helv. Chim. Acta* 1981, *64*, 2617.
12. Setzer, W. N.; Schleyer, P. v. R. *Adv. Organometal. Chem.* 1985, *24*, 353.
13. Rathke, M. W.; Lindirt, A. *J. Am. Chem. Soc.* 1971, *93*, 2318.
14. Seebach, D.; Ehrig, V.; Teschner, M. *Annalen* 1976, 1357.
15. Pfeffer, P. E.; Silbert, L. S. *J. Org. Chem.* 1970, *35*, 262.
16. Evans, D. A. In *Asymmetric Synthesis*; J. D. Morrison, Ed.; Academic Press: Orlando, FL, 1984; Vol. 3; pp 1.
17. d'Angelo, J. *Tetrahedron* 1976, *32*, 2979.
18. Jackman, L. M.; Lange, B. C. *Tetrahedron* 1977, *33*, 2737.
19. Rathke, M. W.; Sullivan, D. F. *Synth. Commun.* 1973, *3*, 67.
20. Stork, G.; Kraus, G. A.; Garcia, G. A. *J. Org. Chem.* 1974, *39*, 3459.
21. House, H. O.; Czuba, L. J.; Gall, M.; Olmstead, H. D. *J. Org. Chem.* 1969, *34*, 2324.
22. Seebach, D.; Ehrig, V.; Teschner, M. *Liebigs Ann. Chem.* 1976, 1357.
23. Corey, E. J.; Gross, A. W. *Tetrahedron Lett.* 1984, *25*, 495.
24. Stork, G.; Hudrlik, P. F. *J. Am. Chem. Soc.* 1968, *90*, 4462.
25. Dau, M.-E. T. H.; Fetizon, M.; Anh, N. T. *Tetrahedron Lett.* 1973, 851.
26. Kobayashi, S.; Nishio, K. *J. Org. Chem.* 1993, *58*, 2647.
27. Stork, G.; Hudrlik, P. F. *J. Am. Chem. Soc.* 1968, *90*, 4464.
28. Kuwajima, I.; Nakamura, E. *J. Am. Chem. Soc.* 1975, *97*, 3257.
29. Kuwajima, I.; Nakamura, E. *Acc. Chem. Res.* 1985, *18*, 181.
30. Noyori, R.; Nishida, I.; Sakata, J. *J. Am. Chem. Soc.* 1983, *105*, 1598.
31. Kuwajima, I.; Nakamura, E.; Shimizu, M. *J. Am. Chem. Soc.* 1982, *104*, 1025.
32. Gariboldi, P.; Jommi, G.; Sisti, M. *J. Org. Chem.* 1982, *47*, 1961.
33. Binkley, E. S.; Heathcock, C. H. *J. Org. Chem.* 1975, *40*, 2156.
34. Chan, T. H.; Zheng, G. Z. *Tetrahedron Lett.* 1993, *34*, 3095.
35. Paolo, G.; Bortolotti, M.; Tagliavini, E.; Trombini, C.; Umni-Ronchi, A. *Tetrahedron Lett.* 1991, *32*, 1229.

36. Whiting, A. *Tetrahedron Lett.* 1991, *32*, 1503.
37. Haner, R.; Laube, T.; Seebach, D. *J. Am. Chem. Soc.* 1985, *107*, 5396.
38. Tidwell, T. T. *Tetrahedron Lett.* 1979, 4615.
39. Lenoir, D.; Seikaly, H. R.; Tidwell, T. T. *Tetrahedron Lett.* 1982, *23*, 4987.
40. Posner, G. H.; Sterling, J. J. *J. Am. Chem. Soc.* 1973, *95*, 3076.
41. Weiss, M. J.; Schaub, R. E.; Poletto, J. F.; Allen, G. R.; Coscia, C. J. *Chem. Ind. (London)* 1963, 118.
42. Petragnani, N.; Yonashiro, M. *Synthesis* 1982, 521.
43. Rasmussen, J. K. *Synthesis* 1977, 91.
44. Rubottom, G. M.; Mott, R. C.; Krueger, D. S. *Synth. Commun.* 1977, *7*, 327.
45. Narula, A. S. *Tetrahedron Lett.* 1982, *22*, 4119.
46. Nakamura, E.; Hashimoto, K.; Kuwajima, I. *Tetrahedron Lett.* 1978, 2079.
47. Zimmerman, H. E.; Chang, W.-H. *J. Am. Chem. Soc.* 1959, *81*, 3634.
48. Ireland, R. E.; Mueller, R. H.; Willard, A. K. *J. Am. Chem. Soc.* 1976, *98*, 2868.
49. Bruice, P. Y. *J. Am. Chem. Soc.* 1990, *112*, 7361.
50. Boche, G.; Langlotz, I.; Marsch, M.; Harms, K.; Nudelman, N. E. S. *Angew. Chem., Int. Ed. Engl.* 1992, *31*, 1205.
51. Olsher, U.; Izatt, R. M.; Bradshaw, J. S.; Dalley, N. K. *Chem. Rev.* 1991, *91*, 137.
52. Mulvey, R. E. *Chem. Soc. Rev.* 1991, *20*, 167.
53. Weiss, E. *Angew. Chem., Int. Ed. Engl.* 1993, *32*, 1501.
54. Sakuma, K.; Gilchrist, J. H.; Romesberg, F. E.; Cajthaml, C. E.; Collum, D. B. *Tetrahedron Lett.* 1993, *34*, 5213.
55. Juaristi, E.; Beck, A. K.; Hansen, J.; Matt, T.; Mukhopadhyay, T.; Simson, M.; Seebach, D. *Synthesis* 1993, 1271.
56. Edwards, A. J.; Hockey, S.; Mair, F. S.; Raithby, P. R.; Snaith, R.; Simpkins, N. S. *J. Org. Chem.* 1993, *58*, 6942.
57. Williard, P. G.; Salvino, J. M. *J. Org. Chem.* 1993, *58*, 1.
58. Mair, F. S.; Clegg, W.; O'Neil, P. A. *J. Am. Chem. Soc.* 1993, *115*, 3388.
59. Romesberg, F. E.; Bernstein, M. P.; Gilchrist, J. H.; Harrison, A. T.; Fuller, D. J.; Collum, D. B. *J. Am. Chem. Soc.* 1993, *115*, 3475.
60. Alberts, A. H.; Wynberg, H. *J. Am. Chem. Soc.* 1989, *111*, 7265.
61. Pollack, R. M. *Tetrahedron* 1989, *45*, 4913.
62. Fleming, I. *Chem. Soc. Rev.* 1981, *10*, 83.
63. Kruger, C. R.; Rochow, E. *J. Organometal. Chem.* 1964, *1*, 476.
64. Reich, H. J.; Holtan, R. C.; Bolm, C. *J. Am. Chem. Soc.* 1990, *112*, 5609.
65. Garst, M. E.; Bonfiglo, J. N.; Grudoski, D. A.; Marks, J. *J. Org. Chem.* 1980, *45*, 2307.
66. Reich, H. J.; Rusek, J. J.; Olson, R. E.; *J. Am. Chem. Soc.* 1979, *101*, 2225.
67. Sato, S.; Matsuda, I.; Izumi, Y. *Tetrahedron Lett.* 1983, *24*, 3855.
68. Hudrlik, P. F.; Hudrlik, A. M.; Kulkarni, A. K. *J. Am. Chem. Soc.* 1985, *107*, 4260.
69. Horiguchi, Y.; Matsuzawa, S.; Nakamura, E.; Kuwajima, I. *Tetrahedron Lett.* 1986, *27*, 4025.
70. Caron, G.; Lessard, J. *Can. J. Chem.* 1973, *51*, 981.
71. Noyori, R.; Nishida, I.; Sakato, J. *Tetrahedron Lett.* 1980, *21*, 2085.
72. Matsuda, I.; Takahashi, K.; Sato, S. *Tetrahedron Lett.* 1990, *31*, 5331.
73. Davis, F. A.; Lal, G. S.; Wei, J. *Tetrahedron Lett.* 1988, *29*, 4269.
74. Panek, E. J.; Kaiser, L. R.; Whitesides, G. M. *J. Am. Chem. Soc.* 1977, *99*, 3708.
75. Kuwajima, I.; Kato, M. *J. Chem. Soc., Chem. Commun.* 1979, 708.
76. Larson, G. L.; Berrios, R.; Prieto, J. A. *Tetrahedron Lett.* 1989, *30*, 283.
77. Nakajima, T.; Segi, M.; Sugimoto, F.; Hioki, R.; Yokota, S.; Miyashita, K. *Tetrahedron* 1993, *49*, 8343.
78. Coates, R. M.; Sandefur, L. O.; Smillie, R. D. *J. Am. Chem. Soc.* 1975, *97*, 1619.
79. Mulzer, J.; Pointner, A.; Chuckolowski, A.; Bruntrup, G. *J. Chem. Soc., Chem. Commun.* 1979, 52.
80. Hudrlik, P. F.; Misra, R. N.; Withers, G. F.; Hudrlik, A. M.; Rona, R. J.; Arcoleo, J. P. *Tetrahedron Lett.* 1976, 1453.
81. Obayashi, M.; Utimoto, K.; Nozaki, H. *Bull. Chem. Soc., Jpn.* 1979, *52*, 2646.
82. Hudrlik, P. F.; Schwartz, R. H.; Kulkarni, A. K. *Tetrahedron Lett.* 1979, 2233.
83. Hudrlik, P. F.; Wan, C.-N.; Withers, G. P. *Tetrahedron Lett.* 1976, 1449.
84. Koreeda, M.; Koo, S. *Tetrahedron Lett.* 1990, *31*, 831.
85. Hippeli, C.; Basso, N.; Dammast, F.; Reißig, H.-U. *Synlett* 1990, 26.
86. Wilcox, C. S.; Babston, R. E. *J. Org. Chem.* 1984, *49*, 1451.
87. Kaye, P. T.; Learmouth, R. A. *Synth. Commun.* 1989, *19*, 2337.
88. Walkup, R. E. *Tetrahedron Lett.* 1987, *28*, 511.
89. Walkup, R. E.; Obeyesekere, N. H. *J. Org. Chem.* 1988, *53*, 920.
90. Joshi, G. C.; Pande, L. M. *Synthesis* 1975, 450.

91. Kaye, P. T.; Learmonth, R. A. *Synth. Commun.* 1990, *20*, 1333.
92. Ireland, R. E.; Willard, A. K. *Tetrahedron Lett.* 1975, 3975.
93. Seebach, D. A., R.; Laube, T.; Schweizer, W. B.; Dunitz, J. D. *J. Am. Chem. Soc.* 1985, *107*, 5403.
94. Ireland, R. E.; Wilcox, C. S. *Tetrahedron Lett.* 1977, 2839.
95. Pereira, S.; Srebnik, M. *Aldrichimica Acta* 1993, *26*, 17.
96. Ireland, R. E.; Wipf, P.; Armstrong, J. D. *J. Org. Chem.* 1991, *56*, 650.
97. Seebach, D.; Mukhopadhyay *Helv. Chim. Acta* 1982, *65*, 385.
98. Herradón, B.; Seebach, D. *Helv. Chim. Acta* 1989, *72*, 690.
99. Masamune, S.; Ellingboe, J. W.; Choy, W. *J. Am. Chem. Soc.* 1982, *104*, 5526.
100. Fataftah, Z. A.; Kopka, I. E.; Rathke, M. W. *J. Am. Chem. Soc.* 1980, *102*, 3959.
101. Klebshick, W. A.; Buse, C. T.; Heathcock, C. H. *J. Am. Chem. Soc.* 1977, *99*, 247.
102. Davenport, K. G.; Eichenauer, H.; Enders, D.; Newcomb, M.; Bergbreiter, D. E. *J. Am. Chem. Soc.* 1979, *101*, 5654.
103. Meyers, A. I.; Brich, Z.; Erickson, G. W.; Traynor, S. G. *J. Chem. Soc., Chem. Commun.* 1979, 566.
104. Hoobler, M. A.; Bergbreiter, D. E.; Newcomb, M. *J. Am. Chem. Soc.* 1978, *100*, 8182.
105. Murai, A.; Abiko, A.; Shimada, N.; Masamune, T. *Tetrahedron Lett.* 1984, *25*, 4951.
106. Gaudemar, M.; Bellassoued, M. *Tetrahedron Lett.* 1989, *30*, 2779.
107. Brown, H. C.; Dhar, R. K.; Bakshi, R. K.; Pandiarajan, P. K.; Singaram, B. *J. Am. Chem. Soc.* 1989, *111*, 3441.
108. Van Horn, D. E.; Masamune, S. *Tetrahedron Lett.* 1979, 2229.
109. Brown, H. C.; Ganesan, K.; Dhar, R. K. *J. Org. Chem.* 1993, *58*, 147.
110. Evans, D. A.; Bartroli, J. *Tetrahedron Lett.* 1982, *23*, 807.
111. Pasto, D. J.; Wojtkowski, P. W. *Tetrahedron Lett.* 1970, 215.
112. Suzuki, A.; Arase, A.; Matsumoto, H.; Itoh, M.; Brown, H. C.; Rogic, M. M.; Rathke, M. W. *J. Am. Chem. Soc.* 1967, *89*, 5708.
113. Boldrini, G. P.; Mancini, F.; Tagliavini, E.; Trombini, C.; Umani-Ronchi, A. *J. Chem. Soc., Chem. Commun.* 1990, 1680.
114. Evans, D. A.; Takacs, J. M.; McGee, L. R.; Ennis, M. D.; Mathre, D. J.; Bartroli, J. *Pure Appl. Chem.* 1981, *53*, 1109.
115. Reetz, M. T.; Maier, W. F.; Heimbach, H.; Giannis, A.; Anastassious, G. *Chem. Ber.* 1980, *113*, 3734.
116. Reetz, M. T.; Maier, W. F.; Chatziiosifidis, I.; Giannis, A.; Heimbach, H.; Lowe, U. *Chem. Ber.* 1980, *113*, 3741.
117. Reetz, M. T.; Schwellnus, K. *Tetrahedron Lett.* 1978, 1455.
118. Chan, T. H.; Paterson, I.; Pinsonnault, J. *Tetrahedron Lett.* 1977, 4183.
119. Paterson, I. *Tetrahedron Lett.* 1979, 1519.
120. Clark, R. D.; Heathcock, C. H. *J. Org. Chem.* 1976, *41*, 1396.
121. Evans, D. A.; Ellman, J. A.; Dorow, R. L. *Tetrahedron Lett.* 1987, *28*, 1123.
122. Lee, S.-H.; Hulce, M. *Synlett* 1992, 485.
123. Askin, D.; Volante, R. P.; Ryan, K. M.; Reamer, R. A.; Shinkai, I. *Tetrahedron Lett.* 1988, *29*, 4245.
124. Lindermann, R. J.; Graves, D. M.; Kwochka, W. R.; Ghannam, A. F.; Anklekar, T. V. *J. Am. Chem. Soc.* 1990, *112*, 7438.
125. Ager, D. J.; East, M. B. *Tetrahedron* 1992, *48*, 2803.
126. Giese, B.; Damm, W.; Witzel, T.; Zeitz, H.-G. *Tetrahedron Lett.* 1993, *34*, 7053.
127. Giese, B.; Damm, W.; Wetterich, F.; Zeitz, H. G.; Rancourt, J.; Guindon, Y. *Tetrahedron Lett.* 1993, *34*, 5885.
128. Evans, D. A. *Aldrichimica Acta* 1982, *15*, 23.
129. Stork, G.; Rosen, P.; Goldman, N.; Coombs, R. V.; Tsuji, J. *J. Am. Chem. Soc.* 1965, *87*, 275.
130. Hubbard, J. S.; Harris, T. M. *J. Am. Chem. Soc.* 1980, *102*, 2110.
131. Ashby, E. C.; Argyropoulos, J. N. *Tetrahedron Lett.* 1984, *25*, 7.
132. MacPhee, J. A.; Dubois, J.-E. *J. Chem. Soc., Perkin Trans. 1* 1977, 694.
133. House, H. O. *Modern Synthetic Reactions*; Benjamin: New York, 1965.
134. Conia, J. M. *Rec. Chem. Prog.* 1963, *24*, 43.
135. Rathke, M. W.; Lindert, A. *Synth. Commun.* 1978, *8*, 9.
136. Faller, J. W.; Tokunga, M. *Tetrahedron Lett.* 1993, *34*, 7359.
137. Morita, Y.; Suzuki, Y.; Noyori, R. *J. Org. Chem.* 1989, *54*, 1785.
138. Masamune, S.; Mori, S.; van Horn, D.; Brook, D. W. *Tetrahedron Lett.* 1979, 1665.
139. Fleming, I.; Lee, T. V. *Tetrahedron Lett.* 1981, *22*, 705.
140. Fleming, I.; Goldhill, J.; Paterson, I. *Tetrahedron Lett.* 1979, 3205.
141. Fleming, I.; Goldhill, J.; Paterson, I. *Tetrahedron Lett.* 1979, 3209.
142. Fleming, I.; Iqbal, J. *Tetrahedron Lett.* 1983, *24*, 327.
143. Paterson, I.; Fleming, I. *Tetrahedron Lett.* 1979, 993.
144. Paterson, I.; Fleming, I. *Tetrahedron Lett.* 1979, 995.
145. Paterson, I.; Fleming, I. *Tetrahedron Lett.* 1979, 2179.

146. Reetz, M. T.; Huttenhaim, S.; Walz, P.; Lowe, U. *Tetrahedron Lett.* 1979, 4971.
147. Khan, H. A.; Paterson, I. *Tetrahedron Lett.* 1982, *23*, 2399.
148. Lee, T. V.; Okonkwo, J. O. *Tetrahedron Lett.* 1983, *24*, 323.
149. Ager, D. J. *Tetrahedron Lett.* 1983, *24*, 419.
150. Iqbal, J.; Mohan, R. *Tetrahedron Lett.* 1989, *30*, 239.
151. Bos, W.; Pabon, H. J. J. *Recueil* 1980, *99*, 141.
152. Cregge, R. J.; Herrmann, J. L.; Lee, C. S.; Richman, J. E.; Schlessinger, R. H. *Tetrahedron Lett.* 1973, 2425.
153. Mulzer, J.; Steffen, U.; Zorn, L.; Schneider, C.; Weinhold, E.; Munch, W.; Rudert, R.; Luger, P.; Harth, H. *J. Am. Chem. Soc.* 1988, *110*, 4640.
154. Takano, S. *Pure Appl. Chem.* 1987, *59*, 353.
155. McGarvey, G. J.; Williams, J. M. *J. Am. Chem. Soc.* 1985, *107*, 1435.
156. Fleming, I.; Lewis, J. J. *J. Chem. Soc., Chem. Commun.* 1985, 149.
157. Fleming, I.; Lawrence, N. J. *Tetrahedron Lett.* 1990, *31*, 3645.
158. Fleming, I.; Hill, J. H. M.; Parker, D.; Waterson, D. *J. Chem. Soc., Chem. Commun.* 1985, 318.
159. Larson, G. L.; Sandoval, S.; Cartledge, F.; Froonczek, F. R. *Organometallics* 1983, *2*, 810.
160. Chan, T.; Wang, D. *Chem. Rev.* 1992, *92*, 995.
161. Dehmlow, E. V. *Angew. Chem., Int. Ed. Engl.* 1977, *16*, 493.
162. Nerinckx, W.; Vandewalle, M. *Tetrahedron: Asymmetry* 1990, *1*, 265.
163. Dehmlow, E. V. *Angew. Chem., Int. Ed. Engl.* 1974, *13*, 170.
164. Seebach, D.; Miller, D. D.; Muller, S.; Weber, T. *Helv. Chim. Acta* 1985, *68*, 949.
165. O'Donnell, M. J.; Bennett, W. D.; Wu, S. *J. Am. Chem. Soc.* 1989, *111*, 2353.
166. O'Donnell, M. J.; Bennett, W. D.; Bruder, W. D.; Jacobsen, W. N.; Knuth, K.; LeClef, B.; Polt, R. L.; Bordwell, F. G.; Mrozack, S. R.; Cripe, T. A. *J. Am. Chem. Soc.* 1988, *110*, 8520.
167. Seebach, D.; Juaristi, E.; Miller, D. D.; Schickli, C.; Weber, T. *Helv. Chim. Acta* 1987, *70*, 237.
168. Schollkopf, U. *Top. Curr. Chem.* 1983, *109*, 65.
169. Schollkopf, U. *Tetrahedron* 1983, *39*, 2085.
170. McIntosh, J. M.; Mishra, P. *Can. J. Chem.* 1986, *64*, 726.
171. McIntosh, J. M.; Leavitt, R. K. *Tetrahedron Lett.* 1986, *27*, 3839.
172. Yamada, S.-i.; Ogurai, T.; Shioiri, T. *J. Chem. Soc., Chem. Commun.* 1976, 136.
173. Kolb, M.; Barth, J. *Tetrahedron Lett.* 1979, 2999.
174. Duhamel, L.; Fouquay, S.; Plaquevent, J.-C. *Tetrahedron Lett.* 1986, *27*, 4975.
175. Duhamel, P.; Eddine, J. J.; Valnot, J.-Y. *Tetrahedron Lett.* 1987, *28*, 3801.
176. El Achqar, A.; Roumestant, M.-L.; Viallefont, P. *Tetrahedron Lett.* 1988, *29*, 2441.
177. Yaozhong, J.; Changyou, Z.; Lan, G.; Guilan, L. *Synth. Commun.* 1989, *19*, 1297.
178. Yaozhong, J.; Peng, G.; Guilan, L. *Synth. Commun.* 1990, *20*, 15.
179. Aiqiao, M.; Jing, W.; Yuanwei, C.; Guishu, Y.; Yaozhong, J. *Synth. Commun.* 1989, *19*, 3337.
180. Dolling, U.-H.; Davis, P.; Grabowski, E. J. J. *J. Am. Chem. Soc.* 1984, *106*, 446.
181. Dolling, U.-H.; Ryan, K. M.; Schoenewaldt, E. F.; Grobowskii, E. J. J. *J. Org. Chem.* 1987, *52*, 4745.
182. Dolling, U.-H.; Huges, D. L.; Battacharya, A.; Ryan, K. M.; Karady, S.; Weinstock, L. M.; Grabowski, E. J. J. In *ACS Symposium Series No. 326*; American Chemical Society: Washington, D. C., 1987; pp 67.
183. Evans, D. A.; Ennis, M. D.; Mathre, D. J. *J. Am. Chem. Soc.* 1982, *104*, 1737.
184. Hauck, R. S.; Nau, H. *Pharm. Res.* 1992, *9*, 850.
185. Page, P. C. B.; Slawin, A. M. Z.; Westwood, D.; Williams, D. J. *J. Chem. Soc., Perkin Trans. 1* 1989, 185.
186. Page, P. C. B.; Klair, S. S.; Westwood, D. *J. Chem. Soc., Perkin Trans. 1* 1989, 2441.
187. Gilday, J. P.; Gallucci, J. C.; Paquette, L. A. *J. Org. Chem.* 1989, *54*, 1399.
188. Ahn, K. H.; Lim, A.; Lee, S. *Tetrahedron: Asymmetry* 1993, *4*, 2435.
189. Larcheveque, M.; Ignatova, E.; Cuvigny, T. *Tetrahedron Lett.* 1978, 3961.
190. Oppolzer, W. *Tetrahedron* 1987, *43*, 1969.
191. Fuji, K.; Node, M.; Tanaka, F. *Tetrahedron Lett.* 1990, *31*, 6553.
192. Fuji, K.; Node, M.; Tanaka, F.; Hosoi, S. *Tetrahedron Lett.* 1989, *30*, 2825.
193. Evans, D. A.; Takacs, J. M. *Tetrahedron Lett.* 1980, *21*, 4233.
194. Jeong, K.-S.; Parris, K.; Ballester, P.; Rebek, J. *Angew. Chem., Int. Ed. Engl.* 1990, *29*, 555.
195. Kawanami, Y.; Ito, Y.; Kitagawa, T.; Taniguchi, Y.; Katsuki, T.; Yamaguchi, M. *Tetrahedron Lett.* 1984, *25*, 857.
196. Negrete, G. R.; Konopelski, J. P. 1991, *Tetrahedron Asymmetry*, 1991, *2*, 105.
197. Evans, D. A.; Urpi, F.; Somers, T. C.; Clark, J. S.; Bilodeau, M. T. *J. Am. Chem. Soc.* 1990, *112*, 8215.
198. Fraser, R. R.; Akiyama, F.; Banville, J. *Tetrahedron Lett.* 1979, 3929.
199. Meyers, A. I. *Pure Appl. Chem.* 1979, *51*, 1255.
200. Whitesell, J. K.; Whitesell, M. A. *J. Org. Chem.* 1977, *42*, 377.
201. Meyers, A. I.; Williams, D. R. *J. Org. Chem.* 1978, *43*, 3245.
202. Meyers, A. I.; Williams, D. R.; Erickson, G. W.; White, S.; Druelinger, M. *J. Am. Chem. Soc.* 1981, *103*, 3081.

203. Meyers, A. I.; Knaus, G. *J. Am. Chem. Soc.* 1974, *96*, 6508.
204. Meyers, A. I.; Knaus, G.; Kamata, K. *J. Am. Chem. Soc.* 1974, *95*, 268.
205. Meyers, A. I.; Knaus, G.; Kamata, K.; Ford, M. E. *J. Am. Chem. Soc.* 1976, *98*, 567.
206. Saigo, K.; Kasahara, A.; Ogawa, S.; Nohira, H. *Tetrahedron Lett.* 1983, *24*, 511.
207. Enders, D.; Eichenauer, H. *Angew. Chem., Int. Ed. Engl.* 1979, *15*, 549.
208. Enders, D.; Gatzweiller, W.; Dederichs, E. *Tetrahedron* 1990, *46*, 4757.
209. Oppolzer, W.; Moretti, R.; Thomi, S. *Tetrahedron Lett.* 1989, *30*, 5603.
210. Enders, D.; Zamponi, A.; Raabe, G. *Synlett* 1992, 897.
211. Davies, S. G.; Seeman, J. I. *Tetrahedron Lett.* 1984, *25*, 1845.
212. Davies, S. G. *Pure Appl. Chem.* 1988, *60*, 13.
213. Davies, S. G.; Middlemiss, D.; Naylor, A.; Wills, M. *J. Chem. Soc., Chem. Commun.* 1990, 797.
214. Collingwood, S. P.; Davies, S. G.; Preston, S. C. *Tetrahedron Lett.* 1990, *31*, 4067.
215. Pearson, A. J.; Mortezaei, R. *Tetrahedron Lett.* 1989, *30*, 5049.
216. Pearson, A. J.; Mallik, S.; Mortezaei, R.; Perry, M. W. D.; Shively, R. J.; Youngs, W. J. *J. Am. Chem. Soc.* 1990, *112*, 8034.
217. Miles, W. H.; Smiley, P. M.; Brinkman, H. R. *J. Chem. Soc., Chem. Commun.* 1989, 1897.
218. Duhamel, L.; Duhamel, P.; Launay, J.-C.; Plaquevent, J.-C. *Bull. Soc. Chim., Fr.* 1984, *II*, 421.
219. Potin, D.; Williams, K.; Rebek, J. *Angew. Chem., Int. Ed. Engl.* 1990, *29*, 1420.
220. Petit, Y.; Sanner, C.; Larcheveque, M. *Tetrahedron Lett.* 1990, *31*, 2149.
221. Aizpurua, J. M.; Oiarbide, M.; Palomo, C. *Tetrahedron Lett.* 1987, *28*, 5361.
222. Ager, D. J. In *Umpoled Synthons: A Survey of Sources and Uses in Synthesis*; T. A. Hase, Ed.; Wiley-Interscience: New York, 1987; pp 19.
223. Fadel, A.; Salaun, J. *Tetrahedron Lett.* 1988, *29*, 6257.
224. Holy, N. L.; Wang, Y. F. *J. Am. Chem. Soc.* 1977, *99*, 944.
225. Danishefsky, S.; Kitahara, T.; McKee, R.; Schuda, P. F. *J. Am. Chem. Soc.* 1976, *98*, 6715.
226. Hooz, J.; Bridson, J. N. *J. Am. Chem. Soc.* 1973, *95*, 602.
227. Corey, E. J. *Chem. Soc. Rev.* 1988, *17*, 111.
228. Corey, E. J.; Kang, M.-C.; Desai, M. C.; Ghosh, A. K.; Houpis, I. N. *J. Am. Chem. Soc.* 1988, *110*, 649.
229. Corey, E. J.; Ghosh, A. K. *Tetrahedron Lett.* 1988, *29*, 3205.
230. Davies, F. A.; Chen, B.-C. *Chem. Rev.* 1992, *92*, 919.
231. Rubottom, G. M.; Gruber, J. M.; Boeckman, R. K.; Ramaiah, M.; Medwid, J. B. *Tetrahedron Lett.* 1978, 4603.
232. Troisi, L.; Cassidei, L.; Lopez, L.; Mello, R.; Curci, R. *Tetrahedron Lett.* 1989, *30*, 257.
233. Adam, W.; Hadjiarapoglou, L.; Klicic, J. *Tetrahedron Lett.* 1990, *31*, 6517.
234. Ager, D. J.; East, M. B. *J. Chem. Soc., Chem. Commun.* 1989, 178.
235. Swern, D. In *Organic Peroxides*; D. Swern, Ed.; Wiley-Interscience: New York, Vol. 2; 1971, pp 355.
236. Kritchevsky, T. H.; Gallagher, T. F. *J. Am. Chem. Soc.* 1951, *73*, 184.
237. Kritchevsky, T. H.; Garmaise, D. L.; Gallagher, T. F. *J. Am. Chem. Soc.* 1952, *74*, 483.
238. Shine, H. J.; Hunt, G. E. *J. Am. Chem. Soc.* 1958, *80*, 2434.
239. Williamson, K. L.; Coburn, J. I.; Herr, M. F. *J. Org. Chem.* 1967, *32*, 3934.
240. Sugimura, T.; Nishiyama, N.; Tai, A.; Hakushi, T. *Tetrahedron: Asymmetry* 1993, *4*, 43.
241. Vedejs, E.; Engler, D. A.; Telschow, J. E. *J. Org. Chem.* 1978, *43*, 188.
242. Guertin, K. R.; Chan, T.-H. *Tetrahedron Lett.* 1991, *32*, 715.
243. Aamboni, R.; Mohr, P.; Waespe-Sarcevic, N.; Tamm, C. *Tetrahedron Lett.* 1985, *26*, 203.
244. Vedejs, E. *J. Am. Chem. Soc.* 1974, *96*, 5944.
245. Adam, W.; Hadjiarapoglou, L.; Wang, X. *Tetrahedron Lett.* 1989, *30*, 6497.
246. Chenault, H. K.; Danishefsky, S. J. *J. Org. Chem.* 1989, *54*, 4249.
247. Rubottom, G. M.; Vazquez, M. A.; Pelegrina, D. R. *Tetrahedron Lett.* 1974, 4319.
248. Rubottom, G. M.; Marrero, R. *J. Org. Chem.* 1975, *40*, 3783.
249. Brook, A. G.; Macrae, D. M. *J. Organometal. Chem.* 1974, *77*, C19.
250. Pennanen, S. I. *Tetrahedron Lett.* 1980, *21*, 657.
251. Horiguchi, Y.; Nakamura, E.; Kuwajima, I. *Tetrahedron Lett.* 1989, *30*, 3323.
252. Hassner, A.; Reuss, R. H.; Pinnick, H. W. *J. Org. Chem.* 1975, *40*, 3427.
253. Musser, A. K.; Fuchs, P. L. *J. Org. Chem.* 1982, *47*, 3121.
254. Rubottom, G. M.; Gruber, J. M.; Kincaid, K. *Synth. Commun.* 1976, *6*, 59.
255. Demir, A. S.; Camkerten, N.; Akgun, H.; Tanyeli, C.; Mahasneh, A. S.; Watt, D. S. *Synth. Commun.* 1990, *20*, 2279.
256. Rubottom, G. M.; Gruber, J. M.; Mong, G. M. *J. Org. Chem.* 1976, *41*, 1673.
257. McCormick, J. P.; Tomasik, W.; Johnson, M. W. *Tetrahedron Lett.* 1981, *22*, 607.
258. Moriarty, R. M.; John, L. S.; Du, P. C. *J. Chem. Soc., Chem. Commun.* 1981, 641.
259. Moriarty, R. M.; Prakash, O. *Acc. Chem. Res.* 1986, *19*, 244.
260. Moriarty, R. M.; Vaid, R. K.; Koser, G. F. *Synlett* 1990, 365.

261. Moriarty, R. M.; Hou, K.-C. *Tetrahedron Lett.* 1984, *25*, 691.
262. Moriarty, R. M.; Hu, H. *Tetrahedron Lett.* 1981, *22*, 2747.
263. Moriarty, R. M.; Penmasta, R.; Awasthi, A. K.; Epa, W. R.; Prakash, I. *J. Org. Chem.* 1989, *54*, 1101.
264. Moriarty, R. M.; Epa, W. R.; Penmasta, R.; Awasthi, A. K. *Tetrahedron Lett.* 1989, *30*, 667.
265. Moriarty, R. M.; Prakash, O.; Duncan, M. P.; Vaid, R. K.; Musallam, H. A. *J. Org. Chem.* 1987, *52*, 150.
266. Moriarty, R. M.; Prakash, O.; Thachet, C. T.; Musallam, H. A. *Heterocycles* 1985, *23*, 633.
267. Koser, G. F.; Relenyi, A. G.; Kalos, A. N.; Rebrovic, L.; Wettach, R. H. *J. Org. Chem.* 1982, *47*, 2487.
268. Koser, G. F.; Lodaya, J. S.; Ray, D. G.; Kokil, P. B. *J. Am. Chem. Soc.* 1988, *110*, 2987.
269. Lodaya, J. S.; Koser, G. F. *J. Org. Chem.* 1988, *53*, 210.
270. Hatzigrigoriou, E.; Varvoglis, A.; Bakola-Christianopoulo, M. *J. Org. Chem.* 1990, *55*, 315.
271. Oppolzer, W.; Dudfield, P. *Helv. Chim. Acta* 1985, *68*, 216.
272. Rubottom, G. M.; Gruber, J. M.; Marrero, R.; Juve, H. D.; Kim, C. W. *J. Org. Chem.* 1983, *48*, 4940.
273. Yamada, T.; Narasaka, K. *Chem. Lett.* 1986, *131.*,
274. Clark, R. D.; Heathcock, C. H. *Tetrahedron Lett.* 1974, 2027.
275. Hillers, S.; Niklaus, A.; Resier, O. *J. Org. Chem.* 1993, *58*, 3169.
276. Friedrich, E.; Lutz, W. *Chem. Ber.* 1980, *113*, 1245.
277. Masui, M.; Ando, A.; Shioiri, T. *Tetrahedron Lett.* 1988, *29*, 2835.
278. Iqbal, J.; Pandey, A. *Tetrahedron Lett.* 1990, *31*, 575.
279. Davis, F. A.; Vishwakarma, L. C.; Billmers, J. M.; Finn, J. *J. Org. Chem.* 1984, *49*, 3241.
280. Davis, F. A.; Sheppard, A. C. *Tetrahedron* 1989, *45*, 5703.
281. Davis, F. A.; Wei, J.; Sheppard, A. C.; Grubernick, S. *Tetrahedron Lett.* 1987, *28*, 5115.
282. Kim, D.; Han, G. H.; Kim, K. *Tetrahedron Lett.* 1989, *30*, 1579.
283. Boschelli, D.; Smith, A. B.; Stringer, O. D.; Jenkins, R. H.; Davis, F. A. *Tetrahedron Lett.* 1981, *22*, 4385.
284. Chen, B.-C.; Weismiller, M. C.; Davis, F. A.; Boschelli, D.; Empfield, J. R.; Smith, A. B. *Tetrahedron* 1991, *47*, 1991.
285. Davis, F. A.; Kumar, A.; Chen, B.-C. *J. Org. Chem.* 1991, *56*, 1143.
286. Davis, F. A.; Haque, M. S.; Przeslawski, R. M. *J. Org. Chem.* 1989, *54*, 2021.
287. Davis, F. A.; Sheppard, A. C.; Chen, B.-C.; Haque, M. S. *J. Am. Chem. Soc.* 1990, *112*, 6679.
288. Davis, F. A.; Weismiller, M. C. *J. Org. Chem.* 1990, *55*, 3715.
289. Davis, F. A.; Kumar, A.; Chen, B.-C. *Tetrahedron Lett.* 1991, *32*, 867.
290. Davis, F. A.; Chen, B.-C. *Tetrahedron Lett.* 1990, *31*, 6823.
291. Evans, D. E.; Morrissey, M. M.; Dorow, R. L. *J. Am. Chem. Soc.* 1985, *107*, 4346.
292. Djuric, S. W.; Miyahiro, J. M.; Penning, T. D. *Tetrahedron Lett.* 1988, *29*, 3459.
293. Davis, F. A.; Sheppard, A. C.; Lal, G. S. *Tetrahedron Lett.* 1989, *30*, 779.
294. Wei, Y.; Bakhavatchalam, R.; Jin, X.-M.; Murphy, C. K.; Davis, F. A. *Tetrahedron Lett.* 1993, *34*, 3715.
295. Davis, F. A.; Kumar, A. *J. Org. Chem.* 1992, *57*, 3337.
296. Enders, D.; Bhushan, V. *Tetrahedron Lett.* 1988, *29*, 2437.
297. Lohray, B. B.; Enders, D. *Helv. Chim. Acta* 1989, *72*, 980.
298. Enders, D.; Nakai, S. *Chem. Ber.* 1991, *124*, 219.
299. Enders, D.; Nakai, S. *Helv. Chim. Acta* 1990, *73*, 1833.
300. Taschner, M. J.; Aminbhavi, A. S. *Tetrahedron Lett.* 1989, *30*, 1029.
301. Guy, A.; Lemor, A.; Imbert, D.; Lemaire, M. *Tetrahedron Lett.* 1989, *30*, 327.
302. Drian, C. L.; Greene, A. E. *J. Am. Chem. Soc.* 1982, *104*, 5473.
303. Wasserman, H. H.; Lipshultz, B. H. *Tetrahedron Lett.* 1975, 1731.
304. Selikson, S. J.; Watt, D. S. *J. Org. Chem.* 1975, *40*, 267.
305. Iwata, C. I.; Takemoto, Y.; Nakamura, A.; Imanishi, T. *Tetrahedron Lett.* 1985, *26*, 3227.
306. Ley, S. V.; Santafianos, D.; Blaney, W. M.; Simmonds, M. S. J. *Tetrahedron Lett.* 1987, *28*, 221.
307. Dembech, P.; Guerrini, A.; Ricci, A.; Seconi, G.; Taddei, M. *Tetrahedron* 1990, *46*, 2999.
308. Adam, W.; Cueto, O. *J. Org. Chem.* 1977, *42*, 38.
309. Anderson, J. C.; Ley, S. V. *Tetrahedron Lett.* 1990, *31*, 3437.
310. Meyers, A. I.; Slade, J. *J. Org. Chem.* 1980, *45*, 2785.
311. Corey, P. F. *Tetrahedron Lett.* 1987, *28*, 2801.
312. Terashima, S.; Jew, S.-S.; Koga, K. *Tetrahedron Lett.* 1978, 4937.
313. Enders, D.; Lotter, H. *Nouv. J. Chim.* 1984, *8*, 747.
314. Craig, D.; Daniels, K.; MacKenzie, A. R. *Tetrahedron Lett.* 1990, *31*, 6441.
315. Sharpless, K. B.; Lauer, R. F.; Teranishi, A. Y. *J. Am. Chem. Soc.* 1973, *95*, 6173.
316. Holmes, A. B.; Nadin, A.; O'Hanlon, P. J.; Pearson, N. D. *Tetrahedron: Asymmetry* 1992, *3*, 1289.
317. Trost, B. M. *Acc. Chem. Res.* 1978, *11*, 453.
318. Pflienger, P.; Mioskowski, C.; Salaun, J. P.; Weissbart, D.; Durst, F. *Tetrahedron Lett.* 1989, *30*, 2791.
319. Kende, A. S.; Mendoza, J. S. *J. Org. Chem.* 1990, *55*, 1125.
320. Aggarwal, V. K.; Warren, S. *Tetrahedron Lett.* 1986, *27*, 101.

321. Oppolzer, W.; Dudfield, P. *Tetrahedron Lett.* 1985, *26*, 5037.
322. Hiroi, K.; Yamada, S.-I. *Chem. Pharm. Bull.* 1973, *21*, 54.
323. Evans, D. A.; Britton, T. C.; Ellman, J. A.; Dorow, R. L. *J. Am. Chem. Soc.* 1990, *112*, 4011.
324. Oppolzer, W.; Pedrosa, R.; Moretti, R. *Tetrahedron Lett.* 1986, *27*, 831.
325. Giordano, C.; Castaldi, G.; Cavicchioli, S.; Villa, M. *Tetrahedron* 1989, *45*, 4243.
326. Giordano, C.; Coppi, L.; Restelli, A. *J. Org. Chem.* 1990, *55*, 5400.
327. Oppolzer, W.; Tamura, O. *Tetrahedron Lett.* 1990, *31*, 991.
328. Loreta, M. A.; Pellacani, L.; Tardella, P. A. *Tetrahedron Lett.* 1989, *30*, 2975.
329. Evans, D. A.; Britton, T. C.; Dorow, R. L.; Dellaria, J. F. *J. Am. Chem. Soc.* 1986, *108*, 6395.
330. Gennari, C.; Columbo, L.; Bertolini, G. *J. Am. Chem. Soc.* 1986, *108*, 6394.
331. Trimble, L. A.; Vederas, C. J. *J. Am. Chem. Soc.* 1986, *108*, 6397.
332. Tamura, Y.; Minamikawa, J.; Ikeda, M. *Synthesis* 1977, 1.
333. Schmitz, E. *Russ. Chem. Rev. (trans.)* 1976, *45*, 16.
334. Wallace, R. G. *Aldrichimica Acta* 1980, *13*, 3.
335. Magnus, P. D.; Mugrage, B. *J. Am. Chem. Soc.* 1990, *112*, 463.
336. Scott, J. W.; Valentine, D. *Science* 1974, *184*, 943.
337. Enders, D. In *Asymmetric Synthesis*; J. D. Morrison, Ed.; Academic Press: Orlando, FL, 1984; Vol. 3; pp 275.
338. Hickmott, P. W. *Tetrahedron* 1982, *38*, 1975.
339. Hickmott, P. W. *Tetrahedron* 1982, *38*, 3363.
340. Kobayashi, Y.; Umeyama, K.; Sato, F. *J. Chem. Soc., Chem. Commun.* 1984, 621.
341. Enders, D.; Eichenauer, H. *Angew. Chem., Int. Ed. Engl.* 1976, *15*, 549.
342. Enders, D.; Eichenauer, H. *Tetrahedron Lett.* 1977, 191.
343. Meyers, A. I.; Williams, D. R.; Erickson, G. W.; White, S.; Druelinger, M. *J. Am. Chem. Soc.* 1981, *103*, 3051.
344. Cardillo, G.; Orena, M.; Romero, M.; Sandri, S. *Tetrahedron* 1989, *45*, 1501.
345. Collum, D. B.; Kahne, D.; Gut, S. A.; DePue, R. T.; Mohamadi, F.; Wanat, R. A.; Clardy, J.; Duyne, G. V. *J. Am. Chem. Soc.* 1984, *106*, 4865.
346. Kitamoto, M.; Hiroi, K.; Terashima, S.; Yamada, S.-I. *Chem. Pharm. Bull.* 1974, *22*, 459.
347. Johnson, F. *Chem. Rev.* 1968, *68*, 375.
348. Hiroi, K.; Yamada, S.-I. *Chem. Pharm. Bull.* 1973, *21*, 47.
349. Meyers, A. I.; Williams, D. R.; Druelinger, M. *J. Am. Chem. Soc.* 1976, *98*, 3032.
350. Knackmuss, H.-J.; Beckmann, W.; Otting, W. *Angew. Chem., Int. Ed. Engl.* 1976, *15*, 549.
351. Wanat, R. A.; Collum, D. B. *J. Am. Chem. Soc.* 1985, *107*, 2078.
352. Enders, D.; Eichenauer, H. *Chem. Ber.* 1979, *112*, 2933.
353. Enders, D.; Eichenauer, H. *Angew. Chem., Int. Ed. Engl.* 1979, *18*, 397.
354. Corey, E. J.; Enders, D. *Chem. Ber.* 1978, *111*, 1362.
355. Pennanen, S. I. *Acta Chem. Scand. (B)* 1981, 555.
356. Corey, E. J.; Enders, D. *Tetrahedron Lett.* 1976, 3.
357. Galiano-Roth, A. S.; Collum, D. B. *J. Am. Chem. Soc.* 1989, *111*, 6772.
358. Meyers, A. I.; Williams, D. R.; White, S.; Erickson, G. W. *J. Am. Chem. Soc.* 1981, *103*, 3088.
359. Whitesell, J. K.; Felman, S. W. *J. Org. Chem.* 1977, *42*, 1663.
360. Hiroi, K.; Achiwa, K.; Yamada, S.-I. *Chem. Pharm. Bull.* 1972, *20*, 246.
361. Stork, G.; Brizzolara, A.; Landesman, H.; Szmuskovicz, J.; Terrell, R. *J. Am. Chem. Soc.* 1963, *85*, 207.
362. Stork, G.; Dowd, S. *J. Am. Chem. Soc.* 1963, *85*, 2178.
363. Méa-Jacheet, D.; Horeau, A. *Bull. Soc. Chim. Fr.* 1968, 4571.
364. Houk, K.; Strozier, R. W.; Rondan, N. G.; Fraser, R. R.; Chuaqui-Offermanns, N. *J. Am. Chem. Soc.* 1980, *102*, 1426.
365. Enders, D.; Karl, W. *Synlett* 1992, 895.
366. Duhamel, L.; Duhamel, P.; Enders, D.; Karl, W.; Leger, F.; Poirier, J. M.; Raale, G. *Synthesis* 1991, 649.
367. Harada, K.; Matsumoto, K. *J. Org. Chem.* 1967, *32*, 1794.
368. Harada, K.; Iwasaki, T.; Okawara, T. *Bull. Chem. Soc., Jpn.* 1973, *46*, 1901.
369. Harada, K. *Nature* 1966, *212*, 1571.
370. Vigneron, J. P.; Kagan, H.; Horeau, A. *Bull. Soc. Chim., Fr.* 1972, 3836.
371. Yamada, S.-I.; Hashimoto, S.-I. *Tetrahedron Lett.* 1976, 997.
372. Yamada, S.-I.; Ikota, N.; Achiwa, K. *Tetrahedron Lett.* 1976, 1001.
373. Duhamel, L.; Normant, H. *C. R. Acad. Sci., Paris (C)* 1976, *282*, 125.
374. Matsushita, H.; Noguchi, M.; Yoshikawa, S. *Chem. Lett.* 1975, 1313.
375. Kober, R.; Papadopoulos, K.; Miltz, W.; Enders, D.; Steglich, W.; Reuter, H.; Puff, H. *Tetrahedron* 1985, *41*, 1693.
376. Hashimoto, S.-I.; Koga, K. *Tetrahedron Lett.* 1978, 573.
377. Hashimoto, S.-I.; Koga, K. *Chem. Pharm. Bull.* 1979, *27*, 2760.
378. Ludwig, J. W.; Newcomb, M.; Bergbreiter, D. E. *Tetrahedron Lett.* 1986, *27*, 2731.

379. Oppolzer, W.; Rodriguez, I.; Starkemann, C.; Walther, E. *Tetrahedron Lett.* 1990, *31*, 5019.
380. Oppolzer, W.; Wills, M.; Starkemann, C.; Bernardinelli, G. *Tetrahedron Lett.* 1990, *31*, 4117.
381. Genet, J. P.; Kopola, N.; Juge, S.; Ruiz-Montes, J.; Antunes, O. A. C.; Tanier, S. *Tetrahedron Lett.* 1990, *31*, 3133.
382. Solladie-Cavallo, A.; Simon, M. C. *Tetrahedron Lett.* 1990, *31*, 6011.
383. Oppolzer, W.; Moretti, R.; Thomi, S. *Tetrahedron Lett.* 1990, *31*, 6009.
384. Meyers, A. I.; Poindexter, G. S.; Birch, Z. *J. Org. Chem.* 1978, *43*, 892.
385. Fraser, R. R.; Banville, J. *J. Chem. Soc., Chem. Commun.* 1979, 47.
386. Meyers, A. I.; Mihelich, E. D. *Angew. Chem., Int. Ed. Engl.* 1976, *15*, 270.
387. Meyers, A. I.; Whitten, C. E. *J. Am. Chem. Soc.* 1975, *97*, 6266.
388. Meyers, A. I.; Mazzu, A.; Whitten, C. E. *Heterocycles* 1977, *6*, 971.
389. Meyers, A. I.; Snyder, E. S.; Ackerman, J. J. H. *J. Am. Chem. Soc.* 1978, *100*, 8186.
390. Meyers, A. I.; Kamata, K. *J. Org. Chem.* 1974, *39*, 1603.
391. Inoue, I.; Shindo, M.; Koga, K.; Tomioka, K. *Tetrahedron: Asymmetry* 1993, *4*, 1603.
392. Whitesell, J. K. *Chem. Rev.* 1989, *89*, 1581.
393. Cain, C. M.; Cousins, R. P. C.; Coumbarides, G.; Simpkins, N. S. *Tetrahedron* 1990, *46*, 523.
394. Simpkins, N. S. *Chem. Soc. Rev.* 1990, *19*, 335.
395. Cox, P. J.; Simpkins, N. S. *Tetrahedron: Asymmetry* 1991, *2*, 1.
396. Simpkins, N. S. *Chem. Ind. (London)* 1988, 387.
397. Barr, D.; Berrisford, D. J.; Jones, R. V. H.; Slawin, A. M. Z.; Snaith, R.; Stoddart, J. F.; Williams, D. J. *Angew. Chem., Int. Ed. Engl.* 1989, *28*, 1044.
398. Eleveld, M. B.; Hogeveen, H. *Tetrahedron Lett.* 1986, *27*, 631.
399. Simpkins, N. S. *J. Chem. Soc., Chem. Commun.* 1986, 88.
400. Shirai, R.; Tanaka, M.; Koga, K. *J. Am. Chem. Soc.* 1986, *108*, 543.
401. Hogeveen, H.; Menge, W. M. P. B. *Tetrahedron Lett.* 1986, *27*, 2767.
402. Ando, A.; Shioirio, T. *J. Chem. Soc., Chem. Commun.* 1987, 656.
403. Murakata, M.; Nakajima, M.; Koga, K. *J. Chem. Soc., Chem. Commun.* 1990, 1657.
404. Leonard, J.; Hewitt, J. D.; Ouali, D.; Rahman, S. K.; Simpson, S. J.; Newton, R. F. *Tetrahedron: Asymmetry* 1990, *1*, 699.
405. Cousins, R. P. C.; Simpkins, N. S. *Tetrahedron Lett.* 1990, *31*, 7241.
406. Kim, H.-D.; Kawasaki, H.; Nakajima, M.; Koga, K. *Tetrahedron Lett.* 1989, *30*, 6537.
407. Sobukawa, M.; Nakajima, M.; Koga, K. *Tetrahedron: Asymmetry* 1990, *1*, 295.
408. Cousins, R. P. C.; Simpkins, N. S. *Tetrahedron Lett.* 1989, *30*, 7241.
409. Yanagisawa, A.; Kuribayashi, T.; Kikuchi, T.; Yamamoto, H. *Angew. Chem., Int. Ed. Engl.* 1994, *33*, 107.
410. Aoki, K.; Noguchi, H.; Tomioka, K.; Koga, K. *Tetrahedron Lett.* 1993, *34*, 5105.
411. Hasegawa, Y.; Kawasaki, H.; Koga, K. *Tetrahedron Lett.* 1993, *34*, 1963.
412. Bunn, B. J.; Cox, P. J.; Simpkins, N. S. *Tetrahedron* 1993, *49*, 207.
413. Honda, T.; Kimura, N.; Tsubuki, M. *Tetrahedron: Asymmetry* 1993, *4*, 21.
414. Honda, T.; Kimura, N.; Tsubuki, M. *Tetrahedron: Asymmetry* 1993, *4*, 1475.
415. Bunn, B. J.; Simpkins, N. S. *J. Org. Chem.* 1993, *58*, 533.
416. Cox, P. J.; Persad, A.; Simpkins, N. S. *Synlett* 1992, 194.
417. Duhamel, L.; Ravard, A.; Plaquevent, J.-C.; Davoust, D. *Tetrahedron Lett.* 1987, *28*, 5517.
418. Muraoka, M.; Kawasaki, H.; Koga, K. *Tetrahedron Lett.* 1988, *29*, 337.
419. Ando, A.; Shioiri, T. *Tetrahedron* 1989, *45*, 4969.
420. Coggins, P.; Simpkins, N. S. *Synlett* 1992, 313.
421. Schwerdtfeger, J.; Hoppe, D. *Angew. Chem., Int. Ed. Engl.* 1992, *31*, 1505.
422. Knochel, P. *Angew. Chem., Int. Ed. Engl.* 1992, *31*, 1459.
423. Whitesell, J. K.; Felma, S. W. *J. Org. Chem.* 1980, *45*, 755.
424. Leonard, J.; Hewitt, J. D.; Ouali, D.; Simpson, S. J.; Newton, R. F. *Tetrahedron Lett.* 1990, *31*, 6703.
425. Asami, M. *Tetrahedron Lett.* 1985, *26*, 5803.
426. Yamashita, T.; Mitsui, H.; Watanabe, H.; Nakamura, N. *Bull. Chem. Soc., Jpn.* 1982, *55*, 961.
427. Sobukawa, M.; Koga, K. *Tetrahedron Lett.* 1993, *34*, 5101.
428. Viteva, L.; Stefanovsky, Y. *Tetrahedron Lett.* 1989, *30*, 4565.
429. McArthur, C. R.; Jiang, J. L.; Leznoff, C. C. *Can. J. Chem.* 1982, *60*, 2984.
430. Matsushita, H.; Noguchi, M.; Saburi, M.; Yoshikawa, S. *Bull. Chem. Soc., Jpn.* 1975, *48*, 3715.
431. Duhamel, L. *C. R. Acad. Sci., Paris (C)* 1976, *282*, 125.
432. Duhamel, L.; Plaquevent, J.-C. *J. Am. Chem. Soc.* 1978, *100*, 7415.
433. Fuji, K.; Tanaka, K.; Miyamoto, H. *Tetrahedron: Asymmetry* 1993, *4*, 247.
434. Yasukata, T.; Koga, K. *Tetrahedron: Asymmetry* 1993, *4*, 35.
435. Pracejus, H. *Annalen* 1960, *634*, 23.
436. Proceus, H.; Tille, A. *Chem. Ber.* 1963, *96*, 854.

437. Winter, S.; Pracejus, H. *Chem. Ber.* 1966, *99*, 151.
438. Salz, U.; Ruechardt, C. *Tetrahedron Lett.* 1982, *23*, 4017.
439. Jaehme, J.; Ruechardt, C. *Angew. Chem., Int. Ed. Engl.* 1981, *20*, 885.
440. Bellucci, G.; Berti, G.; Bianchini, R.; Vecchiani, S. *Gazz. Chim. Ital.* 1988, *118*, 451.
441. Larsen, R. D.; Corley, E. G.; Davis, P.; Reider, P. J.; Grabowski, E. J. J. *J. Am. Chem. Soc.* 1989, *111*, 7650.
442. Seebach, D.; Beck, A. K.; Lehr, F.; Weller, T.; Colvin, E. W. *Angew. Chem., Int. Ed. Engl.* 1981, *20*, 397.
443. Seebach, D.; Beck, A. K.; Mukhopadhyay, T.; Thomas, E. *Helv. Chim. Acta* 1982, *65*, 101.
444. Takano, S.; Uchida, W.; Hatakeyama, S.; Ogasawara, K. *Chem. Lett.* 1982, 733.
445. Takano, S.; Masuda, K.; Hatakeyama, S.; Ogasawara, K. *Heterocycles* 1982, *19*, 1407.
446. Cavelier, F.; Gomez, S.; Jacquier, R.; Verducci, J. *Tetrahedron: Asymmetry* 1993, *4*, 2501.
447. Duhamel, L.; Launay, J.-C. *Tetrahedron Lett.* 1983, *24*, 4209.

Chapter 6

# REACTIONS OF FUNCTIONALIZED CARBONYL SYSTEMS

This chapter discusses the reactions of functionalized carbonyl compounds, such as α,β-unsaturated ketones and β-diketones, where the chemistry of the carbonyl group is responsible for the transformation observed. For reactions where the other functional group is the major participant, as in conjugate additions, Chapter 9 should be consulted. The current chapter is divided into sections that discuss each of the substitution patterns in turn.

Aldol products, as well as β-dicarbonyl systems, can undergo stereoselective reductions or alkylations.[1] Thus, these reactions augment those already described.

As with other areas of asymmetric synthesis, reducing agents have been modified by the addition of chiral ligands, but many have not provided a high degree of stereoselection, although useful insights into reaction mechanisms and pathways may have been gained. Exhaustive lists of these reagents have not been included.[2-4]

## 6.1. ALKYLATIONS OF 1,2-FUNCTIONALIZED CARBONYL COMPOUNDS

The addition of a nucleophile to a chiral ester of an α-keto acid is the foundation of Prelog's generalization [*2.1.4*] (vide infra).[5] A stereogenic center, such as a sulfoxide, adjacent to the carbonyl group can also be used to afford stereoselective reaction.[6]

Other 1,2-dicarbonyl compounds can undergo selective additions. For example, an α-keto amide derived from a chiral amine provides reasonable selectivity with allyltrimethylsilane,[7] while chiral α-keto acetals also show high selectivity in carbonyl alkylation reactions (Scheme 2.8).[8-12] Glyoxal derivatives that have been used and incorporate chirality are based on aminals[13] or carbohydrate-derived hydrazone acetals.[14]

### 6.1.1. ALKYLATIONS OF GLYCOLATES AND GLYOXYLATES

The asymmetric alkylation of a glycolate enolate provides a useful entry to α-hydroxy carboxylic acids.[15] The ketone moiety can be protected as a thioketal. This facilitates anion formation and subsequent reaction with a wide variety of electrophiles.[16] Of course, an alternative methodology is reaction of an α-keto ester with an organometallic reagent (*cf.* Scheme 2.15).[17-22] Indeed, this additional functional group can afford increased stereoselectivity compared to "simple" carbonyl compounds.[23]

Addition of allylmetallic compounds to a chiral glyoxylate affords the α-hydroxy esters. The selectivity is dependent upon the solvent and metal counterion (Scheme 6.1);[17-19,24-27] chiral complexes also provide selectivity with achiral glyoxalate esters.[28]

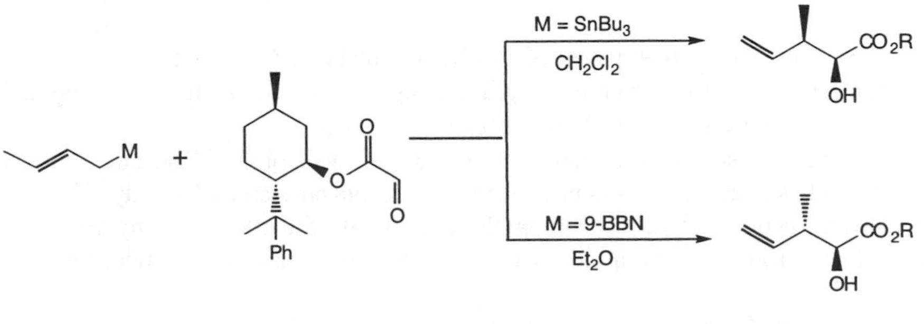

**SCHEME 6.1.**

Organozinc reagents with chiral glyoxalate esters can also give good stereochemical yields of the resultant mandelic acid derivatives,[29] while chiral ligands for Grignard and reducing agents can also provide high selectivity.[30,31]

Other chiral auxiliaries that have been used to mask α-keto acids are based on oxazolines,[32,33] while hydrazones allow entry to 3-substituted 2-keto esters.[34]

## 6.1.2. OTHER 1,2-FUNCTIONALIZED CARBONYL COMPOUNDS

Of course, if the α-substituent is large, this can ensure facial selectivity through enforcement of Houk's rule [*2.2*], as, for example, with α-silyl ketones.[35,36] This applies to both nucleophilic and electrophilic reactions, and it is not necessary to form a chelate to obtain a high degree of selectivity.

### 6.1.2.1. α-Hydroxy Carbonyl Compounds

The presence of an α-hydroxy or other electron-rich group allows for Anh-Felkin and chelation controlled additions, and numerous examples have already been given [*2.1.1 and 2.1.2*].[37-40]

The oxazoline **6.1** provided one such application of this methodology, but some difficulty was encountered in the hydrolytic removal of the chiral auxiliary (Scheme 6.2).[41]

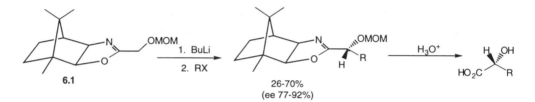

6.1 OMOM — 1. BuLi 2. RX → OMOM R H, 26-70% (ee 77-92%) — $H_3O^+$ → H OH $HO_2C$ R

**SCHEME 6.2.**

Use of a functionalized amide system provides high diastereoselectivites for the preparation of α-hydroxy carboxylic acid derivatives (Scheme 6.3).[42]

TBDMSO O N OMe — 1. LDA 2. $R^1X$ → TBDMSO O N OMe $R^1$, 85-97% (ds >72:28)

**SCHEME 6.3.**

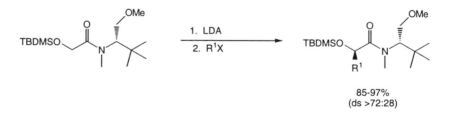

An alternative approach is provided by the dioxolanone **6.2**, but the preparation of the starting material does require a chromatographic separation (Scheme 6.4).[43] Thioglycolic acid provides an analogous system and chemistry.[44,45]

Other means of masking a carbonyl group are also available.[46,47] The direct alkylation of ketals derived from tartaric acid esters does provide reasonable stereoselectivity.[48,49] The nucleophilic substitution of an α-hydroxy carboxylic acid derivative to afford an α-hydroxy ketone can be a useful approach, but this requires that the stereochemistry is already established [*2.5.2.2*].[50]

### 6.1.2.2. α-Amino Carbonyl Compounds

α-Amino aldehydes can undergo high selective additions (e.g., Scheme 6.5),[51-54] as can α-amino ketones.[55]

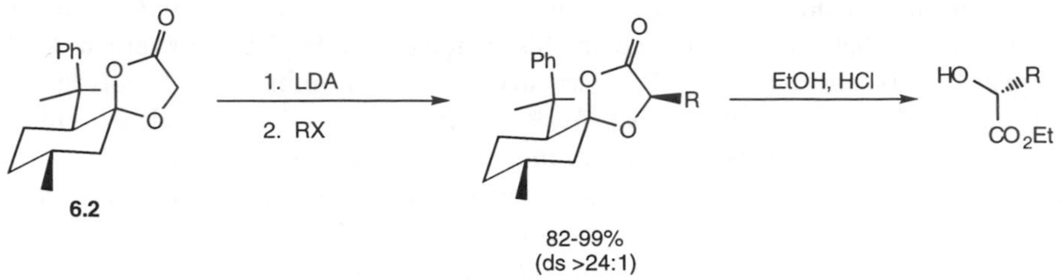
Ph
O
6.2
1. LDA
2. RX
R
82-99%
(ds >24:1)
EtOH, HCl
HO
R
$CO_2Et$

**SCHEME 6.4.**

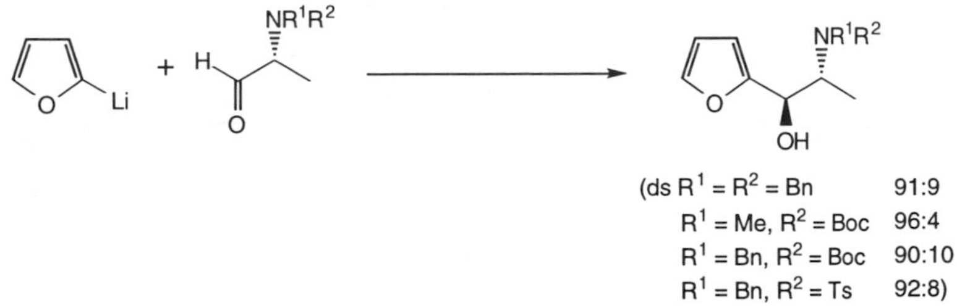
Li
H
O
$NR^1R^2$
OH
(ds $R^1 = R^2 = Bn$ 91:9
$R^1 = Me, R^2 = Boc$ 96:4
$R^1 = Bn, R^2 = Boc$ 90:10
$R^1 = Bn, R^2 = Ts$ 92:8)

**SCHEME 6.5.**

Like carbonyl compounds, imines can react with organometallic reagents and, if the appropriate α-substituent is present, with chelation control (Scheme 6.6).[56-59]

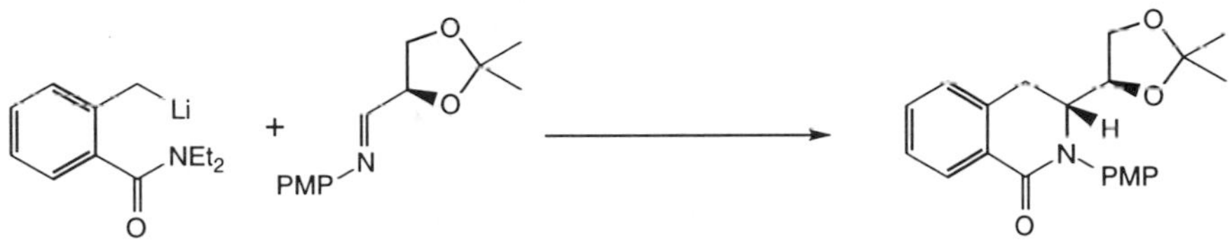
Li
$NEt_2$
O
PMP
N
H
PMP

**SCIIEME 6.6.**

α-Amino esters can be converted to the corresponding imines; a popular choice is the imine derived from benzophenone. Reactions can then be performed at the ester group, such as reduction to the aldehyde and subsequent reaction with an organometallic reagent (Scheme 6.7).[60]

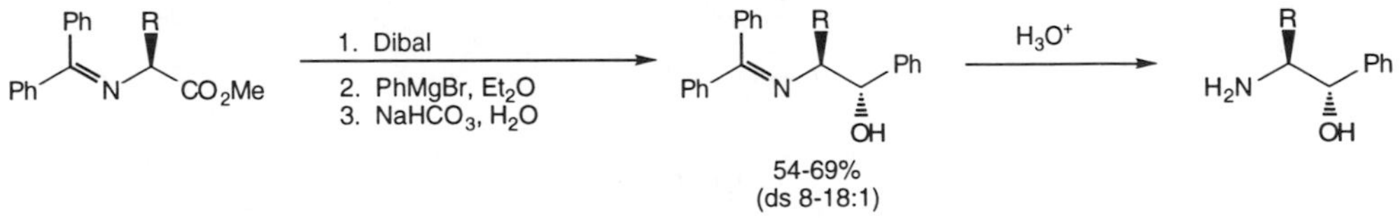
Ph
R
N
$CO_2Me$
1. Dibal
2. PhMgBr, $Et_2O$
3. $NaHCO_3$, $H_2O$
OH
54-69%
(ds 8-18:1)
$H_3O^+$
$H_2N$

**SCHEME 6.7.**

In addition, imines derived from amino acids can be alkylated — this also provides the methodology to modify the stereogenic center of α-amino acids. For the benzophenone derived imines, asymmetric induction has been observed through the use of a chiral alcohol, to form the chiral ester,[61] or a chiral phase transfer catalyst.[62,63] This imine approach also allows aryl groups to be introduced onto glycine through the use of triphenylbismuth carbonate.[64,65] The anions derived from the imines of α-amino esters can undergo reactions other than alkylations, such as Michael additions [*9.5.1*].[66]

Imines of amino acids can also be modified to incorporate a chiral auxiliary at the acid end;[67] auxiliaries that have been used in this context include: 2-hydroxypinan-3-one,[68-73] camphor,[71,74-76] 10-substituted camphor derivatives,[77-79] other amino acids,[80] polyacrylic resins,[81] β-lactams,[82] and phenyloxazinones.[59,83-89] This last case involves a cyclic system so both the amino and acid ends of the amino acid interact with the chiral auxiliary.

Systems have been used where both the nitrogen and ester bear chiral auxiliaries allowing for the potential for double asymmetric induction.[71,76,90,91]

The amino acid can be used as a chiral auxiliary in an imine to allow chirality transfer in a Barbier-type alkylation (Scheme 6.8).[92]

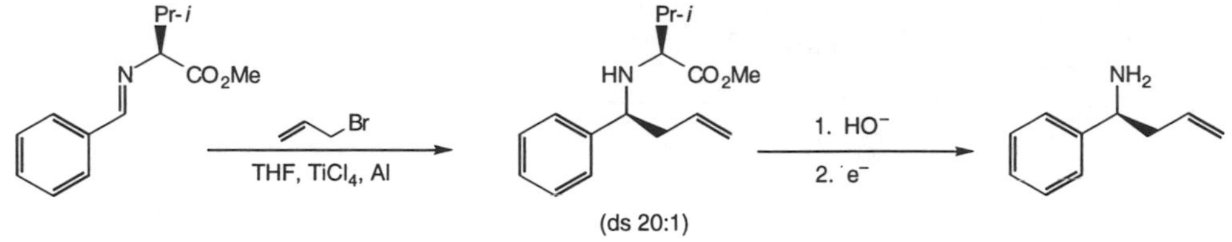

Pr-*i*, N, $CO_2Me$; Br, THF, $TiCl_4$, Al; HN, $CO_2Me$, (ds 20:1); 1. $HO^-$, 2. $e^-$; $NH_2$

**SCHEME 6.8.**

α-Amino esters can be alkylated with a wide degree of diastereoselection[93-95] and the products can be used to prepare a wide range of compounds.[96,97] This alkylation can be accomplished with retention of configuration by formation of a tricyclic system (Scheme 6.9).[98]

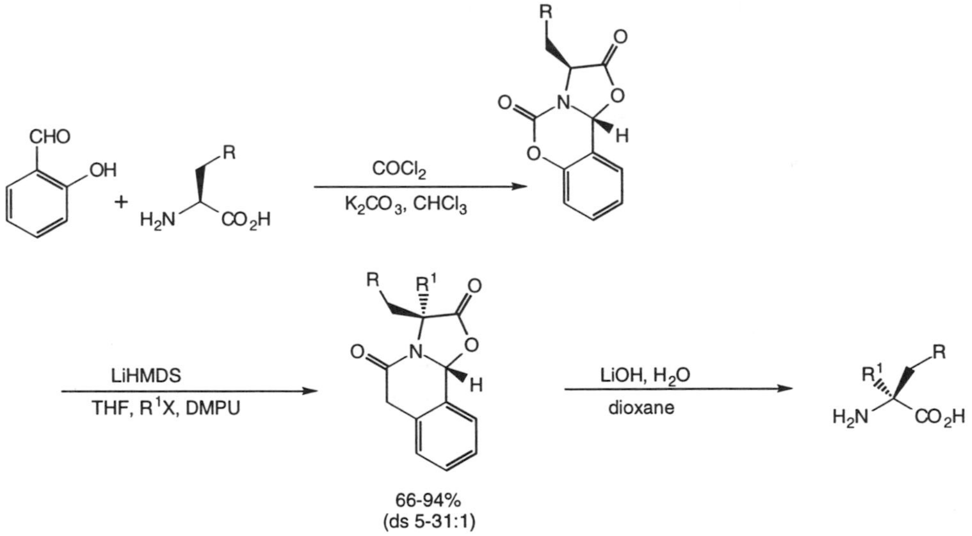

CHO, OH + $H_2N$, R, $CO_2H$; $COCl_2$, $K_2CO_3$, $CHCl_3$; LiHMDS, THF, $R^1X$, DMPU; 66-94% (ds 5-31:1); LiOH, $H_2O$, dioxane; $R^1$, R, $H_2N$, $CO_2H$

**SCHEME 6.9.**

Addition to an α-amino imine can provide high face selectivity through Anh-Felkin control (Scheme 6.10).[99]

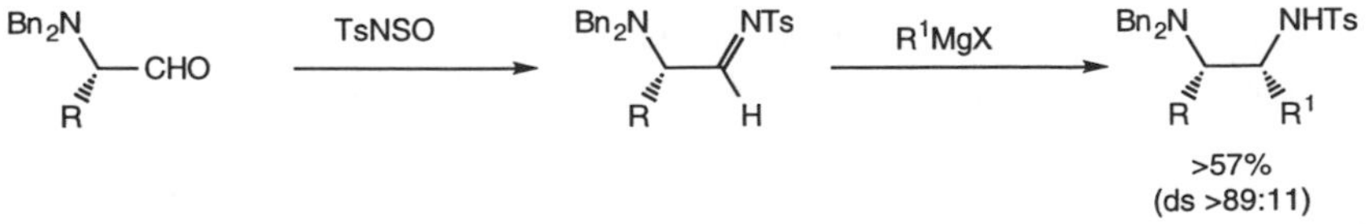

$Bn_2N$, CHO, R; TsNSO; $Bn_2N$, NTs, R, H; $R^1MgX$; $Bn_2N$, NHTs, R, $R^1$, >57% (ds >89:11)

**SCHEME 6.10.**

Reaction of cyclohexylmethylmagnesium bromide with the imine **6.3** gave no addition products unless the Grignard reagent was first treated with cerium(III), then only one amine resulted; the selectivity could be reversed by use of a copper reagent (Scheme 6.11).[100]

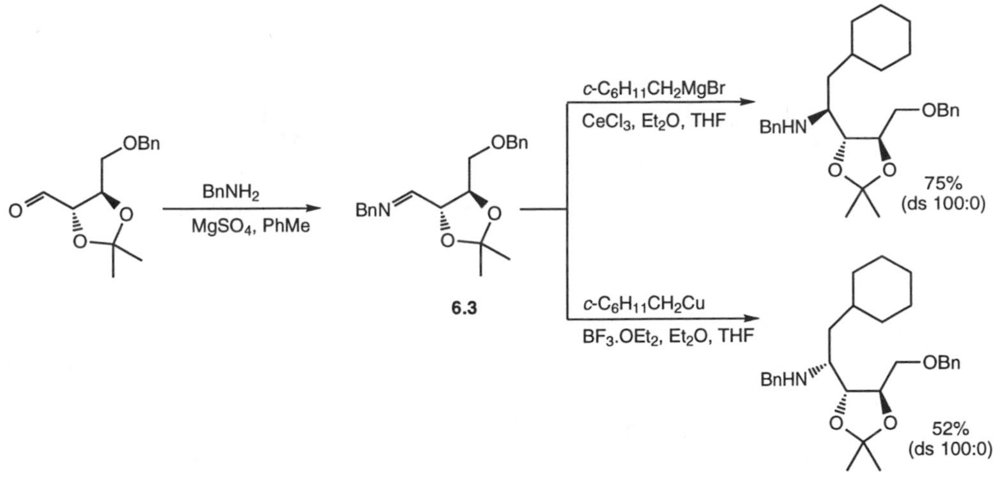

OBn BnNH2 MgSO4, PhMe BnN OBn 6.3 c-C6H11CH2MgBr CeCl3, Et2O, THF BnHN OBn 75% (ds 100:0) c-C6H11CH2Cu BF3.OEt2, Et2O, THF BnHN OBn 52% (ds 100:0)

**SCHEME 6.11.**

### 6.1.2.3. Other α-Functionalized Carbonyl Compounds

Dialkylzinc reagents add to α,β-unsaturated aldehydes in a 1,2-manner on the presence of a chiral amino alcohol; the degree of induction can be good.[101] Of course, additions to enones can be influenced by the steric requirements of both the electrophilic and nucleophilic species.[102,103]

The 1,3-dithiane 1-oxide chiral auxiliary juxtaposed to a carbonyl group allows for both nucleophilic addition to the carbonyl group and enolate alkylations with high diastereofacial selectivity.[104-107]

An example of an α-thio ketone in a face selective reaction is given in Scheme 6.12.[108]

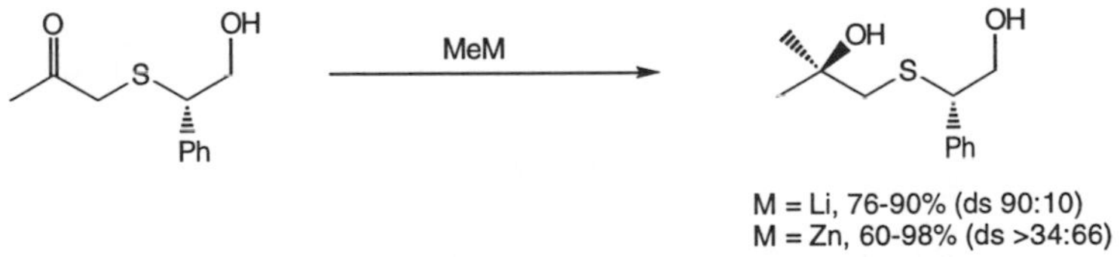

O S OH Ph MeM OH S OH Ph

M = Li, 76-90% (ds 90:10)
M = Zn, 60-98% (ds >34:66)

**SCHEME 6.12.**

Enone alkylation can be achieved by a protection methodology [*9.5.1.3*].[109,110]

## 6.2. REDUCTIONS OF 1,2-FUNCTIONALIZED CARBONYL COMPOUNDS

### 6.2.1. 1,2-DICARBONYL COMPOUNDS

The reduction of α-dicarbonyl systems has often been used as a model system to investigate the utility of new reduction systems.[2,3,111] As a consequence, there are numerous examples in the literature; only reactions that have been used with more than one substrate, show high selectivity, and seem to have promise for general usage have been included in this section.

α-Keto esters are reduced by chiral boranes, such as Eapine-Borane, with good induction (Scheme 6.13).[23,112,113] Of course, a chiral group can be incorporated into the ester moiety.[114] Hydrogenation conditions with chiral catalysis have not given such high degrees of induction,[115-120] except when ruthenium-BINAP is used.[121]

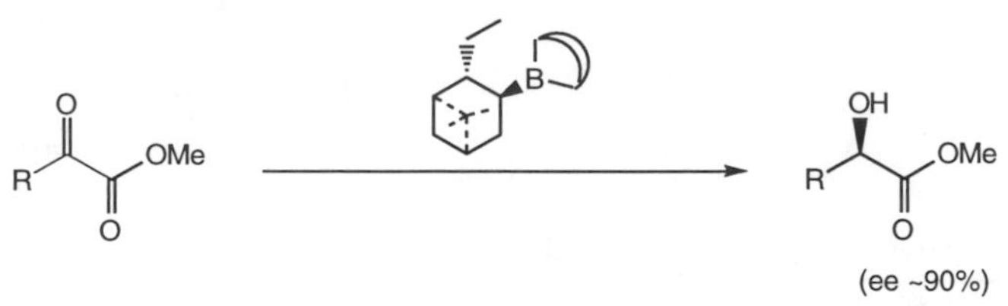

(ee ~90%)

SCHEME 6.13.

Of the many chiral modifiers used with metal hydride reductants, the sodium borohydride–L-tartaric acid system seems to give reasonable ee's (65 to 86%) for the reduction of α-keto acids and esters,[122] although other systems do show promise.[123,124]

The reduction of α-keto acid derivatives has also been the focus of a number of studies that mimic the NADH model.[125-135] These compounds can be reduced by biological methods with good asymmetric induction [*14.2*].[136-138]

A hydrosilylation approach with α-keto esters gave the α-hydroxyester and double asymmetric induction was observed.[139-141] The use of selectride with the ester derived from an inositol showed a reversal of diastereofacial selectivity on addition of 18-crown-6, allowing either alcohol to be accessed.[21]

In a 1,2-dicarbonyl system, one of the carbonyl groups can be protected; this allows for the introduction of asymmetry into the auxiliary unit (Scheme 6.14).[142-145]

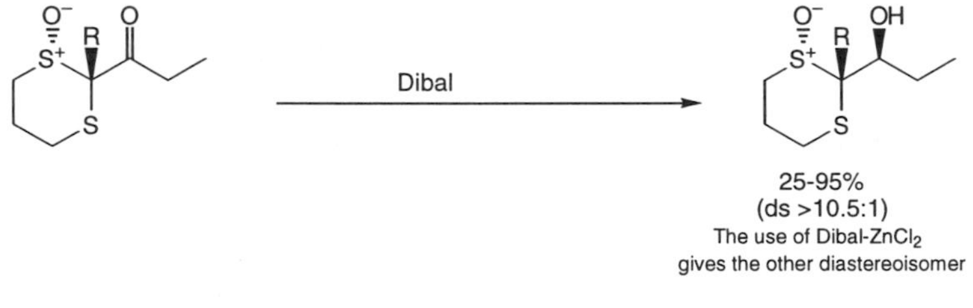

Dibal

25-95%
(ds >10.5:1)
The use of Dibal-$ZnCl_2$ gives the other diastereoisomer

SCHEME 6.14.

Alternatively, a group that allows for the choice between chelation or Anh-Felkin addition can also be incorporated [*2.1*]. The use of a dithiane-protected α-dicarbonyl system provides such a system and allows for good induction with an oxazaborolidine catalyst **6.4** (Scheme 6.15).[146]

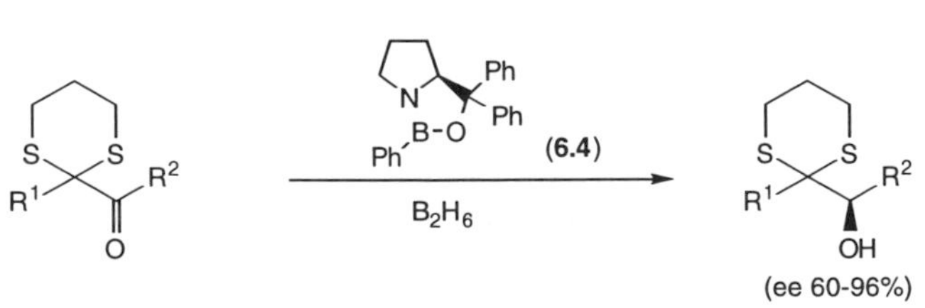

**(6.4)**
$B_2H_6$

(ee 60-96%)

SCHEME 6.15.

Reductive aminations of α-keto esters, especially by a transamination methodology, provides a route to α-amino acid derivatives, the face selectivity being derived from a cyclic intermediate.[147-149] The methodology can also be used to prepare protected α-amino aldehydes (Scheme 6.16).[150]

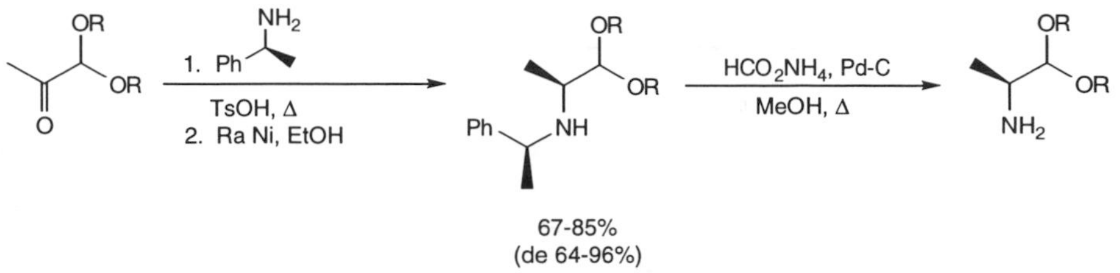
OR
OR
O
$NH_2$
1. Ph
TsOH, Δ
2. Ra Ni, EtOH
OR
OR
Ph
NH
67-85%
(de 64-96%)
$HCO_2NH_4$, Pd-C
MeOH, Δ
OR
OR
$NH_2$

SCHEME 6.16.

## 6.2.2. REDUCTIONS OF OTHER 1,2-DICARBONYL SYSTEMS

The reductions of conjugated systems, as with enones, are discussed in this section. However, as α,β-epoxy carbonyl chemistry falls between 1,2- and 1,3-substitution, it has been covered in the section on epoxide chemistry [*11.3.1.2*], as the resultant compound is also an epoxide.

Biological processes also provide a method for high induction with the reduction of α-functionalized carbonyl compounds [*14.2*].[136]

### 6.2.2.1. α-Hydroxy Carbonyl Compounds

The presence of an α-substituent that has lone pairs and can participate in the formation of a chelate allows for control over a reduction.[151] Thus, reductions of α-hydroxy ketones with zinc borohydride show preference for the *ancat* product (Scheme 6.17).[152] In contrast, use of a large protection group on the oxygen and Red-Al as reductant gives the *syncat* isomer as the preferred product (Scheme 6.18).[152]

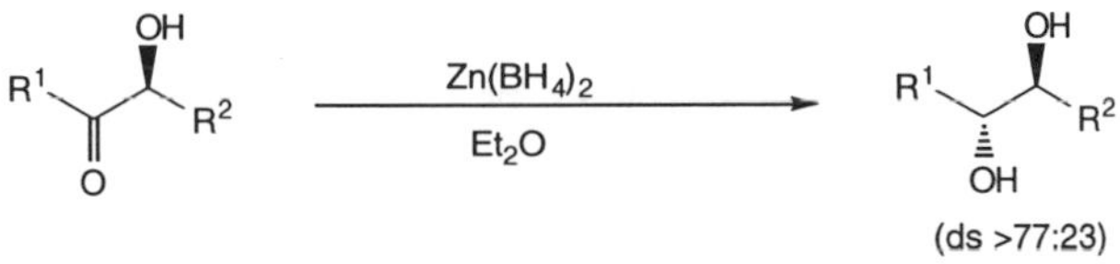
OH
$R^1$
$R^2$
O
$Zn(BH_4)_2$
$Et_2O$
OH
$R^1$
$R^2$
OH
(ds >77:23)

SCHEME 6.17.

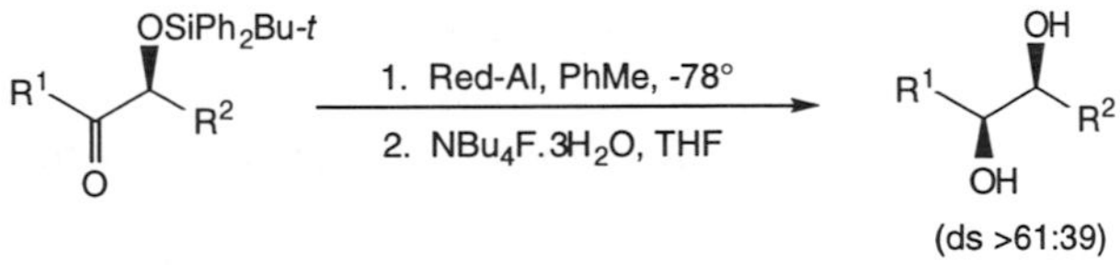
$OSiPh_2Bu$-*t*
$R^1$
$R^2$
O
1. Red-Al, PhMe, -78°
2. $NBu_4F.3H_2O$, THF
OH
$R^1$
$R^2$
OH
(ds >61:39)

SCHEME 6.18.

α-Alkoxy carbonyl compounds are reduced stereoselectively by the careful choice of a reductant and a protective group (Scheme 6.19).[153-156] The observed stereoselection correlates directly to Anh-Felkin or chelation control, as governed by the relevant variables.[157,158] The reactions shown in Scheme 6.19 also illustrate related reductions of β-functionalized carbonyl compounds.

Zn(BH)$_4$, Et$_2$O
(ds 25:1)

1. Zn(BH$_4$)$_2$, Et$_2$O
2. LiAlH$_4$, Et$_2$O
(ds 25:1)

1. TrCl, pyr
2. O$_3$, MeOH
3. Me$_2$S

1. TBDMSCl, imid, DMF
2. Red-Al, PhMe

Zn(BH$_4$)$_2$, Et$_2$O
(ds 25:1)

**SCHEME 6.19.**

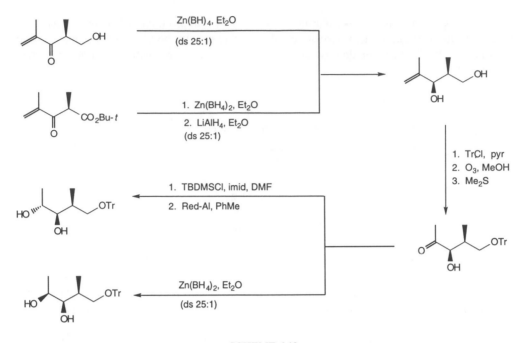

α-Alkoxy esters can be converted to either the *syncat* or *ancat* 1,2-diols through control of the reaction parameters (Scheme 6.20).[159]

1. Dibal, Et$_2$O
2. R$^2$MgX
72-87%
(ds >6:1)

1. R$^2$MgX, THF
2. LiBH$_4$
51-63%
(ds >11:1)

**SCHEME 6.20.**

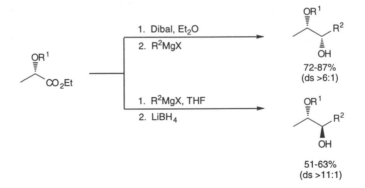

Imines, generated from nitriles, can also provide useful selection in the analogous reaction to Scheme 6.20.[160]

α-Hydroxy and amino carbonyl compounds are also reduced to the corresponding alcohols by hydrogenation in the presence of Ru-BINAP catalyst;[121] a rhodium catalyst did not give as high a degree of induction.[161]

For the reduction of α-hydroxy aryl ketones, substituents on the aryl group can influence the outcome of chelation controlled reductions.[158]

The carbonyl group can also be transformed prior to reduction, as illustrated by the formation of the aminoalcohol **6.5** (Scheme 6.21).[162-164]

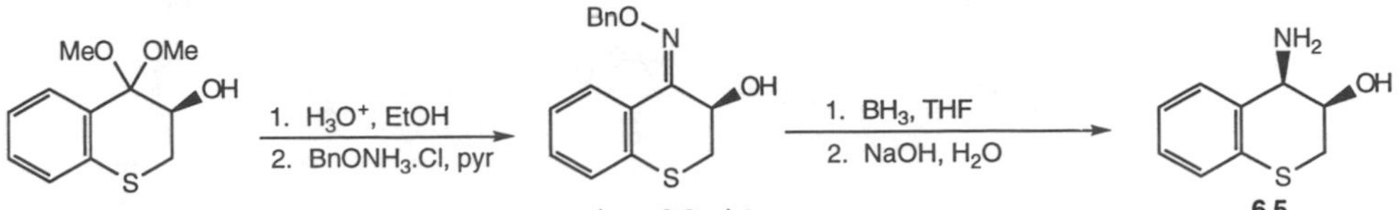

MeO OMe OH S — 1. $H_3O^+$, EtOH 2. $BnONH_3.Cl$, pyr → BnO N OH S (as a 3:2 mixture of isomers) — 1. $BH_3$, THF 2. NaOH, $H_2O$ → $NH_2$ OH S **6.5** 86% (ds 90:10)

**SCHEME 6.21**

### 6.2.2.2. α-Amino Carbonyl Compounds

The presence of a substituent α to the carbonyl group is not limited to oxygen; α-amino ketones can be reduced to α-amino alcohols (e.g., Scheme 6.22).[165-168]

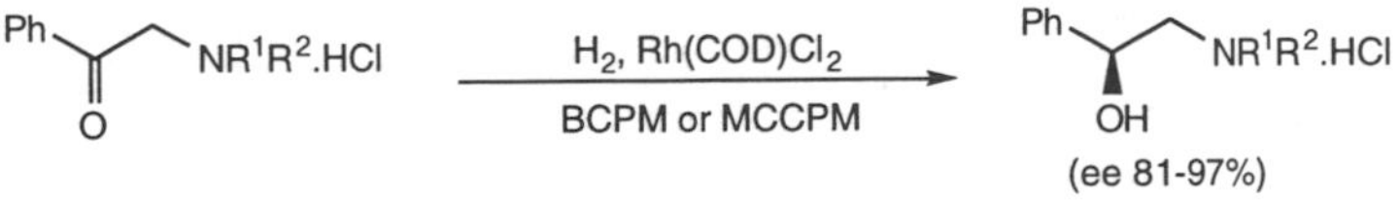

Ph $NR^1R^2$.HCl O — $H_2$, Rh(COD)$Cl_2$ BCPM or MCCPM → Ph $NR^1R^2$.HCl OH (ee 81-97%)

**SCHEME 6.22.**

Reduction of α-amino ketones with sodium borohydride proceeds with good face selectivity and no chelation control (Scheme 6.23).[169]

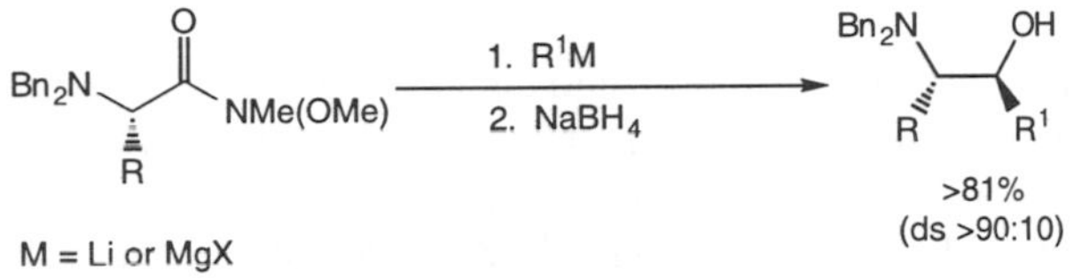

$Bn_2N$ O NMe(OMe) R — 1. $R^1M$ 2. $NaBH_4$ → $Bn_2N$ OH R $R^1$ >81% (ds >90:10)

M = Li or MgX

**SCHEME 6.23.**

This face selectivity allows for product control depending on whether a carbonyl reduction, for example of an α-amino ketone or nucleophilic addition, as with an α-amino aldehyde, is the chosen method, as the two often provide complimentary product stereochemistry.[167] An alternative is to use a different counterion; *syncat* selectivity is observed for lithium aluminum hydride,[170,171] especially in the presence of lithium iodide,[172,173] but is reversed in the presence of titanium.[174]

### 6.2.2.3. Other α-Functionalized Carbonyl Compounds

Similar selectivity to the reduction of α-hydroxy carbonyl compounds is observed for α-thio ketones; L-Selectride provides the *syncat* product while zinc borohydride gives the *ancat* isomer.[175] If the sulfur is converted to a sulfonium salt, the *ancat* product is formed independent of the nature of the metal hydride.[176]

Another substituent that has proven useful to invoke asymmetric transformations is a sulfoxide. The reduction of an α-keto sulfoxide can be affected by the use of a Lewis acid (e.g., Scheme 6.24),[177] and has been extended to acyclic examples.[178]

Thus, the use of α-keto sulfoxides — in some ways these are analogs of β-dicarbonyl systems — allows for control through chelation or Anh-Felkin delivery of a hydride (Schemes 6.24 and 6.25).[178-183]

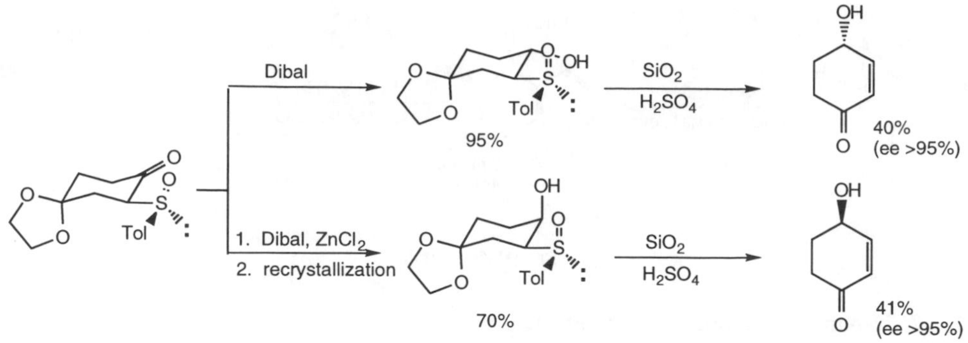

**SCHEME 6.24.**

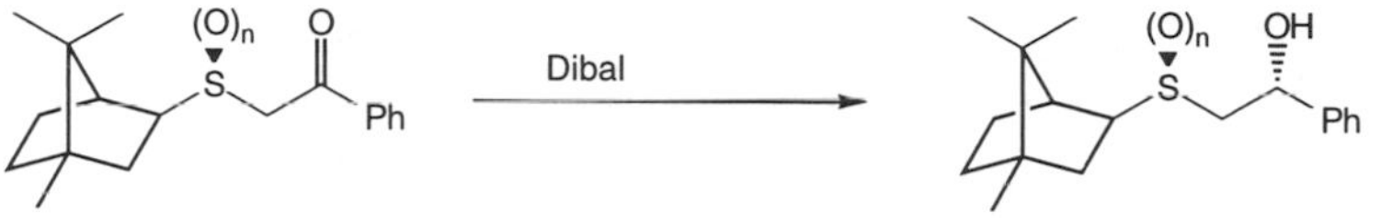

| | | ds |
|---|---|---|
| n = 0 | 90% | 75:25 |
| with $ZnCl_2$ | 83 | 85:15 |
| n = 1 | 91 | 75:25 |
| with $ZnCl_2$ | 99 | 0:100 |

**SCHEME 6.25.**

Of course, the 1,2-reduction of enones and ynones is also a regiochemical problem[184] that can be forced in the 1,2-direction by use of acetoxyborohydrides.[185] Some reagents that provide selectivity for 1,2-reductions of enones have been modified by use of chiral auxiliaries and do provide some induction.[2,3,186] A chiral reducing agent has been developed for the 1,2-reduction of enones; this methodology was paramount in a synthesis of ginkgolide.[187-189] The reduction is achieved with the oxazaborolidine catalysis of Corey. It proceeds in high yield and follows the same model as for "simple" carbonyl compounds [*4.5.5*] (Scheme 6.26).[188-195]

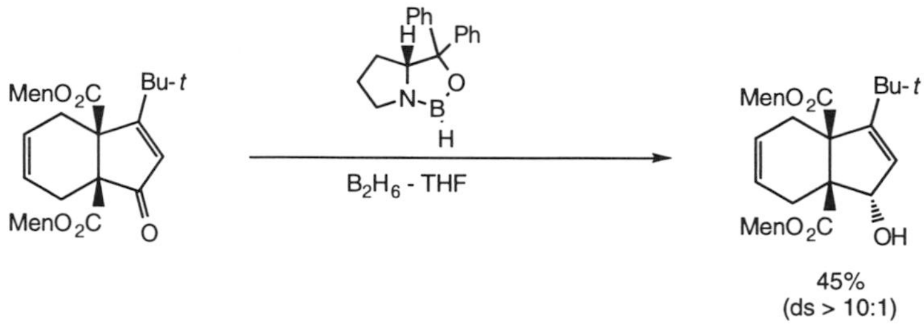

**SCHEME 6.26.**

Other functionalized carbonyl compounds, such as α-chloro ketones, are also effectively reduced by this system.[190,196,197]

The steric environment of the carbonyl group of an enone can influence stereoselectivity. For example, the use of a large sterically-demanding protecting group (**6.6**, R = *p*–$C_6H_5C_6H_4NHCO$) with a bulky borohydride reagent allowed for a high degree of selectivity.[198] However, higher selectivity is observed for a chiral reductant (Scheme 6.27);[199,200] complex formation with the hydroxy group (**6.6**; R = H) can also be used to effect stereoselection, but control over the absolute configuration at the reduced center is then lost.[201]

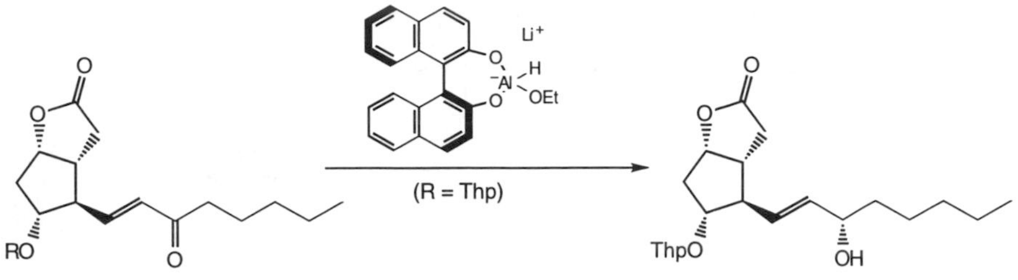

**SCHEME 6.27.**

A 1,2-addition can also be coupled with a 1,4-reduction. Thus, a conjugated enone can undergo 1,2-reduction followed by a hydroalumination to effect the 1,4-reduction (Scheme 6.28).[202-205]

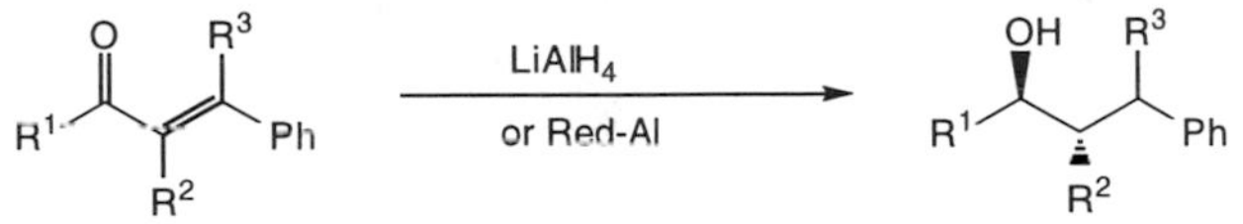

**SCHEME 6.28.**

A number of methods have been developed to provide asymmetric 1,2-reductions but none are as efficient as the Corey method (vide supra).[2,3,206] Regioselectivity can be accomplished for a hydrosilylation of a conjugated enone; use of a di- or trihydrosilane affords 1,2-reduction, while a monohydrosilane provides 1,4-reduction [*9.2.2*].[207] The use of a chiral catalyst does provide some degree of asymmetric induction for 1,2-reductions.[208]

The use of binaphthol-derived aluminum hydrides allows for high asymmetric induction in the reduction of ynones to propargyl alcohols.[200] Complexes derived from lithium aluminum hydride or boranes can also provide some useful induction.[23,155,209-212] Conjugated ynones can also be reduced selectively by biological agents [*14.2*].[213]

Other elements adjacent to carbonyl groups can also provide high selectivity for Anh-Felkin addition just due to their size, such as with silicon.[214]

# 6.3. ALKYLATIONS OF 1,3-FUNCTIONALIZED CARBONYL COMPOUNDS

## 6.3.1. 1,3-DICARBONYL ALKYLATIONS

The regioselective alkylation of β-dicarbonyl compounds can be controlled through the use of the mono- or dianion. The dianion usually undergoes reaction at the distal carbon atom,[215-222] while the monoanion provides substitution at the central carbon atom; diastereoselection is usually poor.[223-226] The enolate geometry is very dependent on substituents.[227] Phase transfer catalysis can also be used for these alkylations.[228,229] Asymmetric induction can be achieved through incorporation of a chiral auxiliary. Variations in the conditions used for the alkylation of the enamine derived from valine allows for facial selectivity; attack from the top face is preferred in toluene in the presence of hexamethylphosphoric triamide, whereas bottom face attack takes place in the presence of THF, dioxane, or an amine (Scheme 6.29).[230,231]

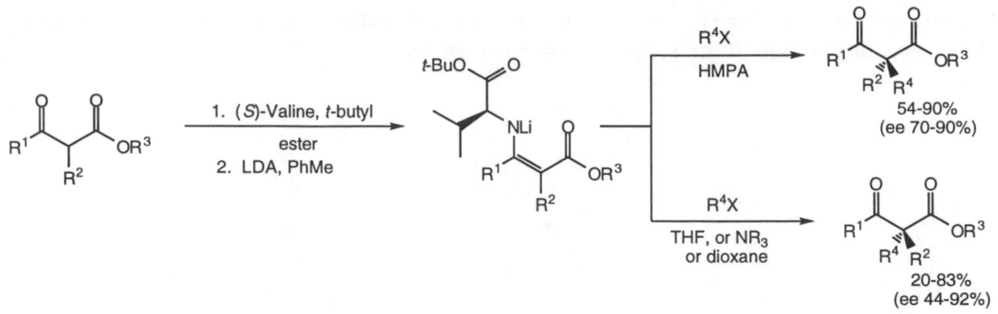

SCHEME 6.29.

An alternative strategy to chiral β-dicarbonyl systems is found through acylation of an enolate. The adducts can then undergo nucleophilic addition, including reduction (Scheme 6.30).[232]

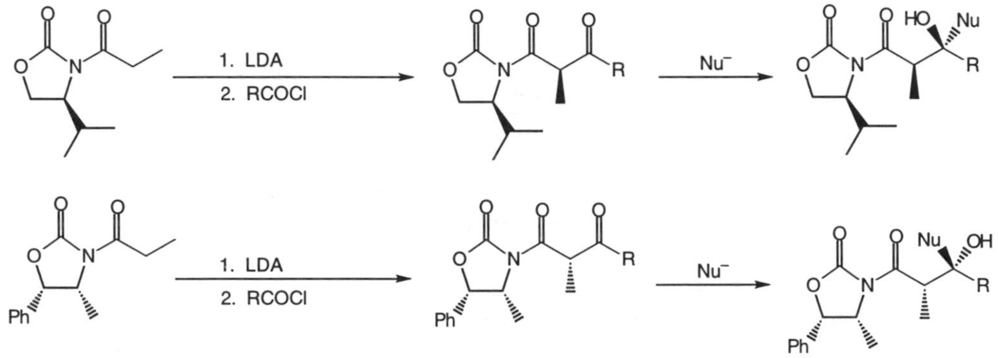

SCHEME 6.30.

The chirality can also be derived from the electrophilic species.[233]

Acetals of β-keto esters, as ketene acetals, also provide an opportunity for the incorporation of a chiral moiety that can control face selectivity for alkylations and reductions (Scheme 6.31). Acetals of β-dicarbonyl systems provide similar opportunities.[234]

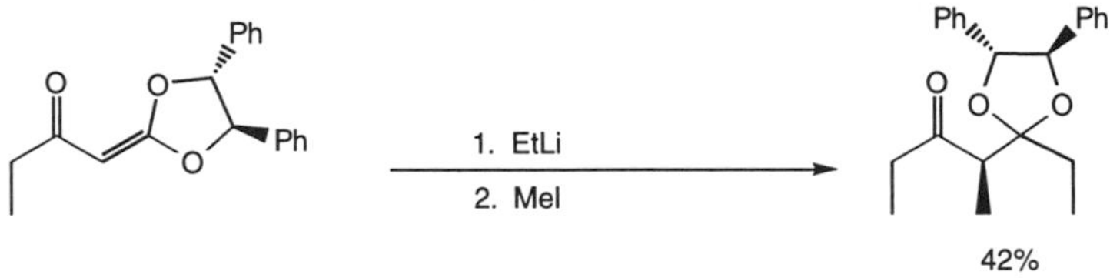

SCHEME 6.31.

The masking of one carbonyl group with a chiral group also can allow for facial selectivity during nucleophilic addition to the unprotected carbonyl group, and generation of a new stereogenic center.[235-237] Malonic acid derivatives have been successfully alkylated in an asymmetric manner through the incorporation of a chiral auxiliary in a monoester.[238]

Another alternative for the introduction of chirality is the use of a chiral alkylation agent, but induction is usually low.[239]

## 6.3.2. OTHER 1,3-FUNCTIONALIZED CARBONYL COMPOUNDS

The use of masked carbonyl functionality provides opportunities for both the incorporation of a chiral auxiliary and the use of umpolung chemistry.[240]

### 6.3.2.1. β-Hydroxy and Amino Carbonyl Compounds

β-Hydroxy esters can be alkylated with a high degree of selectivity;[241,242] β-alkoxy and hydroxy carbonyl compounds, available by an aldol protocol or yeast reduction, provide a good opportunity for asymmetric alkylations, as a chiral center is already present within the substrate (Scheme 6.32).[242-249] The chirality at the hydroxy stereogenic center is not compromised.[250] The affect is attributed to the chelation effect that shields one face of the enolate.[251-254] A cyclic substrate can also provide the means for face selectivity.[255,256] As with other enolate reactions, a wide variety of electrophiles can be employed to provide a wide range of highly functionalized products.[257,258]

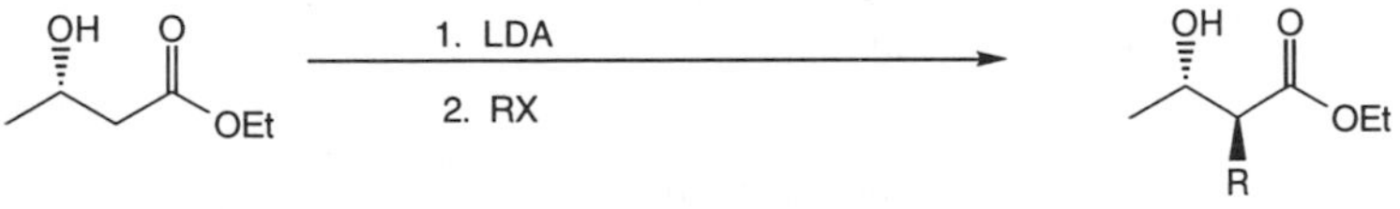

OH O OEt — 1. LDA / 2. RX → OH O OEt, R

SCHEME 6.32.

β-Hydroxy esters can be converted to the corresponding β-chloro esters with inversion of configuration by reaction with aluminum trichloride of the mesylate prepared from the alcohol; this reaction is also applicable to α-hydroxy esters.[259]

A silyl group can be used as a latent hydroxy group to provide the equivalent transformation [*9.4.1*].[260]

β-Amino esters can be alkylated with a wide degree of diastereoselection. The products can also be used to prepare a wide range of compounds.[96] The reactions of β-alkoxy carbonyl compounds with carbon nucleophiles has already been discussed in the context of Anh-Felkin control [*2.1*].[261-263]

### 6.3.2.2. Other β-Functionalized Carbonyl Compounds

Although not strictly functionalized carbonyl compounds, the anion chemistry of β-hydroxy sulfoxides is very similar to that of β-hydroxy ketones. Alkylation of these sulfoxides occurs with high *ancat*¶ selectivity (Scheme 6.33).[264]

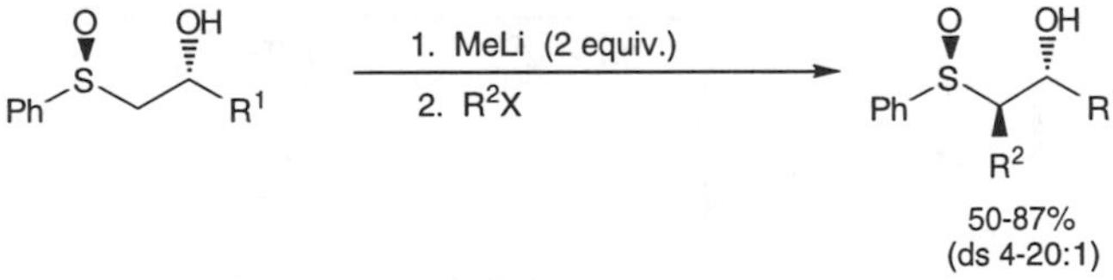

O OH Ph S $R^1$ — 1. MeLi (2 equiv.) / 2. $R^2X$ → O OH Ph S $R^1$, $R^2$

50-87%
(ds 4-20:1)

SCHEME 6.33.

Indeed, sulfoxides provide a convenient method for performing stereoselective alkylations and reductions (Schemes 6.24 and 6.33).[264-268] The products can be converted into a wide variety of other compounds including epoxides, β-hydroxy esters, and lactones.

Incorporation of a large group, as in a 3-silyl-alkyl-3-enal, can provide high facial selectivity for Anh-Felkin reactions at the carbonyl center; the vinylsilane is a masked carbonyl group, so this is an entry to β-hydroxy ketones (Scheme 6.34).[269,270]

¶ With respect to the hydroxy group.

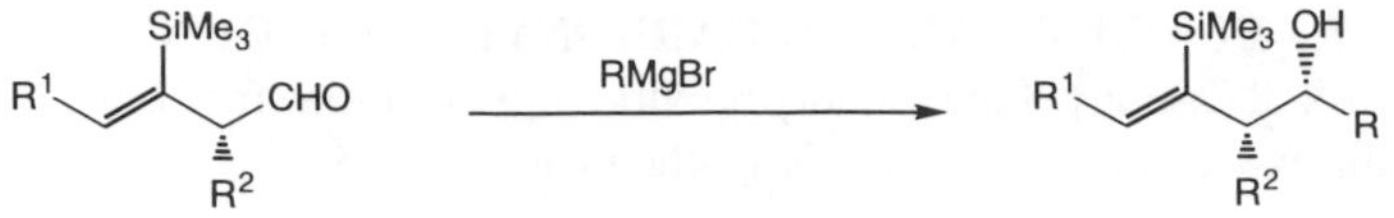

SCHEME 6.34.

The use of pyrrolidine chiral auxiliary allows for the asymmetric alkylation of α-cyanoacetic acid.[271]

## 6.4. REDUCTIONS OF 1,3-FUNCTIONALIZED CARBONYL COMPOUNDS

### 6.4.1. 1,3-DICARBONYL REDUCTIONS

β-Dicarbonyl compounds have been reduced selectively, allowing an alternative to stereoselective aldol reactions (Scheme 6.35).[272-275]

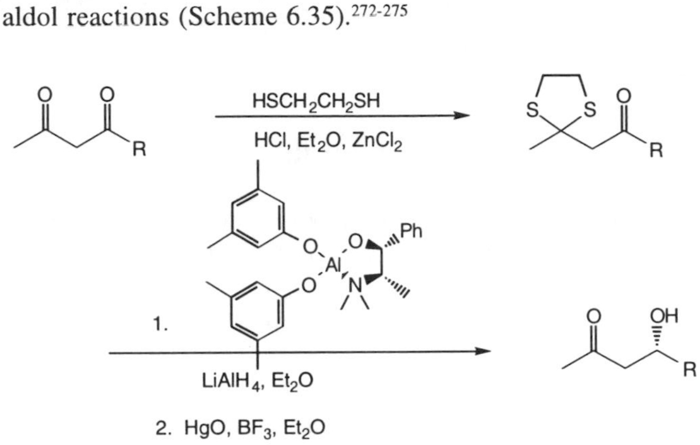

SCHEME 6.35.

β-Keto esters can also be reduced by an asymmetric hydrogenation.[121,276-281] This methodology is extremely powerful when a kinetic resolution occurs, and the β-keto ester can equilibrate (Scheme 6.36).[277,278,282-287]

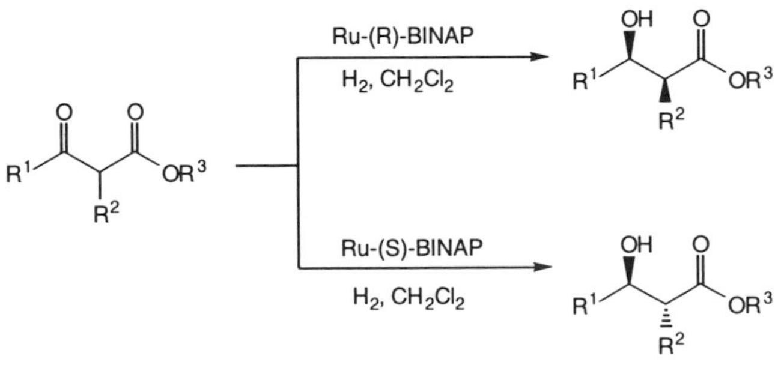

SCHEME 6.36.

The Ru(BINAP) system can also be used for the reduction of 1,3-diketones to *ancat* 1,3-diols.[121,280] The methodology has also been extended to the dioxohexenoate system, **6.7**. The *ancat* product can then be used for further transformations (Scheme 6.37).[288]

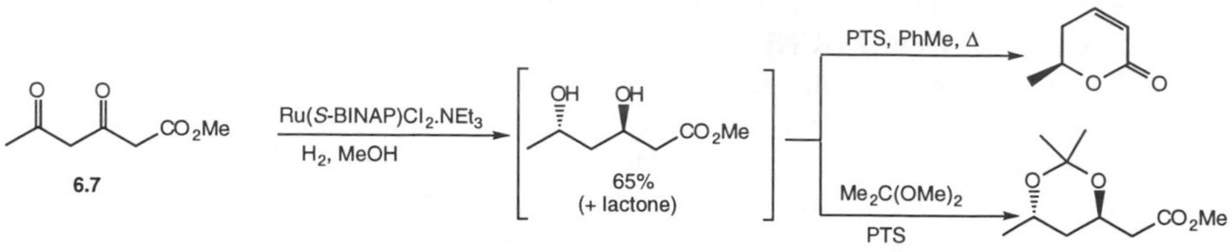

**SCHEME 6.37.**

In contrast, use of a hydrosilylation approach for the reduction of β-keto esters has usually led to a low degree of asymmetric induction.[116,140,289]

Reactions of β-keto esters with zinc borohydride provide the *ancat* product.[156,290-292]

β-Keto esters can also be reduced by modified metal hydrides — again a wide variety of ligands has been proposed but selectivity can still be variable.[2,3,273,275,293-295] The presence of a sulfur group at the α-position of a β-keto ester can provide the *syncat*[§] isomer on reduction with calcium borohydride.[274] For alkyl-substituted β-keto esters, a chiral sultam has been employed to control the new stereogenic center; the resultant stereochemistry is controlled by reagent choice (Scheme 6.38).[296]

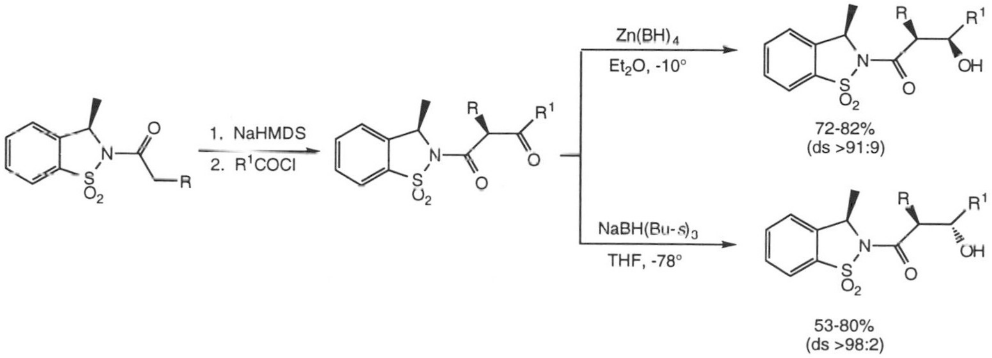

**SCHEME 6.38.**

Other derivatives of 1,3-dicarbonyl systems can be reduced, some by what can be considered a conjugate reduction [*9.5.1.2*]. Hydrogenation of enamines **6.8** with a heterogeneous catalyst provided the β-amino esters, **6.9**, with good stereoselection (Scheme 6.39).[293,297]

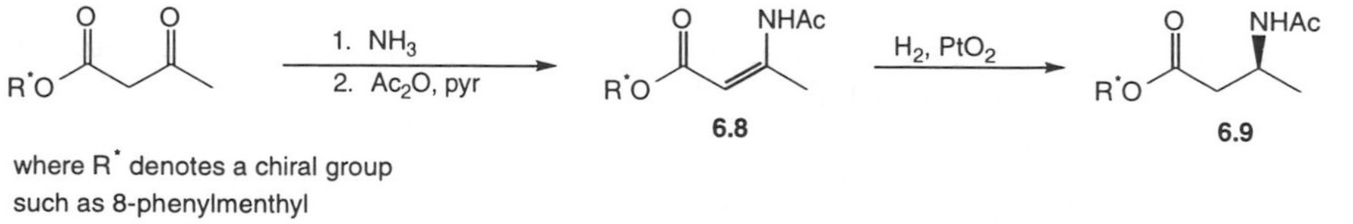

**SCHEME 6.39.**

Hydrosilylations of β-keto α-substituted amides by the use of a silane in the presence of a fluoride ion can provide the *ancat* hydroxy amide with high selectivity.[298]

Of course, biological methods are also available for the reduction of β-keto carbonyl systems [*14.2*].

§ With respect to the hydroxy and sulfur groups.

## 6.4.2. REDUCTIONS OF OTHER 1,3-FUNCTIONALIZED CARBONYL COMPOUNDS

Reactions of β-keto esters with zinc borohydride provide the *ancat* product (vide supra).[154,290] This has been expanded so that the *ancat* selectivity for the reduction of α,β-epoxy carbonyl compounds observed with sodium borohydride[299] is increased significantly through the use of zinc borohydride [*11.3.1.2*].[300] Reactions of α,β-epoxy carbonyl compounds are discussed in Chapter 11, including reductions at the carbonyl center.[§]

### 6.4.2.1. Reductions of β-Hydroxy and Amino Carbonyl Compounds

β-Amino and β-hydroxy ketones have been reduced selectively (Scheme 6.40),[301-305] (Scheme 6.41)[174,306] — the selectivity being much higher than when just lithium aluminum hydride was used.[170-173,307,308]

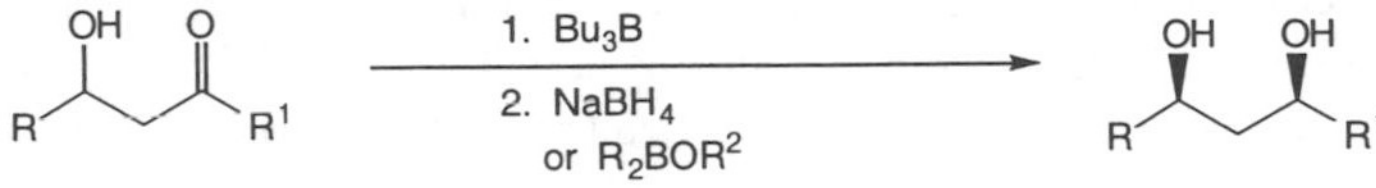

OH
O
R
R1
1. Bu3B
2. NaBH4
or R2BOR2
OH
OH
R
R1


SCHEME 6.40.

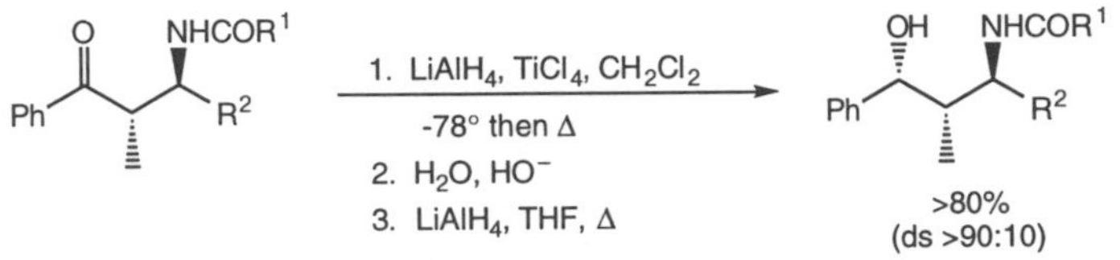

O
NHCOR1
Ph
R2
1. LiAlH4, TiCl4, CH2Cl2
-78° then Δ
2. H2O, HO−
3. LiAlH4, THF, Δ
OH
NHCOR1
Ph
R2
>80%
(ds >90:10)


SCHEME 6.41.

The *syn*-diols are available by a borane, Dibal, or tin hydride reduction,[302,304,309-311] but the size of an α-(C-2)-substituent can have a significant stereochemical effect.[312,313] Of course, these systems are ideal for the practice of a chelation-controlled reduction (Scheme 6.42).[172,173]

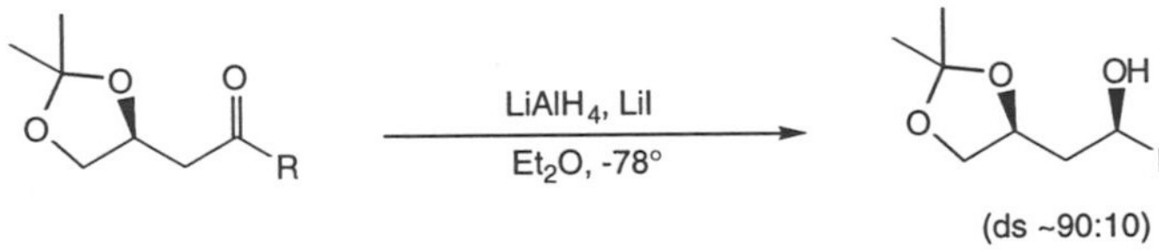

O
O
O
R
LiAlH4, LiI
Et2O, -78°
O
O
OH
R
(ds ~90:10)


SCHEME 6.42.

The mild reagent, tetramethylammonium triacetoxyborohydride, provides the *ancat*-diol with high selectivity (Scheme 6.43).[314,315]

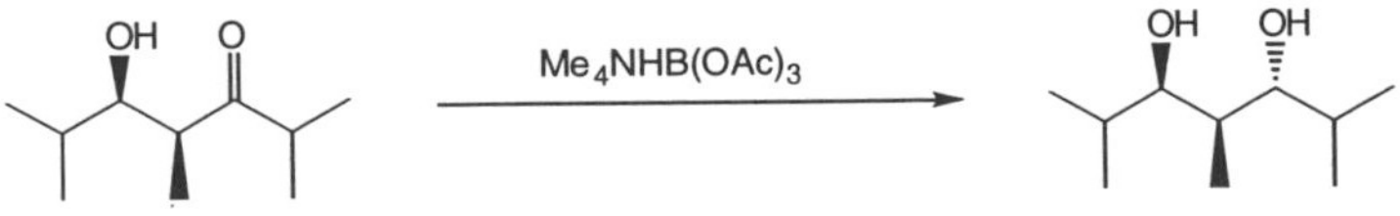

OH
O
Me4NHB(OAc)3
OH
OH


SCHEME 6.43.

§ This chapter deals with the reactions of epoxides. As the reduction of an epoxy carbonyl compound at the carbonyl center results in the formation of another functionalized epoxide, all of the material has been presented in this later chapter.

With silyl protection of an α-alkyl β-hydroxy ketone, lithium aluminum hydride provides the *ancat* product, while lithium triethylborohydride affords the *syncat* 1,3-diol.[157,312]

The reduction of β-amino ketones can also be controlled through the choice of reductant; lithium triethylborohydride provides the *ancat* product while zinc borohydride gives the *syncat* isomer.[303]

### 6.4.2.2. Other β-Functionalized Carbonyl Compounds

Other substituents can be incorporated into the substrate and allow Anh-Felkin addition or chelation control. A striking example is provided by the reduction of the vinylsilane, **6.10**, obtained by Lewis acid catalyzed rearrangement of the 2,3-epoxy alcohol **6.11**.[316,317] The use of lithium triethylborohydride affords excellent selectivity through Anh-Felkin addition that is augmented by the large silyl group, whereas Dibal resulted in poor stereoselectivity (Scheme 6.44).[312]

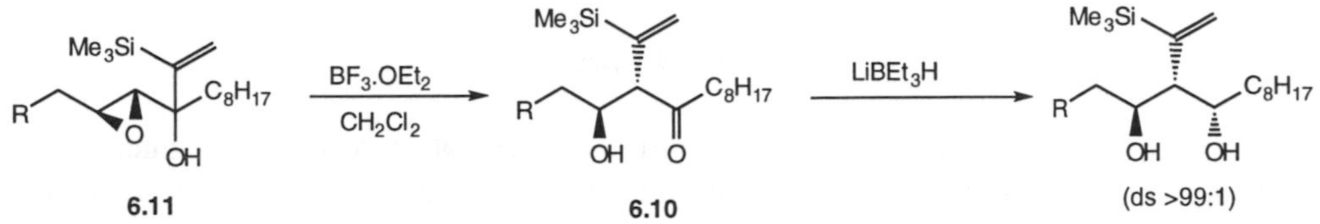

SCHEME 6.44.

A similar increase in selectivity is observed by the introduction of a silyl group on to the cyclopropyl ketones **6.12** (Scheme 6.45).[318]

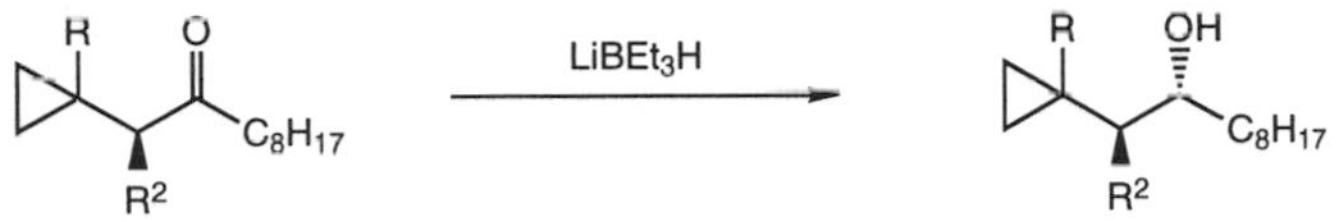

|  |  |  |
|---|---|---|
| $R^2 = Me$ | R = H | ds 78:22 |
| | TMS | >99:1 |
| $R^2 = CH_2OH$ | R = H | ds 60:40 |
| | TMS | >99:1 |
| $R^2 = CH(OH)CH_2OBn$ | R = H | ds 86:14 |
| | TMS | >99:1 |

SCHEME 6.45.

Reduction of α-methyl-β,γ-unsaturated ketones by L-Selectride provided the *ancat* isomer with a high degree of selectivity (Scheme 6.46).[319]

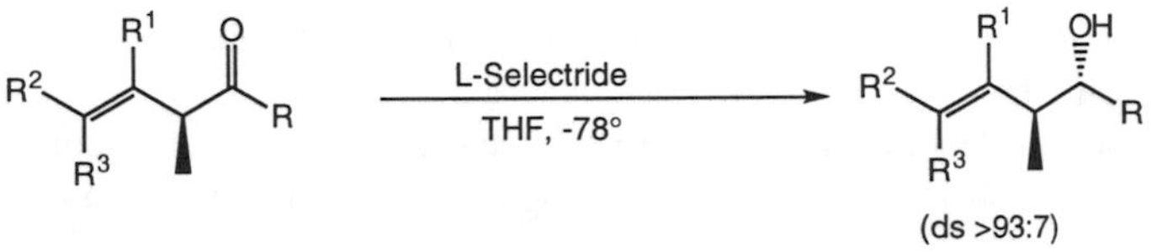

SCHEME 6.46.

## 6.5. OTHER SYSTEMS

Of course, as the two functional groups of functionalized carbonyl compounds become further apart, their interactions become fewer, and they tend to react independently. However, the formation of cyclic complexes, especially with organometallic reagents, can still occur, but control of facial selectivity becomes increasingly more difficult.[158,320,321]

Polyfunctional compounds[1,322,323] can often be alkylated stereoselectively through the use of chelation and face selectivity,[324,325] as, for example, with the enolates of tartaric acid and 2-hydroxy-1,4-diesters.[48,49] In a similar vein, γ-keto acids are available for an alkylation protocol (Scheme 6.47).[326,327] A second alkyl group can also be introduced into the amide **6.13**; again, the second alkyl group is introduced from the bottom face.[328,329] Other chiral auxiliaries have been used to bring about these regio- and stereoselective alkylations.[330-332]

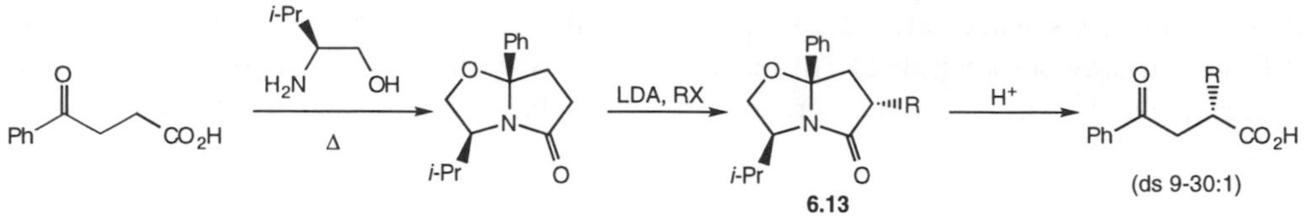

i-Pr, $H_2N$, OH, Δ; Ph; LDA, RX; Ph, R; 6.13; $H^+$; O, R, Ph, $CO_2H$; (ds 9-30:1)

**SCHEME 6.47.**

Treatment of the prochiral 3-methylglutaric acid derivative **6.14** with a nucleophile shows good regioselectivity with secondary amines that amounts to asymmetric induction (Scheme 6.48).[333]

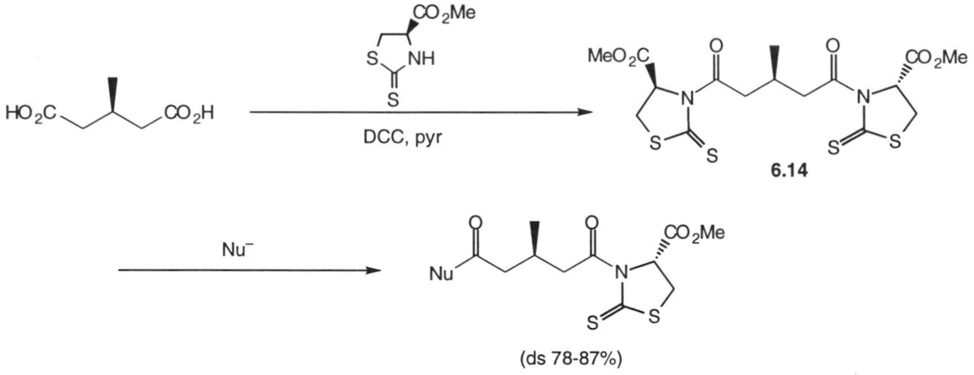

$HO_2C$, $CO_2H$; $CO_2Me$, S, NH, S; DCC, pyr; $MeO_2C$, $CO_2Me$; 6.14; $Nu^-$; Nu, $CO_2Me$; (ds 78-87%)

**SCHEME 6.48.**

### 6.5.1. REACTIONS OF ANHYDRIDES AND RELATED SYSTEMS

Anhydrides can undergo asymmetric ring opening reactions;[334] many examples are catalyzed by biological systems [*14.1*]. They can function as electrophiles to afford ω-keto acids after reaction with the organometallic species.[335]

Cyclic anhydrides can be reduced with moderate asymmetric induction by a chiral ruthenium(II) complex.[336]

The addition of a nucleophile to a lactol can be influenced by a substituent on the cyclic system [*14.1*] (e.g., Scheme 6.49). The selectivity can be interpreted in terms of a cyclic chelate.[337]

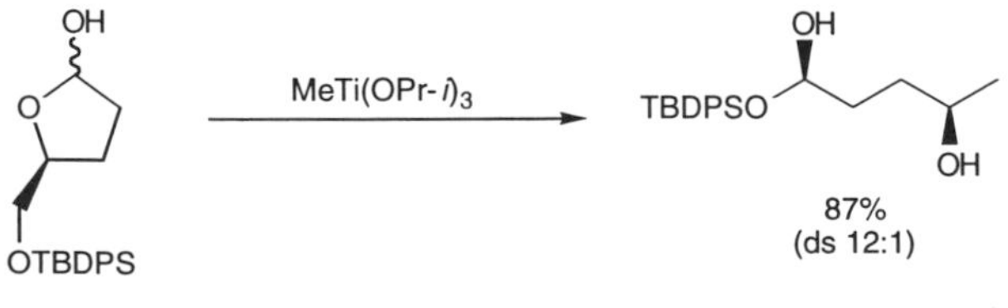

OH, O, OTBDPS; MeTi(OPr-i)$_3$; TBDPSO, OH, OH; 87% (ds 12:1)

**SCHEME 6.49.**

## 6.6. SUMMARY

Functionalized carbonyl compounds allow for a high degree of control over the stereochemistry of a nucleophilic addition, alkylation, or reduction. For systems where the additional functional group also imparts stability to an enolate, as with β-dicarbonyl compounds, the resultant anion is relatively stable, and this reduces many of the problems associated with the scale-up of "simple" enolate reactions.

The stereochemical outcome of most reactions can be predicted, which makes the reactions in this chapter very powerful. If plausible, the use of a functionalized carbonyl compound should be used in a synthesis design, as the advantages invariably outweigh any extra steps. Normal process development will show if use of a simpler system is going to be feasible.

The wide range of reactions available has only allowed a brief survey in this chapter; the reader should consult the primary literature to ensure that precedence exists for the specific system of interest before embarking on a synthesis.

## 6.7. REFERENCES

1. Evans, D. A. *Aldrichimica Acta* 1982, *15*, 23.
2. ApSimon, J. W.; Seguin, R. P. *Tetrahedron* 1979, *35*, 2797.
3. ApSimon, J. W.; Collier, T. E. *Tetrahedron* 1986, *42*, 5157.
4. Kabuto, K.; Ziffer, H. *J. Org. Chem.* 1975, *40*, 3467.
5. Prelog, V. *Helv. Chim. Acta* 1953, *36*, 308.
6. Trost, B. M.; Mallart, S. *Tetrahedron Lett.* 1993, *34*, 8025.
7. Soai, K., Ishizaki, M. *J. Chem. Soc., Chem. Commun.* 1984, 1016.
8. Tamura, Y.; Ko, T.; Kondo, H.; Annoura, H.; Fuji, M.; Takeuchi, R.; Fujioka, H. *Tetrahedron Lett.* 1986, *27*, 2117.
9. Heitz, M. P.; Gellibert, F.; Mioskowski, C. *Tetrahedron Lett.* 1986, *27*, 3859.
10. Tanaka, N.; Suemune, H.; Sakai, K. *Tetrahedron: Asymmetry* 1992, *3*, 1075.
11. Maglioli, P.; De Lucchi, O.; Delogu, G.; Valle, G. *Tetrahedron: Asymmetry* 1992, *3*, 365.
12. Page, P. C. B.; Prodger, J. C.; Westwood, D. *Tetrahedron* 1993, *49*, 10355.
13. Alexakis, A.; Lensen, N.; Mangeney, P. *Tetrahedron Lett.* 1991, *32*, 1171.
14. Thiam, M.; Chastrette, F. *Tetrahedron Lett.* 1990, *31*, 1429.
15. Huet, F.; Pellet, M.; Conia, J. M. *Synthesis* 1979, 33.
16. Bates, G. S.; Ramaswamy, S. *Can. J. Chem.* 1980, *58*, 716.
17. Whitesell, J. K.; Bhattacharya, A.; Henke, K. *J. Chem. Soc., Chem. Commun.* 1982, 988.
18. Whitesell, J. K.; Deyo, D.; Bhattacharya, A. *J. Chem. Soc., Chem. Commun.* 1983, 802.
19. Yamamoto, Y.; Maeda, N.; Maruyama, K. *J. Chem. Soc., Chem. Commun.* 1983, 774.
20. Boireau, G.; Deberly, A.; Abenhaim, D. *Tetrahedron Lett.* 1988, *29*, 2175.
21. Akiyama, T.; Nishimoto, H.; Ozaki, S. *Tetrahedron Lett.* 1991, *32*, 1335.
22. Hamon, D. P. G.; Holman, J. W.; Massy-Westropp, R. A. *Tetrahedron* 1993, *49*, 9593.
23. Brown, H. C.; Ramachandran, P. V.; Weissman, S. A.; Swaminathan, S. *J. Org. Chem.* 1990, *55*, 6328.
24. Ojima, I.; Miyazawa, Y.; Kumagai, M. *J. Chem. Soc., Chem. Commun.* 1976, 927.
25. Whitesell, J. K.; Nabona, K.; Deyo, D. *J. Org. Chem.* 1989, *54*, 2258.
26. Yamamoto, Y.; Komatsu, T.; Maruyama, K. *J. Chem. Soc., Chem. Commun.* 1983, 191.
27. Yamamoto, Y.; Yatagai, H.; Ishihara, Y.; Maeda, N.; Maruyama, K. *Tetrahedron* 1984, *40*, 2239.
28. Aoki, S.; Mikami, K.; Terada, M.; Nakai, T. *Tetrahedron* 1993, *49*, 1783.
29. Boireau, G.; Deberly, A.; Abenhaim, D. *Tetrahedron* 1989, *45*, 5837.
30. Abenhaim, D.; Boireau, G.; Sabourault, B. *Tetrahedron Lett.* 1980, *21*, 3043.
31. Mukaiyama, T.; Sakito, Y.; Asami, M. *Chem. Lett.* 1979, 705.
32. Meyers, A. I.; Slade, J. *J. Org. Chem.* 1980, *45*, 2785.
33. Meyers, A. I.; Slade, J. *Synth. Commun.* 1976, *6*, 601.
34. Enders, D.; Dyker, H.; Raabe, G. *Angew. Chem., Int. Ed. Engl.* 1992, *31*, 618.
35. Utimoto, K.; Obayashi, M.; Nozaki, H. *J. Org. Chem.* 1976, *41*, 2940.
36. Gilday, J. P.; Gallucci, J. C.; Paquette, L. A. *J. Org. Chem.* 1989, *54*, 1399.

37. Cardillo, G.; Orena, M.; Romero, M.; Sandri, S. *Tetrahedron* 1989, *45*, 1501.
38. Underwood, R.; Fraser-Reid, B. *J. Chem. Soc., Perkin Trans. 1* 1990, 731.
39. Stolz, F.; Strazewski, P.; Tamm, C.; Neuberger, M.; Zehnder, M. *Angew. Chem., Int. Ed. Engl.* 1992, *31*, 193.
40. Castellino, S.; Volk, D. E. *Tetrahedron Lett.* 1993, *34*, 967.
41. Kelly, T. R.; Arvanitis, A. *Tetrahedron Lett.* 1984, *25*, 39.
42. Ludwig, J. W.; Newcomb, M.; Bergbreiter, D. E. *Tetrahedron Lett.* 1986, *27*, 2731.
43. Pearson, W. H.; Cheng, M.-C. *J. Org. Chem.* 1986, *51*, 3746.
44. Po, S.-Y.; Lui, H.-H.; Uang, B.-J. *Tetrahedron: Asymmetry* 1990, *1*, 143.
45. Liu, H.-H.; Chen, E.-N.; Uang, B.-J.; Wang, S.-L. *Tetrahedron Lett.* 1990, *31*, 257.
46. Dondoni, A.; Franco, S.; Merchan, F. L.; Merino, P. *Tetrahedron Lett.* 1993, *34*, 5475.
47. Enders, D.; Jegelka, U. *Tetrahedron Lett.* 1993, *34*, 2453.
48. Naef, R.; Seebach, D. *Angew. Chem., Int. Ed. Engl.* 1981, *20*, 1030.
49. Seebach, D.; Aebi, J.; Wasmuth, D. *Org. Synth.* 1984, *63*, 109.
50. Brown, J. D. *Tetrahedron: Asymmetry* 1992, *3*, 1551.
51. Raczko, J.; Golebiowski, A.; Krajewski, J. W.; Glunzinski, P.; Jurczak, J. *Tetrahedron Lett.* 1990, *31*, 3797.
52. Reetz, M. T.; Drewes, M. W.; Schmitz, A. *Angew. Chem., Int. Ed. Engl.* 1987, *26*, 1141.
53. Reetz, M. T.; Reif, W.; Holdgrün, X. *Heterocycles* 1989, *28*, 707.
54. Heneghan, M.; Procter, G. *Synlett* 1992, 489.
55. Audoye, P.; Gaset, A.; Lattes, A. *J. Appl. Chem. Biotech.* 1975, *25*, 19.
56. Clark, R. D.; Jahangir; Souchet, M.; Kern, J. R. *J. Chem. Soc., Chem. Commun.* 1989, 930.
57. Clark, R. D.; Souchet, M. *Tetrahedron Lett.* 1990, *31*, 193.
58. Brussee, J.; Dofferhoff, F.; Kruse, C. G.; Gen, A. V. D. *Tetrahedron* 1990, *46*, 1653.
59. Williams, R. M.; Hendrix, J. A. *J. Org. Chem.* 1990, *55*, 3703.
60. Polt, R.; Peterson, M. A. *Tetrahedron Lett.* 1990, *31*, 4985.
61. Genet, J. P.; Kopola, N.; Juge, S.; Ruiz-Montes, J.; Antunes, O. A. C.; Tanier, S. *Tetrahedron Lett.* 1990, *31*, 3133.
62. O'Donnell, M. J.; Bennett, W. D.; Wu, S. *J. Am. Chem. Soc.* 1989, *111*, 2353.
63. O'Donnell, M. J.; Wu, S. *Tetrahedron: Asymmetry* 1992, *3*, 591.
64. O'Donnell, M. J.; Bennett, W. D.; Jacobsen, W. N.; Ma, Y.-A.; Huffman, J. C. *Tetrahedron Lett.* 1989, *30*, 3909.
65. O'Donnell, M. J.; Bennett, W. D.; Jacobsen, W. N.; Ma, Y.-A. *Tetrahedron Lett.* 1989, *30*, 3913.
66. Kanemasa, S.; Uchida, O.; Wada, E. *J. Org. Chem.* 1990, *55*, 4411.
67. Hamon, D. P. G.; Massy-Westropp, R. A.; Razzino, P. *Tetrahedron* 1993, *49*, 6419.
68. Yamada, S.-I.; Ogurai, T.; Shioiri, T. *J. Chem. Soc., Chem. Commun.* 1976, 136.
69. El Achqar, A.; Roumestant, M.-L.; Viallefont, P. *Tetrahedron Lett.* 1988, *29*, 2441.
70. Aiqiao, M.; Jing, W.; Yuanwei, C.; Guishu, Y.; Yaozhong, J. *Synth. Commun.* 1989, *19*, 3337.
71. Yaozhong, J.; Changyou, Z.; Huri, P. *Synth. Commun.* 1989, *19*, 881.
72. Solladié-Cavallo, A.; Simon, M. C. *Tetrahedron Lett.* 1989, *30*, 6011.
73. Yuanwei, C.; Aiqiao, M.; Xun, X.; Yaozhong, J. *Synth. Commun.* 1989, *19*, 1423.
74. McIntosh, J. M.; Mishra, P. *Can. J. Chem.* 1986, *64*, 726.
75. Yaozhong, J.; Jingen, D.; Wenhao, H.; Giulan, L.; Aiqiao, M. *Synth. Commun.* 1990, *20*, 3077.
76. Yaozhong, J.; Guilan, L.; Changyou, Z.; Huri, P.; Lanjun, W.; Aiqiao, M. *Synth. Commun.* 1991, *21*, 1087.
77. Oppolzer, W.; Moretti, R.; Thomi, S. *Tetrahedron Lett.* 1989, *30*, 6009.
78. Yaozhong, J.; Peng, G.; Guilan, L. *Synth. Commun.* 1990, *20*, 15.
79. Vedejs, E.; Fields, S. C.; Schrimpf, M. R. *J. Am. Chem. Soc.* 1993, *115*, 11612.
80. Seebach, D.; Bossler, H.; Gründler, H.; Shoda, S.-I. *Helv. Chim. Acta* 1991, *74*, 197.
81. Calmes, M.; Daunis, J.; Ismaili, H.; Jacquier, R.; Koudou, J.; Nkusi, G.; Zouanate, A. *Tetrahedron* 1990, *46*, 6021.
82. Ojima, I.; Komata, T.; Qiu, X. *J. Am. Chem. Soc.* 1990, *112*, 770.
83. Williams, R. M.; Zhai, W. *Tetrahedron* 1988, *44*, 5425.
84. Zhai, D.; Zhai, W.; Williams, R. M. *J. Am. Chem. Soc.* 1988, *110*, 2501.
85. Williams, R. M.; Sinclair, P. J.; Zhai, D.; Chen, D. *J. Am. Chem. Soc.* 1988, *110*, 1547.
86. Williams, R. M.; Sinclair, P. J.; Zhai, W. *J. Am. Chem. Soc.* 1988, *110*, 482.
87. Sinclair, P. J.; Zhai, D.; Reibenspies, J.; Williams, R. M. *J. Am. Chem. Soc.* 1986, *108*, 1103.
88. Dellaria, J. F.; Santarsiero, B. D. *J. Org. Chem.* 1989, *54*, 3916.
89. Baldwin, J. E.; Lee, V.; Schofield, C. J. *Synlett* 1992, 249.
90. Fiaud, J.-C.; Kagan, H. B. *Tetrahedron Lett.* 1971, 1019.
91. Fiaud, J.-C.; Kagan, H. B. *Tetrahedron Lett.* 1970, 1813.
92. Tanaka, H.; Inoue, K.; Pokorski, U.; Taniguchi, M.; Torii, S. *Tetrahedron Lett.* 1990, *31*, 3023.
93. McIntosh, J. M.; Thangarasa, R.; Ager, D. J.; Zhi, B. *Tetrahedron* 1992, *48*, 6219.
94. McIntosh, J. M.; Thangarasa, R.; Foley, N.; Ager, D. J.; Froen, D. E.; Klix, R. C. *Tetrahedron* 1994, *50*, 1967.

95. Ager, D. J.; Froen, D. E.; Klix, R. C.; Zhi, B.; McIntosh, J. M.; Thangarasa, R. *Tetrahedron* 1994, *50*, 1975.
96. Estermann, H.; Seebach, D. *Helv. Chim. Acta* 1988, *71*, 1824.
97. Harding, K. E.; Davis, C. S. *Tetrahedron Lett.* 1988, *29*, 1891.
98. Zydowsky, T. M.; de Lara, E.; Spanton, S. G. *J. Org. Chem.* 1990, *55*, 5437.
99. Reetz, M. T.; Jaeger, R.; Drewlies, R.; Hubel, M. *Angew. Chem., Int. Ed. Engl.* 1991, *30*, 103.
100. Matsumoto, T.; Kobayashi, Y.; Takemoto, Y.; Ito, Y.; Kamijo, T.; Harada, H.; Terashima, S. *Tetrahedron Lett.* 1990, *31*, 4175.
101. Hayashi, M.; Kaneko, T.; Oguni, N. *J. Chem. Soc., Perkin Trans. 1* 1991, 25.
102. Hoffmann, R. W. *Angew. Chem., Int. Ed. Engl.* 1982, *21*, 555.
103. Zair, T.; Santelli-Rouvier, C.; Santelli, M. *J. Org. Chem.* 1993, *58*, 2686.
104. Page, P. C. B.; Westwood, D.; Slawin, A. M. Z.; Williams, D. J. *J. Chem. Soc., Perkin Trans. 1* 1989, 1158.
105. Page, P. C. B.; Slawin, A. M. Z.; Westwood, D.; Williams, D. J. *J. Chem. Soc., Perkin Trans. 1* 1989, 185.
106. Page, P. C. B.; Klair, S. S.; Westwood, D. *J. Chem. Soc., Perkin Trans. 1* 1989, 2441.
107. Pyne, S. G.; Boche, G. *J. Org. Chem.* 1989, *54*, 2663.
108. Fujisawa, T.; Takemura, I.; Ukaji, Y. *Tetrahedron Lett.* 1990, *31*, 5479.
109. Fleming, I.; Lee, T. V. *Tetrahedron Lett.* 1981, *22*, 705.
110. Stork, G.; Benaim, J. *J. Am. Chem. Soc.* 1971, *93*, 5938.
111. Chauvin, R. *Tetrahedron: Asymmetry* 1990, *1*, 737.
112. Brown, H. C.; Pai, G. G.; Jadhav, P. K. *J. Am. Chem. Soc.* 1984, *106*, 1531.
113. Flynn, G. A.; Beight, D. W. *Tetrahedron Lett.* 1988, *29*, 423.
114. Hamon, D. P. G.; Holman, J. W.; Massy-Westropp, R. A. *Tetrahedron: Asymmetry* 1992, *3*, 1533.
115. Hatat, C.; Kokel, N.; Mortreux, A.; Petit, F. *Tetrahedron Lett.* 1990, *31*, 4139.
116. Ojima, I.; Kogure, T.; Achiwa, K. *J. Chem. Soc., Chem. Commun.* 1977, 428.
117. Garland, M.; Blaser, H.-U. *J. Am. Chem. Soc.* 1990, *112*, 7048.
118. Ojima, I.; Kogure, T. *J. Organometal. Chem.* 1980, *195*, 239.
119. Inoguchi, K.; Saruraba, S.; Achiwa, K. *Synlett* 1992, 169.
120. Augustine, R. L.; Tanielyan, S. K.; Doyle, L. K. *Tetrahedron: Asymmetry* 1993, *4*, 1803.
121. Kitamura, M.; Ohkuma, T.; Inoue, S.; Sayo, N.; Kumobayashi, H.; Akutagawa, S.; Ohta, T.; Takaya, H.; Noyori, R. *J. Am. Chem. Soc.* 1988, *110*, 629.
122. Yatagai, Y.; Ohnuki, T. *J. Chem. Soc., Chem. Commun.* 1990, 1826.
123. Gamez, P.; Fache, F.; Mangeney, P.; Lemaire, M. *Tetrahedron Lett.* 1993, *34*, 6897.
124. Xiang, Y. B.; Snow, K.; Belley, M. *J. Org. Chem.* 1993, *58*, 993.
125. Baba, N.; Oda, J.; Inouye, Y. *J. Chem. Soc., Chem. Commun.* 1980, 815.
126. Davies, S. G.; Skerlj, R. T.; Whittaker, M. *Tetrahedron: Asymmetry* 1990, *1*, 725.
127. Burgess, V. A.; Davies, S. G.; Skerlj, R. T. *J. Chem. Soc., Chem. Commun.* 1990, 1759.
128. Davies, S. G.; Skelj, R. T.; Whittaker, M. *Tetrahedron Lett.* 1990, *31*, 3213.
129. Nishiyama, K.; Baba, N.; Oda, J.; Inouye, Y. *J. Chem. Soc., Chem. Commun.* 1976, 101.
130. Ohno, A.; Yamamoto, H.; Kimura, T.; Oka, S.; Ohnishi, Y. *Tetrahedron Lett.* 1976, 4585.
131. Ohnishi, Y.; Kagami, M.; Ohno, A. *J. Am. Chem. Soc.* 1975, *97*, 4766.
132. Versleijen, J. P.; Sanders-Hovens, M. S.; Vanhommerig, S. A.; Vekemans, J. A.; Meijer, E. M. *Tetrahedron* 1993, *49*, 7793.
133. Beckett, R. P.; Burgess, V. A.; Davies, S. G.; Whittaker, M. *Tetrahedron Lett.* 1993, *34*, 3617.
134. Burgess, V. A.; Davies, S. G.; Skerlj, R. T.; Whittaker, M. *Tetrahedron: Asymmetry* 1992, *3*, 871.
135. Combret, Y.; Duflos, J.; Dupas, G.; Bourguignon, J.; Quéguiner, G. *Tetrahedron* 1993, *49*, 5237.
136. Ridley, D. D.; Stralow, M. *J. Chem. Soc., Chem. Commun.* 1975, 400.
137. Casy, G.; Lee, T. V.; Lovell, H. *Tetrahedron Lett.* 1992, *33*, 817.
138. Pirrung, M. C.; Holmes, C. P.; Horowitz, D. M.; Nunn, D. S. *J. Am. Chem. Soc.* 1991, *113*, 1020.
139. Ojima, I.; Kogure, T.; Nagai, Y. *Tetrahedron Lett.* 1974, 1889.
140. Ojima, I.; Kogure, T.; Kumagai, M. *J. Org. Chem.* 1977, *42*, 1671.
141. Corriu, R. J. P.; Lanneau, G. F.; Yu, Z. *Tetrahedron* 1993, *49*, 9019.
142. Page, P. C. B.; Prodger, J. C. *Synlett* 1990, 460.
143. Tamura, Y.; Annoura, H.; Fujioka, H. *Tetrahedron Lett.* 1987, *28*, 5681.
144. Page, P. C. B.; Gareh, M. T.; Porter, R. A. *Tetrahedron Lett.* 1993, *34*, 5159.
145. Poli, G.; Belvisi, L.; Manzoni, L.; Scolastico, C. *J. Org. Chem.* 1993, *58*, 3165.
146. DeNinno, M. P.; Perneer, R. J.; Lijewski, L. *Tetrahedron Lett.* 1990, *31*, 7415.
147. Bycroft, B. W.; Lee, G. R. *J. Chem. Soc., Chem. Commun.* 1975, 988.
148. Corey, E. J.; Sachdev, H. S.; Gougoutas, J. Z.; Saenger, W. *J. Am. Chem. Soc.* 1970, *92*, 2488.
149. Harada, K.; Kataoka, Y. *Tetrahedron Lett.* 1978, 2103.
150. Bringmann, G.; Geisler, J.-P. *Synthesis* 1989, 608.
151. Yamaguchi, S.; Kabuto, K. *Bull. Chem. Soc., Jpn.* 1977, *50*, 3033.
152. Nakata, T.; Tanaka, T.; Oishi, T. *Tetrahedron Lett.* 1983, *24*, 2653.

153. Nakata, T.; Fukui, M.; Ohtsuka, H.; Oishi, T. *Tetrahedron Lett.* 1983, *24*, 2661.
154. Nakata, T.; Fukui, M.; Ohtsuka, H.; Oishi, T. *Tetrahedron* 1984, *40*, 2225.
155. Midland, M. M. *Chem. Rev.* 1989, *89*, 1553.
156. Oishi, T.; Nakata, T. *Acc. Chem. Res.* 1984, *17*, 338.
157. Bloch, R.; Gilbert, L.; Girard, C. *Tetrahedron Lett.* 1988, *29*, 1021.
158. Samuels, W. D.; Nelson, D. A.; Hallen, R. T. *Tetrahedron Lett.* 1986, *27*, 3091.
159. Burke, S. D.; Deaton, D. N.; Olsen, R. J.; Armistead, D. M.; Blough, B. E. *Tetrahedron Lett.* 1987, *28*, 3905.
160. Jackson, W. R.; Jacobs, H. A.; Matthews, B. R.; Jayatilake, G. S. *Tetrahedron Lett.* 1990, *31*, 1447.
161. Takeda, H.; Tachinami, T.; Aburatani, M. *Tetrahedron Lett.* 1989, *30*, 367.
162. Ghosh, A. K.; Mckee, S. P.; Sanders, W. M. *Tetrahedron Lett.* 1991, *32*, 711.
163. Williams, D. R.; Osterhout, M. H.; Reddy, J. P. *Tetrahedron Lett.* 1993, *34*, 3271.
164. Khiar, N.; Fernández, I.; Alcudia, F.; Hua, D. H. *Tetrahedron Lett.* 1993, *34*, 699.
165. Takahashi, H.; Morimoto, T.; Achiwa, K. *Tetrahedron Lett.* 1989, *30*, 363.
166. Takahashi, H.; Sakuraba, S.; Takeda, H.; Achiwa, K. *J. Am. Chem. Soc.* 1990, *112*, 5876.
167. Duhamel, P.; Duhamel, L.; Gralak, J. *Tetrahedron Lett.* 1972, 2329.
168. Cho, B. T.; Chun, Y. S. *Tetrahedron: Asymmetry* 1992, *3*, 341.
169. Reetz, M. T.; Drewes, M. W.; Lennick, K.; Schmitz, A.; Holdgrün, X. *Tetrahedron: Asymmetry* 1990, *1*, 375.
170. Fouquey, C.; Jacques, J.; Angiolini, L.; Tramontini, M. *Tetrahedron* 1974, *30*, 2801.
171. Briene, M.-J.; Fouquey, C.; Jacques, J. *Bull. Soc. Chim. Fr.* 1969, 2395.
172. Mori, Y.; Kuhara, M.; Takeuchi, A.; Suzuki, M. *Tetrahedron Lett.* 1988, *29*, 5419.
173. Mori, Y.; Takeuchi, A.; Kageyama, H.; Suzuki, M. *Tetrahedron Lett.* 1988, *29*, 5423.
174. Barluenga, J.; Aguilar, E.; Olano, B.; Fustero, S. *Synlett* 1990, 463.
175. Shimagaki, M.; Maeda, T.; Matsuzaki, Y.; Hori, I.; Nakata, T.; Oishi, T. *Tetrahedron Lett.* 1984, *25*, 4775.
176. Shimagaki, M.; Matsuzaki, Y.; Hori, I.; Nakata, T.; Oishi, T. *Tetrahedron Lett.* 1984, *25*, 4779.
177. Carreño, M. C.; Ruano, J. L. G.; Garrido, M.; Ruiz, M. P.; Solladié, G. *Tetrahedron Lett.* 1990, *31*, 6653.
178. Carreño, M. C.; Ruano, J. L. G.; Martín, A. M.; Pedregal, C.; Rodriguez, J. H.; Rubio, A.; Sanchez, J.; Solladié, G. *J. Org. Chem.* 1990, *55*, 2120.
179. Hung, S.-M.; Lee, D.-S.; Yang, T.-K. *Tetrahedron: Asymmetry* 1990, *1*, 873.
180. Solladié, G.; Hutt, J.; Frechou, C. *Tetrahedron Lett.* 1987, *28*, 61.
181. Guanti, G.; Narisano, E.; Pero, F.; Banfi, L.; Scolastico, C. *J. Chem. Soc., Perkin Trans. 1* 1984, 189.
182. Solladié-Cavallo, A.; Suffert, J.; Adib, A.; Solladié, G. *Tetrahedron Lett.* 1990, *31*, 6649.
183. Solladié, G.; Demailly, G.; Creck, C. *Tetrahedron Lett.* 1985, *26*, 435.
184. Marshall, J. A.; Carroll, R. D. *J. Org. Chem.* 1965, *30*, 2748.
185. Nutaitis, C. F.; Bernardo, J. E. *J. Org. Chem.* 1989, *54*, 5629.
186. Giacomelli, G.; Caporusso, A. M.; Lardicci, L. *Tetrahedron Lett.* 1981, *22*, 3663.
187. Corey, E. J. *Chem. Soc. Rev.* 1988, *17*, 111.
188. Corey, E. J.; Gavai, A. V. *Tetrahedron Lett.* 1988, *29*, 3201.
189. Corey, E. J.; Su, W.-G. *Tetrahedron Lett.* 1988, *29*, 3423.
190. Corey, E. J.; Bakshi, R. K.; Shibata, S.; Chen, C.-P.; Singh, V. K. *J. Am. Chem. Soc.* 1987, *109*, 7925.
191. Nevalainen, V. *Tetrahedron: Asymmetry* 1992, *3*, 921.
192. Nevalainen, V. *Tetrahedron: Asymmetry* 1992, *3*, 1563.
193. Jones, D. K.; Liotta, D. C.; Shinkai, I.; Mathre, D. J. *J. Org. Chem.* 1993, *58*, 799.
194. Nevalainen, V. *Tetrahedron: Asymmetry* 1992, *3*, 1441.
195. Nevalainen, V. *Tetrahedron: Asymmetry* 1992, *3*, 933.
196. Corey, E. J.; Bakshi, R. K.; Shibata, S. *J. Am. Chem. Soc.* 1987, *109*, 5551.
197. Corey, E. J.; Bakshi, R. K. *Tetrahedron Lett.* 1990, *31*, 611.
198. Corey, E. J.; Becker, K. B.; Varma, R. K. *J. Am. Chem. Soc.* 1972, *94*, 8616.
199. Noyori, R.; Tomino, I.; Nishizawa, M. *J. Am. Chem. Soc.* 1979, *101*, 5844.
200. Noyori, R. *Pure Appl. Chem.* 1981, *53*, 2315.
201. Iguchi, S.; Nakai, H.; Hayashi, M.; Yamamoto, H. *J. Org. Chem.* 1979, *44*, 1363.
202. Gammill, R. B.; Nash, S. A.; Bell, L. T.; Watt, W.; Mizsak, S. A.; Scahill, T. A.; Sobieray, D. *Tetrahedron Lett.* 1990, *31*, 5303.
203. Johnson, M. R.; Rickborn, B. *J. Org. Chem.* 1969, *35*, 1041.
204. Iqbal, K.; Jackson, W. R. *J. Chem. Soc. (C)* 1968, 616.
205. Jackson, W. R.; Zurqiyah, A. *J. Chem. Soc.* 1965, 5280.
206. Terashima, S.; Tanno, N.; Koga, K. *J. Chem. Soc., Chem. Commun.* 1980, 1026.
207. Ojima, I.; Hirai, K. In *Asymmetric Synthesis*; J. D. Morrison, Ed.; Academic Press: Orlando, FL, 1985; Vol. 5; pp 103.
208. Kogure, T.; Ojima, I. *J. Organometal. Chem.* 1982, *234*, 249.
209. Brown, H. C.; Jadhaw, P. K.; Mandal, A. K. *Tetrahedron* 1981, *37*, 3547.
210. Brinkmeyer, R. S.; Kapoor, V. M. *J. Am. Chem. Soc.* 1977, *99*, 8339.

211. Huguet, J.; Reyes, M. d. C. *Tetrahedron Lett.* 1990, *31*, 4279.
212. Midland, M. M.; McDowell, D. C.; Hatch, R. L.; Tramontano, A. *J. Am. Chem. Soc.* 1980, *102*, 86.
213. Amici, M. D.; Micheli, C. D.; Carrea, G.; Spezia, S. *J. Org. Chem.* 1989, *54*, 2646.
214. Hudrlik, P. F.; Peterson, D. *Tetrahedron Lett.* 1974, 1133.
215. Sum, F. W.; Weiler, L. *J. Am. Chem. Soc.* 1979, *101*, 4401.
216. Huckin, S. N.; Weiler, L. *J. Am. Chem. Soc.* 1974, *96*, 1082.
217. Huckin, S. N.; Weiler, L. *Can. J. Chem.* 1974, *52*, 2157.
218. Huckin, S. N.; Wieler, L. *Tetrahedron Lett.* 1971, 4835.
219. Masamune, S.; Imperiali, B.; Garvey, B. *J. Am. Chem. Soc.* 1982, *104*, 5528.
220. Meyer, H.; Seebach, D. *Liebigs Ann. Chem.* 1975, 2261.
221. Kongkathip, B.; Kongkathip, N. *Tetrahedron Lett.* 1984, *25*, 2175.
222. White, J. D.; Kawasaki, M. *J. Am. Chem. Soc.* 1990, *112*, 4991.
223. Danishefsky, S.; Zamboni, R.; Kahn, M.; Etheredge, S. J. *J. Am. Chem. Soc.* 1981, *103*, 3460.
224. Huang, H. C.; Rehmann, J. K.; Gray, G. R. *J. Org. Chem.* 1982, *47*, 4018.
225. Trost, B. M.; Lautens, M. *J. Am. Chem. Soc.* 1987, *109*, 1469.
226. Kreiser, W.; Below, P. *Tetrahedron Lett.* 1981, *22*, 429.
227. Krause, V.; Lauer, W.; Meier, H. *Chem. Ber.* 1989, *122*, 1719.
228. Winkler, J. D.; Finck-Estes, M. *Tetrahedron Lett.* 1989, *30*, 7293.
229. Fiaud, J.-C. *Tetrahedron Lett.* 1975, 3495.
230. Tomioka, K.; Ando, K.; Takemasa, Y.; Koga, K. *J. Am. Chem. Soc.* 1984, *106*, 2718.
231. Tomioka, K.; Seo, W.; Ando, K.; Koga, K. *Tetrahedron Lett.* 1987, *28*, 6637.
232. Evans, D. A.; Ennis, M. D.; Le, T.; Mandel, N.; Mandel, G. *J. Am. Chem. Soc.* 1984, *106*, 1154.
233. Hoffman, R. V.; Kim, H.-O. *Tetrahedron Lett.* 1993, *34*, 2051.
234. Eid, C. N.; Konopelski, J. P. *Tetrahedron* 1991, *47*, 975.
235. Hoppe, I.; Hoppe, D.; Wolff, C.; Egert, E.; Herbst, R. *Angew. Chem., Int. Ed. Engl.* 1989, *28*, 67.
236. Cativiela, C.; Diaz-de-Villegas, M. D.; Gálvez, J. A. *Tetrahedron: Asymmetry* 1993, *4*, 229.
237. Cativiela, C.; Diaz-de-Villegas, M. D.; Gálvez, J. A. *Tetrahedron: Asymmetry* 1993, *4*, 1445.
238. Ihara, M.; Takahashi, T.; Niitsuma, H.; Taniguchi, N.; Yasui, K.; Fukumoto, K. *J. Org. Chem.* 1989, *54*, 5413.
239. Umemura, K.; Matsuyama, H.; Watanabe, N.; Kobayashi, M.; Kamigata, N. *J. Org. Chem.* 1989, *54*, 2374.
240. Rychnovsky, S. D. *J. Org. Chem.* 1989, *54*, 4982.
241. Seebach, D.; Wasmuth, D. *Helv. Chim. Acta* 1980, *63*, 197.
242. Frater, G. *Helv. Chim. Acta* 1979, *62*, 2829.
243. Brooks, D. W.; Kellogg, R. P. *Tetrahedron Lett.* 1982, *23*, 4991.
244. Meyer, H. H. *Liebigs Ann. Chem.* 1984, 791.
245. Frater, G.; Muller, U.; Gunther, W. *Tetrahedron* 1984, *40*, 1269.
246. Imamoto, T.; Matsumoto, T.; Yokoyama, H.; Yamaguchi, K.-I. *J. Org. Chem.* 1984, *49*, 1105.
247. Trost, B. M.; Metz, P.; Hane, J. T. *Tetrahedron Lett.* 1986, *27*, 5691.
248. Mori, K.; Takikawa, H. *Tetrahedron* 1990, *46*, 4473.
249. Ikeda, N.; Omori, K.; Yamamoto, H. *Tetrahedron Lett.* 1986, *27*, 1175.
250. Frater, G. *Tetrahedron Lett.* 1981, *22*, 425.
251. Chamberlin, A. R.; Dezube, M. *Tetrahedron Lett.* 1982, *23*, 3055.
252. Michel, J.; Canonne, P. *Can. J. Chem.* 1971, *49*, 4084.
253. Martin, V. A.; Murray, D. H.; Pratt, N. E.; Zhao, Y.-b.; Albizati, K. F. *J. Am. Chem. Soc.* 1990, *112*, 6965.
254. Knochel, P.; Brieden, P.; Rozema, M. J.; Eisenberg, C. *Tetrahedron Lett.* 1993, *34*, 5881.
255. Sato, M.; Hisamichi, H.; Kitazawa, N.; Kaneko, C.; Furuya, T.; Suzaki, N.; Inukai, N. *Tetrahedron Lett.* 1990, *31*, 3605.
256. Fráter, G.; Günther, W.; Müller, U. *Helv. Chim. Acta* 1989, *72*, 1846.
257. Genet, J. P.; Juge, S.; Mallart, S. *Tetrahedron Lett.* 1988, *29*, 6765.
258. Guanti, G.; Banfi, L.; Narisano, E. *Tetrahedron Lett.* 1989, *30*, 5515.
259. Azzena, U.; Delogu, G.; Melloni, G.; Piccolo, O. *Tetrahedron Lett.* 1989, *30*, 4555.
260. Fleming, I.; Hill, J. H. M.; Parker, D.; Waterson, D. *J. Chem. Soc., Chem. Commun.* 1985, 318.
261. Still, W. C.; Schneider, J. A. *Tetrahedron Lett.* 1980, *21*, 1035.
262. Reetz, M. T.; Jung, A. *J. Am. Chem. Soc.* 1983, *105*, 4833.
263. Baker, R.; O'Mahoney, M. J.; Swain, C. J. *Tetrahedron Lett.* 1986, *27*, 3059.
264. Sato, T.; Itoh, T.; Fujisawa, T. *Tetrahedron Lett.* 1987, *28*, 5677.
265. Solladié, G.; Matloubi-Maghadam, F. *J. Org. Chem.* 1982, *47*, 91.
266. Solladié, G.; Frechon, C.; Demailly, G. *Nouv. J. Chim.* 1985, *9*, 21.
267. Carreño, M. C.; Ruano, J. L. C.; Maestro, M. C.; González, M. P. *Tetrahedron* 1993, *49*, 11009.
268. Barros, D.; Carreño, M. C.; Ruano, J. L. G.; Maestro, M. C. *Tetrahedron Lett.* 1992, *33*, 2733.
269. Sato, F.; Takahashi, O.; Kato, T.; Kobayashi, Y. *J. Chem. Soc., Chem. Commun.* 1985, 1638.
270. Uchiyama, H.; Kobayashi, Y.; Sato, F. *Chem. Lett.* 1985, 467.

271. Hanamoto, T.; Katsuki, T.; Yamaguchi, M. *Tetrahedron Lett.* 1986, *27*, 2463.
272. Ager, D. J.; East, M. B. *Tetrahedron* 1993, *49*, 5683.
273. Stahl, I.; Wrabletz, F.; Gosselck, J. *Chem. Ber.* 1989, *122*, 371.
274. Shimagaki, M.; Shiokawa, M.; Sugai, K.; Terenaka, T.; Nakata, T.; Oishi, T. *Tetrahedron Lett.* 1988, *29*, 659.
275. Soai, K.; Yamanoi, T.; Hikima, H.; Oyamada, H. *J. Chem. Soc., Chem. Commun.* 1985, 138.
276. Noyori, R.; Ohkuma, T.; Kitamura, M.; Takaya, H.; Sayo, N.; Kumobayashi, H.; Akutagawa, S. *J. Am. Chem. Soc.* 1987, *109*, 5856.
277. Kukuda, N.; Mashima, K.; Matsumura, Y.-I.; Takaya, H. *Tetrahedron Lett.* 1990, *31*, 7185.
278. Murata, M.; Morimoto, T.; Achiwa, K. *Synlett* 1991, 827.
279. Heiser, B.; Broger, E. A.; Crameri, Y. *Tetrahedron: Asymmetry* 1991, *2*, 51.
280. Kawano, H.; Ishii, Y.; Saburi, M.; Uchida, Y. *J. Chem. Soc., Chem. Commun.* 1988, 87.
281. Shao, L.; Kawano, H.; Saburi, M.; Uchida, Y. *Tetrahedron* 1993, *49*, 1997.
282. Noyori, R.; Ikeda, T.; Ohkuma, T.; Widhalm, M.; Kitamura, M.; Takaya, H.; Akutagawa, S.; Sayo, N.; Saito, T.; Taketomi, T.; Kumobayashi, H. *J. Am. Chem. Soc.* 1989, *111*, 9134.
283. Kitahara, M.; Ohkuma, T.; Tokunaga, M.; Noyori, R. *Tetrahedron: Asymmetry* 1990, *1*, 1.
284. Noyori, R. *Chem. Soc. Rev.* 1989, *18*, 187.
285. Takaya, H.; Ohta, T.; Mashima, K.; Noyori, R. *Pure Appl. Chem.* 1990, *62*, 1135.
286. Akutagawa, S. In *Chirality in Industry*; A. N. Collins; G. N. Sheldrake and J. Crosby, Ed.; John Wiley & Sons: New York, 1992; pp 325.
287. King, S. A.; Thompson, A. S.; King, A. O.; Verhoeven, T. R. *J. Org. Chem.* 1992, *57*, 6689.
288. Shao, L.; Seki, T.; Kawano, H.; Saburi, M. *Tetrahedron Lett.* 1991, *32*, 7699.
289. Valentine, D.; Scott, J. W. *Synthesis* 1978, 329.
290. Nakata, T.; Oishi, T. *Tetrahedron Lett.* 1980, *21*, 1641.
291. Nakata, T.; Fukui, M.; Oishi, T. *Tetrahedron Lett.* 1988, *29*, 2219.
292. Nakata, T.; Kuwabara, T.; Tani, Y.; Oishi, T. *Tetrahedron Lett.* 1982, *23*, 1015.
293. Furukawa, M.; Okawara, T.; Noguchi, Y.; Terawaki, Y. *Chem. Pharm. Bull.* 1979, *27*, 2223.
294. Yatagai, M.; Ohnuki, T. *J. Chem. Soc., Perkin Trans. 1* 1990, 1826.
295. Taniguchi, M.; Fujii, H.; Oshima, K.; Utimoto, K. *Tetrahedron* 1993, *49*, 11169.
296. Oppolzer, W.; Rodriguez, I.; Starkemann, C.; Walther, E. *Tetrahedron Lett.* 1990, *31*, 5019.
297. Potin, D.; Dumas, F.; d'Angelo, J. *J. Am. Chem. Soc.* 1990, *112*, 3483.
298. Fujita, M.; Hiyama, T. *J. Org. Chem.* 1988, *53*, 5405.
299. Chautemps, P.; Pierre, J.-L. *Tetrahedron* 1976, *32*, 549.
300. Nakata, T.; Tanaka, T.; Oishi, T. *Tetrahedron Lett.* 1981, *22*, 4723.
301. Andrisano, R.; Angeloni, A. S.; Mazocchi, S. *Tetrahedron* 1973, *29*, 913.
302. Narasaka, K.; Pai, F.-C. *Tetrahedron* 1984, *40*, 2233.
303. Pilli, R. A.; Russowsky, D.; Dias, L. C. *J. Chem. Soc., Perkin Trans. 1* 1990, 1213.
304. Evans, D. A.; Hoveyda, A. H. *J. Org. Chem.* 1990, *55*, 5190.
305. Guanti, G.; Banfi, L.; Riva, R.; Zannetti, M. T. *Tetrahedron Lett.* 1993, *34*, 5483.
306. Barluenga, J.; Aguilar, E.; Fustero, S.; Olano, B.; Viado, A. L. *J. Org. Chem.* 1992, *57*, 1219.
307. Barluenga, J.; Olano, B.; Fustero, S. *J. Org. Chem.* 1985, *55*, 4052.
308. Barluenga, J.; Olano, B.; Fustero, S.; Foces-Foces, M. C.; Hernández, F. *J. Chem. Soc., Chem. Commun.* 1988, 410.
309. Chen, K.-M.; Hardtmann, G. E.; Pasad, K.; Repic, O.; Shapiro, M. J. *Tetrahedron Lett.* 1987, *28*, 155.
310. Narasaka, K.; Pai, H. C. *Chem. Lett.* 1980, 1415.
311. Kiyooka, S.-I.; Kuroda, H.; Shimasaki, Y. *Tetrahedron Lett.* 1986, *27*, 3009.
312. Suzuki, K.; Shimazaki, M.; Tsuchihashi, G.-I. *Tetrahedron Lett.* 1986, *27*, 6233.
313. Suzuki, K.; Miyazawa, M.; Shimazaki, M.; Tsuchihashi, G.-I. *Tetrahedron Lett.* 1986, *27*, 6237.
314. Evans, D. A.; Chapman, K. T.; Carreira, E. M. *J. Am. Chem. Soc.* 1988, *110*, 3560.
315. Evans, D. A.; Chapman, K. T. *Tetrahedron Lett.* 1986, *27*, 5939.
316. Maruoka, K.; Hasegawa, M.; Yamomoto, H.; Suzuki, K.; Shimazaki, M.; Tsuchihashi, G. *J. Am. Chem. Soc.* 1986, *108*, 3827.
317. Suzuki, K.; Katayama, E.; Tsuchihashi, G. *Tetrahedron Lett.* 1984, *25*, 1817.
318. Shimazaki, M.; Hara, H.; Suzuki, K. *Tetrahedron Lett.* 1989, *30*, 5447.
319. Suzuki, K.; Katayama, E.; Tsuchihashi, G.-I. *Tetrahedron Lett.* 1984, *25*, 2479.
320. Ramón, D. J.; Yus, M. *Tetrahedron Lett.* 1990, *31*, 3763.
321. Ramón, D. J.; Yus, M. *Tetrahedron Lett.* 1990, *31*, 3767.
322. Petragnani, N.; Yonashiro, M. *Synthesis* 1982, 521.
323. Evans, D. A. In *Asymmetric Synthesis*; J. D. Morrison, Ed.; Academic Press: Orlando, FL, 1984; Vol. 3; pp 1.
324. Ager, D. J.; East, M. B. *Tetrahedron* 1992, *48*, 2803.
325. Molander, G. A.; Haar, J. P. *J. Am. Chem. Soc.* 1993, *115*, 40.
326. Meyers, A. I.; Harre, M.; Garland, R. *J. Am. Chem. Soc.* 1984, *106*, 1146.

327. Meyers, A. I.; Westrum, L. J. *Tetrahedron Lett.* 1993, *34*, 7701.
328. Meyers, A. I.; Wallace, R. H.; Harre, M.; Garland, R. *J. Org. Chem.* 1990, *55*, 3137.
329. Meyers, A. I.; Wanner, K. T. *Tetrahedron Lett.* 1985, *26*, 2047.
330. Bashiardes, G.; Collingwood, S. P.; Davies, S. G.; Preston, S. C. *J. Chem. Soc., Perkin Trans. 1* 1989, 1162.
331. Masamune, S.; Kennedy, R. M.; Petersen, J. S.; Houk, K. N.; Wu, Y.-D. *J. Am. Chem. Soc.* 1986, *108*, 7404.
332. Bravo, P.; Piovosi, E.; Resnati, G.; Fronza, G. *J. Org. Chem.* 1989, *54*, 5171.
333. Nagao, Y.; Ikeda, T.; Yagi, M.; Fujita, E.; Shiro, M. *J. Am. Chem. Soc.* 1982, *104*, 2079.
334. Aitken, R. A.; Gopal, J. *Tetrahedron: Asymmetry* 1990, *1*, 517.
335. Santelli-Rouvier, C. *Tetrahedron Lett.* 1984, *25*, 4371.
336. Osakada, K.; Obana, M.; Ikariya, T.; Saburi, M.; Yoshikawa, S. *Tetrahedron Lett.* 1981, *21*, 4297.
337. Toomoka, K.; Okinaga, T.; Suzuki, K.; Tsuchihashi, G.-I. *Tetrahedron Lett.* 1987, *28*, 6335.

Chapter 7

# ALDOL AND RELATED REACTIONS

This chapter only covers aldol-type reactions that can, or could, provide control over one of the new stereogenic centers.† Thus, reactions that control only regiochemistry, as for the preparation of β-keto esters,[1,2] are not discussed.

As described below, a wide variety of metals has been used in conjunction with aldol-type chemistry. More "exotic" metals, such as the late transition metals,[3] have to show considerable advantages to be considered useful for synthetic transformations when compared to more common types of reagents. For example, α-mercurio ketones, in the presence of a Lewis acid, provide for *syncat* aldol reactions,[4] but seem to hold no significant advantages over alternative strategies that employ less toxic reagents. Only metals that provide useful selectivity are discussed below.

To circumvent some of the problems associated with the aldol reaction, alternative strategies have been employed, including the use of allyl anions [*4.1.5*], the Claisen rearrangement [*12.2*],[5] and reductions of β-dicarbonyl compounds [*6.4.1*]. The aldol approach is, however, very powerful; nature also uses this reaction to form carbon–carbon bonds [*14.3*].

This chapter has been organized into sections based on the metal counterion with an introductory section covering the various transition state models and how the reaction outcome can be manipulated. This introduction also covers ways that can be used to control the stereochemical outcome of an aldol approach through the use of various metal counterions and enolate geometries. In addition, a section on the formose reaction has been included, as this allows coupling of carbonyl compounds as well, although in a 1,2-manner.

The large number of variables available to control the aldol reaction and its analogs have spurned a number of models, explanations, and rationalizations.[6-12] We have used the current, widely accepted models that allow for relatively simple predictions to be made and avoided the more complex arguments.¶

For the preparation of cyclic compounds through an intramolecular aldol reaction, Baldwin's rules provide guidelines to predict whether a reaction pathway will be favorable [*2.4.1*].

## 7.1. THE ALDOL REACTION: METHODS OF CONTROL

The aldol reaction,[3,13-26] condensation of a nucleophilic carbonyl species with a electrophilic carbonyl moiety, together with its many derivatives, has proven to be one of the most versatile methods for the formation of carbon–carbon bonds. The addition has, until recently, been plagued by dehydration of the initial aldol adduct and by regiochemical problems. Even with these problems circumvented, four different stereoisomers can be formed.[27-29] Most of the methods for stereochemical control of the aldol reaction rely on the use of chiral auxiliaries or the use of chiral organometallic reagents.[20,30] The use of functionalized enolates in an aldol type reaction[3,13,14,16-18,20-24,31] with controlled addition allows for the construction of several centers with stereochemical control and a convergent approach.[32-47] Indeed, functionality or chiral auxiliaries can be incorporated into one or more of the substrates.[1,2,48-52] For high diastereoselection, however, matched pairs must be used to achieve double asymmetric induction.[25,53-55] For example, with nucleophilic additions to carbonyl compounds (e.g., Scheme 2.7) [Chapter 4], the acetonide of L-glyceraldehyde has proven to be a useful building block in aldol methodology.[56-61] An alternative approach is to use a metal complex is an integral part of the nucleophilic moiety, such as an acetyl iron complex, as this provides for high induction.[62-66]

† For the purposes of this chapter, all aldol-type reactions have been considered together, whether the nucleophilic species is derived from a ketone, ester, etc. Thus, specifc references to reactions, such as the Claisen condensation, will not be found.

¶ This approach was adopted to keep in line with the use of this book as a reference work. However, the reader is cautioned that these simple models may not hold for all cases, and the original citations should be consulted.

Nitrogen analogs of carbonyl compounds have been utilized to effect the equivalent of aldol reactions,[67-70] as have thionium ions.[71]

Many studies have addressed the issue of a stereoselective aldol reaction. The major variables are the metal counterion, together with its associated ligands, and the reaction conditions.[§] [72-77] High stereoselectivity correlates with the enolate geometry [*5.1*], and the steric influences within this moiety, while the steric constraints associated with the electrophilic carbonyl moiety tend to play a minor role.[31,78] Anh-Felkin or chelation control can also be important (vide infra).[79]

The enolate anion or enol ether necessary to preform an aldol reaction can be generated by a wide variety of methods [*5.1*]. Indeed, enolate generation followed by subsequent reaction with a carbonyl compound is the normal experimental methodology of an aldol reaction, as it ensures that the first carbonyl compound is converted to the nucleophilic species and alleviates problems associated with self-condensation reactions. As with enolate alkylations [*5.2.1*], especially those with lithium as the counterion, aggregates could play a key role in the stereochemical outcome of the aldol reaction.[65,80]

The observed stereochemistry can usually be interpreted in terms of a chair-like transition state. These principles are summarized in Scheme 7.1.[†] [12,24,81-87]

**SCHEME 7.1.**

Thus, the *E*-enolate would be expected to give the *ancat* (*threo*) product as the transition state **7.1** has less serious steric interactions than in Scheme 7.2.[88-90] In many cases, the preferred transition state may not be a chair, or the energy difference between the two pathways is very small.[24,91] For enolates with a tetrasubstituted double bond, the Zimmerman-Traxler model tends to be followed whatever the metal counterion.[92]

In some of the early studies it was found that a large alkyl group in the nucleophilic moiety affords good stereoselection (Scheme 7.2).[88,93-97] Silicon can also play the role of a large group, as in acylsilanes (*cf.* Schemes 2.3 and 2.12).[98]

The use of a metal counterion, such as lithium or magnesium, that can form a chelate does offer selectivity through the use of either Anh-Felkin or chelation controlled additions.[99-102] The use of titanium can also prove useful,[103-105] and if chiral ligands are present then an enantioselective transformation becomes available (vide infra). Boron enolates have also found widespread usage. Some of the systems that provide selectivity are summarized in Table 7.1.[¶]

§ Both theoretical and experimental studies to determine how these variables influence the stereochemical outcome of the reaction are being performed.[6-12]

† This is often called the Zimmerman-Traxler model.

¶ Unlike other tables in this book, examples with <90% ee or de have been included to allow some of the subtle nuances of the reaction to be seen.

**TABLE 7.1**
**Systems that Have Been Used to Control Selectivity in Aldol and Related Reactions**

| System | Metal Counterion | Product Stereo-chemistry[a] | Reference | System | Metal Counterion | Product Stereo-chemistry | Reference |
|---|---|---|---|---|---|---|---|
| 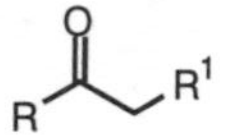 | Sn or Zr[b] | S | 106,107 | 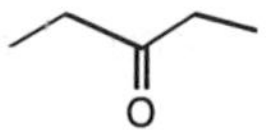 | B[c] | S | 108 |
| | Ti | A | 109 | 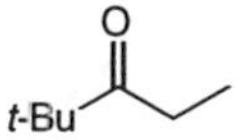 | Li, Mg, or B | S | 88,90,110 |
| 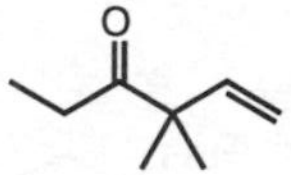 | Li | S | 111 | 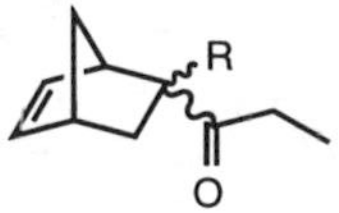 | Li | S | 97 |
| | Ti | A | 111 | 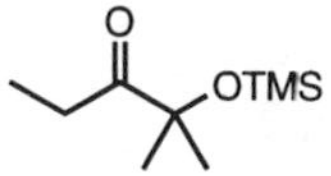 | Li | S | 94,112–114 |
| 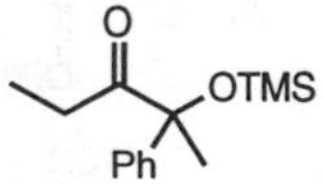 | Li | S | 115 | | Si[d] | A | 111 |
| 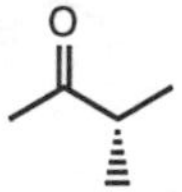 | Li | — | 116 | 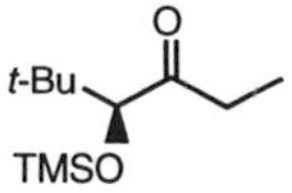 | B. Li, or Ti | S[e] | 93,96,113,117 |
| 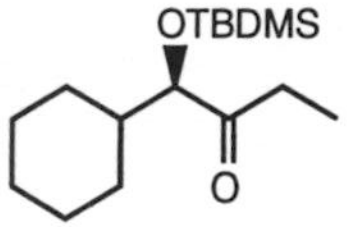 | B, Li, or Ti | S[f] | 32,96,118–121 | | Mg | A | 26,117 |

**TABLE 7.1 (continued)**

| System | Metal Counterion | Product Stereo-chemistry[a] | Reference | System | Metal Counterion | Product Stereo-chemistry | Reference |
|---|---|---|---|---|---|---|---|
| 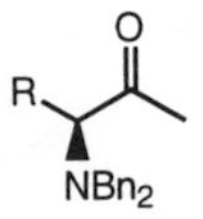 | Li | — | 122 | 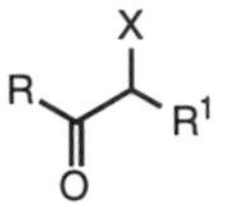 | Sn | S[g] | 123 |
| 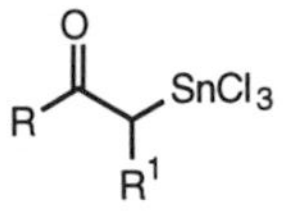 | — | A | 124 | 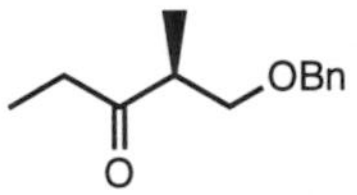 | B | S[h] | 125 |
| 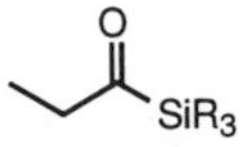 | Li | S | 98 | 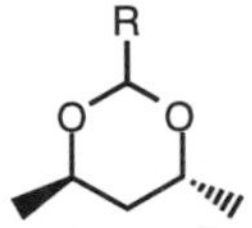 | Ti | S | 126 |
| 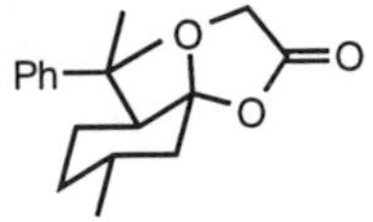 | Li or Mg | A | 127,128 | 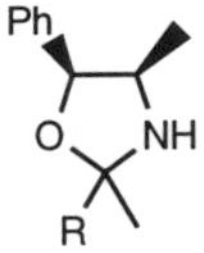 | Sn | — | 129,130 |
| | Zr | S | 127 |  | Li | A | 131,132 |
| 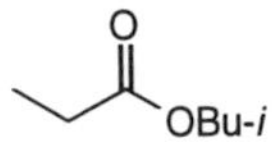 | B[h,i] | A | 73,133,134 | 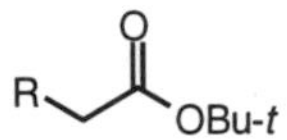 | Ti[b] | — | 135,136 |
| 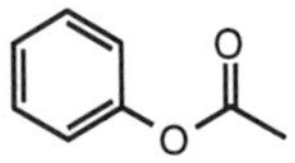 | Ti[b] | S | 137 | 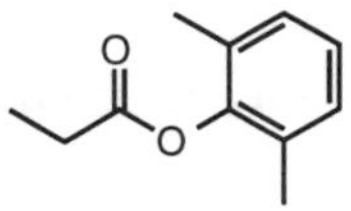 | Li | A | 138-141 |

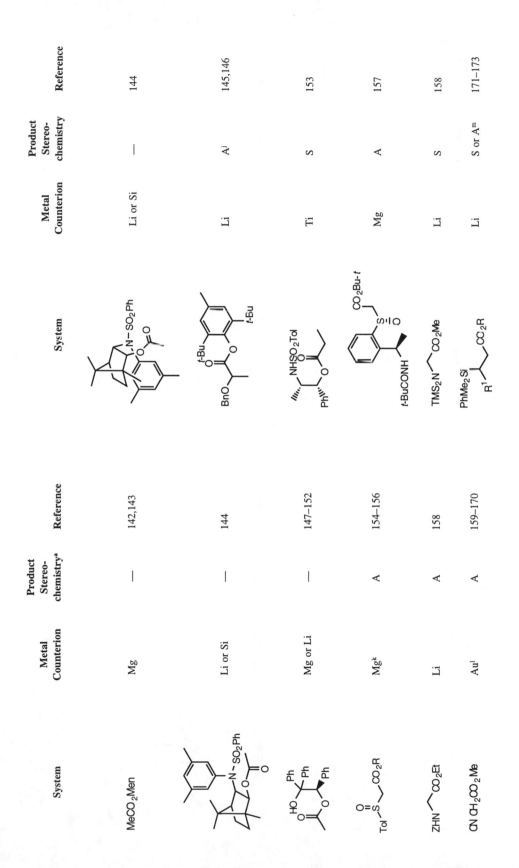

| System | Metal Counterion | Product Stereo-chemistry[a] | Reference | System | Metal Counterion | Product Stereo-chemistry | Reference |
|---|---|---|---|---|---|---|---|
| $MeCO_2Men$ | Mg | — | 142,143 | N–$SO_2$Ph, O, O | Li or Si | — | 144 |
| N–$SO_2$Ph, O, O | Li or Si | — | 144 | BnO, O, O, *t*-Bu, *t*-Bu | Li | A[j] | 145,146 |
| HO, Ph, Ph, Ph, O, O | Mg or Li | — | 147–152 | $NHSO_2$Tol, O, Ph, O | Ti | S | 153 |
| Tol–S(=O)–$CH_2CO_2R$ | Mg[k] | A | 154–156 | $CO_2$Bu-*t*, S, O, *t*-BuCONH | Mg | A | 157 |
| ZHN–$CH_2$–$CO_2Et$ | Li | A | 158 | $TMS_2N$–$CH_2$–$CO_2Me$ | Li | S | 158 |
| $CN\ CH_2CO_2Me$ | Au[l] | A | 159–170 | $PhMe_2Si$, $R^1$, $CO_2R$ | Li | S or A[m] | 171–173 |

## TABLE 7.1 (continued)

| System | Metal Counterion | Product Stereo-chemistry[a] | Reference | System | Metal Counterion | Product Stereo-chemistry | Reference |
|---|---|---|---|---|---|---|---|
| 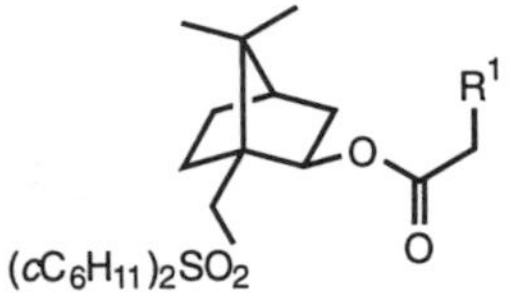 $(cC_6H_{11})_2SO_2$, O, $R^1$ | Si | A | 174 | 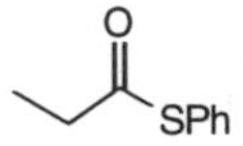 O, SPh | B[b] | S | 73,133,134 |
| 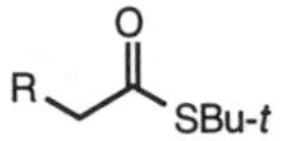 O, R, SBu-*t* | Al or B | A | 90,110,175–178 | 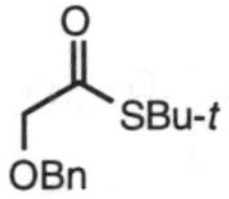 O, SBu-*t*, OBn | Ti | A | 179,180 |
|  | Li or Zr | S | 177,181 | 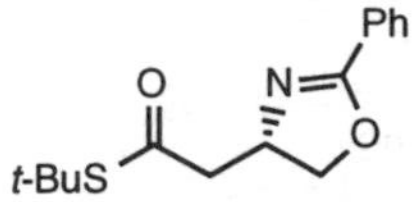 O, *t*-BuS, N, O, Ph | Li | A | 48 |
| 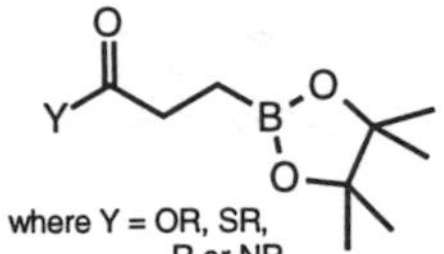 O, Y, B, O, O; where Y = OR, SR, R or $NR_2$ | Li | S | 182,183 | 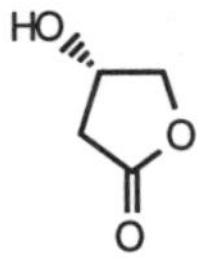 HO, O, O | Li | — | 184 |
| 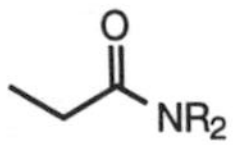 O, $NR_2$ | Zr | S | 181 | 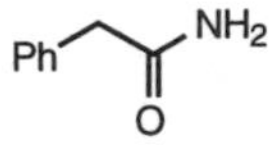 Ph, $NH_2$, O | Na or Li | A | 185 |
| 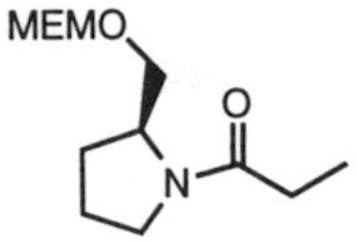 MEMO, N, O | Li or Zr | S | 186 | 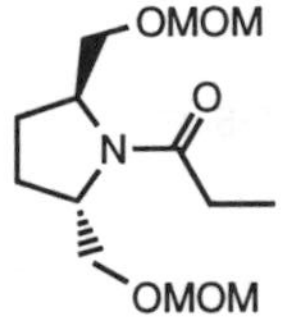 OMOM, N, O, OMOM | Zr | S | 187 |

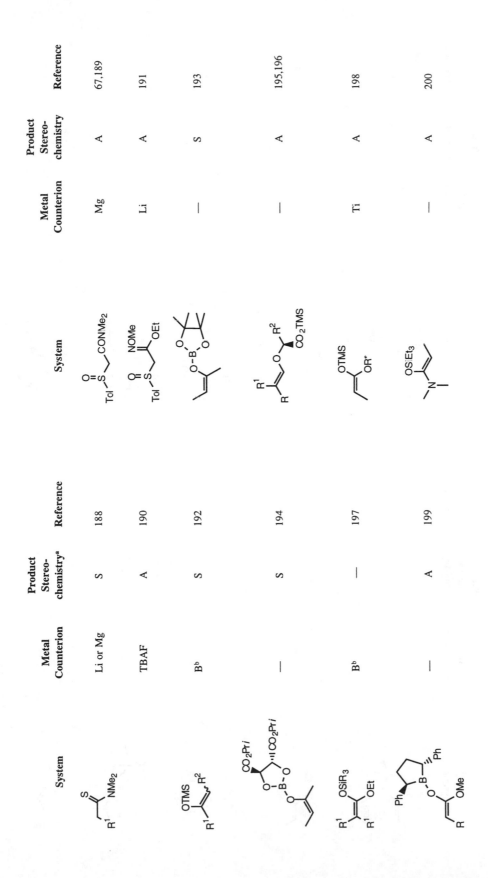

| System | Metal Counterion | Product Stereo-chemistry[a] | Reference | System | Metal Counterion | Product Stereo-chemistry | Reference |
|---|---|---|---|---|---|---|---|
| 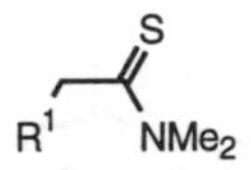 | Li or Mg | S | 188 | 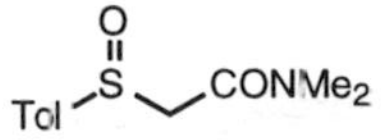 | Mg | A | 67,189 |
| | TBAF | A | 190 | 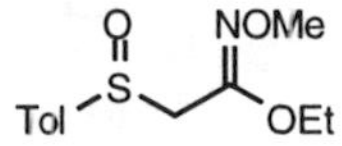 | Li | A | 191 |
| 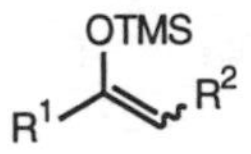 | B[b] | S | 192 | 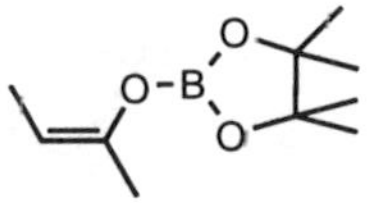 | — | S | 193 |
| 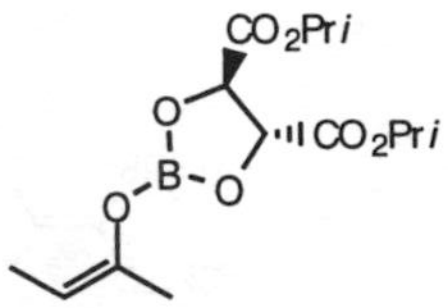 | — | S | 194 | 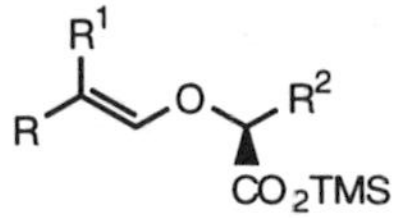 | — | A | 195,196 |
| 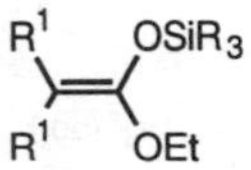 | B[b] | — | 197 | 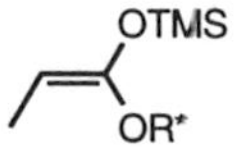 | Ti | A | 198 |
| 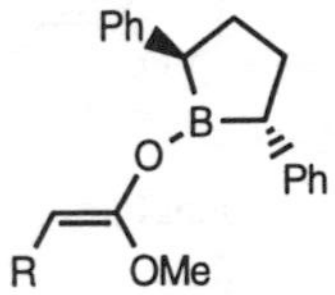 | — | A | 199 | 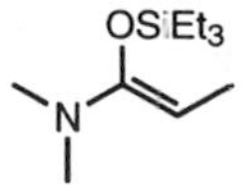 | — | A | 200 |

**TABLE 7.1 (continued)**

| System | Metal Counterion | Product Stereo-chemistry[a] | Reference | System | Metal Counterion | Product Stereo-chemistry | Reference |
|---|---|---|---|---|---|---|---|
| 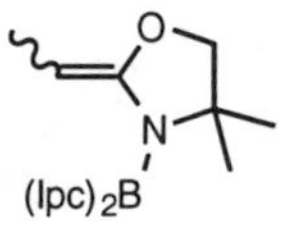 O, N, $(Ipc)_2B$ | — | A | 201 | 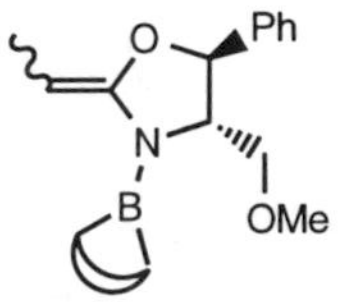 O, Ph, N, B, OMe | — | S | 201 |
| 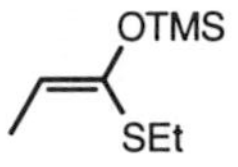 OTMS, SEt | —[b] | S | 202 | 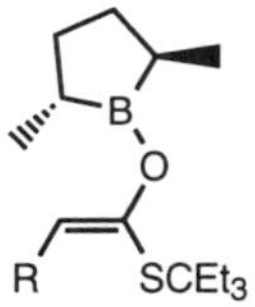 B, O, R, $SCEt_3$ | — | A | 175,203 |
| 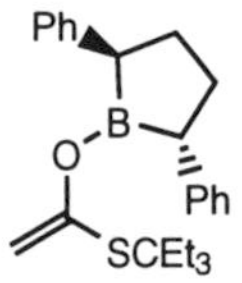 Ph, B, O, Ph, $SCEt_3$ | — | — | 204 | 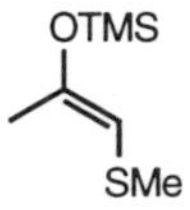 OTMS, SMe | — | S | 204 |
| 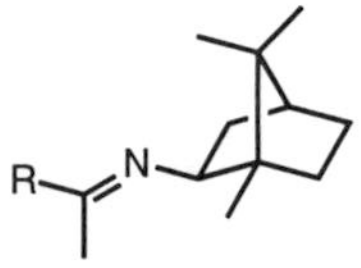 R, N | B | —[n] | 69 | 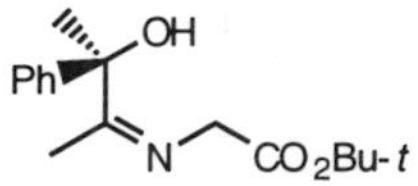 OH, Ph, N, $CO_2Bu$-*t* | Mg | — | 23,205 |
| 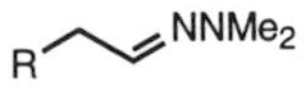 R, $NNMe_2$ | Ti | S | 68 | 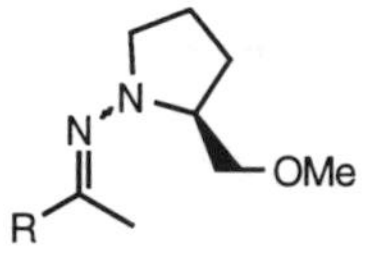 N, N, OMe, R | Li | — | 206,207 |

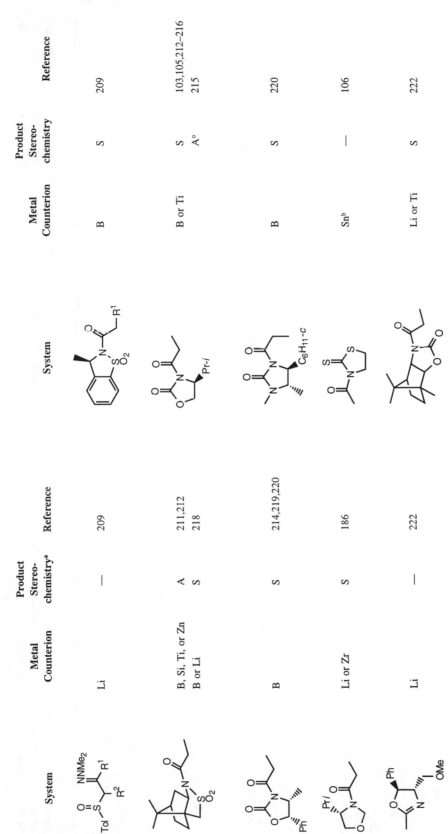

| System | Metal Counterion | Product Stereo-chemistry[a] | Reference | System | Metal Counterion | Product Stereo-chemistry | Reference |
|---|---|---|---|---|---|---|---|
| 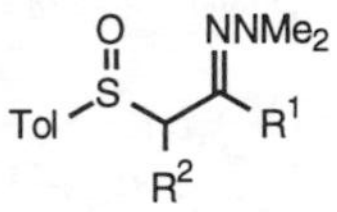 | Li | — | 209 | 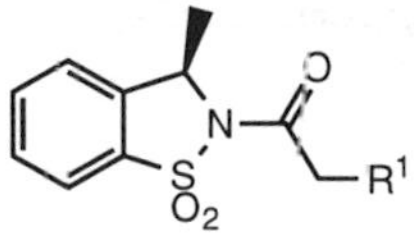 | B | S | 209 |
| | B, Si, Ti, or Zn | A | 211,212 | | B or Ti | S | 103,105,212–216 |
| | B or Li | S | 218 | | | A[c] | 215 |
| 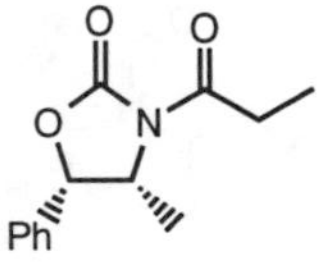 | B | S | 214,219,220 | 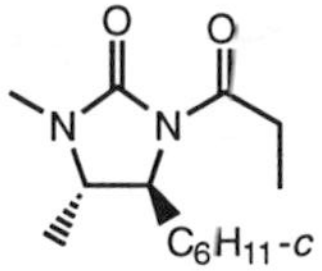 | B | S | 220 |
| 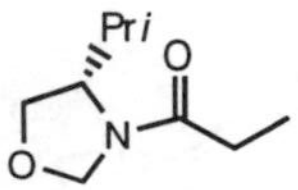 | Li or Zr | S | 186 | 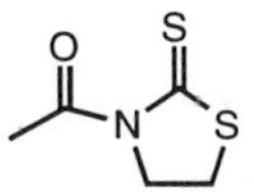 | Sn[b] | — | 106 |
| 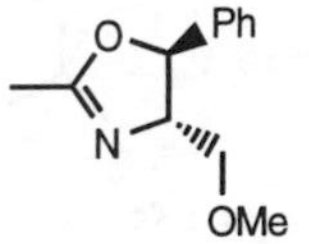 | Li | — | 222 | 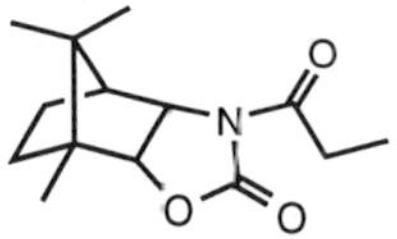 | Li or Ti | S | 222 |

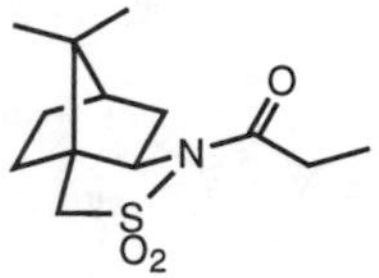

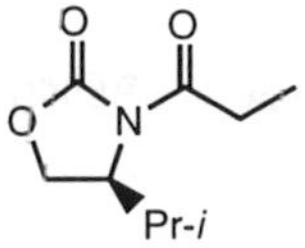

**TABLE 7.1 (continued)**

| System | Metal Counterion | Product Stereo-chemistry[a] | Reference | System | Metal Counterion | Product Stereo-chemistry | Reference |
|---|---|---|---|---|---|---|---|
| 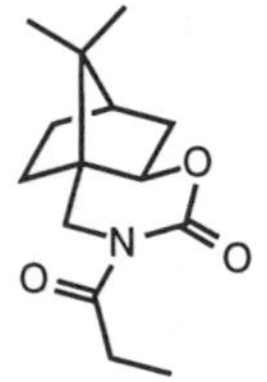 | Ti, B | S | 224,225 | 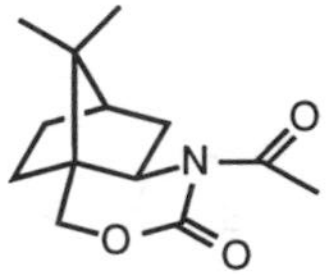 | Li or Ti | S | 225 |
| 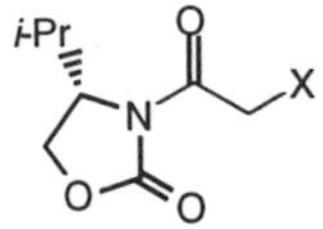 i-Pr, X; X = Br or Cl | B, Li, Sn, or Zn | S[p] | 227–229 | 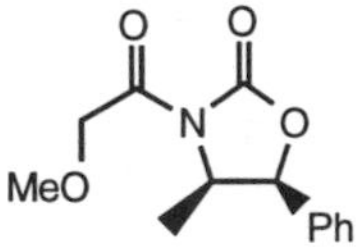 MeO, Ph | B | S | 229 |
| 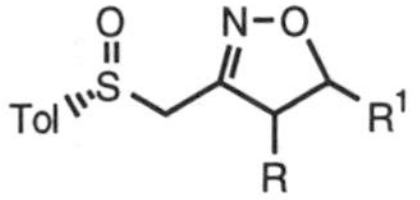 Tol, R, $R^1$ | Mg | A | 231 |  $Cp(Ph_3P)(CO)Fe$ | Al or Li | — | 62,231–233 |
| 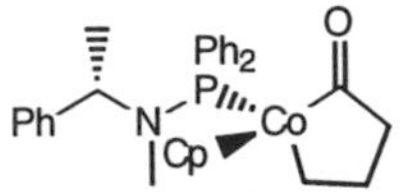 Ph, $Ph_2$, Cp, Co | Li | A[q] | 235 | 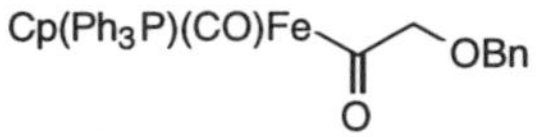 $Cp(Ph_3P)(CO)Fe$, OBn | Li | — | 235 |

[a] Relative stereochemistry of the new stereogenic centers is given (A = *ancat*, S = *syncat*). The use of a chiral ligand or auxiliary is noted where appropriate although the product stereochemistry has not. [b] In the presence of a chiral ligand or reagent. [c] As $(Ipc)_2B$. [d] Denotes that a silyl enol ether is the precursor even if a transmetallation is conducted. [e] With respect to the new stereocenters. The trimethylsiloxy group is *ancat*. [f] The *tert*-butyldimethylsiloxy group is also *syncat*. [g] As the epoxy ketone. [h] The stereochemistry, with respect to the stereogenic center within the substrate, can be controlled through the use of chiral boron agents. [i] The α-stannyl ester can also be used as a precursor. [j] With respect to the hydroxy and methyl groups. [k] The sulfoxide can be chiral. [l] In the presence of $[Au(c\text{-HexNC})_2]BF_4$ and a chiral ferrocene-derived ligand. [m] The product outcome is determined by enolate geometry. [n] Not applicable. Reaction occurs at the methyl group. [o] Two equivalents of borane are required. [p] Aromatic aldehydes give the *ancat* isomer. [q] This serves as a cyclobutanone equivalent.

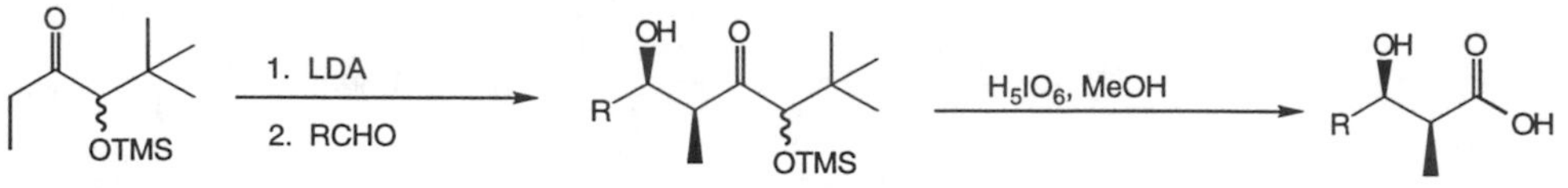

SCHEME 7.2.

The use of chiral catalysts to bring about an asymmetric aldol reaction[237,238] was exploited in a synthesis of erythromycin (Scheme 7.3).[239-241] Use of L-proline gave racemic products, whereas D-proline led to good asymmetric induction — an example of double asymmetric induction.[242]

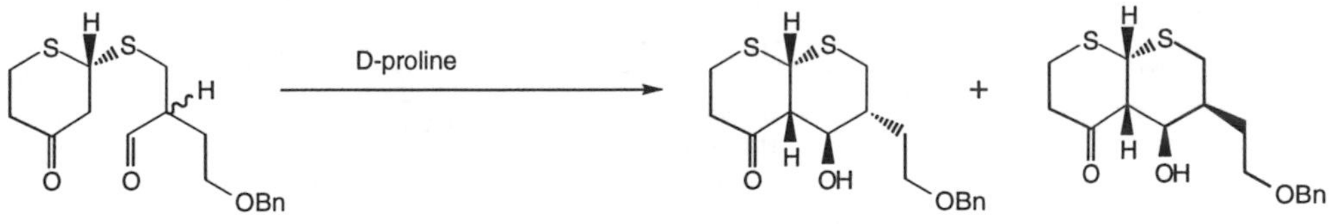

SCHEME 7.3.

Removal of the sulfur atoms provided the acyclic aldol product [*13.4*].[239,240,243]

## 7.2. THE ALDOL REACTION: THE ROLE OF ENOLATE GEOMETRY AND METAL COUNTERION

To access all of the four possible aldol products, various metals have to be employed so that the geometry of the transition state can be manipulated. These subtleties are demonstrated in Scheme 7.4. The key starting material for this protocol is the α-siloxy ketone **7.3**, prepared as shown in Scheme 7.5.[‡ 117]

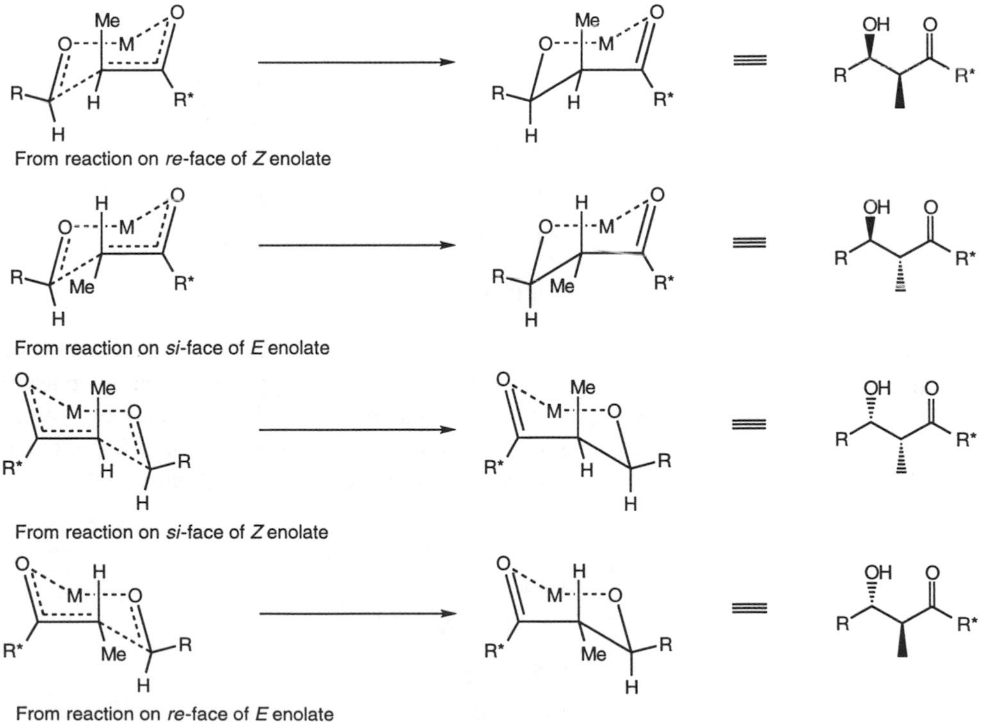

From reaction on *re*-face of *Z* enolate

From reaction on *si*-face of *E* enolate

From reaction on *si*-face of *Z* enolate

From reaction on *re*-face of *E* enolate

SCHEME 7.4.

‡ Contamination of the silyl ether by alcohol from the organometallic step necessitates the hydrolysis, resilylation procedure.

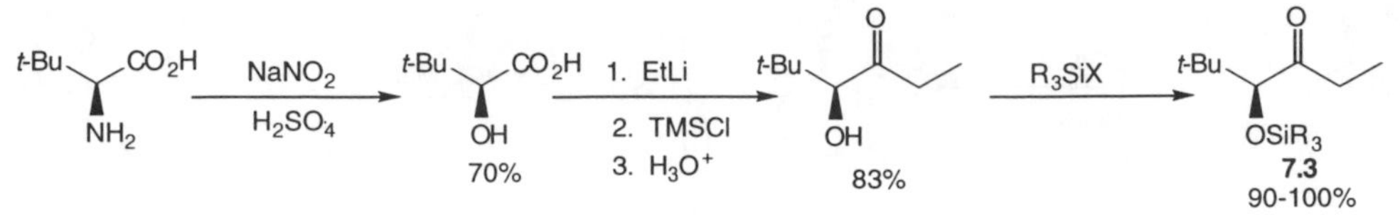

SCHEME 7.5.

The reactions to obtain the various geometric products are summarized in Scheme 7.6.[117]

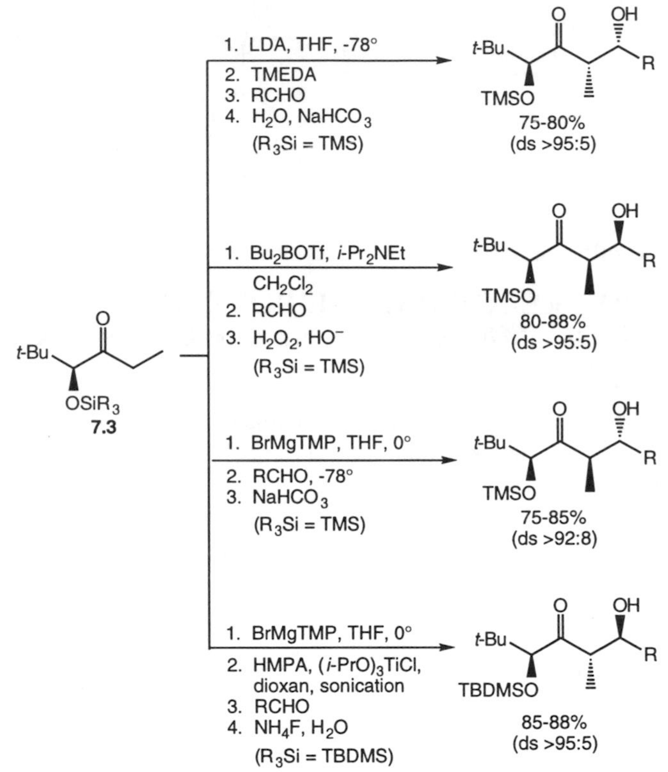

SCHEME 7.6.

This selectivity can be rationalized in terms of the transition state and how the metal complexes. Lithium (Figure 7.1) and magnesium form a complex with three oxygen atoms in the transition state (Figure 7.1),[244] while boron (Figure 7.2) and titanium can coordinate to only two oxygen atoms [*5.1*], so the siloxy group must be released for reaction to occur.[117,245] In addition, dipolar repulsions of the adjacent carbon–oxygen bonds provide the orientation (Figure 7.2).[26,246]

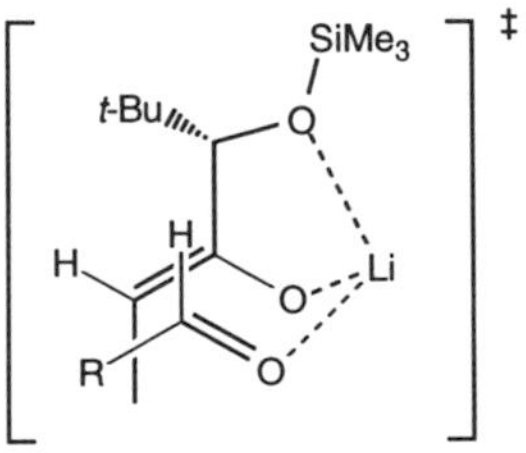

**FIGURE 7.1.** Aldol transition state with lithium as counterion.

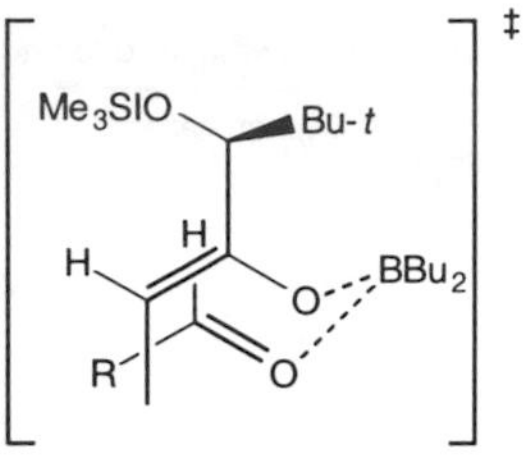

**FIGURE 7.2.** Aldol transition state with boron as counterion.

Thus, the *Z* lithium and *Z* boron enolates of the α-siloxy ketones **7.3** (R = Me) and **7.4** provide the complimentary diastereoisomeric products (Schemes 7.7 and 7.8).[26,93,117,118]

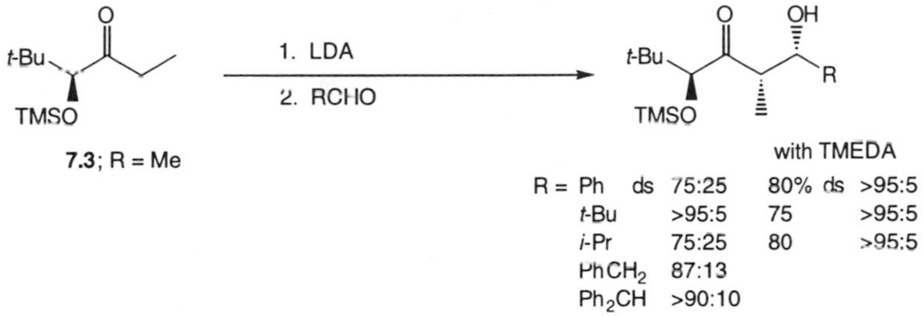

**7.3**; R = Me

1. LDA
2. RCHO

| R = | ds | with TMEDA | |
|---|---|---|---|
| Ph | 75:25 | 80% | ds >95:5 |
| *t*-Bu | >95:5 | 75 | >95:5 |
| *i*-Pr | 75:25 | 80 | >95:5 |
| $PhCH_2$ | 87:13 | | |
| $Ph_2CH$ | >90:10 | | |

**SCHEME 7.7.**

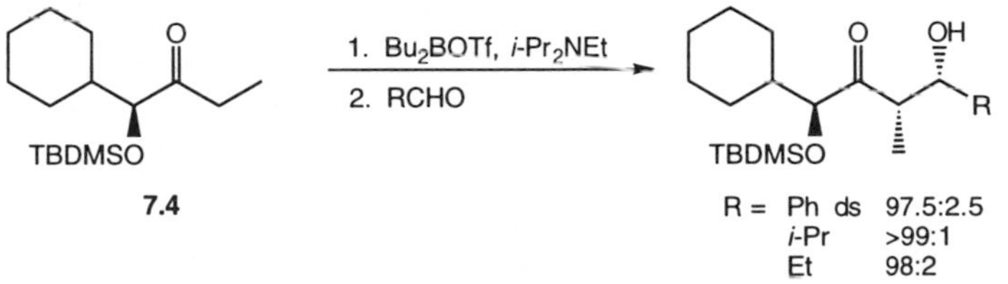

**7.4**

1. $Bu_2BOTf$, $i$-$Pr_2NEt$
2. RCHO

| R = | ds |
|---|---|
| Ph | 97.5:2.5 |
| *i*-Pr | >99:1 |
| Et | 98:2 |

**SCHEME 7.8.**

The use of magnesium provides the *E*-enolate that, in turn, provides the *ancat* product (Scheme 7.9).[26]

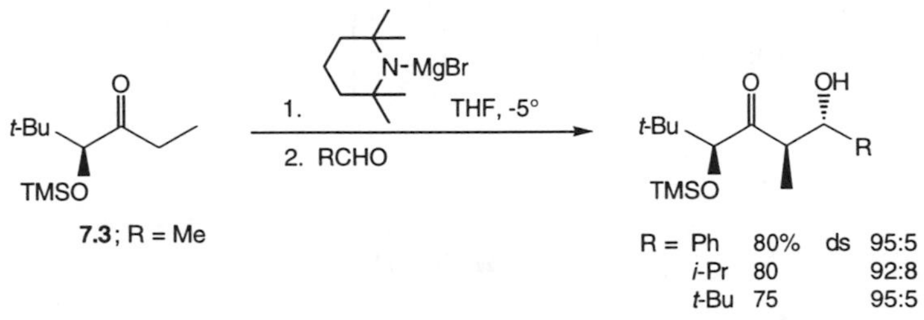

**7.3**; R = Me

1. N-MgBr, THF, -5°
2. RCHO

| R = | | ds |
|---|---|---|
| Ph | 80% | 95:5 |
| *i*-Pr | 80 | 92:8 |
| *t*-Bu | 75 | 95:5 |

**SCHEME 7.9.**

To obtain the last stereoisomer, it was necessary to use a titanium enolate (Scheme 7.10).[26]

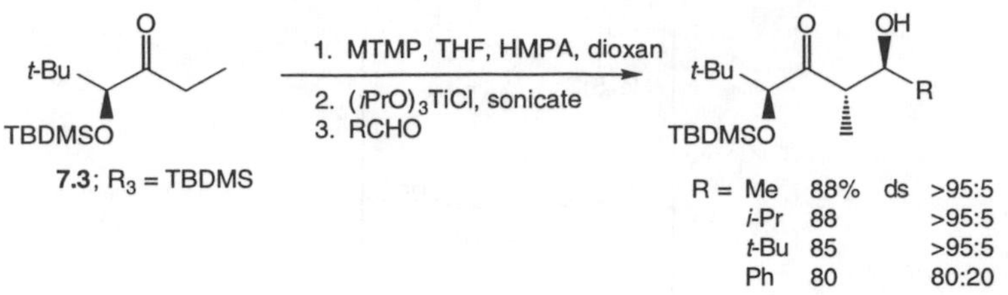

| R | Yield | ds |
|---|---|---|
| Me | 88% | >95:5 |
| *i*-Pr | 88 | >95:5 |
| *t*-Bu | 85 | >95:5 |
| Ph | 80 | 80:20 |

**SCHEME 7.10.**

Again, only two oxygen atoms can complex to the metal in the transition state with dipolar interactions providing the stereochemical control (Figure 7.3).

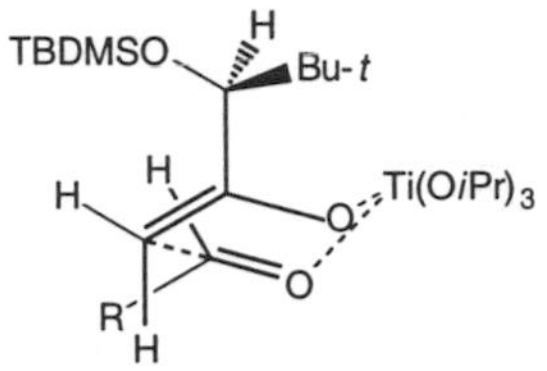

**FIGURE 7.3.** Aldol transition state with titanium as counterion.

It is noteworthy that with oxazolidinones, titanium reverses the stereochemistry of the new stereogenic centers, while still being *syncat*, compared to boron. This was rationalized by titanium being able to participate in chelation control (*cf.* Table 7.1).[217]

With the same enolate geometry, boron and titanium mediated aldol reactions provide the same *syncat* product;[247] isolated yields are often higher with titanium. The transition state requirements can, therefore, be summarized for both of these metals by Scheme 7.11.[248]

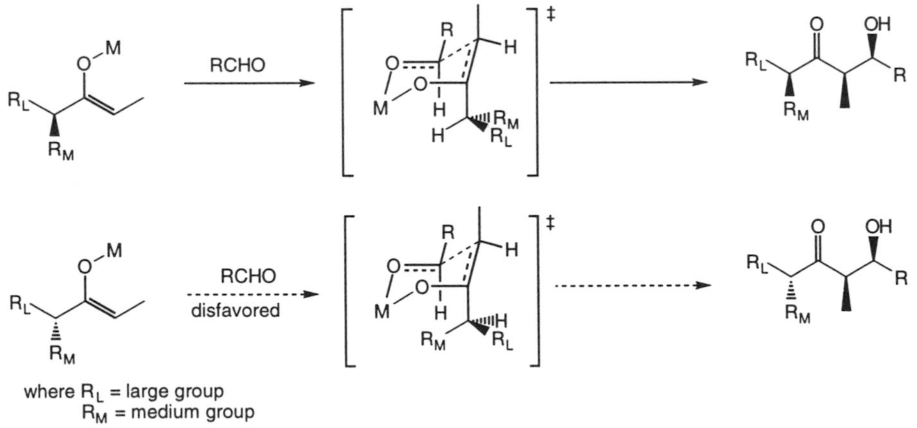

where $R_L$ = large group
$R_M$ = medium group

**SCHEME 7.11.**

## 7.2.1. BORON ENOLATES

Boron enolates have been used extensively to control the stereochemistry during an aldol reaction. Boron's small size does not allow coordination to other oxygen atoms present within the reaction parameters. In addition to high selectivity (Scheme 7.12),[90,118,249] a reversal of that seen with a lithium counterion can also be observed.[31,90,110,181,216]

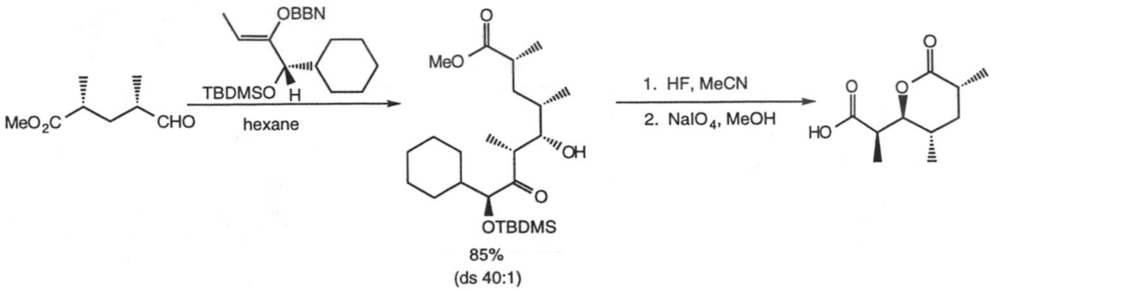

SCHEME 7.12.

In early studies, alkylboranes were added to preformed lithium enolates to generate the boron enolate.[250] This technique has been superseded by treatment of the ketone with an appropriately substituted borane in the presence of the base,[251-253] as this allows the use of a chiral borane, but other approaches are available [*5.1*].[47,90,108,125,133,194,204,254-257] A silyl enol ether can also be used as a boron enolate precursor.[258] A suitable choice of chiral reagent overcomes the low selectivity associated with mismatched pairs.[125] Indeed, some very high enantioselectivities have been observed (Scheme 7.13).[134,192,259]

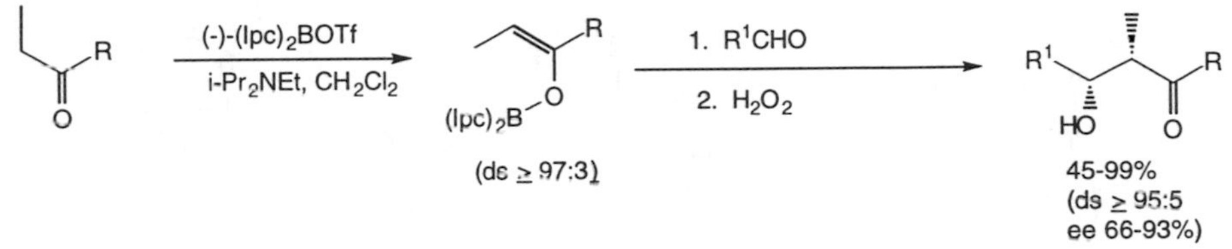

SCHEME 7.13.

The use of a chiral boron enolate allows for a choice in the product stereochemistry (Scheme 7.14),[259,260] as does the β-keto imide **7.5** (Scheme 7.15).[261]

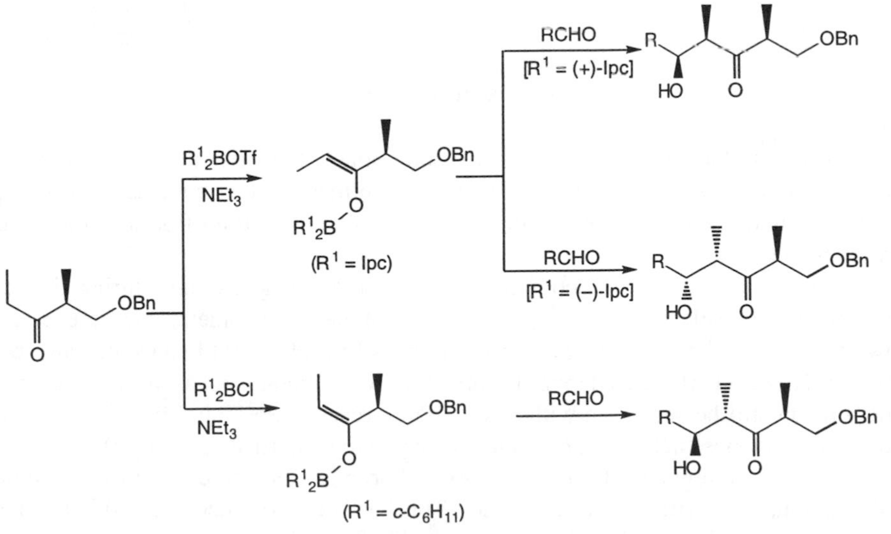

SCHEME 7.14.

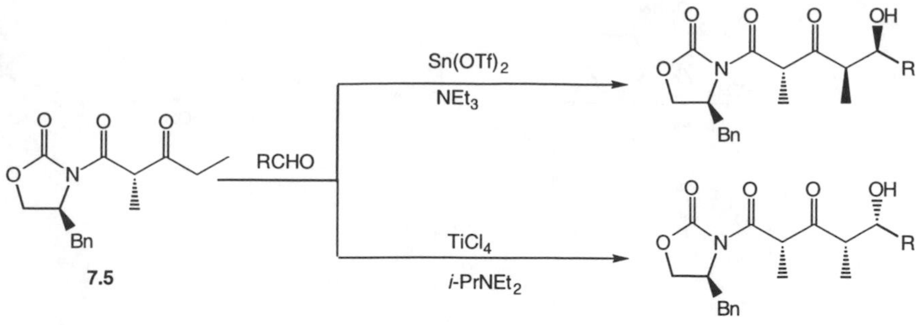

SCHEME 7.15.

To observe high selectivity with oxazolidinone auxiliaries, an α-substituent should be present. This observation has been rationalized in terms of the Zimmerman-Traxler model where the interactions between the α-substituent (Y) and the control group drive the two possible reaction pathways. The approach of the aldehyde is away from the bulky control group to favor intermediate **7.6**, although the alternative, **7.7**, is still viable when Y = H (Scheme 7.16).[24,31,52]

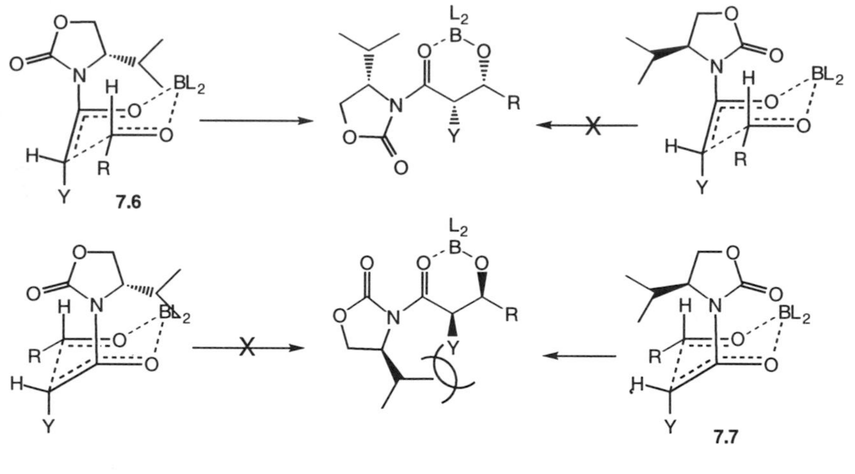

SCHEME 7.16.

For boron enolates based on Evans' oxazolidinone chemistry, the use of triethylamine, rather than Hunig's base, has been advocated, as the former provides high diastereoselectivities. This implies that the resultant ammonium salt plays a role in the transition state for the reaction with aldehydes.[216,220]

As noted above, boron can coordinate to only two oxygen atoms during the reaction. However, as face selectivity is still high, control of enolate geometry can be exercised and allows access to either the *syncat* or *ancat* products [*5.1*];[262] chiral ligands can also be borne by boron. However, if the nucleophilic carbonyl group contains an asymmetric center, good induction can still be seen even with achiral boron reagents.[256] Chiral ligands are, of course, required for ketones that do not contain a stereogenic center (Scheme 7.17).[108]

Thus, there is a tendency for the chirality of boron reagents to control the outcome of an aldol reaction, even with mismatched cases. This has been exploited to provide methodology to *syncat,ancat* (**7.8** and 7.**9**) and *syncat,syncat* (**7.10** and **7.11**) diastereoisomers with >93% de and >99% ee (Schemes 7.14 and 7.18).[125]

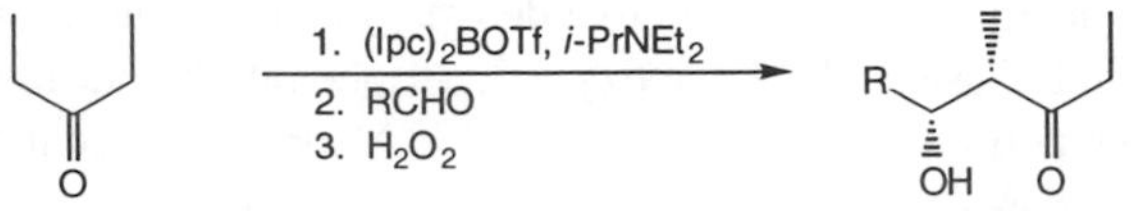

**SCHEME 7.17.**

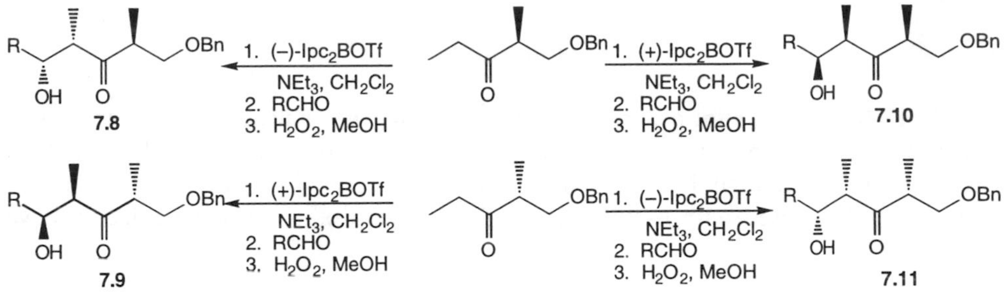

**SCHEME 7.18.**

The *syncat* selectivity is a consequence of the chiral boron enolate.[108]

## 7.2.2. LITHIUM ENOLATES

Lithium was the first metal to provide useful selectivity in aldol-type reactions. It has been the subject of many studies. Perhaps one of the most important findings, from early studies, was a strong correlation between product stereochemistry and enolate geometry.[54,83,88] As an example, stereocontrolled aldol strategy has been employed in a synthesis of ristosamine (**7.12**) (Scheme 7.19).[263,264]

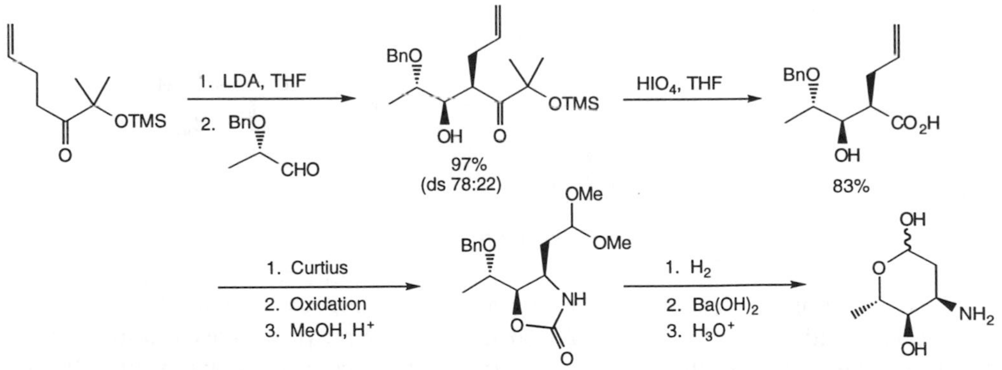

**SCHEME 7.19.**

Lithium still provides a plethora of methods and our understanding of the control factors continues to expand. For α-alkoxy ketones, and other α-substituted carbonyl compounds, there is a marked rate of acceleration for the aldol reaction compared to unsubstituted examples.[265] Interestingly, reaction of an ester enolate with an aldehyde in the solid state gave the same diastereoselectivity as the solution reaction.[266]

The addition of an acetic acid equivalent to aldehydes can be achieved with a high degree of stereoselection through the use of (*R*)-2-acetoxy-1,2,2-triphenylethyl acetate (**7.13**),[147-149] in turn available from (*R*)-mandelic acid (Scheme 7.20).[149,267,268] The (*R*)-reagent adds to the aldehyde predominately from the *Re*-face. The corresponding (*S*)-reagent gives *Si*-face attack.[24]

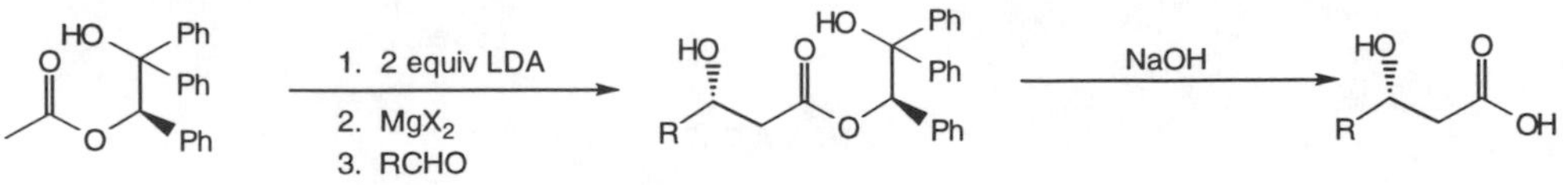

1. 2 equiv LDA
2. $MgX_2$
3. RCHO

NaOH

**7.13**

**SCHEME 7.20.**

With aldehydes containing a chiral center, the problem of mismatched pairs can arise; when the facial selectivity works in opposition, a smaller diastereomeric excess is obtained.[53] The selectivity can still be high, however, with β-alkoxy aldehydes (Scheme 7.21).[24] This chiral auxiliary has been used with α-amino aldehydes and respectable diastereoselectivity was obtained.[150,151]

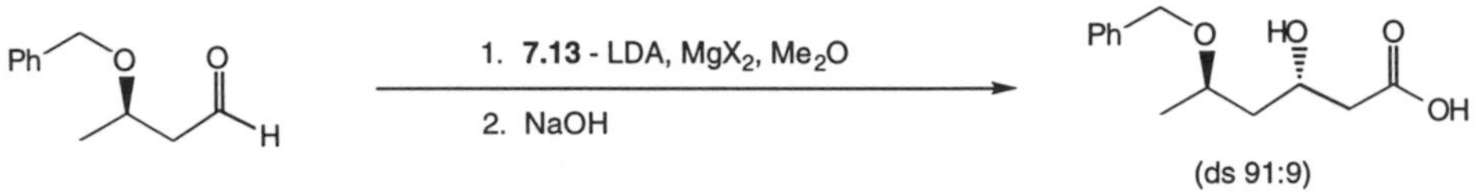

1. **7.13** - LDA, $MgX_2$, $Me_2O$
2. NaOH

(ds 91:9)

**SCHEME 7.21.**

The use of a chiral base [*5.4.1*] can provide reasonable enantioselectivity (Scheme 7.22),[269,270] but it is limited to methyl ketones where the second group does not contain an acidic α-hydrogen.[269,270]

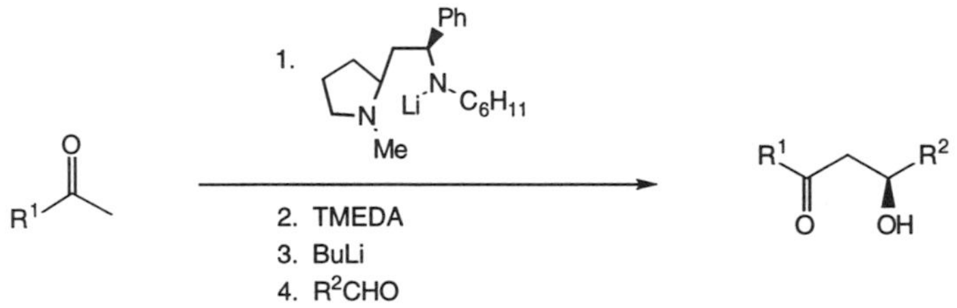

1.
2. TMEDA
3. BuLi
4. $R^2CHO$

**SCHEME 7.22.**

This type of approach allows for the preparation of contiguous asymmetric centers.[159-161]

Special mention must be made of 2-oxazolines, which provide routes to β-hydroxy and β-alkoxy acids. Although the enantiomeric excesses are usually not high (*ca.* 20 to 25%),[271] the adducts are useful precursors to 1,4-addition methodology [*9.5.2*].[222,272-275] Indeed, many of the chiral auxiliaries used for the α-alkylation of carbonyl compounds, such as hydrazones, have been used to effect stereoselective aldol reactions with varying degrees of success (Table 7.1) [*5.2.1*].[207]

Mention must also be made of carboxylic dianions that result in an *ancat* selective reaction with aldehydes;[276,277] the dianion of propanedithioic acid can provide some facial selectivity.[278] However, as any stereogenic center must be an integral part of the molecule, the use of chiral auxiliaries, other than metal ligands, is rather difficult.[131,132]

### 7.2.3. MAGNESIUM ENOLATES

The most important reactions of magnesium enolates have been discussed in the introductory sections (see also Scheme 7.20).

Condensation of chiral esters with aldehydes usually results in low levels of asymmetric induction.[143] Incorporation of a chiral sulfoxide, however, as a chiral auxiliary allows for the synthesis of β-hydroxy acids in good chemical and optical yields (Scheme 7.23).[154,155,189,191,209,279-281]

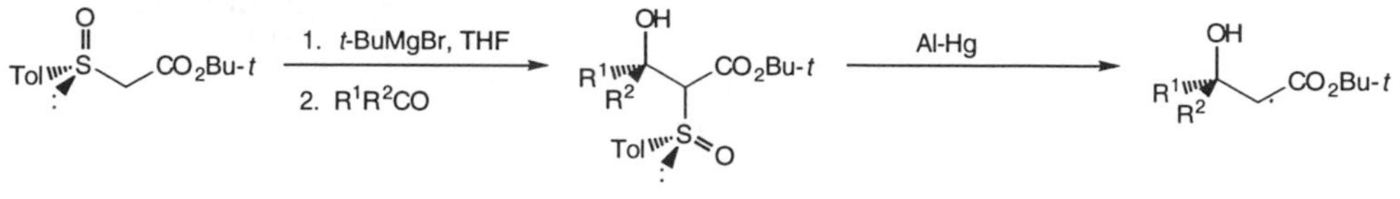
Tol
S
O
CO2Bu-t
1. t-BuMgBr, THF
2. R1R2CO
OH
R1
R2
CO2Bu-t
Tol
S
O
Al-Hg
OH
R1
R2
CO2Bu-t

SCHEME 7.23.

### 7.2.4. SILYL ENOL ETHERS

Silyl enol ethers react, in the presence of a Lewis acid catalyst,¶ with carbonyl compounds or acetals (the Mukaiyama reaction)[282,283] to afford aldol-type products (Scheme 7.24).[126,284-292] Although most examples do not show high stereoselection, those derived from *tert*-alkyl ketones do (Scheme 7.25).[111,293] The use of enoxysilacyclobutyl and other silicon derivatives that do not require the presence of a Lewis acid has been advocated.[294,295]

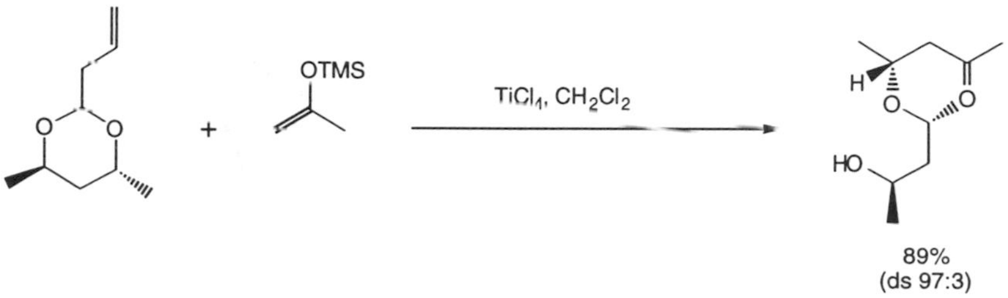
O
O
+
OTMS
TiCl4, CH2Cl2
H
O
O
HO
89%
(ds 97:3)

SCHEME 7.24.

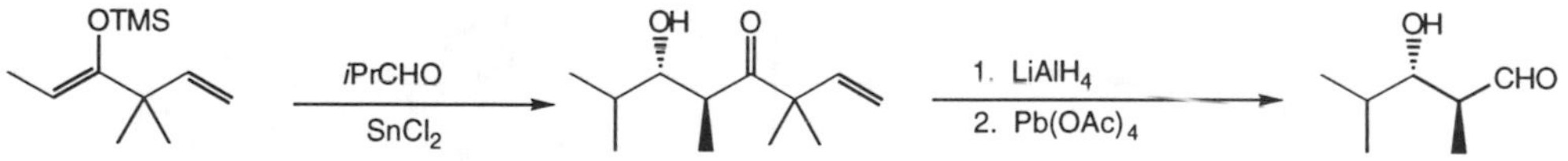
OTMS
iPrCHO
SnCl2
OH
O
1. LiAlH4
2. Pb(OAc)4
OH
CHO

SCHEME 7.25.

Silyl enol ethers often provide excellent diastereofacial selectivity with aldehydes to afford the Anh-Felkin product.[100,101] Stereoselectivity can arise from the stereochemical requirements of one of the reactants, such as the acetal structure (*cf.* Scheme 7.24),[296-300] the silyl enol ether,[117,144,211,301,302] or from the incorporation of a chiral ligand into the catalyst (Scheme 7.26).[202,303,304]

Indeed, a wide variety of Lewis acid catalysts has been proposed for the Mukaiyama reaction to try and exercise control over the product's relative stereochemistry.[290,305] Although an acetal can provide good diastereoselectivity, the chiral auxiliary is destroyed during its removal, by a β-elimination (*cf.* Scheme 7.24), and several steps are required to ensure that this elimination does not occur within the desired product's framework.[126]

¶ This can result in a transmetallation. For convenience, all reactions that employ a silyl enol ether as precursor have been included in this section.

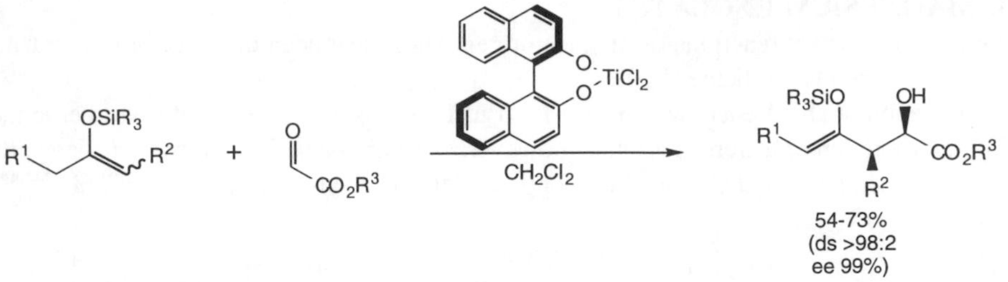

SCHEME 7.26.

In contrast to lithium enolates,[306] reaction of a silyl enol ether or ketene acetal with a chiral aldehyde can provide high stereoselection.[26] This effect has been rationalized in terms of the angles of approach of the nucleophile [*2.1*].[100] For the addition of a nucleophile to a chiral aldehyde, the trajectory of the nucleophile will, most likely, be away from the bulky group; this is away from the stereogenic center (Scheme 7.27). This minimizes the effect of this center. Lewis acids, such as boron trifluoride etherate, complex to the carbonyl oxygen atom of the aldehyde *cis* to the hydrogen, allowing the bulky Lewis acid group to counteract the normal steric bias and increase diastereofacial preference compared to the uncomplexed aldehyde (Scheme 7.28).[26,100,307]

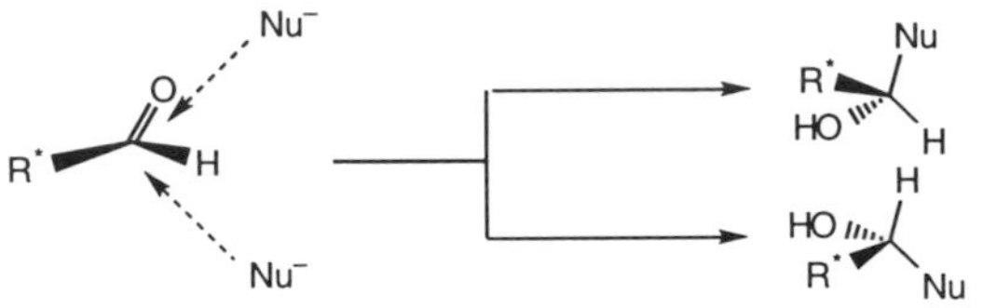

SCHEME 7.27.

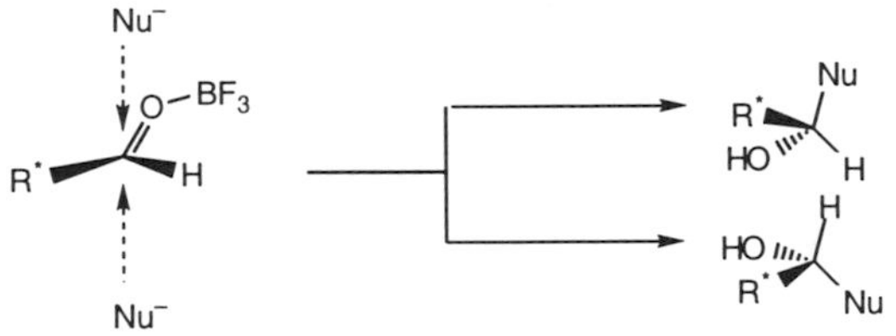

SCHEME 7.28.

This argument also holds for the reaction of a silyl enol ether with an acetal.[297] In addition, the use of a silyl enol ether can "reverse" the stereochemical outcome of an aldol reaction compared to a lithium-based approach.[211,218]

Functionalized carbonyl compounds, such as menthyl pyruvate can also be used with silyl enol ethers to provide some degree (<68%) of asymmetric induction.[308] The use of a functionalized group on the enol oxygen allows for the incorporation of a stereogenic center as illustrated in Scheme 7.29; the starting material **7.14** is derived from an aldehyde and an α-hydroxy acid.[195]

Ketene acetals can undergo reaction with chiral α-thio- or α-alkoxy aldehydes to give either the *syncat* or *ancat* product; the Lewis acid is the key control factor (Scheme 7.30).[309-311] α-Amino aldehydes can be generated and used *in situ* from α-amino esters.[122,312]

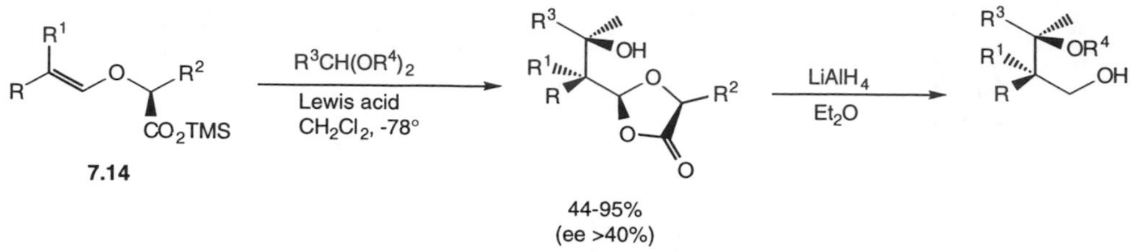

R¹ R O R² CO₂TMS 7.14 → R³CH(OR⁴)₂ Lewis acid CH₂Cl₂, -78° → R³ OH R¹ R O R² O O 44-95% (ee >40%) → LiAlH₄ Et₂O → R³ OR⁴ R¹ R OH

SCHEME 7.29.

SMe i-Pr CHO + OTMS OMe → TiCl₄ → SMe i-Pr CO₂Me OH 76% (ds 93:7); → BF₃•OEt₂ → SMe i-Pr CO₂Me OH 90% (ds ≥98:2)

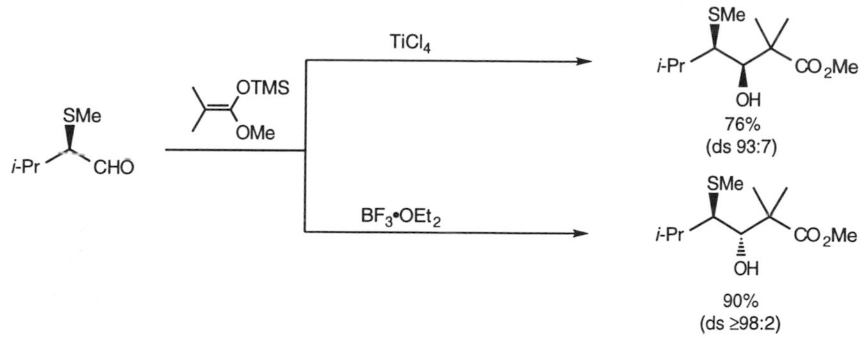

SCHEME 7.30.

Oxazolidines (e.g., **7.15**) provide an alternative to acetals for this type of approach (Scheme 7.31).[312-314] The stereoselection is not, however, high across a broad range of substrates. The silyl enol ether approach can be extended to silyl ketene acetals where the additional alkoxy moiety is a useful handle for the incorporation of a chiral unit.[198]

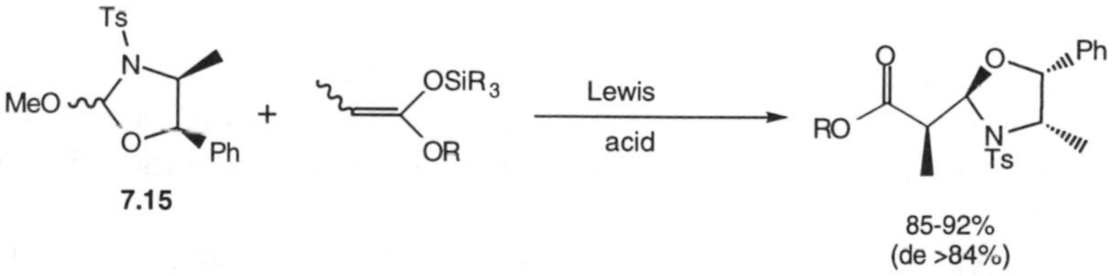

Ts N MeO O Ph 7.15 + OSiR₃ OR → Lewis acid → O RO O Ph N Ts 85-92% (de >84%)

SCHEME 7.31.

The use of a ketene acetal with titanium tetrachloride has enhanced selectivity for the *ancat* isomer if triphenylphosphine is added.[302]

In addition to reactions that are designed for the preparation of β-hydroxy carbonyl compounds, other functionality is available as illustrated by the preparation of 1,3-diols (Scheme 7.32).[315]

The use of a lactone enol ether **7.16** provided just the diastereoisomer **7.17** with the titanium chloride reaction of acetaldehyde. The Peterson elimination then provided the unsaturated lactone (Scheme 7.33).[316,317]

Chirality transfer can occur between a chiral silane, as part of a silyl enol ether, in the reaction with acetals. However, the degree of induction is not high (<26% ee).[317]

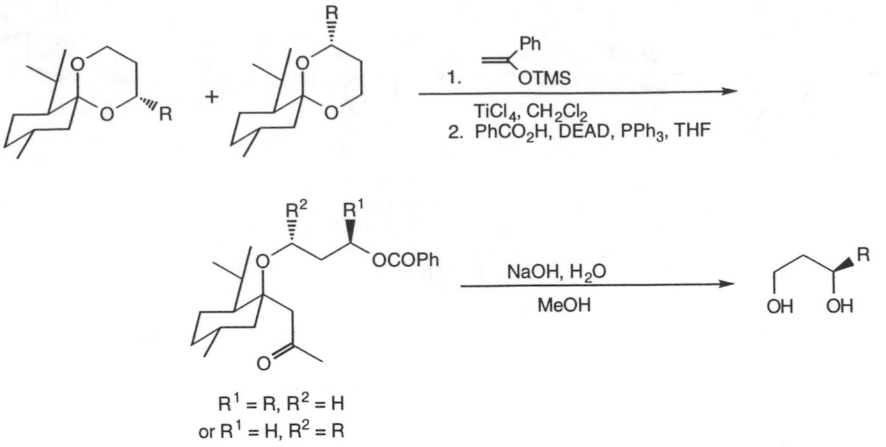

1. Ph, OTMS; $TiCl_4$, $CH_2Cl_2$
2. $PhCO_2H$, DEAD, $PPh_3$, THF

OCOPh

$NaOH$, $H_2O$, MeOH

OH OH

$R^1 = R$, $R^2 = H$
or $R^1 = H$, $R^2 = R$

SCHEME 7.32.

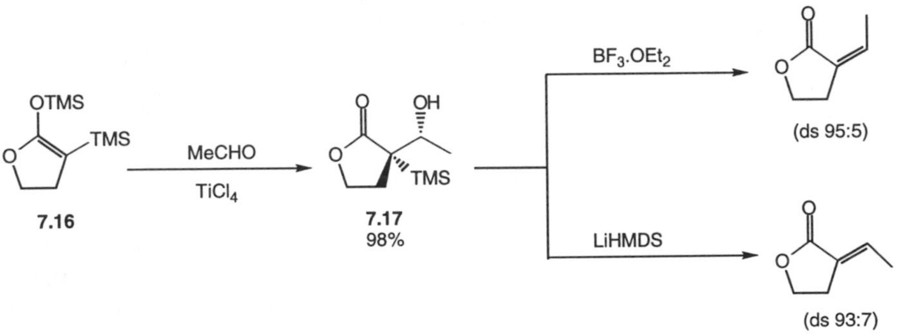

OTMS, TMS, **7.16**; MeCHO, $TiCl_4$; OH, TMS, **7.17** 98%; $BF_3.OEt_2$ (ds 95:5); LiHMDS (ds 93:7)

SCHEME 7.33.

As with allylic anions [*4.1.5*], carbon–carbon unsaturation can be masked through complex formation with a transition metal;[318] otherwise, the unsaturation can influence the reaction's outcome.[319] One face of an aryl group can be masked by use of a transition metal–aryl complex.[320]

The use of silyl enol ethers has been invoked for metal exchange reactions, such as with lithium or titanium, the resultant chemistry then being a function of that metal's enolate chemistry (vide supra).[258,288,321-323] The fluoride-catalyzed aldol reaction seems to follow the "non-chelate" transition state even when the enol ether is not sterically demanding. The use of tetrabutylammonium as the counterion provides very similar selectivity to the use of tris(dimethylamino)sulfonium,[324-327] while chiral ammonium salts are also accessible.[328]

## 7.2.5. TIN COMPOUNDS

In addition to silyl enol ethers, tin enolates can also be used — without the addition of a Lewis acid — to afford *ancat* selectivity.[329-331] An example is seen in the synthesis of the branch chain sugar, 2-*C*-methyl-D,L-*lyxo*-furanoside (**7.18**) (Scheme 7.34).[106,332]

Many control groups, including chiral auxiliaries, have been exercised in the aldol reaction. One simple example is the use of tin to provide the *ancat* β-hydroxy ester **7.19** (Scheme 7.35); the dienolate of the α,β-unsaturated carboxylic acid derivative **7.20** without a tin group gave a mixture of compounds.[33]

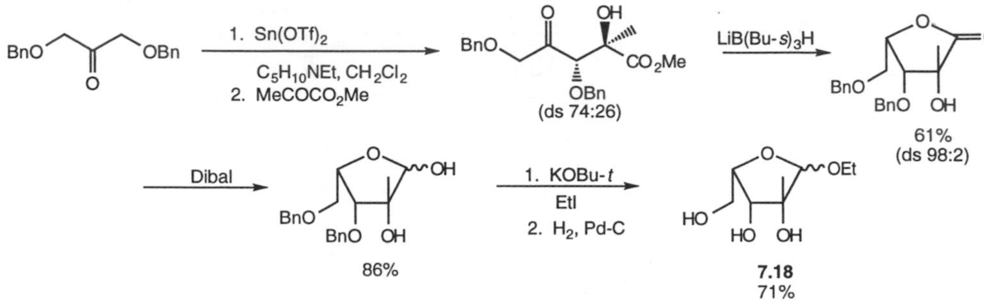

SCHEME 7.34.

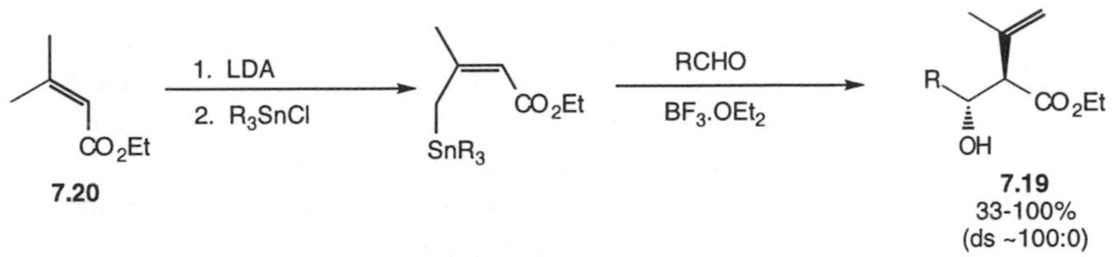

SCHEME 7.35.

The use of chiral ligands with tin enolates has led to selective methodology with respect to geometrical isomers and asymmetric induction (e.g., Scheme 7.36).[106] The highest selectivity is seen when R is bulky, such as *tert*-butyl.

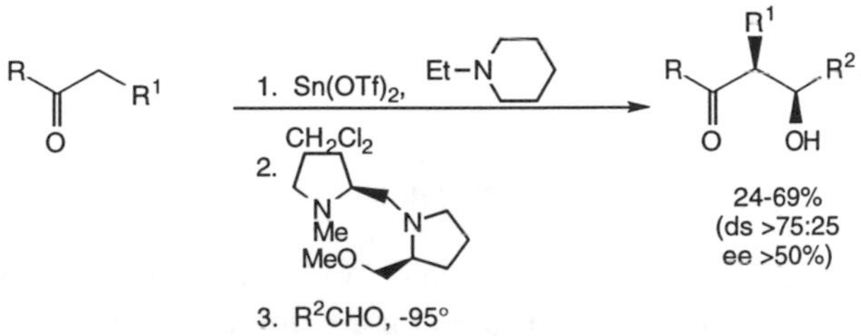

SCHEME 7.36.

### 7.2.6. TITANIUM ENOLATES

In addition to lithium and boron, many other metals have been proposed to control the stereoselection of the aldol reaction (vide supra).[13,22,24,31] A high degree of success is being realized with titanium enolates, especially when used in conjunction with a chiral auxiliary.[120,217,223,248] One of the problems associated with the use of titanium enolates was the need to preform the enolate either by reaction with a silyl enol ether or lithium enolate, but they are available through the use of titanium(IV) chloride in the presence of Hunig's base.[333,334]

High enantioselectivity can be achieved by use of a chiral ligand for titanium (Scheme 7.37).[135-137,335,336]

### 7.2.7. ZINC ENOLATES AND THE REFORMATSKY REACTION

The Reformatsky reaction is closely related to the aldol reaction; the zinc enolate is formed by reaction of an α-bromocarbonyl compound.[337-343] The Reformatsky reagent has been shown to exist as a cyclic dimer in the solid state,[344] but the exact mechanism is still a subject of discussion.[342,345]

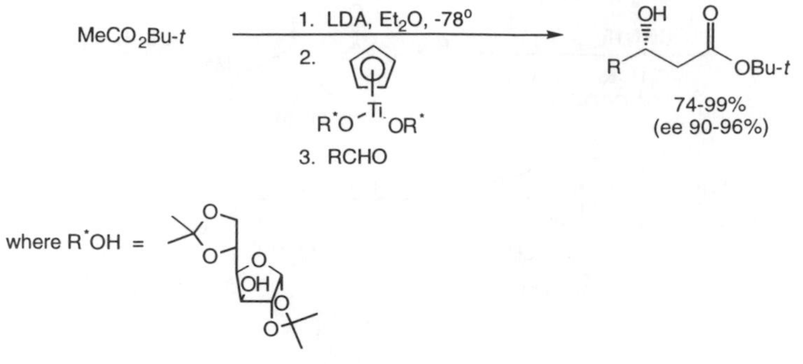

SCHEME 7.37.

Diastereoselectivity can be observed with zinc enolate addition to carbonyl compounds.[81,340,346,347] The Reformatsky reaction with imines has been studied extensively, as β-lactams are the result of this condensation.[348-360] The major product is usually the *trans* β-lactam,[350,354,361] although this can be influenced by substituents.[347,349,351-353]

One of the major problems associated with the Reformatsky approach to β-lactams is the stereoselectivity of the reaction with a zinc counterion. In some cases, low selectivity is observed, while in others there is a high solvent dependency.[337,353,362] The use of a lithium enolate (vide infra), as it does not require an α-bromo ester for its generation, can remove some of these limitations.[337,363-368] Although the stereochemical outcome for these lithium reactions is still solvent dependent, it can be rationalized in terms of the ester enolate geometry [*5.1*];[337] an example is given in Scheme 7.38.[31]

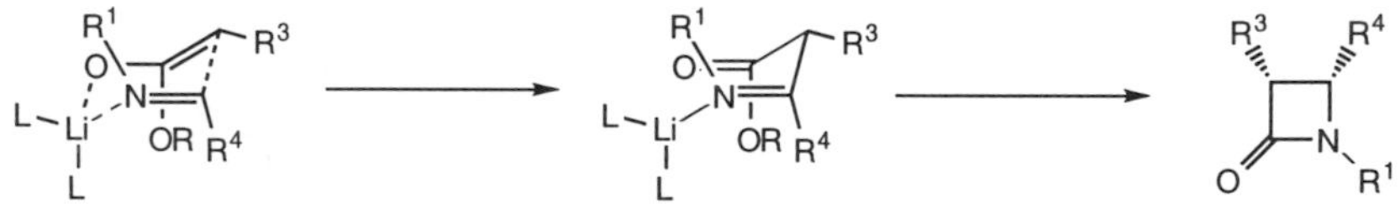

SCHEME 7.38.

In addition to zinc, a zinc-diethylaluminium chloride reagent provides for *syncat* selectivity.[369-371] Even aqueous conditions can be used for a Reformatsky reaction.[372] Imines can also be generated *in situ* from suitable precursors,[356] or alternatives such as 2-substituted 3-benzyl-1,3-oxazinanes can also be used as substrates.[373]

Asymmetric induction, as with other aldol-derived reactions, has been attempted by incorporation of a chiral auxiliary into the enolate moiety as an ester or other derivative.[353,362,363,374-378] Some success can be achieved through application of Prelog's rule [*2.1.4*]. The induction can also be derived through the use of a chiral imine.[362,364,379,380] Both of the approaches for induction can be employed with ketene acetals (e.g., Scheme 7.39),[381] and can also include the use of a chiral Lewis acid.[382]

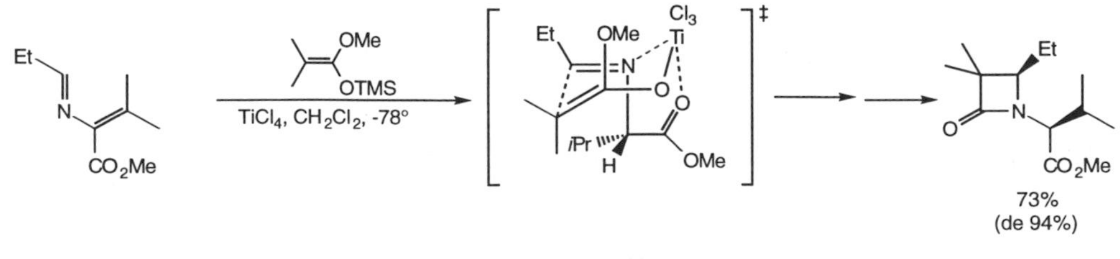

SCHEME 7.39.

The use of a chiral imine with a ketene acetal does provide some diastereoselectivity (72 to 89%) and asymmetric induction (44 to 78%).[381] Reaction with an aldehyde, rather than an imine, does provide some selectivity, but it is not useful compared to alternative approaches.[346]

Asymmetric induction occurs if the reaction is conducted with ketones or imines in the presence of a chelation agent, such as sparteine,[362,383-385] or if a chiral ester is incorporated.[386-388] Chiral auxiliaries can also be employed, the steric requirements of the auxiliary system determining the stereochemistry, and degree of induction in the product.[271]

The use of acetals, or equivalents, with a Reformatsky reagent, in the presence of a Lewis acid catalyst, again allows for enantioselectivity (Scheme 7.40).[285,389,390]

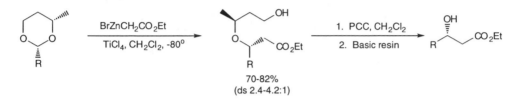

$BrZnCH_2CO_2Et$; $TiCl_4$, $CH_2Cl_2$, -80°; 70-82% (ds 2.4-4.2:1); 1. PCC, $CH_2Cl_2$ 2. Basic resin

**SCHEME 7.40.**

In addition to zinc enolates, other metal enolates react with imines to provide β-amino carbonyl compounds [*cf. 4.3*] or β-lactams.[298-300,356,363] Such an example is provided by β-hydroxy esters where the presence of lithium chloride was found to improve the stereoselection for the reaction with imines (Scheme 7.41).[70,391-393] Similar observations can be made with β-amino esters.[394]

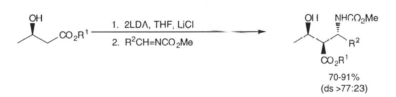

1. 2LDA, THF, LiCl 2. $R^2CH{=}NCO_2Me$; 70-91% (ds >77:23)

**SCHEME 7.41.**

Tin can also be used in place of zinc and provides the *syncat* product (Scheme 7.42); the action of the tin(II) triflate on the parent ketone provides similar results.[23,253,395,396]

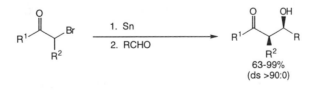

1. Sn 2. RCHO; 63-99% (ds >90:0)

**SCHEME 7.42.**

The Reformatsky reaction can also be used in an intramolecular version. One version is promoted by samarium rather than zinc, and provides a methodology to functionalized valerolactones (Scheme 7.43).[397]

Returning to zinc chemistry, this provides methods for the use of homoenolates, where the stereochemistry at the α-position is preserved (Scheme 7.44); alkylations are also available with this organometallic reagent.[398]

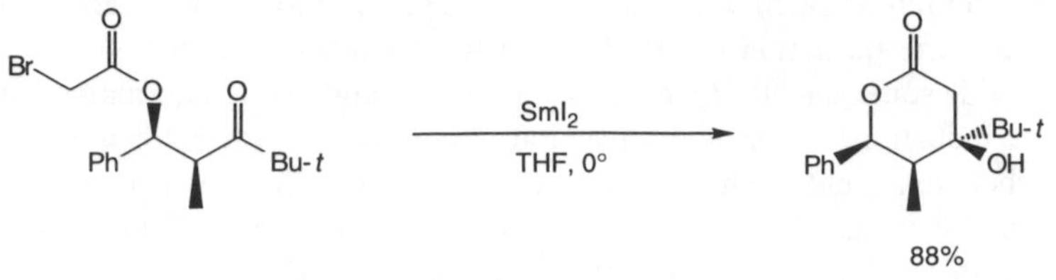

SCHEME 7.43.

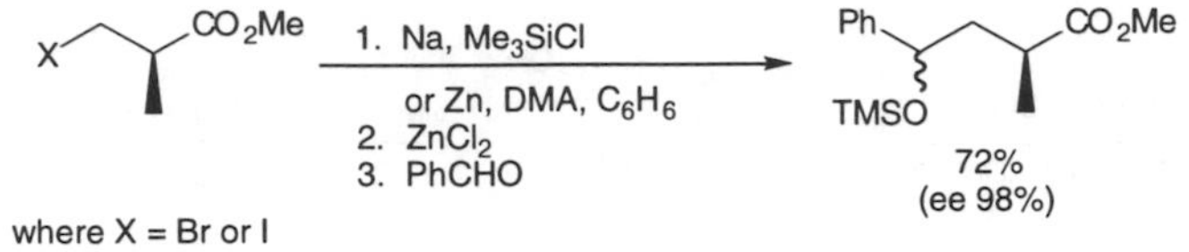

where X = Br or I

SCHEME 7.44.

## 7.3. THE FORMOSE REACTION

The formose reaction[399,400] affords a useful method for the formation of poly-1,2-diol units from formaldehyde.[5,401,402] The use of this reaction, however, has practical problems of execution on a small scale, including isolation of the product,[403] and is more amenable to an industrial scale.[400,402]

The formose reaction is also catalyzed by thiazolium salts, such as thiamin; these catalysts have provided a selective method for the conversion of formaldehyde to 1,3-dihydroxyacetone.[404-408]

An iterative process that adds a one carbon unit to a carbonyl compound is better suited in the laboratory to the use of a formyl anion equivalent.[409] The addition of the formyl anion equivalent has to be diastereoselective; 2-trimethylsilylthiazole has been advocated [*4.1.3*].

In a related reaction, condensation of an enone with an aldehyde under hydrosilylation conditions provides a route to β-siloxy carbonyl compounds (Scheme 7.45).[410]

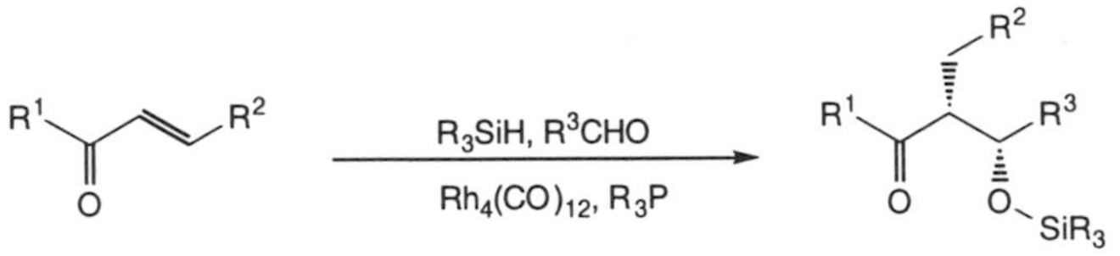

SCHEME 7.45.

An alternative strategy to 1,2-functionality of this type is the addition of a formyl anion equivalent to a carbonyl compound with diastereofacial control (e.g., Scheme 4.7). The resultant masked α-hydroxy carbonyl compound can then be manipulated in a wide variety of ways. Many umpolung reagents are available to fulfill this role, but the control over the new stereogenic centers lies in face selection [*2.1*].[409,411-413] Reaction of an α-alkoxy organometallic reagent [*3.2*] with an aldehyde can also provide an entry to 1,2-diols.[414] This approach to vicinal diols has been extended to a wide variety of functionalized allyl anions [*4.1.5*].[415-418] This is illustrated by the allylstannane **7.21** that allows preparation of either the *syncat* or *ancat* products with an additional change in regiochemical control (Scheme 7.46).[419]

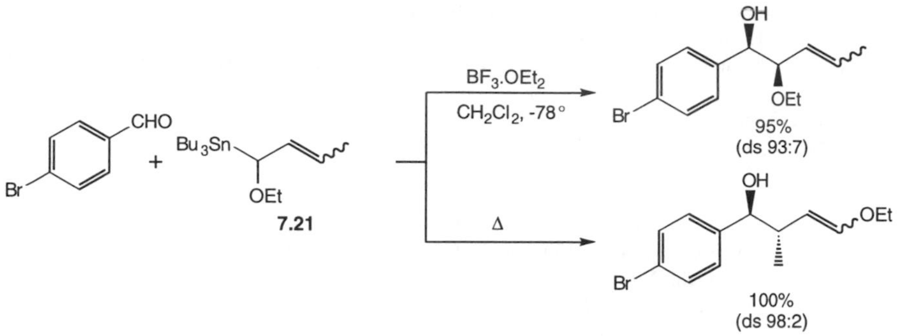

SCHEME 7.46.

## 7.4. SUMMARY

The aldol and closely related reactions are very powerful weapons in the armory of synthetic methodology. Not only is a carbon–carbon bond made, but control over the new stereogenic centers can be exercised by careful choice of reaction conditions and reagents employed. The major variables are control of the enolate geometry and metal counterion. All four of the possible stereoisomers are accessible, although some of these isomers do require very careful control of the reaction conditions — alternative approaches may be more successful in these cases.

With the potential for chirality to be incorporated into either of the carbonyl substrates, as well as ligands for the metal, the absolute control of asymmetry is also possible. The use of boron enolates has found wide acceptance in this regard.

The formose reaction allows for the preparation of 1,2-functionalized systems, as opposed to the aldol's 1,3-systems, with some control over the relative stereochemistry.

## 7.5. REFERENCES


1. Huckin, S. N.; Weiler, L. *Can. J. Chem.* 1974, *52*, 2157.
2. Huckin, S. N.; Wieler, L. *Tetrahedron Lett.* 1971, 4835.
3. Burkhardt, E. R.; Doney, J. J.; Slough, G. A.; Stack, J. M.; Heathcock, C. H.; Bergman, R. G. *Pure Appl. Chem.* 1988, *60*, 1.
4. Yamamoto, Y.; Maruyama, K. *J. Am. Chem. Soc.* 1982, *104*, 2323.
5. Ager, D. J.; East, M. B. *Tetrahedron* 1993, *49*, 5683.
6. Goodman, J. M.; Kahn, S. D.; Paterson, I. *J. Org. Chem.* 1990, *55*, 3295.
7. Bernardi, A.; Capelli, A. M.; Gennari, C.; Goodman, J. M.; Paterson, I. *J. Org. Chem.* 1990, *55*, 3576.
8. Denmark, S. E.; Henke, B. R. *J. Am. Chem. Soc.* 1991, *113*, 2177.
9. Denmark, S. E.; Henke, B. R. *J. Am. Chem. Soc.* 1989, *111*, 8032.
10. Li, Y.; Paddon-Row, N.; Houk, K. N. *J. Org. Chem.* 1990, *55*, 481.
11. Corey, E. J.; Lee, D.-H. *Tetrahedron Lett.* 1993, *34*, 1737.
12. Vulpetti, A.; Bernardi, A.; Gennari, C.; Goodman, J. M.; Paterson, I. *Tetrahedron* 1993, *49*, 685.
13. Masamune, S.; Choy, W. *Aldrichimica Acta* 1982, *15*, 47.
14. Evans, D. A. *Aldrichimica Acta* 1982, *15*, 23.
15. Evans, D. A. In *Asymmetric Synthesis*; J. D. Morrison, Ed.; Academic Press: Orlando, FL, 1984; Vol. 3; pp 1.
16. Davies, S. G. *Pure Appl. Chem.* 1988, *60*, 13.
17. Rasmussen, J. K. *Synthesis* 1977, 91.
18. Fleming, I. *Chem. Soc. Rev.* 1981, *10*, 83.
19. Kahn, S. D.; Pau, C. F.; Hehre, W. J. *J. Am. Chem. Soc.* 1986, *108*, 7396.

20. Nielsen, A. T.; Houlihan, W. J. *Org. Reactions* 1968, *16*, 1.
21. Heathcock, C. H. *Science* 1981, *214*, 395.
22. Heathcock, C. H. In *Asymmetric Synthesis*; J. D. Morrison, Ed.; Academic Press: Orlando, FL, 1984; Vol. 3; pp 111.
23. Mukaiyama, T. *Pure Appl. Chem.* 1982, *54*, 2455.
24. Braun, M. *Angew. Chem., Int. Ed. Engl.* 1987, *26*, 24.
25. Masamune, S. *Heterocycles* 1984, *21*, 107.
26. Heathcock, C. H. *Aldrichimica Acta* 1990, *23*, 99.
27. Stork, G.; Kraus, G. A.; Garcia, G. A. *J. Org. Chem.* 1974, *39*, 3459.
28. Kelleher, R. G.; McKervey, M. A.; Vibuljan, P. *J. Chem. Soc., Chem. Commun.* 1980, 486.
29. Nakamura, E.; Horiguchi, Y.; Shimada, J.-I.; Kuwajima, I. *J. Chem. Soc., Chem. Commun.* 1983, 796.
30. Mukaiyama, T. *Org. Reactions* 1982, *28*, 203.
31. Evans, D. A.; Nelson, J. V.; Taber, T. R. *Top. Stereochem.* 1982, *13*, 1.
32. Boschelli, D.; Ellingboe, J. W.; Masamune, S. *Tetrahedron Lett.* 1984, *25*, 3395.
33. Yamamoto, Y.; Hatsuya, S.; Yamada, J.-I. *J. Chem. Soc., Chem. Commun.* 1987, 561.
34. Evans, D. A.; Kaldor, S. W.; Jones, T. K.; Clardy, J.; Stout, T. J. *J. Am. Chem. Soc.* 1990, *112*, 7001.
35. Martin, S. F.; Pacofsky, G. J.; Gist, R. P.; Lee, W.-C. *J. Am. Chem. Soc.* 1989, *111*, 7634.
36. Evans, D. A.; Rieger, D. L.; Jones, T. K.; Kaldor, S. W. *J. Org. Chem.* 1990, *55*, 6260.
37. Evans, D. A.; Sheppard, G. S. *J. Org. Chem.* 1990, *55*, 5192.
38. Evans, D. A.; Dow, R. L.; Shih, T. L.; Takacs, J. M.; Zahler, R. *J. Am. Chem. Soc.* 1990, *112*, 5290.
39. Roy, R.; Rey, A. W. *Synlett* 1990, 448.
40. Ojima, I.; Pei, Y. *Tetrahedron Lett.* 1990, *31*, 977.
41. Kende, A. S.; Kawamura, K.; Orwat, M. J. *Tetrahedron Lett.* 1989, *30*, 5821.
42. Paterson, I.; McClure, C. K.; Schumann, R. C. *Tetrahedron Lett.* 1989, *30*, 1293.
43. Trost, B. M.; Urabe, H. *J. Org. Chem.* 1990, *55*, 3982.
44. Walkup, R. D.; Kane, R. R.; Boatman, P. D.; Cunningham, R. T. *Tetrahedron Lett.* 1990, *31*, 7587.
45. Duplantier, A. J.; Nantz, M. H.; Roberts, J. C.; Short, R. P.; Somfai, P.; Masamune, S. *Tetrahedron Lett.* 1989, *30*, 7357.
46. Mukai, C.; Nagami, K.; Hanaaoka, M. *Tetrahedron Lett.* 1989, *30*, 5623.
47. Deloux, L.; Srebnik, M. *Chem. Rev.* 1993, *93*, 763.
48. McGarvey, G. J.; Hiner, R. N.; Williams, J. M.; Matasubara, Y.; Poarch, J. W. *J. Org. Chem.* 1986, *51*, 3742.
49. Genet, J. P.; Juge, S.; Mallart, S. *Tetrahedron Lett.* 1988, *29*, 6765.
50. Nakata, T.; Suenaga, T.; Oishi, T. *Tetrahedron Lett.* 1989, *30*, 6525.
51. Blanchette, M. A.; Malamas, M. S.; Nantz, M. H.; Roberts, J. C.; Somfai, P.; Whritenour, D. C.; Masamune, S.; Kageyama, M.; Tamura, T. *J. Org. Chem.* 1989, *54*, 2817.
52. Evans, D. A.; Takacs, J. M.; McGee, L. R.; Ennis, M. D.; Mathre, D. J.; Bartroli, J. *Pure Appl. Chem.* 1981, *53*, 1109.
53. Masamune, S.; Choy, W.; Petersen, J. S.; Sita, L. R. *Angew. Chem., Int. Ed. Engl.* 1985, *24*, 1.
54. Heathcock, C. H.; White, C. T.; Morrison, J. J.; VanDerveer, D. *J. Org. Chem.* 1981, *46*, 1296.
55. Heathcock, C. H.; White, C. T. *J. Am. Chem. Soc.* 1979, *101*, 7076.
56. Mukaiyama, T. In *Trends in Synthetic Carbohydrate Chemistry*; D. Horton and L. D. Hawkins, Eds.; American Chemical Society: Washington, D. C., 1989; pp 278.
57. McGarvey, G. J.; Kimura, M.; Oh, T. *J. Carbohydr. Chem.* 1984, *3*, 125.
58. Mukaiyama, T.; Tsuzuki, R.; Kato, J. *Chem. Lett.* 1985, 837.
59. Lopez-Herrera, F. J.; Valpuesta-Fernandez, M.; Garcia-Claros, S. *Tetrahedron* 1990, *46*, 7165.
60. Mukaiyama, T.; Miwa, T.; Nakatsuka, T. *Chem. Lett.* 1982, 145.
61. Dondoni, A.; Fantin, G.; Fogagnolo, M. *Tetrahedron Lett.* 1989, *30*, 6063.
62. Davies, S. G.; Dordor-Hedgecock, I. M.; Warner, P.; Jones, R. H.; Prout, K. *J. Organometal. Chem.* 1985, *285*, 213.
63. Davies, S. G.; Easton, R. J. C.; Walker, J. C.; Warner, P. *Tetrahedron* 1986, *42*, 175.
64. Lin, S.-H.; Vong, W.-J.; Cheng, C.-Y.; Wang, S.-L.; Lui, R.-S. *Tetrahedron Lett.* 1990, *31*, 7645.
65. Davies, S. G.; Shipton, M. R. *Synlett* 1991, 25.
66. Uemura, M.; Minami, T.; Hayashi, Y. *Tetrahedron Lett.* 1989, *30*, 6383.
67. Colombo, L.; Gennari, C.; Poli, G.; Scolastico, C.; Annunziata, R.; Cinquini, M.; Cozzi, F. *J. Chem. Soc., Chem. Commun.* 1983, 403.
68. Reetz, M. T.; Steinbach, R.; Kesseler, K. *Angew. Chem., Int. Ed. Engl.* 1982, *21*, 864.
69. Sugasawa, T.; Toyoda, T. *Tetrahedron Lett.* 1979, 1423.
70. Cainelli, G.; Panunzio, M.; Giacomini, D.; Martelli, G.; Spunta, G. *J. Am. Chem. Soc.* 1988, *110*, 6879.
71. Mori, I.; Bartlett, P. A.; Heathcock, C. H. *J. Org. Chem.* 1990, *55*, 5966.
72. Seebach, D. *Angew. Chem., Int. Ed. Engl.* 1988, *27*, 1624.
73. Corey, E. J.; Imwinkelried, R.; Pikul, S.; Xiang, Y. B. *J. Am. Chem. Soc.* 1989, *111*, 5493.

74. Dubois, J. E.; Dubois, M. *J. Chem. Soc., Chem. Commun.* 1968, 1567.
75. Arnett, E. M.; Fisher, F. J.; Nichols, M. A.; Ribeiro, A. A. *J. Am. Chem. Soc.* 1989, *111*, 748.
76. House, H. O.; Crumrine, D. S.; Teranishi, A. Y.; Olmstead, H. D. *J. Am. Chem. Soc.* 1973, *95*, 3310.
77. Juaristi, E.; Beck, A. K.; Hansen, J.; Matt, T.; Mukhopadhyay, T.; Simson, M.; Seebach, D. *Synthesis* 1993, 1271.
78. Dubois, J.-E.; Dubois, M. *Bull. Soc. Chim., Fr.* 1969, 3120.
79. Heathcock, C. H.; Young, S. D.; Hagen, J. P.; Pirrung, M. C.; White, C. T.; Van Derveer, D. *J. Org. Chem.* 1980, *45*, 3846.
80. Toda, F.; Tanaka, K.; Hamai, K. *J. Chem. Soc., Perkin Trans. 1* 1990, 3207.
81. Zimmerman, H. E.; Traxler, M. D. *J. Am. Chem. Soc.* 1957, *79*, 1920.
82. Mulzer, J.; Segner, J.; Bruntrup, G. *Tetrahedron Lett.* 1977, 4651.
83. Mulzer, J.; Bruntrup, G.; Finke, J.; Zippel, M. *J. Am. Chem. Soc.* 1979, *101*, 7723.
84. Dubois, J.-E.; Dubois, M. *Tetrahedron Lett.* 1967, 4215.
85. Dubois, J.-E.; Fort, J.-F. *Tetrahedron* 1972, *28*, 1653.
86. Dubois, J.-E.; Fort, J.-F. *Tetrahedron* 1972, *28*, 1665.
87. Fellmann, P.; Dubois, J.-E. *Tetrahedron* 1978, *34*, 1349.
88. Heathcock, C. H.; Buse, C. T.; Kleschick, W. A.; Pirrung, M. C.; Sohn, J. E.; Lampe, J. *J. Org. Chem.* 1980, *45*, 1066.
89. Dubois, J. E.; Fellmann, P. *Tetrahedron Lett.* 1975, 1225.
90. Evans, D. A.; Nelson, J. V.; Vogel, E.; Taber, T. R. *J. Am. Chem. Soc.* 1981, *103*, 3099.
91. Dougherty, D. A. *Tetrahedron Lett.* 1982, *23*, 4891.
92. Yamago, S.; Machii, D.; Nakamura, E. *J. Org. Chem.* 1991, *56*, 2098.
93. Heathcock, C. H.; Pirrung, M. C.; Buse, C. T.; Hagen, J. P.; Young, S. D.; Sohn, J. E. *J. Am. Chem. Soc.* 1979, *101*, 7077.
94. Buse, C. T.; Heathcock, C. H. *J. Am. Chem. Soc.* 1977, *99*, 8109.
95. Kleischick, W. A.; Buse, C. T.; Heathcock, C. H. *J. Am. Chem. Soc.* 1977, *99*, 247.
96. Panyachotipun, C.; Thornton, E. R. *Tetrahedron Lett.* 1990, *31*, 6001.
97. Bloch, R.; Gilbert, L. *Tetrahedron Lett.* 1986, *27*, 3511.
98. Schinzer, D. *Synthesis* 1989, 179.
99. Dubois, J.-E.; Fellman, P. *C. R. Acad. Sci., Paris (C)* 1972, *274*, 1307.
100. Heathcock, C. H.; Flippin, L. A. *J. Am. Chem. Soc.* 1983, *105*, 1667.
101. Evans, D. A.; Gage, J. R. *Tetrahedron Lett.* 1990, *31*, 6129.
102. Roush, W. R.; Bannister, T. D.; Wendt, M. D. *Tetrahedron Lett.* 1993, *34*, 8387.
103. Nerz-Stormes, M.; Thornton, E. R. *Tetrahedron Lett.* 1986, *27*, 897.
104. Panek, J. S.; Bula, O. A. *Tetrahedron Lett.* 1988, *29*, 1661.
105. Shirodkar, S.; Nerz-Stormes, M.; Thornton, E. R. *Tetrahedron Lett.* 1990, *31*, 4699.
106. Mukaiyama, T.; Iwasawa, N.; Stevens, R. W.; Haga, T. *Tetrahedron* 1984, *40*, 1381.
107. Yamamoto, Y.; Maruyama, K. *Tetrahedron Lett.* 1980, *21*, 4607.
108. Paterson, I.; Lister, M. A.; McClure, C. K. *Tetrahedron Lett.* 1986, *27*, 4787.
109. Reetz, M. T.; Peter, R. *Tetrahedron Lett.* 1981, *22*, 4691.
110. Evans, D. A.; Vogel, E.; Nelson, J. V. *J. Am. Chem. Soc.* 1979, *101*, 6120.
111. Mori, I.; Ishihara, H.; Heathcock, C. H. *J. Org. Chem.* 1990, *55*, 1114.
112. Buse, C. T.; Heathcock, C. H. *J. Am. Chem. Soc.* 1977, *99*, 2337.
113. Heathcock, C. H.; Pirrung, M. C.; Lampe, J.; Buse, C. T.; Young, S. D. *J. Org. Chem.* 1981, *46*, 2290.
114. White, C. T.; Heathcock, C. H. *J. Org. Chem.* 1981, *46*, 191.
115. Masamune, S.; Ali, S. A.; Snitman, D. L.; Garvey, D. S. *Angew. Chem., Int. Ed. Engl.* 1980, *19*, 557.
116. Seebach, D.; Ehrig, V.; Teschner, M. *Liebigs Ann. Chem.* 1976, 1357.
117. Van Draanen, N. A.; Arseniyadis, S.; Crimmins, M. T.; Heathcock, C. H. *J. Org. Chem.* 1991, *56*, 2499.
118. Masamune, S.; Choy, W.; Kerdesky, F. A. J.; Imperali, B. *J. Am. Chem. Soc.* 1981, *103*, 1566.
119. Masamune, S.; Kaiho, T.; Garvey, D. S. *J. Am. Chem. Soc.* 1982, *104*, 5521.
120. Siegel, C.; Thornton, E. R. *J. Am. Chem. Soc.* 1989, *111*, 5722.
121. Siegel, C.; Thornton, E. R. *Tetrahedron Lett.* 1986, *27*, 457.
122. Lagu, B. R.; Crane, H. M.; Liotta, D. C. *J. Org. Chem.* 1993, *58*, 4191.
123. Shibata, I.; Yamasaki, H.; Baba, A.; Matsuda, H. *Synlett* 1990, 490.
124. Nakamura, E.; Kuwajima, I. *Tetrahedron Lett.* 1983, *24*, 3347.
125. Paterson, I.; Lister, M. A. *Tetrahedron Lett.* 1988, *29*, 585.
126. Johnson, W. S.; Edington, C.; Elliott, J. D.; Silverman, I. R. *J. Am. Chem. Soc.* 1984, *106*, 7588.
127. Pearson, W. H.; Hines, J. V. *J. Org. Chem.* 1989, *54*, 4235.
128. Pearson, W. H.; Cheng, M.-C. *J. Org. Chem.* 1987, *52*, 3176.
129. Narasaka, K.; Miwa, T. *Chem. Lett.* 1985, 1217.
130. Narasaka, K.; Miwa, T.; Hayashi, H.; Ohta, M. *Chem. Lett.* 1984, 1399.

131. Mulzer, J.; Zippel, M.; Brüntrup, G.; Segner, J.; Finke, J. *Liebigs Ann. Chem.* 1980, 1108.
132. Mulzer, J.; de Lasalle, P.; Chuckolowski, A.; Blaschek, U.; Brüntrup, G.; Jibril, I.; Huttner, G. *Tetrahedron* 1984, *40*, 2211.
133. Corey, E. J.; Kim, S. S. *Tetrahedron Lett.* 1990, *31*, 3715.
134. Corey, E. J.; Kim, S. S. *J. Am. Chem. Soc.* 1990, *112*, 4976.
135. Duthaler, R. O.; Herold, P.; Lottenbach, W.; Oertle, K.; Riediker, M. *Angew. Chem., Int. Ed. Engl.* 1989, *28*, 495.
136. Bold, G.; Duthaler, R. O.; Riediker, M. *Angew. Chem., Int. Ed. Engl.* 1989, *28*, 497.
137. Duthaler, R. O.; Herold, P.; Wyler-Helfer, S.; Riediker, M. *Helv. Chim. Acta* 1990, *73*, 659.
138. Aggarwal, V. K.; Coldham, I.; McIntyre, S.; Sansbury, F. H.; Villa, M.-J.; Warren, S. *Tetrahedron Lett.* 1988, *29*, 4885.
139. Montgomery, S. H.; Pirrung, M. C.; Heathcock, C. H. *Org. Synth.* 1990, *Coll. Vol. 7*, 190.
140. Pirrung, M. C.; Heathcock, C. H. *J. Org. Chem.* 1980, *45*, 1727.
141. Heathcock, C. H.; Pirrung, M. C.; Montgomery, S. H.; Lampe, J. *Tetrahedron* 1981, *37*, 4087.
142. Dongala, E. B.; Dull, D. L.; Mioskowski, C.; Solladié, G. *Tetrahedron Lett.* 1973, 4983.
143. Kudo, Y.; Iwasawa, M.; Kobayashi, M.; Senda, Y.; Mitsui, S. *Tetrahedron Lett.* 1972, 2125.
144. Helmchen, G.; Leikauf, U.; Taufer-Knopfel, I. *Angew. Chem., Int. Ed. Engl.* 1985, *24*, 874.
145. Heathcock, C. H.; Hagen, J. P.; Jarvi, E. T.; Pirrung, M. C.; Young, S. D. *J. Am. Chem. Soc.* 1981, *103*, 4972.
146. Heathcock, C. H.; Pirrung, M. C.; Young, S. D.; Hagen, J. P.; Jarvi, E. T.; Badertscher, U.; Märki, H.-P.; Montgomery, S. *J. Am. Chem. Soc.* 1984, *106*, 8161.
147. Braun, M.; Devant, R. *Tetrahedron Lett.* 1984, *25*, 5031.
148. Prasad, K.; Chen, K.-M.; Repic, O.; Hardtmann, G. E. *Tetrahedron: Asymmetry* 1990, *1*, 703.
149. Devant, R.; Mahler, U.; Braun, M. *Chem. Ber.* 1988, *121*, 397.
150. Braun, M.; Waldmuller, D. *Synthesis* 1989, 856.
151. Wuts, P. G. M.; Putt, S. R. *Synthesis* 1989, 951.
152. Braun, M.; Sacha, H. *Angew. Chem., Int. Ed. Engl.* 1991, *30*, 1318.
153. Xiang, Y.; Olivier, E.; Ouimet, N. *Tetrahedron Lett.* 1992, *33*, 457.
154. Mioskowski, C.; Solladié, G. *J. Chem. Soc., Chem. Commun.* 1977, 162.
155. Mioskowski, C.; Solladié, G. *Tetrahedron* 1980, *36*, 227.
156. Solladié, G.; Frechon, C.; Demailly, G. *Nouv. J. Chim.* 1985, *9*, 21.
157. Wills, M.; Butlin, R. J.; Linney, I. D. *Tetrahedron Lett.* 1992, *33*, 5427.
158. Shanzer, A.; Somekh, L.; Butina, D. *J. Org. Chem.* 1979, *44*, 3967.
159. Ito, Y.; Sawamura, M.; Hayashi, T. *Tetrahedron Lett.* 1988, *29*, 239.
160. Ito, Y.; Sawamura, M.; Kobayashi, M.; Hayashi, T. *Tetrahedron Lett.* 1988, *29*, 6321.
161. Ito, Y.; Sawamura, M.; Hayashi, T. *J. Am. Chem. Soc.* 1986, *108*, 6405.
162. Pastor, S. D.; Togni, A. *Tetrahedron Lett.* 1990, *31*, 839.
163. Pastor, S. D.; Togni, A. *J. Am. Chem. Soc.* 1989, *111*, 2333.
164. Sawamura, M.; Hamashima, H.; Ito, Y. *J. Org. Chem.* 1990, *55*, 5935.
165. Togni, A.; Pastor, S. D. *Tetrahedron Lett.* 1989, *30*, 1071.
166. Togni, A.; Pastor, S. D. *Helv. Chim. Acta* 1989, *72*, 1038.
167. Sawamura, M.; Ito, Y.; Hayashi, T. *Tetrahedron Lett.* 1989, *30*, 2247.
168. Togni, A.; Pastor, S. D.; Rihs, G. *Helv. Chim. Acta* 1989, *72*, 1471.
169. Sawamura, M.; Ito, Y.; Hayashi, T. *Tetrahedron Lett.* 1990, *31*, 2723.
170. Ito, Y.; Sawamura, M.; Hamashima, H.; Emura, T.; Hayashi, T. *Tetrahedron Lett.* 1989, *30*, 4681.
171. Fleming, I.; Kilburn, J. D. *J. Chem. Soc., Chem. Commun.* 1986, 1198.
172. Fleming, I.; Kilburn, J. D. *J. Chem. Soc., Chem. Commun.* 1986, 305.
173. Fleming, I.; Sarkar, A. K. *J. Chem. Soc., Chem. Commun.* 1986, 1199.
174. Oppolzer, W.; Marco-Contelles, J. *Helv. Chim. Acta* 1986, *69*, 1699.
175. Hirama, M.; Garvey, D. S.; Lu, L. D.-L.; Masamune, S. *Tetrahedron Lett.* 1979, 3937.
176. Hirama, H.; Masamune, S. *Tetrahedron Lett.* 1979, 2225.
177. Iwasaki, G.; Shibasaki, M. *Tetrahedron Lett.* 1987, *28*, 3257.
178. Gennari, C.; Moresca, D.; Vieth, S.; Vulpetti, A. *Angew. Chem., Int. Ed. Engl.* 1993, *32*, 1618.
179. Mukai, C.; Lim, I. J.; Hanaoka, M. *Tetrahedron: Asymmetry* 1992, *3*, 1007.
180. Mukai, C.; Kim, I. J.; Furu, E.; Hanaoka, M. *Tetrahedron* 1993, *49*, 8323.
181. Evans, D. A.; McGee, L. R. *Tetrahedron Lett.* 1980, *21*, 3975.
182. Mears, R. J.; Whiting, A. *Tetrahedron* 1993, *49*, 177.
183. Curtis, A. D. M.; Mears, R. J.; Whiting, A. *Tetrahedron* 1993, *49*, 187.
184. Shieh, H.-M.; Prestwich, G. D. *J. Org. Chem.* 1981, *46*, 4319.
185. von Schriltz, D. M.; Kaiser, E. M.; Hauser, C. R. *J. Org. Chem.* 1967, *32*, 2610.
186. Evans, D. A.; McGee, L. R. *J. Am. Chem. Soc.* 1981, *103*, 2876.
187. Katsuki, T.; Yamaguchi, M. *Tetrahedron Lett.* 1985, *26*, 5807.

188. Tamaru, Y.; Harada, T.; Nishi, S.-I.; Mizutami, M.; Hioki, T.; Yoshida, Z.-I. *J. Am. Chem. Soc.* 1980, *102*, 7806.
189. Annunziata, R.; Cinquinin, M.; Cozzi, F.; Montanari, F.; Restelli, A. *J. Chem. Soc., Chem. Commun.* 1983, 1138.
190. Goasdoue, C.; Goasdoue, N.; Gaudemar, M. *Tetrahedron Lett.* 1983, *24*, 4001.
191. Bernardi, A.; Colombo, L.; Gennari, C.; Prati, L. *Tetrahedron* 1984, *41*, 3769.
192. Furuta, K.; Maruyama, T.; Yamamoto, H. *J. Am. Chem. Soc.* 1991, *113*, 1041.
193. Hoffmann, R. W.; Ditrich, K. *Tetrahedron Lett.* 1984, *25*, 1781.
194. Basile, T.; Biondi, S.; Boldrini, G. P.; Tagliavini, E.; Trombini, C.; Umani-Ronchi, A. *J. Chem. Soc., Perkin Trans. 1* 1989, 1025.
195. Faunce, J. A.; Grisso, B. A.; Mackenzie, P. B. *J. Am. Chem. Soc.* 1991, *113*, 3418.
196. Faunce, J. A.; Friebe, T. L.; Grisso, B. A.; Losey, E. N.; Sabat, M.; Mackenzie, P. B. *J. Am. Chem. Soc.* 1989, *111*, 4508.
197. Kiyooka, S.-i.; Kaneko, Y.; Komura, M.; Matsuo, H.; Nakano, M. *J. Org. Chem.* 1991, *56*, 2276.
198. Gennari, C.; Molinari, F.; Cozzi, P.; Oliva, A. *Tetrahedron Lett.* 1989, *30*, 5163.
199. Reetz, M. T.; Kunisch, F.; Heitmann, P. *Tetrahedron Lett.* 1986, *27*, 4721.
200. Meyers, A. G.; Widdowson, K. L. *J. Am. Chem. Soc.* 1990, *112*, 9672.
201. Meyers, A. I.; Yamamoto, Y. *J. Am. Chem. Soc.* 1981, *103*, 4278.
202. Mukaiyama, T.; Kobayashi, S.; Sano, T. *Tetrahedron* 1990, *46*, 4653.
203. Short, R. P.; Masamune, S. *Tetrahedron Lett.* 1987, *28*, 2841.
204. Reetz, M. T.; Rivadeneira, E.; Niemeyer, C. *Tetrahedron Lett.* 1990, *31*, 3863.
205. Uenishi, J.-i.; Tomozane, H.; Yamato, M. *J. Chem. Soc., Chem. Commun.* 1985, 717.
206. Nakatsuka, T.; Miwa, T.; Mukaiyama, T. *Chem. Lett.* 1981, 279.
207. Eichenauer, H.; Friewdrich, E.; Lutz, W.; Enders, D. *Angew. Chem., Int. Ed. Engl.* 1978, *17*, 206.
208. Enders, D.; Eichenauer, H.; Pieter, R. *Chem. Ber.* 1979, *112*, 3703.
209. Annunziata, R.; Cozzi, F.; Cinquini, M.; Colombo, L.; Gennari, C.; Poli, G.; Scolastico, C. *J. Chem. Soc., Perkin Trans. 1* 1985, 251.
210. Oppolzer, W.; Rodriguez, I.; Starkemann, C.; Walther, E. *Tetrahedron Lett.* 1990, *31*, 5019.
211. Oppolzer, W.; Starkemann, C.; Rodriguez, I.; Bernardinelli, G. *Tetrahedron Lett.* 1991, *32*, 61.
212. Oppolzer, W.; Lienard, P. *Tetrahedron Lett.* 1993, *34*, 4321.
213. Evans, D. A.; Bartroli, J.; Shih, T. L. *J. Am. Chem. Soc.* 1981, *103*, 2127.
214. Evans, D. A.; Sjorgren, E. B., Bartroli, J.; Dow, R. L. *Tetrahedron Lett.* 1986, *27*, 4957.
215. Meyers, A. I.; Yamamoto, Y. *Tetrahedron* 1984, *40*, 2309.
216. Danda, H.; Hansen, M. M.; Heathcock, C. H. *J. Org. Chem.* 1990, *55*, 173.
217. Nerz-Stormes, M.; Thornton, E. R. *J. Org. Chem.* 1991, *56*, 2489.
218. Oppolzer, W.; Blagg, J.; Rodriguez, I.; Walther, E. *J. Am. Chem. Soc.* 1990, *112*, 2767.
219. Evans, D. A.; Dow, R. L. *Tetrahedron Lett.* 1986, *27*, 1007.
220. Baker, R.; Castro, J. L.; Swain, C. J. *Tetrahedron Lett.* 1988, *29*, 2247.
221. Drewes, S. E.; Mallissar, D. G. S.; Roos, G. H. P. *Tetrahedron: Asymmetry* 1992, *3*, 515.
222. Meyers, A. I.; Knaus, G. *Tetrahedron Lett.* 1974, 1333.
223. Bonner, M. P.; Thornton, E. R. *J. Am. Chem. Soc.* 1991, *113*, 1299.
224. Yan, T.-H.; Lee, H.-C.; Tan, C. W. *Tetrahedron Lett.* 1993, *34*, 3559.
225. Yan, T.-H.; Tan, C.-W.; Lee, H.-C.; Lo, H.-C.; Huang, T.-Y. *J. Am. Chem. Soc.* 1993, *115*, 2613.
226. Ahn, K. H.; Lee, S.; Lim, A. *J. Org. Chem.* 1992, *57*, 5065.
227. Abdel-Magid, A.; Pridgen, L. N.; Eggleston, D. S.; Lantos, I. *J. Am. Chem. Soc.* 1986, *108*, 4595.
228. Abdel-Magid, A.; Pridgen, L. N.; Eggleston, D. S.; Lantos, I. *Tetrahedron Lett.* 1984, *25*, 3273.
229. Pridgen, L. N.; Abdel-Magid, A.; Lantos, I. *Tetrahedron Lett.* 1989, *30*, 5539.
230. Ku, T. W.; Kondrad, K. H.; Gleason, J. G. *J. Org. Chem.* 1989, *54*, 3487.
231. Annunziata, R.; Cinquini, M.; Cozzi, F.; Restelli, A. *J. Chem. Soc., Chem. Commun.* 1984, 1253.
232. Davies, S. G.; Dordor, I. M.; Warner, P. *J. Chem. Soc., Chem. Commun.* 1984, 956.
233. Liebeskind, L. S.; Welker, M. E. *Tetrahedron Lett.* 1984, *24*, 4341.
234. Ojima, I.; Kwon, H. B. *J. Am. Chem. Soc.* 1988, *110*, 5617.
235. Theopold, K. H.; Becker, P. N.; Bergman, R. G. *J. Am. Chem. Soc.* 1982, *104*, 5250.
236. Davies, S. G.; Middlemass, D.; Naylor, A.; Wills, M. *Tetrahedron Lett.* 1989, *30*, 2971.
237. Martens, J. *Chemiker-Zeitung* 1986, *110*, 169.
238. Martens, J. *Top. Curr. Chem.* 1984, *125*, 165.
239. Woodward, R. B.; Logusch, E.; Nambiar, K. P.; Sakan, K.; Ward, D. E.; Au-Yeung, B.-W.; Balaram, P.; Browne, L. J.; Card, P. J.; Chen, C. H.; Chenevert, R. B.; Fliri, A.; Frobel, K.; Gais, H.-J.; Garratt, D. G.; Hayakawa, K.; Heggie, W.; Hesson, D. P.; Hoppe, D.; Hoppe, I. et al. *J. Am. Chem. Soc.* 1981, *103*, 3210.
240. Woodward, R. B.; Logusch, E.; Nambiar, K. P.; Sakan, K.; Ward, D. E.; Au-Yeung, B.-W.; Balaram, P.; Browne, L. J.; Card, P. J.; Chen, C. H.; Chenevert, R. B.; Fliri, A.; Frobel, K.; Gais, H.-J.; Garratt, D. G.; Hayakawa, K.; Heggie, W.; Hesson, D. P.; Hoppe, D.; Hoppe, I. et al. *J. Am. Chem. Soc.* 1981, *103*, 3213.

241. Woodward, R. B.; Logusch, E.; Nambiar, K. P.; Sakan, K.; Ward, D. E.; Au-Yeung, B.-W.; Balaram, P.; Browne, L. J.; Card, P. J.; Chen, C. H.; Chenevert, R. B.; Fliri, A.; Frobel, K.; Gais, H.-J.; Garratt, D. G.; Hayakawa, K.; Heggie, W.; Hesson, D. P.; Hoppe, D.; Hoppe, I. et al. *J. Am. Chem. Soc.* 1981, *103*, 3215.
242. Agami, C.; Puchot, C.; Sevestre, H. *Tetrahedron Lett.* 1986, *27*, 1501.
243. Vedejs, E.; Krafft, G. A. *Tetrahedron* 1982, *38*, 2857.
244. Nakata, T.; Fukui, M.; Oishi, T. *Tetrahedron Lett.* 1988, *29*, 2223.
245. Roush, W. R.; Adam, M. A.; Walts, A. E.; Harris, D. J. *J. Am. Chem. Soc.* 1986, *108*, 3422.
246. Heathcock, C. H. In *Asymmetric Reactions and Processes in Chemistry*; E. L. Eliel and S. Otsuka, Eds.; American Chemical Society: Washington, D. C., 1982.
247. Paterson, I.; Bower, S.; Tillyer, R. D. *Tetrahedron Lett.* 1993, *34*, 4393.
248. Evans, D. A.; Rieger, D. L.; Bilodeau, M. T.; Urpi, F. *J. Am. Chem. Soc.* 1991, *113*, 1047.
249. Masamune, S.; Hirama, M.; Mori, S.; Ali, S. A.; Garvey, D. S. *J. Am. Chem. Soc.* 1981, *103*, 1568.
250. Yamamoto, Y.; Yatagai, H.; Maruyama, K. *Tetrahedron Lett.* 1982, *23*, 2387.
251. Fenzl, W.; Köster, R. *Liebigs Ann. Chem.* 1975, 1322.
252. Fenzl, W.; Kosfeld, H.; Köster, R. *Liebigs Ann. Chem.* 1976, 1370.
253. Inoue, T.; Mukaiyama, T. *Bull. Chem. Soc., Jpn.* 1980, *53*, 174.
254. Corey, E. J. *Pure Appl. Chem.* 1990, *62*, 1209.
255. Mukaiyama, T.; Murakami, M.; Oriyama, T.; Yamaguchi, M. *Chem. Lett.* 1981, 1193.
256. Paterson, I.; McClure, C. K. *Tetrahedron Lett.* 1987, *28*, 1229.
257. Masamune, S.; Mori, S.; van Horn, D.; Brook, D. W. *Tetrahedron Lett.* 1979, 1665.
258. Kuwajima, I.; Kato, M.; Mori, A. *Tetrahedron Lett.* 1980, *21*, 4291.
259. Paterson, I.; Goodman, J. M.; Isaka, M. *Tetrahedron Lett.* 1990, *31*, 7121.
260. Brown, H. C.; Dhar, R. K.; Bakshi, R. K.; Pandiarajan, P. K.; Singaram, B. *J. Am. Chem. Soc.* 1989, *111*, 3441.
261. Evans, D. A.; Clark, J. S.; Metternich, R.; Novack, V. J.; Sheppard, G. S. *J. Am. Chem. Soc.* 1990, *112*, 866.
262. Van Horn, D. E.; Masamune, S. *Tetrahedron Lett.* 1979, 2229.
263. Heathcock, C. H.; Montgomery, S. H. *Tetrahedron Lett.* 1983, *24*, 4637.
264. Koft, E. R.; Dorff, P.; Kullnig, R. *J. Org. Chem.* 1989, *54*, 2936.
265. Das, G.; Thornton, E. R. *J. Am. Chem. Soc.* 1990, *112*, 5360.
266. Wei, Y.; Bakthavatchalam *Tetrahedron Lett.* 1991, *32*, 1535.
267. McKenzie, A.; Wren, H. *J. Chem. Soc.* 1910, *97*, 473.
268. Roger, R.; McKay, W. B. *J. Chem. Soc.* 1931, 2229.
269. Muraoka, M.; Kawasaki, H.; Koga, K. *Tetrahedron Lett.* 1988, *29*, 337.
270. Ando, A.; Shioiri, T. *Tetrahedron* 1989, *45*, 4969.
271. Ito, Y.; Terashima, S. *Tetrahedron Lett.* 1987, *28*, 6629.
272. Meyers, A. I.; Mihelich, E. D. *Angew. Chem., Int. Ed. Engl.* 1976, *15*, 270.
273. Meyers, A. I.; Whitten, C. E. *Heterocycles* 1976, *4*, 1687.
274. Meyers, A. I.; Whitten, C. E. *Tetrahedron Lett.* 1976, 1947.
275. Evans, D. A.; Sacks, C. E.; Kleschick, W. A.; Taber, T. R. *J. Am. Chem. Soc.* 1979, *101*, 6789.
276. Petragnani, N.; Yonashiro, M. *Synthesis* 1982, 521.
277. Kuo, Y.; Yahner, J. A.; Ainsworth, C. A. *J. Am. Chem. Soc.* 1971, *93*, 6321.
278. Beslin, P.; Houtteville, M.-C. *Tetrahedron* 1989, *45*, 4445.
279. Hoye, T. R.; Kurth, M. J. *J. Org. Chem.* 1980, *45*, 3549.
280. Kingsbury, C. A. *J. Org. Chem.* 1972, *37*, 102.
281. Abushanab, E.; Reed, D.; Suzuki, F.; Sih, C. J. *Tetrahedron Lett.* 1978, 3415.
282. Mukaiyama, T.; Banno, H.; Narasaka, K. *J. Am. Chem. Soc.* 1974, *96*, 7503.
283. Murata, S.; Suzuki, M.; Noyori, R. *Tetrahedron* 1988, *44*, 4259.
284. Alexakis, A.; Mangeney, P.; Ghribi, A.; Marek, I.; Sedrani, R.; Guir, C.; Normant, J. *Pure Appl. Chem.* 1988, *60*, 49.
285. Basile, T.; Tagliavini, E.; Trombini, C.; Umani-Ronchi, A. *J. Chem. Soc., Chem. Commun.* 1989, 596.
286. Murata, S.; Suzuki, M.; Noyori, R. *Tetrahedron Lett.* 1980, *21*, 2527.
287. Saigo, K.; Osaki, M.; Mukaiyama, T. *Chem. Lett.* 1975, 989.
288. Chan, T. H.; Aida, T.; Lau, P. W. K.; Gorys, V.; Harpp, D. N. *Tetrahedron Lett.* 1979, 4029.
289. Heathcock, C. H.; Hug, K. T.; Flippin, L. A. *Tetrahedron Lett.* 1984, *25*, 5973.
290. Gong, L.; Streitwieser, A. *J. Org. Chem.* 1990, *55*, 6235.
291. Reetz, M. T.; Fox, D. N. A. *Tetrahedron Lett.* 1993, *34*, 1119.
292. Hong, Y.; Norris, D. J.; Collins, S. *J. Org. Chem.* 1993, *58*, 3591.
293. Heathcock, C. H.; Davidson, S. K.; Hug, K. T.; Flippin, L. A. *J. Org. Chem.* 1986, *51*, 3027.
294. Kobayashi, S.; Nishio, K. *J. Org. Chem.* 1993, *58*, 2647.
295. Denmark, S. E.; Griedel, B. D.; Coe, D. M. *J. Org. Chem.* 1993, *58*, 988.
296. Ishihara, K.; Yamamoto, H.; Heathcock, C. H. *Tetrahedron Lett.* 1989, *30*, 1825.

297. Mori, I.; Ishihara, K.; Flippin, L. A.; Nozaki, K.; Yamamoto, H.; Bartlett, P. A.; Heathcock, C. H. *J. Org. Chem.* 1990, *55*, 6107.
298. Ojima, I.; Inaba, S.-I.; Yoshida, K. *Tetrahedron Lett.* 1977, 3643.
299. Ojima, I.; Inaba, S.-I. *Tetrahedron Lett.* 1980, *21*, 2081.
300. Furukawa, M.; Okawara, T.; Noguchi, H.; Terawaki, Y. *Heterocycles* 1977, *6*, 1323.
301. Gennari, C.; Bernardi, A.; Colombo, L.; Scolastico, C. *J. Am. Chem. Soc.* 1985, *107*, 5812.
302. Palazzi, C.; Colombo, L.; Gennari, C. *Tetrahedron Lett.* 1986, *27*, 1735.
303. Mikami, K.; Matsukawa, S. *J. Am. Chem. Soc.* 1993, *115*, 7039.
304. Kobayashi, S.; Uchiro, H.; Shiina, I.; Mukaiyama, T. *Tetrahedron* 1993, *49*, 1761.
305. Ranu, B. C.; Chakraborty, R. *Tetrahedron* 1993, *49*, 5333.
306. Seebach, D.; Ehrig, V.; Teschner, M. *Annalen* 1976, 1357.
307. Reetz, M. T.; Hüllman, M.; Massa, W.; Berger, S.; Rademacher, P.; Heymann, P. *J. Am. Chem. Soc.* 1986, *108*, 2405.
308. Ojima, I.; Yoshida, K.; Inaba, S.-I. *Chem. Lett.* 1977, 429.
309. Annunziata, R.; Cinquini, M.; Cozzi, F.; Cozzi, P. G. *Tetrahedron Lett.* 1990, *31*, 6733.
310. Reetz, M. T.; Kesseler, K.; Jung, A. *Tetrahedron Lett.* 1984, *25*, 729.
311. Kudo, K.; Hashimoto, Y.; Sukegawa, M.; Hasegawa, M.; Saigo, K. *J. Org. Chem.* 1993, *58*, 579.
312. Kiyooka, S.-I.; Suzuki, K.; Shirouchi, M.; Kaneko, Y.; Tanimori, S. *Tetrahedron Lett.* 1993, *34*, 5729.
313. Bernardi, A.; Cardani, S.; Carugo, O.; Colombo, L.; Scolastico, C.; Villa, R. *Tetrahedron Lett.* 1990, *31*, 2779.
314. Palazzi, C.; Poli, G.; Scolastico, C.; Villa, R. *Tetrahedron Lett.* 1990, *31*, 4223.
315. Harada, T.; Kurokawa, H.; Oku, A. *Tetrahedron Lett.* 1987, *28*, 4847.
316. Yamamoto, K.; Tomo, Y.; Suzuki, S. *Tetrahedron Lett.* 1980, *21*, 2861.
317. Jung, M. E.; Hogan, K. T. *Tetrahedron Lett.* 1988, *29*, 6199.
318. Ju, J.; Reddy, B. R.; Khan, M.; Nicholas, K. M. *J. Org. Chem.* 1989, *54*, 5426.
319. Chow, K.; Danishefsky, S. *J. Org. Chem.* 1989, *54*, 6016.
320. Mukai, C.; Cho, W. J.; Hanaoka, M. *Tetrahedron Lett.* 1989, *30*, 7435.
321. Nakamura, E.; Shimada, J.-I.; Horiguchi, Y.; Kuwajima, I. *Tetrahedron Lett.* 1983, *24*, 3341.
322. Nakamura, E.; Kuwajima, I. *Tetrahedron Lett.* 1983, *24*, 3343.
323. Wada, M. *J. Chem. Soc., Chem. Commun.* 1981, 153.
324. Nakamura, E.; Yamago, S.; Machii, D.; Kuwajima, I. *Tetrahedron Lett.* 1988, *29*, 2207.
325. Noyori, R.; Nishida, I.; Sakata, J.; Nishizawa, M. *J. Am. Chem. Soc.* 1980, *102*, 1223.
326. Noyori, R.; Nishida, I.; Sakata, J. *J. Am. Chem. Soc.* 1983, *105*, 1598.
327. Nakamura, E.; Shimizu, M.; Kuwajima, I. *Tetrahedron Lett.* 1976, 1699.
328. Ando, A.; Miura, T.; Tatematsu, T.; Shioiri, T. *Tetrahedron Lett.* 1993, *34*, 1507.
329. Shenvi, S.; Stille, J. K. *Tetrahedron Lett.* 1982, *23*, 627.
330. Shenvi, S.; Stille, J. K. *Tetrahedron* 1984, *40*, 2329.
331. Yamamoto, Y.; Yatagai, H.; Maruyama, K. *J. Chem. Soc., Chem. Commun.* 1981, 162.
332. Stevens, R. W.; Iwasawa, N.; Mukaiyama, T. *Chem. Lett.* 1982, 1459.
333. Harrison, C. R. *Tetrahedron Lett.* 1987, *28*, 4135.
334. Evans, D. A.; Urpi, F.; Somers, T. C.; Clark, J. S.; Bilodeau, M. T. *J. Am. Chem. Soc.* 1990, *112*, 8215.
335. Oertle, K.; Beyeler, H.; Duthaler, R. O.; Lottenbach, W.; Riediker, M.; Steiner, E. *Helv. Chim. Acta* 1990, *73*, 353.
336. Duthaler, R. O.; Hafner, A. *Chem. Rev.* 1992, *92*, 807.
337. Hart, D. J.; Ha, D.-C. *Chem. Rev.* 1989, *89*, 1447.
338. Furstner, A. *Synthesis* 1989, 571.
339. Gaudemar, M. *Organometal. Chem. Rev. A.* 1972, *8*, 183.
340. Rathke, M. W. *Org. Reactions* 1975, *22*, 423.
341. Shriner, R. L. *Org. Reactions* 1942, *1*, 1.
342. Dewar, M. J. S.; Merz, K. M. *J. Am. Chem. Soc.* 1987, *109*, 6553.
343. Knochel, P.; Singer, R. D. *Chem. Rev.* 1993, *93*, 2117.
344. Dekker, J.; Boersma, J.; van der Kerk, G. J. M. *J. Chem. Soc., Chem. Commun.* 1981, 553.
345. Maiz, J.; Arrieta, A.; Lopez, X.; Ugalde, J. M.; Cossio, F. P. *Tetrahedron Lett.* 1993, *34*, 6111.
346. Matsumoto, T.; Hosoda, Y.; Mori, K.; Fukui, K. *Bull. Chem. Soc., Jpn.* 1972, *45*, 3156.
347. Bellassould, M.; Gaudemar, M. *J. Organometal. Chem.* 1975, *102*, 1.
348. Gilman, H.; Speeter, M. *J. Am. Chem. Soc.* 1943, *65*, 2255.
349. Kagan, H. B.; Basselier, J.-J.; Luche, J.-L. *Tetrahedron Lett.* 1964, 941.
350. Dardoize, F.; Moreau, J.-L.; Gaudemar, M. *Bull. Soc. Chim., Fr.* 1973, 1668.
351. Dardoize, F.; Gaudemar, M. *Bull. Soc. Chim., Fr.* 1974, 939.
352. Dardoize, F.; Gaudemar, M. *Bull. Soc. Chim., Fr.* 1976, 1561.
353. Luche, J. L.; Kagan, H. B. *Bull. Soc. Chim., Fr.* 1969, 3500.

354. Luche, J. L.; Kagan, H. B. *Bull. Soc. Chim., Fr.* 1971, 2260.
355. Dardoize, F.; Moreau, J.-L.; Gaudemar, M. *C. R. Acad. Sci., Paris (C)* 1970, *270*, 233.
356. Shono, T.; Kise, N.; Sanda, F.; Ohi, S.; Yoshioka, K. *Tetrahedron Lett.* 1989, *30*, 1253.
357. Curé, J.; Gaudemar, G. *Bull. Soc. Chim. Fr.* 1973, 2418.
358. Cainelli, G.; Panunzio, M.; Andreoli, P.; Martelli, G.; Spunta, G.; Giacomini, D.; Bandini, E. *Pure Appl. Chem.* 1990, *62*, 605.
359. Robinson, A. J.; Wyatt, P. B. *Tetrahedron* 1993, *49*, 11329.
360. van Maanen, H. L.; Jastrzebski, J. T. B. H.; Verweij, J.; Kieboom, A. P. G.; Spek, A. L.; van Koten, G. *Tetrahedron: Asymmetry* 1993, *4*, 1441.
361. Dardoize, F.; Moreau, J.-L.; Gaudemar, M. *CRAS* 1970, *270*, 233.
362. Furukawa, M.; Okawara, T.; Naguchi, Y.; Terawaki, Y. *Chem. Pharm. Bull.* 1978, *26*, 260.
363. Gluchouski, C.; Cooper, L.; Bergbreiter, D. E.; Newcomb, M. *J. Org. Chem.* 1980, *45*, 3413.
364. Overman, L. E.; Osawa, T. *J. Am. Chem. Soc.* 1985, *107*, 1698.
365. Bose, A. K.; Khajavi, M. S.; Mankas, M. S. *Synthesis* 1982, 407.
366. Ha, D.-C.; Hart, D. J.; Yang, T. K. *J. Am. Chem. Soc.* 1984, *106*, 4819.
367. Ha, D.-C.; Hart, D. J. *J. Antibiot.* 1987, *40*, 309.
368. Guanti, G.; Banfi, L.; Narisano, E.; Scolastico, C.; Bosone, E. *Synthesis* 1985, 609.
369. Noyori, R.; Nishida, I.; Sakata, J. *J. Am. Chem. Soc.* 1981, *103*, 2106.
370. Maruoka, K.; Hashimoto, S.; Kitagawa, Y.; Yamamoto, H.; Nozaki, H. *J. Am. Chem. Soc.* 1977, *99*, 7705.
371. Tsuboniwa, N.; Marsubara, S.; Morizawa, Y.; Oshima, K.; Nozaki, H. *Tetrahedron Lett.* 1984, *25*, 2569.
372. Chan, T. H.; Li, C. J.; Wei, Z. Y. *J. Chem. Soc., Chem. Commun.* 1990, 505.
373. Alberola, A.; Alvarez, M. A.; Andrés, C.; González, A.; Pedrosa, R. *Synthesis* 1990, 1057.
374. Hart, D. J.; Lee, C. S.; Pirkle, W. H.; Hyon, H. M.; Tsipouras, A. *J. Am. Chem. Soc.* 1986, *108*, 6054.
375. Pirkle, W. H.; Tsipouras, A.; Hyan, M. H.; Hart, D. J.; Lee, C. S. *J. Chromatogrpah.* 1986, *358*, 377.
376. van der Steen, F. H.; Kleijn, H.; Jastrebski, T. B. H.; van Koten, G. *Tetrahedron Lett.* 1989, *30*, 765.
377. Iimori, T.; Ishida, Y.; Shibasaki, M. *Tetrahedron Lett.* 1986, *27*, 2153.
378. Iimori, T.; Shibasaki, M. *Tetrahedron Lett.* 1985, *26*, 1523.
379. Yamada, T.; Suzuki, H.; Mukaiyama, T. *Chem. Lett.* 1987, 293.
380. Shibasaki, M.; Ishida, Y.; Iwasaki, G.; Iimori, T. *J. Org. Chem.* 1987, *52*, 3489.
381. Ojima, I.; Inaba, S. *Tetrahedron Lett.* 1980, *21*, 2077.
382. Hattori, K.; Miyata, M.; Yamamoto, H. *J. Am. Chem. Soc.* 1993, *115*, 1151.
383. Guette, M.; Guette, J.-P.; Capillon, J. *Tetrahedron Lett.* 1971, 2863.
384. Guette, M.; Capillon, J.; Guette, J.-P. *Tetrahedron* 1973, *29*, 3659.
385. Soai, K.; Hirose, Y.; Sakata, S. *Tetrahedron: Asymmetry* 1992, *3*, 677.
386. Basavaiah, D.; Bharathi, T. K. *Synth. Commun.* 1989, *19*, 2035.
387. Palmer, M. H.; Reid, J. A. *J. Chem. Soc.* 1960, 931.
388. Palmer, M. H.; Reid, J. A. *J. Chem. Soc.* 1962, 1762.
389. Basile, T.; Tagliavini, E.; Trombini, C.; Umani-Ronchi, A. *Synthesis* 1990, 305.
390. Andrés, C.; González, A.; Pedrosa, R.; Pérez-Encabo, A. *Tetrahedron Lett.* 1992, *33*, 2895.
391. Hatanaka, M.; Park, O.-S.; Ueda, I. *Tetrahedron Lett.* 1990, *31*, 7631.
392. Ha, D.-C.; Hart, D. J. *Tetrahedron Lett.* 1987, *28*, 4489.
393. Gallucci, J. C.; Ha, D.-C.; Hart, D. J. *Tetrahedron* 1989, *45*, 1283.
394. Alcaide, B.; Plumet, J.; Rodriguez-López, J.; Sánchez-Cantalljo, Y. M. *Tetrahedron Lett.* 1990, *31*, 2493.
395. Harada, T.; Mukaiyama, T. *Chem. Lett.* 1982, 467.
396. Mukaiyama, T.; Stevens, R. W.; Iwasawa, N. *Chem. Lett.* 1982, 353.
397. Molander, G.; Etter, J. B. *J. Am. Chem. Soc.* 1987, *109*, 6556.
398. Nakamura, E.; Sekiya, K.; Kuwajima, I. *Tetrahedron Lett.* 1987, *28*, 337.
399. Zamojski, A.; Grynkiewicz, G. In *The Synthesis of Natural Products*; J. ApSimon, Ed.; John Wiley & Sons: New York, 1984; Vol. 6; pp 141.
400. Mizuno, T.; Weiss, A. H. A. *Carbohydr. Res.* 1974, *28*, 173.
401. Weiss, A. H.; LaPierre, R. B.; Shapira, J. *J. Catal.* 1970, *16*, 332.
402. Weiss, A. H.; Tambawala, H.; Partridge, R. D.; Shapira, J. In *Dechema-Monographen #1264-1291*Verlag Chemie: Weinhelm/Berstrasse, 1971; Vol. 68; pp 239.
403. Weiss, A. H.; Shapira, J. *Hydrocarbon Process.* 1970, *40*, 119.
404. Saimoto, H.; Kotani, K.; Shigemasa, Y.; Suzuki, M.; Harada, K.-I. *Tetrahedron Lett.* 1989, *30*, 2553.
405. Matsumoto, T.; Inoue, S. *J. Chem. Soc., Chem. Commun.* 1983, 171.
406. Castells, J.; Geijo, F.; Lopez-Calahorra, F. *Tetrahedron Lett.* 1980, *21*, 4517.
407. Shigemasa, Y.; Sasaki, Y.; Ueda, N.; Nakashima, R. *Bull. Chem. Soc., Japan* 1984, *57*, 2761.
408. Matumoto, T.; Yamamoto, H.; Inoue, S. *J. Am. Chem. Soc.* 1984, *106*, 4829.
409. Ager, D. J. In *Umpoled Synthons: A Survey of Sources and Uses in Synthesis*; T. A. Hase, Ed.; Wiley-Interscience: New York, 1987; pp 19.

410. Matsuda, I.; Takahashi, K.; Sato, S. *Tetrahedron Lett.* 1990, *31*, 5331.
411. Walba, D. M.; Ward, D. M. *Tetrahedron Lett.* 1982, *23*, 4995.
412. Paulsen, H.; Schuller, M.; Nashed, M. A.; Heitman, A.; Redlich, H. *Tetrahedron Lett.* 1985, *26*, 3689.
413. Enders, D.; Lotter, H. *Angew. Chem., Int. Ed. Engl.* 1981, *20*, 795.
414. Yamada, J.-i.; Abe, H.; Yamamoto, Y. *J. Am. Chem. Soc.* 1990, *112*, 6118.
415. Corey, E. J.; Erickson, B. W.; Noyori, R. *J. Am. Chem. Soc.* 1971, *93*, 1724.
416. Still, W. C.; Macdonald, T. L. *J. Org. Chem.* 1976, *41*, 3620.
417. Yamamoto, Y.; Yatagai, H.; Maruyama, K. *Chem. Lett.* 1979, 385.
418. Evans, D. A.; Andrews, G. C.; Buckwalter, B. *J. Am. Chem. Soc.* 1974, *96*, 5560.
419. Dumartin, G.; Pereyre, M.; Quintard, J.-P. *Tetrahedron Lett.* 1987, *28*, 3935.

Chapter 8

# PREPARATION AND REACTIONS OF ALKENES

This chapter is concerned with the preparation of simple and functionalized alkenes together with reactions of carbon–carbon unsaturation. However, only the reactions of simple alkenes have been covered. The plethora of methods that relates to the chemistry of functionalized alkenes is discussed elsewhere [Chapter 9]. Due to the importance of the oxidations of carbon–carbon unsaturation, this topic is covered separately. The oxidations of simple alkenes will be found in Chapter 10, while those of functionalized alkenes are covered in Chapter 11.

## 8.1. ALKENE SYNTHESIS

Many of the asymmetric methods cited in this book rely on the addition of various reagents to an alkene in a stereoselective manner.[†] [5-8] To provide good stereoselection for asymmetric transformations, only one isomer of the alkene must be present in the substrate. It is pertinent, therefore, to address reactions that can be used for the stereoselective preparation of alkenes.[§] These methods fall into four main categories: reduction of an acetylene; condensation of an organometallic species with either a carbonyl or vinyl compound; and elimination reactions.[9-20]

### 8.1.1. ALKYNE REDUCTIONS AND ADDITIONS

Alkynes provide access to a wide variety of alkenes. It should be noted, that to avoid a mixture of diastereomers, many of the examples in this section rely on the use of terminal or symmetrical alkynes.

#### 8.1.1.1. Simple Alkenes

Acetylenes can provide *E*-alkenes by reduction with lithium aluminum hydride[21,22] or alkali metals (Scheme 8.1).[23] The use of hydrogenation or diisobutylaluminum hydride with an alkyne provides the *Z*-isomer. Hydroboration gives access to either *E*- or *Z*-alkenes (Scheme 8.2).[17,24-39] 1-Bromo-1-alkynes can also be used as substrates to access either alkene isomer.[24,25,33] Trisubstituted alkenes can also be obtained by hydroboration of internal alkynes with dialkyl or alkylbromoboranes.[27,32,40]

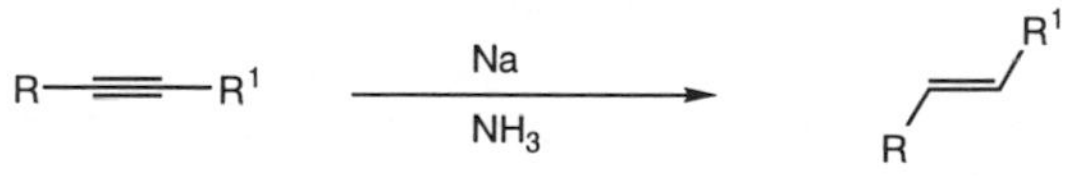

R—≡—R$^1$ —(Na / $NH_3$)→ R–CH=CH–R$^1$ (*E*)

**SCHEME 8.1.**

H—≡—R$^1$ + $R_2BX$ —($LiAlH_4$)→ $R_2B$–CH=CH–R$^1$ —(NaOMe / $I_2$)→ R–CH=CH–R$^1$ (*Z*)

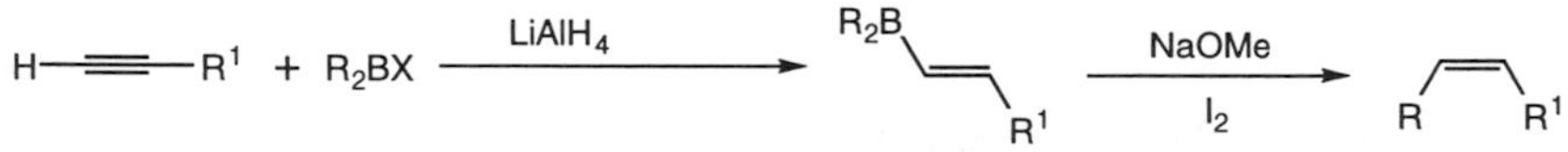

**SCHEME 8.2.**

Symmetrical acetylenes can be mono-alkylated with an organoaluminum reagent that adds in a *syn* manner (Scheme 8.3).[41-44] Hydroalumination of 1-chloro-1-alkynes furnishes a stereoselective route to both di- and trisubstituted alkenes.[45]

† Indeed, many of these methods have been employed in the synthesis of insect pheromones.[1-4]

§ Many examples have been incorporated into other reaction sequences throughout this book. The reader is urged to find these examples by use of the index.

Organocuprates and organocopper–magnesium reagents have also been used in analogous reactions.[46-50] Indeed, two different alkyl groups can be introduced by subsequent reaction of the vinyl organometallic intermediate (vide infra) (Scheme 8.4).[49,51,52]

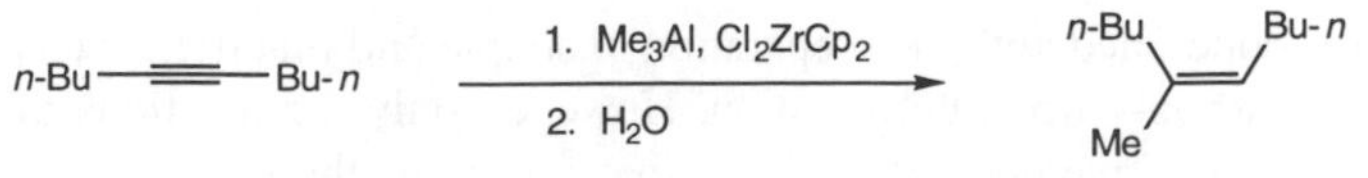

n-Bu—≡—Bu-n → 1. $Me_3Al$, $Cl_2ZrCp_2$; 2. $H_2O$ → (n-Bu)(Me)C=CH(Bu-n)

**SCHEME 8.3.**

$R^1$—≡—H → 1. $R^2Cu.MgX_2$; 2. $R^3X$ → $R^1$, $R^2$, $R^3$ alkene

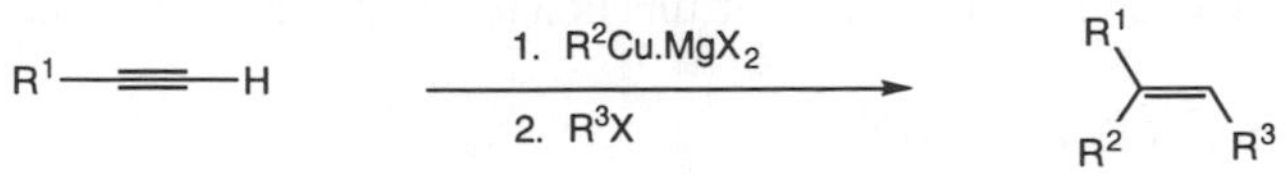

**SCHEME 8.4.**

### 8.1.1.2. Conjugated Alkenes

High *E*-stereoselectivity has been observed for the synthesis of α,β-unsaturated esters from the reaction of alkoxy or siloxy acetylenes with aldehydes (Scheme 8.5).[53-56] Haloboration of terminal or alkoxy acetylenes also provides a useful protocol for the synthesis of *E*-α,β-unsaturated esters.[57,58] α,β-Unsaturated acids, nitriles, and amides are also available from terminal alkynes through the *syn* addition of organocopper–magnesium reagents (Scheme 8.6).[59,60]

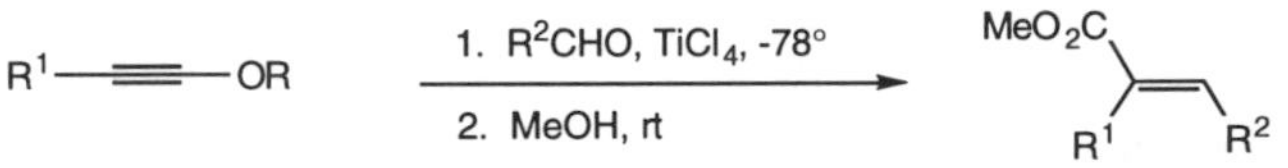

$R^1$—≡—OR → 1. $R^2CHO$, $TiCl_4$, -78°; 2. MeOH, rt → $MeO_2C$, $R^1$, $R^2$ alkene

**SCHEME 8.5.**

$R^1$—≡—H → 1. $R^2CuMgBr_2$; 2. $CO_2$, $P(OEt)_3$, HMPT → $R^1$, $R^2$, $CO_2H$ alkene

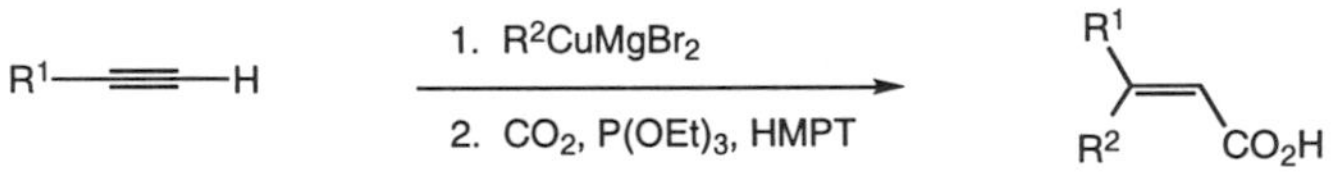

**SCHEME 8.6.**

High regio- and stereoselectivity are also seen in the addition of organocuprates to acetylenic esters.[61-64] The resultant anion can also react with a variety of electrophiles, allowing entry to functionalized alkenes (vide infra).

Hydroboration of alkynes provides a stereoselective route to various isomers of conjugated dienes.[28,65,66]

### 8.1.1.3. Heteroatom Substituted Alkenes

Hydroboration of 1-halo-1-alkynes and acetylenes provides a stereoselective synthesis of *Z*- and *E*-1-halo-1-alkenes.[33,67-72] The stereoselective synthesis of 1-bromo-1-(trimethylsilyl)-1-alkenes provides a carbonyl anion and cation equivalent together with the potential for further elaboration (Scheme 8.7).[73-75] Disubstituted acetylenes react with Grignard reagents, in the presence of a titanium catalyst, to give *E*-alkenyl Grignard reagents and iodides.[76]

Hydroboration of silylacetylenes is both regio- and stereoselective.[77-83] Vinylsilanes are also available from silylacetylenes through nickel, platinum, or palladium catalyzed *syn* additions.[77,84-87] This mode of addition is also seen for other approaches to vinylsilanes, such as the addition of silylcuprates to acetylenes,[88] and the hydrosilylation of alkynes in the presence of a catalyst, such as chloroplatinic acid.[89-91] Although the regiospecificity is dependent upon the alkyne substrate in the latter case (Scheme 8.8),[89,92] stereoselective examples are limited to symmetrical and terminal alkynes.[89,93]

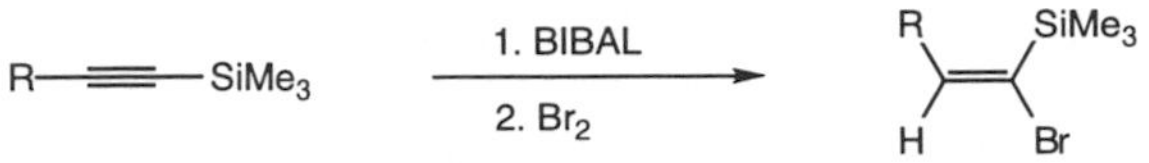

$$R{-}C{\equiv}C{-}SiMe_3 \xrightarrow[2.\ Br_2]{1.\ BIBAL} (R)(H)C{=}C(SiMe_3)(Br)$$

**SCHEME 8.7.**

$$R^1{-}C{\equiv}C{-}H \xrightarrow[H_2PtCl_6]{R_3SiH} (R^1)(H)C{=}C(H)(SiR_3)$$

$$R^1{-}C{\equiv}C{-}SiMe_3 \xrightarrow[H_2PtCl_6]{R_3SiH} (R^1)(R_3Si)C{=}C(SiMe_3)(H)$$

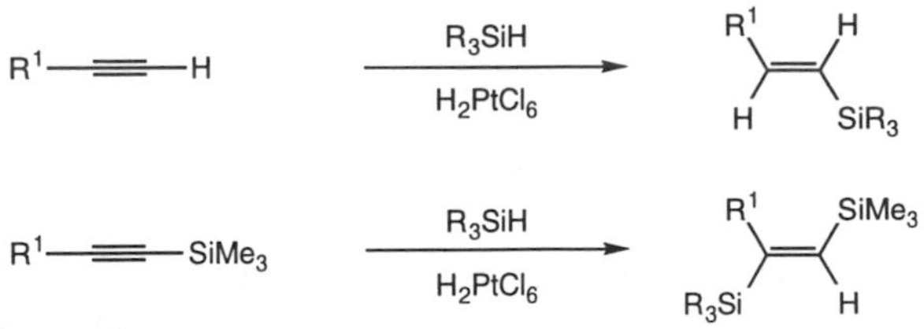

**SCHEME 8.8.**

Silyl sulfonyl acetylenes can be stereoselectively reduced to form either *trans* or *cis* vinyl silyl sulfones.[94]

The hydroboration of 1-phenylseleno-1-alkynes provides a stereoselective route to *Z*-1-phenylseleno-1-alkenes.[95] The reaction of organocuprate or aluminum reagents with terminal acetylenes results in *syn* addition; treatment of the intermediate with a variety of electrophiles leads to the stereospecific synthesis of functionalized alkenes.[74,79,96-106]

Although alkynes undergo addition with acids, the resultant stereochemistry is dependent on the nature of the substrate and reaction conditions.[107-113] Terminal and symmetrical acetylenes react with hydrogen iodide with complete *cis* selectivity through a *syn*-addition, in a Markovnikov fashion, to generate *E*-alkenyl iodides.[114]

#### 8.1.1.4. Allyl Alcohols and Related Compounds

Allyl alcohols may be prepared from the partial reduction of propargyl alcohols.[115,116] An alternative approach is the reaction of organometallic reagents with an alkyne that results in the formation of a vinyl anion; this can then be treated with a carbonyl compound.[117-120] Hydrolysis of alkenyl boranes derived from alkynes also provides access to allyl alcohols.[121]

*Z*-Allylsilanes are readily available from propargylsilanes by hydroboration or reduction with diisobutylaluminum hydride.[122] Allylsilanes can also be accessed by the reaction of propargyl acetates with silylcuprates.[123] The silyl group is a masked hydroxy group [*9.4.1*].

### 8.1.2. REACTION OF ORGANOMETALLICS WITH CARBONYL COMPOUNDS

#### 8.1.2.1. Simple Alkenes

There are three main protocols for the synthesis of alkenes from carbonyl compounds: the Wittig, Peterson, and Julia reactions, although many heteroatoms and metals can be employed in this context (vide infra).[124]

##### *8.1.2.1.1. The Wittig Reaction*

The Wittig reaction has been investigated extensively, and it is possible to control the stereochemical outcome of this condensation to a large degree through manipulation of the reaction parameters (Scheme 8.9).[125-138]

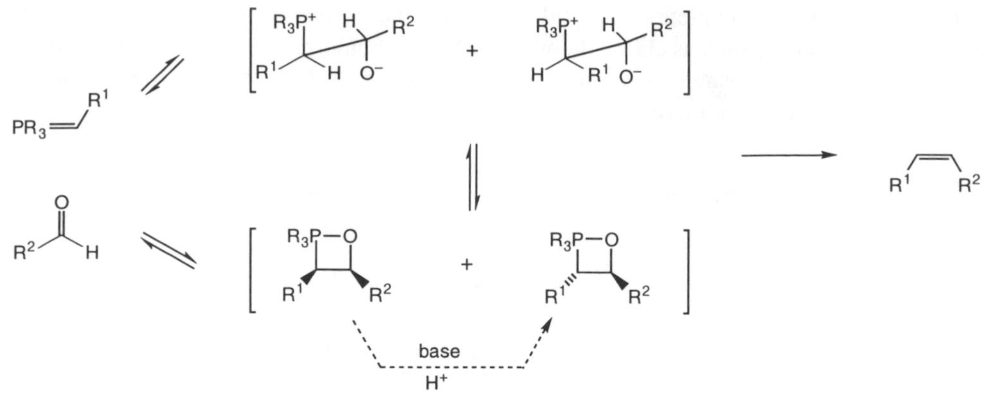

SCHEME 8.9.

Although the Z-alkene is usually the major isomer from nonstabilized ylides, stabilized ylides usually provide the E-alkene. The selectivity for Z-alkenes increases under "salt free" conditions with the use of sodium rather than lithium bases.[125,139-143] The Schlosser modification of the Wittig reaction provides access to E-alkenes from nonstabilized ylides by low temperature deprotonation of the intermediate betaine and subsequent warming (Scheme 8.10).[142,144-146] The major limitations of the Wittig reaction are the low reactivity towards ketones and the absence of a satisfactory route to Z-alkenes bearing electron-withdrawing substituents.[147]

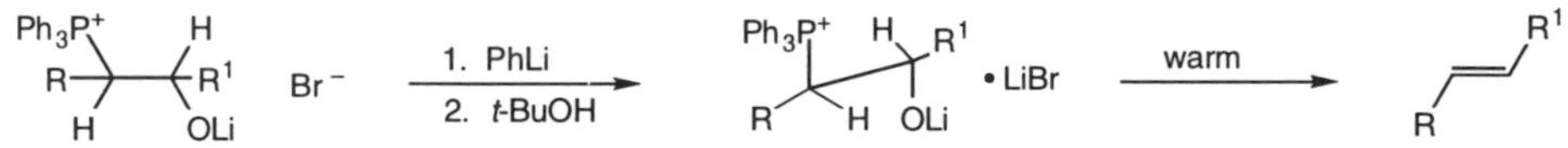

SCHEME 8.10.

In addition to the lack of stereoselectivity sometimes observed in the traditional Wittig reaction, the removal of the triphenylphosphine oxide by-product can also be problematic. The use of the Horner-Wittig (Scheme 8.11) or Horner-Wadsworth-Emmons reactions (Scheme 8.12) allows for the removal of phosphorous as a water soluble by-product.[147,148] The Horner-Wittig, with a careful choice of solvent, followed by the separation of the β-hydroxy phosphinyl intermediate, gives either the pure Z- or E-diastereoisomer in high yield.[149,150] The E- or Z-isomers are also available in a modified Horner-Wittig reaction by the stereoselective reduction of a phosphinoyl ketone (Scheme 8.13).[151,152]

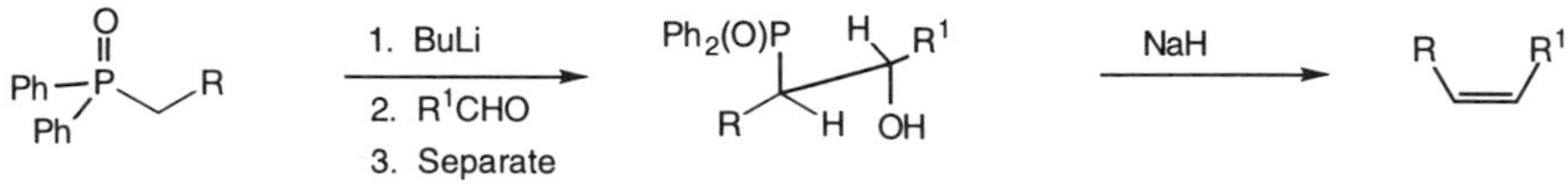

SCHEME 8.11.

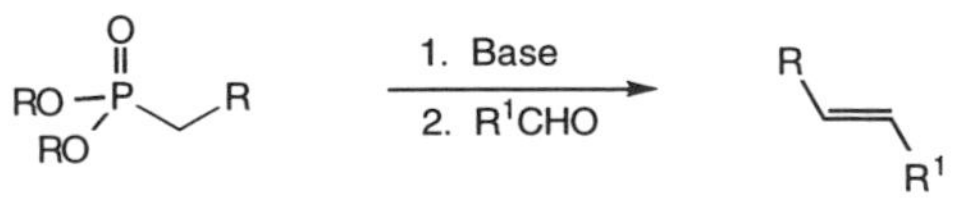

SCHEME 8.12.

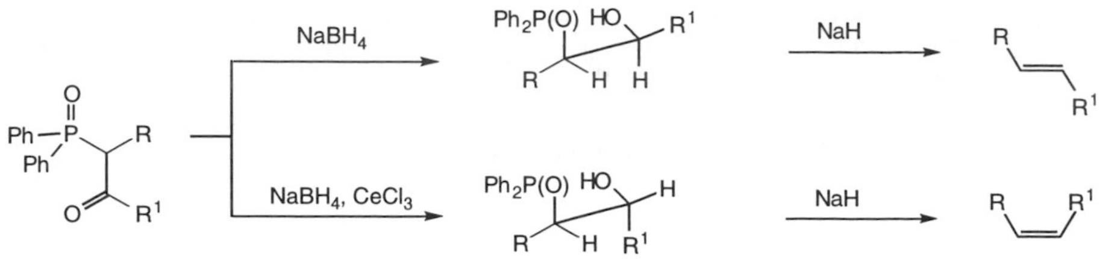

**SCHEME 8.13.**

### *8.1.2.1.2. The Peterson Reaction*

The silicon analog of the Wittig reaction, the Peterson reaction, can also be used for the stereoselective preparation of alkenes, although separation of the diastereoisomeric intermediate β-hydroxy silanes may be necessary if a single alkene isomer is the desired product.[124,153-155] However, treatment of one isomer of the hydroxy silane with base, and the other with acid, results in the same product diastereoisomer (Scheme 8.14). The Anh-Felkin controlled reduction of α-silyl ketones provides an attractive alternative to the separation of these diastereoisomeric β-hydroxy silanes, and impressive stereoselectivity can result (Scheme 8.15).[156,157] High *E*-stereoselectivity has been observed in the Peterson-type reaction of bis(trimethylsilyl)methyl derivatives in the presence of fluoride ion.[158]

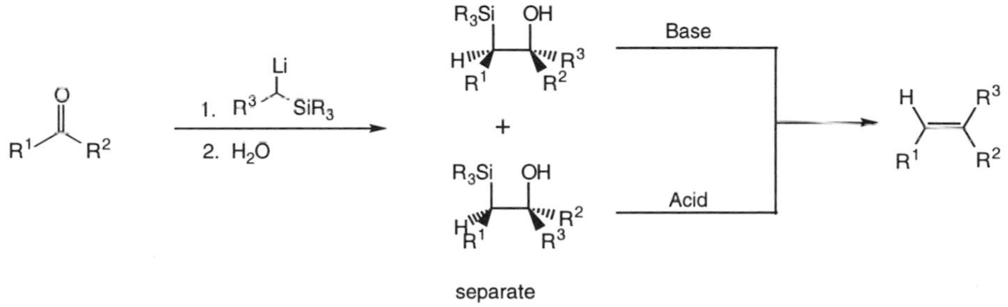

**SCHEME 8.14.**

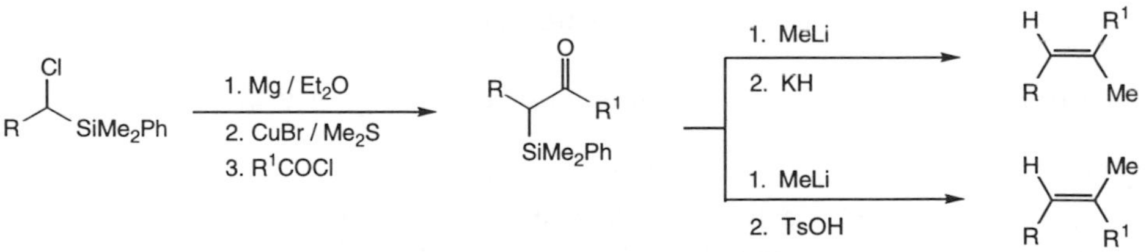

**SCHEME 8.15.**

### *8.1.2.1.3. The Julia Reaction*

Olefins are also available by the Julia method where a sulfone is reacted with a carbonyl compound.[159-161] Subsequent conversion of the hydroxy to a better leaving group, such as acetate, followed by reductive elimination, usually affords the *E*-alkene (Scheme 8.16).[161-167] A modification of the Julia reaction, that employs β-hydroxy imidazolyl sulfones, allows for the use of milder conditions (Scheme 8.17).[161] An analogous reaction with sulfides to provide *Z*-alkenes is known.[168] Selenium can also be a useful alternative to sulfur.[169,170]

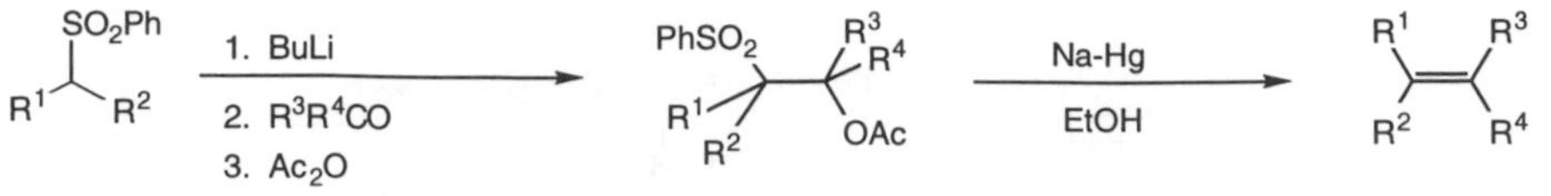

SCHEME 8.16.

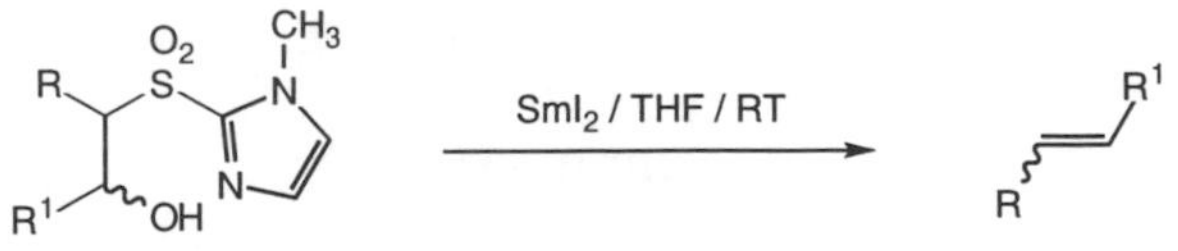

SCHEME 8.17.

*8.1.2.1.4. Trisubstituted Alkenes*

The stereoselective synthesis of trisubstituted alkenes presents additional difficulties.[171] The Wittig reaction often fails due to poor control of the relative configurations, or because of the reversibility of the reaction. However, the Horner-Wittig variant has been more successful.[138]

The use of cyclic phosphonamidates in a modified Horner-Wadsworth-Emmons reaction allows for the stereospecific construction of substituted alkenes (Scheme 8.18). Alkylation of the β-keto phosphonamidates **8.1** occurs with excellent diastereoselectivity. The alkenes are isolated by reduction followed by thermal elimination.[172]

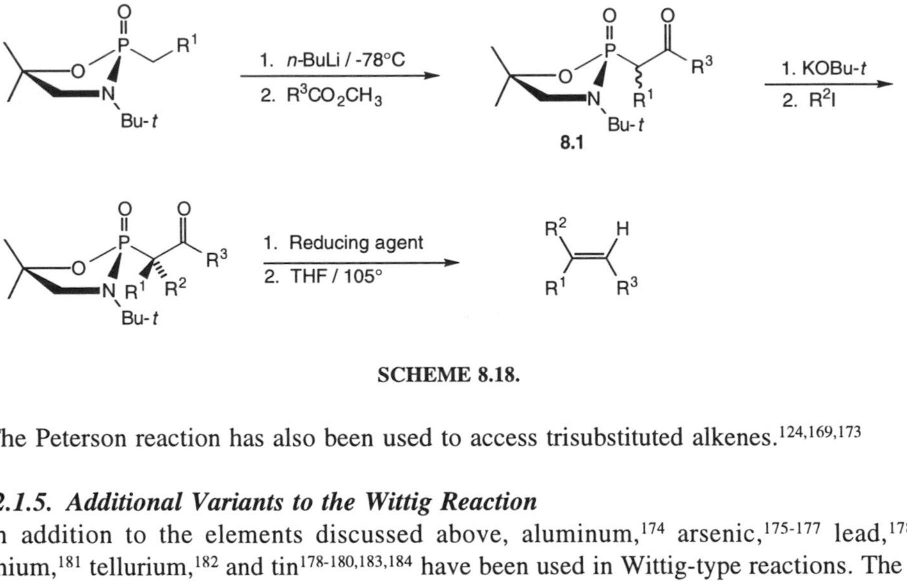

SCHEME 8.18.

The Peterson reaction has also been used to access trisubstituted alkenes.[124,169,173]

*8.1.2.1.5. Additional Variants to the Wittig Reaction*

In addition to the elements discussed above, aluminum,[174] arsenic,[175-177] lead,[178-180] selenium,[181] tellurium,[182] and tin[178-180,183,184] have been used in Wittig-type reactions. The use of tri-*n*-butylarsine or dibutyl telluride, in a catalytic Horner-Wittig-type reaction yields *E*-alkenes with high stereoselectivity.[175,185]

**8.1.2.1.5.1. The Boron-Wittig Reaction**

The Boron-Wittig reaction provides a useful route to both *E*- and *Z*-alkenes.[186-191] Condensation of dimesitylboryl stabilized carbanions with aldehydes followed by reaction of the intermediate betaine with trifluoroacetic anhydride or trimethylsilyl chloride provides a route to the *E*-and *Z*-isomers respectively (Scheme 8.19).[187]

*8.1.2.1.6. Alternate Routes*

The hydroboration of enamines formed from aldehydes and ketones allows for the preparation of either isomer of an alkene.[192,193] The conversion of α-halo sulfones, the Ramberg-Bäcklund reaction, to alkenes usually results in an isomeric mixture, although the *Z*-isomer usually predominates.[194,195] In addition, the reductive coupling of carbonyl compounds with titanium, the McMurry reaction, has also proved useful in the synthesis of simple alkenes.[196-200] Carbonyl compounds with bulky substituents give excellent yields of sterically hindered alkenes, predominately as the *trans* isomer.[196,200] In addition, mixed carbonyl compounds can be used to synthesize unsymmetrical alkenes, often giving impressive yields.[196,200-202]

Hydrazone derivatives have been converted stereoselectively to di- and trisubstituted alkenes.[203-205]

ArCHO + $Mes_2BCHLiR$ —(-78°)→ 1. $Me_3SiCl$, -78°; 2. aq. HF, MeCN → (Z)-ArCH=CHR; TFAA, -110° → (E)-ArCH=CHR

**SCHEME 8.19.**

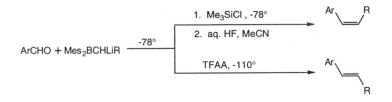

### 8.1.2.2. Conjugated Alkenes

The Wittig and Horner-Wittig reactions have proved to be classical methods to prepare these compounds, although the *E*-isomer usually results.[206] Variation of the cation, temperature, or solvent can alter the isomeric product ratio.[207] The Horner-Wadsworth-Emmons reaction is a popular method for the synthesis of α,β-unsaturated esters.[147] Although the methodology usually provides the *E*-isomer, increased *Z*-selectivity can be obtained by use of bis(trifluoroethyl)phosphonoacetates (Scheme 8.20).[208]

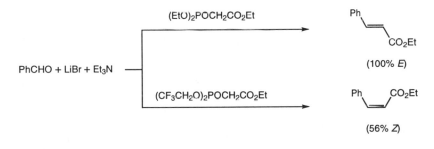

PhCHO + LiBr + $Et_3N$ — $(EtO)_2POCH_2CO_2Et$ → PhCH=CH$CO_2Et$ (100% *E*); $(CF_3CH_2O)_2POCH_2CO_2Et$ → PhCH=CH$CO_2Et$ (56% *Z*)

**SCHEME 8.20.**

The *Z*-isomer has been found to predominate in the Horner-Wadsworth-Emmons reaction with the use of low temperatures with cyclic phosphonocarboxylates derived from *N,N′*-dimethylethylenediamine (Scheme 8.21).[206] However, stereoselection is often not observed for the synthesis of trisubstituted alkenes. The use of less expensive and milder bases together with lithium or magnesium salts increases the utility of this approach.[208] The Horner-Wadsworth-Emmons reaction has also been used in the stereoselective construction of polyunsaturated compounds and α,β-unsaturated acylsilanes.[209,210]

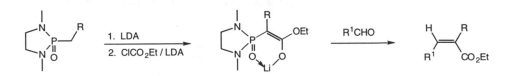

1. LDA; 2. $ClCO_2Et$ / LDA → (OEt, O–Li chelate intermediate) —$R^1CHO$→ H, R, $R^1$, $CO_2Et$

**SCHEME 8.21.**

The main problem associated with a Wittig-type approach for the synthesis of α,β-unsaturated compounds is the low reactivity with ketones.[211] An imino type ylide has shown increased reactivity in this context and can be used to prepare α,β-unsaturated carbonyl compounds (Scheme 8.22).[212]

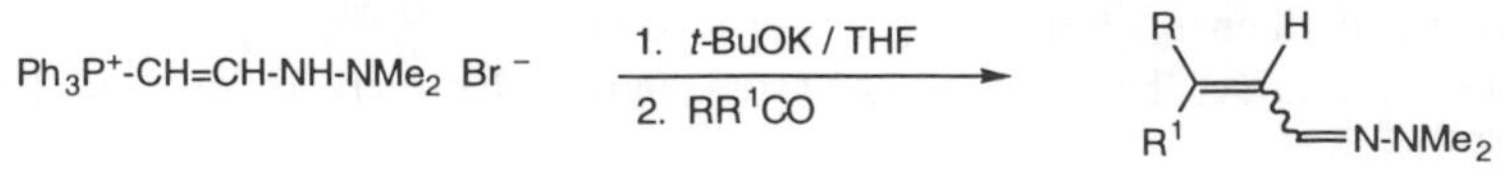

$Ph_3P^+$-CH=CH-NH-$NMe_2$ $Br^-$ → (1. t-BuOK / THF; 2. $RR^1CO$) → R($R^1$)C=CH-CH=N-$NMe_2$

SCHEME 8.22.

Both β-mono and β,β-disubstituted α,β-unsaturated esters can be prepared, with excellent diastereoselectivity, from 2-bromo-1-alkenylboranes (Scheme 8.23).[29] The Peterson reaction has also been used extensively for the synthesis of α,β-unsaturated compounds.[124] Tosyl substituted α,β-unsaturated esters have been converted to their corresponding alkyl, alkenyl, or alkynyl derivatives with high stereoselection from just one diastereoisomer.[213] Ketene thioacetals can also be used to prepare *E*-α,β-unsaturated carbonyl compounds.[214]

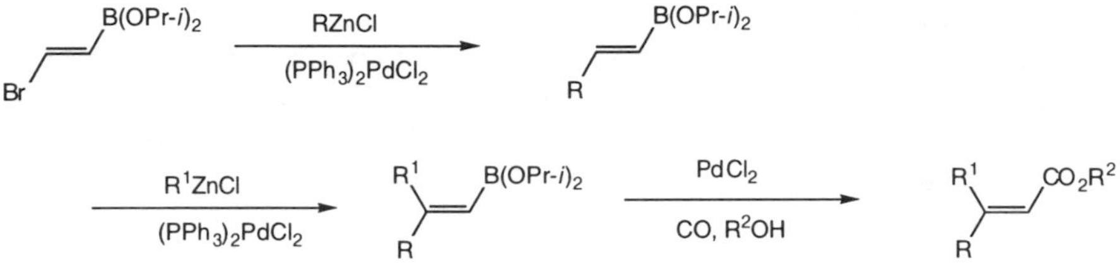

Br-CH=CH-B(OPr-$i$)$_2$ → (RZnCl, $(PPh_3)_2PdCl_2$) → R-CH=CH-B(OPr-$i$)$_2$ → ($R^1$ZnCl, $(PPh_3)_2PdCl_2$) → R($R^1$)C=CH-B(OPr-$i$)$_2$ → ($PdCl_2$; CO, $R^2$OH) → R($R^1$)C=CH-$CO_2R^2$

SCHEME 8.23.

A number of chiral auxiliaries have been used to control product selection (e.g., Scheme 8.24).[215-218]

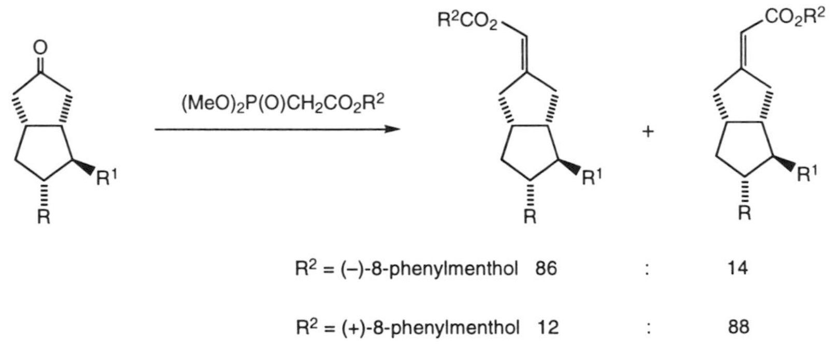

(MeO)$_2$P(O)CH$_2$CO$_2$R$^2$

| | | |
|---|---|---|
| $R^2$ = (–)-8-phenylmenthol | 86 : | 14 |
| $R^2$ = (+)-8-phenylmenthol | 12 : | 88 |

SCHEME 8.24.

*Z*-Conjugated nitriles are available by a highly stereoselective addition of tris-(trimethylsilyl)ketenimine to aldehydes (Scheme 8.25),[219] or through a Reformatsky–Peterson reaction.[220]

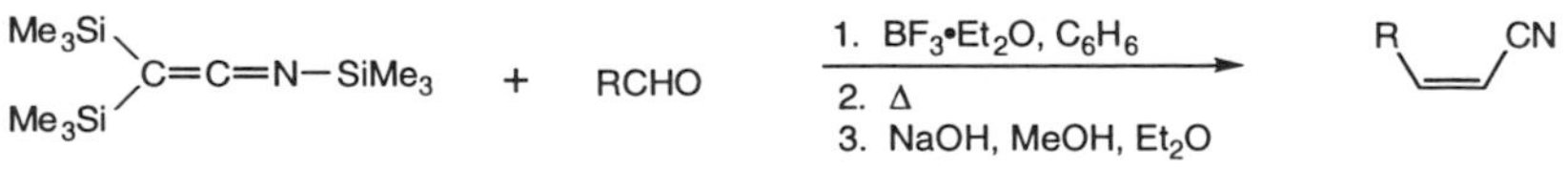

$(Me_3Si)_2$C=C=N–$SiMe_3$ + RCHO → (1. $BF_3$•$Et_2O$, $C_6H_6$; 2. Δ; 3. NaOH, MeOH, $Et_2O$) → R-CH=CH-CN

SCHEME 8.25.

#### 8.1.2.3. Heteroatom Substituted Alkenes

Vinylsilanes, vinyl halides, and vinyl sulfides offer a variety of methods for the preparation of alkenes. The vinylsilanes are, in turn, available from a number of routes including the reduction of acetylenes, silylacetylenes, and carbonyl condensation reactions. The silyl group can then be replaced by a wide variety of electrophilic species (vide infra).[79-83,86,219,221-228]

The *E*-isomers of vinyl sulfones are readily available through the Horner-Wittig reaction by phosphorylation of methyl phenyl sulfone.[229] Vinyl sulfides have been used to synthesize either *Z*- or *E*-alkenes.[230,231] The reaction of a vinyl sulfide with a Grignard reagent in the presence of a nickel catalyst allows for the stereoselective replacement of the sulfide moiety.[231-235] Vinyl sulfones react with mercury amalgam to yield the simple alkene.[236]

Vinyl halides have also proved useful in the synthesis of alkenes,[73,74,173,231,237,238] and are available by a wide variety of methods, including the Peterson (e.g., Scheme 8.26)[173,237] and Wittig reactions.[238-241] The halogen can then be transmetallated to provide a vinyl organometallic species (vide infra), or used as an electrophilic component.

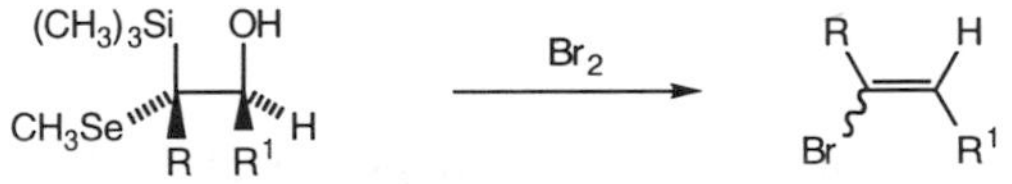

$(CH_3)_3Si$ OH
$CH_3Se$ H
R $R^1$
$Br_2$
R H
Br $R^1$

**SCHEME 8.26.**

#### 8.1.2.4. Allylic Compounds

Allylic alcohols are substrates for a wide variety of important stereoselective reactions. They are available by an extension of the Schlosser modification of the Wittig reaction. Condensation of a β-oxido ylide, formed from the deprotonation of the betaine, with an aldehyde results in the synthesis of *Z*-allylic alcohols (Scheme 8.27).[242-251] Some asymmetric induction has been observed, such as with the use of a metal catalyst [*9.6*].

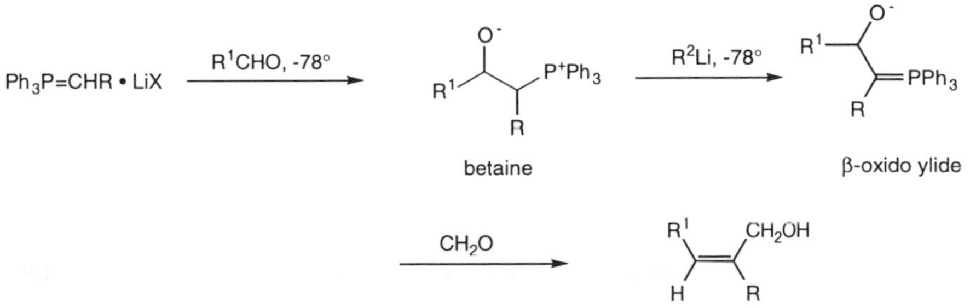

$Ph_3P{=}CHR \bullet LiX$
$R^1CHO$, -78°
$O^-$ $R^1$ $P^+Ph_3$ R
betaine
$R^2Li$, -78°
$O^-$ $R^1$ $=PPh_3$ R
β-oxido ylide
$CH_2O$
$R^1$ $CH_2OH$
H R

**SCHEME 8.27.**

A regioselective route to allylsilanes involves the treatment of α-halo carbonyl compounds with trimethylsilylmethylmagnesium chloride.[252]

### 8.1.3. REACTION OF HETEROALKENES

There are a number of stereoselective routes to heteroalkenes (vide supra). Although the presence of a heteroatom allows for further elaboration, the heteroatom can be exchanged for a proton to yield a simple alkene.

#### 8.1.3.1. Simple Alkenes

Many metals are available to catalyze the condensation of vinyl anions with a wide variety of electrophiles providing a versatile preparation of alkenes.[231,253] The vinyl anions are available by transmetallation procedures, as well as from acetylenes (vide supra).[254,255]

Metal–halogen exchange of vinyl halides often results in retention of configuration.[255-257] Vinyl anions derived from vinyl halides can be quenched with electrophiles,[254,258] including alkyl halides to provide a stereoselective route to trisubstituted alkenes.[258] Organocuprates, including vinyl derivatives, react with alkyl halides in high yield and good stereoselectivity.[259-261]

Vinylsilanes can be stereospecifically converted to alkenes.[262-264] The silyl group can be displaced in a regioselective, and often stereoselective, manner by a wide variety of electrophiles, including protiodesilylation (Scheme 8.28).[222,255,263,265]

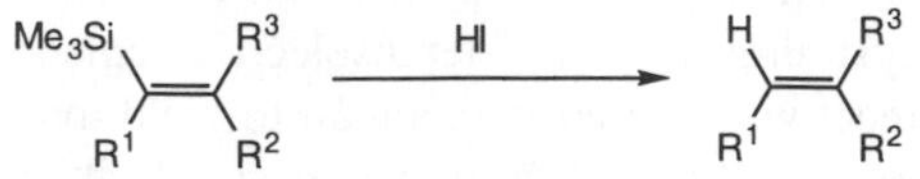

**SCHEME 8.28.**

In addition, Grignard reagents and organolithium compounds add to vinylsilanes and vinyl sulfur compounds regioselectively [*9.5*].[266]

### 8.1.3.2. Conjugated Alkenes

Vinylsilanes react under Friedel–Craft conditions to yield α,β-unsaturated ketones.[267,268] An alternative approach to conjugated alkenes is by heteroatom–metal exchange to form the vinyl anion;[257,258,269,270] subsequent reaction with an electrophile then provides compounds such as α,β-unsaturated aldehydes and acids.[254]

### 8.1.3.3. Heteroatom Substitution

Vinylsilanes can be converted to either *E*- or *Z*-vinyl halides depending on the reagents used (Scheme 8.29);[82,271] the stereoselectivity can also be controlled by the concentration of the Lewis acid.[265]

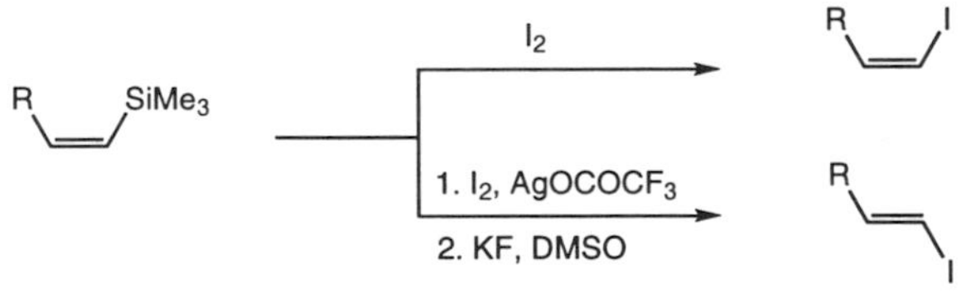

**SCHEME 8.29.**

Vinylsilyl phosphates, readily available from α-halo ketones, react with organocuprates to form vinylsilanes.[272,273]

### 8.1.3.4. Allylic Compounds

Allylsilanes can be synthesized by a number of routes from vinyl halides, silyl enol ethers, enol phosphates, and vinyl triflates.[274,275] In a similar manner, allyl alcohols can be prepared by a number of methods including the reaction of vinyl anions with carbonyl compounds.[254]

## 8.1.4. ELIMINATIONS

Elimination of a wide variety of functional groups, either through an $E_2$ reaction[276] or by thermal elimination, provides stereospecific routes to alkenes.[277-279] In general, the former reactions proceed with *anti*-stereochemistry, while the latter are *syn* (Scheme 8.30).[6,280] However, both approaches can destroy two stereogenic centers to afford one alkene isomer.[281-284]

Sulfoxide elimination, along with the selenoxide analog, have been used to prepare alkenes (Scheme 8.31),[285-289] but the stereoselectivity of these *syn*-eliminations can be compromised by the problems associated with the asymmetric introduction of the heteroatom.[287,290,291]

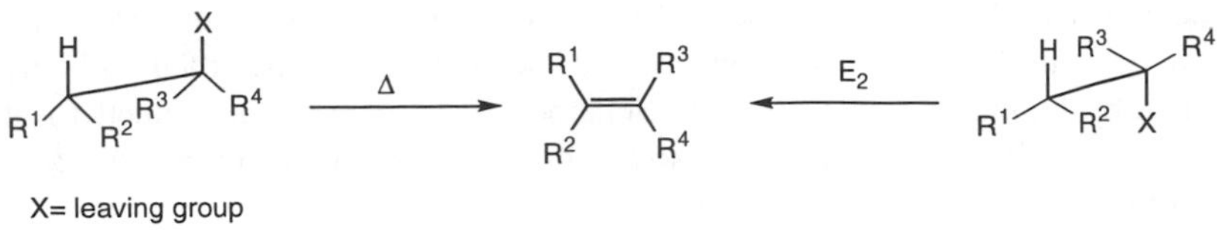

SCHEME 8.30.

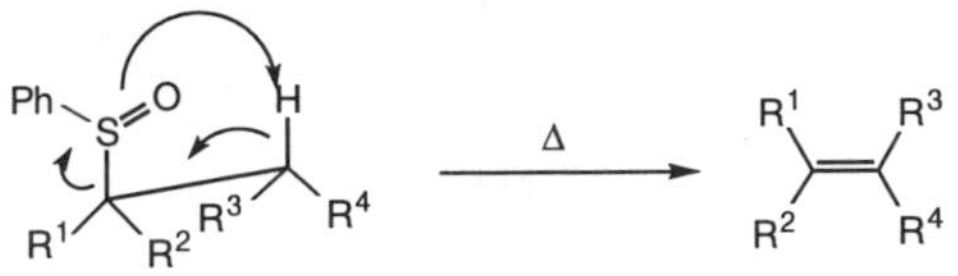

SCHEME 8.31.

The deoxygenation of diols also relies on *syn* elimination, and a number of methodologies are available.[292-299] Some of these methods form intermediates, such as a thiocarbonate (Scheme 8.32),[293] or a 2-dimethylamino-1,3-dioxolane,[292] that undergoes the elimination reaction, while others form an intermediate that spontaneously converts to the desired alkene under the reaction conditions (e.g., Scheme 8.33).[299]

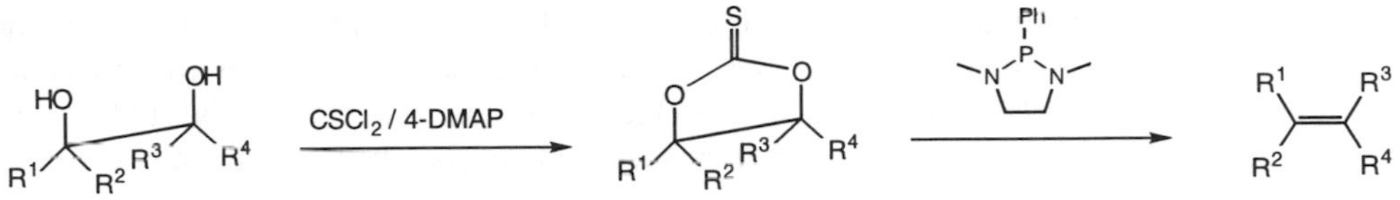

SCHEME 8.32.

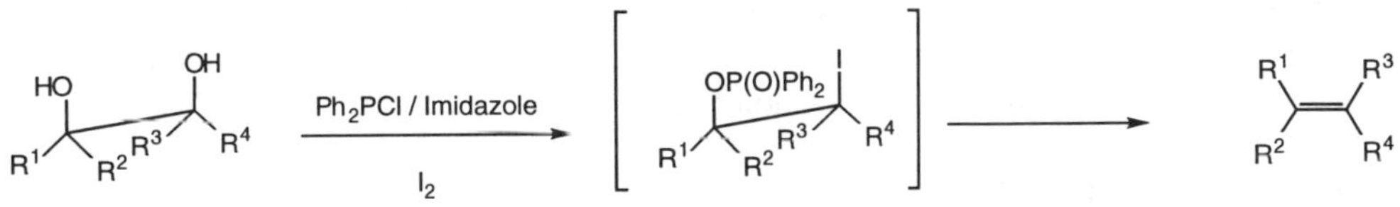

SCHEME 8.33.

Alkenes have also been obtained by the stereoselective dehydrative decarboxylation of β-hydroxy carboxylic acids.[300]

Other alkenes are also available by deoxygenation of epoxides,[301-307] thermolysis of esters,[278] lactones,[284] ketones,[308] and episulfides[298,301,309] together with episulfoxides or episulfones.[310,311]

Classical reactions, such as the Claisen condensation, Stobbe condensation, Knoevenagel condensation, and Reformatsky reaction, can also be used to prepare conjugated alkenes.[312-314] The retro Diels-Alder reaction has also proved valuable in the synthesis of alkenes [*12.1.11*].

## 8.1.5. INVERSION

Many methods also exist for the inversion of olefin stereochemistry.[301,315-319] A classic photochemical reaction is the isomerization of alkenes. Although the *trans*-isomer of disubstituted alkenes is thermodynamically more stable, *cis* tends to predominate in the photostationary state.[6,315] Sensitizers have also been used to selectively excite the *trans*-isomer, further increasing the amount of the *cis*-isomer.[320] However, the need to separate the isomers, together with by-product formation, detracts from this methodology.

Treatment of an alkene with iodine and the silver salt of an acid, the Prévost reaction, results in a *trans* diol.[301,321] Conversion back to the opposite isomer of the alkene can be accomplished by deoxygenation of the diol (Scheme 8.34) (vide supra). Adducts formed from a *syn,* or net *syn,* addition to alkenes can also be converted back to the opposite alkene isomer by an *anti* elimination.[301,322]

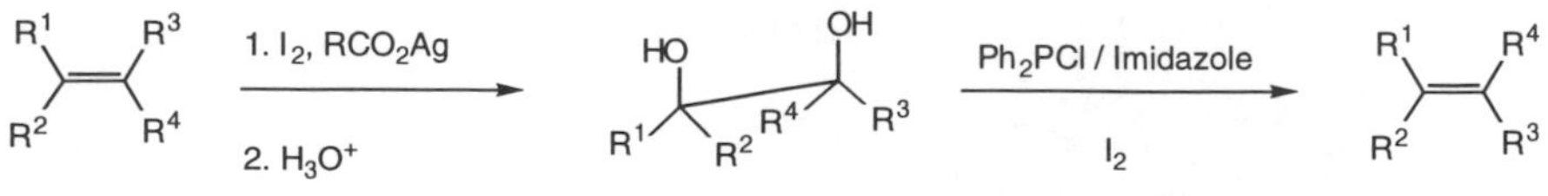

1. $I_2$, $RCO_2Ag$ / 2. $H_3O^+$ ; $Ph_2PCl$ / Imidazole / $I_2$

SCHEME 8.34.

## 8.2. REACTIONS OF ALKENES

### 8.2.1. ADDITION REACTIONS

Many reagents add stereoselectively to alkenes, such as hydrogen halides,[323,324] and halogens.[6,113,325-327] Such an addition can be useful to control relative stereochemistry, especially if regiochemical control is accomplished through application (or violation!) of Markovnikov addition.[6,328-331] In contrast, very few methods are currently available to control absolute stereoselectivity during these additions;[332] however, advances in this area have been rapid (vide infra).[333]

Addition to a double bond can take place by electrophilic, nucleophilic,[333] free radical, or cyclic mechanisms.[6,331] Selection of the appropriate starting material, together with reagents and reaction conditions can determine the mechanism followed. Electrophilic addition to unsymmetrical alkenes proceeds by attack of the electrophile from the least substituted alkene carbon to give the more stable carbocation — Markovnikov addition. In addition to being regiospecific, the product is usually *anti* except where an ion pair is formed. Nucleophilic addition, such as the Michael reaction, requires the presence of an anion stabilizing group on the alkene [*9.5*]. Free radical mechanisms result in *anti*-Markovnikov addition as the radical attaches to the least hindered alkene carbon. Cyclic mechanisms result in simultaneous *syn* addition such as in the Diels-Alder reaction [*12.1*].

#### 8.2.1.1. Addition of Acids

Strong acids add directly to alkenes, while weaker acids, water, and alcohols require the presence of a strong acid.[327,334] Markovnikov addition is observed for hydrogen halides except for hydrogen bromide under conditions that favor radical formation. Protic solvents can lead to high levels of solvolysis products. Although the major product is usually *pref,* low temperature and steric effects can reverse this (Scheme 8.35).[335] Hydrogen cyanide adds stereospecifically to alkenes in the presence of a palladium, nickel, or platinum catalyst through *syn* addition (Scheme 8.36).[323,336]

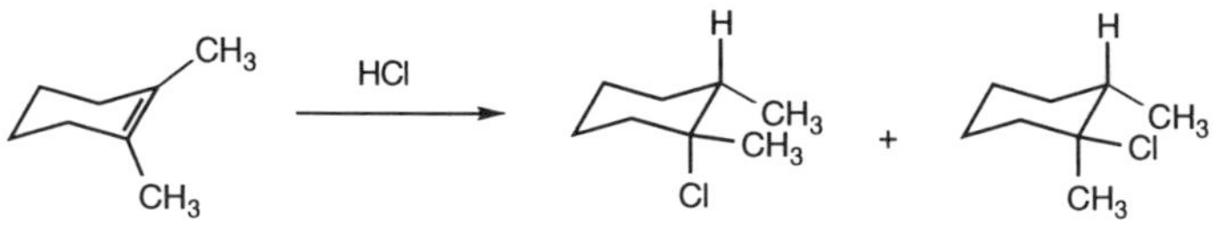

HCl

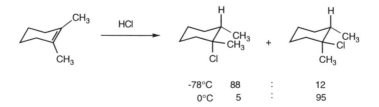

| | | | |
|---|---|---|---|
| -78°C | 88 | : | 12 |
| 0°C | 5 | : | 95 |

SCHEME 8.35.

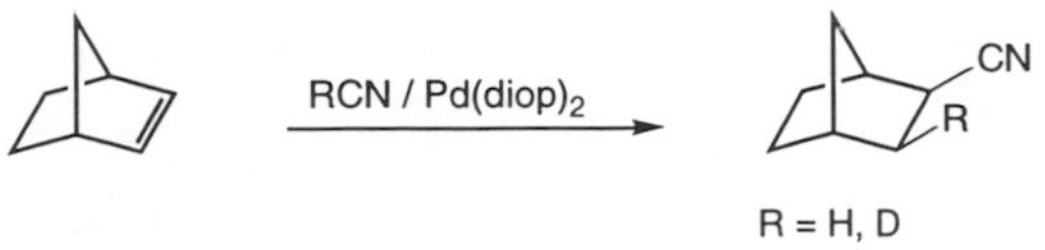

RCN / Pd(diop)$_2$

R = H, D

SCHEME 8.36.

#### 8.2.1.2. Addition of Halogens

All of the halogens can react with alkenes to give addition products. However, both fluorine and iodine require special reaction conditions.[6] The normal mode of addition is *trans* except for fluorine. Cation stabilizing substituents together with polar solvents, lead to an increase in *cis*-addition.[337,338] Bromine, due to its larger size and higher polarizability, generally leads to a higher percentage of *trans*-addition than chlorine.[339] Hypohalous acids also lead to an intermediate halonium ion that can react with other nucleophiles in the reaction medium (e.g., Scheme 8.37).[340,341]

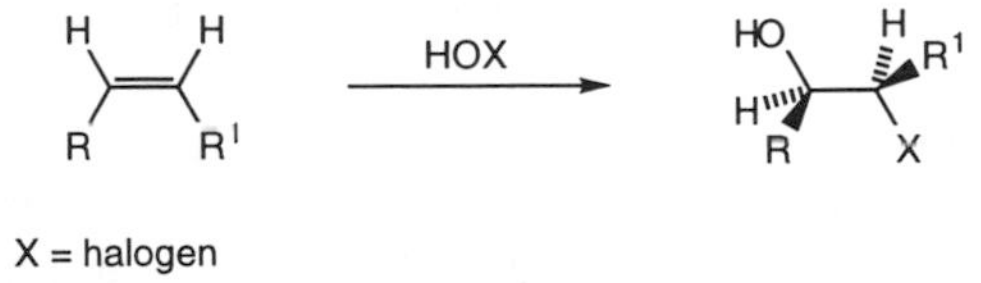

HOX

X = halogen

SCHEME 8.37.

#### 8.2.1.3. Oxymercuration

Oxymercuration provides excellent regioselectivity leading to the Markovnikov product.[330,342,343] Terminal alkenes are more reactive; hence, selectivity can be obtained with compounds that contain more than one double bond. The removal of mercury with sodium borohydride has been shown to proceed by a radical process, consequently leading to a mixture of *pref* and *parf* isomers (Scheme 8.38).[344]

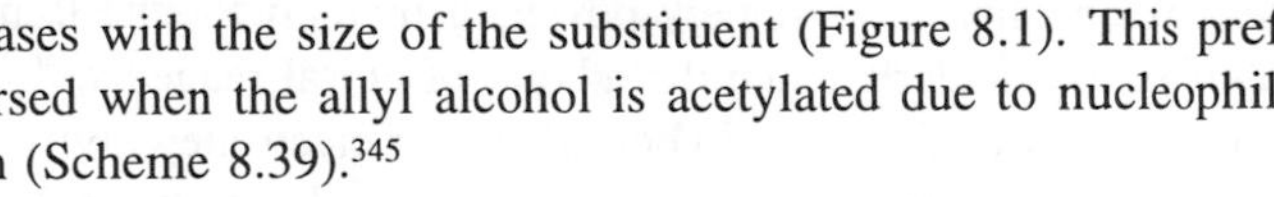

1. Hg(OAc)$_2$
2. NaBH$_4$

SCHEME 8.38.

The oxymercuration of terminal allyl alcohols exhibits reasonable diastereoselectivity with a preference for *anti*-addition. The hydroxy group directs the approach of the mercuric ion and selectivity increases with the size of the substituent (Figure 8.1). This preference for *anti*-addition is reversed when the allyl alcohol is acetylated due to nucleophilic attack by the carbonyl oxygen (Scheme 8.39).[345]

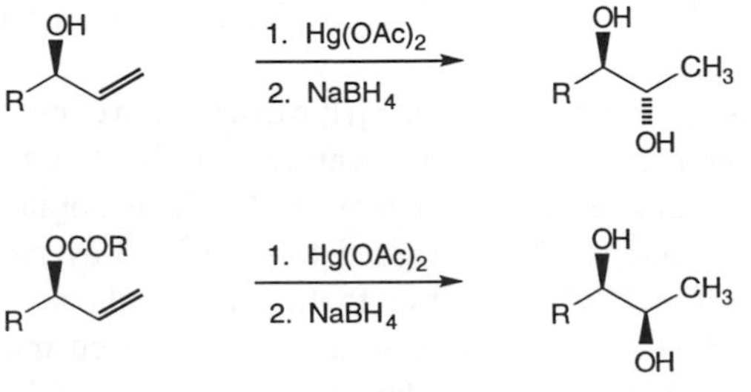

1. Hg(OAc)$_2$
2. NaBH$_4$

1. Hg(OAc)$_2$
2. NaBH$_4$

SCHEME 8.39.

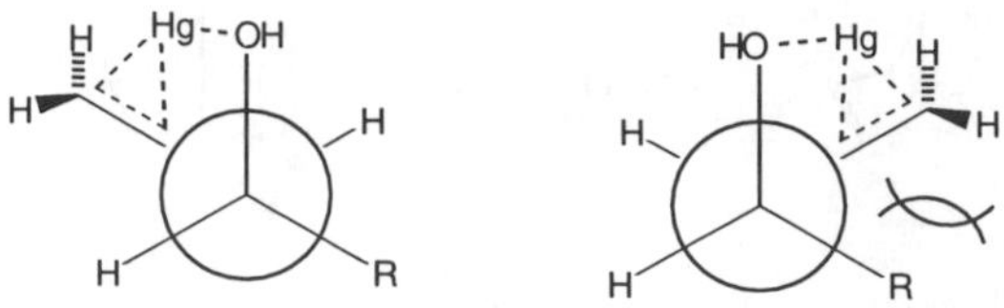

FIGURE 8.1. Possible conformations for the oxymercuration of allyl alcohols.

## 8.2.2. REDUCTIONS

### 8.2.2.1. Hydrogenations

Despite the ubiquitous use of hydrogenation in organic synthetic methodology, asymmetric hydrogenation of simple alkenes still remains elusive.[346-350] The elegance of such an approach has, however, spurred many attempts, but the asymmetric induction with simple alkenes remains low.[351-353] In contrast, considerable success has been achieved when a functional group near to the alkene can act as a ligand for the metal catalyst [*9.1*].[353,354] Of course, hydrogenation usually provides *cis*-delivery of the hydrogen that can still be very useful.[355,356]

### 8.2.2.2. Hydrosilylations

In contrast, hydrosilylation of an alkene usually provides a higher degree of asymmetric induction than hydrogenation.[357-361] The methodology introduces a silyl group into the substrate, which must be removed by subsequent manipulation (Scheme 8.40).[362] Hydrosilylation has been used to reduce a number of conjugated functional groups;[361,363-365] the use of allyl alcohols allows a stereoselective synthesis of 1,3-diols (e.g., Scheme 9.17).[366,367]

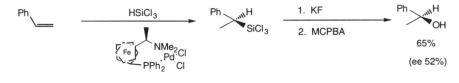

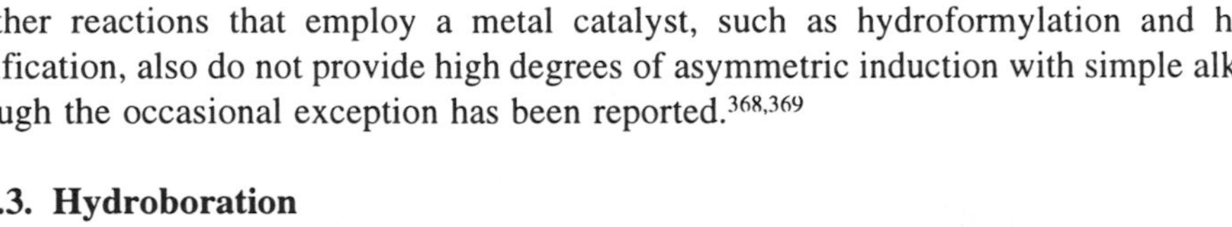

SCHEME 8.40.

Other reactions that employ a metal catalyst, such as hydroformylation and hydroesterification, also do not provide high degrees of asymmetric induction with simple alkenes, although the occasional exception has been reported.[368,369]

### 8.2.2.3. Hydroboration

Hydroboration has become an extremely powerful method for the transformation of an alkene into an alcohol, particularly since the advent of chiral reagents.[370-372] Hydroboration results in *anti*-Markovnikov addition due to boron's lower electronegativity and steric factors. Like many organometallic reagents, *syn*-addition occurs, allowing for a highly regio- and stereoselective reaction.[19,373,374] Increased regioselectivity can be obtained by the use of dialkylboranes and haloboranes.[19,375]

The synthesis of diisopinocampheylborane [$(Ipc)_2BH$] from commercially impure α-pinene by resolution was a major milestone for the synthesis of chiral compounds, as high enantiomeric purities of the reagent can now be achieved.[376,377] This borane shows a high degree of selectivity toward *cis*-alkenes,[26,378] while good levels of 1,3-asymmetric induction are observed with terminal olefins.[379,380] In addition to the traditional conversion of the borane to an alcohol (e.g., Scheme 8.41),[381] a borane can also be converted into many other functional groups including amines, alkyl halides, aldehydes, carboxylic acids, ketones, esters, nitriles, acetylenes, alkenes, and allenes.[19,370,376,382,383]

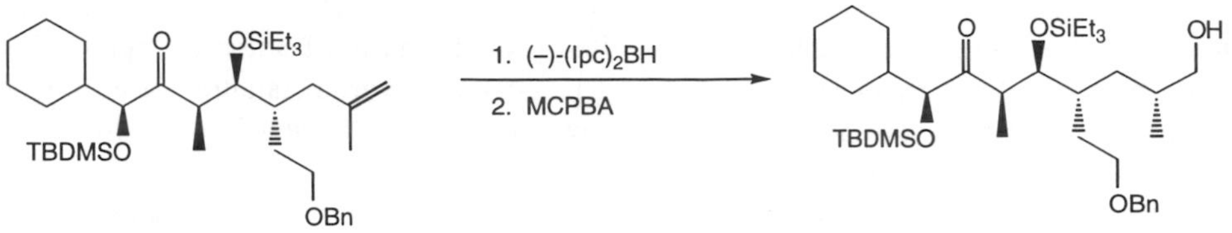

SCHEME 8.41.

The use of borolanes has been advocated over the use of chiral boranes (Scheme 8.42) for diastereoselective synthesis through double asymmetric induction.[384]

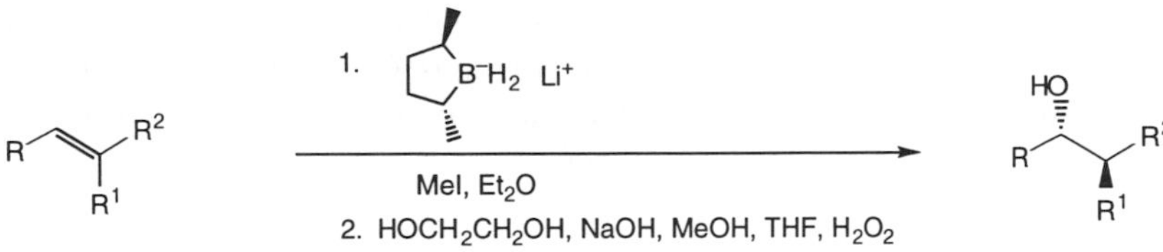

SCHEME 8.42.

Some asymmetric induction has been observed for the addition of catecholborane to an alkene in the presence of a chiral rhodium(I) catalyst.[385-387] The improvement of reagents continues,[388-390] not only through theoretical investigations into the reaction pathway and the factors influencing the outcome, but by synthetic approaches.[391] For allyl alcohol substrates, other factors can play an important role to provide excellent stereochemical control [*9.3.1.*].

### 8.2.3. ALLYLIC FUNCTIONALIZATION

The formation of a free radical, cation, or anion α to a double bond is facilitated by conjugation with the π-electrons. Halogenation can be accomplished with a number of reagents, the most popular being *N*-bromosuccinimide.[392] Although halogenation does occur at the allylic position, sometimes rearrangements can result in a mixture of products.[393]

Allylic oxidation of an alkene provides a useful entry to allyl alcohols. Selenium dioxide can provide high stereoselectivity, especially with trisubstituted alkenes (Scheme 8.43).[242,394] However, this approach can be troublesome due to over-oxidation to unsaturated aldehydes and ketones or the formation of rearranged products.[394,395]

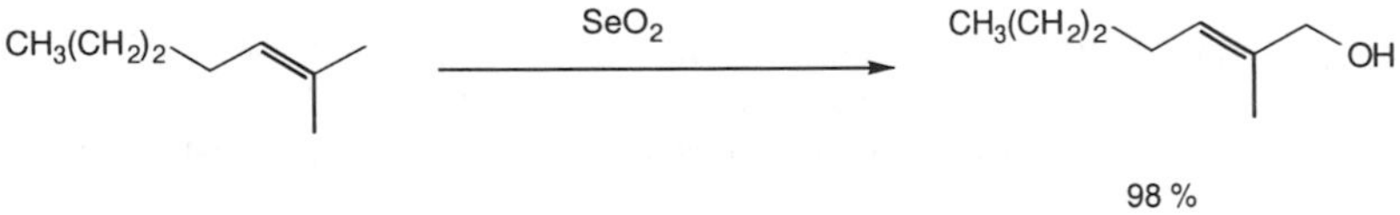

SCHEME 8.43.

## 8.3. SUMMARY

A plethora of approaches is available for the stereoselective synthesis of alkenes. This has been, in part, due to the lack of a general route. Of these, carbonyl-based approaches, especially the Wittig and related reactions, provide the largest number of examples as well as being the most-used approaches. However, the Wittig and related reactions are generally unreactive towards ketones. The removal of triphenylphosphine oxide can also prove problematic in the Wittig reaction. Often a specific route will lead to an unacceptable amount of the undesired isomer. Separation of alkene isomers is often far from trivial. Despite of these detractions, the Wittig and related reactions often provide the most attractive approach to alkenes due to the directness and availability of precursors. Hydrometallation of acetylenes often provides an attractive alternative. However, this latter approach is often limited to terminal or symmetrical acetylenes.

Most methods for the preparation of alkenes, however, require the use of organometallic reagents, and they may not be amenable to large scale due to the problems associated with the handling and use of these types of reagents. The use of a stabilized phosphorus ylide, where a "mild" base can be employed, offers many advantages in this regard, even though reaction times may be increased.

Alkenes do react predictably with a number of reagents often in a Markovnikov fashion. However, control of stereoselectivity is limited to a few reactions. Hydrogenations, hydrosilylations, and oxymercuration often provide excellent stereoselectivity. Heavy metals are not very practicable particularly in large-scale reactions. Control of stereoselectivity is best achieved by hydroboration. The range of functional groups that can be incorporated by hydroboration also adds to its utility in synthesis.

## 8.4. REFERENCES


1. Rossi, R. *Synthesis* 1978, 413.
2. Mori, K. In *The Total Synthesis of Natural Products*; J. ApSimon, Ed.; John Wiley & Sons: New York, 1981; Vol. 4; pp 1.
3. Mori, K. *Tetrahedron* 1989, *45*, 3233.
4. Henrick, C. A. *Tetrahedron* 1977, *33*, 1845.
5. Ager, D. J.; East, M. B. *Tetrahedron* 1992, *48*, 2803.
6. March, J. *Advanced Organic Chemistry: Reactions Mechanisms and Structure;* 3rd ed.; John Wiley & Sons: New York, 1985.
7. Reucroft, J.; Sammes, P. G. *Quart. Rev.* 1971, *25*, 135.
8. Ager, D. J.; East, M. B. *Tetrahedron* 1993, *49*, 5683.
9. Hayashi, T.; Tajika, M.; Tamao, K.; Kumada, M. *J. Am. Chem. Soc.* 1976, *98*, 3718.
10. Fouquet, G.; Schlosser, M. *Angew. Chem., Int. Ed. Engl.* 1974, *13*, 82.
11. Neumann, H.; Seebach, D. *Tetrahedron Lett.* 1976, 4839.
12. Brunner, H.; Probster, M. *J. Organometal. Chem.* 1981, *209*, C1.
13. Fiandanese, V.; Marchese, G.; Naso, F.; Ronzini, L. *J. Chem. Soc., Chem. Commun.* 1982, 647.
14. Wenkert, E.; Ferreira, T. W. *J. Chem. Soc., Chem. Commun.* 1982, 840.
15. Schwartz, J.; Arvanitis, G. M.; Smegel, J. A.; Meier, I. K.; Clift, S. M.; Engen, D. V. *Pure Appl. Chem.* 1988, *60*, 65.
16. Kataoka, Y.; Takai, K.; Oshima, K.; Utimoto, K. *Tetrahedron Lett.* 1990, *31*, 365.
17. Daeuble, J. F.; McGettigan, C.; Stryker, J. M. *Tetrahedron Lett.* 1990, *31*, 2397.
18. Sandler, S. R.; Karo, W. *Organic Functional Group Preparations;* 2nd ed.; Academic Press: Orlando, FL, 1983.
19. Pelter, A.; Smith, K.; Brown, H. C. *Borane Reagents*; Academic Press: London, 1988.
20. Choudary, B. M.; Sharma, G. V. M.; Bharathi, P. *Angew. Chem., Int. Ed. Engl.* 1989, *28*, 465.
21. Magoon, E. F.; Slaugh, L. H. *Tetrahedron* 1967, *23*, 4509.
22. Huang, H. C.; Rehmann, J. K.; Gray, G. R. *J. Org. Chem.* 1982, *47*, 4018.
23. Schwartz, M.; Waters, R. M. *Synthesis* 1972, 567.
24. Brown, H. C.; Basavaiah, D. *J. Org. Chem.* 1982, *47*, 3806.
25. Brown, H. C.; Bakshi, R. K.; B., S. *J. Am. Chem. Soc.* 1988, *110*, 1529.
26. Brown, H. C.; Yoon, N. M. *Isr. J. Chem.* 1977, *15*, 12.
27. Brown, H. C.; Basavaiah, D. *J. Org. Chem.* 1982, *47*, 5407.
28. Negishi, E.-I.; Yoshida, T.; Abramovitch, A.; Lew, G.; Williams, R. M. *Tetrahedron* 1991, *47*, 343.
29. Yamashina, N.; Hyuga, S.; Hara, S.; Suzuki, A. *Tetrahedron Lett.* 1989, *30*, 6555.
30. Hyuga, N.; Yamashina, S.; Hara, S.; Suzuki, A. *Chem. Lett.* 1988, 809.

31. Hyuga, S.; Chiba, Y.; Yamashina, S.; Hara, S.; Suzuki, A. *Chem. Lett.* 1987, 1757.
32. Zweifel, G.; Arzoumanian, H.; Whitney, C. C. *J. Am. Chem. Soc.* 1967, *89*, 3652.
33. Zweifel, G.; Arzoumanian, H. *J. Am. Chem. Soc.* 1967, *89*, 5086.
34. Wilke, G.; Müller, H. *Chem. Ber.* 1956, *89*, 444.
35. Gensler, W. J.; Bruno, J. J. *J. Org. Chem.* 1963, *28*, 1254.
36. Cram, D. J.; Allinger, N. L. *J. Am. Chem. Soc.* 1956, *78*, 2518.
37. Zweifel, G.; Clark, G. M.; Polston, N. L. *J. Am. Chem. Soc.* 1971, *93*, 3395.
38. Brown, H. C.; Gupta, S. K. *J. Am. Chem. Soc.* 1972, *94*, 4370.
39. Brown, H. C.; Zweifel, G. *J. Am. Chem. Soc.* 1959, *81*, 1512.
40. Zweifel, G.; Fisher, R. P.; Whitney, C. C. *J. Am. Chem. Soc.* 1971, *93*, 6309.
41. Yoshida, T.; Negishi, E. *J. Am. Chem. Soc.* 1981, *103*, 4985.
42. Negishi, E. *Pure Appl. Chem.* 1981, *53*, 2333.
43. Zhang, Y.; Miller, J. A.; Negishi, E. *J. Org. Chem.* 1989, *54*, 2043.
44. Molander, G. A.; Mautner, K. *J. Org. Chem.* 1989, *54*, 4042.
45. Miller, J. A. *J. Org. Chem.* 1989, *54*, 998.
46. Normant, J.-F.; Cahiez, G.; Bourgain, M.; Chuit, C.; Villieras, J. *Bull. Chem. Soc. Fr.* 1974, 1656.
47. Alexakis, A.; Cahiez, G.; Normant, J.-F. *Org. Synthesis* 1984, *62*, 1.
48. Alexakis, A.; Normant, J.-F.; Villieras, J. *Tetrahedron Lett.* 1976, 3461.
49. Normant, J.-F.; Alexakis, A. *Synthesis* 1981, 841.
50. Miller, J. G. *Tetrahedron Lett.* 1989, *30*, 4913.
51. McGuirk, P. R.; Marfat, A.; Helquist, P. *Tetrahedron Lett.* 1978, 2465.
52. Iyer, R. S.; Helquist, P. *Org. Synth.* 1985, *64*, 1.
53. Kowalski, C. J.; Sakdarat, S. *J. Org. Chem.* 1990, *55*, 1977.
54. Vieregge, H.; Schmidt, H. M.; Renema, J.; Bos, H. J. T.; Arens, J. F. *Recueil* 1966, *85*, 929.
55. Pornet, J.; Khouz, B.; Miginiac, L. *Tetrahedron Lett.* 1985, *26*, 1861.
56. Pornet, J.; Rayadh, A.; Miginiac, L. *Tetrahedron Lett.* 1986, *27*,
57. Satoh, M.; Tayano, T.; Suzuki, A. *Tetrahedron Lett.* 1989, *30*, 5153.
58. Yamahina, N.; Hyuga, S.; Hara, S.; Suzuki, A. *Tetrahedron Lett.* 1989, *30*, 6555.
59. Normant, J.-F.; Cahiez, G.; Chuit, C.; Villieras, J. *J. Organomet. Chem.* 1973, *54*, C53.
60. Westmijze, H.; Vermeer, P. *Synthesis* 1977, 784.
61. Anderson, R. J.; Corbin, V. L.; Cotterrell, G.; Cox, G. R.; Henrick, C. A.; Schaub, F.; Siddall, J. B. *J. Am. Chem. Soc.* 1975, *97*, 1197.
62. Bowlus, S. B.; Katzenellenbogen, J. A. *J. Org. Chem.* 1973, *38*, 2733.
63. Liedtke, R. J.; Djerassi, C. *J. Org. Chem.* 1972, *37*, 2111.
64. Michelot, D.; Linstrumelle, G. *Tetrahedron Lett.* 1976, 275.
65. Zweifel, G.; Polston, N. L. *J. Am. Chem. Soc.* 1970, *92*, 4086.
66. Zweifel, G.; Polston, N. L.; Whitney, C. C. *J. Am. Chem. Soc.* 1968, *90*, 6243.
67. Brown, H. C.; Blue, C. D.; Nelson, D. J.; Bhat, N. G. *J. Org. Chem.* 1989, *54*, 6064.
68. Brown, H. C.; Subrahmanyam, C.; Hamaoka, T.; Ravindran, N.; Bowman, D. H.; Misumi, S.; Unni, M. K.; Somayaji, V.; Bhat, N. G. *J. Org. Chem.* 1989, *54*, 6068.
69. Brown, H. C.; Hamaoka, T.; Ravindran, N.; Subrahmanyam, C.; Somayaji, V.; Bhat, N. G. *J. Org. Chem.* 1989, *54*, 6075.
70. Brown, H. C.; Larock, R. C.; Gupta, S. K.; Rajagopalan, S.; Bhat, N. G. *J. Org. Chem.* 1989, *54*, 6079.
71. Brown, H. C.; Hamaoka, T.; Ravindran, N. *J. Am. Chem. Soc.* 1973, *95*, 6456.
72. Brown, H. C.; Hamaoka, T.; Ravindran, N. *J. Am. Chem. Soc.* 1973, *95*, 5786.
73. Miller, R. B.; Al-Hassan, M. I.; McGarvey, G. *Syn. Commun.* 1983, *13*, 969.
74. Zweifel, G.; Lewis, W. *J. Org. Chem.* 1978, *43*, 2739.
75. Miller, R. B.; Al-Hassan, M. I. *J. Org. Chem.* 1984, *49*, 725.
76. Sato, F.; Ishikawa, H.; Sato, M. *Tetrahedron Lett.* 1981, *22*, 85.
77. Soderquist, J. A.; Léon-Colón, G. *Tetrahedron Lett.* 1991, *32*, 43.
78. Soderquist, J. A.; Colberg, J. C.; Del Valle, L. *J. Am. Chem. Soc.* 1989, *111*, 4873.
79. Uchida, K.; Utimoto, K.; Nozaki, H. *J. Org. Chem.* 1976, *41*, 2941.
80. Hoshi, M.; Masuda, Y.; Arase, A. *J. Chem. Soc., Perkin Trans. 1* 1990, 3237.
81. Soderquist, J. A.; Santiago, B. *Tetrahedron Lett.* 1990, *31*, 5113.
82. Miller, R. B.; Reichenbach, T. *Tetrahedron Lett.* 1974, 543.
83. Uchida, K.; Utimoto, K.; Nozaki, H. *Tetrahedron* 1977, *33*, 2987.
84. Arcadi, A.; Cacchi, S.; Marinelli, F. *Tetrahedron Lett.* 1986, *27*, 6397.
85. Arcadi, A.; Cacchi, C.; Marinelli, F. *Tetrahedron* 1985, *41*, 5121.
86. Snider, B. B.; Karras, M.; Conn, R. S. E. *J. Am. Chem. Soc.* 1978, *100*, 4624.
87. Snider, B. B.; Conn, R. S. E.; Karras, M. *Tetrahedron Lett.* 1979, 1679.
88. Fleming, I.; Newton, T. W.; Roesseler, F. *J. Chem. Soc., Perkin Trans. 1* 1981, 2527.

89. Benkeser, R. A. *Pure Appl. Chem.* 1966, *13*, 133.
90. Seyferth, D.; Vaughan, L. G.; Suzuki, R. *J. Organometal. Chem.* 1964, *1*, 437.
91. Benkeser, R. A.; Cunico, R. F.; Dunny, S.; Jones, P. R.; Nerlekar, P. G. *J. Org. Chem.* 1967, *32*, 2634.
92. Dunogues, J.; Bourgeois, P.; Pillot, J. P.; Merault, G.; Calas, R. *J. Organometal. Chem.* 1975, *87*, 169.
93. Ryan, J. W.; Speier, J. L. *J. Org. Chem.* 1966, *31*, 2698.
94. Paquette, L. A.; Williams, R. V. *Tetrahedron Lett.* 1981, *22*, 4643.
95. Raucher, S.; Hansen, M. R.; Colter, M. A. *J. Org. Chem.* 1978, *43*, 4885.
96. Chou, S. S.; Kuo, H. L.; Wang, C. J.; Tsai, C. Y.; Sun, C. M. *J. Org. Chem.* 1989, *54*, 868.
97. Zweifel, G.; Whitney, C. C. *J. Am. Chem. Soc.* 1967, *89*, 2753.
98. Newman, H. *J. Am. Chem. Soc.* 1973, *95*, 4098.
99. Newman, H. *Tetrahedron Lett.* 1971, 4571.
100. Baba, S.; van Horn, D. E.; Negishi, E. *Tetrahedron Lett.* 1976, 1927.
101. Palei, B. A.; Gavrilenko, V. V.; Zakharkin, L. I. *Bull. Acad. Sci. USSR* 1969, 2590.
102. Stork, G.; Jung, M. E.; Colvin, E.; Noel, Y. *J. Am. Chem. Soc.* 1974, *96*, 3684.
103. Uchida, K.; Utimoto, K.; Nozaki, H. *J. Org. Chem.* 1976, *41*, 2215.
104. Miller, R. B.; McGarvey, G. *J. Org. Chem.* 1978, *43*, 4424.
105. Obayashi, M.; Utimoto, K.; Nozaki, H. *Tetrahedron Lett.* 1977, 1805.
106. Westmijze, H.; Meijer, J.; Vermeer, P. *Tetrahedron Lett.* 1977, 1823.
107. Yates, K.; Go, T. A. *J. Org. Chem.* 1980, *45*, 2385.
108. Pincock, J. A.; Yates, K. *J. Am. Chem. Soc.* 1968, *90*, 5643.
109. Cousseau, J. *Synthesis* 1980, 805.
110. Cousseau, J.; Gouin, L. *J. Chem. Soc., Perkin Trans. 1* 1977, 1797.
111. Fahey, R. C.; Payne, M. T.; Lee, D. J. *J. Org. Chem.* 1974, *39*, 1124.
112. Pincock, J. A.; Yates, K. *Can. J. Chem.* 1970, *48*, 3332.
113. Fahey, R. C. *Top. Stereochem.* 1968, *3*, 237.
114. Kamiya, N.; Chikami, Y.; Ishii, Y. *Synlett* 1990, 675.
115. Roush, W. R.; Spada, A. P. *Tetrahedron Lett.* 1982, *23*, 3773.
116. Rao, A. V. R.; Reddy, E. R. *Tetrahedron Lett.* 1986, *27*, 2279.
117. Piers, E.; Morton, H. E. *J. Org. Chem.* 1980, *45*, 4263.
118. Attenburrow, J.; Cameron, A. F. B.; Chapman, J. H.; Evans, R. M.; Hems, B. A.; Jansen, A. B. A.; Walker, T. *J. Chem. Soc.* 1952, 1094.
119. Corey, E. J.; Wollenberg, R. H. *J. Org. Chem.* 1975, *40*, 2265.
120. Corey, E. J.; Hopkins, P. B.; Munroe, J. E.; Marfat, A.; Hashimoto, S.-I. *J. Am. Chem. Soc.* 1980, *102*, 7986.
121. Corey, E. J.; Ravindranathan, T. *J. Am. Chem. Soc.* 1972, *94*, 4013.
122. Rajagopalan, S.; Zweifel, G. *Synthesis* 1984, 111.
123. Fleming, I.; Terrett, N. K. *J. Organometal. Chem.* 1984, *264*, 99.
124. Ager, D. J. *Org. React.* 1990, *38*, 1.
125. Schlosser, M. *Top. Stereochem.* 1970, *5*, 1.
126. Reitz, A. B.; Mutter, M. S.; Maryanoff, B. E. *J. Am. Chem. Soc.* 1984, *106*, 1873.
127. Maryanoff, B. E.; Reitz, A. B.; Duhl-Emswiler, B. A. *J. Am. Chem. Soc.* 1985, *107*, 217.
128. Maryanoff, B. E.; Reitz, A. B. *Chem. Rev.* 1989, *89*, 863.
129. McKenna, E. G.; Walker, B. J. *Tetrahedron Lett.* 1988, *29*, 485.
130. Mackenzie, P. B.; Whelan, J.; Bosnich, B. *J. Am. Chem. Soc.* 1985, *107*, 2046.
131. Vedejs, E.; Marth, C. F. *J. Am. Chem. Soc.* 1988, *110*, 3948.
132. Maryanoff, B. E.; Reitz, A. B.; Graden, D. W.; Almond, H. R. *Tetrahedron Lett.* 1989, *30*, 1361.
133. Liu, Z.-I.; Schlosser, M. *Tetrahedron Lett.* 1990, *31*, 5753.
134. Ward, W. J.; McEwen, W. E. *J. Org. Chem.* 1990, *55*, 493.
135. Vedejs, E.; Marth, C. F. *J. Am. Chem. Soc.* 1989, *111*, 1519.
136. Vedejs, E.; Marth, C. F. *J. Am. Chem. Soc.* 1990, *112*, 3905.
137. Maercker, A. *Org. React.* 1965, *14*, 270.
138. Ayrey, P. M.; Warren, S. *Tetrahedron Lett.* 1989, *30*, 4581.
139. Liu, Y. Y.; Thom, E.; Liebman, A. A. *J. Heterocycl. Chem.* 1979, *16*, 799.
140. Koreeda, M.; Patel, P. D.; Brown, L. *J. Org. Chem.* 1985, *50*, 5912.
141. Schlosser, M.; Christmann, K. F. *Justus Liebigs, Ann. Chem.* 1967, *708*, 1.
142. Schlosser, M.; Christmann, K. F.; Piskala, A. *Chem. Ber.* 1970, *103*, 2814.
143. Schlosser, M.; Schaub, B. *Chimica* 1982, *36*, 396.
144. Maryanoff, B. E.; Duhl-Emswiler, B. A. *Tetrahedron Lett.* 1981, *22*, 4185.
145. Corey, E. J.; Ulrich, P.; Venkateswarlu, A. *Tetrahedron Lett.* 1977, 3229.
146. Corey, E. J.; Kang, J. *J. Am. Chem. Soc.* 1982, *104*, 4742.
147. Wadsworth, W. S. *Org. React.* 1977, *25*, 73.
148. Boutagy, J.; Thomas, R. *Chem. Rev.* 1974, *74*, 87.

149. Buss, A. D.; Warren, S. *Tetrahedron Lett.* 1983, *24*, 3931.
150. Buss, A. D.; Warren, S. *J. Chem. Soc., Perkin Trans. 1* 1985, 2307.
151. Elliott, J.; Warren, S. *Tetrahedron Lett.* 1986, *27*, 645.
152. Greeves, N.; Warren, S. *Tetrahedron Lett.* 1986, *27*, 259.
153. Hudrlik, P. F.; Peterson, D. *Tetrahedron Lett.* 1974, 1133.
154. Hudrlik, P. F.; Peterson, D. *J. Am. Chem. Soc.* 1975, *97*, 1464.
155. Ager, D. J. *Synthesis* 1984, 384.
156. Barrett, A. G. M.; Hill, J. M.; Wallace, E. M. *J. Org. Chem.* 1992, *57*, 386.
157. Barrett, A. G. M.; Flygare, J. A. *J. Org. Chem.* 1991, *56*, 638.
158. Palomo, C.; Aizpurua, J. M.; García, J. M.; Ganboa, I.; Cossio, F. P.; Lecea, B.; López, C. *J. Org. Chem.* 1990, *55*, 2498.
159. Julia, M. *Pure Appl. Chem.* 1985, *57*, 763.
160. Magnus, P. D. *Tetrahedron* 1977, *33*, 2019.
161. Kende, A. S.; Mendoza, J. S. *Tetrahedron Lett.* 1990, *31*, 7105.
162. Kocienski, P. J. *Chem. Int. (London)* 1981, 548.
163. Julia, M.; Paris, J.-M. *Tetrahedron Lett.* 1973, 4833.
164. Kocienski, P. J.; Lythgoe, B.; Ruston, S. *J. Chem. Soc., Perkin Trans. 1* 1978, 829.
165. Yamada, S.; Shiraishi, M.; Ohmori, M.; Takayama, H. *Tetrahedron Lett.* 1984, *25*, 3347.
166. Street, S. D. A.; Yeates, C.; Kocienski, P.; Campbell, S. F. *J. Chem. Soc., Chem. Commun.* 1985, 1386.
167. Baudin, J. B.; Hareau, G.; Julia, S. A.; Ruel, O. *Tetrahedron Lett.* 1991, *32*, 1175.
168. Shimagaki, M.; Matsuzaki, Y.; Hori, I.; Nakata, T.; Oishi, T. *Tetrahedron Lett.* 1984, *25*, 4779.
169. Sachdev, K. *Tetrahedron Lett.* 1976, 4041.
170. Lucchetti, J.; Krief, A. *Tetrahedron Lett.* 1978, 2693.
171. Etemad-Moghadam, G.; Seyden-Penne, J. *Tetrahedron* 1984, *40*, 5153.
172. Denmark, S. E.; Amburgey, J. *J. Am. Chem. Soc.* 1993, *115*, 10386.
173. Dumont, W.; Van Ende, D.; Krief, A. *Tetrahedron Lett.* 1979, 485.
174. Cainelli, G.; Bertini, F.; Grasselli, P.; Zubiani, G. *Tetrahedron Lett.* 1967, 1581.
175. Shi, L.; Wang, W.; Wang, Y.; Huang, Y.-Z. *J. Org. Chem.* 1989, *54*, 2027.
176. Boubia, B.; Mioskowski, C.; Bellamy, F. *Tetrahedron Lett.* 1989, *30*, 5263.
177. Boubia, B.; Mann, A.; Bellamy, F. D.; Mioskowski, C. *Angew. Chem., Int. Ed. Engl.* 1990, *29*, 1454.
178. Davis, D.; Gray, C. E. *J. Org. Chem.* 1970, *35*, 1303.
179. Kauffmann, T.; Ahler, H.; Joussen, R.; Kriegsmann, R.; Vahrenhorst, A.; Woltermann, A. *Tetrahedron Lett.* 1978, 4399.
180. Tilhard, H.; Ahlers, H.; Hauffmann, T. *Tetrahedron Lett.* 1980, *21*, 2803.
181. Reich, H. J.; Chow, F. *J. Chem. Soc., Chem. Commun.* 1975, 790.
182. Peteagnani, N.; Comasseto, J. V. *Synthesis* 1986, 1.
183. Kauffmann, T.; Kreigsmann, R. A. *Angew. Chem., Int. Ed. Engl.* 1977, *16*, 862.
184. Seebach, D.; Willert, I.; Beck, A. K.; Grobl, B. *Helv. Chim. Acta* 1978, *61*, 2510.
185. Huang, Y.; Shi, L.; Shi, S.; Wen, X. *J. Chem. Soc., Perkin Trans. 1* 1989, 2397.
186. Cainelli, G.; Bello, G. D.; Zubiaini, G. *Tetrahedron Lett.* 1966, 4315.
187. Pelter, A.; Buss, D.; Colclough, E. *J. Chem. Soc., Chem. Commun.* 1987, 297.
188. Pelter, A.; Smith, K.; Elgendy, S.; Rowlands, M. *Tetrahedron Lett.* 1989, *30*, 5647.
189. Pelter, A.; Singaram, B.; Wilson, J. W. *Tetrahedron Lett.* 1983, *24*, 635.
190. Pelter, A.; Buss, D.; Pitchford, A. *Tetrahedron Lett.* 1985, *26*, 5093.
191. Pelter, A.; Smith, K.; Jones, K. D. *J. Chem. Soc., Perkin Trans. 1* 1992, 747.
192. Singaram, B.; Goralski, C. T.; Rangaishenvi, M. V.; Brown, H. C. *J. Am. Chem. Soc.* 1989, *111*, 384.
193. Singaram, B.; Rangaishenvi, M. V.; Brown, H. C.; Goralski, C. T.; Hasha, D. L. *J. Org. Chem.* 1991, *56*, 1543.
194. Paquette, L. A. *Org. React.* 1977, *25*, 1.
195. Braverman, S. *The Chemistry of Sulphoxides and Sulphones*; John Wiley & Sons: New York, 1988.
196. McMurry, J. E.; Fleming, M. P.; Kees, K. L.; Krepski, L. R. *J. Org. Chem.* 1978, *43*, 3255.
197. McMurry, J. E.; Fleming, M. P. *J. Am. Chem. Soc.* 1974, *96*, 4708.
198. McMurry, J. E. *Acc. Chem. Res.* 1974, *7*, 281.
199. McMurry, J. E. *Acc. Chem. Res.* 1983, *16*, 405.
200. Lenoir, D. *Synthesis* 1989, 883.
201. McMurry, J. E.; Krepski, L. R. *J. Org. Chem.* 1976, *41*, 3929.
202. Nishida, S.; Kataoka, F. *J. Org. Chem.* 1978, *43*, 1612.
203. Myers, A. G.; Kukkola, P. J. *J. Am. Chem. Soc.* 1990, *112*, 8208.
204. Adlington, R. M.; Barrett, A. G. M. *Acc. Chem. Res.* 1983, *16*, 55.
205. Shapiro, R. H. *Org. React.* 1976, *23*, 405.
206. Patois, C.; Savignac, P. *Tetrahedron Lett.* 1991, *32*, 1317.
207. Thomson, S. K.; Heathcock, C. H. *J. Org. Chem.* 1990, *55*, 3386.

208. Rathke, M. W.; Bouhlel, E. *Synth. Commun.* 1990, *20*, 869.
209. Kann, N.; Rein, T.; Björn, A.; Helquist, P. *J. Org. Chem.* 1990, *55*, 5312.
210. Nowick, J. S.; Danheiser, R. L. *J. Org. Chem.* 1989, *54*, 2798.
211. Martin, S. F. *Synthesis* 1979, 633.
212. Cristau, H.-J.; Gasc, M.-B. *Tetrahedron Lett.* 1990, *31*, 341.
213. Nájera, C.; Yus, M. *Tetrahedron Lett.* 1989, *30*, 173.
214. Seebach, D.; Kolb, M.; Gröbel, B.-T. *Tetrahedron Lett.* 1974, 3171.
215. Gais, H. J. *Tetrahedron Lett.* 1988, *29*, 1773.
216. Rehwinkel, H.; Skupsch, J.; Vorbruggen, H. *Tetrahedron Lett.* 1988, *29*, 1775.
217. Hanessian, S.; Delorme, D.; Beaudoin, S.; Leblanc, Y. *J. Am. Chem. Soc.* 1984, *106*, 5754.
218. Erdelmeier, I.; Gais, H.-J.; Lindner, H. J. *Angew. Chem., Int. Ed. Engl.* 1986, *25*, 935.
219. Sato, Y.; Niinomi, Y. *J. Chem. Soc. Chem. Commun.* 1982, 56.
220. Palomo, C.; Aizpurua, J. M.; Aurrekoetxea, N. *Tetrahedron Lett.* 1990, *31*, 2209.
221. Eisch, J. J.; Damasevitz, G. A. *J. Org. Chem.* 1976, *41*, 2214.
222. Chan, T. H.; Fleming, I. *Synthesis* 1979, 761.
223. Murphy, P. J.; Spencer, J. L.; Procter, G. *Tetrahedron Lett.* 1990, *31*, 1051.
224. Soderquist, J. A.; Leon-Colon, G. *Tetrahedron Lett.* 1991, *32*, 43.
225. Maruyama, K.; Ishihara, Y.; Yamamoto, Y. *Tetrahedron Lett.* 1981, *22*, 4235.
226. Hosomi, A. *Acc. Chem. Res.* 1988, *21*, 200.
227. Ironside, M. D.; Murray, A. W. *Tetrahedron Lett.* 1989, *30*, 1691.
228. Lucast, D. H.; Wemple, J. *Tetrahedron Lett.* 1977, 1103.
229. Lee, J. W.; Oh, D. Y. *Synth. Commun.* 1989, *19*, 2209.
230. Kwon, T. W.; Smith, M. B. *J. Org. Chem.* 1989, *54*, 4250.
231. Naso, F. *Pure Appl. Chem.* 1988, *60*, 79.
232. Casey, C. P.; Marten, D. F. *Tetrahedron Lett.* 1974, 925.
233. Kobayashi, S.; Mukaiyama, T. *Chem. Lett.* 1974, 405.
234. Kobayashi, S.; Mukaiyama, T. *Chem. Lett.* 1974, 1425.
235. Corey, E. J.; Chen, R. H. K. *Tetrahedron Lett.* 1973, 387.
236. Azzena, U.; Cossu, S.; De Lucchi, O.; Melloni, G. *Tetrahedron Lett.* 1989, *30*, 1845.
237. Seyferth, D.; Lefferts, J. L.; Lambert, R. L. *J. Organomet. Chem.* 1977, *142*, 39.
238. Stork, G.; Zhao, K. *Tetrahedron Lett.* 1989, *30*, 2173.
239. Bestmann, H. J.; Rippel, H. C.; Dostalek, R. *Tetrahedron Lett.* 1989, *30*, 5261.
240. Corey, E. J.; Shulman, J. I.; Yamamoto, H. *Tetrahedron Lett.* 1970, 447.
241. Schlosser, M.; Christmann, K. F. *Synthesis* 1969, 38.
242. Bhalerao, U. T.; Rapoport, H. *J. Am. Chem. Soc.* 1971, *93*, 4835.
243. Corey, E. J.; Yamamoto, H.; Herron, D. K.; Achiwa, K. *J. Am. Chem. Soc.* 1970, *92*, 6635.
244. Corey, E. J.; Yamamoto, H. *J. Am. Chem. Soc.* 1970, *92*, 6636.
245. Corey, E. J.; Yamamoto, H. *J. Am. Chem. Soc.* 1970, *92*, 6637.
246. Corey, E. J.; Yamamoto, H. *J. Am. Chem. Soc.* 1970, *92*, 226.
247. Schlosser, M.; Christmann, K. F.; Piskala, A.; Coffinat, D. *Synthesis* 1971, 29.
248. Schlosser, M.; Christmann, K. F. *Synthesis* 1971, 380.
249. Schlosser, M.; Christmann, K. F. *Synthesis* 1972, 575.
250. Schlosser, M.; Tuong, H. B.; Respondek, J.; Schaub, B. *Chimica* 1983, *37*, 10.
251. Corey, E. J.; Ulrich, P.; Venkateswarlu, A. *Tetrahedron Lett.* 1977, 3231.
252. Barluenga, J.; Fernández-Simón, J. L.; Concellón, J. M.; Miguel, Y. *Tetrahedron Lett.* 1989, *30*, 5927.
253. Hayashi, T.; Yamamoto, A.; Hojo, M.; Ito, Y. *J. Chem. Soc., Chem. Commun.* 1989, 495.
254. Cahiez, G.; Bernard, D.; Normant, J. F. *Synthesis* 1976, 245.
255. Miller, R. B.; McGarvey, G. *J. Org. Chem.* 1979, *44*, 4623.
256. Corey, E. J.; Ulrich, P. *Tetrahedron Lett.* 1975, 3685.
257. Neumann, N.; Seebach, D. *Tetrahedron Lett.* 1976, 4839.
258. Millon, J.; Lorne, R.; Linstrumelle, G. *Synthesis* 1975, 434.
259. Linstrumelle, G.; Kreiger, J. K.; Whitesides, G. M. *Org. Synth.* 1976, *55*, 103.
260. Vig, O. P.; Kapur, J. C.; Sharma, S. D. *J. Int. Chem. Soc.* 1968, *45*, 1026.
261. Posner, G. H. *Org. React.* 1975, *22*, 253.
262. Utimoto, K.; Kitai, M.; Naruse, M.; Nozaki, H. *Tetrahedron Lett.* 1976, 4233.
263. Utimoto, K.; Kitai, M.; Nozaki, H. *Tetrahedron Lett.* 1975, 2825.
264. Chan, T. H.; Mychajlowskij, W.; Ong, B. S.; Harpp, D. N. *J. Organometal. Chem.* 1976, *107*, C1.
265. Chan, T. H.; Koumaglo, K. *Tetrahedron Lett.* 1986, *27*, 883.
266. Buell, G. R.; Corriu, R.; Guerin, C.; Spialter, L. *J. Am. Chem. Soc.* 1970, *92*, 7424.
267. Fleming, I.; Pearce, A. *J. Chem. Soc., Chem. Commun.* 1975, 633.
268. Pillot, J.-P.; Dunogues, J.; Calas, R. *Bull. Soc. Chim. Fr.* 1975, 2143.

269. Neumann, H.; Seebach, D. *Chem. Ber.* 1978, *111*, 2785.
270. Kauffmann, T. *Top. Curr. Chem.* 1986, *92*, 109.
271. Koenig, K. E.; Weber, W. P. *Tetrahedron Lett.* 1973, 2533.
272. Koerwitz, F. L.; Hammond, G. B.; Weimer, D. F. *J. Org. Chem.* 1989, *54*, 738.
273. Koerwitz, F. L.; Hammond, G. B.; Weimer, D. F. *J. Org. Chem.* 1989, *54*, 743.
274. Sarkar, T. K. *Synthesis* 1990, 969.
275. Sarkar, T. K. *Synthesis* 1990, 1101.
276. Bach, R. D.; Badger, R. C.; Lang, T. J. *J. Am. Chem. Soc.* 1979, *101*, 2845.
277. Spencer, H. K.; Hill, R. K. *J. Org. Chem.* 1976, *41*, 2485.
278. King, R. W.; DePuy, C. H. *Chem. Rev.* 1960, *60*, 431.
279. Maccoll, A. *Chem. Rev.* 1969, *69*, 33.
280. Cram, D. J.; Elhafez, F. A. A. *J. Am. Chem. Soc.* 1952, *74*, 5828.
281. Mulzer, J.; Lammer, O. *Angew. Chem., Int. Ed. Engl.* 1983, *22*, 628.
282. Mulzer, J.; Pointner, A.; Chuckolowski, A.; Bruntrup, G. *J. Chem. Soc., Chem. Commun.* 1979, 52.
283. Hara, S.; Taguchi, H.; Yamaoto, H.; Nozaki, H. *Tetrahedron Lett.* 1975, 1545.
284. Adam, W.; Baeza, J.; Liu, J.-C. *J. Am. Chem. Soc.* 1972, *94*, 2000.
285. Trost, B. M. *Chem. Rev.* 1978, *78*, 363.
286. Ochiai, M.; Tada, S.-I.; Sumi, K.; Fujita, E. *J. Chem. Soc., Chem. Commun.* 1982, 281.
287. Trost, B. M. *Acc. Chem. Res.* 1978, *11*, 453.
288. Fleming, I.; Perry, D. A. *Tetrahedron Lett.* 1981, *22*, 5095.
289. Grieco, P. A.; Boxler, D.; Pogonowski, C. S. *J. Chem. Soc., Chem. Commun.* 1974, 497.
290. Solladie, G.; Zimmerman, R.; Bartsch, R. *Tetrahedron Lett.* 1983, *24*, 755.
291. Liotta, D.; Saindane, M.; Brothers, D. *J. Org. Chem.* 1982, *47*, 1600.
292. King, J. L.; Posner, B. A.; Mak, K. T.; Yang, N.-C. C. *Tetrahedron Lett.* 1987, *28*, 3919.
293. Corey, E. J.; Hopkins, P. B. *Tetrahedron Lett.* 1982, *23*, 1979.
294. Beels, C. M. D.; Coleman, M. J.; Taylor, R. J. K. *Synlett* 1990, 479.
295. Tanata, S.; Yasuda, A.; Yamamoto, H.; Nozaki, H. *J. Am. Chem. Soc.* 1975, *97*, 3251.
296. Bargiotti, A.; Hanessian, S.; LaRue, M. *Tetrahedron Lett.* 1978, *8*, 737.
297. Barrett, A. G. M.; Barton, D. H. R.; Bielski, R.; McCombie, S. *J. Chem. Soc., Chem. Commun.* 1977, 866.
298. Block, E. *Org. React.* 1984, *30*, 457.
299. Liu, Z.; Classon, B.; Samuelsson, B. *J. Org. Chem.* 1990, *55*, 4273.
300. Mulzer, J.; Lammer, O. *Angew. Chem., Int. Ed. Engl.* 1983, *22*, 628.
301. Sonnet, P. E. *Tetrahedron* 1980, *36*, 557.
302. Sonnet, P. E. *Synthesis* 1980, 828.
303. Schobert, R.; Hohlein, U. *Synlett* 1990, 465.
304. Suzuki, H.; Fuchita, T.; Iwasa, A.; Mishina, T. *Synthesis* 1978, 905.
305. Martin, M. G.; Ganem, B. *Tetrahedron Lett.* 1984, *25*, 251.
306. Kaufmann, T.; Bisling, M. *Tetrahedron Lett.* 1984, *25*, 293.
307. Caputo, R.; Mangoni, L.; Neri, O.; Palumbo, G. *Tetrahedron Lett.* 1981, *22*, 3551.
308. Motherwell, W. B. *J. Chem. Soc., Chem. Commun.* 1973, 935.
309. Neureiter, N. P.; Bordwell, F. G. *J. Am. Chem. Soc.* 1959, *81*, 578.
310. Aalbersberg, W. G. L.; Vollhardt, K. P. C. *J. Am. Chem. Soc.* 1977, *99*, 2792.
311. Opitz, G.; Ehlis, T.; Rieth, K. *Tetrahedron Lett.* 1989, *30*, 3131.
312. Johnson, W. S.; Daub, G. H. *Org. React.* 1951, *6*, 1.
313. Jones, G. *Org. React.* 1967, *15*, 204.
314. Rathke, M. W. *Org. React.* 1975, *22*, 423.
315. Kagan, J. *Organic Photochemistry: Principles and Applications*; Academic Press: San Diego, 1993.
316. Lyons, J. E. *J. Chem. Soc., Chem. Commun.* 1971, 562.
317. Hubert, A. J.; Reimlinger, H. *Synthesis* 1969, 97.
318. Whitesides, G. M.; Coe, G. L.; Cope, A. C. *J. Am. Chem. Soc.* 1969, *91*, 2608.
319. Radhakrishna, A. S.; Suri, S. K.; Prasad Rao, K. R. K.; Sivaprakash, K.; Singh, B. B. *Synth. Commun.* 1990, *20*, 345.
320. Hammond, G. S.; Saltiel, J.; Lamola, A. A.; Turro, N. J.; Bradshaw, J. S.; Cowan, D. O.; Counsell, R. C.; Vogt, V.; Dalton, C. *J. Am. Chem. Soc.* 1964, *86*, 3197.
321. Wilson, C. V. *Org. React.* 1957, *9*, 332.
322. Wittig, G.; Haag, W. *Chem. Ber.* 1955, *88*, 1654.
323. Jackson, W. R.; Lovel, C. G. *Aust. J. Chem.* 1982, *35*, 2053.
324. Elmes, P. S.; Jackson, W. R. *Aust. J. Chem.* 1982, *35*, 2041.
325. Freeman, F. *Chem. Rev.* 1975, *75*, 439.
326. V'yunov, K. A.; Ginak, A. I. *Russ. Chem. Rev.* 1981, *50*, 151.
327. Sergeev, G. B.; Smirnov, V. V.; Rostovshchikova, T. N. *Russ. Chem. Rev.* 1983, *52*, 259.

328. Isenberg, N.; Grdinic, M. *J. Chem. Ed.* 1969, *46*, 601.
329. Dubois, J.-E.; Chretien, J. R. *J. Am. Chem. Soc.* 1978, *100*, 3506.
330. Kitching, W. *Organomet. Reactions* 1972, *3*, 319.
331. Deslongchamps, P. *Stereoelectronic Effects in Organic Chemistry*; Pergamon Press: Oxford, 1983.
332. Bodner, G. S.; Fernandez, J. M.; Arif, A. M.; Gladysz, J. A. *J. Am. Chem. Soc.* 1988, *110*, 4082.
333. Bernasconi, C. F. *Tetrahedron* 1989, *45*, 4017.
334. Sharts, C. M.; Sheppard, W. A. *Org. React.* 1974, *21*, 125.
335. Becker, K. B.; Grob, C. A. *Synthesis* 1973, 789.
336. Elmes, P. S.; Jackson, R. *Aust. J. Chem.* 1982, *35*, 2041.
337. Fahey, R. C. *J. Am. Chem. Soc.* 1966, *88*, 4681.
338. Fahey, R. C.; Schubert, C. *J. Am. Chem. Soc.* 1965, *87*, 5172.
339. Abraham, R. J.; Monasterios, J. R. *J. Chem. Soc., Perkin Trans. 1* 1973, 1446.
340. House, H. O. *J. Am. Chem. Soc.* 1955, *77*, 5083.
341. Boguslavskaya, L. S. *Russ. Chem. Rev.* 1972, *41*, 740.
342. Brown, H. C.; Geoghegan, P. J. *J. Org. Chem.* 1970, *35*, 1844.
343. Moon, S.; Waxman, B. H. *J. Org. Chem.* 1969, *34*, 1157.
344. Hill, C. L.; Whitesides, G. M. *J. Am. Chem. Soc.* 1974, *96*, 870.
345. Giese, B.; Bartmann, D. *Tetrahedron Lett.* 1985, *26*, 1197.
346. Ojima, I.; Clos, N.; Bastos, C. *Tetrahedron* 1989, *45*, 6901.
347. Marko, L.; Heil, B. *Catalysis Rev.* 1973, *8*, 269.
348. Bogdanovic, B. *Angew. Chem., Int. Ed. Engl.* 1973, *12*, 954.
349. Harmon, R. E.; Gupta, S. K.; Brown, D. J. *Chem. Rev.* 1973, *73*, 21.
350. Matteoli, U.; Frediani, P.; Bianchi, M.; Botteghi, C.; Gladiali, S. *J. Mol. Catal.* 1981, *12*, 265.
351. Horner, L.; Siegler, H.; Buthe, H. *Angew. Chem., Int. Ed. Engl.* 1968, *7*, 942.
352. Izumi, Y. *Angew. Chem., Int. Ed. Engl.* 1971, *10*, 871.
353. Knowles, W. S. *Acc. Chem. Res.* 1983, *16*, 106.
354. Martens, J. *Top. Curr. Chem.* 1984, *125*, 165.
355. Osborn, J. A.; Jardine, F. H.; Young, J. F.; Wilkinson, G. *J. Chem. Soc. (A)* 1966, 1711.
356. Eliel, E. L. *Stereochemistry of Carbon Compounds*; McGraw-Hill Kogakusha: Tokyo, 1962.
357. Sommer, L. H.; Lyons, J. E.; Fujimoto, H. *J. Am. Chem. Soc.* 1969, *91*, 7051.
358. Chalk, A. J. *J. Organometal. Chem.* 1970, *21*, 207.
359. Balavoine, G.; Clinet, J. C.; Lellouche, I. *Tetrahedron Lett.* 1989, *30*, 5141.
360. Yamamoto, K.; Hayashi, T.; Uramoto, Y.; Ito, R.; Kumada, M. *J. Organometal. Chem.* 1976, *118*, 331.
361. Brunner, H. *Angew. Chem., Int. Ed. Engl.* 1983, *22*, 897.
362. Hayashi, T.; Tamao, K.; Katsuro, Y.; Nakae, I.; Kumada, M. *Tetrahedron Lett.* 1980, *21*, 1871.
363. Ojima, I.; Kumagai, M.; Nagai, Y. *J. Organometal. Chem.* 1976, *111*, 43.
364. Kogure, T.; Ojima, I. *J. Organometal. Chem.* 1982, *234*, 249.
365. Keinan, E. *Pure Appl. Chem.* 1989, *61*, 1737.
366. Tamao, K.; Nakagawa, Y.; Arai, H.; Higuchi, N.; Ito, Y. *J. Am. Chem. Soc.* 1988, *110*, 3712.
367. Tamao, K.; Tohma, T.; Inui, N.; Nakayama, O.; Ito, Y. *Tetrahedron Lett.* 1990, *31*, 7333.
368. Ojima, I.; Hirai, K. In *Asymmetric Synthesis*; J. D. Morrison, Ed.; Academic Press: Orlando, FL, 1985; Vol. 5; pp 103.
369. Botteghi, C.; Consiglio, G.; Pino, P. *Liebigs Ann. Chem.* 1974, 864.
370. Brown, H. C.; Jadhaw, P. K.; Mandal, A. K. *Tetrahedron* 1981, *37*, 3547.
371. Brown, H. C.; Jadhav, P. K. In *Asymmetric Synthesis*; J. D. Morrison, Ed.; Academic Press: Orlando, FL, 1983; Vol. 2; pp 1.
372. Midland, M. M. *Chem. Rev.* 1989, *89*, 1553.
373. Brown, H. C.; Zweifel, G. *J. Am. Chem. Soc.* 1959, *81*, 247.
374. Brown, H. C.; Zweifel, G. *J. Am. Chem. Soc.* 1961, *83*, 2544.
375. Brown, H. C.; Kulkarni, S. U. *J. Organomet. Chem.* 1982, *239*, 23.
376. Brown, H. C.; Singaram, B. *J. Am. Chem. Soc.* 1984, *106*, 1797.
377. Riediker, M.; Duthaler, R. O. *Angew. Chem., Int. Ed. Engl.* 1989, *28*, 494.
378. Brown, H. C.; Yoon, N. M. *J. Am. Chem. Soc.* 1977, *99*, 5514.
379. Evans, D. A.; Bartrolli, J.; Godel, T. *Tetrahedron Lett.* 1982, *23*, 4577.
380. Houk, K. N.; Rondan, N. G.; Wu, Y.-D.; Metz, J. T.; Padden-Row, M. N. *Tetrahedron* 1984, *40*, 2257.
381. Masamune, S.; Lu, L. D.-L.; Jackson, W. P.; Kaiko, T.; Toyoda, T. *J. Am. Chem. Soc.* 1982, *104*, 5523.
382. Brown, H. C. *Hydroboration*; Benjamin: New York, 1962.
383. Brown, H. C.; Pai, G. G.; Jadhav, P. K. *J. Am. Chem. Soc.* 1984, *106*, 1531.
384. Masamune, S.; Kim, B. M.; Petersen, J. S.; Sato, T.; Veenstra, S. J.; Imai, T. *J. Am. Chem. Soc.* 1985, *107*, 4549.
385. Sato, M.; Miyaura, N.; Suzuki, A. *Tetrahedron Lett.* 1990, *31*, 231.

386. Evans, D. A.; Fu, G. C. *J. Org. Chem.* 1990, *55*, 2280.
387. Hayashi, T.; Matsumoto, Y.; Ito, Y. *J. Am. Chem. Soc.* 1989, *111*, 3426.
388. Brown, H. C.; Joshi, N. N.; Pyun, C.; Singaram, B. *J. Am. Chem. Soc.* 1989, *111*, 1754.
389. Richter, B. K. D.; Bonato, M.; Follet, M.; Kamenka, J.-M. *J. Org. Chem.* 1990, *55*, 2855.
390. Brown, H. C.; Weissman, S. A.; Perumal, P. T.; Dhokte, U. P. *J. Org. Chem.* 1990, *55*, 1217.
391. Wang, X.; Li, Y.; Wu, Y.-D.; Paddon-Row, M. N.; Rondan, N. G.; Houk, K. N. *J. Org. Chem.* 1990, *55*, 2601.
392. Nechvatal, A. *Adv. Free Radical Chem.* 1972, *4*, 175.
393. Bateman, L.; Cunneen, J. I. *J. Chem. Soc.* 1950, 941.
394. Rabjohn, N. *Org. React.* 1976, *24*, 261.
395. Umbreit, M. A.; Sharpless, K. B. *J. Am. Chem. Soc.* 1977, 5526.

Chapter 9

# REACTIONS OF FUNCTIONALIZED ALKENES

Many of the reactions discussed in the last chapter have been extended to alkenes bearing a functional group either at an $sp^2$ carbon or in the allylic position. We will not be concerned with the stereoselectivity arising from the inherent steric requirements of a substituent in this chapter; we will concentrate on differences and similarities in stereoselection and reactivity arising from interactions between the functional group and reagent. Oxidations of these systems are discussed in Chapter 11.

There is a degree of overlap with the reactions of functionalized carbonyl systems, especially if an enol or enamine derivative is involved. These reactions are discussed in Chapter 6. Transformations that involve the alkene functionality of a conjugated system as the primary reaction site, as with conjugate (1,4-) addition, are described within this chapter, while 1,2-reductions of enones will be found in Chapter 6.

## 9.1. HYDROGENATIONS

Unlike the situation for isolated alkenes, considerable success has been realized for the reductions of functionalized alkenes with hydrogen.[1-8] Unfortunately, there is no "magic" catalyst that will ensure stereospecific reduction of a wide range of substrates, even within a specific class of compounds. The stringent requirements of functional group arrangement within the substrate detract greatly from the general application of asymmetric hydrogenations, although a plethora of ligands has been investigated to overcome this limitation (vide infra).[9,10] As our understanding of the mechanisms of these organometallic reactions increases, the knowledge can be used to design new, or modify existing ligands and catalysts. This could not only improve stereoselectivity but allow "general" reactions to be developed.[10-13]

The majority of catalysts that have been developed employ reduction of α,β-unsaturated esters or acids.[9,14-18] Introduction of an amide group then allows for the asymmetric synthesis of α-amino acids (Scheme 9.1).[2,9,14,19-21]

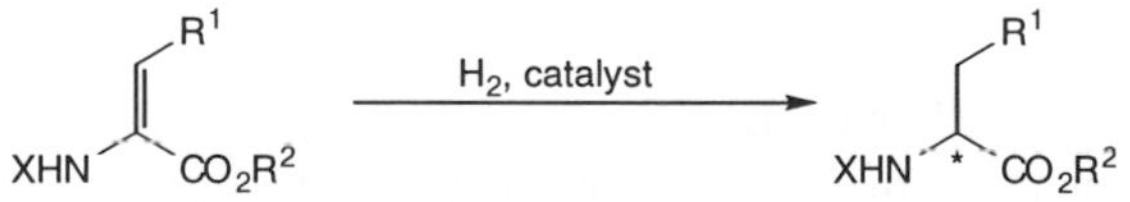

R1
XHN CO2R2
H2, catalyst
R1
XHN * CO2R2

**SCHEME 9.1.**

Most of the useful catalysts have chiral phosphorus ligands bound to rhodium; although other systems have been investigated, none has provided the scope for a diverse variety of substrates that rhodium has. The possible exception is the use of Ru(BINAP) catalysts (vide infra). In addition, these catalysts are homogeneous with the substrate acting as a ligand, ensuring intimate contact during the key steps, and do not rely on surface phenomenon for stereoselection.[5-8,11,22] A mechanism for the hydrogenation shown in Scheme 9.1 with [Rh(I)dipamp] as catalyst has been proposed. As with the related reduction of ethyl *Z*-α-acetamidocinnamate with Rh(chiraphos),[23,24] the major reaction product is derived from a minor, less stable organometallic intermediate, due to the latter's higher reactivity to hydrogen (Scheme 9.2).[21]

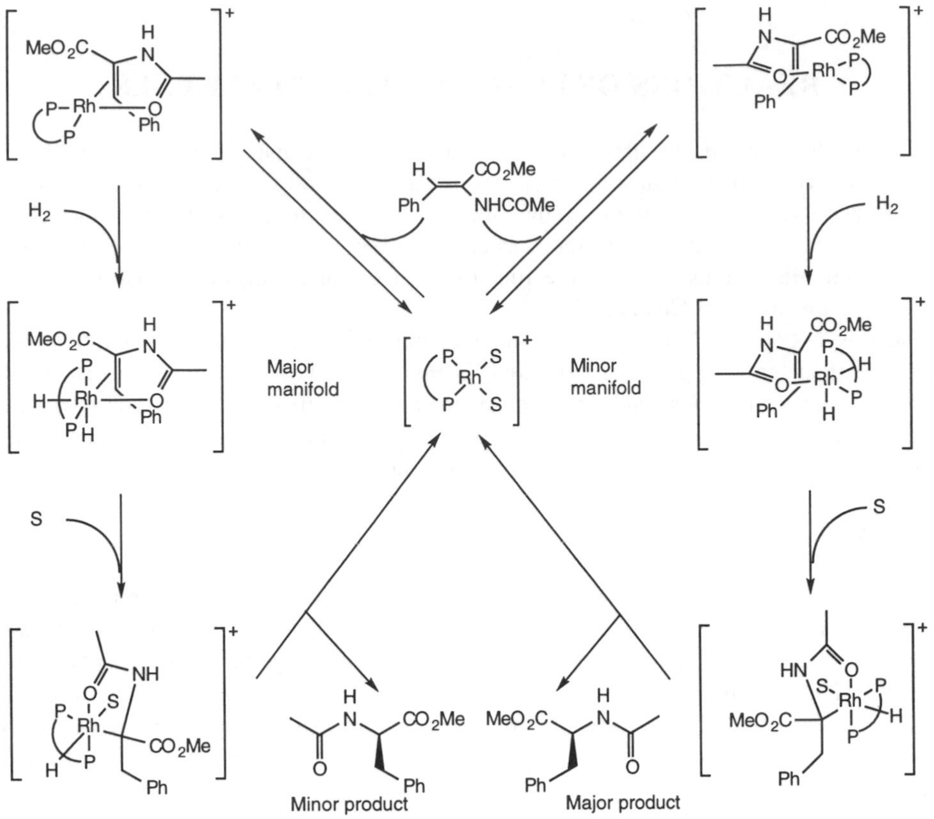

**SCHEME 9.2.**

The amide ligand is essential for selectivity. Of course, this reaction encompasses one of the first asymmetric syntheses and is used for the commercial preparation of α-amino acids,[19] in particular L-Dopa.[20] Kinetic resolutions in asymmetric reductions, such as with an α-amidoalkyl acrylate, have been observed (see Table 9.1).[25]

## 9.1.1. FUNCTIONALIZED ALKENES

The literature abounds with examples of catalysts and ligands to perform asymmetric hydrogenations with functionalized alkenes.[9,26-29] To help locate a system that may be considered in a synthetic sequence, examples of catalysts to perform asymmetric reductions for a number of functionalized alkenes are given in Tables 9.1 and 9.2. For inclusion, ee's of greater than 90% must have been reported with more than one example for that class of compounds.[¶ †]

Examination of α,β-unsaturated carbonyl substrates has allowed some structure activity relationships to be developed. For example, the first successful substrate for an asymmetric reduction was *Z*-α-amidocinnamic acid (**9.1**) (Figure 9.1).[4,20,26,83]

The aryl region can accommodate a wide variety of functionality ranging from substituted phenyl groups to alkyl to hydrogen; however, a halide cannot be placed at this position.[26] The hydrogen region seems only to accommodate a hydrogen when the aryl region is filled with

¶ More comprehensive lists are given in references 9, 26–28.

† In these tables, references are given to examples that fall below the 90% ee criterion — this has been done to allow the scope and limitations of that particular catalyst system to be determined. In addition, some of the citations may be concerned with the other antipode!

**TABLE 9.1**
**Reductions of Dehydroamino Acid Derivatives**

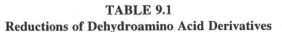

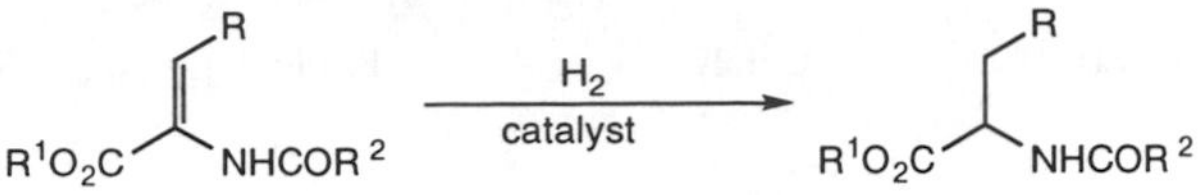

$$R^1O_2C\text{–}C(\text{NHCOR}^2)\text{=CHR} \xrightarrow[\text{catalyst}]{H_2} R^1O_2C\text{–}CH(\text{NHCOR}^2)\text{–}CH_2R$$

$R^1$ = H or R

| Catalyst[a] | Product Configuration | Reference |
|---|---|---|
| (S,S)-CHIRAPHOS | R | 30–33 |
| (S)-PROLOPHOS | S | 34 |
| (R)-PROPHOS | S | 31–33,35–38 |
| (R)-PHE-PHOS | R | 39 |
| (R,R)-DIPAMP[b] | S | 20,26,32,40–42 |
| (S,S)-NORPHOS | S | 43 |
| PHELLANPHOS | R | 44 |
| 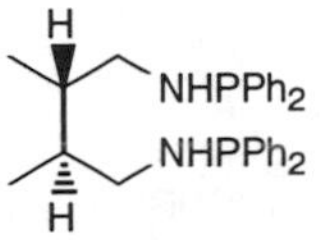 H, NHPPh$_2$, NHPPh$_2$, H | S | 45 |
| (S,S)-SKEWPHOS | R | 31 |
| 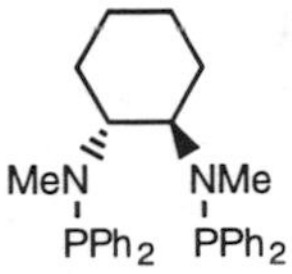 MeN, NMe, PPh$_2$, PPh$_2$ | S | 45–48 |
| (S,S)-BPPM[b] | R | 32,33,41,49–51 |
| (R)-PROPHOS | S | 31–33,35–38 |
| (S,S)-R-CAPP | S | 52 |
| (R)-CYCPHOS | S | 36–38 |
| 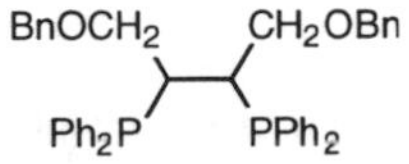 BnOCH$_2$, CH$_2$OBn, Ph$_2$P, PPh$_2$ | S | 35,53 |
| (S)-BINAP | R | 54-56 |
| (R,R)-DIOP | R | 57-64 |
| (R,R)-PPCP | S | 65 |
| 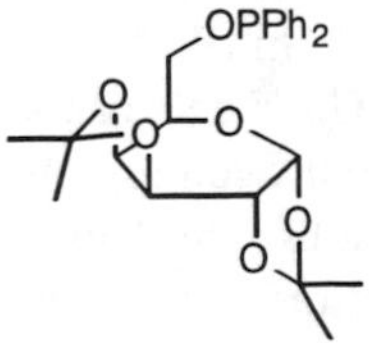 OPPh$_2$, O, O, O, O | R | 66 |
| 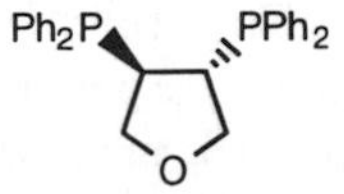 Ph$_2$P, PPh$_2$, O | S | 67 |

[a] With Rh as the metal, unless otherwise noted. [b] Triethylamine may be an additive in some examples.

**TABLE 9.2**
**Other Substrates for Asymmetric Hydrogenations**

| Substrate | Catalyst[a] | Product | Reference |
|---|---|---|---|
| 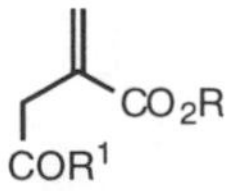 $CO_2R$, $COR^1$; $R^1$ = OH, OMe, or $NHR^1$ | BPPM | 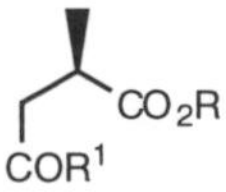 $CO_2R$, $COR^1$ | 68–71 |
| | BINAP[b] | | 72–74 |
| | DIPAMP[c] | | 75,76 |
| | MOD-DIOP | | 77 |
| | CAPP | | 52 |
| 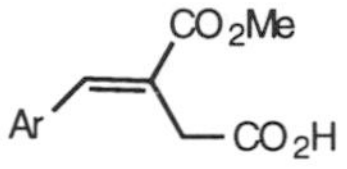 $CO_2Me$, Ar, $CO_2H$ | MOD-DIO5 | 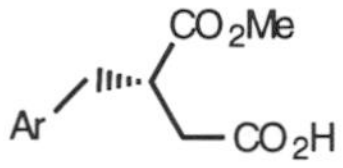 $CO_2Me$, Ar, $CO_2H$ | 78 |
| 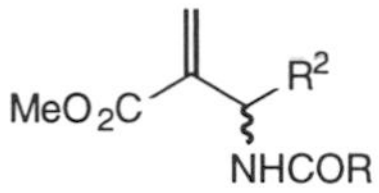 $MeO_2C$, $R^2$, NHCOR | DIPAMP[c] | 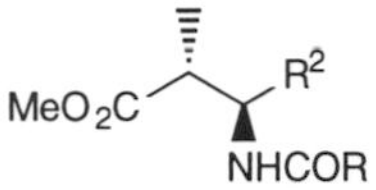 $MeO_2C$, $R^2$, NHCOR | 79 |
| 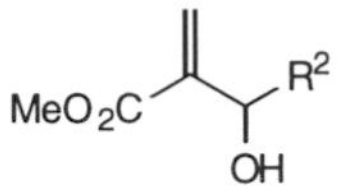 $MeO_2C$, $R^2$, OH | DIPAMP | 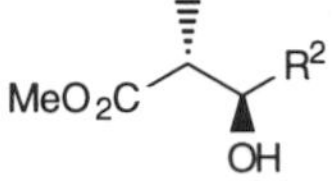 $MeO_2C$, $R^2$, OH | 80 |
| 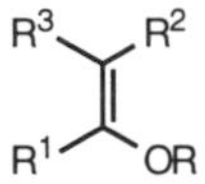 $R^3$, $R^2$, $R^1$, OR | DIPAMP | 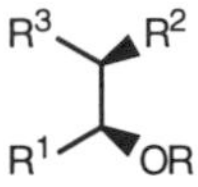 $R^3$, $R^2$, $R^1$, OR | 26,81 |
| 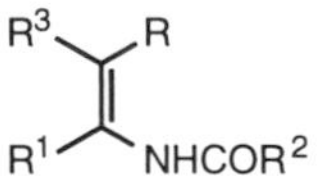 $R^3$, R, $R^1$, $NHCOR^2$ | DIOP | 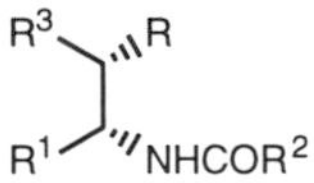 $R^3$, R, $R^1$, $NHCOR^2$ | 59,82 |

[a] With Rh as the metal unless otherwise stated. [b] With Rh or Ru. [c] Kinetic resolution can be observed.

a bulky substituent.[84] The size of the substituents on the amide group seems to have little impact on the degree of induction.[57,85-87] The acid region can accommodate other functionality provided that it is electron-withdrawing, such as an amide or nitrile;[81,83] an ester functionality can be employed without any substantial penalty in ee.[88] Good ee's can still be observed for compounds that have aryl or alkyl substituents in the "hydrogen region," but only when a hydrogen is present in the "aryl region."[32] Reductions of substrates lacking the amide group, acrylic acids, tend to give much lower degrees of induction;[76] it has been proposed that a β-dicarbonyl system is necessary to form a rigid complex with the metal, as illustrated by the high selectivities observed with itaconic acid derivatives (Table 9.2).[76,89-91]

The *Z*-enamide **9.2** is reduced with high asymmetric induction in the presence of Ru(BINAP); the *E*-isomer is not reduced under similar reaction conditions. The approach allows entry to a number of alkaloid classes (Scheme 9.3).[92,93]

It is possible to reduce appropriately substituted α,β-unsaturated acids with good optical yields.[17,72,75,77,79,80,94] Good asymmetric induction is also seen with Ru(BINAP) for the reduction of butyrolactones and cyclopentenones.[95]

Although many examples do not fulfill our imposed >90% ee criterion, the use of ruthenium BINAP catalysts is worthy of mention. A family of catalysts has been developed, differing mainly in the counterion, that can be used to reduce conjugated unsaturated carboxylic acids.[17,95,96] The asymmetric induction is not as high for acyclic examples as it is for aryl analogs, where it can be impressive (e.g., Scheme 9.4). The sense of induction is dependent upon the substrate structure and reaction conditions, particularly hydrogen pressure.[7,97,98]

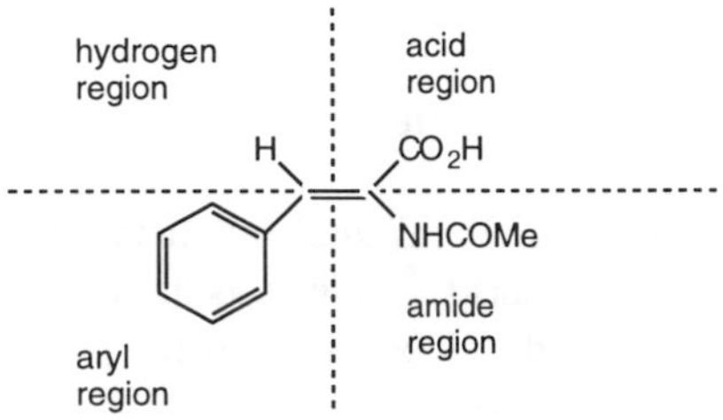

**FIGURE 9.1.** Regions of amidocinnamic acids for hydrogenations.

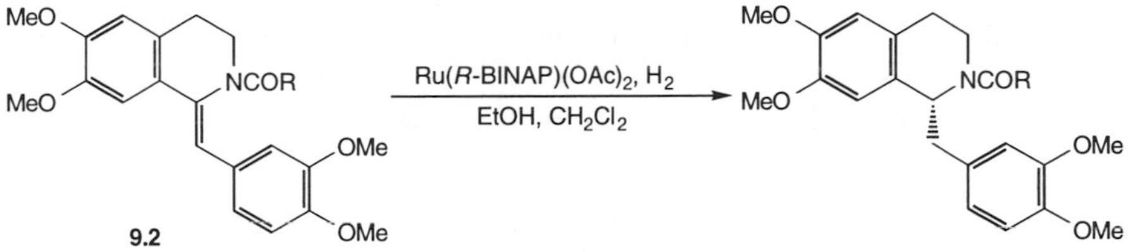

**SCHEME 9.3.**

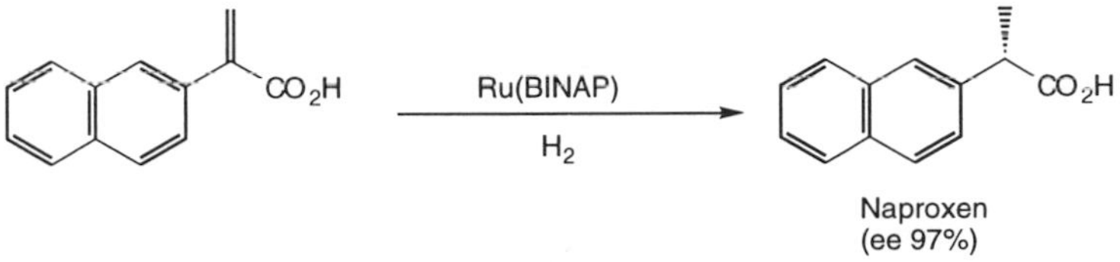

**SCHEME 9.4.**

In this reaction, the hydrogen delivered to the double bond is not completely derived from the molecular hydrogen. The hydrogen β to the carbonyl group is derived from the protic solvent.[99]

β-Amino acids have not been available by asymmetric catalytic hydrogenation approaches with a high degree of enantioselectivity;[100] although not a complete solution, BINAP now provides reasonable (60 to 96%) ee's.[101] No doubt improvements will continue to be made in this area as selectivity is increased through the use of BINAP analogs.[102]

An alternative catalyst system to accomplish similar transformations of acrylic acid derivatives is based on a ferrocene derivative.[94] Effective kinetic resolutions have been achieved with the asymmetric hydrogenations of α-(hydroxyalkyl)acrylate esters.[80]

To alleviate some of the problems associated with the preparation of chiral ligands in high enantiomeric purity, alternative strategies, such as chiral poisoning, are being pursued.[103,104]

So far, we have considered hydrogenations of functionalized alkenes by chiral reagents. Asymmetric induction can also be achieved by use of a chiral substrate, especially when one face is hindered.[105,106] An example is given in Scheme 9.5, where a homogeneous catalyst did provide useful stereofacial discrimination.[107]

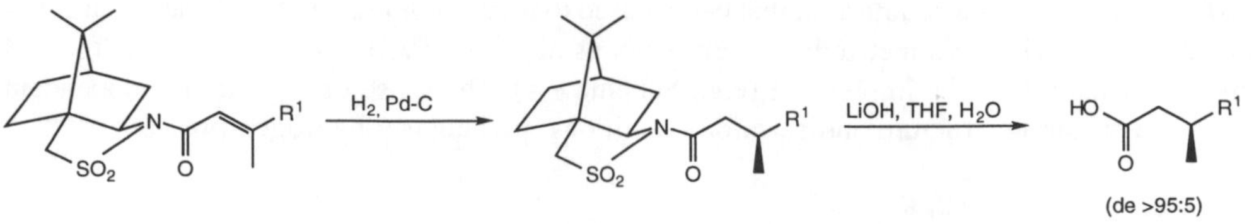

$H_2$, Pd-C

LiOH, THF, $H_2O$

(de >95:5)

SCHEME 9.5.

Of course, until the development of chiral catalysts, face selectivity of an achiral catalyst on a chiral substrate was the only means to achieve asymmetric synthesis by this approach. Many ingenious methods have been developed such as the preparation of α-amino acids by a reduction, but invariably the stereoselectivity was low.[2,18,108] In many cases, the constraints of a cyclic system allow for facial selectivity during the reduction. This effect has allowed for the asymmetric synthesis of amino acid derivatives with improved stereoselection.[108-111] With a cyclic dipeptide system, one amino acid derivative is dehydro, and this can be converted to the parent amino acid by reduction, while the other amino acid acts as a chiral template and allows for useful degrees of asymmetric induction on a heterogeneous catalyst surface.[112]

Other cyclic systems lend themselves to face-selective reactions. An example is provided in a synthesis of a protected sub-unit of nonactin (Scheme 9.6).[113]

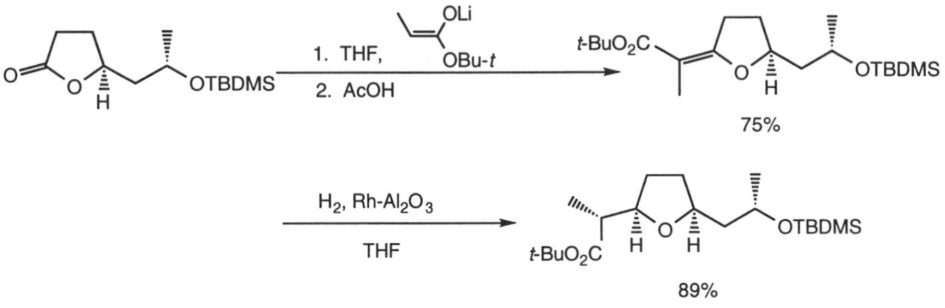

1. THF, (OLi, OBu-*t*)
2. AcOH

75%

$H_2$, Rh-$Al_2O_3$
THF

89%

SCHEME 9.6.

In addition to the control of absolute configuration, hydrogenation can also provide useful methodology for the control of relative stereochemistry; again, a nonactin synthesis provides an example (Scheme 9.7).[114,115]

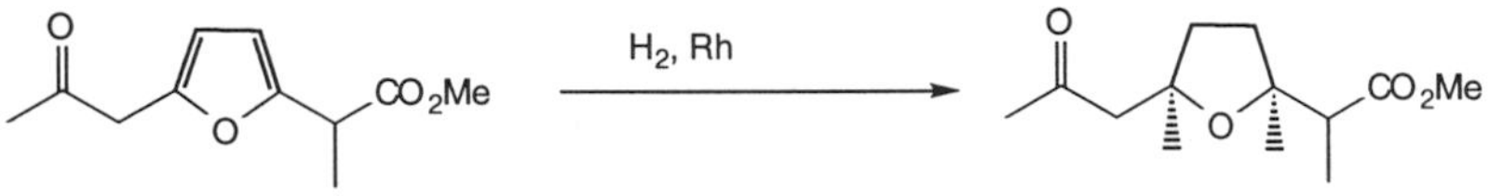

$H_2$, Rh

SCHEME 9.7.

The use of a hydrogen bond to afford a "pseudo-cyclic" system has been used successfully in the chemistry of β-substituted carbonyl compounds [*6.3*].

As with achiral hydrogenations,[116] reagents can be used in place of hydrogen gas for the transformations outlined above. However, the degree of asymmetric induction with a transfer agent is usually lower for a comparable reaction that uses hydrogen gas.[117]

### 9.1.2. ALLYLIC SUBSTRATES

Allyl alcohols can be reduced to either antipode (Scheme 9.8).[118,119] In addition, the methodology can be used to effect a kinetic resolution.[103,120,121]

S-BINAP-Ru, $H_2$

R-BINAP-Ru, $H_2$

S-BINAP-Ru, $H_2$

**SCHEME 9.8.**

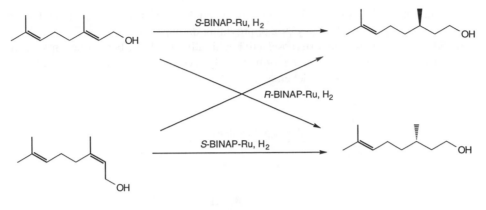

As inferred in Scheme 9.8, Ru(BINAP) catalyzed reductions are very specific for allyl alcohols; isolated alkenes are left intact.[122] The utility of the approach is illustrated by a synthesis of the α-tocopherol sidechain (Scheme 9.9); the isomerization is discussed in Section 9.2.1.[118,123,124]

1. Rh(S-BINAP)$^+$ 2. $H_2O$ — CHO (ee 98%)

$Me_2CO$, NaOH / $H_2O$, $H_2$, Pd-C — MgCl, THF — OH

1. $PCl_3$, pyr 2. NaOAc 3. NaOH, $H_2O$ 4. Fractional distillation — OH (ee 98%) — $H_2$, Ru(S-BINAP) — OH (98% this isomer)

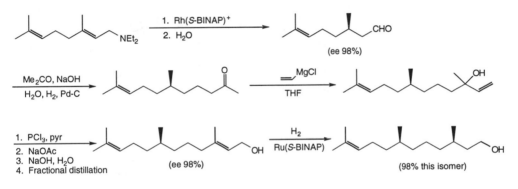

**SCHEME 9.9.**

Substituted allyl alcohols bearing a chiral auxiliary are reduced selectively. However, the extent of diastereoselection depends upon the catalyst-to-substrate stoichiometry (Scheme 9.10). The hydroxy group allows complex formation with the catalysts to ensure high facial selectivity.[125,126]

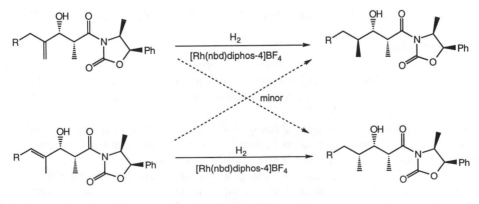

$H_2$, $[Rh(nbd)diphos\text{-}4]BF_4$ — minor — $H_2$, $[Rh(nbd)diphos\text{-}4]BF_4$

**SCHEME 9.10.**

This rhodium catalyst reduces 3-stannyl allyl alcohols to the *syncat* alcohol; the tin group can then be converted to a hydroxy group, providing an entry to 1,3-diols.[127]

An iridium catalyst was less stereoselective with allyl alcohols, but further investigation showed that with cyclic substrates, the iridium catalyst, could function well at lower loadings as the selectivity was increased (Scheme 9.11).[125,128,129]

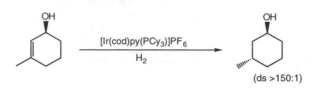

**SCHEME 9.11.**

This effect is demonstrated by reduction of the cyclohexane derivative **9.3**. Heterogeneous catalyzed hydrogenation led to preferential addition from the least hindered face, providing **9.4** as the major product. However, use of a cationic heterogeneous catalyst derived from iridium, or better, rhodium, allowed directed hydrogenation and formation of the desired isomer **9.5** (Scheme 9.12).[130]

**SCHEME 9.12.**

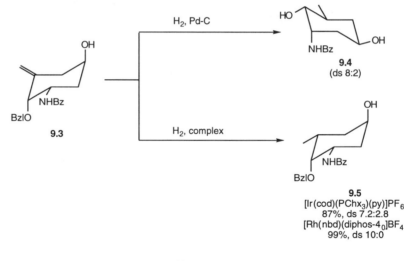

# 9.2. OTHER TRANSITION METAL CATALYZED REACTIONS

## 9.2.1. ISOMERIZATIONS

In addition to reductions, metal catalysts can bring about asymmetric isomerizations of allylic systems. Perhaps the best known example is in the commercial synthesis of menthol where rhodium BINAP catalyzes the asymmetric isomerization of an allyl amine to an enamine (Scheme 9.13).[7,123,131-133]

Both antipodes are available at the newly created center, and this can be controlled by the geometry of the enamine or catalyst (Scheme 9.14).[7,123,134] An additional example of this chemistry is shown in the tocopherol synthesis (Scheme 9.9).

Allyl alcohols can also undergo an analogous rearrangement, but the stereoselectivity is usually not as good.[135] The isomerization product is an enol that rapidly converts to the ketone.[136-138] The approach has been used in a kinetic resolution to obtain the starting material for a prostaglandin synthesis (Scheme 9.15).[139]

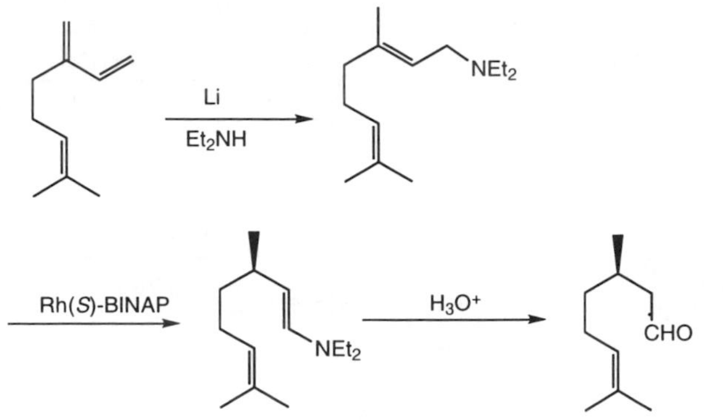

**SCHEME 9.13.**

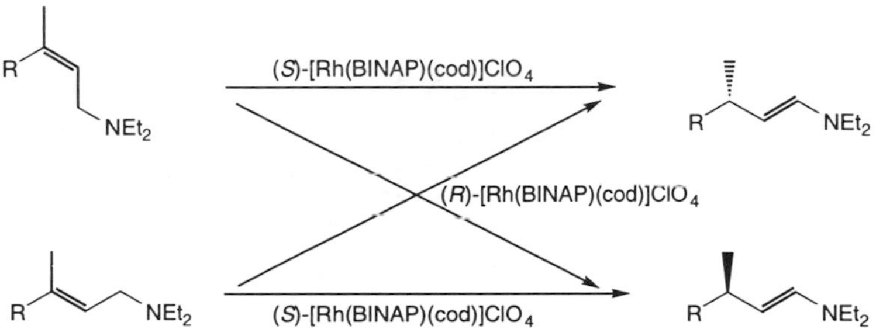

**SCHEME 9.14.**

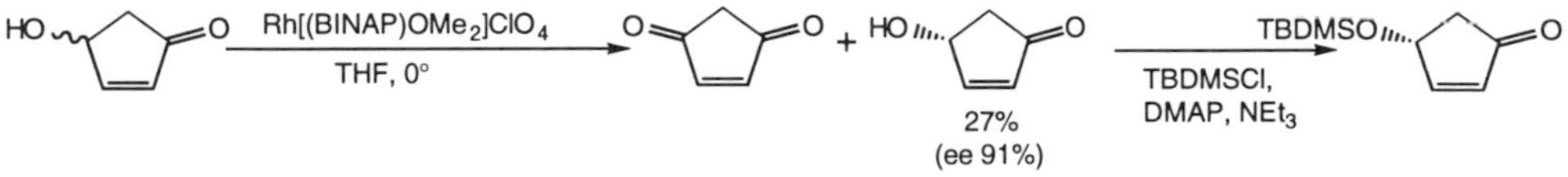

**SCHEME 9.15.**

## 9.2.2. HYDROSILYLATIONS

Hydrosilylation has provided useful methodology for the reduction of simple functional groups [*8.2.2.2*]. The approach can be used to reduce conjugated α,β-unsaturated carbonyl compounds by, what is effectively, a 1,4-reduction.[140,141] 1,2-Reduction is observed if a dihydro- or trihydrosilane is employed as the reagent,[142,143] together with a palladium(0) or molybdenum(0) catalyst.[144,145] Although use of a chiral rhodium catalyst does impart some induction in a 1,4-reduction, the degree is low.[146] Higher selectivity is observed with a chiral substrate (Scheme 9.16).[147]

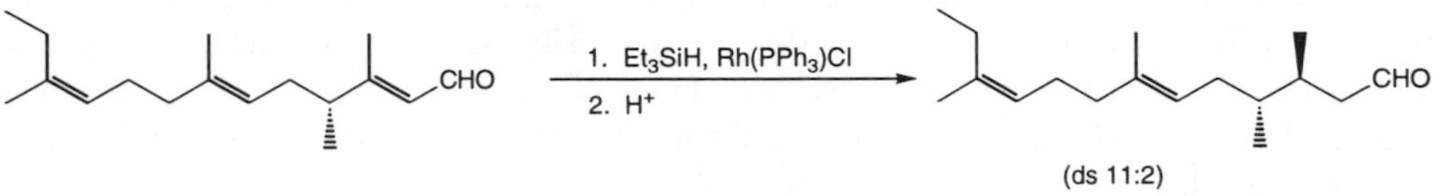

**SCHEME 9.16.**

Although hydrosilylation has been used to reduce a number of conjugated functional groups,[3,140,148-150] the use of allyl alcohols allows a stereoselective synthesis of 1,3-diols (Scheme 9.17).[151-153]

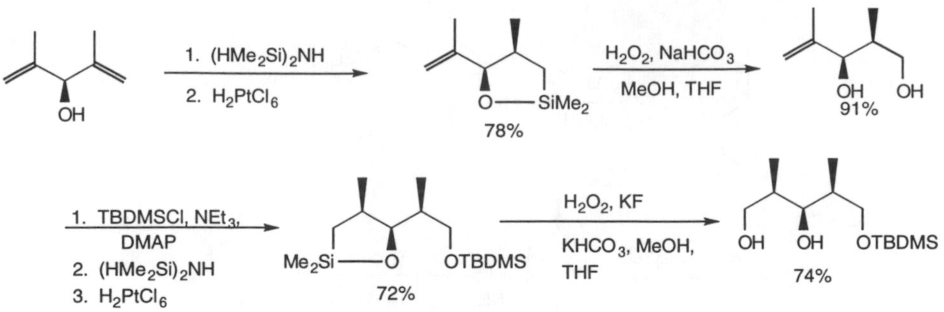

SCHEME 9.17.

The high selectivity has been rationalized through an intramolecular reaction (Schemes 9.18 and 9.19). Both are in accord with Houk's rule.[151,153]

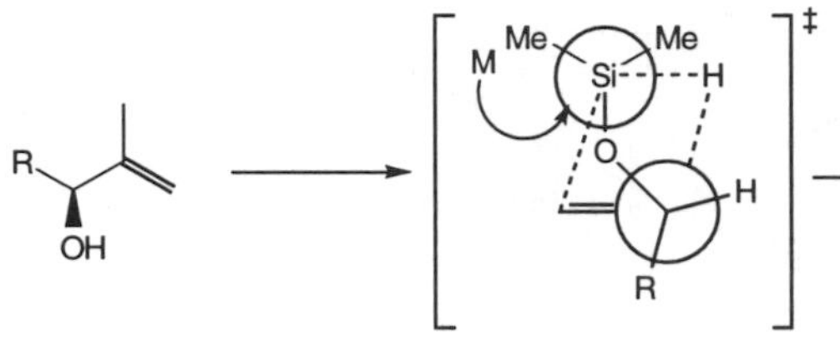

SCHEME 9.18.

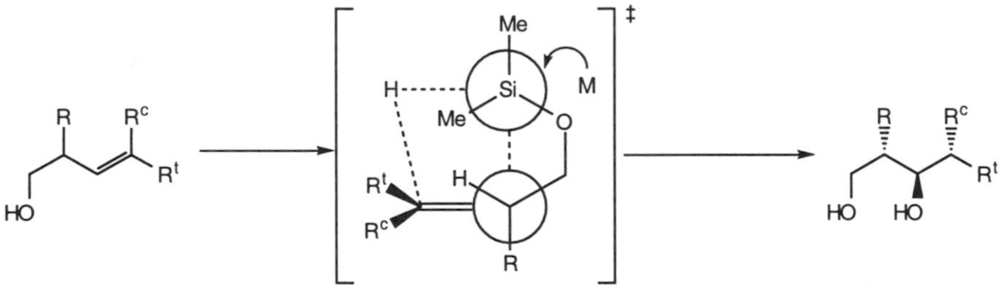

SCHEME 9.19.

As the reaction is intramolecular, the product stereochemistry is *syncat* for an allyl alcohol substrate, whereas an intermolecular hydroboration on the same substrate would provide the *ancat* isomer [*9.3.1*]; this makes the two methodologies complimentary. A useful variant of this intramolecular platinum catalyzed hydrosilylation reaction provides an entry to *syncat*-1,2-amino alcohols.[154]

Although not a true hydrosilylation, as a conjugate addition occurs, palladium catalyzed disilylation of an enone provides a facile entry to β-silyl ketones, and, hence, β-hydroxy ketones (Scheme 9.20).[155] In addition, hydrosilylation of a 1,3-diene can provide an allylsilane with moderate (<66% ee) induction. The resultant allylsilanes can be converted to allyl alcohols [*9.4.1*].[156]

Reductions with silanes are also useful for the conversion of an allylic acetate, or similar ester, to an alkene.[150,157-159] The reaction occurs with inversion of configuration at the original oxygenated center (Scheme 9.21).[159,160]

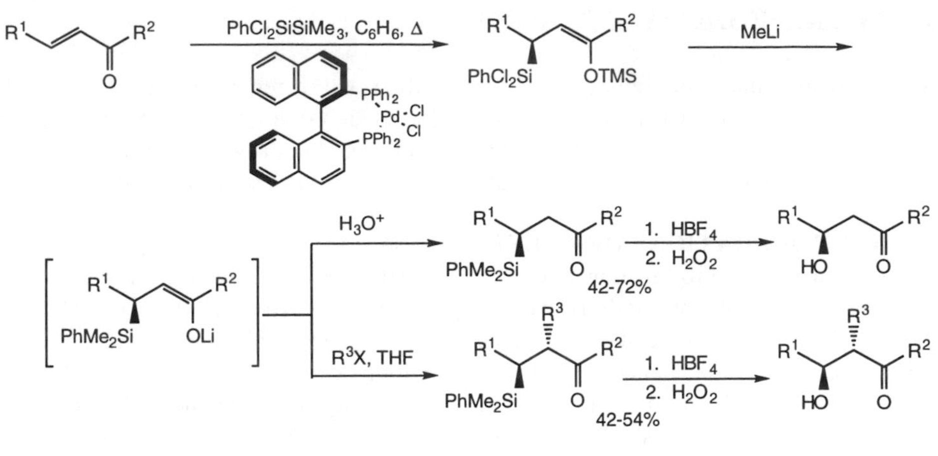

SCHEME 9.20.

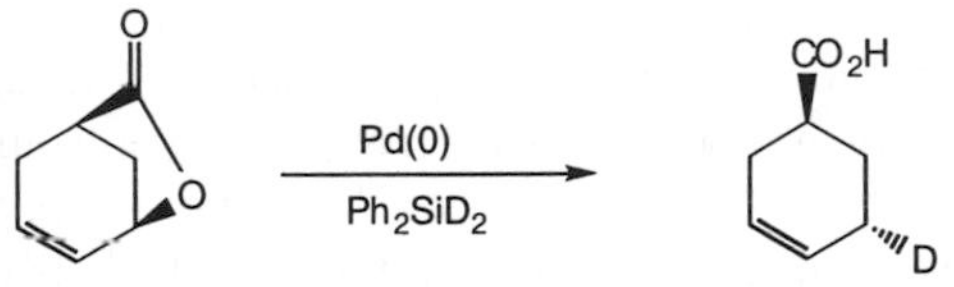

SCHEME 9.21.

### 9.2.3. HYDROFORMYLATIONS

In addition to hydroformylations, hydrocarbonylations are also possible, and have provided methodology to aldehydes and carboxylate derivatives.[161-165] Unfortunately, asymmetric hydrocarboxylations still wait to be developed.[161,166-168] Hydroformylation of simple alkenes can provide some asymmetric induction. The stereoselectivity is higher with functionalized alkenes (e.g., Scheme 9.22),[161,165,169] but still leaves much to be desired.[161,170-174]

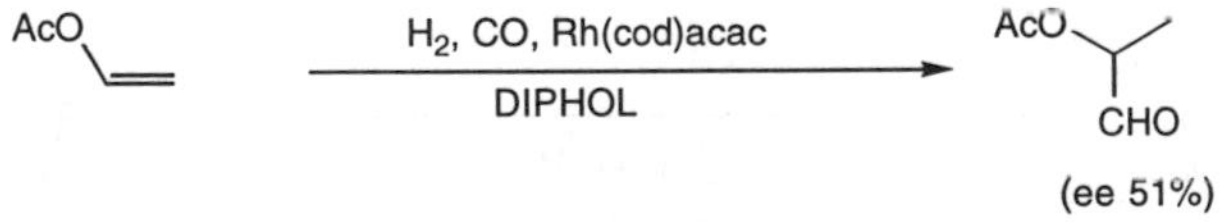

SCHEME 9.22.

### 9.2.4. HYDROCARBOXYLATIONS AND HYDROESTERIFICATIONS

These reactions can provide higher selectivity than hydroformylations (e.g., Scheme 9.23).[168,175-178]

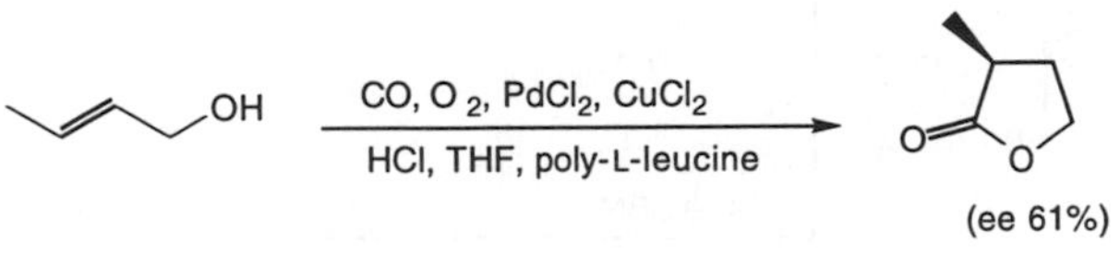

SCHEME 9.23.

However, most examples are limited to simple alkenes.[161]

## 9.3. HYDRIDE DELIVERY AGENTS

The problem with α,β-unsaturated systems is often one of regiochemistry (1,4- vs. 1,2-addition);[179,180] reagents that react with enone and related systems at the carbonyl center are discussed in Chapter 6. Use of a chiral additive can bring about asymmetric induction (vide infra). Complete reductions of enones to saturated alcohols are also covered in Chapter 6.

### 9.3.1. HYDROBORATIONS OF ALLYL SYSTEMS

Boranes can be used to functionalize alkenes or act as reducing agents. The reaction of a borane with an allyl alcohol or derivative can provide control over relative stereochemistry (Scheme 9.24).[181-184] The outcome does depend on the degree of substitution; indeed, in most reactions the (1,2-) *ancat* isomer of the 1,3-diol predominates (Scheme 9.25).[185-187] The most sterically demanding boranes afford better stereoselectivity; the outcome is then consistent with Houk's rule.[188,189] The selectivity may be reversed by the use of a rhodium catalyst (Scheme 9.26),[190-192] or by use of a vinyl ether (Scheme 9.27).[193-197] If the hydroxy group of the allyl alcohol is protected, anti-Markovnikov addition is observed with *syn*-addition control of relative stereochemistry.[198,199] For the rhodium catalyzed additions, the larger the protection, the better the induction observed; the oxygen acceptor group then adopts the *anti*-position.[190] Clearly, an oxygen atom is not a prerequisite for Houk-type addition; indeed, allylsilanes provide a useful entry to 1,3-diols [*9.4.1*].[200] Hydroboration of allyl amines with, or in the absence of, a rhodium catalyst to give 1,3-amino alcohols parallels the reactions of allyl alcohols and ethers.[196,201,202] The use of an enamine with a chiral borane provides an entry to 2-amino alcohols.[203]

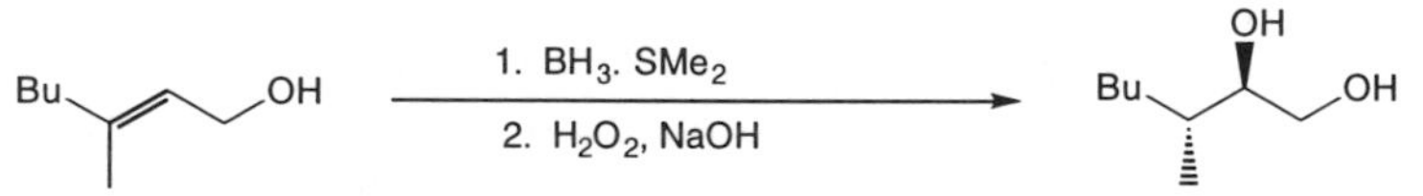

Bu OH — 1. $BH_3 \cdot SMe_2$; 2. $H_2O_2$, NaOH → Bu (OH) OH

**SCHEME 9.24.**

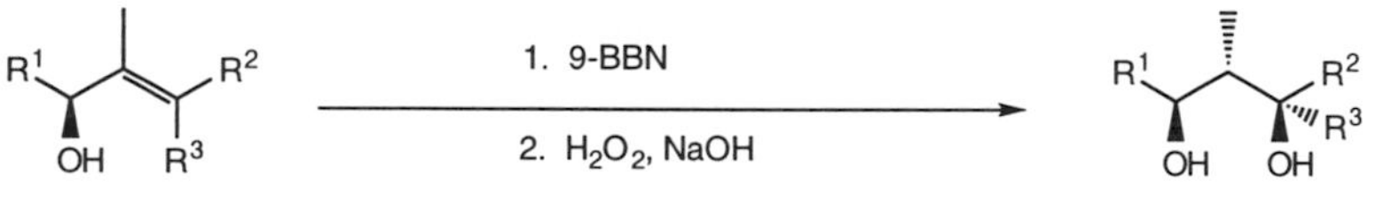

$R^1$ $R^2$ OH $R^3$ — 1. 9-BBN; 2. $H_2O_2$, NaOH → $R^1$ $R^2$ $R^3$ OH OH

**SCHEME 9.25.**

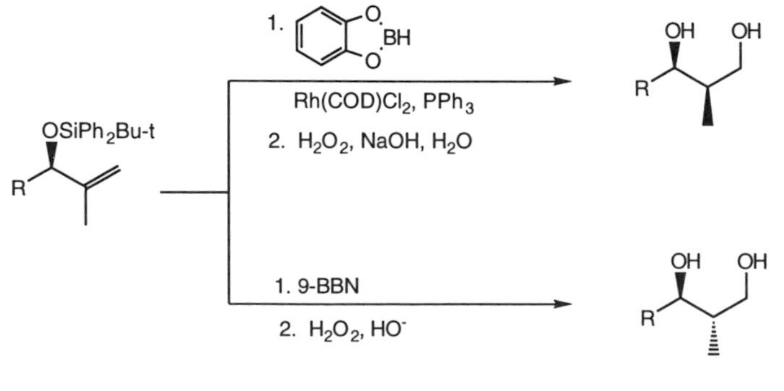

$OSiPh_2Bu$-t R — 1. (catecholborane) O BH O, $Rh(COD)Cl_2$, $PPh_3$; 2. $H_2O_2$, NaOH, $H_2O$ → OH OH R

— 1. 9-BBN; 2. $H_2O_2$, $HO^-$ → OH OH R

**SCHEME 9.26.**

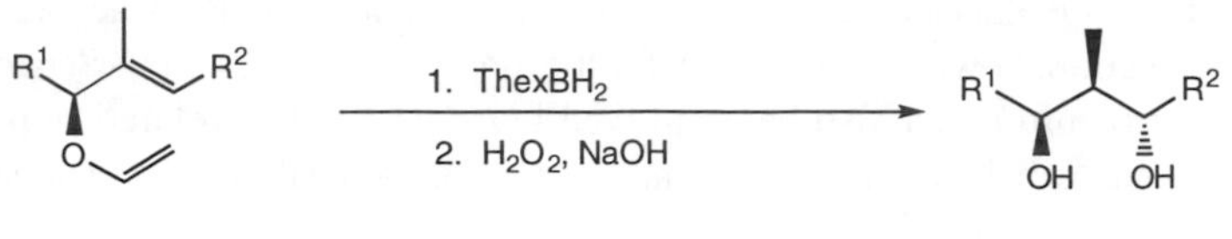
R1
R2
O
1. ThexBH2
2. H2O2, NaOH
R1
R2
OH
OH

SCHEME 9.27.

# 9.4. REACTIONS OF SOME FUNCTIONALIZED ALKENES

## 9.4.1. REACTIONS OF FUNCTIONALIZED SILANES

Organosilicon chemistry allows for a wide variety of transformations, often in a stereoselective manner.[204-207] The use of this chemistry has expanded greatly by the development of methods for conversion of a silyl group to an alcohol.¶ [209-215] Many of the reactions in this section comply with Houk's rule [*2.2*][188] — the silyl group acts as the large substituent.

Allylsilanes allow for the regiochemical introduction of a wide variety of electrophilic species.[165,216-222] An intramolecular proton transfer can provide good asymmetric induction.[223,224] Conversion of the silyl group to hydroxy then provides an allyl alcohol.[225,226]

Reaction of a chiral allylsilane with an aldehyde provides the homoallyl alcohol with good asymmetric induction [*4.1.5.2*]. Oxidative cleavage of the unsaturation affords the β-hydroxy acid (Scheme 4.30).[227,228] The selectivity arises from the *anti*-$S_E{}'$ mode of the reaction. Thus, condensation of a chiral allylsilane with a wide variety of electrophiles proceeds with high asymmetric induction.[229,230] Allylsilanes undergo hydroboration with good regio- and stereochemical control. The resultant alcohols can be converted to 1,3-diols (Scheme 9.28).[200,231]

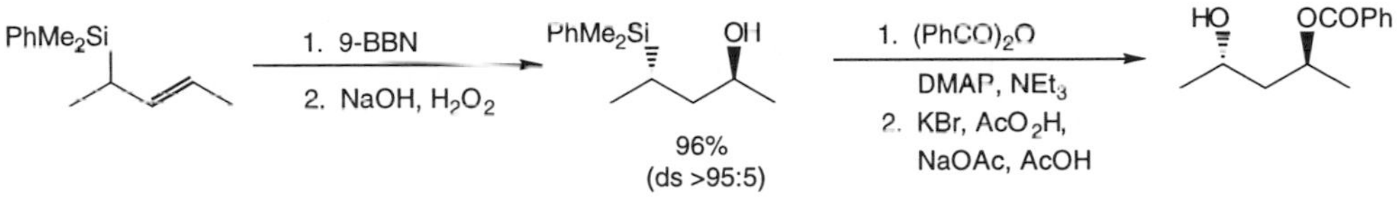
PhMe2Si
1. 9-BBN
2. NaOH, H2O2
PhMe2Si
OH
96%
(ds >95:5)
1. (PhCO)2O
DMAP, NEt3
2. KBr, AcO2H,
NaOAc, AcOH
HO
OCOPh

SCHEME 9.28.

*Ancat*-1,2-diols are available by reaction of an allylborane with an aldehyde followed by oxidative substitution of a silyl group. The use of a chiral tartrate ligand on the boron also allows for enantioselectivity (Scheme 9.29).[232]

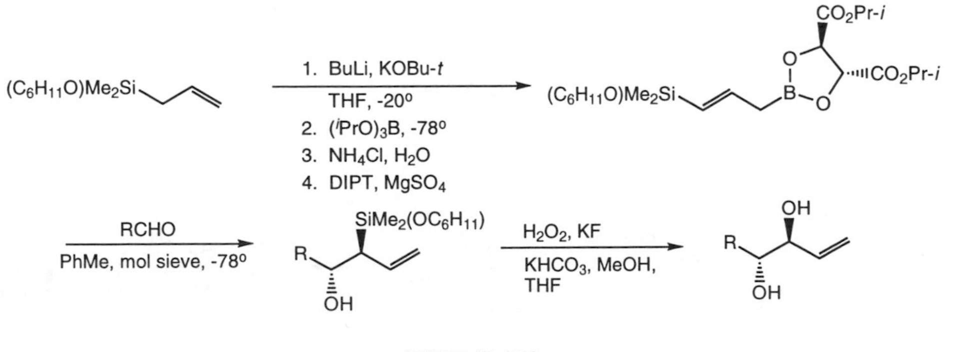
(C6H11O)Me2Si
1. BuLi, KOBu-t
THF, -20°
2. (iPrO)3B, -78°
3. NH4Cl, H2O
4. DIPT, MgSO4
(C6H11O)Me2Si
CO2Pr-i
CO2Pr-i
O
B
O
RCHO
PhMe, mol sieve, -78°
SiMe2(OC6H11)
R
OH
H2O2, KF
KHCO3, MeOH,
THF
OH
R
OH

SCHEME 9.29.

¶ A similar reaction has been performed with tin.[208]

Phenyldimethylallylsilanes react with osmium tetroxide and MCPBA according to Houk's rule; the stereoselection, however, can be low for the methyl series (R = Me) (Scheme 2.17).[233,234] Chiral ligands of osmium can also be employed [*11.4.1*].[235,236] A related sequence uses an α-alkoxysilane and Anh-Felkin controlled addition of an acylsilane [*2.1*] (Scheme 9.30; see also Schemes 2.3 and 2.9).[237,238]

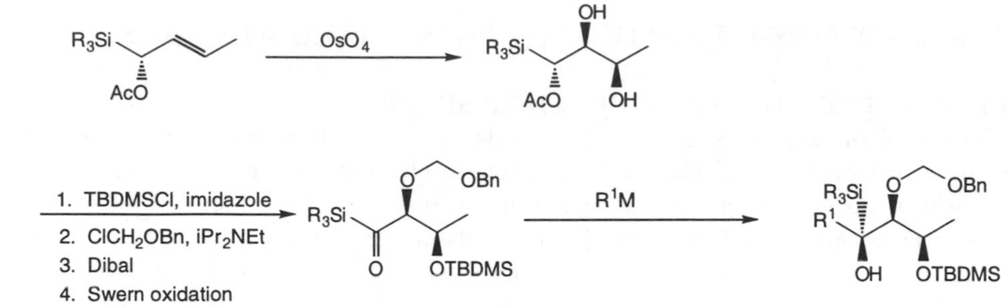

$R_3Si$ AcO OsO$_4$ $R_3Si$ OH AcO OH

1. TBDMSCl, imidazole
2. $ClCH_2OBn$, $iPr_2NEt$
3. Dibal
4. Swern oxidation

$R_3Si$ O OBn O OTBDMS $R^1M$ $R_3Si$ $R^1$ O OBn OH OTBDMS

**SCHEME 9.30.**

β-Hydroxy esters are available through conjugate addition of a silyl cuprate and then alkylation; the sequence of addition controls the relative stereochemistry (Scheme 9.31).[239-244] The selectivity observed is attributed to electronic factors. The scheme can be extended to aldol reactions,[245-247] and can be made asymmetric through employment of a chiral auxiliary (Scheme 9.32).[240]

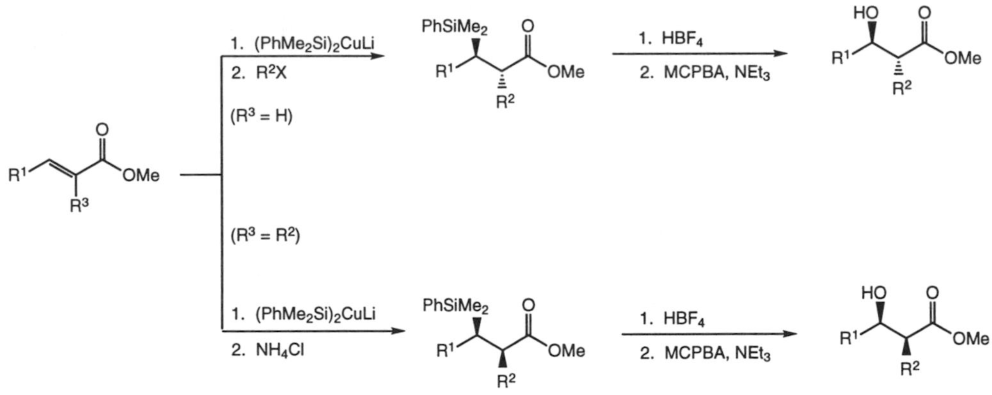

$R^1$ O OMe $R^3$

1. $(PhMe_2Si)_2CuLi$
2. $R^2X$
($R^3$ = H)

PhSiMe$_2$ O $R^1$ OMe $R^2$

1. $HBF_4$
2. MCPBA, $NEt_3$

HO O $R^1$ OMe $R^2$

($R^3$ = $R^2$)

1. $(PhMe_2Si)_2CuLi$
2. $NH_4Cl$

PhSiMe$_2$ O $R^1$ OMe $R^2$

1. $HBF_4$
2. MCPBA, $NEt_3$

HO O $R^1$ OMe $R^2$

**SCHEME 9.31.**

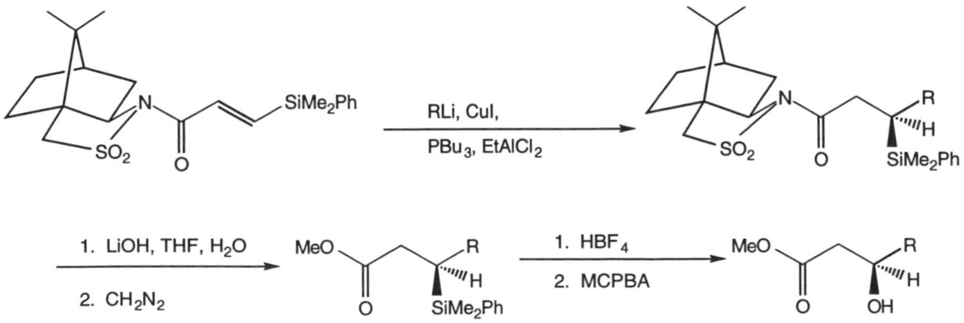

N SO$_2$ O SiMe$_2$Ph

RLi, CuI,
$PBu_3$, $EtAlCl_2$

N SO$_2$ O R H SiMe$_2$Ph

1. LiOH, THF, $H_2O$
2. $CH_2N_2$

MeO O R H SiMe$_2$Ph

1. $HBF_4$
2. MCPBA

MeO O R H OH

**SCHEME 9.32.**

Conjugate introduction of a silyl group to an enone provides an approach to β-hydroxy ketones.[155] Vinyl sulfoxides can also be Michael acceptors for silyl cuprates.[248]

Alkylation of a β-silylamide provides introduction of the electrophile *anti* to the silyl group. This reaction was used to prepare γ-butyrolactones — the chiral starting material **9.6** being derived from a Sharpless kinetic resolution [*11.1.2.1*] (Scheme 9.33).[249]

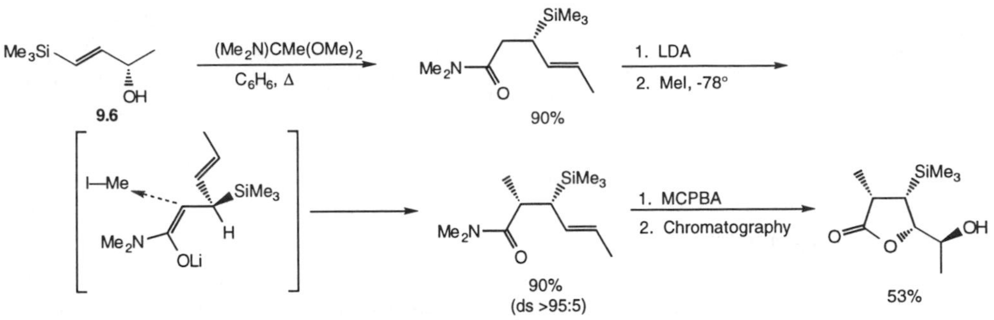

**SCHEME 9.33.**

A chiral propargyl silane provides a convenient entry to chiral propargyl alcohols through the use of an alkylation methodology.[250]

The use of a chiral vinylsilane does allow for the preparation of chiral allyl alcohols, although a separation of the intermediate diastereoisomers is required (Scheme 9.34).[251]

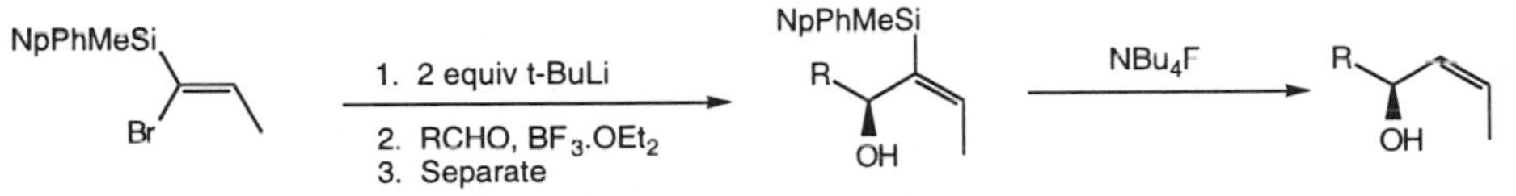

**SCHEME 9.34.**

Hydrosilylations of allyl alcohols provide a stereoselective approach to 1,3-diols (Scheme 9.17).[151]

There are a large number of synthetic methods to introduce a silyl group, including the addition of a silane to an enone (vide supra); this sequence also illustrates the flexibility of the methodology (Scheme 9.20).[155]

## 9.4.2. PROPARGYLIC SUBSTRATES

Functionalized acetylenes are useful substrates for reduction reactions, as a vinyl organometallic species often results that can undergo further reaction to provide functionalized alkenes [*8.1.1*]. An example is provided by the lithium aluminum hydride reduction of a propargyl alcohol; the geometry of the product vinyl iodide is dependent upon whether an intramolecular complex is formed (Scheme 9.35).[252-259]

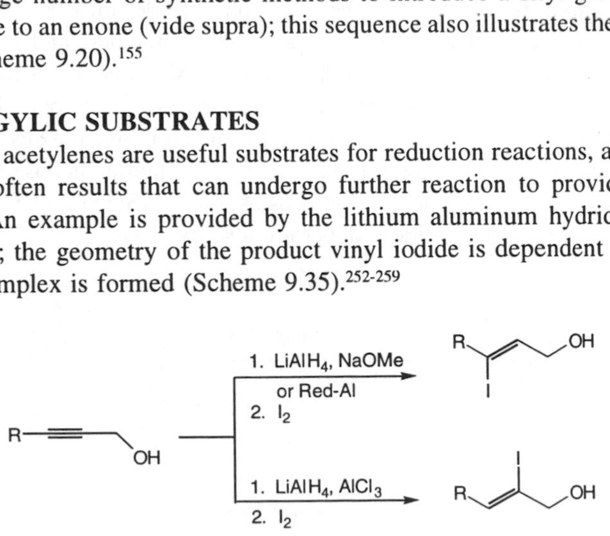

**SCHEME 9.35.**

There are a number of extensions of this methodology, including the use of other metals. As the control is regiochemical in nature, these methods have been discussed in Section 8.1.1. However, the methodology can be applied to chiral propargyl alcohols **9.7** — obtained by a resolution with the Sharpless methodology [*11.1.2.1*] — as shown in the synthesis of butenolides (Scheme 9.36).[260-262]

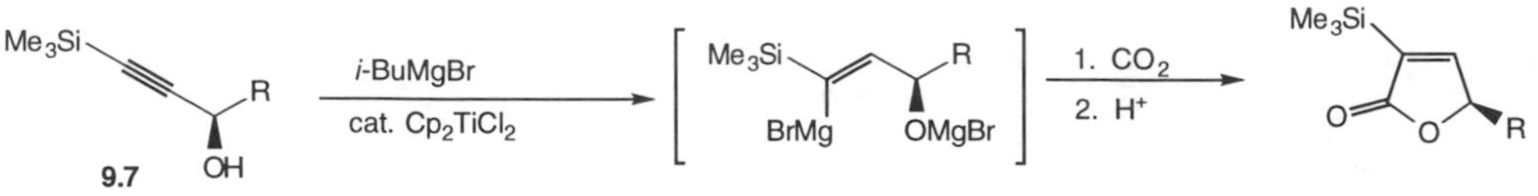

$Me_3Si$–C≡C–CH(OH)R (**9.7**) $\xrightarrow[\text{cat. } Cp_2TiCl_2]{i\text{-BuMgBr}}$ [ $Me_3Si$(BrMg)C=CH–CH(OMgBr)R ] $\xrightarrow[\text{2. } H^+]{\text{1. } CO_2}$ butenolide ($Me_3Si$, R)

SCHEME 9.36.

### 9.4.3. VINYL COMPOUNDS

The presence of additional functionality allows for a wide range of chemical transformations to become available. For example, the functionality can be displaced and substituted by another group. An alternative is that the functionality stabilizes an α-carbanion that can be used in further reactions. As $sp^2$ centers are involved in these examples, control is of a regio- and diastereoisomeric nature and these reactions have been discussed in detail in Section 8.1.3. In addition, reactions that involve addition of a vinyl anion to a carbonyl group are discussed in Section 4.1.4, while those that involve the addition of an organometallic reagent across the alkene are considered conjugate additions (vide infra).

## 9.5. CONJUGATE ADDITIONS

### 9.5.1. α,β-UNSATURATED CARBONYL SYSTEMS

The conjugate addition[263] of a nucleophile to an α,β-unsaturated system involves a question of controlling facial selectivity.[264,265] Of course, the problem of 1,2- vs. 1,4-addition must also be overcome.[266]

#### 9.5.1.1. Carbon Nucleophiles

Conjugate addition is often accomplished with a cuprate reagent,[267-271] although other metals, such as zinc, have been used.[272-279] Conjugate addition is usually observed when a "soft," stabilized anion is employed as the nucleophile (vide infra).[280-292]

The resultant enolate is formed in a regioselective manner, and can undergo further elaboration, such as aldol condensations [Chapter 7], or silyl enol ether formation [*5.1*].[293-296] The methodology also allows for the introduction of functionality in both the nucleophilic and electrophilic moieties.[239,287,297-300] In cyclic cases, the nucleophilic and electrophilic groups are introduced in a relative *trans*-configuration.[301-306] This phenomenon has been used to provide the strategy behind an elegant approach to prostaglandins (Scheme 9.37);[301,302,307,308] extensions have shown that metals besides copper are also effective.[272,303-305,309,310]

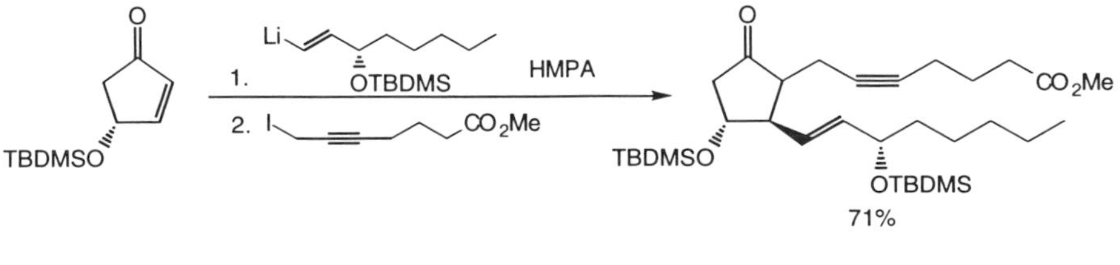

TBDMSO-cyclopentenone $\xrightarrow[\text{2. I–}CH_2C{\equiv}C(CH_2)_3CO_2Me]{\text{1. Li–CH=CH–CH(OTBDMS)}C_5H_{11}\text{, HMPA}}$ prostaglandin precursor (TBDMSO, OTBDMS, $CO_2Me$), 71%

SCHEME 9.37.

The *trans*-addition methodology does translate to acyclic systems, provided that a large nucleophile is employed (Scheme 9.38).[284,288,311]

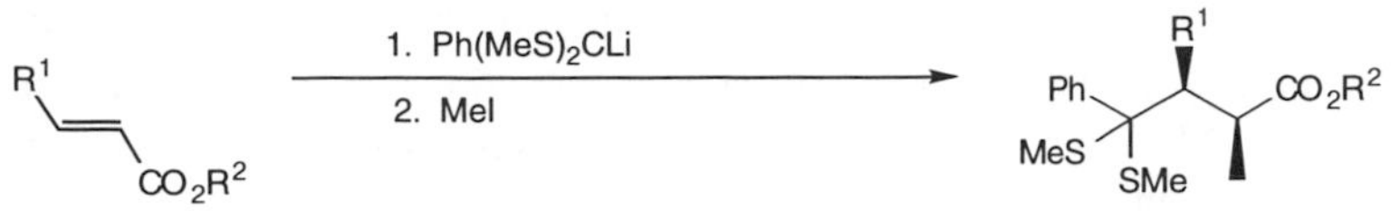
R1
CO2R2
1. Ph(MeS)2CLi
2. MeI
R1
Ph
CO2R2
MeS
SMe

**SCHEME 9.38.**

In cyclic systems, facial selectivity can be dictated by steric effects from appendant substituents.[290,312-315] Control in acyclic systems has centered around the use of chiral auxiliaries (e.g., Schemes 9.39 and 9.40).[240,316-332]

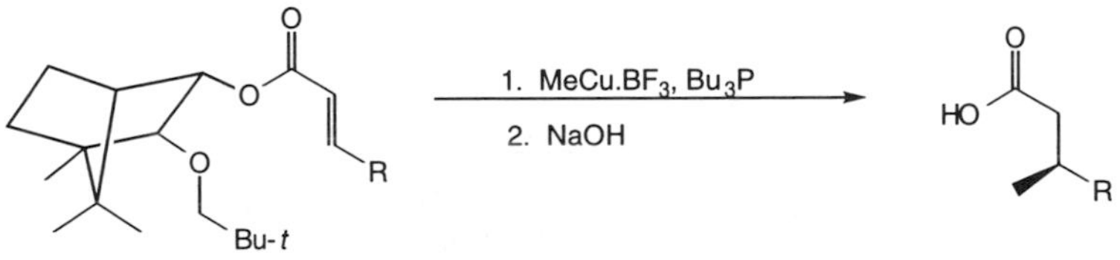
O
O
O
R
Bu-t
1. MeCu.BF3, Bu3P
2. NaOH
O
HO
R

**SCHEME 9.39.**

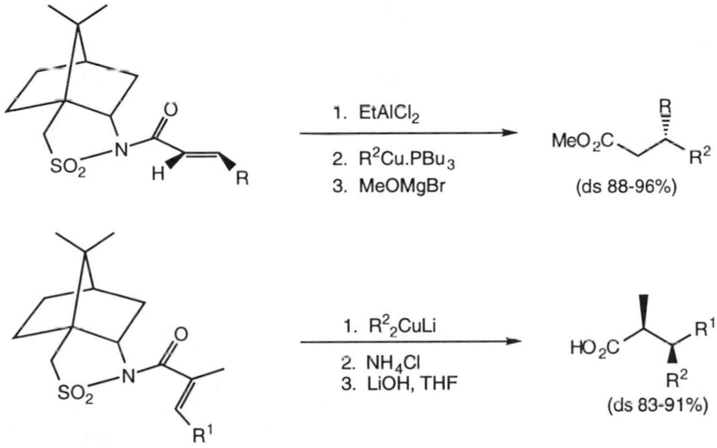
O
SO2
N
H
R
1. EtAlCl2
2. R2Cu.PBu3
3. MeOMgBr
R
MeO2C
R2
(ds 88-96%)
O
SO2
N
R1
1. R22CuLi
2. NH4Cl
3. LiOH, THF
HO2C
R1
R2
(ds 83-91%)

**SCHEME 9.40.**

Asymmetric conjugate additions are observed when a chiral sulfoxide is present on the α-position of the enone. Although most of this work has focused on cyclopentenone derivatives,[333-335] examples are available in the acyclic (Scheme 9.41),[336] and lactone series.[337,338] Selectivity is also observed for simple chiral sulfoxides and sulfoximines (vide infra).[339-344]

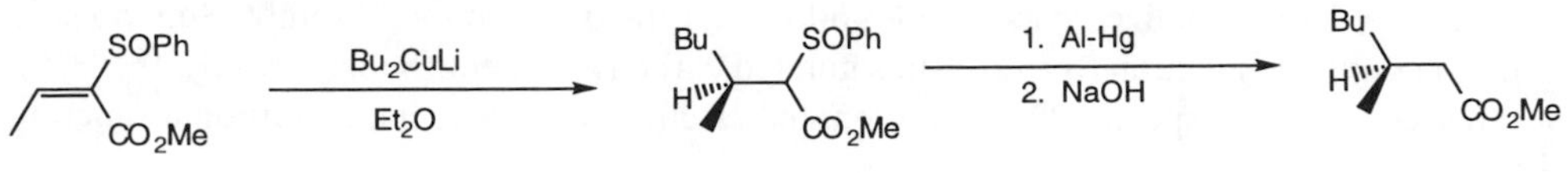
SOPh
CO2Me
Bu2CuLi
Et2O
Bu
SOPh
H
CO2Me
1. Al-Hg
2. NaOH
Bu
H
CO2Me

**SCHEME 9.41.**

With organocuprates, the addition of chiral chelation agents, such as sparteine, results in only low levels of asymmetric induction.[263,345] Higher optical yields were obtained with proline derivatives[346] and chiral copper complexes (vide infra).[347,348] Chiral ligands for organozinc reagents also allow asymmetric 1,4-additions.[349-354]

Michael additions can be catalyzed by chiral phase transfer catalysts, although optical yields are variable.[355-358] Lewis acid catalysis can also be employed, and may even be required, as for the addition of an allylsilane.[359,360]

The nucleophile that partakes in a conjugate addition can be chiral, such as a sulfoxide,[361-371] a β-hydroxy ester,[372] amide,[373] imine,[374-378] α-aminonitrile,[379] oxazinone,[380] or oxazepine.[381,382] As an alternative, a chiral ligand can be used to provide induction (vide supra).[276,346,347,349,351,383-390]

The use of the RAMP/SAMP hydrazone methodology [*5.3*] provides an entry to chiral δ-keto esters (Scheme 9.42).[391-393]

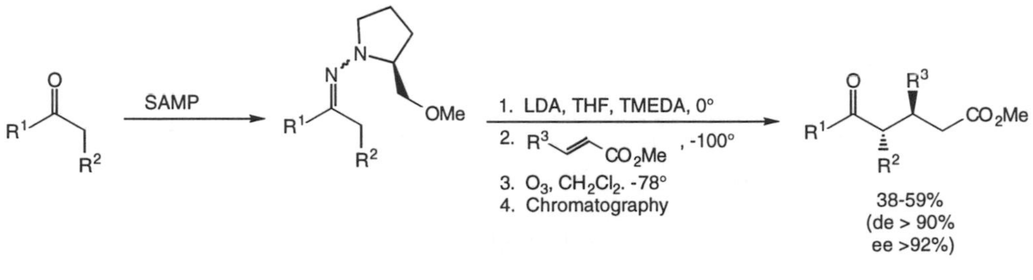

**SCHEME 9.42.**

The chiral ligand can also be an integral part of the substrate molecule (Scheme 9.42), and good stereoselection can be observed.[394,395] This approach has been coupled with an intramolecular alkylation reaction (Scheme 9.43)¶ [396]

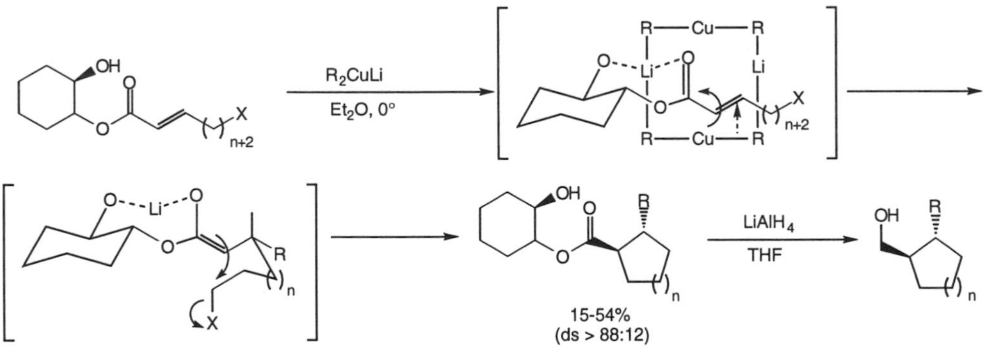

**SCHEME 9.43.**

The use of a chiral nucleophile is adequately illustrated by the synthesis of (+)-γ-pelargonolactone (**9.8**) (Scheme 9.44).[397,398]

The use of functionalized nucleophiles allows for the introduction of remote functionality through conjugate addition to an α,β-unsaturated carbonyl system (Scheme 9.45),[263,397,399-404] or other electrophilic species.[405] In addition, the electrophile can contain additional functionality.[406]

As with other reactions involving carbonyl groups, the degree of stereoselection observed is a function of solvent and counterion. Transition state models are being developed to explain

¶ The cuprate is shown as square planar as it was in the original citation.

the experimental observations.[387,407-411] As yet, no universally applicable model has been found to predict the stereochemical outcome of a 1,4-addition, but significant advances are being made in this area.[401,409,412,413] The use of chiral ligands with cuprates has led to a transition state proposal for use with cyclic enones (vide infra).[387,410,414-417] Michael addition is promoted by the presence of a Lewis acid,[316,317,418-422] or a silyl chloride;[278,279,423-432] these experimental results are providing information about the transition state requirements.[433]

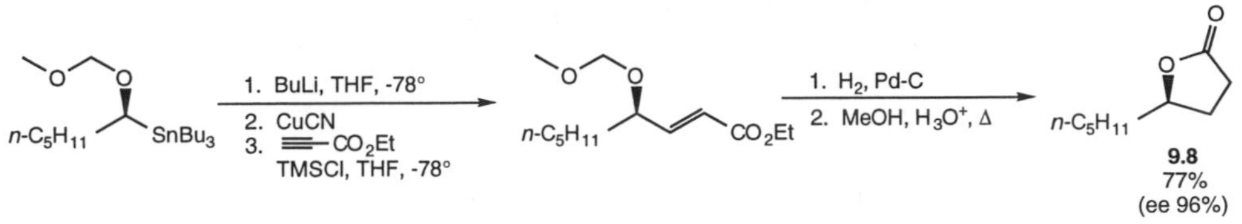

1. BuLi, THF, -78°
2. CuCN
3. ≡—$CO_2Et$
TMSCl, THF, -78°

1. $H_2$, Pd-C
2. MeOH, $H_3O^+$, Δ

**9.8**
77%
(ee 96%)

**SCHEME 9.44.**

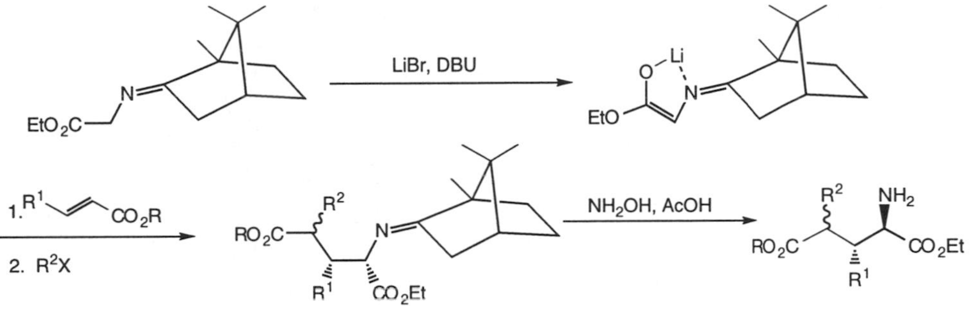

LiBr, DBU

1. $R^1$–CH=CH–$CO_2R$
2. $R^2X$

$NH_2OH$, AcOH

**SCHEME 9.45.**

The development of a model for the conjugate addition of a cuprate to an α,β-unsaturated carbonyl system has been hindered, to some extent, by our lack of knowledge of the exact structure of the cuprate;[270,271,434] higher-order cuprates, in particular.[411,435-437]

Cuprate additions have been the subject of many studies and a number of mechanisms have been proposed, including single electron transfers.[268,438] Lower order cuprates,[†] of the type $R_2CuLi$, exist as dimers.[271,439] However, as with other organometallic reagents, equilibria exist and the major component need not be the most reactive. Based on the data obtained from structural elucidation of the reagents in the solid state,[440,441] coordination of lithium (or magnesium) to the enone carbonyl group in the transition state,[442-445] and kinetic studies,[269,442,443,446-449] a transition state model has been proposed.[410,449] The addition of a cuprate (**9.9**) to an enone proceeds through a d,π*-complex (**9.10**) and an enone-copper(III) adduct (**9.11**) (Scheme 9.46).

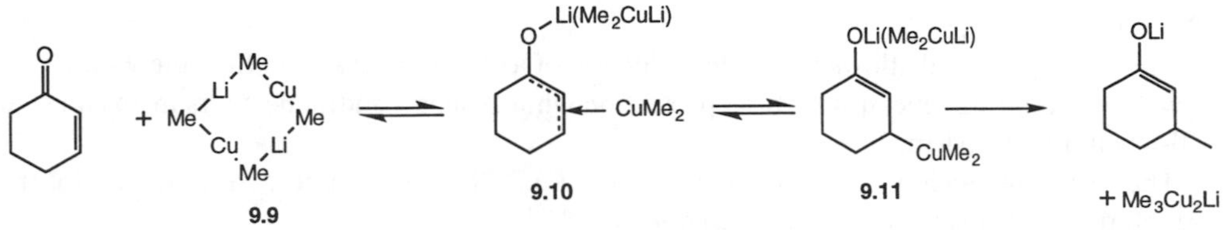

**9.9** **9.10** **9.11** + $Me_3Cu_2Li$

**SCHEME 9.46.**

[†] Also known as Gilman cuprates or reagents.

For 5-substituted 2-cycloalkenones, *trans*-addition occurs; this can be rationalized by the d,$\pi^*$ complex **9.12** being favored over **9.13** (Figure 9.2),[449] due to the unfavorable allylic interactions in the latter.[§]

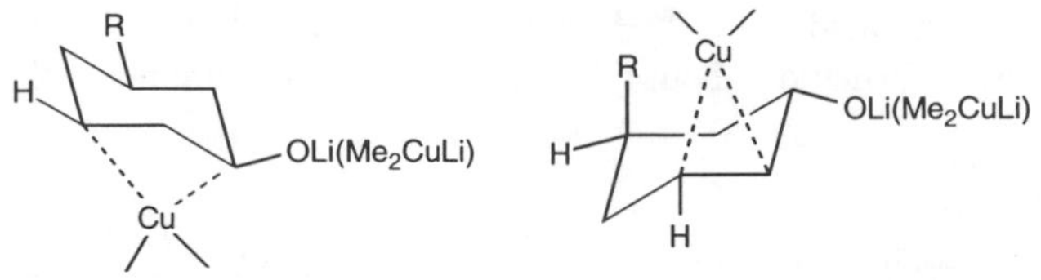

**FIGURE 9.2.** Simple model for the addition of a cuprate to a 5-substituted cyclohexenone.

This simple model of Figure 9.2 has been modified slightly to those of Figure 9.3, where the bonds to copper are not intended to depict a specific bond order, but rather show coordination geometry; the transition state **9.14** is preferred.[410]

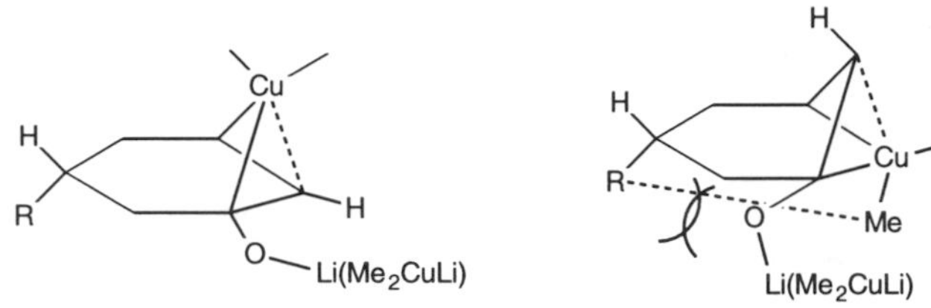

**FIGURE 9.3.** Modified model for the addition of a cuprate to a 5-substituted cyclohexenone.

This transition state model has been used to rationalize and design enones that undergo conjugate additions with high selectivity (Scheme 9.47).[410,414,449]

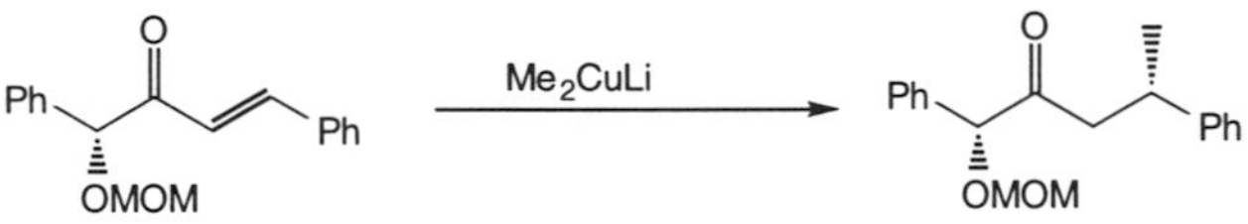

**SCHEME 9.47.**

Models, other than those for cuprate reagents, can also be useful for the stereoselective control of conjugate additions. For example, even with Grignard reagents, Michael addition can be enforced if a large group is present next to the carbonyl group of an α,β-unsaturated carbonyl system.[311,450] In addition, Houk's rule can rationalize subsequent reactions after a 1,4-addition has occurred [*2.2*].

The use of an acetal allows for the introduction of a chiral moiety (e.g., Scheme 9.48),[451,452] as well as providing a means for chiral protection while Michael addition occurs in an adjacent α,β-unsaturated system.[453]

The self-regeneration of a stereogenic center [*1.3.2.3*] provides a powerful method for the preparation of β-hydroxy acids (Scheme 9.49).[424,454-458]

§ Control through allylic interactions means that selectivity in acyclic systems can also be rationalized [*2.3*].

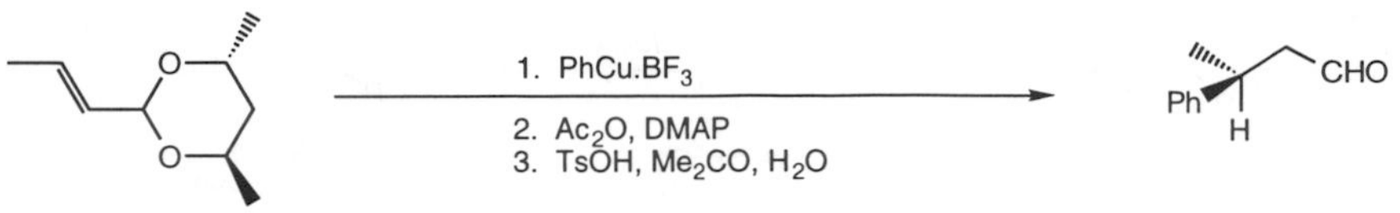
1. PhCu.BF3
2. Ac2O, DMAP
3. TsOH, Me2CO, H2O
Ph
H
CHO

SCHEME 9.48.

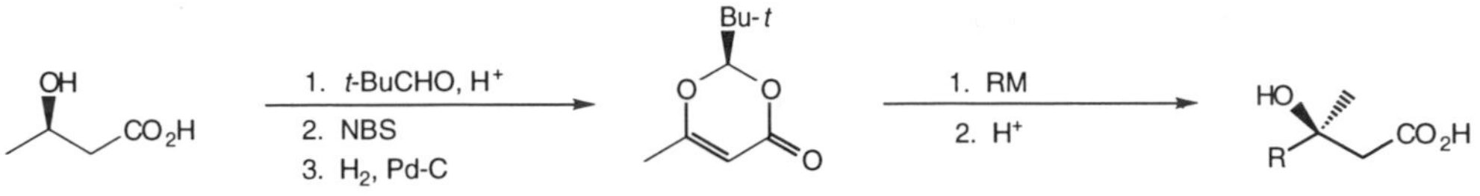
OH
CO2H
1. t-BuCHO, H+
2. NBS
3. H2, Pd-C
Bu-t
O
O
O
1. RM
2. H+
HO
R
CO2H

SCHEME 9.49.

Allylic alkylations can also be achieved by an organocopper reagent, and potential competing reaction pathways can exist, as for the substrates **9.15** and **9.16**. In this example, allylic alkylation is observed [9.6]; reduction of the product **9.17** and cyclization provides the δ-lactone **9.18** (Scheme 9.50).[459-462]

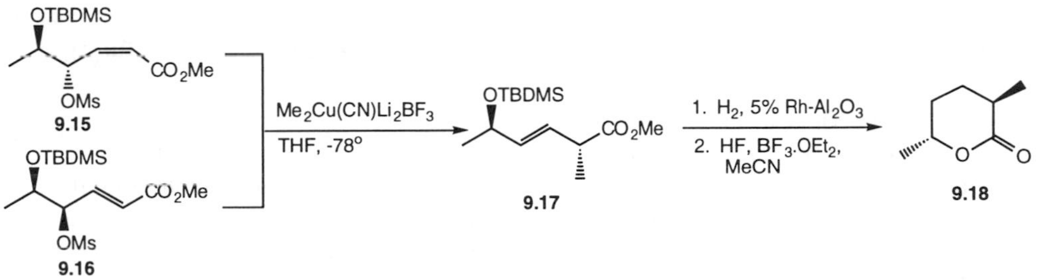
OTBDMS
CO2Me
OMs
9.15
OTBDMS
CO2Me
OMs
9.16
Me2Cu(CN)Li2BF3
THF, -78°
OTBDMS
CO2Me
9.17
1. H2, 5% Rh-Al2O3
2. HF, BF3.OEt2,
MeCN
O
O
9.18

SCHEME 9.50.

1,2-Addition of a cuprate occurs with γ-alkoxy enals in the presence of a silyl chloride, indicating that conjugate addition does not always occur.[463]

#### 9.5.1.2. Conjugate Reductions

The methodology not only allows for the stereoselective 1,4-addition of an alkyl group to an α,β-unsaturated system, but also for conjugate reductions. A wide range of reagents has been employed to promote overall hydride delivery at the β-position of an α,β-unsaturated carbonyl system, and include CuI-LiAlH$_4$,[464] homogeneous or complex metal hydrides,[465-469] magnesium in methanol,[470] lithium with an amine or ammonia,[471] LDA,[472] zinc–copper couple,[473] or aluminum or nickel chloride.[474] As with conjugate delivery of an alkyl group, the enolate derived from conjugate reductions can be used in a wide variety of subsequent reactions.[293] This flexibility can allow for alternative stereochemistry to be available at the β-position compared to methodology based on the delivery of an alkyl group. A number of metal hydride reagents have been developed that not only provide conjugate reduction, but achieve asymmetric reduction on the basis of their chiral auxiliaries (e.g., Scheme 9.51).[9,146,180,475-484]

The most general approach to obtain conjugate reduction with a chiral substrate is to use a borane, such as catecholborane,[485] or a hindered borohydride (Scheme 9.52), the inherent chirality being obtained from the substrate.[194,486,487]

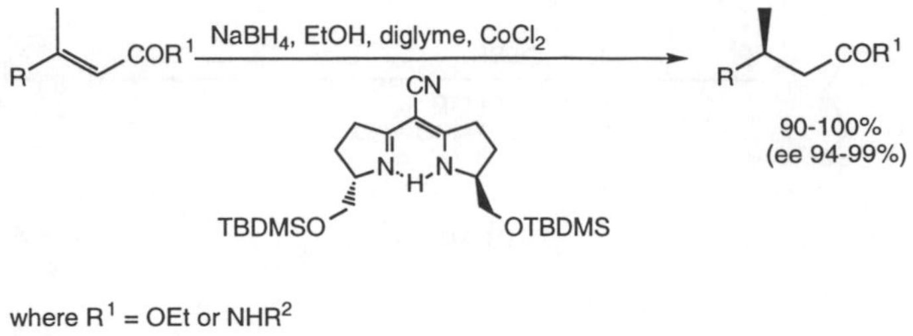
COR¹
R
NaBH₄, EtOH, diglyme, CoCl₂
CN
N
H
N
TBDMSO
OTBDMS
COR¹
R
90-100%
(ee 94-99%)
where R¹ = OEt or NHR²

SCHEME 9.51.

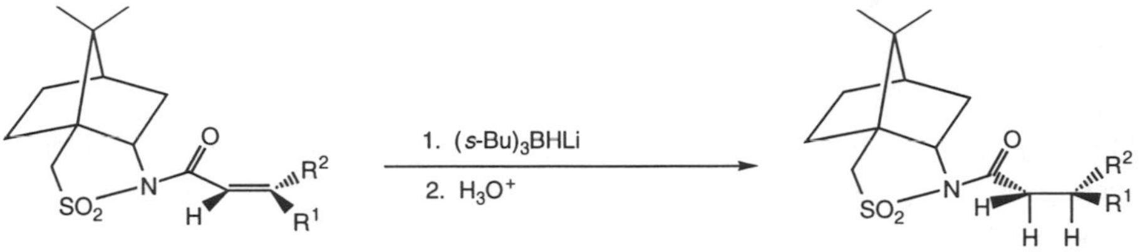
O
N
SO₂
H
R²
R¹
1. (s-Bu)₃BHLi
2. H₃O⁺
H H

SCHEME 9.52.

Reductions can also be achieved through use of a hydrogenation approach [*9.1.1*].

### 9.5.1.3. Other Nucleophiles

A conjugate addition methodology allows functionalization at the β-position of an unsaturated system through the use of a masking group to protect the unsaturation; many of these methods rely on the addition of a heteroatom group, such as silicon or a halogen, that can then be eliminated once the desired reaction has been performed.[488-497] Indeed, some enantioselectivity has been observed for the addition of thiols to enones in the presence of a chiral catalyst.[357,498-501] Use of a chiral enone allows for high induction, as illustrated by the synthesis of the oxathiane **9.19**, which has been employed as an acyl anion equivalent and chiral auxiliary (Scheme 9.53).[498,502]

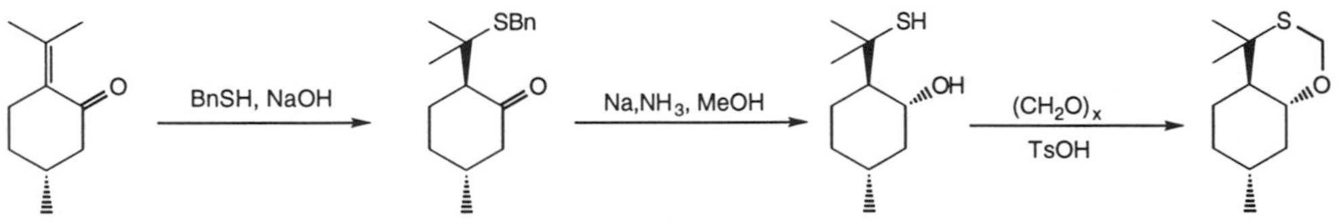
O
BnSH, NaOH
SBn
Na,NH₃, MeOH
SH
OH
(CH₂O)ₓ
TsOH
S
O

9.19

SCHEME 9.53.

The use of a nitrogen nucleophile in a Michael addition provides a rapid entry to β-amino acids; the use of a chiral amine provides asymmetric induction (Scheme 9.54).[383,503-508]

Once again, the resultant enolate can be used in a wide variety of reactions as illustrated by the (achiral) example in Scheme 9.55.[16,509] This sequence reveals that the enolate **9.20** is generated by the conjugate addition, as determined by silylation (not shown), and that it can be converted to the other isomer **9.21** by a protonation-deprotonation manipulation.

As with other nucleophiles, the use of a chiral substrate allows for induction in the reaction (Scheme 9.56).[510]

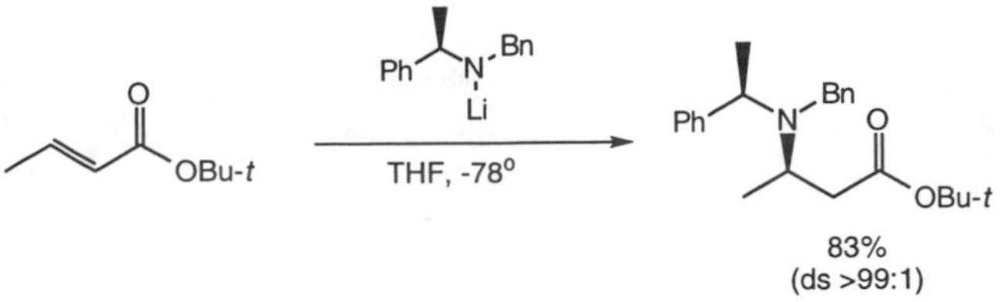

**SCHEME 9.54.**

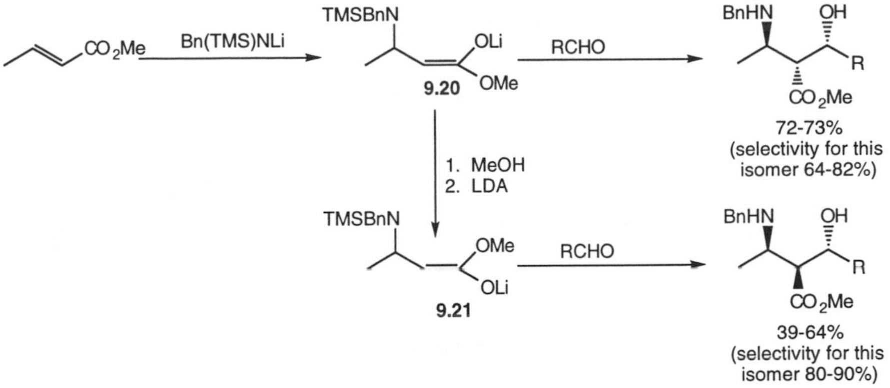

**SCHEME 9.55.**

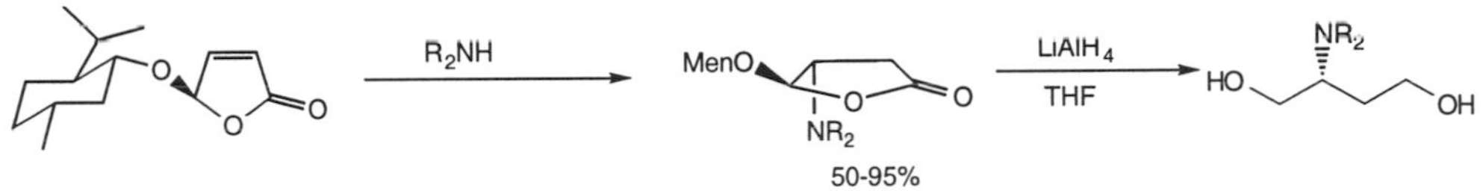

**SCHEME 9.56.**

## 9.5.2. OTHER SYSTEMS

As for α,β-unsaturated carbonyl systems, conjugate additions occur with systems where the resultant carbanion from the addition across carbon–carbon unsaturation is stabilized by a functional group that can also contain chirality. Examples of unsaturated systems that have been used in this context are based on the functional groups: amides,[373,511,512] aminals (Scheme 9.57),[513] enamines,[514-516] hydrazones,[391] imides,[517,518] imines,[517-523] nitro compounds,[524-537] oxazepines,[538,539] oxazinones,[380] oxazolines (Scheme 9.58),[540-546] α-silyl sulfones (Scheme 9.59),[547-554] sulfinimines,[340,555] sulfones,[556-558] and sulfoxides,[248,338,339,341,342,344,368,559-561] among others.[325,385,396]

The nucleophilic species can be carbon-based or a heteroatom with most of the Michael acceptors quoted (vide supra). The functional group on the alkene has to stabilize the resultant carbanion, and this can be used for subsequent chemistry. A chiral auxiliary can be an integral part of the acceptor (e.g., Schemes 9.57 and 9.58), or the functional group of the Michael acceptor can be used for further elaboration (Scheme 9.59). Consider the vinyl sulfur class of compounds. Additions of nucleophiles to these are well known, and they can show enantioselectivity as chirality can be incorporated into the heteroatom moiety. Indeed, selectivity is also observed for simple chiral sulfoxides and sulfoximines.[339-344] Conjugate addition to a vinyl sulfone in an asymmetric manner can be achieved in the presence of an adjacent chiral auxiliary, or asymmetric center (Scheme 9.60).[248,548,549,553]

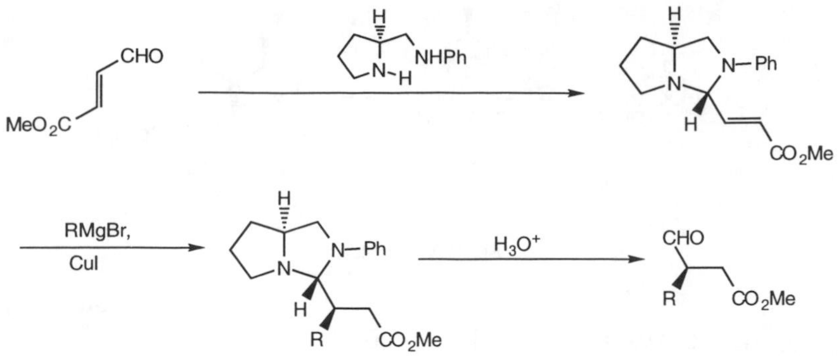

SCHEME 9.57.

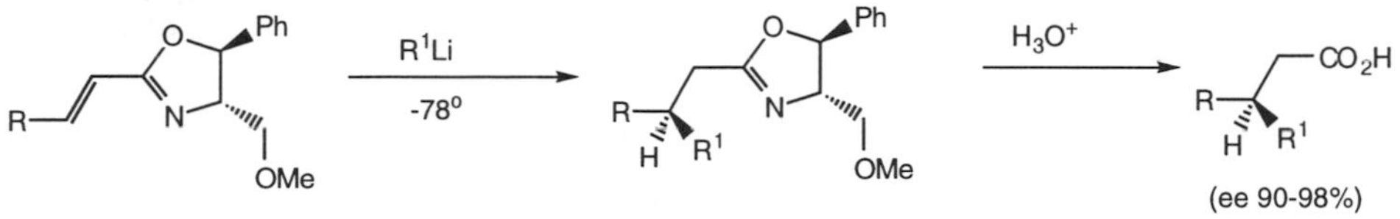

(ee 90-98%)

SCHEME 9.58.

(ds >100:1)

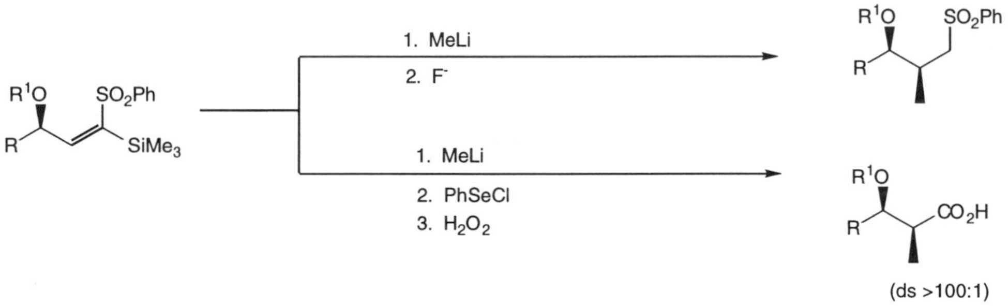

SCHEME 9.59.

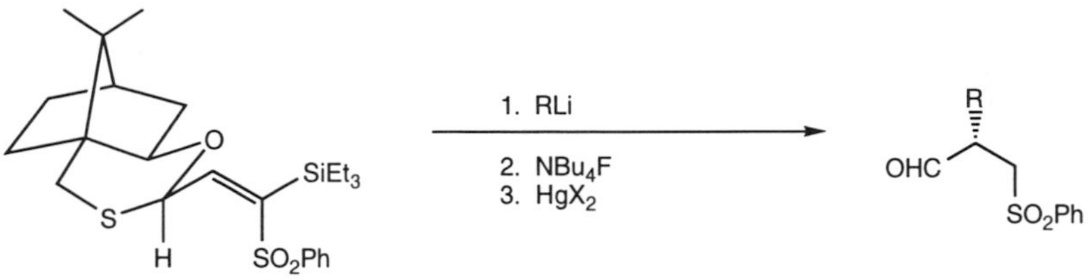

SCHEME 9.60.

## 9.6. ALLYL ALKYLATIONS

Allylic systems can undergo nucleophilic substitution by either an $S_N2$ or $S_N2'$ reaction.[562-565] The $S_N2$ reaction can compete effectively with an $S_N2'$ reaction. In some cases the mechanism cannot be clearly defined as either (*cf.* Scheme 3.8).[421,460,566-569] Regiochemistry can often be controlled, such as by the use of cuprates.[570-576] This methodology has been employed for the regioselective synthesis of allylsilanes; the urethane gives better selectivity than the corresponding acetate.[577] Use of carbamate as a chiral auxiliary does allow for some asymmetric induction (Scheme 9.61).[578]

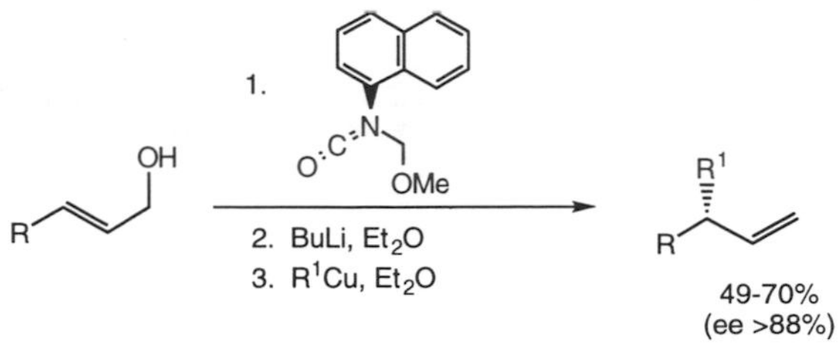

SCHEME 9.61.

The choice between molybdenum or palladium does allow for some control over the regiochemical outcome of an allyl alkylation.[579,580]

The use of a metal catalyst, such as palladium, also provides for some asymmetric induction when an allylic system is treated with a stabilized ester anion (Schemes 9.62 and 9.63),[580-598] or with other nucleophiles.[195,573,583,589,599-614] This approach also allows for the kinetic resolution of allyl acetates.[599,615]

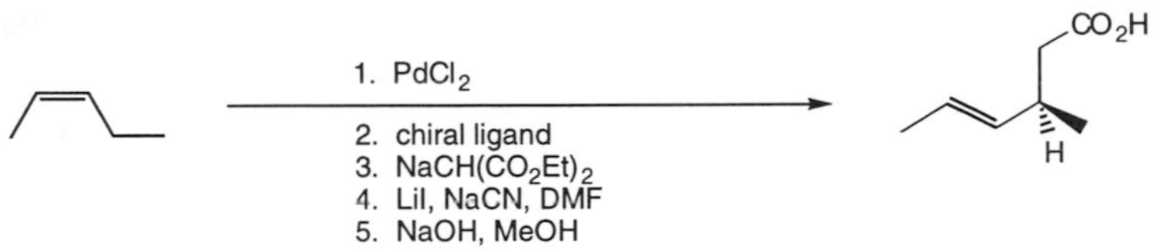

SCHEME 9.62.

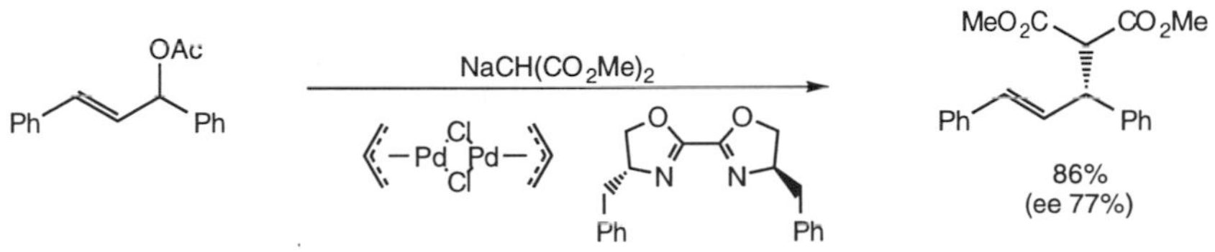

SCHEME 9.63.

The mechanism of allylic alkylation reactions has been the subject of significant investigation.[564,616] The two major approaches were based on the use of a cyclic substrate and of chiral substrates, and have given rise to the model shown in Scheme 9.64. The model also includes the general observations:

1. The allyl complex is generally formed with inversion of configuration.
2. If the leaving group is antiperiplanar to the metal, then formation of the π-allyl intermediate usually follows.[617-619]
3. Attack by a "soft" nucleophile leads to inversion of configuration in the second step.[620-629]
4. Attack by a "hard" nucleophile results in retention of configuration in the second step as the metal center is attacked first (pathway B).[630-633]

Palladium catalyzed allylation reactions proceed with net retention of configuration that is the result of two inversions.[582,600,620-622,634,635] Grignard reactions with nickel or copper catalysis proceed with net inversion.[630,636-638]

The methodology suffers from competition between *syn* and *anti* mechanisms in acyclic cases. Indeed, the pathway may depend on the nature of nucleophile and substrate.[562,639] In addition, there is no clear cut differentiation between "hard" and "soft" nucleophiles in the model described above.

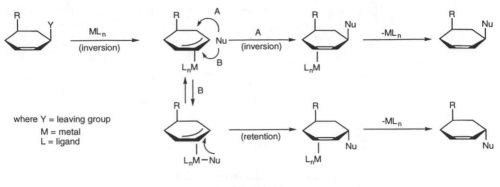

**SCHEME 9.64.**

Cuprates can provide selective $S_N2'$ reactions (Scheme 9.65).[572,575,640]



**SCHEME 9.65.**

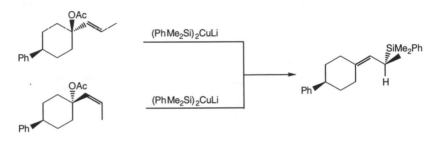

The presence of additional functionality on the allylic substrate can also enhance selectivity, as illustrated by the high diastereoselectivities observed with the reaction of the allyl chloride **9.22** with copper–zinc reagents (Scheme 9.66),[574] and the use of chiral carbamates (Scheme 9.67).[578]



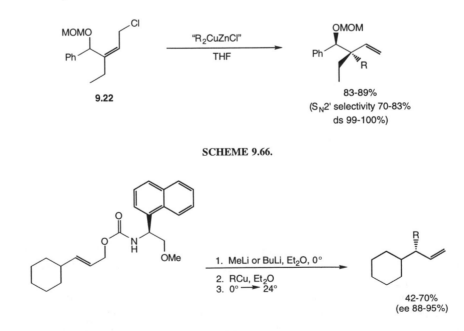

**SCHEME 9.66.**



**SCHEME 9.67.**

In addition to carbon nucleophiles, amines provide an entry to allyl amines by this methodology; chiral ferrocenylphosphine–palladium complexes provide reasonable asymmetric induction (Scheme 9.68).[608]

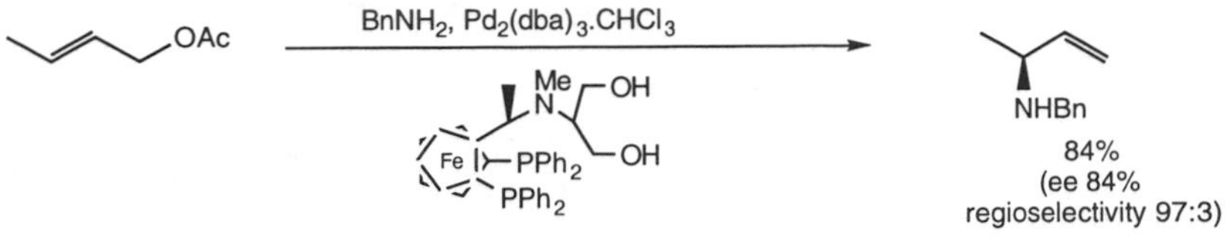

**SCHEME 9.68.**

Allylic systems can also be used as nucleophiles. As with electrophilic reactions, regioselectivity can be a problem with the ambidoselective allyl anions, in addition to the stereochemical integrity of the allyl anion itself.[641] The reactions of these systems with carbonyl electrophiles are discussed in Section 4.1.5, as they provide an alternative methodology to an aldol reaction. Although a wide variety of functionalized allyl anions are known, as are their reactions with a wide range of electrophiles, the regioselectivity problem has overwhelmed asymmetric method developments.[399] The methods that do overcome regioselectivity usually afford stereoselective approaches to alkenes [*8.1*].

An example of an asymmetric alkylation is provided by the chiral amino nitrile **9.23** (Scheme 9.69).[512]

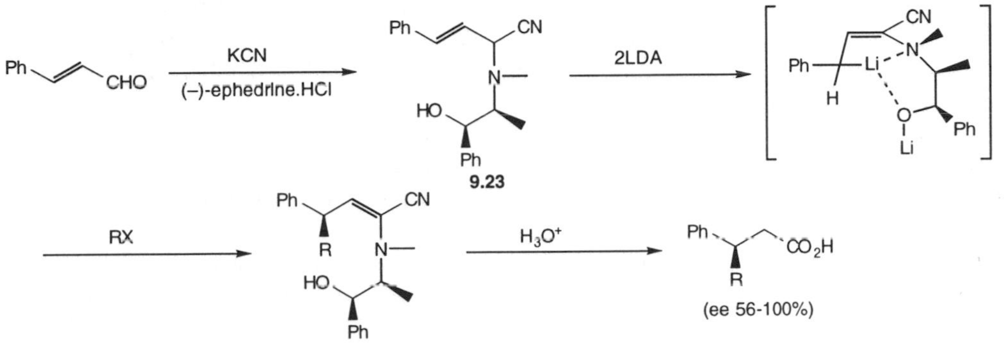

**SCHEME 9.69.**

## 9.7. OTHER REACTIONS

In addition to the general reactions cited above, others have proven useful for the preparation of 1,3-functionality. Homoallyl alcohols provide an entry into 1,3-diols [*13.2.1*] (Scheme 9.70);[642,643] the analogous reaction with allylamines provides 1,2-aminoalcohols as the four-membered cyclic intermediate is formed.[154]

The reaction of Scheme 9.70 can also be achieved by use of a carbonate in place of the phosphate esters.[642,644] The two hydroxy groups of a *meso*-1,3-diol can be differentiated by Lewis-acid promoted ring opening of the corresponding spiroketal derived from menthone.[645,646]

The pivotal role of glycerol has led to the development of many synthetic methods to control both relative and regiochemical problems associated with hydroxy group protection.[647]

Some of the reactions discussed above can be used to provide "remote" asymmetric induction. Such an example is the hydroboration of a diene where a cyclic intermediate can provide for high selectivity (*cf.* Schemes 9.27 and 9.71).[648-650]

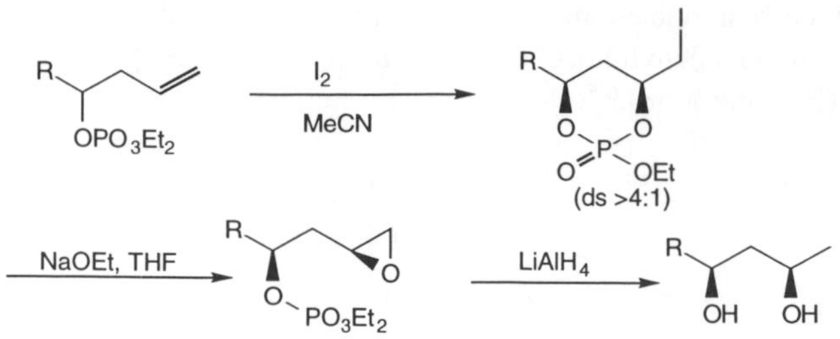

**SCHEME 9.70.**

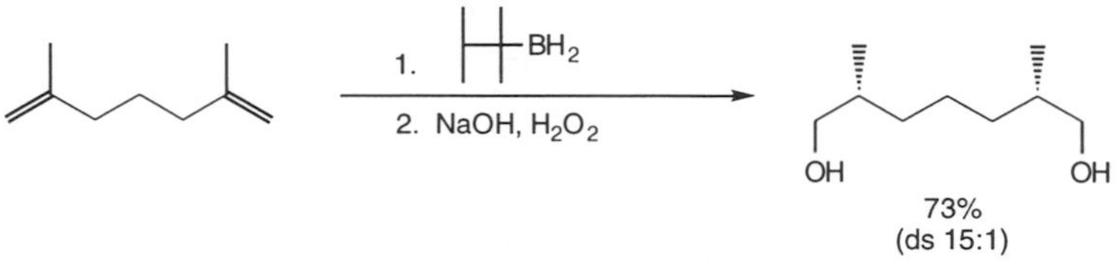

**SCHEME 9.71.**

## 9.8. SUMMARY

An array of almost unrelated reactions has been covered in this chapter; the common theme to connect them is the use of a functionalized alkene as one of the substrates.

Hydrogenation can provide high asymmetric induction with a number of allylic compounds. The reaction is simple to scale up and some of these reactions provide items of commerce. However, there is no single catalyst that can be relied upon to provide high chemical and stereochemical yields with a range of substrates. In addition, many of the catalysts are very sensitive to oxygen and require special precautions in their handling to avoid degradation. As the usage of the catalysts is usually low, asymmetric catalytic hydrogenation can provide a very economical approach to establish a stereogenic center.

A closely related reaction is the transition metal catalyzed rearrangement of an allylic system. Again, induction can be high, and economic considerations certainly allow for the large-scale usage of this methodology.

One other reaction that provides good asymmetric induction is hydrosilylation. As hydrogen is replaced by a silane, costs can increase, but induction can still be high.

Despite a wide array of chiral agents and ligands being used with hydride delivery reagents, only organoboranes provide general methodology by reaction with functionalized alkenes. As the resultant borane can undergo a wide variety of reactions, this methodology holds a plethora of opportunities.

Of the functionalized alkenes, allyl alcohols and conjugated enones are, perhaps, the best known as precursors for the establishment of new stereogenic centers. Silanes, through the oxidative conversion of the silyl group to hydroxy, as well as control of facial selectivity during reactions due to the comparative large size of the silyl group, provide a number of potential opportunities for asymmetric methodologies, although they have not been fully exploited.

Conjugate additions to unsaturated systems provide a simple means for the control of relative stereochemistry as *trans*-addition usually results, the second group being introduced by enolate reaction. The proposal of a model has allowed for some control in enantioselectivity during these reactions, and ligands continue to be developed to improve on ee's.

Allylic alkylations have a number of problems associated with the approach, including regiochemical control as well as face selectivity. Despite these, chiral ligands have been developed that allow for good asymmetry induction.

The chemistry described in this chapter has overcome a number of problems to allow for useful methodology; in the successful cases, the reaction conditions may only apply to a limited number of substrates. In the less successful methods, a number of products can result. In general, therefore, reactions of functionalized alkenes should not be a first choice unless strong literature precedence exists, particularly if a rapid, high yielding solution is sought. In contrast, the potential for elegant, inexpensive methodology also lies in the reactions of functionalized alkenes.

Some of the reactions described in this chapter are currently practiced on a large scale, such as hydrogenations and isomerizations. However, other methodologies that utilize low temperature or large amounts of heavy metals carry this baggage with them when considered for scale-up.

## 9.9. REFERENCES


1. Matteoli, U.; Frediani, P.; Bianchi, M.; Botteghi, C.; Gladiali, S. *J. Mol. Catal.* 1981, *12*, 265.
2. Izumi, Y. *Angew. Chem., Int. Ed. Engl.* 1971, *10*, 871.
3. Ojima, I.; Clos, N.; Bastos, C. *Tetrahedron* 1989, *45*, 6901.
4. Knowles, W. S. *Acc. Chem. Res.* 1983, *16*, 106.
5. Halpern, J. *Inorg. Chim. Acta* 1981, *50*, 11.
6. Brown, J. M. *Chem. Int. (London)* 1982, 737.
7. Noyori, R. *Chem. Soc. Rev.* 1989, *18*, 187.
8. Harmon, R. E.; Gupta, S. K.; Brown, D. J. *Chem. Rev.* 1973, *73*, 21.
9. ApSimon, J. W.; Collier, T. E. *Tetrahedron* 1986, *42*, 5157.
10. Bosnich, B. *Pure Appl. Chem.* 1990, *62*, 1131.
11. Halpern, J. In *Asymmetric Synthesis*; J. D. Morrison, Ed.; Academic Press: Orlando, FL, 1985; Vol. 5; pp 41.
12. Brunner, H. *Pure Appl. Chem.* 1990, *62*, 589.
13. Inoguchi, K.; Saruraba, S.; Achiwa, K. *Synlett* 1992, 169.
14. Valentine, D.; Scott, J. W. *Synthesis* 1978, 329.
15. Achiwa, K. *Tetrahedron Lett.* 1978, 1475.
16. Uyehara, T.; Asao, N.; Yamamoto, Y. *J. Chem. Soc., Chem. Commun.* 1989, 753.
17. Ashby, M. T.; Halpern, J. *J. Am. Chem. Soc.* 1991, *113*, 589.
18. Babievskii, K. K.; Latov, V. K. *Russ. Chem. Rev.* 1969, *38*, 456.
19. Knowles, W. S.; Sabacky, M. J.; Vineyard, B. D. *J. Chem. Soc., Chem. Commun.* 1972, 10.
20. Vineyard, B. D.; Knowles, W. S.; Sabacky, M. J.; Bachman, G. L.; Weinkauff, D. J. *J. Am. Chem. Soc.* 1977, *99*, 5946.
21. Landis, C. R.; Halpern, J. *J. Am. Chem. Soc.* 1987, *109*, 1746.
22. Marko, L.; Heil, B. *Catalysis Rev.* 1973, *8*, 269.
23. Chan, A. S. C.; Pluth, J. J.; Halpern, J. *J. Am. Chem. Soc.* 1980, *102*, 5952.
24. Chua, P. S.; Roberts, N. K.; Bosnich, B.; Okrasinski, S. J.; Halpern, J. *J. Chem. Soc., Chem. Commun.* 1981, 1278.
25. Brown, J. M. *Chem. Int. (London)* 1988, 612.
26. Koenig, K. E. In *Asymmetric Synthesis*; J. D. Morrison, Ed.; Academic Press: Orlando, FL, 1985; Vol. 5; pp 71.
27. Horner, L. *Pure Appl. Chem.* 1980, *52*, 843.
28. ApSimon, J. W.; Seguin, R. P. *Tetrahedron* 1979, *35*, 2797.
29. Selke, R.; Facklam, C.; Foken, H.; Heller, D. *Tetrahedron: Asymmetry* 1993, *4*, 369.
30. Fryzuk, M. D.; Bosnich, B. *J. Am. Chem. Soc.* 1977, *99*, 6262.
31. Bosnich, B.; Roberts, N. K. In *Catalytic Aspects of Metal Phosphine Complexes;* American Chemical Society: Washigton, D. C., 1982; pp 337.
32. Scott, J. W.; Keith, D. D.; Nix, G.; Parrish, D. R.; Remington, S.; Roth, G. P.; Townsend, J. M.; Valentine, D.; Yang, R. *J. Org. Chem.* 1981, *46*, 5086.

33. Hayashi, T.; Mise, T.; Kumada, M. *Tetrahedron Lett.* 1976, 4351.
34. Cesasotti, E.; Chiesa, A. *J. Organometal. Chem.* 1983, *251*, 79.
35. Stille, J. K.; Amma, J. P. *J. Org. Chem.* 1982, *47*, 468.
36. Brown, J. M.; Murrer, B. A. *J. Chem. Soc., Perkin Trans.* 2 1982, 489.
37. Oliver, J. D.; Riley, D. P. *Organometallics* 1983, *2*, 1032.
38. Riley, D. P.; Shumate, R. E. *J. Org. Chem.* 1980, *45*, 5187.
39. Bergstein, W.; Kleemann, A.; Martens, J. *Synthesis* 1981, 76.
40. Vineyard, B. D.; Knowles, W. S.; Sabacky, M. J. *J. Mol. Catal.* 1983, *19*, 159.
41. Hengartner, U.; Valentine, D.; Johnson, K. K.; Larscheid, M. E.; Pigott, F.; Scheidl, F.; Scott, J. W.; Sun, R. C.; Townsend, J. M.; Williams, T. H. *J. Org. Chem.* 1979, *44*, 3741.
42. Knowles, W. S.; Sabacky, M. J.; Vineyard, B. D.; Weinkauff, D. J. *J. Am. Chem. Soc.* 1975, *97*, 2567.
43. Kyba, E. P.; Davies, R. E.; Juri, P. N.; Shirley, K. R. *Inorg. Chem.* 1981, *20*, 3616.
44. Ball, R. G.; Payne, N. C. *Inorg. Chem.* 1977, *16*, 1187.
45. Onuma, K.; Ito, T.; Nakamura, A. *Bull. Chem. Soc., Jpn.* 1980, *53*, 2016.
46. Onuma, K.; Ito, T.; Nakamura, A. *Chem. Lett.* 1979, 905.
47. Hanaki, K.; Kashiwabara, K.; Fujita, J. *Chem. Lett.* 1978, 489.
48. Kashiwabara, K.; Hanaki, K.; Fujita, J. *Bull. Chem. Soc., Jpn.* 1980, *53*, 2275.
49. Achiwa, K. *J. Am. Chem. Soc.* 1976, *98*, 8265.
50. Ojima, I.; Kogure, T.; Yoda, N. *Chem. Lett.* 1979, 495.
51. Achiwa, K.; Chaloner, P.; Parker, D. *J. Organometal. Chem.* 1981, *218*, 249.
52. Ojima, I.; Yoda, N. *Tetrahedron Lett.* 1980, *21*, 1051.
53. Lafont, D.; Sinou, D.; Descotes, G. *J. Chem. Res. (S)* 1982, 117.
54. Miyashita, A.; Yasuda, A.; Takaya, H.; Toriumi, K.; Ito, T.; Souchi, T.; Noyori, R. *J. Am. Chem. Soc.* 1980, *102*, 7932.
55. Ikariya, T.; Ishii, Y.; Kawano, H.; Arai, T.; Saburi, M.; Yoshikawa, S.; Akutagawa, S. *J. Chem. Soc., Chem. Commun.* 1985, 922.
56. Miyashita, A.; Takaya, H.; Souchi, T.; Noyori, R. *Tetrahedron* 1984, *40*, 1245.
57. Glaser, R.; Geresh, S. *Tetrahedron* 1979, *35*, 2381.
58. Gelbard, G.; Kagan, H. B.; Stern, R. *Tetrahedron* 1976, *32*, 233.
59. Sinou, D.; Kagan, H. B. *J. Organometal. Chem.* 1976, *114*, 325.
60. Kagan, H. B.; Dang, T.-P. *J. Am. Chem. Soc.* 1972, *94*, 6429.
61. Detellier, C.; Gelbard, G.; Kagan, H. B. *J. Am. Chem. Soc.* 1978, *100*, 7556.
62. Glaser, R.; Geresh, S.; Twaik, M. *Isr. J. Chem.* 1980, *20*, 102.
63. Glaser, R.; Geresh, S.; Blumenfeld, J.; Vainas, B.; Twaik, M. *Isr. J. Chem.* 1976, *15*, 17.
64. Dang, T.-P.; Poulin, J.-C.; Kagan, H. B. *J. Organometal. Chem.* 1975, *91*, 105.
65. Inoguchi, K.; Achiwa, K. *Synlett* 1991, 49.
66. Yamada, M.; Yamashita, M.; Inokawa, S. *CarRes* 1981, *95*, C9.
67. Terfort, A. *Synthesis* 1992, 951.
68. Yamashita, H., K.; Yamada, M.; Suzuki, N.; Inokawa, S. *Bull. Chem. Soc., Jpn.* 1982, *55*, 2917.
69. Ojima, I.; Kogure, T.; Achiwa, J. *Chem. Lett.* 1978, 567.
70. Achiwa, K. *Tetrahedron Lett.* 1978, 2583.
71. Brunner, H.; Graf, E.; Leitner, W.; Wutz, K. *Synthesis* 1989, 743.
72. Kawano, H.; Ishii, Y.; Ikariya, T.; Saburi, M.; Yoshikawa, S.; Uchida, Y.; Kumobayashi, H. *Tetrahedron Lett.* 1987, *28*, 1905.
73. Brown, J. M.; Brunner, H.; Leitner, W.; Rose, M. *Tetrahedron: Asymmetry* 1991, *2*, 331.
74. Shao, L.; Miyata, S.; Muramatsu, H.; Kawano, H.; Ishii, Y.; Saburi, M.; Uchida, Y. *J. Chem. Soc., Perkin Trans. 1* 1990, 1441.
75. Brown, J. M.; James, A. P. *J. Chem. Soc., Chem. Commun.* 1987, 181.
76. Christopfel, W. C.; Vineyard, B. D. *J. Am. Chem. Soc.* 1979, *101*, 4406.
77. Morimoto, T.; Chiba, M.; Achiwa, K. *Tetrahedron Lett.* 1989, *30*, 735.
78. Morimoto, T.; Chiba, M.; Achiwa, K. *Tetrahedron* 1993, *49*, 1793.
79. Brown, J. M.; James, A. P.; Prior, L. M. *Tetrahedron Lett.* 1987, *28*, 2179.
80. Brown, J. M.; Cutting, I. *J. Chem. Soc., Chem. Commun.* 1985, 578.
81. Koenig, K. E.; Bachman, G. L.; Vineyard, B. D. *J. Org. Chem.* 1980, *45*, 2362.
82. Kagan, H. B.; Langloirs, N.; Dang, T. P. *J. Organometal. Chem.* 1975, *96*, 353.
83. Koenig, K. E. In *Catalysis of Organic Reactions*; J. R. Kosak, Ed.; Marcel Dekker: New York, 1984; pp 63.
84. Koenig, K. E.; Sabacky, M. J.; Bachman, G. L.; Christopfel, W. C.; Barnstorff, H. D.; Friedman, R. B.; Knowles, W. S.; Stults, B. R.; Vineyard, B. D.; Weinkauff, D. J. *Ann. N.Y. Acad. Sci.* 1980, *333*, 16.
85. Lafont, D.; Sinou, D.; Descotes, G.; Glaser, R.; Geresh, S. *J. Mol. Catal.* 1981, *10*, 305.
86. Glaser, R.; Vainas, B. *J. Organometal. Chem.* 1976, *121*, 249.
87. Kreuzfeld, H.-J.; Döbler, C.; Krause, H. W. *Tetrahedron: Asymmetry* 1993, *4*, 2047.

88. Glaser, R.; Blumenfeld, J.; Twaik, M. *Tetrahedron Lett.* 1977, 4639.
89. Chan, A. S. C.; Pluth, J. J.; Halpern, J. *Inorg. Chim. Acta* 1979, *37*, 1477.
90. Brown, J. M.; Chaloner, P. A.; Glaser, R.; Geresh, S. *Tetrahedron* 1980, *36*, 815.
91. Brown, J. M.; Parker, D. *J. Org. Chem.* 1982, *47*, 2722.
92. Noyori, R.; Ohta, M.; Hsiao, Y.; Kitamura, M.; Ohta, T.; Takaya, H. *J. Am. Chem. Soc.* 1986, *108*, 7117.
93. Kitamura, M.; Noyori, R.; Takaya, H. *Tetrahedron Lett.* 1987, *28*, 4829.
94. Hayashi, T.; Kawamura, N.; Ito, Y. *Tetrahedron Lett.* 1988, *29*, 5969.
95. Ohta, T.; Miyake, T.; Seido, N.; Kumobayashi, H.; Akutagawa, S.; Takaya, H. *Tetrahedron Lett.* 1992, *33*, 635.
96. Genet, J. P.; Mallart, S.; Pinel, C.; Juge, S.; Laffitte, J. A. *Tetrahedron: Asymmetry* 1991, *2*, 43.
97. Ohta, T.; Takaya, H.; Kitamura, M.; Nagai, K.; Noyori, R. *J. Org. Chem.* 1987, *52*, 3174.
98. Takaya, H.; Ohta, T.; Mashima, K.; Noyori, R. *Pure Appl. Chem.* 1990, *62*, 1135.
99. Ohta, T.; Takaya, H.; Noyori, R. *Tetrahedron Lett.* 1990, *31*, 7189.
100. Achiwa, K.; Soga, T. *Tetrahedron Lett.* 1978, 1119.
101. Lubell, W. D.; Kitamura, M.; Noyori, R. *Tetrahedron: Asymmetry* 1991, *2*, 543.
102. Zhang, X.; Mashima, K.; Koyana, K.; Sayo, N.; Kumobayashi, H.; Akutagawa, S.; Takaya, H. *Tetrahedron Lett.* 1991, *32*, 7283.
103. Faller, J. W.; Tokunga, M. *Tetrahedron Lett.* 1993, *34*, 7359.
104. Faller, J. W.; Parr, J. *J. Am. Chem. Soc.* 1993, *115*, 804.
105. Reetz, M. T.; Kayser, F. *Tetrahedron: Asymmetry* 1992, *3*, 1377.
106. Cativiela, C.; Diaz-de-Villegas, M. D.; Galvez, J. A. *Tetrahedron: Asymmetry* 1992, *3*, 567.
107. Oppolzer, W.; Mills, R. J.; Reglier, M. *Tetrahedron Lett.* 1986, *27*, 183.
108. Poisel, H.; Schmidt, U. *Chem. Ber.* 1973, *106*, 3408.
109. Caplar, V.; Lisini, A.; Kajfez, F.; Kolbah, D.; Sunjic, V. *J. Org. Chem.* 1978, *43*, 1355.
110. Bycroft, B. W.; Lee, G. R. *J. Chem. Soc., Chem. Commun.* 1975, 988.
111. Vigneron, J. P.; Kagan, H.; Horeau, A. *Tetrahedron Lett.* 1968, 5681.
112. Izumiya, N.; Lee, S.; Kanmera, T.; Aoyagi, H. *J. Am. Chem. Soc.* 1977, *99*, 8346.
113. Barrett, A. G. M.; Sheth, H. G. *J. Chem. Soc., Chem. Commun.* 1982, 170.
114. Arco, M. J.; Trammell, M. H.; White, J. D. *J. Org. Chem.* 1976, *41*, 2075.
115. Gerlach, H.; Wetter, H. *Helv. Chim. Acta* 1974, *57*, 2306.
116. Johnstone, R. A. W.; Wilby, A. H.; Entwistle, I. D. *Chem. Rev.* 1985, *85*, 129.
117. Zassinovich, G.; Mestroni, G.; Gladiali, S. *Chem. Rev.* 1992, *92*, 1051.
118. Takaya, H.; Ohta, T.; Sayo, N.; Kumobayashi, H.; Akutagawa, S.; Inoue, S. I.; Kasahara, I.; Noyori, R. *J. Am. Chem. Soc.* 1987, *109*, 1596.
119. Takaya, H.; Ohta, T.; Sayo, N.; Kumobayashi, H.; Akutagawa, S.; Inoue, S.-I.; Kasahara, I.; Noyori, R. *J. Am. Chem. Soc.* 1987, *109*, 4129.
120. Heiser, B.; Broger, E. A.; Crameri, Y. *Tetrahedron: Asymmetry* 1991, *2*, 51.
121. Kitamura, M.; Kasahara, K.; Noyori, R.; Takaya, H. *J. Org. Chem.* 1988, *53*, 708.
122. Imperiali, B.; Zimmerman, J. W. *Tetrahedron Lett.* 1988, *29*, 5343.
123. Akutagawa, S. In *Chirality in Industry*; A. N. Collins; G. N. Sheldrake and J. Crosby, Eds.; John Wiley & Sons: New York, 1992; pp 325.
124. Inoue, S.; Osada, M.; Koyano, K.; Takaya, H.; Noyori, R. *Chem. Lett.* 1985, 1007.
125. Evans, D. A.; Morrissey, M. M. *Tetrahedron Lett.* 1984, *25*, 4637.
126. Evans, D. A.; Morrissey, M. M. *J. Am. Chem. Soc.* 1984, *106*, 3866.
127. Lautens, M.; Zhang, C.-H.; Crudden, C. M. *Angew. Chem., Int. Ed. Engl.* 1992, *31*, 232.
128. Stork, G.; Kahne, D. E. *J. Am. Chem. Soc.* 1983, *105*, 1072.
129. Crabtree, R. H.; Davis, M. W. *Organometallics* 1983, *2*, 681.
130. Machado, A. S.; Oleseker, A.; Castillion, S.; Lukacs, G. *J. Chem. Soc., Chem. Commun.* 1985, 330.
131. Inoue, S.-I.; Takaya, H.; Tani, K.; Otsuka, S.; Sato, T.; Noyori, R. *J. Am. Chem. Soc.* 1990, *112*, 4897.
132. Noyori, R. *Science* 1990, *248*, 1194.
133. Tani, K.; Yamagata, T.; Akutagawa, S.; Kumolayashi, H.; Taketomi, T.; Takaya, H.; Mujashita, A.; Noyori, R.; Otsuki, S. *J. Am. Chem. Soc.* 1984, *106*, 5208.
134. Rosini, C.; Franzini, L.; Raffaelli, A.; Salvadori, P. *Synlett* 1992, 503.
135. Otsuka, S.; Tani, K. In *Asymmetric Synthesis*; J. D. Morrison, Ed.; Academic Press: Orlando, FL, 1985; Vol. 5; pp 171.
136. Sato, S.; Matsuda, I.; Izumi, Y. *Tetrahedron Lett.* 1983, *24*, 3855.
137. Sato, S.; Okada, H.; Matsuda, I.; Izumi, Y. *Tetrahedron Lett.* 1984, *25*, 769.
138. Bäckvall, J.-E.; Andreasson, U. *Tetrahedron Lett.* 1993, *34*, 5459.
139. Kitamura, M.; Manabe, K.; Noyori, R.; Takaya, H. *Tetrahedron Lett.* 1987, *28*, 4719.
140. Ojima, I.; Kumagai, M.; Nagai, Y. *J. Organometal. Chem.* 1976, *111*, 43.
141. Tour, J. M.; Cooper, J. P.; Pendalwar, S. L. *J. Org. Chem.* 1990, *55*, 3452.

142. Ojima, I.; Kogure, T. *Organometallics* 1982, *1*, 1390.
143. Kogure, T.; Nagai, Y. *Tetrahedron Lett.* 1972, 5035.
144. Kagan, H. B. *Pure Appl. Chem.* 1975, *43*, 401.
145. Ojima, I.; Nihonyanagi, M.; Kogure, T.; Kumagai, M.; Horiuchi, S.; Nakatsugawa, K. *J. Organometal. Chem.* 1975, *94*, 449.
146. Hayashi, T.; Yamamoto, K.; Kumada, M. *Tetrahedron Lett.* 1975, 3.
147. Kobayashi, M.; Koyama, T.; Ogura, K.; Seto, S.; Ritter, F. J.; Brüggemann-Rotgans, I. E. M. *J. Am. Chem. Soc.* 1980, *102*, 6602.
148. Brunner, H. *Angew. Chem., Int. Ed. Engl.* 1983, *22*, 897.
149. Kogure, T.; Ojima, I. *J. Organometal. Chem.* 1982, *234*, 249.
150. Keinan, E. *Pure Appl. Chem.* 1989, *61*, 1737.
151. Tamao, K.; Nakagawa, Y.; Arai, H.; Higuchi, N.; Ito, Y. *J. Am. Chem. Soc.* 1988, *110*, 3712.
152. Tamao, K.; Tohma, T.; Inui, N.; Nakayama, O.; Ito, Y. *Tetrahedron Lett.* 1990, *31*, 7333.
153. Tamao, K.; Nakajima, T.; Sumiya, R.; Arai, H.; Higuchi, N.; Ito, Y. *J. Am. Chem. Soc.* 1986, *108*, 6090.
154. Tamao, K.; Nakagawa, Y.; Ito, Y. *J. Org. Chem.* 1990, *55*, 3438.
155. Hayashi, T.; Matsumoto, Y.; Ito, Y. *J. Am. Chem. Soc.* 1988, *110*, 5579.
156. Tayashi, T.; Matsumoto, Y.; Morikawa, I.; Ito, Y. *Tetrahedron: Asymmetry* 1990, *1*, 151.
157. Keinan, E.; Greenspoon, N. *J. Org. Chem.* 1983, *48*, 3545.
158. Keinan, E.; Greenspoon, N. *Isr. J. Chem.* 1984, *24*, 82.
159. Greenspoon, N.; Keinan, E. *J. Org. Chem.* 1988, *53*, 3723.
160. Keinan, E.; Greenspoon, N. *Tetrahedron Lett.* 1982, *23*, 241.
161. Ojima, I.; Hirai, K. In *Asymmetric Synthesis*; J. D. Morrison, Ed.; Academic Press: Orlando, FL, 1985; Vol. 5; pp 103.
162. Pino, P.; Piacenti, F.; Bianchi, M. In *Organic Syntheses via Metal Carbonyls*; I. Wender and P. Pino, Eds.; Wiley Interscience: New York, 1977; Vol. 2; pp 43.
163. Falbe, J. In *New Syntheses with Carbon Monoxide;* Springer-Verlag: Berlin, 1980; pp 1.
164. Falbe, J. In *New Syntheses with Carbon Monoxide;* Springer-Verlag: Berlin, 1980; pp 243.
165. Doyle, M. M.; Jackson, W. R.; Perlmutter, P. *Tetrahedron Lett.* 1989, *30*, 233.
166. Murai, S.; Sonoda, N. *Angew. Chem., Int. Ed. Engl.* 1979, *18*, 837.
167. Pittman, C. U.; Kawabata, Y.; Flowers, L. I. *J. Chem. Soc., Chem. Commun.* 1982, 473.
168. Alper, H. *Aldrichimica Acta* 1991, *24*, 3.
169. Hobbs, C. F.; Knowles, W. S. *J. Org. Chem.* 1981, *46*, 4422.
170. Becker, Y.; Eisenstadt, A.; Stille, J. K. *J. Org. Chem.* 1980, *45*, 2145.
171. Tanaka, M.; Watanabe, Y.; Mitsudo, T.; Yamamoto, K.; Takegami, Y. *Chem. Lett.* 1972, 483.
172. Botteghi, C. *Gazz. Chim. Ital.* 1975, *105*, 233.
173. Botteghi, C.; Branca, M.; Saba, A. *J. Organometal. Chem.* 1980, *184*, C17.
174. Jackson, W. R.; Perlmutter, P.; Tasdelen, E. E. *Tetrahedron Lett.* 1990, *31*, 2461.
175. Alper, H.; Hamel, N. *J. Chem. Soc., Chem. Commun.* 1990, 135.
176. Alper, H.; Hamel, N. *J. Am. Chem. Soc.* 1990, *112*, 2803.
177. Alper, H.; Leonard, D. *J. Chem. Soc., Chem. Commun.* 1985, 511.
178. Alper, H.; Leonard, N. *Tetrahedron Lett.* 1985, *26*, 5639.
179. Marshall, J. A.; Carroll, R. D. *J. Org. Chem.* 1965, *30*, 2748.
180. Loupy, A.; Seyden-Penne, J. *Tetrahedron* 1980, *36*, 1937.
181. Suzuki, T.; Saimoto, H.; Tomioka, H.; Oshima, K.; Nozaki, H. *Tetrahedron Lett.* 1982, *23*, 3597.
182. Brown, H. C.; Krishnamurthy, S. *Tetrahedron* 1979, *35*, 567.
183. Midland, M. M. *Chem. Rev.* 1989, *89*, 1553.
184. Burgess, K.; Ohlmeyer, M. J. *Chem. Rev.* 1991, *91*, 1179.
185. Johnson, M. R.; Nakata, T.; Kishi, Y. *Tetrahedron Lett.* 1979, 4343.
186. Still, W. C.; Barrish, J. C. *J. Am. Chem. Soc.* 1983, *105*, 2487.
187. Kishi, Y. *Pure Appl. Chem.* 1981, *53*, 1163.
188. Paddon-Row, M. N.; Rondan, N. G.; Houk, K. N. *J. Am. Chem. Soc.* 1982, *104*, 7162.
189. Houk, K. N.; Rondan, N. G.; Wu, Y.-D.; Metz, J. T.; Padden-Row, M. N. *Tetrahedron* 1984, *40*, 2257.
190. Burgess, K.; Cassidy, J.; Ohlmeyer, M. J. *J. Org. Chem.* 1991, *56*, 1020.
191. Evans, D. A.; Fu, G. C.; Hoveyda, A. H. *J. Am. Chem. Soc.* 1988, *110*, 6917.
192. Evans, D. A.; Fu, G. C. *J. Org. Chem.* 1990, *55*, 2280.
193. Harada, T.; Matsuda, U., J.; Oku, A. *J. Chem. Soc., Chem. Commun.* 1989, 1429.
194. Burgess, K.; Ohlmeyer, M. J. *Tetrahedron Lett.* 1989, *30*, 395.
195. Trost, B. M.; Keinan, E. *Tetrahedron Lett.* 1980, *21*, 2591.
196. Burgess, K.; Ohlmeyer, M. J. *Tetrahedron Lett.* 1989, *30*, 5857.
197. Burgess, K.; Ohlmeyer, M. J. *Tetrahedron Lett.* 1989, *30*, 5861.
198. Mori, K.; *Tetrahedron* 1977, *33*, 289.

199. Schmid, G. S.; Fukuyama, T.; Akasaka, K.; Kishi, Y. *J. Am. Chem. Soc.* 1979, *101*, 259.
200. Fleming, I.; Lawrence, N. J. *Tetrahedron Lett.* 1988, *29*, 2073.
201. Burgess, K.; Ohlmeyer, M. J. *J. Org. Chem.* 1991, *56*, 1027.
202. Gurjar, M. K. *Pure Appl. Chem.* 1990, *62*, 1293.
203. Fisher, G. B.; Goralski, C. T.; Nicholson, L. W.; Singaram, B. *Tetrahedron Lett.* 1993, *34*, 7693.
204. Fleming, I. *Pure Appl. Chem.* 1988, *60*, 71.
205. Corey, E. J. *Pure Appl. Chem.* 1990, *62*, 1209.
206. Schinzer, D. *Synthesis* 1988, 263.
207. Fleming, I. *Pure Appl. Chem.* 1990, *62*, 1879.
208. Herndon, J. W.; Wu, C. *Tetrahedron Lett.* 1989, *30*, 6461.
209. Tamao, K.; Iwahara, T.; Kanatani, R.; Kumada, M. *Tetrahedron Lett.* 1984, *25*, 1909.
210. Fleming, I.; Sanderson, P. E. J. *Tetrahedron Lett.* 1987, *28*, 4229.
211. Fleming, I.; Henning, R.; Plaut, H. *J. Chem. Soc., Chem. Commun.* 1984, 29.
212. Tamao, K.; Hayashi, T.; Ito, Y. *Tetrahedron Lett.* 1989, *30*, 6533.
213. Sato, K.; Kira, M.; Sakurai, H. *Tetrahedron Lett.* 1989, *30*, 4375.
214. Boons, G. J. P. H.; Overhand, M.; van der Marel, G. A.; van Boom, J. H. *Angew. Chem., Int. Ed. Engl.* 1989, *28*, 1504.
215. Fleming, I.; Winter, S. B. D. *Tetrahedron Lett.* 1993, *34*, 7287.
216. Hosomi, A. *Acc. Chem. Res.* 1988, *21*, 200.
217. Fleming, I. *Chem. Soc. Rev.* 1981, *10*, 83.
218. Chan, T. H.; Fleming, I. *Synthesis* 1979, 761.
219. Westerlund, C. *Tetrahedron Lett.* 1982, *23*, 4835.
220. Sakurai, H. *Pure Appl. Chem.* 1982, *54*, 1.
221. Kahn, S. D.; Pau, C. F.; Hehre, W. J. *J. Am. Chem. Soc.* 1986, *108*, 7396.
222. Imai, T.; Nishida, S. *J. Org. Chem.* 1990, *55*, 4849.
223. Wilson, S. R.; Price, M. F. *Tetrahedron Lett.* 1983, *24*, 569.
224. Wilson, S. R.; Price, M. F. *J. Am. Chem. Soc.* 1982, *104*, 1124.
225. Lamothe, S.; Chan, T. H. *Tetrahedron Lett.* 1991, *32*, 1847.
226. Nativi, C.; Ravida, N.; Ricci, A.; Seconi, G.; Taddei, M. *J. Org. Chem.* 1991, *56*, 1951.
227. Hayashi, T.; Konishi, M.; Kumada, M. *J. Org. Chem.* 1983, *48*, 281.
228. Hayashi, T.; Konishi, M.; Kumada, M. *J. Am. Chem. Soc.* 1982, *104*, 4963.
229. Hayashi, T.; Konichi, M.; Ito, H.; Kumada, M. *J. Am. Chem. Soc.* 1982, *104*, 4962.
230. Hayashi, T.; Ito, H.; Kumada, M. *Tetrahedron Lett.* 1982, *23*, 4605.
231. Fleming, I.; Lawrence, N. J. *Tetrahedron Lett.* 1988, *29*, 2077.
232. Roush, W. R.; Grover, P. T.; Lin, X. *Tetrahedron Lett.* 1990, *31*, 7563.
233. Fleming, I.; Sarkar, A. K.; Thomas, A. P. *J. Chem. Soc., Chem. Commun.* 1987, 157.
234. Nativi, C.; Palio, G.; Taddei, M. *Tetrahedron Lett.* 1991, *32*, 1583.
235. Soderquist, J. A.; Rane, A. M.; López, C. J. *Tetrahedron Lett.* 1993, *34*, 1893.
236. Okamoto, S.; Tani, K.; Sato, F.; Sharpless, K. B.; Zargarian, D. *Tetrahedron Lett.* 1993, *34*, 2509.
237. Cirillo, P. F.; Panek, J. S. *J. Org. Chem.* 1990, *55*, 6071.
238. Panek, J. S.; Cirillo, P. F. *J. Am. Chem. Soc.* 1990, *112*, 4873.
239. McGarvey, G. J.; Williams, J. M. *J. Am. Chem. Soc.* 1985, *107*, 1435.
240. Oppolzer, W.; Mills, R. J.; Pachinger, W.; Stevenson, T. *Helv. Chim. Acta* 1986, *69*, 1542.
241. Fleming, I.; Hill, J. H. M.; Parker, D.; Waterson, D. *J. Chem. Soc., Chem. Commun.* 1985, 318.
242. Yamamoto, Y.; Maryama, J. *J. Chem. Soc., Chem. Commun.* 1984, 904.
243. Bernhard, W.; Fleming, I.; Waterson, D. *J. Chem. Soc., Chem. Commun.* 1984, 28.
244. Engel, W.; Fleming, I.; H., S. R. *J. Chem. Soc., Perkin Trans. 1* 1986, 1637.
245. Fleming, I.; Kilburn, J. D. *J. Chem. Soc., Chem. Commun.* 1986, 1198.
246. Fleming, I.; Kilburn, J. D. *J. Chem. Soc., Chem. Commun.* 1986, 305.
247. Fleming, I.; Sarkar, A. K. *J. Chem. Soc., Chem. Commun.* 1986, 1199.
248. Takaki, K.; Maeda, T.; Ishikawa, M. *J. Org. Chem.* 1989, *54*, 58.
249. Russell, A. T.; Procter, G. *Tetrahedron Lett.* 1987, *28*, 2041.
250. Hartley, R. C.; Lamothe, S.; Chan, T. H. *Tetrahedron Lett.* 1993, *34*, 1449.
251. Torres, E.; Larson, G. L.; McGarvey, G. J. *Tetrahedron Lett.* 1988, *29*, 1355.
252. Rao, A. V. R.; Reddy, E. R. *Tetrahedron Lett.* 1986, *27*, 2279.
253. Corey, E. J.; Katzenellenbogen, J. A.; Posner, G. H. *J. Am. Chem. Soc.* 1967, *89*, 4245.
254. Denmark, S. E.; Jones, T. K. *J. Org. Chem.* 1982, *47*, 4595.
255. Cowell, A.; Stille, J. K. *Tetrahedron Lett.* 1979, 133.
256. Kim, K. D.; Magriotis, P. A. *Tetrahedron Lett.* 1990, *31*, 6137.
257. Sato, F.; Tomuro, Y.; Ishikawa, H.; Sato, M. *Chem. Lett.* 1980, 99.
258. Corey, E. J.; Achiwa, K.; Katzenellenbogen, J. A. *J. Am. Chem. Soc.* 1969, *91*, 4318.

259. Takai, K.; Kataoka, Y.; Utimoto, K. *J. Org. Chem.* 1990, *55*, 1707.
260. Ito, T.; Okamoto, S.; Sato, F. *Tetrahedron Lett.* 1989, *30*, 7083.
261. Ito, T.; Okamoto, S.; Sato, F. *Tetrahedron Lett.* 1990, *31*, 6399.
262. Ito, T.; Yamakawa, I.; Okamoto, S.; Kobayashi, Y.; Sato, F. *Tetrahedron Lett.* 1991, *32*, 371.
263. Tomioka, K.; Koga, K. In *Asymmetric Synthesis*; J. D. Morrison, Ed.; Academic Press: Orlando, FL, 1983; Vol. 2; pp 201.
264. Stork, G.; Logusch, E. W. *Tetrahedron Lett.* 1979, 3361.
265. Logusch, E. W. *Tetrahedron Lett.* 1979, 3365.
266. Anh, N. T. *Top. Curr. Chem.* 1980, *88*, 145.
267. Posner, G. H. *Org. Reactions* 1972, *19*, 1.
268. House, H. O. *Acc. Chem. Res.* 1976, *9*, 959.
269. Ullenius, C.; Christenson, B. *Pure Appl. Chem.* 1988, *60*, 57.
270. House, H. O.; Fischer, W. F. *J. Org. Chem.* 1968, *33*, 949.
271. House, H. O.; Respess, W. L.; Whitesides, G. M. *J. Org. Chem.* 1966, *31*, 3128.
272. Schwartz, J.; Loots, M. J.; Kosugi, H. *J. Am. Chem. Soc.* 1980, *102*, 1333.
273. Fenzl, W.; Koster, R.; Zimmerman, H.-J. *Annalen* 1975, 2201.
274. Pasto, D. J.; Wojtkowski, P. W. *Tetrahedron Lett.* 1970, 215.
275. Suzuki, A.; Arase, A.; Matsumoto, H.; Itoh, M.; Brown, H. C.; Rogic, M. M.; Rathke, M. W. *J. Am. Chem. Soc.* 1967, *89*, 5708.
276. Soai, K.; Hayashi, T.; Ugajin, S. *J. Chem. Soc., Chem. Commun.* 1989, 516.
277. Cahiez, G.; Alami, M. *Tetrahedron Lett.* 1989, *30*, 7365.
278. Cahiez, G.; Alami, M. *Tetrahedron Lett.* 1990, *31*, 7423.
279. Cahiez, G.; Alami, M. *Tetrahedron Lett.* 1990, *31*, 7425.
280. Myers, M. R.; Cohen, T. *J. Org. Chem.* 1989, *54*, 1290.
281. Ager, D. J.; East, M. B. *J. Org. Chem.* 1986, *51*, 3983.
282. Kreiser, W.; Below, P. *Tetrahedron Lett.* 1981, *22*, 429.
283. Berrada, S.; Metzner, P. *Tetrahedron Lett.* 1987, *28*, 409.
284. Yamaguchi, M.; Hamada, M.; Nakashima, H.; Minami, T. *Tetrahedron Lett.* 1987, *28*, 1785.
285. Yamaguchi, M.; Tsukamoto, Y.; Hayashi, A.; Minami, T. *Tetrahedron Lett.* 1990, *31*, 2423.
286. Kpegba, K.; Metzner, P. *Tetrahedron Lett.* 1990, *31*, 1853.
287. Hirama, M. *Tetrahedron Lett.* 1981, *22*, 1905.
288. Oare, D. A.; Heathcock, C. H. *Tetrahedron Lett.* 1986, *27*, 6169.
289. Tomioka, K.; Koga, K. *Tetrahedron Lett.* 1984, *25*, 1599.
290. Jansen, J. F. G. A.; Feringa, B. L. *Tetrahedron Lett.* 1989, *30*, 5481.
291. Seebach, D.; Bürstinghaus, R. *Angew. Chem., Int. Ed. Engl.* 1975, *14*, 57.
292. Bürstinghaus, R.; Seebach, D. *Chem. Ber.* 1977, *110*, 841.
293. Stork, G.; Rosen, P.; Goldman, N.; Coombs, R. V.; Tsuji, J. *J. Am. Chem. Soc.* 1965, *87*, 275.
294. Binkley, E. S.; Heathcock, C. H. *J. Org. Chem.* 1975, *40*, 2156.
295. Jackman, L. M.; Lange, B. C. *Tetrahedron* 1977, *33*, 2737.
296. Posner, G. H.; Sterling, J. J.; Whitten, C. E.; Lentz, C. M.; Brunelle, D. J. *J. Am. Chem. Soc.* 1975, *97*, 107.
297. Gariboldi, P.; Jommi, G.; Sisti, M. *J. Org. Chem.* 1982, *47*, 1961.
298. Mulzer, J.; Bruntrup, G.; Hartz, G.; Kuhl, U.; Blaschek, U.; Bohrer, G. *Chem. Ber.* 1981, *114*, 3701.
299. Oppolzer, W.; Pitteloud, R. *J. Am. Chem. Soc.* 1982, *104*, 6478.
300. Sih, C. J.; Heather, J. B.; Sood, R.; Price, P.; Peruzzotti, G.; Lee, L. F. H.; Lee, S. S. *J. Am. Chem. Soc.* 1975, *97*, 865.
301. Suzuki, M.; Morita, Y.; Koyano, H.; Koga, M.; Noyori, R. *Tetrahedron* 1990, *46*, 4809.
302. Suzuki, M.; Koyano, H.; Morita, Y.; Noyori, R. *Synlett* 1989, 22.
303. Okamoto, S.; Kobayashi, Y.; Sato, F. *Tetrahedron Lett.* 1989, *30*, 4379.
304. Takahashi, T.; Nakazawa, M.; Kanoh, M.; Yamamoto, K. *Tetrahedron Lett.* 1990, *31*, 7349.
305. Tsujiyama, H.; Ono, N.; Yoshino, T.; Okamoto, S.; Sato, F. *Tetrahedron Lett.* 1990, *31*, 4481.
306. Charonnat, J. A.; Mitchell, A. L.; Keogh, B. P. *Tetrahedron Lett.* 1990, *31*, 315.
307. Miller, J. G.; Kurz, W.; Untch, K. G.; Stork, G. *J. Am. Chem. Soc.* 1974, *96*, 6774.
308. Noyori, R.; Suzuki, M. *Angew. Chem., Int. Ed. Engl.* 1984, *23*, 847.
309. Patterson, J. W. *J. Org. Chem.* 1990, *55*, 5528.
310. Babiak, K. A.; Behling, J. R.; Dygos, J. H.; McLaughlin, K. T.; Ng, J. S.; Kalish, V. J.; Kramer, S. W.; Shone, R. L. *J. Am. Chem. Soc.* 1990, *112*, 7441.
311. Kawasaki, H.; Tomioka, K.; Koga, K. *Tetrahedron Lett.* 1985, *26*, 3031.
312. Paulsen, H.; Koebernick, W.; Koebernick, H. *Tetrahedron Lett.* 1976, 2297.
313. Fitzsimmons, B. J.; Fraser-Reid, B. *J. Am. Chem. Soc.* 1979, *101*, 6123.
314. Liotta, C. L. *Tetrahedron Lett.* 1975, 519.
315. Baudouy, R.; Crabbe, P.; Greene, A. E.; Le Drian, C.; Orr, A. F. *Tetrahedron Lett.* 1977, 2973.

316. Oppolzer, W.; Moretti, R.; Godel, T.; Meunier, A.; Loher, H. *Tetrahedron Lett.* 1983, *24*, 4971.
317. Oppolzer, W.; Dudfield, P.; Stevenson, T.; Godel, T. *Helv. Chim. Acta* 1985, *68*, 212.
318. Oppolzer, W.; Loher, H. J. *Helv. Chim. Acta* 1981, *64*, 2808.
319. Oppolzer, W.; Stevenson, T. *Tetrahedron Lett.* 1986, *27*, 1139.
320. Oppolzer, W.; Kingma, A. J.; Poli, G. *Tetrahedron* 1989, *45*, 479.
321. Oppolzer, W.; Rodriguez, I.; Starkemann, C.; Walther, E. *Tetrahedron Lett.* 1990, *31*, 5019.
322. Oppolzer, W.; Schneider, P. *Helv. Chim. Acta* 1986, *69*, 1817.
323. Oppolzer, W.; Kingma, A. J. *Helv. Chim. Acta* 1989, *72*, 1337.
324. Rück, K.; Kunz, H. *Angew. Chem., Int. Ed. Engl.* 1991, *30*, 694.
325. Tomioka, K.; Suenaga, T.; Koga, K. *Tetrahedron Lett.* 1986, *27*, 369.
326. Larchevêque, M.; Tamagnan, G.; Petit, Y. *J. Chem. Soc., Chem. Commun.* 1989, 31.
327. Li, G.; Jarosinski, M. A.; Hruby, V. J. *Tetrahedron Lett.* 1993, *34*, 2561.
328. Li, G.; Russell, K. C.; Jarosinski, M. A.; Hruby, V. J. *Tetrahedron Lett.* 1993, *34*, 2565.
329. Giese, B.; Hoffmann, U.; Roth, M.; Veit, A.; Wyss, C.; Zehnder, M.; Zipse, H. *Tetrahedron Lett.* 1993, *34*, 2445.
330. Corey, E. J.; Houpis, I. N. *Tetrahedron Lett.* 1993, *34*, 2421.
331. Rück, K.; Kunz, H. *Synthesis* 1993, 1018.
332. Sato, M.; Murakami, M.; Kaneko, C.; Furuya, T. *Tetrahedron* 1993, *49*, 8529.
333. Posner, G. H.; Kogan, T. P.; Hulce, M. *Tetrahedron Lett.* 1984, *25*, 383.
334. Posner, G. H.; Hulce, M. *Tetrahedron Lett.* 1984, *25*, 379.
335. Posner, G. H.; Mallamo, J. P.; Hulce, M.; Frye, L. L. *J. Am. Chem. Soc.* 1982, *104*, 4180.
336. Posner, G. H.; Mallamo, J. P.; Muira, K. *J. Am. Chem. Soc.* 1981, *103*, 2886.
337. Posner, G. H.; Kogan, T. P.; Haines, S. R.; Frye, L.-L. *Tetrahedron Lett.* 1984, *25*, 2627.
338. Posner, G. H.; Mallamo, J. P.; Muira, K.; Hulce, M. *Pure Appl. Chem.* 1981, *53*, 2307.
339. Posner, G. H. In *Asymmetric Synthesis*; J. D. Morrison, Ed.; Academic Press: Orlando, FL, 1983; Vol. 2; pp 225.
340. Pyne, S. G. *Tetrahedron Lett.* 1986, *27*, 1691.
341. Pyne, S. G.; Griffith, R.; Edwards, M. *Tetrahedron Lett.* 1988, *29*, 2089.
342. Abbot, D. J.; Colonna, S.; Stirling, C. J. M. *J. Chem. Soc., Chem. Commun.* 1971, 471.
343. Abbot, D. J.; Colonna, S.; Stirling, C. J. M. *J. Chem. Soc., Perkin Trans. 1* 1976, 49.
344. Pyne, S. G.; Bloem, P.; Chapman, S. L.; Dixon, C. E.; Griffith, R. *J. Org. Chem.* 1990, *55*, 1086.
345. Zweig, J. S.; Fuche, J. L.; Barreiro, E.; Crabbe, P. *Tetrahedron Lett.* 1975, 2355.
346. Leyendecker, F.; Laucher, D. *Tetrahedron Lett.* 1983, *24*, 3517.
347. Desimoni, G.; Quadrelli, P.; Righetti, P. P. *Tetrahedron* 1990, *46*, 2927.
348. Tanaka, K.; Suzuki, H. *J. Chem. Soc., Chem. Commun.* 1991, 101.
349. Soai, K.; Okudo, M.; Okamoto, M. *Tetrahedron Lett.* 1991, *32*, 95.
350. Morita, Y.; Suzuki, Y.; Noyori, R. *J. Org. Chem.* 1989, *54*, 1785.
351. Jansen, J. F. G. A.; Feringa, B. L. *J. Org. Chem.* 1990, *55*, 4168.
352. Corma, A.; Iglesias, M.; Martín, M. V.; Rubio, J.; Sánchez, F. *Tetrahedron: Asymmetry* 1992, *3*, 845.
353. Uemura, M.; Miyake, R.; Nakayama, Y. *Tetrahedron: Asymmetry* 1992, *3*, 713.
354. Jansen, J. F. G. A.; Feringa, B. L. *Tetrahedron: Asymmetry* 1992, *3*, 581.
355. Dehmlow, E. V.; Singh, P.; Heider, J. *J. Chem. Research (S)* 1981, 292.
356. Wynberg, H.; Helder, R. *Tetrahedron Lett.* 1975, 4057.
357. Wynberg, H.; Greijdanus, B. *J. Chem. Soc., Chem. Commun.* 1978, 427.
358. Latvala, A.; Stanchev, S.; Linden, A.; Hesse, M. *Tetrahedron: Asymmetry* 1993, *4*, 173.
359. Toroyama, T.; Pan, L.-R. *Tetrahedron Lett.* 1989, *30*, 197.
360. Herndon, J. W.; Wu, C. *Tetrahedron Lett.* 1989, *30*, 5745.
361. Hiroi, K.; Yamada, S.-I. *Chem. Pharm. Bull.* 1973, *21*, 47.
362. Otani, G.; Yamada, S.-I. *Chem. Pharm. Bull.* 1973, *21*, 2130.
363. Otani, G.; Yamada, S.-I. *Chem. Pharm. Bull.* 1973, *21*, 2112.
364. Otani, G.; Yamada, S. *Chem. Pharm. Bull.* 1973, *21*, 2125.
365. Casey, M.; Manage, A. C.; Nezhat, L. *Tetrahedron Lett.* 1988, *29*, 5821.
366. Sone, T.; Hiroi, K.; Yamada, S.-I. *Chem. Pharm. Bull.* 1973, *21*, 2331.
367. Matloubi, F.; Solladie, G. *Tetrahedron Lett.* 1979, 2141.
368. Tsuchihashi, G.-I.; Mitamura, S.; Inoue, S.; Ogura, K. *Tetrahedron Lett.* 1973, 323.
369. Casey, M.; Manage, A. C.; Gairns, R. S. *Tetrahedron Lett.* 1989, *30*, 6919.
370. Swindell, C. S.; Blase, F. R.; Eggleston, D. S.; Krause, J. *Tetrahedron Lett.* 1990, *31*, 5409.
371. Hua, D. H.; Bharathi, S. N.; Takusagawa, F.; Tsujimoto, A.; Panangadan, J. A. K.; Hung, M.-H.; Bravo, A. A.; Ernelding, A. M. *J. Org. Chem.* 1989, *54*, 5659.
372. Zuger, M.; Weller, T.; Seebach, D. *Helv. Chim. Acta* 1980, *63*, 2005.
373. Yamaguchi, M.; Hasebe, K.; Tanaka, S.; Minami, T. *Tetrahedron Lett.* 1986, *27*, 959.

374. Jianguo, C.; Lingchong, Y. *Synth. Commun.* 1990, *20*, 2887.
375. Jianguo, C.; Lingchong, Y. *Synth. Commun.* 1990, *20*, 2895.
376. Tomioka, K.; Masumi, F.; Yamashita, T.; Koga, K. *Tetrahedron* 1989, *45*, 643.
377. Kogen, H.; Tomioka, K.; Hashimoto, S.-I.; Koga, K. *Tetrahedron Lett.* 1980, *21*, 4005.
378. d'Angelo, J.; Desmaële, D.; Dumas, F.; Guingant, A. *Tetrahedron: Asymmetry* 1992, *3*, 459.
379. Enders, D.; Gerdes, P.; Kepphardt, H. *Angew. Chem., Int. Ed. Engl.* 1990, *29*, 179.
380. Meyers, A. I.; Leonard, W. R.; Romine, J. L. *Tetrahedron Lett.* 1991, *32*, 597.
381. Mukaiyama, T.; Hirako, I. Y.; Takeda, T. *Chem. Lett.* 1978, 461.
382. Mukaiyama, T.; Fujimoto, K.; Hirose, T.; Takeda, T. *Chem. Lett.* 1980, 635.
383. Davies, S. G.; Ichihara, O. *J. Chem. Soc., Chem. Commun.* 1990, 1554.
384. Aoki, S.; Sasaki, S.; Koga, K. *Tetrahedron Lett.* 1990, *31*, 7229.
385. Davies, S. G. *Pure Appl. Chem.* 1988, *60*, 13.
386. Dieter, R. K.; Lagu, B.; Deo, N.; Dieter, J. W. *Tetrahedron Lett.* 1990, *31*, 4105.
387. Rossiter, B. E.; Eguchi, M. *Tetrahedron Lett.* 1990, *31*, 965.
388. Duerr, B. F.; Czarnik, A. W. *Tetrahedron Lett.* 1989, *30*, 6951.
389. Alonso-Lopez, M.; Martín-Lomas, M.; Panadés, S. *Tetrahedron Lett.* 1986, *27*, 3551.
390. Yamaguchi, M.; Shiraishi, T.; Hirama, M. *Angew. Chem., Int. Ed. Engl.* 1993, *32*, 1176.
391. Enders, D.; Papadopoulos, K.; Rendenbach, B. E. M. *Tetrahedron Lett.* 1986, *27*, 3491.
392. Enders, D.; Scherer, H. J.; Raabe, G. *Angew. Chem., Int. Ed. Engl.* 1991, *30*, 1664.
393. Enders, D.; Papadopoulis, K.; Herdtweck, E. *Tetrahedron* 1993, *49*, 1821.
394. Datta, A.; Schmidt, R. R. *Tetrahedron Lett.* 1993, *34*, 4161.
395. Leonard, J.; Mohialdin, S.; Reed, D.; Jones, M. F. *Synlett* 1992, 741.
396. Fang, C.; Suemune, H.; Sakai, K. *Tetrahedron Lett.* 1990, *31*, 4751.
397. Linderman, R. J.; Griedel, B. D. *J. Org. Chem.* 1990, *55*, 5428.
398. Lindermann, R. J.; Godfrey, A.; Horne, K. *Tetrahedron* 1989, *45*, 495.
399. Ager, D. J. In *Umpoled Synthons: A Survey of Sources and Uses in Synthesis*; T. A. Hase, Ed.; Wiley-Interscience: New York, 1987; pp 19.
400. Kanemasa, S.; Tatsukawa, A.; Wada, E. *J. Org. Chem.* 1991, *56*, 2875.
401. Bernardi, A.; Capelli, A. M.; Comotti, A.; Gennari, C.; Scolastico, C. *Tetrahedron Lett.* 1991, *32*, 823.
402. Lipschutz, B. H.; Ellsworth, E. L.; Dimock, S. H.; Smith, R. A. J. *J. Am. Chem. Soc.* 1990, *112*, 4404.
403. Lipshutz, B. H.; Crow, R.; Dimock, S. H.; Ellsworth, E. L.; Smith, R. A. J.; Behling, J. R. *J. Am. Chem. Soc.* 1990, *112*, 4063.
404. Jones, R. C. F.; Hirst, S. C. *Tetrahedron Lett.* 1989, *30*, 5361.
405. Pyne, S. P.; Spellmeyer, D. C.; Chen, S.; Fuchs, P. L. *J. Am. Chem. Soc.* 1982, *104*, 5728.
406. Leonard, J.; Ryan, G.; Swain, P. A. *Synlett* 1990, 613.
407. Oare, D. A.; Heathcock, C. H. *J. Org. Chem.* 1990, *55*, 157.
408. Oare, D. A.; Henderson, M. A.; Sanner, M. A.; Heathcock, C. H. *J. Org. Chem.* 1990, *55*, 132.
409. Dorigo, A. E.; Morokuma, K. *J. Am. Chem. Soc.* 1989, *111*, 6524.
410. Corey, E. J.; Hannon, F. J. *Tetrahedron Lett.* 1990, *31*, 1393.
411. Lipshutz, B. H.; Kayser, F.; Siegmann, K. *Tetrahedron Lett.* 1993, *34*, 6693.
412. Bernardi, A.; Capelli, A. M.; Gennari, C.; Scolastico, C. *Tetrahedron: Asymmetry* 1990, *1*, 21.
413. Sevin, A.; Masure, D.; Giessner-Prettre, C.; Pfau, M. *Helv. Chim. Acta* 1990, *73*, 552.
414. Rossiter, B. E.; Swingle, N. M. *Chem. Rev.* 1992, *92*, 771.
415. Zhou, Q.-L.; Pfaltz, A. *Tetrahedron Lett.* 1993, *34*, 7725.
416. Rossiter, B. E.; Miao, G.; Swingle, N. M.; Eguchi, M.; Hernández, A. E.; Patterson, R. G. *Tetrahedron: Asymmetry* 1992, *3*, 231.
417. Rossiter, B. M.; Eguchi, M.; Miao, G.; Swingle, N. M.; Hernández, A. E.; Vickers, D.; Fluckiger, E.; Patterson, R. G.; Reddy, K. V. *Tetrahedron* 1993, *49*, 965.
418. Cabral, J.; Laszlo, P.; Mahe, L.; Montaufier, M.-T.; Randriamahefa, S. L. *Tetrahedron Lett.* 1989, *30*, 3969.
419. Hara, S.; Hyaga, S.; Aoyama, M.; Sato, M.; Suzuki, A. *Tetrahedron Lett.* 1990, *31*, 247.
420. Lipshutz, B. H.; Ellsworth, E. L.; Siahaan, T. J. *J. Am. Chem. Soc.* 1989, *111*, 1351.
421. Ibuka, T.; Yamamoto, Y. *Synlett* 1992, 769.
422. Bernardi, A.; Cavicchioli, M.; Scolastico, C. *Tetrahedron* 1993, *49*, 10913.
423. Horiguchi, Y.; Komatsu, M.; Kuwajima, I. *Tetrahedron Lett.* 1989, *30*, 7087.
424. Amberg, W.; Seebach, D. *Chem. Ber.* 1990, *123*, 2439.
425. Alexakis, A.; Sedrani, R.; Mangeney, P. *Tetrahedron Lett.* 1990, *31*, 345.
426. Sakata, H.; Aoki, Y.; Kuwajima, I. *Tetrahedron Lett.* 1990, *31*, 1161.
427. Matsuzawa, S.; Horiguchi, Y.; Nakamura, E.; Kuwajima, I. *Tetrahedron* 1989, *45*, 349.
428. Tamura, R.; Tamai, S.; Katayama, H.; Suzuki, H. *Tetrahedron Lett.* 1989, *30*, 3685.
429. Aoki, Y.; Kuwajima, I. *Tetrahedron Lett.* 1990, *31*, 7457.
430. Bergdahl, M.; Lindstedt, E.-L.; Nilsson, M.; Olsson, T. *Tetrahedron* 1989, *45*, 535.

431. Lipshutz, B. H.; James, B. *Tetrahedron Lett.* 1993, *34*, 6689.
432. Lipshutz, B. H.; Dimock, S. H.; James, B. *J. Am. Chem. Soc.* 1993, *115*, 9283.
433. Viteva, L.; Stefanovsky, Y. *Tetrahedron Lett.* 1990, *31*, 5649.
434. Bertz, S. H.; Smith, R. A. J. *J. Am. Chem. Soc.* 1989, *111*, 8276.
435. Bertz, S. H. *J. Am. Chem. Soc.* 1990, *112*, 4031.
436. Lipshultz, B. H.; Sharma, S.; Ellsworth, E. L. *J. Am. Chem. Soc.* 1990, *112*, 4032.
437. Lipshultz, B. H.; Ellsworth, E. L.; Dimock, S. H. *J. Am. Chem. Soc.* 1990, *112*, 5869.
438. House, H. O.; Wilkins, J. M. *J. Org. Chem.* 1978, *43*, 2443.
439. van Koten, G.; Noltes, J. G. *J. Chem. Soc., Chem. Commun.* 1972, 940.
440. van Goten, G.; Jastzebski, J. T. B. H.; Muller, F.; Stam, C. H. *J. Am. Chem. Soc.* 1985, *107*, 697.
441. Hope, H.; Olmstead, M. M.; Powers, P. O.; Sandell, J.; Yu, X. *J. Am. Chem. Soc.* 1985, *107*, 4337.
442. Kraus, S. R.; Smith, S. G. *J. Am. Chem. Soc.* 1981, *103*, 141.
443. Hallnemo, G.; Olsson, T.; Ullenius, T. *J. Organometal. Chem.* 1985, *282*, 133.
444. Ouannes, G.; Dessaire, G.; Langlais, Y. *Tetrahedron Lett.* 1977, 815.
445. Hallnemo, G.; Ullenius, C. *Tetrahedron Lett.* 1986, *27*, 395.
446. Corey, E. J.; Boaz, N. W. *Tetrahedron Lett.* 1984, *25*, 3063.
447. Corey, E. J.; Boaz, N. W. *Tetrahedron Lett.* 1985, *26*, 6015.
448. Corey, E. J.; Boaz, N. W. *Tetrahedron Lett.* 1985, *26*, 6019.
449. Corey, E. J.; Hannon, F. J.; Boaz, N. W. *Tetrahedron* 1989, *45*, 545.
450. Seebach, D.; Locher, R. *Angew. Chem., Int. Ed. Engl.* 1979, *18*, 957.
451. Mangeney, P.; Alexakis, A.; Normant, J.-F. *Tetrahedron Lett.* 1986, *27*, 3143.
452. Yanagisawa, A.; Habaue, S.; Yamamoto, H. *J. Am. Chem. Soc.* 1989, *111*, 366.
453. Alexakis, A.; Sedrani, R.; Mangeney, P.; Normant, J. F. *Tetrahedron Lett.* 1988, *29*, 4411.
454. Seebach, D.; Zimmerman, J.; Gysel, U.; Ziegler, R.; Ha, T.-K. *J. Am. Chem. Soc.* 1988, *110*, 4763.
455. Seebach, D.; Zimmerman, J. *Helv. Chim. Acta* 1986, *69*, 1147.
456. Amberg, W.; Seebach, D. *Chem. Ber.* 1990, *123*, 2413.
457. Amberg, W.; Seebach, D. *Chem. Ber.* 1990, *123*, 2429.
458. Konopelski, J. P.; Chu, K. S.; Negrete, G. R. *J. Org. Chem.* 1991, *56*, 1355.
459. Ibuka, T.; Akimoto, N.; Tanaka, M.; Nishii, S.; Yamamoto, Y. *J. Org. Chem.* 1989, *54*, 4055.
460. Yamamoto, Y.; Asao, N. *J. Org. Chem.* 1990, *55*, 5303.
461. Ibuka, T.; Tanaka, M.; Nishii, S.; Yamamoto, Y. *J. Am. Chem. Soc.* 1989, *111*, 4864.
462. Ibuka, T.; Habashita, H.; Funakoshi, S.; Fujii, N.; Oguchi, Y.; Uychara, T.; Yamamoto, Y. *Angew. Chem., Int. Ed. Engl.* 1990, *29*, 801.
463. Arai, M.; Nemoto, T.; Ohashi, Y.; Nakamura, E. *Synlett* 1992, 309.
464. Tsunda, T.; Fujii, T.; Kawasaki, K.; Salgasa, T. *J. Chem. Soc., Chem. Commun.* 1980, 1014.
465. Boger, D. L.; Habibi, M. M.; Fauth, D. J. *J. Org. Chem.* 1980, *45*, 3860.
466. Fuzuzawa, S. I.; Fujinami, T. T.; Yameuchi, S.; Sakai, S. *J. Chem. Soc., Perkin Trans. 1* 1986, 1929.
467. Brestensky, D. M.; Stryker, J. M. *Tetrahedron Lett.* 1989, *30*, 5677.
468. Koenig, T. M.; Daeuble, J. F.; Brestensky, D. M.; Stryker, J. M. *Tetrahedron Lett.* 1990, *31*, 3237.
469. Lipshultz, B. H.; Ung, C. S.; Sengupta, S. *Synlett* 1989, 64.
470. Brettle, R.; Shibib, S. M. *J. Chem. Soc., Perkin Trans. 1* 1981, 2912.
471. Burgstahler, A. W.; Sanders, M. E. *Synthesis* 1980, 400.
472. Malemed, U.; Feit, B. A. *J. Chem. Soc., Perkin Trans. 1* 1980, 1267.
473. Sondngam, B. L.; Formum, Z. T.; Sharles, G.; Akam, T. M. *J. Chem. Soc., Perkin Trans. 1* 1983, 1219.
474. Hazarika, M. J.; Barua, N. C. *Tetrahedron Lett.* 1989, *30*, 6567.
475. Leutenegger, U.; Modin, A.; Pfalz, A. *Angew. Chem., Int. Ed. Engl.* 1989, *28*, 60.
476. Cabaret, D.; Welvart, Z. *J. Organometal. Chem.* 1978, *78*, 295.
477. von Matt, P.; Pfaltz, A. *Tetrahedron: Asymmetry* 1991, *2*, 691.
478. Pfaltz, A. *Chimia* 1990, *44*, 202.
479. Brinkmeyer, R. S.; Kapoor, V. M. *J. Am. Chem. Soc.* 1977, *99*, 8339.
480. Fouquey, C.; Jacques, J.; Angiolini, L.; Tramontini, M. *Tetrahedron* 1974, *30*, 2801.
481. Ojima, I.; Kogure, T.; Nagai, Y. *Tetrahedron Lett.* 1972, 5085.
482. Nakata, T.; Fukui, M.; Oishi, T. *Tetrahedron Lett.* 1988, *29*, 2219.
483. Giacomelli, G.; Caporusso, A. M.; Lardicci, L. *Tetrahedron Lett.* 1981, *22*, 3663.
484. Makino, T.; Baba, N.; Oda, J.; Inouye, Y. *Chem. Int. (London)* 1977, 277.
485. Evans, D. A.; Fu, G. C. *J. Org. Chem.* 1990, *55*, 5678.
486. Oppolzer, W.; Poli, G. *Tetrahedron Lett.* 1986, *27*, 4717.
487. Oppolzer, W.; Poli, G.; Starkemann, C.; Bernardinelli, G. *Tetrahedron Lett.* 1988, *29*, 3559.
488. Schroder, M. *Chem. Rev.* 1980, *80*, 187.
489. Itoh, A.; Ozawa, S.; Oshima, K.; Nozaki, H. *Tetrahedron Lett.* 1980, *21*, 361.
490. Guaciaro, M. A.; Wovkulich, P. M.; Smith, A. B. *Tetrahedron Lett.* 1978, 4661.

491. Ager, D. J.; Fleming, I. *J. Chem. Soc., Chem. Commun.* 1978, 177.
492. Ager, D. J.; Fleming, I.; Patel, S. K. *J. Chem. Soc., Perkin Trans. 1* 1981, 2520.
493. Barbero, A.; Cuadrado, P.; Fleming, I.; González, A. M.; Pulido, F. J. *J. Chem. Res. (S)* 1990, 297.
494. Lipshultz, B. H.; Reuter, D. C.; Ellsworth, E. L. *J. Org. Chem.* 1989, *54*, 4975.
495. Sharma, S.; Oehlschlager, A. C. *J. Org. Chem.* 1989, *54*, 5383.
496. Mohrig, J. R.; Fu, S. S.; King, R. W.; Warnet, R.; Gustafson, G. *J. Am. Chem. Soc.* 1990, *112*, 3665.
497. Suzuki, M.; Kawagishi, T.; Noyori, R. *Tetrahedron Lett.* 1981, *22*, 1809.
498. Suzuki, K.; Ikegawa, A.; Mukaiyama, T. *Bull. Chem. Soc., Jpn.* 1982, *55*, 3277.
499. Bhole, S. I.; Gogte, V. N. *Ind. J. Chem.* 1981, *20B*, 218.
500. Bhole, S. I.; Gogte, V. N. *Ind. J. Chem.* 1981, *20B*, 222.
501. Ahuja, R. R.; Natu, A. A.; Gogte, V. N. *Tetrahedron Lett.* 1980, *21*, 4743.
502. Eliel, E. L.; Lynch, J. E. *Tetrahedron Lett.* 1981, *22*, 2855.
503. Davies, S. G.; Ichihara, O. *Tetrahedron: Asymmetry* 1991, *2*, 183.
504. d'Angelo, J.; Maddaluno, J. *J. Am. Chem. Soc.* 1986, *108*, 8112.
505. Lange, B. D.; Bolhuis, F. V.; Feringa, B. L. *Tetrahedron* 1989, *45*, 6799.
506. Hawkins, J. M.; Fu, G. C. *J. Org. Chem.* 1986, *51*, 2820.
507. Rudolf, K.; Hawkins, J. M.; Loncharich, R. J.; Houk, K. N. *J. Org. Chem.* 1988, *53*, 3879.
508. Juaristi, E.; Quintana, D.; Escalante, J. *Aldrichimica Acta* 1994, *27*, 3.
509. Basavaiah, D.; Gowriswari, V. V. L.; Sarma, P. K. S.; Rao, P. D. *Tetrahedron Lett.* 1990, *31*, 1621.
510. Lange, B. D.; Bolhuis, F. V.; Feringa, B. L. *Tetrahedron* 1989, *45*, 6799.
511. Mukaiyama, T.; Iwasawa, N. *Chem. Lett.* 1981, 913.
512. Fang, J.-M.; Chang, C.-J. *J. Chem. Soc., Chem. Commun.* 1989, 1787.
513. Asami, M.; Mukaiyama, T. *Chem. Lett.* 1979, 579.
514. Tomioka, K.; Ando, K.; Yasuda, K.; Koga, K. *Tetrahedron Lett.* 1986, *27*, 959.
515. Tomioka, K.; Yasuda, K.; Koga, K. *Tetrahedron Lett.* 1986, *27*, 4611.
516. Tomioka, K.; Seo, W.; Ando, K.; Koga, K. *Tetrahedron Lett.* 1987, *28*, 6637.
517. Hashimoto, S.-I.; Yamada, S.-I.; Koga, K. *J. Am. Chem. Soc.* 1976, *98*, 7450.
518. Hashimoto, S.-I.; Komeshima, N.; Yamada, S.-I.; Koga, K. *Tetrahedron Lett.* 1977, 2907.
519. Hashimoto, S.-I.; Kogen, H.; Tomioka, K.; Koga, K. *Tetrahedron Lett.* 1979, 3009.
520. Hashimoto, S.-I.; Yamada, S.-I.; Koga, K. *Chem. Pharm. Bull.* 1979, *27*, 771.
521. Hashimoto, S.; Komeshima, N.; Yamada, S.; Koga, K. *Chem. Pharm. Bull.* 1979, *27*, 2437.
522. d'Angelo, J.; Guingant, A. *Tetrahedron Lett.* 1988, *29*, 2667.
523. Boga, C.; Savoia, D.; Umani-Ronchi, A. *Tetrahedron: Asymmetry* 1990, *1*, 291.
524. Mateos, A. F.; de la Fuente Blanco, J. A. *J. Org. Chem.* 1990, *55*, 1349.
525. Kamimura, A.; Ono, N. *Tetrahedron Lett.* 1989, *30*, 731.
526. Yoshikoshi, A.; Miyashita, M. *Acc. Chem. Res.* 1985, *18*, 284.
527. Stevens, R. W.; Mukaiyama, T. *Chem. Lett.* 1985, 855.
528. Seebach, D.; Golinski, J. *Helv. Chim. Acta* 1981, *64*, 1413.
529. Seebach, D.; Brook, M. A. *Helv. Chim. Acta* 1985, *68*, 319.
530. Pecunioso, A.; Menicagli, R. *J. Org. Chem.* 1988, *53*, 45.
531. Ochiai, M.; Arimoto, M.; Fujita, E. *Tetrahedron Lett.* 1981, *22*, 1115.
532. Uno, H.; Fujiki, S.; Suzuki, H. *Bull. Chem. Soc., Jpn.* 1986, *59*, 1267.
533. Uno, H.; Watanabe, N.; Fujiki, S.; Suzuki, H. *Synthesis* 1987, 471.
534. Yamamoto, Y.; Nishii, S. *J. Org. Chem.* 1988, *53*, 3597.
535. Brook, M. A.; Seebach, D. *Can. J. Chem.* 1987, *65*, 836.
536. Fuji, K.; Node, M.; Nagasawa, H.; Naniwa, Y.; Taga, T.; Machida, K.; Snatzke, G. *J. Am. Chem. Soc.* 1989, *111*, 7921.
537. Morris, M. L.; Sturgess, M. A. *Tetrahedron Lett.* 1993, *34*, 43.
538. Mukaiyama, T.; Takeda, T.; Osaki, M. *Chem. Lett.* 1977, 1165.
539. Mukaiyama, T.; Takeda, T.; Fujimoto, K. *Bull. Chem. Soc., Jpn.* 1978, *51*, 3368.
540. Meyers, A. I.; Whitten, C. E. *J. Am. Chem. Soc.* 1975, *97*, 6266.
541. Meyers, A. I.; Whitten, C. E. *Heterocycles* 1976, *4*, 1687.
542. Meyers, A. I.; Whitten, C. E. *Tetrahedron Lett.* 1976, 1947.
543. Meyers, A. I.; Smith, R. K.; Whitten, C. E. *J. Org. Chem.* 1979, *44*, 2250.
544. Mangeney, P.; Alexakis, A.; Normant, J. F. *Tetrahedron Lett.* 1983, *24*, 373.
545. Huche, M.; Aubouet, J.; Pourcelot, G.; Berlan, J. *Tetrahedron Lett.* 1983, *24*, 585.
546. Langlois, N.; Dahuron, N. *Tetrahedron Lett.* 1990, *31*, 7433.
547. Isobe, M.; Ichikawa, Y.; Funabashi, Y.; Mio, S.; Goto, T. *Tetrahedron* 1986, *42*, 2863.
548. Isobe, M.; Hirose, Y.; Shimokawa, K.-I.; Nishikawa, T.; Goto, T. *Tetrahedron Lett.* 1990, *31*, 5499.
549. Alcaraz, C.; Carretero, J. C.; Dominguez, E. *Tetrahedron Lett.* 1991, *32*, 1385.
550. Isobe, M.; Kitamura, M.; Goto, T. *J. Am. Chem. Soc.* 1982, *104*, 4997.

551. Isobe, M.; Kitamura, M.; Goto, T. *Tetrahedron Lett.* 1980, *21*, 4797.
552. Isobe, M.; Funabashi, Y.; Ichikawa, Y.; Mio, S.; Goto, T. *Tetrahedron Lett.* 1984, *25*, 2021.
553. Isobe, M.; Obeyama, J.; Funabashi, Y.; Goto, T. *Tetrahedron Lett.* 1988, *29*, 4773.
554. Isobe, M.; Ichikawa, Y.; Goto, T. *Tetrahedron Lett.* 1981, *22*, 4287.
555. Bund, J.; Gais, H.-J.; Erdelmeier, I. *J. Am. Chem. Soc.* 1991, *113*, 1442.
556. Simpkins, N. S. *Tetrahedron* 1990, *46*, 6951.
557. Fuchs, P. L.; Braish, T. F. *Chem. Rev.* 1986, *86*, 903.
558. Dominguez, E.; Carretero, J. C. *Tetrahedron Lett.* 1993, *34*, 5803.
559. Mikolajczyk, M.; Drabowicz, J. *Top. Stereochem.* 1982, *13*, 333.
560. Kahn, S. D.; Hehre, W. J. *J. Am. Chem. Soc.* 1986, *108*, 7399.
561. Craig, D.; Daniels, K.; MacKenzie, A. R. *Tetrahedron Lett.* 1990, *31*, 6441.
562. Magid, R. M. *Tetrahedron* 1980, *36*, 1901.
563. Tsuji, J. *Pure Appl. Chem.* 1989, *61*, 1673.
564. Consiglio, G.; Waymouth, R. M. *Chem. Rev.* 1989, *89*, 257.
565. Uguen, D. *Tetrahedron Lett.* 1984, *25*, 541.
566. Kang, S.-K.; Kim, S.-G.; Lee, J.-S. *Tetrahedron: Asymmetry* 1992, *3*, 1139.
567. Kang, S.-K.; Park, Y.-W.; Lee, D.-H.; Sim, H.-S.; Jeon, J.-H. *Tetrahedron: Asymmetry* 1992, *3*, 705.
568. Kang, S.-K.; Lee, D.-H.; Sim, H.-S.; Lim, J.-S. *Tetrahedron Lett.* 1993, *34*, 91.
569. Girard, C.; Mandville, G.; Bloch, R. *Tetrahedron: Asymmetry* 1993, *4*, 613.
570. Fouquet, G.; Schlosser, M. *Angew. Chem., Int. Ed. Engl.* 1974, *13*, 82.
571. Tanigawa, Y.; Kanamura, H.; Sonada, A.; Murahashi, S. *J. Am. Chem. Soc.* 1977, *99*, 2361.
572. Posner, G. H. *Org. Reactions* 1975, *22*, 253.
573. Muruyama, K.; Yamamoto, Y. *J. Am. Chem. Soc.* 1977, *99*, 8068.
574. Nakamura, E.; Sekiya, K.; Arai, M.; Aoki, S. *J. Am. Chem. Soc.* 1989, *111*, 3091.
575. Backvall, J.-E.; Sellen, M.; Grant, B. *J. Am. Chem. Soc.* 1990, *112*, 6615.
576. Anderson, R. J.; Henrick, C. A.; Siddall, J. B. *J. Am. Chem. Soc.* 1970, *92*, 735.
577. Fleming, I.; Thomas, A. P. *J. Chem. Soc., Chem. Commun.* 1985, 411.
578. Denmark, S. E.; Marble, L. K. *J. Org. Chem.* 1990, *55*, 1984.
579. Trost, B. M.; Lautens, M. *J. Am. Chem. Soc.* 1987, *109*, 1469.
580. Frost, C. G.; Howarth, J.; Williams, J. M. J. *Tetrahedron: Asymmetry* 1992, *3*, 1089.
581. Trost, B. M.; Dietsche, T. J. *J. Am. Chem. Soc.* 1973, *95*, 8200.
582. Trost, B. M.; Strege, P. E. *J. Am. Chem. Soc.* 1977, *99*, 1649.
583. Trost, B. M. *Acc. Chem. Res.* 1980, *13*, 385.
584. Hirao, T.; Enda, J.; Ohshiro, Y.; Agawa, T. *Tetrahedron Lett.* 1981, *22*, 3079.
585. Trost, B. M.; Strege, P. E. *J. Am. Chem. Soc.* 1975, *97*, 2534.
586. Trost, B. M.; Fullerton, T. J. *J. Am. Chem. Soc.* 1973, *95*, 292.
587. Okada, Y.; Minami, T.; Sasaki, Y.; Umezu, Y.; Yamaguchi, M. *Tetrahedron Lett.* 1990, *31*, 3905.
588. Muller, D.; Umbricht, G.; Weber, B.; Pfaltz, A. *Helv. Chim. Acta* 1991, *74*, 232.
589. Auburn, P. R.; Mackenzie, P. B.; Bosnich, B. *J. Am. Chem. Soc.* 1985, *107*, 2033.
590. Leutenegger, U.; Umbricht, G.; Fahrni, C.; von Matt, P.; Pfaltz, A. *Tetrahedron* 1992, *48*, 2143.
591. von Matt, P.; Pfaltz, A. *Angew. Chem., Int. Ed. Engl.* 1993, *32*, 566.
592. Reiser, O. *Angew. Chem., Int. Ed. Engl.* 1993, *32*, 547.
593. Sprinz, J.; Helmchen, G. *Tetrahedron Lett.* 1993, *34*, 1769.
594. Frost, C. G.; Williams, J. M. J. *Tetrahedron Lett.* 1993, *34*, 2015.
595. Kubota, H.; Nakajima, M.; Koga, K. *Tetrahedron Lett.* 1993, *34*, 8135.
596. Dawson, G. J.; Frost, C. G.; Williams, J. M. J.; Coote, S. J. *Tetrahedron Lett.* 1993, *34*, 3149.
597. Yamazaki, A.; Morimoto, T.; Achiwa, K. *Tetrahedron: Asymmetry* 1993, *4*, 2287.
598. Frost, C. G.; Williams, J. M. J. *Tetrahedron: Asymmetry* 1993, *4*, 1785.
599. Sawamura, M.; Ito, Y. *Chem. Rev.* 1992, *92*, 857.
600. Mackenzie, P. B.; Whelan, J.; Bosnich, B. *J. Am. Chem. Soc.* 1985, *107*, 2046.
601. Trost, B. M. *Pure Appl. Chem.* 1979, *51*, 787.
602. Trost, B. M. *Tetrahedron* 1977, *33*, 2615.
603. Ziegler, F. E.; Kneisley, A.; Wester, R. T. *Tetrahedron Lett.* 1986, *27*, 1221.
604. Genet, J. P.; Ferroud, D.; Juge, S.; Montes, J. R. *Tetrahedron Lett.* 1986, *27*, 4573.
605. Hayashi, T. *Pure Appl. Chem.* 1988, *60*, 7.
606. Hiroi, K.; Abe, J.; Suya, K.; Sato, S. *Tetrahedron Lett.* 1989, *30*, 1543.
607. Yamaguchi, M.; Shima, T.; Yamagishi, T.; Hida, M. *Tetrahedron Lett.* 1990, *31*, 5049.
608. Hayashi, T.; Kishi, K.; Yamamoto, A.; Ito, Y. *Tetrahedron Lett.* 1990, *31*, 1743.
609. Hayashi, T.; Yamamoto, A.; Ito, Y.; Nishioka, E.; Miura, H.; Yanagi, K. *J. Am. Chem. Soc.* 1989, *111*, 6301.
610. Fotiadu, F.; Cros, P.; Faure, B.; Buono, G. *Tetrahedron Lett.* 1990, *31*, 77.
611. Jumnah, R.; Williams, J. M. J.; Williams, A. C. *Tetrahedron Lett.* 1993, *34*, 6619.

612. Inami, H.; Ito, T.; Urabe, H.; Sato, F. *Tetrahedron Lett.* 1993, *34*, 5919.
613. Fujii, N.; Nakai, K.; Habashita, H.; Yoshizawa, H.; Ibuka, T.; Garrido, F.; Mann, A.; Chounan, Y.; Yamamoto, Y. *Tetrahedron Lett.* 1993, *34*, 4227.
614. Tsuji, Y.; Yamada, N.; Tanaka, S. *J. Org. Chem.* 1993, *58*, 16.
615. Hayashi, T.; Yamamoto, A.; Ito, Y. *J. Chem. Soc., Chem. Commun.* 1986, 1090.
616. Trost, B. M.; Van Vranken, D. L.; Bingel, C. *J. Am. Chem. Soc.* 1992, *114*, 9327.
617. Fiaud, J.-C.; Aribi-Zouioneche, L. *J. Chem. Soc., Chem. Commun.* 1986, 390.
618. Fiaud, J.-C.; Legros, J.-Y. *J. Org. Chem.* 1987, *52*, 1907.
619. Faller, J. W.; Linebarrier, D. *Organometallics* 1988, *7*, 1670.
620. Trost, B. M.; Verhoeven, T. R. *J. Org. Chem.* 1976, *41*, 3215.
621. Trost, B. M.; Verhoeven, T. R. *J. Am. Chem. Soc.* 1980, *102*, 4730.
622. Trost, B. M.; Weber, L. *J. Am. Chem. Soc.* 1975, *97*, 1611.
623. Keinan, E.; Sahai, M.; Roth, Z.; Nudelman, A.; Herzig, J. *J. Org. Chem.* 1985, *50*, 3558.
624. Trost, B. M.; Keinan, E. *J. Am. Chem. Soc.* 1978, *100*, 7779.
625. Fiaud, J.-C. *J. Chem. Soc., Chem. Commun.* 1983, 1055.
626. Auburn, P. R.; Whelan, J.; Bosnich, B. *J. Chem. Soc., Chem. Commun.* 1986, 186.
627. Trost, B. M.; Schmuff, N. R. *J. Am. Chem. Soc.* 1985, *107*, 396.
628. Trost, B. M.; Verhoeven, T. R.; Fortunak, J. M. *Tetrahedron Lett.* 1979, 2301.
629. Bäckvall, J.-E.; Nordberg, R. E. *J. Am. Chem. Soc.* 1981, *103*, 4959.
630. Consiglio, G.; Morandini, F.; Piccolo, O. *J. Am. Chem. Soc.* 1981, *103*, 1845.
631. Sheffy, F. K.; Stille, J. K. *J. Am. Chem. Soc.* 1983, *105*, 7173.
632. Matsushita, H.; Negishi, E. *J. Chem. Soc., Chem. Commun.* 1982, 150.
633. Hayashi, T.; Yamamoto, A.; Hagihara, T. *J. Org. Chem.* 1986, *51*, 723.
634. Trost, B. M.; Schmuff, N. R.; Miller, M. J. *J. Am. Chem. Soc.* 1980, *102*, 5979.
635. Tanikaga, R.; Jun, T. X.; Kaji, A. *J. Chem. Soc., Perkin Trans. 1* 1990, 1185.
636. Felkin, H.; Joly-Goudket, M.; Davies, S. G. *Tetrahedron Lett.* 1981, *22*, 1157.
637. Gendreau, Y.; Normant, J. F. *Tetrahedron Lett.* 1979, 1617.
638. Consiglio, G.; Morandini, F.; Piccolo, O. *J. Chem. Soc., Chem. Commun.* 1983, 112.
639. Hayashi, T.; Yamamoto, A.; Hayihara, T.; Ito, Y. *Tetrahedron Lett.* 1986, *27*, 191.
640. Fleming, I.; Terrett, N. K. *Tetrahedron Lett.* 1984, *24*, 5103.
641. Thompson, T. B.; Ford, W. T. *J. Am. Chem. Soc.* 1979, *101*, 5459.
642. Bartlett, P. A.; Meadows, J. D.; Brown, E. G.; Morimoto, A.; Jernstedt, K. K. *J. Org. Chem.* 1982, *47*, 4013.
643. Bartlett, P. A.; Jernstedt, K. K. *Tetrahedron Lett.* 1980, *21*, 1607.
644. Bongini, A.; Cardillo, G.; Orena, M.; Porzi, G.; Sandri, S. *J. Org. Chem.* 1982, *47*, 4626.
645. Harada, T.; Sakamoto, K.; Ikemura, Y.; Oku, A. *Tetrahedron Lett.* 1988, *29*, 3097.
646. Harada, T.; Kurokawa, H.; Oku, A. *Tetrahedron Lett.* 1987, *28*, 4847.
647. Takano, S.; Akiyama, M.; Ogasawara, K. *Chem. Pharm. Bull.* 1984, *32*, 791.
648. Still, W. C.; Darst, K. P. *J. Am. Chem. Soc.* 1980, *102*, 7385.
649. Still, W. C.; Shaw, K. R. *Tetrahedron Lett.* 1981, *22*, 3725.
650. Morgans, D. J. *Tetrahedron Lett.* 1981, *22*, 3721.

Chapter 10

# ASYMMETRIC OXIDATIONS OF ISOLATED ALKENES

The reactions of isolated alkenes with a wide variety of reagents are discussed in Chapter 8. This chapter is concerned with oxidations of isolated alkenes; functionalized alkenes are covered in Chapter 11. For simplicity, this chapter has been divided into two major sections: oxidations and reactions of epoxides, and closely related compounds, making up the first part. The second half deals with reactions that involve 1,2-diols and related compounds. Oxidative cleavage reactions of alkenes are not discussed as they do not involve asymmetric methodology.

## 10.1. EPOXIDATIONS

Epoxidation of an alkene can be achieved through a halohydrin, or by reaction with an oxidant which effects overall delivery of an oxygen atom,[1] such as a peroxy acid,[2] or a porphyrin.[3-9] In the former situation, the chirality of the halohydrin must be controlled prior to epoxide formation.[10-12] The key to asymmetric induction with an oxygen donor is discrimination between the two faces of the alkene. This has been achieved by use of chiral peroxy acids and chiral oxaziridines, although the magnitude of the asymmetric induction is generally not high.

### 10.1.1. PEROXY ACID EPOXIDATION

An alkene reacts with a peroxy acid to afford an epoxide in one step. The very nature of the epoxidation reaction, however, ensures *cis*-addition and the relative stereochemistry at two adjacent centers is controlled.[13-22] When prior association between the peroxy acid and alkene is possible, as in the case of an allyl alcohol [*11.1.3*], stereoselection can be good. In the case of a simple alkene, peroxy acid epoxidation is sensitive to steric effects caused by the degree of substitution on the alkene,[13,23] while for cyclic alkenes, the steric constraints of the ring system can impart facial selectivity.[13,18,20] The solvent can also have an effect.[24] The stereoselectivity series for the rate of epoxidation (Figure 10.1) indicates that there is little selectivity in reaction rates between *cis*, *trans*, and 1,1-disubstituted alkenes.[25] Facial selectivity, therefore, is difficult to achieve with a simple alkene.

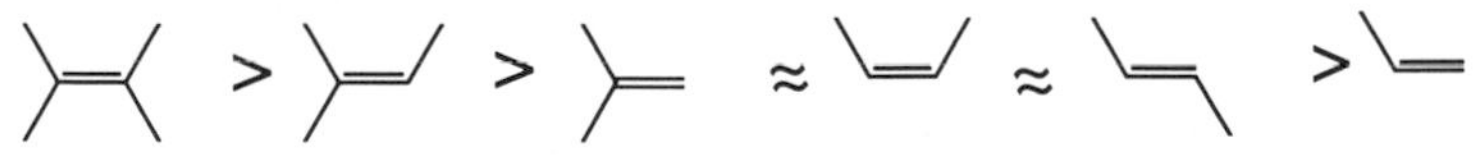

**FIGURE 10.1.** Reactivity of alkene isomers with peroxy acid epoxidations.

A number of achiral peroxy acids have been advocated as advantageous to perform the transformation, but the main reason is invariably the stability of the peroxy acid itself.[2,26,27] The current fashionable peroxy acid, replacing MCPBA, is magnesium monoperoxyphthalate (MMPP).[28]

Some progress has been made toward geometrical differentiation for disubstituted alkenes.[29] Reaction of the peroxy acid **10.1** with an alkene shows that as the size of the alkyl group of the ester increases, the *Z*-alkene reacts preferentially to the *E*-isomer (Figure 10.2).[25,30,31]

Use of a chiral peroxy acid with alkenes results in very low asymmetric induction (<8% ee's) with chirality transfer from the peroxy acid to the "major" product enantiomer; the model follows Houk's rule.[23,24,30-33]

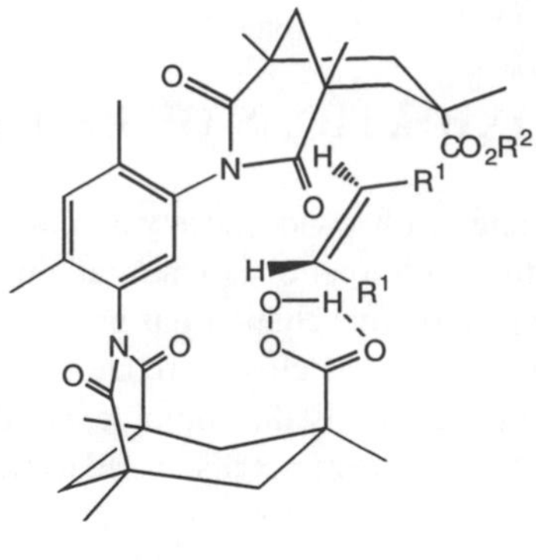

10.2

FIGURE 10.2. Reaction of peroxy acid **10.2** with a Z-alkene.

## 10.1.2. OXAZIRIDINE EPOXIDATION

Chiral oxaziridines **10.3** and **10.4** have been used to oxidize alkenes, and have shown much greater enantioselectivity than when a chiral peroxy acid was employed.[34,35] The configuration of the three-membered heterocycle determines the configuration of the epoxide product. Despite the increase in enantioselectivity over peroxy acids, the degree of asymmetric induction is not yet high enough for general synthetic application.[34,36] This includes alkene oxidations (ee's 17 to 61%) with chiral *N*-sulfonyloxaziridines derived from 3-substituted 1,2-benzoisothiazole-1,1-dioxide oxides.[37] The transition state geometry has been calculated,[38] and is borne out by experimental observations, to be analogous to the parallel peroxy acid mechanism (Figure 10.3).[36,39,40]

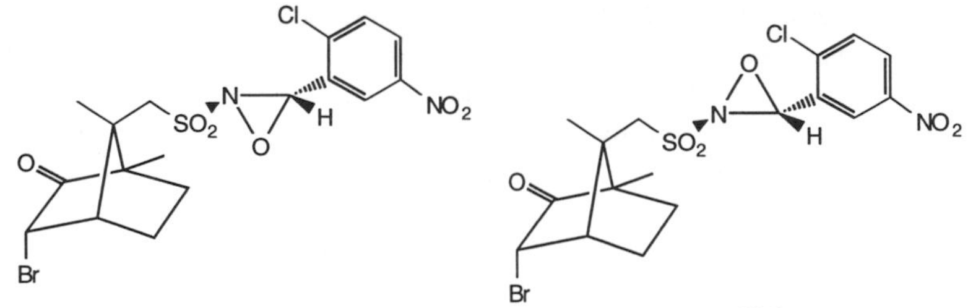

10.3 10.4

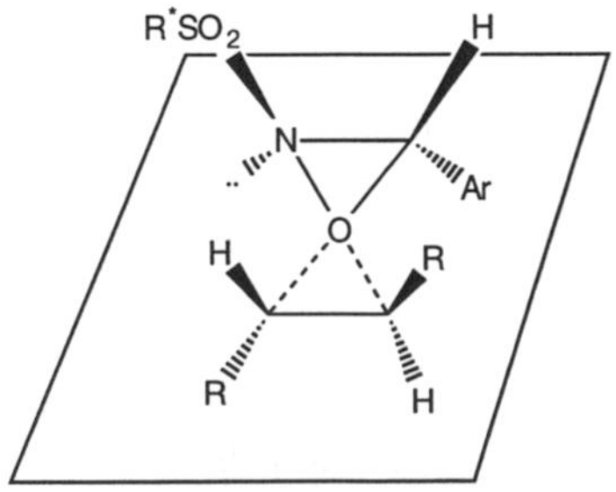

FIGURE 10.3. Transition state for epoxide formation with an aziridine.

## 10.1.3. METAL CATALYZED EPOXIDATIONS

Some metals such as ruthenium, tungsten, vanadium, chromium, and molybdenum catalyze epoxide formation in addition to diol formation from an alkene.[41-58] Use of a molybdenum(VI) catalyst in the presence of diisopropyl tartrate led to low (<11% ee) asymmetric induction.[59]

The use of (*S*)-*N,N*-dimethyllactamide as a ligand led to slightly better optical yields (16 to 35% ee's),[59,60] which, in turn, has been improved upon by a platinum(II) complex (18 to 41% ee's).[61] Although the facial selectivity has not been optimized, manganese complexes provide sufficient induction for synthetic utility (Scheme 10.1).[56,62-66] This manganese(III) salen complex **10.5** can also use bleach as the oxidant rather than an iodosylarene.[67,68]

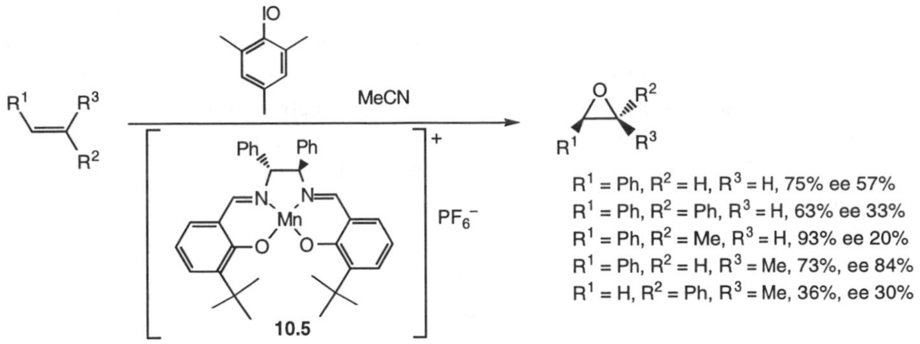

IO

MeCN

Ph Ph

Mn

$PF_6^-$

10.5

$R^1$ = Ph, $R^2$ = H, $R^3$ = H, 75% ee 57%
$R^1$ = Ph, $R^2$ = Ph, $R^3$ = H, 63% ee 33%
$R^1$ = Ph, $R^2$ = Me, $R^3$ = H, 93% ee 20%
$R^1$ = Ph, $R^2$ = H, $R^3$ = Me, 73%, ee 84%
$R^1$ = H, $R^2$ = Ph, $R^3$ = Me, 36%, ee 30%

SCHEME 10.1.

In effect the metal is planar, but the ligands are slightly buckled and direct the approach of the alkene to the metal reaction site in a particular orientation. The enantioselectivity is highest for *Z*-alkenes (Scheme 10.1).[64] The use of blocking groups other than *t*-butyl provides similar selectivities.[63,67,69,70] Epoxide formation based on mimics of cytochrome P-450 oxidation (vide infra) of alkenes can provide high chemical yields.[9]

Thus, despite numerous efforts, a reliable method for the epoxidation of a simple alkene with high enantioselectivity still has to be found, although significant advances have been made for *Z*-alkenes. The use of a kinetic resolution could provide an alternative strategy.[48]

It should be noted that chiral epoxides are available by an allyl alcohol epoxidation protocol [*11.1.1.1*] (Scheme 10.2).[71,72]

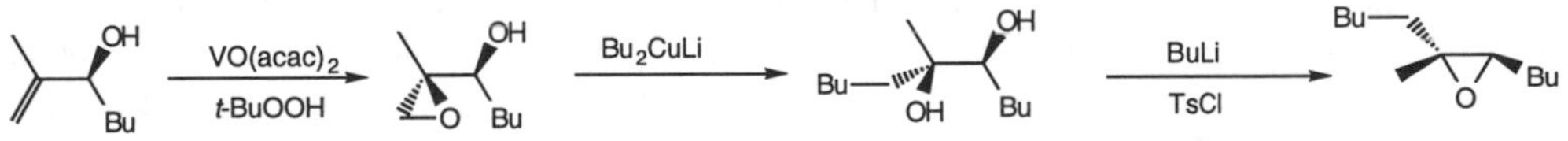

$VO(acac)_2$, *t*-BuOOH; $Bu_2CuLi$; BuLi, TsCl

SCHEME 10.2.

Studies have been conducted with iron, or other metal, porphyrins in the presence of iodosylbenzene or other oxidants. Enantiomeric excesses are moderate, although chemical yields are good.[3-8,73-85] Similarities between the porphyrin and salen systems have been noted.[86] Disubstitute alkenes undergo epoxidation with high enantioselectivity in the presence of chloroperoxidase.[87]

## 10.1.4. OTHER EPOXIDATION METHODS

Epoxides can still be accessed by the classical method of an intramolecular displacement of a leaving group by an oxygen on the adjacent carbon atom (Scheme 10.3).[88,89]

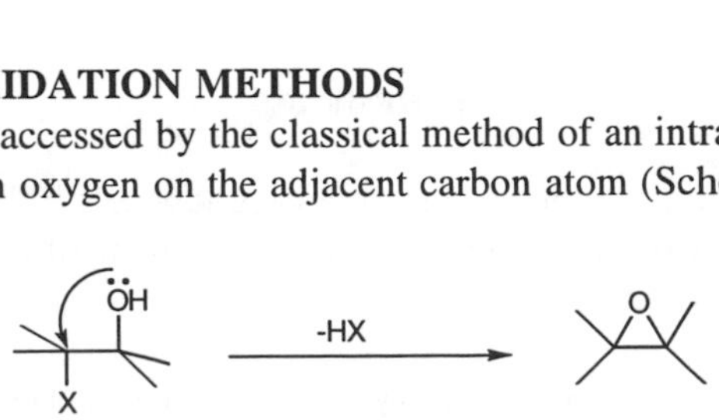

-HX

SCHEME 10.3.

The substitution can proceed under acid or base conditions, which in turn can effect the stereochemical outcome of the reaction [*cf.* Chapter 2], although inversion at the reaction center usually occurs.[90,91] The problem of this approach is that the stereochemistry already has to be in place at the oxygen-bearing carbon center, and, usually, the other center as well. A wide variety of groups has been used as the leaving group.[92]

If the substituted alcohol is not available in chiral form, then a kinetic resolution during epoxide formation has to be accomplished. Cobalt salen complexes allow such an approach (ee <67%) for the asymmetric cyclization of chlorohydrins.[93] A kinetic resolution during epoxide formation from chlorohydrins under phase transfer conditions has been claimed.[¶] [95]

The use of a camphor chiral auxiliary allows for the asymmetric preparation of a halohydrin and, consequently, an epoxide (Scheme 10.4).[96] An alternative procedure relies on the enantioselective reduction of trichloromethyl ketones with Corey's oxazaborolidine catalyst system [*4.5.5*] followed by dechlorination and cyclization.[97]

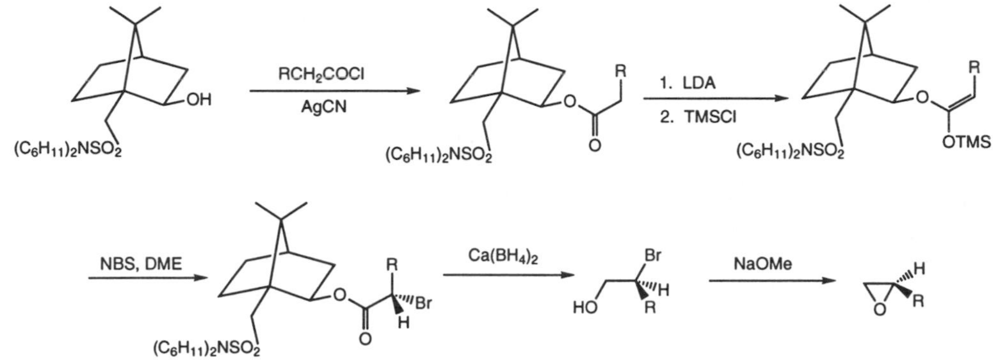

**SCHEME 10.4.**

Epoxides are available through condensation of various nucleophilic species that contain a heteroatom leaving group, such as sulfur,[89,98-101] selenium,[102] and arsenic or tellurium[103] with an aldehyde or ketone.[104-108] Chiral epoxides are available from chiral sulfoxides and derivatives by a number of strategies (e.g., Scheme 10.5).[109-119] This methodology is an extension of the sulfide variant introduced by Corey.[98] The addition of (–)-*N,N*-dimethylephedrinium bromide to the standard Corey reaction with trimethylsulfonium iodide resulted in ee's of 25 to 61%.[120] Another variant of this approach with sulfonium salts is derived from a $C_2$-symmetric thiolane; it has been used to prepare stilbene oxides with ee's up to 83% (Scheme 10.6).[116,121]

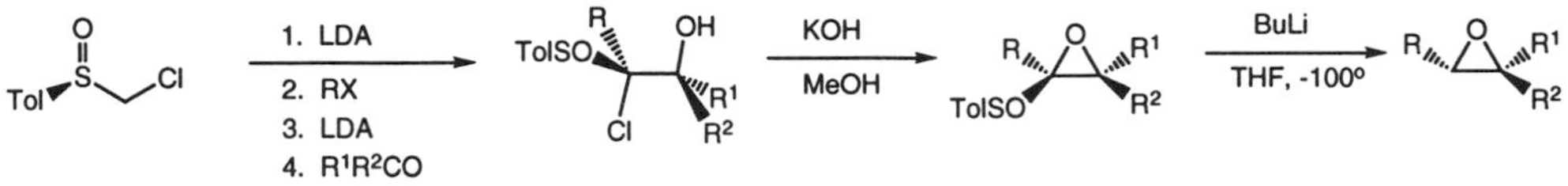

**SCHEME 10.5.**

A similar sequence with a camphoryl sulfide gave low ee's.[116,122]

The resolution of epoxides has been accomplished by use of a variant of the sulfonium methodology (Scheme 10.7).[115]

¶ Reactions that involve chiral induction by chiral β-hydroxyammonium catalysts under basic conditions have been questioned with regard to the source of the optical activity observed.[94]

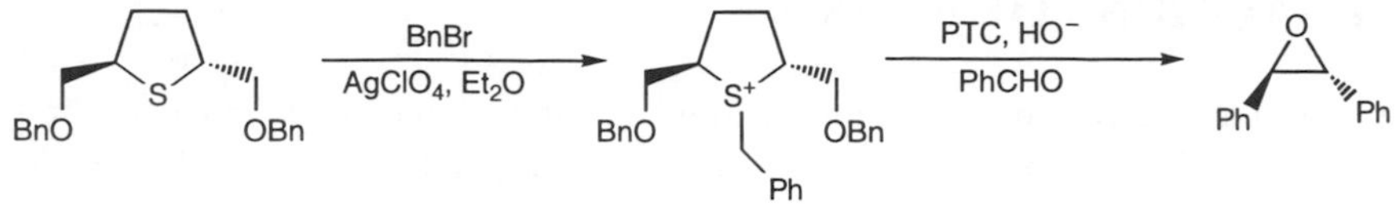
BnBr
$AgClO_4$, $Et_2O$
BnO
OBn
S
$S^+$
Ph
PTC, $HO^-$
PhCHO
O
Ph
Ph

**SCHEME 10.6.**

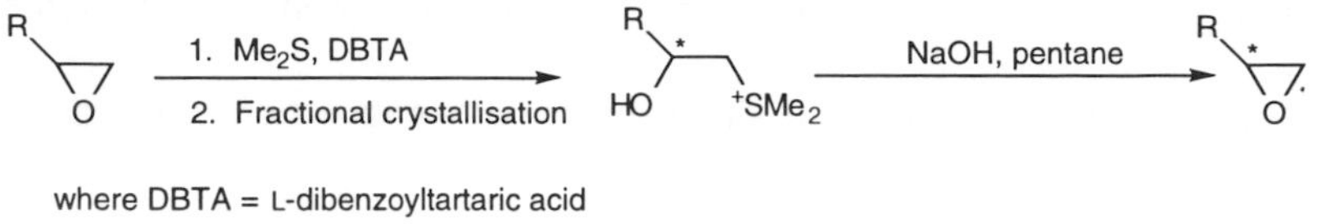
R
O
1. $Me_2S$, DBTA
2. Fractional crystallisation
HO
$^+SMe_2$
NaOH, pentane
where DBTA = L-dibenzoyltartaric acid

**SCHEME 10.7.**

α-Thio ketones provide a useful approach to *trans*-epoxides through control of relative stereochemistry (Scheme 10.8).[98,123]

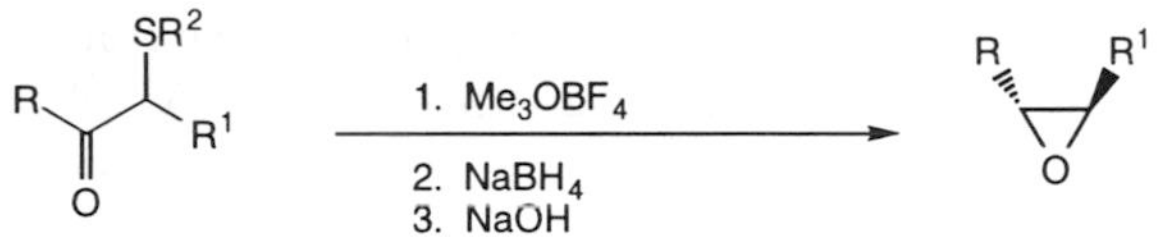
$SR^2$
R
$R^1$
O
1. $Me_3OBF_4$
2. $NaBH_4$
3. NaOH

**SCHEME 10.8.**

Schemes 10.2 and 10.5 are illustrations of functionalized epoxides being converted to "simple" ones. Functionalized alkene epoxidation is discussed in the following chapter. An epoxy alcohol can be used as a precursor to an epoxide [*11.2*] through nucleophilic substitution. This is illustrated by the synthesis of disparlure (**10.6**) (Scheme 10.9).[72,124-127]

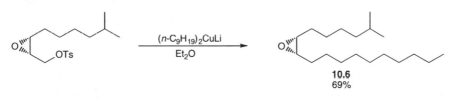

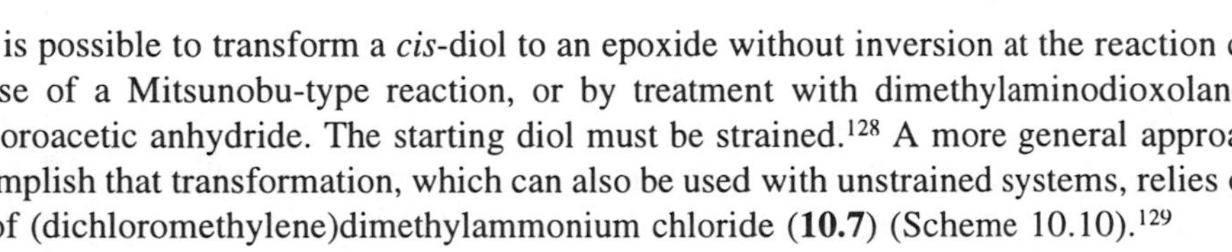
O
OTs
$(n\text{-}C_9H_{19})_2CuLi$
$Et_2O$
10.6
69%

**SCHEME 10.9.**

It is possible to transform a *cis*-diol to an epoxide without inversion at the reaction center by use of a Mitsunobu-type reaction, or by treatment with dimethylaminodioxolane and trifluoroacetic anhydride. The starting diol must be strained.[128] A more general approach to accomplish that transformation, which can also be used with unstrained systems, relies on the use of (dichloromethylene)dimethylammonium chloride (**10.7**) (Scheme 10.10).[129]

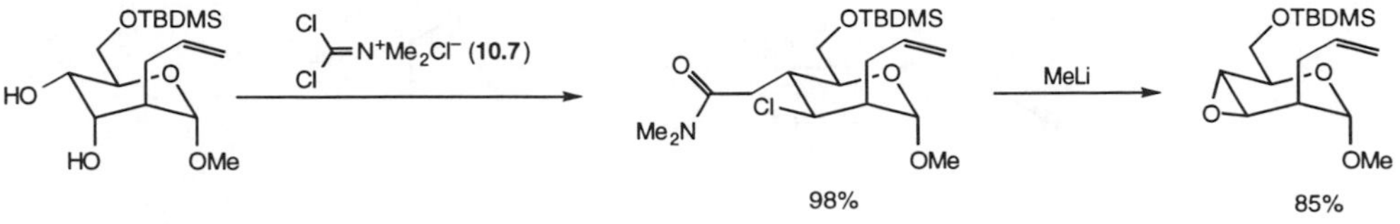
OTBDMS
HO
HO
OMe
Cl
Cl
$N^+Me_2Cl^-$ (10.7)
$Me_2N$
98%
MeLi
85%

**SCHEME 10.10.**

## 10.2. REACTIONS OF EPOXIDE

The regioselective control for the nucleophilic opening of an epoxide in an acyclic system is well known.[90,130,131] Under basic conditions, the nucleophile usually attacks the sterically less encumbered site, while under acidic conditions, the sterically more hindered site is favored.[90,131-134] The product invariably contains the functional groups in an *ancat*-disposition, when an $S_N2$ pathway is followed.[90,131,135-142] Epoxides react with a wide variety of nucleophiles.[91] Some examples of reactions that employ mild reaction conditions are summarized in Table 10.1.

A procedure that allows inversion of styrene oxide relies on reaction at the benzylic center (Scheme 10.11).[189]

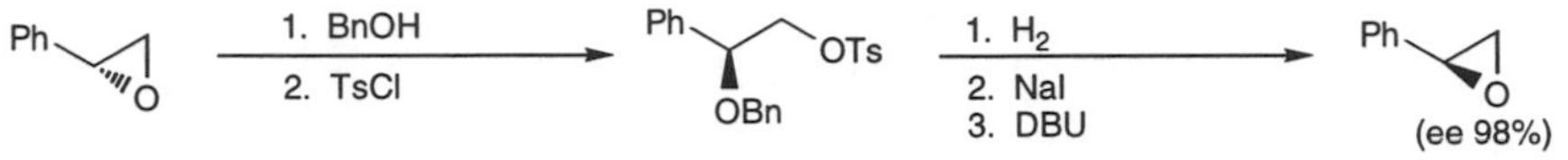
Ph
O
1. BnOH
2. TsCl
Ph
OTs
OBn
1. H2
2. NaI
3. DBU
Ph
O
(ee 98%)

**SCHEME 10.11.**

A related process that, in effect, isomerizes the parent alkene from which the epoxide is derived, relies on the use of a silylmetal reagent (Scheme 10.12).[190,191]

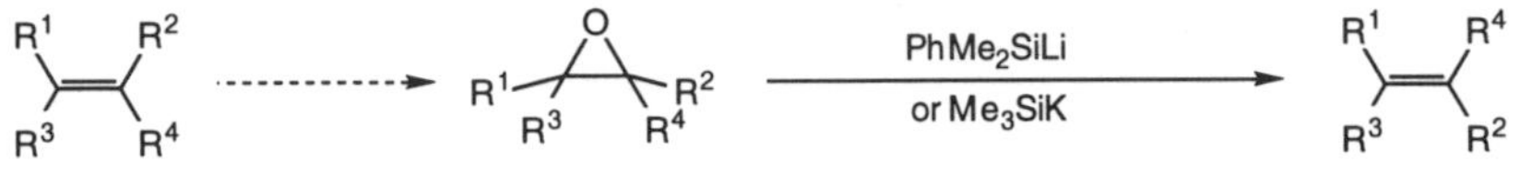
R1 R2
R3 R4
O
R1 R2
R3 R4
PhMe2SiLi
or Me3SiK
R1 R4
R3 R2

**SCHEME 10.12.**

Triphenylphosphine–hydrogen iodide can also be used to convert an epoxide to the alkene through an intermediate iodohydrin.[192,193]

If the epoxide is chiral, then an optically active product can result from ring opening (e.g., Scheme 10.13).[178,194,195] The use of metal salts can provide useful catalysis for nitrogen, among other nucleophiles, to attack at the least hindered end of an epoxide.[149,156,157,161,167,168,172,179,189]

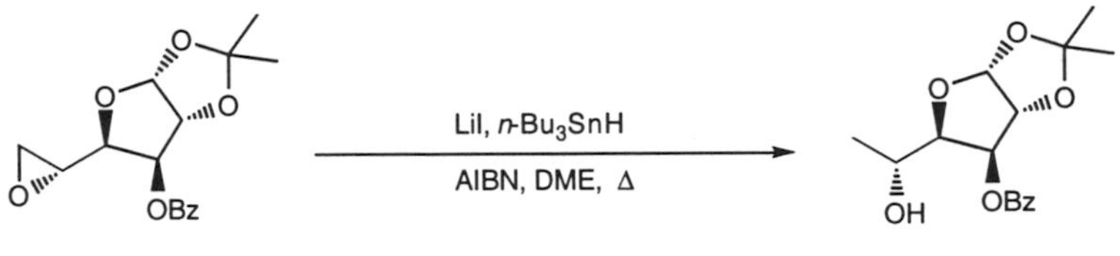
O
O
O
O
OBz
LiI, n-Bu3SnH
AIBN, DME, Δ
O
O
O
OH
OBz

**SCHEME 10.13.**

The ring opening of a *meso*-epoxide with *B*-halodiisopinocampheylborane provides an enantioselective method to the halohydrin (Scheme 10.14).[196]

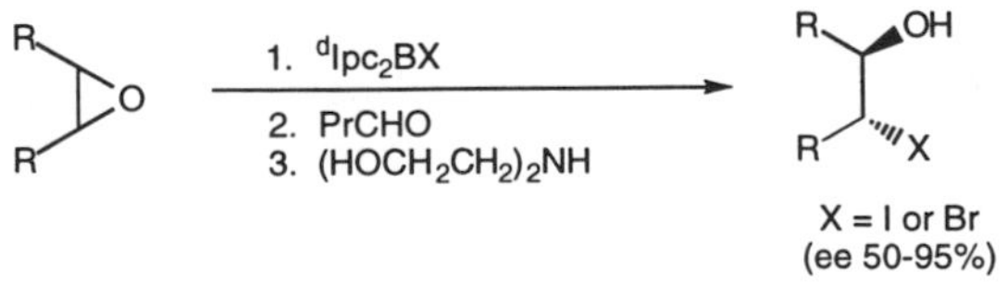
R
O
R
1. dIpc2BX
2. PrCHO
3. (HOCH2CH2)2NH
R OH
R X
X = I or Br
(ee 50-95%)

**SCHEME 10.14.**

**TABLE 10.1**
**Reactions of Epoxides[90]**

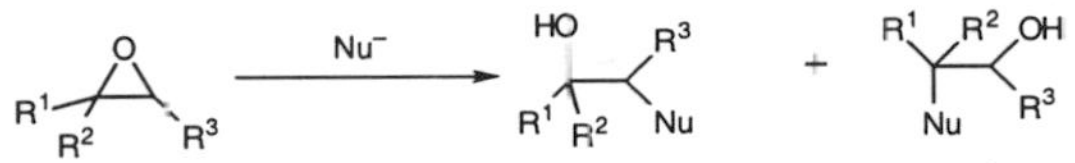

| Nucleophile | Reagent | Major Product[a] | Ref. |
|---|---|---|---|
| $RO^-/ROH$ | RONa | 1 | 143–148 |
| | ROH, $H^+$ | 1[b] | 90 |
| | ROH, CAN | 1 | 149 |
| | ROH, OPC[c] | 1[d] | 150,151 |
| | | 2 | 150,151 |
| | ROH, DDQ | 2 | 152 |
| $H_2O$ | $H_2O$[e] | —[f] | 153,154 |
| $RNH^-/RNH_2$ | $RNH_2$ | 1 | 155 |
| | $RNH_2$, $CoCl_2$ | 1 | 156 |
| | $RNH_2$, $LiClO_4$ | 1 | 157 |
| | RNHLi | 1 | 158 |
| | RNHNa | 1 | 144,159 |
| | RNHMgX | 1 | 160 |
| | $RNHPbR_3$ | 1 | 161 |
| | $RNHSnMe_3$ | 2 | 162,163 |
| | $RNHSbR_3$ | 2 | 139 |
| | $RNHSiR^1_3$[g] | 2 | 163 |
| | $RNHAlR_2$ | 2 | 164 |
| $R_2N^-/R_2NH$ | $R_2NH$ | 1 | 144 |
| | $R_2NMgBr$ | 1 | 165 |
| $N_3^-$ | $NaN_3$ | 1 | 166 |
| | $NaN_3$, $NH_4Cl$ | 1 | 167 |
| | $NaN_3$, $LiClO_4$ | 1 | 167 |
| | $Me_3SiN_3$, $Al(OPr\text{-}i)_3$ | 2 | 168 |
| $CN^-$ | $Me_3SiCN$, $Pd(CN)_2$ | 2 | 169,170 |
| | $Me_3SiCN$, $SnCl_2$ | 2 | 169,170 |
| | $Me_3SiCN$, $R_nAlX_{(3-n)}$ | 1 | 171 |
| | $Me_3SiCN$, $Yb(CN)_3$ | 1 | 172 |
| $RS^-$ | RSNa | 1 | 144,173 |
| $Cl^-/HCl$ | HCl | 1[b] | 90 |
| | $Li_2CuCl_4$ | 1 | 136 |
| | $R_3SiCl$ | 1[h] | 174 |
| $Br^-$ | $NiBr_4^{2-}$ | 1 | 175 |
| $H^-$ | $LiAlH_4$ | 1 | 176 |
| | $LiBH_4$ | 1 | 177 |
| $H^-$ (cont.) | Ra Ni | 2 | 144,176 |
| | NaI, AIBN, $Bu_3SnH$ | 1 | 178 |
| | $Zn(BH_4)_2$, $SiO_2$ | 2 | 179 |
| $R^-$ | $NaAlR_4$ | 1 | 132 |
| | $R_2Cu(CN)Li_2$ | 1 | 137 |
| | $R_2CuLi$ | 1 | 138,180 |
| | $MeC(ONa)=CHCO_2Et$ | 1 | 181 |
| | RMgX | 1 | 182,183 |
| | | 3[i] | 133,184 |
| $Me^-$ | $Me_3Al$ | 2 | 185 |
| $Allyl^-$ | AllylMgX, CuI | 1 | 12,186 |
| | $(Allyl)Ti(OR_4)_3$ | 2 | 134 |
| $RCH=CH^-$ | $RCH=CHAlR_2$ | 1 | 141 |
| $Ar^-$ | ArH, $AlCl_3$ | 1 | 187 |
| | ArLi | 1 | 188 |

[a] The representation is used to denote *trans*-addition at the least hindered carbon atom (**1**) or more hindered center (**2**). [b] Significant amounts of **2** can be formed and may even be the major product; the outcome of the reaction depends on the epoxide structure. [c] Organotin phosphate condensates. [d] When $R^2 = R^3 = H$. [e] The epoxide can be prepared *in situ*. [f] Not applicable. [g] In the presence of a Lewis acid. [h] The hydroxy group of the product is silylated. [i] The magnesium salt catalyzes rearrangement of the epoxide to an aldehyde ($R^3 = H$), which, in turn, reacts with the nucleophile to give 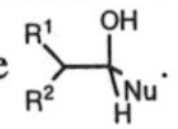 .

Treatment of an epoxide with strong base provides a route to allyl alcohols.[197-199] Chiral bases have also been employed in this reaction [*5.4.1*],[200] as have silicon reagents.[199,201,202] Kinetic resolutions of epoxides are also possible, giving access to 1,2-functionality.[203]

Epoxides can be transformed to substituted cyclopropanes by use of a functionalized sulfur-stabilized carbanion (Scheme 10.15). The reaction most likely involves a carbene, an alkene being generated by silicon-based chemistry.[204]

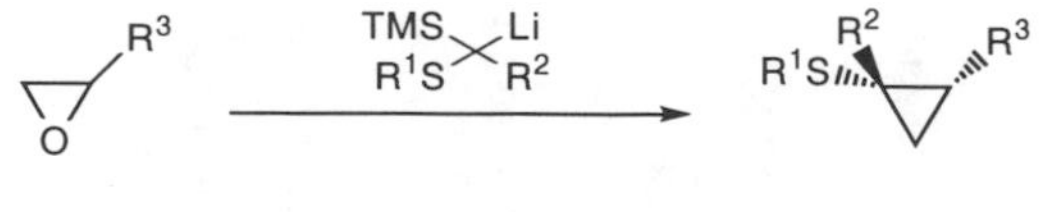

**SCHEME 10.15.**

Epoxides react with a sulfoxide stabilized anion. The resultant diastereoisomers can be separated by conventional resolution means. Reduction of the sulfur group then provides a methyl group equivalent (Scheme 10.16).[135]

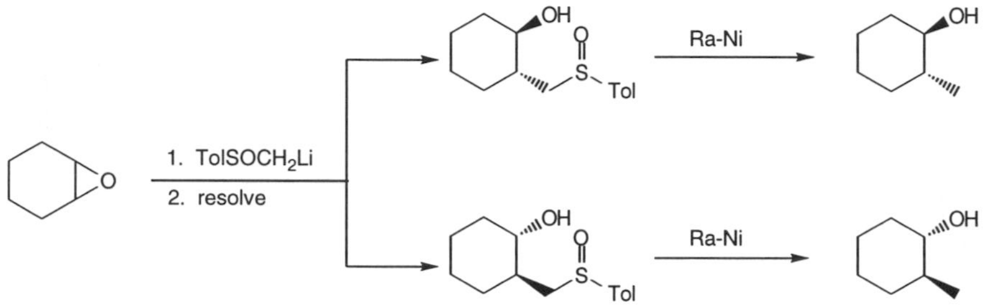

**SCHEME 10.16.**

Other heteroatom nucleophiles that allow further elaboration of the resultant alcohols have also been used effectively.[205-207]

## 10.3. 1,2-DIOLS

1,2-Diols are available by a number of methods. The sections below describe the one-step conversion of an alkene to a *cis*-1,2-diol. This methodology complements the epoxide approach to *trans*-1,2-diols where the *trans*-isomer is formed.

### 10.3.1. METAL OXIDE ADDITIONS

Many metal oxides have been employed to convert an alkene to a *cis*-1,2-diol.[208] This approach is complemented by nucleophilic opening of an epoxide, which invariably results in the net formation of a *trans*-1,2-diol.[153]

In both the osmium and manganese oxidations, an intermediate cyclic ester accounts for the *cis*-stereochemistry.[209-219] Reaction conditions have to be carefully controlled to avoid oxidative cleavage of the diol product.[220]

The hypervalent iodine reagent, [hydroxy(tosyloxy)iodo]benzene (**10.8**) oxidizes alkenes to the *syncat* tosylate by a *syn*-addition (Scheme 10.17);[221-223] however, the exact mechanism has yet to be determined.

### 10.3.2. OSMIUM REAGENTS

Osmium reagents can be used reliably to form *cis*-vicinal diols[224,225] from the less hindered face of a carbon–carbon double bond.[209,226-230]

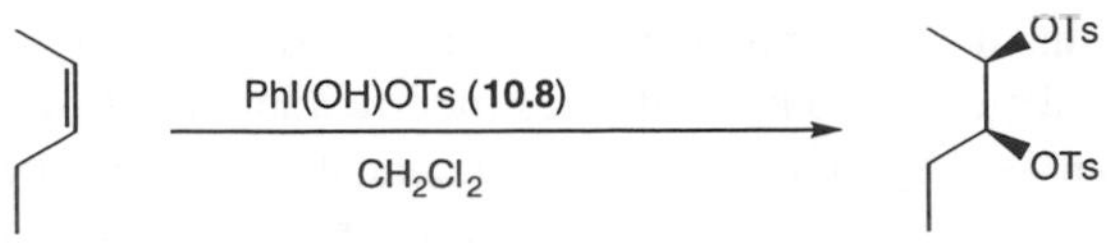

**SCHEME 10.17.**

Oxidation of an alkene by osmium tetroxide,[209,224,225,231] or alkaline potassium permanganate[220,232] occurs by *syn*-addition from the less hindered face of the double bond to form a cyclic ester.[217,228,233] This steric effect is amplified in cyclic substrates (Scheme 10.18).[234,235] Such an approach has been employed in the preparation of many natural products, including carbohydrate derivatives.[234-239]

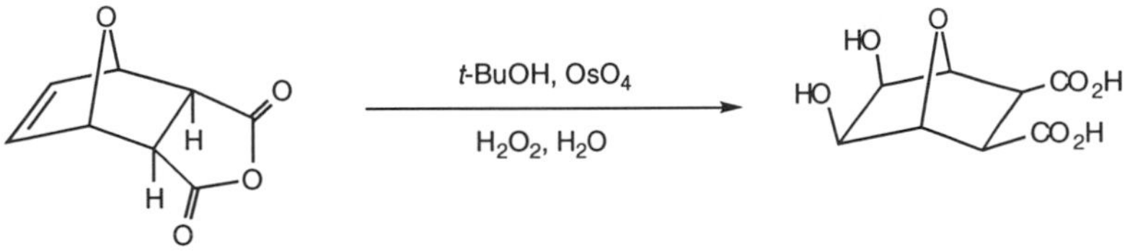

**SCHEME 10.18.**

Significant advances have been made towards an asymmetric transformation through the use of chiral ligands.[240-245] Early work on the oxidation of an alkene by osmium tetroxide in the presence of a chiral ligand, such as dihydroquinine acetate (**10.9**; R = Ac) or dihydroquinidine acetate (**10.10**; R = Ac), led to diol formation with some enantiomeric excess.[246] However, this asymmetric dihydroxylation problem has now been solved by the use of cinchona alkaloid esters (**10.9** and **10.10**; R = *p*-$ClC_6H_4$) together with a catalytic amount of osmium tetroxide. The alkaloid esters act as pseudoenantiomeric ligands (Scheme 10.19).[247-251] They can also be supported on a polymer.[252]

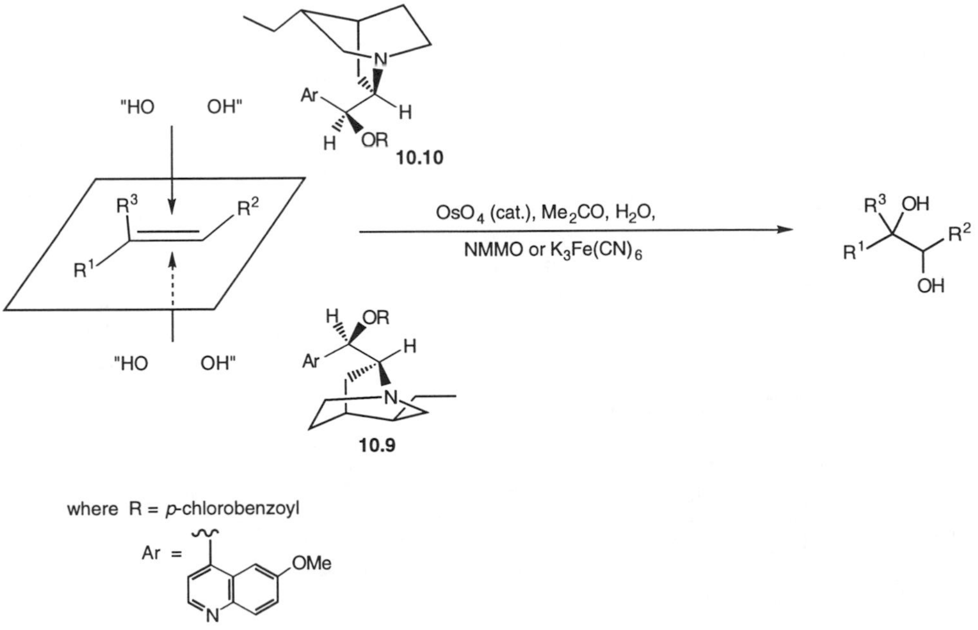

**SCHEME 10.19.**

The original procedure has been modified by the use of a slow addition of the alkene to afford the diol in higher optical purity, and ironically this modification results in a faster reaction. This behavior can be rationalized by consideration of two catalytic cycles operating for the alkene (Scheme 10.20); the use of low alkene concentrations effectively removes the second, low enantioselective cycle.[248,253] The use of potassium ferricyanide in place of *N*-methylmorpholine-*N*-oxide (NMMO) as oxidant also improves the level of asymmetric induction.[254,255]

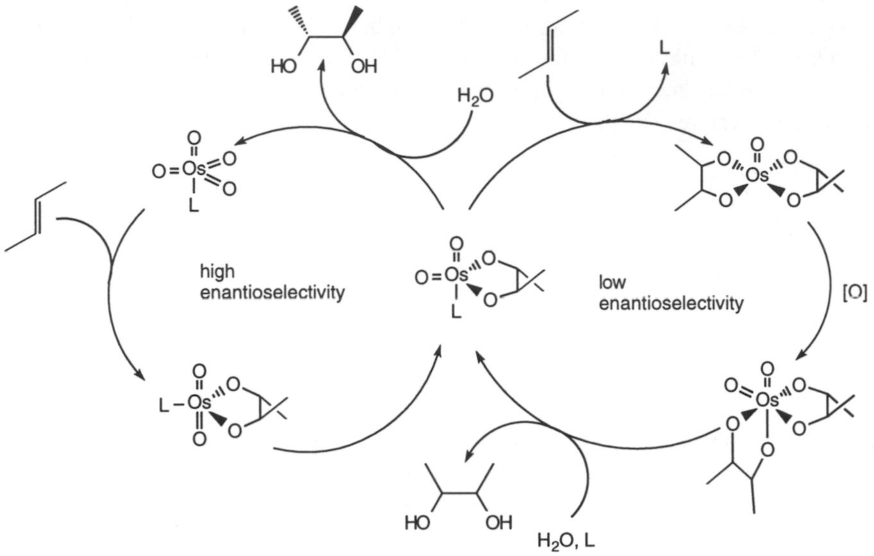

**SCHEME 10.20.**

X-ray, NMR, kinetic analyses, and theoretical approaches have provided insight into the mechanism of the oxidation.[243,248,251,256-261]

Although the use of cinchona alkaloids as chiral ligands does provide high asymmetric induction with a number of types of alkene, the search for better systems has continued.[245,262-267] Thus, the selectivities and scope continue to expand as illustrated in Scheme 10.21.[268-272]

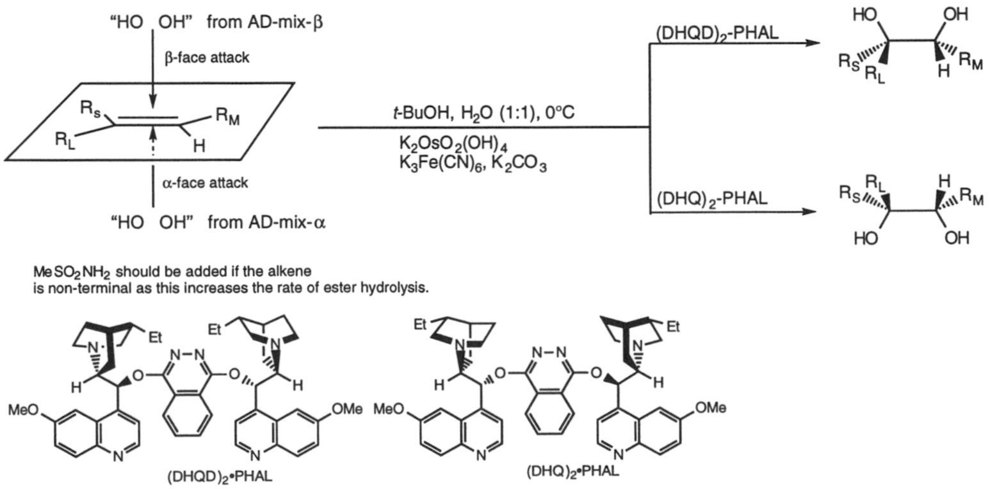

**SCHEME 10.21.**

The use of a mnemonic allows the prediction of the reaction outcome, and the reactivity. This model also works for other ligands (aside from those shown in Scheme 10.21). The steric requirements of the system indicate that a hydrogen (i.e., a trisubstituted alkene is the best substrate) is necessary for good asymmetric induction (Figure 10.4).[268]

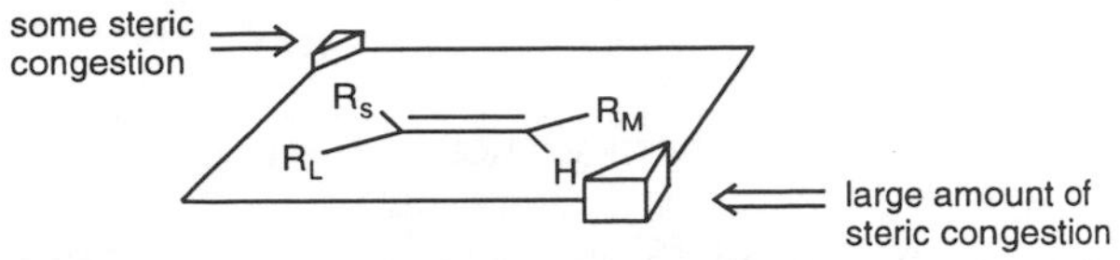

some steric congestion
$R_S$ $R_M$ $R_L$ H
large amount of steric congestion

**FIGURE 10.4.** Steric requirements for the asymmetric dihydroxylation procedure.

Kinetic resolutions can be achieved by the dihydroxylation approach.[273]

The 1,2-diols formed by the asymmetric oxidation can be used as substrates in a wide variety of transformations. Conversion of the hydroxy groups to *p*-toluenesulfonates then allows nucleophilic displacement by azide at both centers with inversion of configuration (Scheme 10.22).[274]

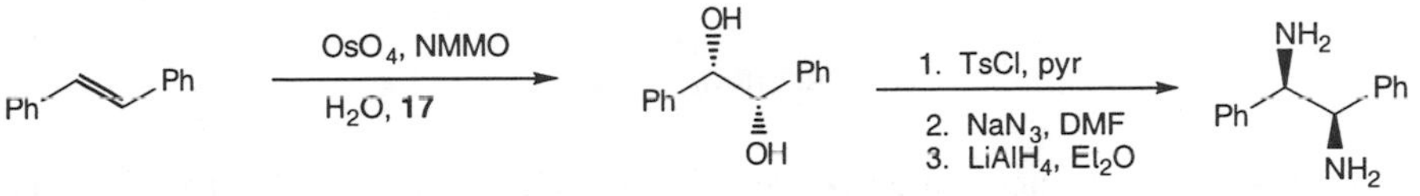

$OsO_4$, NMMO; $H_2O$, **17**
1. TsCl, pyr; 2. $NaN_3$, DMF; 3. $LiAlH_4$, $Et_2O$

**SCHEME 10.22.**

Reaction of 1,2-diols with *p*-toluenesulfonyl isocyanate, in the presence of a Pd(0) catalyst, gives the oxazolidinone with retention at both stereogenic centers.[275]

Allyl ethers can be used as substrates for this oxidative methodology, providing alternative strategies to the epoxidation of allyl alcohols [*11.1*].[276]

## 10.4. FORMATION AND REACTIONS OF CYCLIC SULFATES

Cyclic sulfates provide a useful alternative to epoxides now that production of a chiral diol from an alkene is viable. These cyclic compounds are prepared by reaction of the diol with thionyl chloride, followed by ruthenium catalyzed oxidation of the sulfur (Scheme 10.23).[277] This oxidation has the advantage over previous procedures, as it only uses a small amount of the transition metal catalyst.[278,279]

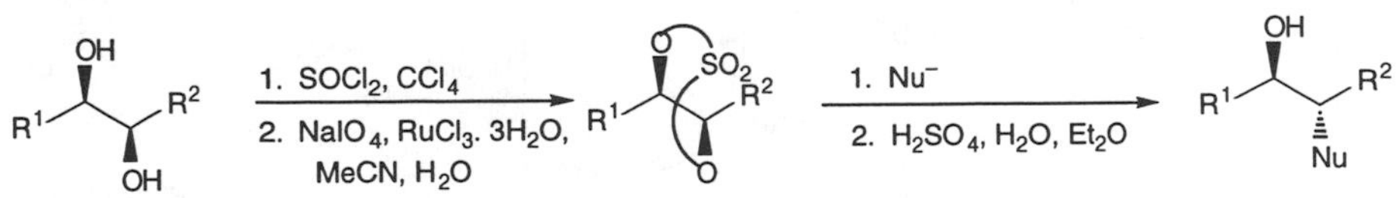

1. $SOCl_2$, $CCl_4$; 2. $NaIO_4$, $RuCl_3$. $3H_2O$, MeCN, $H_2O$
1. $Nu^-$; 2. $H_2SO_4$, $H_2O$, $Et_2O$

**SCHEME 10.23.**

The cyclic sulfates undergo ring opening with a wide variety of nucleophiles, such as hydride, azide, fluoride, benzoate, amines, and Grignard reagents. The reaction of an amidine with a cyclic sulfate provides an expeditious entry to chiral imidazolines (**10.11**) and 1,2-diamines (Scheme 10.24).[280]

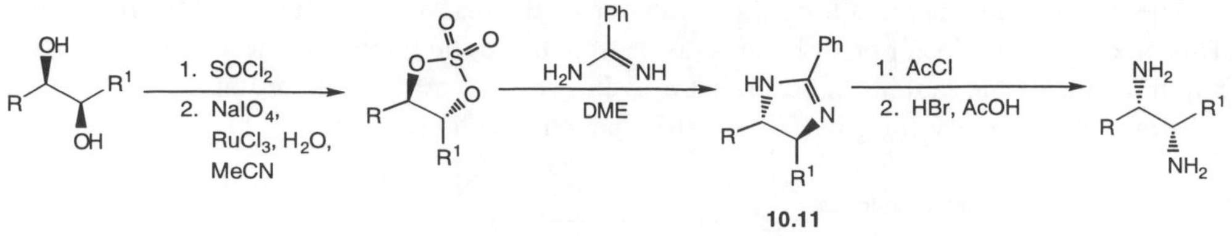

SCHEME 10.24.

In the case of an ester ($R^2 = CO_2Me$), the addition occurs exclusively at C-2 (Scheme 10.25); the analogous epoxide does not demonstrate such selectivity.[277,281-283] Terminal cyclic sulfates (Scheme 10.25; $R^2 = H$) open in a manner completely analogous to the corresponding epoxide.[281]

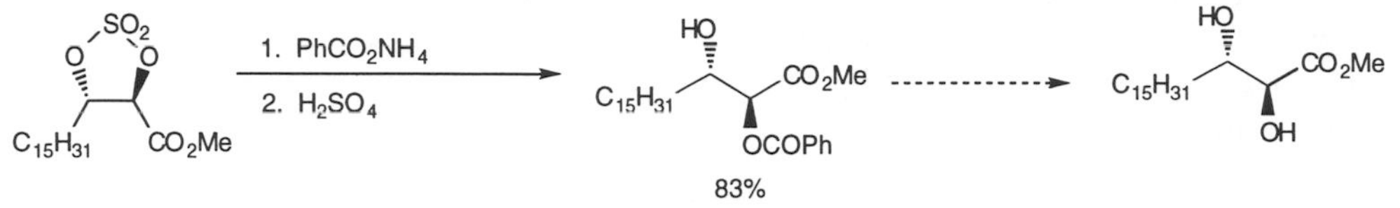

SCHEME 10.25.

The resultant sulfate ester can be converted to the alcohol by acid hydrolysis. If an acid sensitive group is present, this hydrolysis is still successful through use of a catalytic amount of sulfuric acid in the presence of 0.5 to 1.0 equivalents of water with tetrahydrofuran as solvent. The use of base in the formation of the cyclic sulfates themselves can also alleviate problems associated with acid sensitive groups.[256,284] 1,2-Cyclic sulfates provide methodology to aziridine derivatives (vide infra).[282,285]

A synthesis of (*R*)-reticuline included a comparison of epoxide and cyclic sulfate chemistry (Scheme 10.26).[286]

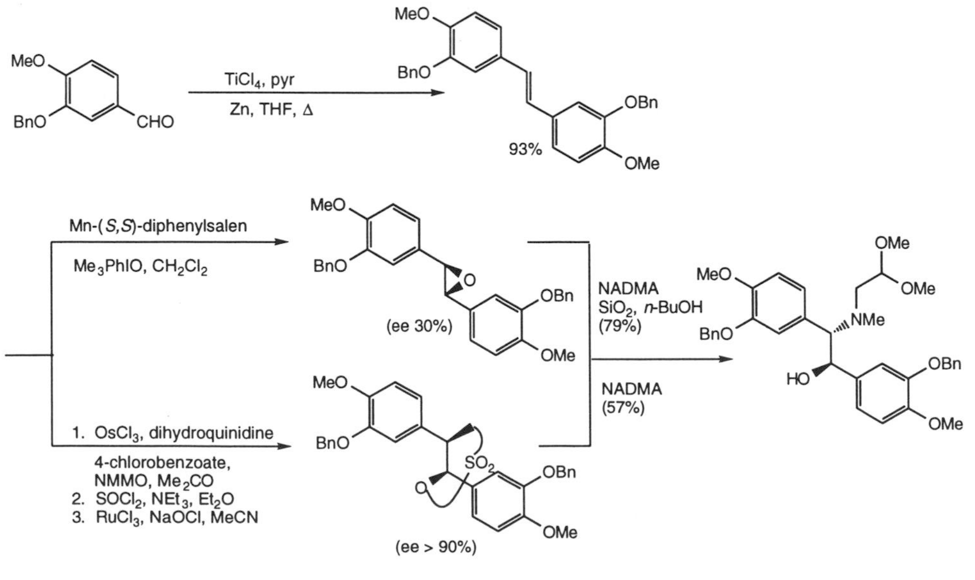

SCHEME 10.26.

The sulfites, obtained by reaction of the 1,2-diol with thionyl chloride (*cf.* Scheme 10.23), also undergo facile ring opening with concurrent inversion at the reaction center when treated with azide.[285]

## 10.5. PREPARATION AND REACTIONS OF AZIRIDINES

These nitrogen analogs of epoxides are of growing importance in organic synthesis.[287] They are available by a nitrene reaction with alkenes or by a ring closure approach analogous to epoxide formation.[287,288] For asymmetric preparations, the latter method holds potential as does methodology starting from amino acids;[289-293] *N*-tosyl aziridines are available from α-amino acids by a tosylation, reduction, cyclization sequence.[294] Some selectivity has been observed for the reaction of a nitrene to an alkene, the very reactive type of intermediate by incorporation of a chiral auxiliary.[295,296] Aziridines are also available through the chemistry of 1,2-cyclic sulfates (Scheme 10.27).[282,285]

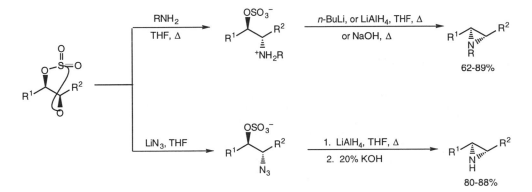

$RNH_2$, THF, Δ; $n$-BuLi, or $LiAlH_4$, THF, Δ or NaOH, Δ — 62-89%

$LiN_3$, THF; 1. $LiAlH_4$, THF, Δ 2. 20% KOH — 80-88%

**SCHEME 10.27.**

Variations on the manganese salen approach to epoxides also show promise for the preparation of chiral aziridines.[70,297] In addition to alternative ligands and metals, the nitrogen source can be [*N*-(*p*-toluenesulfonyl)imino]phenyliodinane.[70,298]

Aziridines can be used as precursors to β-lactams by use of a stereoselective ring expansion catalyzed by a rhodium(I) complex (Scheme 10.28.).[299]

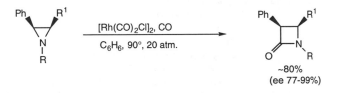

$[Rh(CO)_2Cl]_2$, CO; $C_6H_6$, 90°, 20 atm. — ~80% (ee 77-99%)

**SCHEME 10.28.**

In many respects, aziridines act in an analogous manner to epoxides.[300-318] An example is provided by the catalyzed opening with trimethylsilyl cyanide (Scheme 10.29).[319]

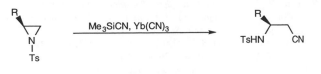

$Me_3SiCN$, $Yb(CN)_3$

**SCHEME 10.29.**

## 10.6. SUMMARY

Despite the plethora of useful reactions that employ epoxides, and the ability to control the position of ring opening, the usefulness of chiral epoxides from simple alkenes is blighted by the indirect approaches that have to be employed to obtain them. This problem has been alleviated, to a certain extent, by the development of asymmetric dihydroxylation methodology, that in turn, led to the manganese-salen reagents. Both of these approaches can provide useful ee's for the oxidation of simple alkenes. The dihydroxylation procedure is more promiscuous and, when coupled with the chemistry of cyclic sulfates, provides some useful general methods. For synthetic design, the use of cyclic sulfates offers the more attractive approach, unless other functionality is present within the substrate that allows for epoxidation formation without significant deviation from the synthetic strategy.

As low temperatures are not customary for this type of oxidation, the potential for a straightforward scale-up exists. This also carries over to the reactions of epoxides [*cf.* Chapter 11].

## 10.7. REFERENCES


1. Rebek, J.; Wolf, S.; Mossman, A.; *J. Org. Chem.* 1978, *43*, 180.
2. Emmons, W. D.; Pagano, A. S. *J. Am. Chem. Soc.* 1955, *77*, 89.
3. Groves, J. T.; Nemo, T. E. *J. Am. Chem. Soc.* 1983, *105*, 5786.
4. Groves, J. T.; Myers, R. S. *J. Am. Chem. Soc.* 1983, *105*, 5791.
5. Ostovic, D.; Bruice, T. C. *J. Am. Chem. Soc.* 1988, *110*, 6906.
6. Castellino, A. J.; Bruice, T. C. *J. Am. Chem. Soc.* 1988, *110*, 7512.
7. Castellino, A. J.; Bruice, T. C. *J. Am. Chem. Soc.* 1988, *110*, 1313.
8. Castellino, A. J.; Bruice, T. C. *J. Am. Chem. Soc.* 1988, *110*, 158.
9. Higuchi, T.; Ohtake, H.; Hirobe, M. *Tetrahedron Lett.* 1989, *30*, 6545.
10. Seuring, B.; Seebach, D. *Helv. Chim. Acta* 1977, *60*, 1175.
11. Kirby, A. J. *Adv. Phys. Org. Chem.* 1980, *17*, 183.
12. Johnston, B. D.; Slessor, K. N. *Can. J. Chem.* 1979, *57*, 233.
13. Swern, D. In *Organic Peroxides*; D. Swern, Ed.; Wiley-Interscience: New York, Vol. 2; 1971, pp 355.
14. Kishi, Y. *Aldrichimica Acta* 1980, *13*, 23.
15. Katsuki, T. *J. Syn. Org. Chem., Jpn.* 1987, *45*, 90.
16. Bartlett, P. D. *Rec. Chem. Prog.* 1950, 47.
17. Ogata, Y.; Tabushi, I. *J. Am. Chem. Soc.* 1961, *83*, 3444.
18. Henbest, H. B. *Chem. Soc., Spec. Publ.* 1965, *19*, 83.
19. Kwart, H.; Hoffman, D. M. *J. Org. Chem.* 1966, *31*, 419.
20. Berti, G. *Top. Stereochem.* 1973, *7*, 93.
21. Dryuk, V. G. *Tetrahedron* 1976, *32*, 2855.
22. Plesnicar, B.; Tasevski, M.; Azman, A. *J. Am. Chem. Soc.* 1978, *100*, 743.
23. Montanari, F.; Moretti, I.; Torre, G. *Gazz. Chim. Ital.* 1974, *104*, 7.
24. Ewins, R. C.; Henbest, H. B.; McKervey, M. A. *J. Chem. Soc., Chem. Commun.* 1967, 1085.
25. Rebek, J.; Marshall, L.; Wolak, R.; McMannis, J. *J. Am. Chem. Soc.* 1984, *106*, 1170.
26. Rastetter, W. H.; Richard, T. J.; Lewis, M. D. *J. Org. Chem.* 1978, *43*, 3163.
27. Fringuelli, F.; Germani, R.; Pizzo, F.; Savelli, G. *Tetrahedron Lett.* 1989, *30*, 1427.
28. Brougham, P.; Cooper, M. S.; Cummerson, D. A.; Heaney, H.; Thompson, N. *Synthesis* 1987, 1015.
29. Rebek, J.; McReady, R. *Tetrahedron Lett.* 1979, 1001.
30. Montanari, F.; Moretti, I.; Torre, G. *J. Chem. Soc., Chem. Commun.* 1969, 135.
31. Yonezawa, T.; Kato, H.; Yamaoto, O. *Bull. Chem. Soc., Jpn.* 1967, *40*, 307.
32. Bowman, R. M.; Collins, J. F.; Grundon, M. F. *J. Chem. Soc., Chem. Commun.* 1967, 1131.
33. Bowman, R. M.; Grundon, M. F. *J. Chem. Soc. (C)* 1967, 2368.
34. Davis, F. A.; Harakal, M. E.; Awad, S. B. *J. Am. Chem. Soc.* 1983, *105*, 3123.
35. Davis, F. A.; Sheppard, A. C. *Tetrahedron* 1989, *45*, 5703.
36. Davis, F. A.; Chattopadhyay, S. *Tetrahedron Lett.* 1986, *27*, 5079.

37. Davis, F. A.; ThimmaReddy, R.; McCauley, J. P.; Przeslawski, R. M.; Harakal, M. E.; Carroll, P. J. *J. Org. Chem.* 1991, *56*, 809.
38. Davis, F. A.; Vishwakarma, L. C.; Billmers, J. M.; Finn, J. *J. Org. Chem.* 1984, *49*, 3241.
39. Davis, F. A.; Billmers, J. M.; Gosciniak, D. J.; Towson, J. C.; Bach, R. D. *J. Org. Chem.* 1986, *51*, 4240.
40. Bach, R. D.; Wolber, G. J. *J. Am. Chem. Soc.* 1984, *106*, 1410.
41. Jorgensen, K. A. *Chem. Rev.* 1989, *89*, 431.
42. Mimoun, H.; Roch, I. S. D.; Sajus, L. *Tetrahedron* 1970, *26*, 37.
43. Sheng, M. N.; Zajacek, J. G. *J. Org. Chem.* 1970, *35*, 1839.
44. Breslow, R.; Maresca, L. M. *Tetrahedron Lett.* 1977, 623.
45. Sharpless, K. B.; Verhoeven, T. R. *Aldrichimica Acta* 1979, *12*, 63.
46. Sheldon, R. A. *J. Mol. Catal.* 1980, *7*, 107.
47. Sharpless, K. B.; Teranishi, A. Y.; Backvall, J. *J. Am. Chem. Soc.* 1977, *99*, 3120.
48. Schurig, V.; Betschinger, F. *Chem. Rev.* 1992, *92*, 873.
49. Mimoun, H.; Saussine, L.; Daire, E.; Postel, M.; Fischer, J.; Weiss, R. *J. Am. Chem. Soc.* 1983, *105*, 3101.
50. Tolstikov, G. A.; Yur'ev, V. P.; Dzhemilev, U. M. *Russ. Chem. Rev. (Engl. Trans.)* 1975, *44*, 319.
51. Bortolini, O.; Furia, F. D.; Scrimin, P.; Modena, G. *J. Mol. Catal.* 1980, *7*, 59.
52. Balavoine, G.; Eskenazi, C.; Meunier, F.; Riviere, H. *Tetrahedron Lett.* 1984, *25*, 3187.
53. Prandi, J.; Kagan, H. B.; Mimoun, H. *Tetrahedron Lett.* 1986, *27*, 2617.
54. Quenard, M.; Bonmarin, V.; Gelbard, G. *Tetrahedron Lett.* 1987, *28*, 2237.
55. Kato, J.-i.; Ota, H.; Matsukawa, K.; Endo, T. *Tetrahedron Lett.* 1988, *29*, 2843.
56. Okamoto, Y.; Still, W. C. *Tetrahedron Lett.* 1988, *29*, 971.
57. Kureshy, R. I.; Khan, N. H.; Abdi, S. H. R.; Bhatt, K. N. *Tetrahedron: Asymmetry* 1993, *4*, 1693.
58. Colleti, S. L.; Halterman, R. L. *Tetrahedron Lett.* 1992, *33*, 1005.
59. Tani, K.; Hanafusa, M.; Otsuka, S. *Tetrahedron Lett.* 1979, 3017.
60. Kagan, H. B.; Mimoun, H.; Mark, C.; Schurig, V. *Angew. Chem., Int. Ed. Engl.* 1979, *18*, 485.
61. Sinigalia, R.; Michelin, R. A.; Pinna, F.; Strukul, G. *Organometal.* 1987, *6*, 728.
62. Zhang, W.; Loebach, J. L.; Wilson, S. R.; Jacobsen, E. N. *J. Am. Chem. Soc.* 1990, *112*, 2801.
63. Irie, R.; Noda, K.; Ito, Y.; Matsumoto, N.; Katsuki, T. *Tetrahedron Lett.* 1990, *31*, 7345.
64. Zhang, W.; Loebach, J. L.; Wilson, S. R.; Jacobsen, E. J. *J. Am. Chem. Soc.* 1990, *112*, 2801.
65. Van Draanen, N. A.; Arseniyadis, S.; Crimmins, M. T.; Heathcock, C. H. *J. Org. Chem.* 1991, *56*, 2499.
66. Schwenkreis, T.; Berkessel, A. *Tetrahedron Lett.* 1993, *34*, 4785.
67. Zhang, W.; Jacobsen, E. N. *J. Org. Chem.* 1991, *56*, 2296.
68. Deng, L.; Jacobsen, E. N. *J. Org. Chem.* 1992, *57*, 4320.
69. Irie, R.; Noda, K.; Ito, Y.; Katsuki, T. *Tetrahedron Lett.* 1991, *32*, 1055.
70. O'Connor, K. J.; Wey, S.-J.; Burrows, C. J. *Tetrahedron Lett.* 1992, *33*, 1001.
71. Yasuda, A.; Tanaka, S.; Yamamoto, H.; Nozaki, H. *Bull. Chem. Soc., Jpn.* 1979, *52*, 1701.
72. Bell, T. W.; Clacclo, J. A. *Tetrahedron Lett.* 1988, *29*, 865.
73. Bruice, T. C. *Aldrichimica Acta* 1988, *21*, 87.
74. Traylor, T. G.; Xu, F. *J. Am. Chem. Soc.* 1988, *110*, 1953.
75. Groves, J. T.; Viski, P. *J. Org. Chem.* 1990, *55*, 3628.
76. Nee, M. W.; Bruice, T. C. *J. Am. Chem. Soc.* 1982, *104*, 6123.
77. Traylor, T. G.; Miksztal, A. R. *J. Am. Chem. Soc.* 1989, *111*, 7443.
78. Traylor, T. G.; Fann, W.-P.; Bandyopadhyay, D. *J. Am. Chem. Soc.* 1989, *111*, 8009.
79. Collman, J. P.; Zhang, X.; Hembre, R.-T.; Brauman, J. I. *J. Am. Chem. Soc.* 1990, *112*, 5356.
80. Ledon, H. J.; Durbut, P.; Varlscon, F. *J. Am. Chem. Soc.* 1981, *103*, 3601.
81. Adam, W.; Richter, M. J. *Acc. Chem. Res.* 1994, *27*, 57.
82. Ostovic, D.; Bruice, T. C. *Acc. Chem. Res.* 1992, *25*, 314.
83. Saalfrank, R. W.; Reihs, S.; Hug, M. *Tetrahedron Lett.* 1993, *34*, 6033.
84. Traylor, T. G.; Tsuchiya, S.; Byun, Y.-S.; Kim, C. *J. Am. Chem. Soc.* 1993, *115*, 2775.
85. Hirao, T.; Moriuchi, T.; Mikami, S.; Ikeda, I.; Ohshiro, Y. *Tetrahedron Lett.* 1993, *34*, 1031.
86. Breslow, R.; Brown, A. B.; McCullough, R. D.; White, P. W. *J. Am. Chem. Soc.* 1989, *111*, 4517.
87. Allain, E. J.; Hager, L. P.; Deng, L.; Jacobsen, E. N. *J. Am. Chem. Soc.* 1993, *115*, 4415.
88. March, J. *Advanced Organic Chemistry: Reactions Mechanisms and Structure;* 3rd ed.; John Wiley & Sons, New York, 1985.
89. Townsend, J. M.; Sharpless, K. B. *Tetrahedron Lett.* 1972, 3313.
90. Parker, R. E.; Isaacs, N. S. *Chem. Rev.* 1959, *59*, 737.
91. Williams, N. R. *Adv. Carbohydr. Chem. Biochem.* 1970, *25*, 109.
92. Fujiwara, M.; Hitomi, K.; Baba, A.; Matsuda, H. *Synthesis* 1990, 106.
93. Takeichi, T.; Arihara, M.; Ishimori, M.; Tsuruta, T. *Tetrahedron* 1980, *36*, 3391.
94. Dehmlow, E. V.; Singh, P.; Heider, J. *J. Chem. Res. (S)* 1981, 292.

95. Hummelen, J. C.; Wynberg, H. *Tetrahedron Lett.* 1978, 1089.
96. Oppolzer, W.; Dudfield, P. *Tetrahedron Lett.* 1985, *26*, 5037.
97. Corey, E. J.; Helal, C. J. *Tetrahedron Lett.* 1993, *34*, 5227.
98. Corey, E. J.; Chaykovsky, M. *J. Am. Chem. Soc.* 1965, *87*, 1353.
99. Johnson, C. R.; Schroeck, C. W. *J. Am. Chem. Soc.* 1971, *93*, 5303.
100. Trost, B. M.; Melvin, L. S. *Sulfur Ylides*; Academic Press: New York, 1975.
101. Shanklin, J. R.; Johnson, C. R.; Ollinger, J.; Coates, R. M. *J. Am. Chem. Soc.* 1973, *95*, 3429.
102. Takaki, K.; Yasumura, M.; Negoro, K. *Angew. Chem., Int. Ed. Engl.* 1981, *20*, 671.
103. Shi, L.-L.; Zhou, Z.-L.; Huang, Y.-Z. *Tetrahedron Lett.* 1990, *31*, 4173.
104. Still, W. C.; Novack, N. J. *J. Am. Chem. Soc.* 1981, *103*, 1283.
105. Allen, D. G.; Roberts, N. K.; Wild, S. B. *J. Chem. Soc., Chem. Commun.* 1978, 346.
106. Lloyd, D.; Gosney, I.; Ormistead, R. A. *Chem. Rev.* 1987, *16*, 45.
107. Gosney, I.; Lillie, T. J.; Lloyd, D. *Angew. Chem., Int. Ed. Engl.* 1977, *16*, 487.
108. Trippett, S.; Walker, M. A. *J. Chem. Soc. (C)* 1971, 1114.
109. Satoh, T.; Oohara, T.; Yamakawa, K. *Tetrahedron Lett.* 1988, *29*, 2851.
110. Johnson, C. R.; Schroelk, C. W. *J. Am. Chem. Soc.* 1973, *95*, 7418.
111. Kano, S.; Yokomatsu, T.; Shibuya, S. *J. Chem. Soc., Chem. Commun.* 1978, 785.
112. Solladie, G.; Demailly, G.; Creck, C. *Tetrahedron Lett.* 1985, *26*, 435.
113. Yeates, C.; Street, S. D. A.; Kocienski, P.; Campbell, S. F. *J. Chem. Soc., Chem. Commun.* 1985, 1388.
114. Williams, D. R.; Phillips, J. G. *Tetrahedron* 1986, *42*, 3013.
115. Cimetiere, B.; Jacob, L.; Julia, M. *Tetrahedron Lett.* 1986, *27*, 6329.
116. Trost, B. M.; Hammen, R. F. *J. Am. Chem. Soc.* 1973, *95*, 962.
117. Ohta, H.; Matsumoto, S.; Sugai, T. *Tetrahedron Lett.* 1990, *31*, 2895.
118. Satoh, T.; Oohara, T.; Ueda, Y.; Yamakawa, K. *J. Org. Chem.* 1989, *54*, 3130.
119. Satoh, T.; Kaneko, Y.; Yamakawa, K. *Tetrahedron Lett.* 1986, *27*, 2379.
120. Hiyama, T.; Mishima, T.; Sawada, H.; Nozaki, H. *J. Am. Chem. Soc.* 1975, *97*, 1626.
121. Breau, L.; Ogilvie, W. W.; Durst, T. *Tetrahedron Lett.* 1990, *31*, 35.
122. Furukawa, N. F.; Sugihara, Y.; Fujihara, H. *J. Org. Chem.* 1989, *54*, 4222.
123. Shimagaki, M.; Matsuzaki, Y.; Hori, I.; Nakata, T.; Oishi, T. *Tetrahedron Lett.* 1984, *25*, 4779.
124. Mori, K.; Kbata, T. *Tetrahedron* 1986, *42*, 3471.
125. Kurth, M. J.; Abreo, M. A. *Tetrahedron* 1990, *46*, 5085.
126. Marczak, S.; Masnyk, M.; Wicha, J. *Tetrahedron Lett.* 1989, *30*, 2845.
127. Kang, S.-K.; Kim, Y.-S.; Lim, J.-S.; Kim, K.-S.; Kim, S.-G. *Tetrahedron Lett.* 1991, *32*, 363.
128. Palomino, E.; Schaap, A. P.; Heeg, M. J. *Tetrahedron Lett.* 1989, *30*, 6797.
129. Sunay, U.; Mootoo, D.; Molino, B.; Fraser-Reid, B. *Tetrahedron Lett.* 1986, *27*, 4697.
130. Rao, A. S.; Paknikar, S. K.; Kirtane, J. G. *Tetrahedron* 1983, *39*, 2323.
131. Bartók, M.; Láng, K. L. *Chem. Heterocycl. Comp.* 1984, *42 (pt.3)*, 1.
132. Boireau, G.; Abenhaim, D.; Bernardon, C.; Henry-Basch, E.; Sabourault, B. *Tetrahedron Lett.* 1975, 2521.
133. Schauder, J. R.; Krief, A. *Tetrahedron Lett.* 1982, *23*, 4389.
134. Tanaka, T.; Inoue, T.; Kamei, K.; Murakami, K.; Iwata, C. *J. Chem. Soc., Chem. Commun.* 1990, 906.
135. Tsuchihashi, G.-I.; Iriuchijima, S.; Ishibashi, M. *Tetrahedron Lett.* 1972, 4605.
136. Ciaccio, J. A.; Addess, K. J.; Bell, T. W. *Tetrahedron Lett.* 1986, *27*, 3697.
137. Lipshutz, B. H.; Kozlowski, J.; Wilhelm, R. S. *J. Am. Chem. Soc.* 1982, *104*, 2305.
138. Herr, R. W.; Wieland, D. M.; Johnson, C. R. *J. Am. Chem. Soc.* 1970, *92*, 3813.
139. Fujiwara, M.; Imada, M.; Baba, A.; Matsuda, H. *Tetrahedron Lett.* 1989, *30*, 739.
140. Mori, K.; Tamada, S. *Tetrahedron* 1979, *35*, 1279.
141. Alexakis, A.; Jachiet, D. *Tetrahedron* 1989, *45*, 6197.
142. Flippin, L. A.; Brown, P. A.; Jalali-Araghi, K. *J. Org. Chem.* 1989, *54*, 3588.
143. Chitwood, H. C.; Freure, B. T. *J. Am. Chem. Soc.* 1946, *68*, 680.
144. Graham, A. R.; Millidge, A. F.; Young, D. P. *J. Chem. Soc.* 1954, 2180.
145. Petrov, A. A.; Gantseva, B. V.; Kiseleva, O. A. *Zh. Obsh. Khim.* 1953, *23*, 737.
146. McSweeney, G. P.; Wiggins, L. F.; Wood, D. J. C. *J. Chem. Soc.* 1952, 37.
147. Sexton, A. R.; Britton, E. C. *J. Am. Chem. Soc.* 1948, *70*, 3606.
148. Winstein, S.; Ingraham, L. L. *J. Am. Chem. Soc.* 1952, *74*, 1160.
149. Iranpoor, N.; Baltork, I. M. *Synth. Commun.* 1990, *20*, 2789.
150. Otera, J.; Yoshida, Y.; Hirakawa, K. *Tetrahedron Lett.* 1985, *32*, 3219.
151. Otera, J.; Niibo, Y.; Tatsumi, N.; Nozaki, H. *J. Org. Chem.* 1988, *53*, 275.
152. Iranpoor, N.; Baltork, I. M. *Tetrahedron Lett.* 1990, *31*, 735.
153. Fringuelli, F.; Germani, R.; Pizzo, F.; Savelli, G. *Synth. Commun.* 1989, *19*, 1939.
154. Kotsuki, H.; Kataoka, M.; Nishizawa, H. *Tetrahedron Lett.* 1993, *34*, 4031.
155. Coote, S. J.; Davies, S. G.; Middlemus, D.; Naylor, A. *J. Chem. Soc., Perkin Trans. 1* 1989, 2223.

156. Iqbal, J.; Pandey, A. *Tetrahedron Lett.* 1990, *31*, 575.
157. Chini, M.; Crotti, P.; Macchia, F. *Tetrahedron Lett.* 1990, *31*, 4661.
158. Kissel, C. L.; Richborn, B. *J. Org. Chem.* 1972, *37*, 2060.
159. Birch, A. J. *J. Proc. Roy. Soc., N. S. Wales* 1949, *83*, 245.
160. Carre, M. C.; Houmouou, J. P.; Caubere, P. *Tetrahedron Lett.* 1985, *26*, 3107.
161. Yamada, J.-I.; Yumoto, M.; Yamamoto, Y. *Tetrahedron Lett.* 1989, *30*, 4255.
162. Fiorenza, M.; Ricci, A.; Taddei, M.; Tassi, D. *Synthesis* 1983, 640.
163. Papini, A.; Ricci, A.; Taddei, M.; Secondi, G.; Dembech, P. *J. Chem. Soc., Perkin Trans. 1* 1984, 2261.
164. Overman, L. E.; Flippin, L. A. *Tetrahedron Lett.* 1981, *22*, 195.
165. Chadha, A.; Goergens, U.; Schneider, M. P. *Tetrahedron: Asymmetry* 1993, *4*, 1449.
166. VanderWerf, C. A.; Heisler, R. Y.; McEwan, W. E. *J. Am. Chem. Soc.* 1954, *76*, 1231.
167. Chini, M.; Crotti, P.; Macchia, F. *Tetrahedron Lett.* 1990, *31*, 5641.
168. Sutowardoyo, K.; Emziane, M.; Sinou, D. *Tetrahedron Lett.* 1989, *30*, 4673.
169. Gassman, P. G.; Guggenheim, T. L. *J. Am. Chem. Soc.* 1982, *104*, 1849.
170. Gassman, P. G.; Guggenheim, T. L. *Org. Synth.* 1985, *64*, 39.
171. Imi, K.; Yanagihara, N.; Utimoto, K. *J. Org. Chem.* 1987, *52*, 1013.
172. Matsubara, S.; Onishi, H.; Utimoto, K. *Tetrahedron Lett.* 1990, *31*, 6209.
173. Schuetz, R. D. *J. Am. Chem. Soc.* 1951, *73*, 1881.
174. Andrews, G. C.; Crawford, T. C.; Contello, L. G. *Tetrahedron Lett.* 1981, *22*, 3803.
175. Dawe, R. D.; Molinski, T. F.; Turner, J. V. *Tetrahedron Lett.* 1984, *25*, 2061.
176. Newman, M. S.; Underwood, G.; Renoll, M. W. *J. Am. Chem. Soc.* 1949, *71*, 3362.
177. Fuchs, R.; VanderWerf, C. A. *J. Am. Chem. Soc.* 1954, *76*, 1631.
178. Bonini, C.; Fabio, R. D.; Sotgiu, G.; Cavagnero, S. *Tetrahedron* 1989, *45*, 2895.
179. Ranu, B. C.; Das, A. R. *J. Chem. Soc., Chem. Commun.* 1990, 1334.
180. Hanamoto, T.; Katsuki, T.; Yamaguchi, M. *Tetrahedron Lett.* 1987, *28*, 6191.
181. Adams, R. M.; VanderWerf, C. A. *J. Am. Chem. Soc.* 1950, *72*, 4368.
182. Blaise, E. E. *C. R. Acad. Sci.* 1902, *134*, 551.
183. van der Louw, J.; Out, G. J. J.; van der Baan, J. L.; Bichelhaupt, F.; Klumpp, G. W. *Tetrahedron Lett.* 1989, *30*, 4863.
184. Koreeda, M.; Koizumi, N. *Tetrahedron Lett.* 1978, 1641.
185. Fukumasa, M.; Furuhashi, K.; Umezawa, J.; Takahashi, O.; Hirai, T. *Tetrahedron Lett.* 1991, *32*, 1059.
186. Derguini-Boumechal, F.; Lome, R.; Linstrumelle, G. *Tetrahedron Lett.* 1977, 1181.
187. Likhterov, V. R.; Etlis, V. S. *Zh. Obsh. Khim.* 1957, *27*, 2867.
188. Cristol, S. J.; Douglass, J. R.; Meek, J. S. *J. Am. Chem. Soc.* 1951, *73*, 816.
189. Niibo, Y.; Nakata, T.; Otera, J.; Nozaki, H. *Synlett* 1991, 97.
190. Reetz, M. T.; Plachky, M. *Synthesis* 1976, 199.
191. Dervan, P. B.; Shippey, M. A. *J. Am. Chem. Soc.* 1976, *98*, 1265.
192. Sonnet, P. E. *Synthesis* 1980, 828.
193. Inokuchi, T.; Kawafuchi, H.; Torii, S. *Synlett* 1992, 510.
194. Mori, K.; Tamada, S.; Matsui, M. *Tetrahedron Lett.* 1978, 901.
195. Corey, E. J.; Mehrotra, M. M. *Tetrahedron Lett.* 1986, *27*, 5173.
196. Joshi, N. N.; Srebnik, M.; Brown, H. C. *J. Am. Chem. Soc.* 1988, *110*, 6246.
197. Yandovskii, V. N.; Ershov *Russ. Chem. Rev.* 1972, *41*, 403.
198. Mosset, P.; Manna, S.; Vials, J.; Falck, J. R. *Tetrahedron Lett.* 1986, *27*, 299.
199. Murata, S.; Suzuki, M.; Noyori, R. *J. Am. Chem. Soc.* 1979, *101*, 2738.
200. Leonard, J.; Hewitt, J. D.; Ouali, D.; Simpson, S. J.; Newton, R. F. *Tetrahedron Lett.* 1990, *31*, 6703.
201. Detty, M. R. *J. Org. Chem.* 1980, *45*, 924.
202. Sakurai, H.; Sasaki, K.; Hosomi, A. *Tetrahedron Lett.* 1980, *21*, 2329.
203. Naruse, Y.; Esaki, T.; Yamamoto, H. *Tetrahedron Lett.* 1988, *29*, 1417.
204. Schaumann, E.; Friese, C. *Tetrahedron Lett.* 1989, *30*, 7033.
205. Rücker, C. *Tetrahedron Lett.* 1984, *25*, 4349.
206. Kabat, M. M.; Wicha, J. *Tetrahedron Lett.* 1991, *32*, 1073.
207. Fleming, I.; Floyd, C. *J. Chem. Soc., Perkin Trans. 1* 1981, 969.
208. Mimoun, H.; Roch, I. S. D.; Sajus, L. *Bull. Soc. Chim., Fr.* 1969, 1481.
209. Schroder, M. *Chem. Rev.* 1980, *80*, 187.
210. Lee, D. G.; Brownridge, J. R. *J. Am. Chem. Soc.* 1973, *95*, 3033.
211. Lee, D. G.; Brownridge, J. R. *J. Am. Chem. Soc.* 1974, *96*, 5517.
212. Wiberg, K. B.; Deutsch, C. J.; Rocek, J. *J. Am. Chem. Soc.* 1973, *95*, 3034.
213. Simandi, L. I.; Jaky, M. *J. Am. Chem. Soc.* 1976, *98*, 1995.
214. Lee, D. G.; Brown, K. C. *J. Am. Chem. Soc.* 1982, *104*, 5076.
215. Ogino, T.; Mochizuki, T. *Chem. Lett.* 1979, 433.

216. Ogino, T. *Tetrahedron Lett.* 1980, *21*, 177.
217. Sivik, M. R.; Gallucci, J. C.; Paquette, L. A. *J. Org. Chem.* 1990, *55*, 391.
218. Göbel, T.; Sharpless, K. B. *Angew. Chem., Int. Ed. Engl.* 1993, *32*, 1329.
219. Stewart, R. *Oxidation Mechanisms*; Benjamin: New York, 1964.
220. Ogino, T.; Mochizuki, K. *Chem. Lett.* 1979, 443.
221. Moriarty, R. M.; Vaid, R. K.; Koser, G. F. *Synlett* 1990, 365.
222. Koser, G. F.; Rebrovic, L.; Wettach, R. H. *J. Org. Chem.* 1981, *46*, 4324.
223. Rebrovic, L.; Koser, G. F. *J. Org. Chem.* 1984, *49*, 2462.
224. Milas, N. A.; Sussman, S. *J. Am. Chem. Soc.* 1936, *58*, 1302.
225. Criegee, R. *Annalen* 1936, *522*, 75.
226. Vanhheenan, V.; Kelly, R. C.; Cha, D. Y. *Tetrahedron Lett.* 1976, 1973.
227. Corey, E. J.; Pan, B.-C.; Hua, D. H.; Deardorff, D. R. *J. Am. Chem. Soc.* 1982, *104*, 6816.
228. Ray, R.; Matteson, D. S. *Tetrahedron Lett.* 1980, *21*, 449.
229. Cainelli, G.; Contento, M.; Manescalchi, F.; Plessi, L. *Synthesis* 1989, 45.
230. Sharpless, K. B.; Akashi, K. *J. Am. Chem. Soc.* 1976, *98*, 1986.
231. Akashi, K.; Palermo, R. E.; Sharpless, K. B. *J. Org. Chem.* 1978, *43*, 2063.
232. Dockx, J. *Synthesis* 1973, 441.
233. Paaren, H.; Schnoes, H. K.; DeLuca, H. F. *J. Org. Chem.* 1983, *48*, 3819.
234. Ager, D. J.; East, M. B. *Tetrahedron* 1993, *49*, 5683.
235. Daniels, R.; Fischer, J. L. *J. Org. Chem.* 1963, *28*, 320.
236. Ager, D. J.; East, M. B. *Tetrahedron* 1992, *48*, 2803.
237. Ager, D. J.; East, M. B. *J. Chem. Soc., Chem. Commun.* 1989, 178.
238. Kelly, R. C.; Schletter, I. *J. Am. Chem. Soc.* 1973, *95*, 7156.
239. Just, G.; Martel, A. *Tetrahedron Lett.* 1973, 1517.
240. Tokles, M.; Snyder, J. K. *Tetrahedron Lett.* 1986, *27*, 3951.
241. Yamada, T.; Narasaka, K. *Chem. Lett.* 1986, *131*.,
242. Hirama, M.; Oishi, T.; Ito, S. *J. Chem. Soc., Chem. Commun.* 1989, 665.
243. Corey, E. J.; Lotto, G. I. *Tetrahedron Lett.* 1990, *31*, 2665.
244. Tomioka, K.; Nakajima, M.; Koga, K. *Tetrahedron Lett.* 1990, *31*, 1741.
245. Oishi, T.; Hirama, M. *J. Org. Chem.* 1989, *54*, 5834.
246. Hentges, S. G.; Sharpless, K. B. *J. Am. Chem. Soc.* 1980, *102*, 4263.
247. Jacobsen, E. N.; Markó, I.; Mungall, W. S.; Schröder, G.; Sharpless, K. B. *J. Am. Chem. Soc.* 1988, *110*, 1968.
248. Jacobsen, E. N.; Markó, I.; France, M. B.; Svendsen, J. S.; Sharpless, K. B. *J. Am. Chem. Soc.* 1989, *111*, 737.
249. Lohray, B. B.; Kalantar, T. H.; Kim, B. M.; Park, C. Y.; Shibata, T.; Wai, J. S. M.; Sharpless, K. B. *Tetrahedron Lett.* 1989, *30*, 2041.
250. Shibata, T.; Gilheany, D. G.; Blackburn, B. K.; Sharpless, K. B. *Tetrahedron Lett.* 1990, *31*, 3817.
251. Jorgensen, K. A. *Tetrahedron Lett.* 1990, *31*, 6417.
252. Kim, B. M.; Sharpless, K. B. *Tetrahedron Lett.* 1990, *31*, 3003.
253. Wai, J. S. M.; Marko, I.; Svendsen, J. S.; Finn, M. G.; Jacobsen, E. N.; Sharpless, K. B. *J. Am. Chem. Soc.* 1989, *111*, 1123.
254. Kwong, H.-L.; Sorato, C.; Ogino, Y.; Chen, H.; Sharpless, K. B. *Tetrahedron Lett.* 1990, *31*, 2999.
255. Minato, M.; Yamamoto, K.; Tsuji, J. *J. Org. Chem.* 1990, *55*, 766.
256. Pearlstein, R. M.; Blackburn, B. K.; Davis, W. M.; Sharpless, K. B. *Angew. Chem., Int. Ed. Engl.* 1990, *29*, 639.
257. Dijkstra, G. D. H.; Kellogg, R. M.; Wynberg, H.; Svendsen, J. S.; Marko, I.; Sharpless, K. B. *J. Am. Chem. Soc.* 1989, *111*, 8069.
258. Svendsen, J. S.; Marko, I.; Jacobsen, E.; Rao, C. P.; Bott, S.; Sharpless, K. B. *J. Org. Chem.* 1989, *54*, 2263.
259. Corey, E. J.; Noe, M. C.; Sarshar, S. *J. Am. Chem. Soc.* 1993, *115*, 3828.
260. Kolb, H. C.; Andersson, P. G.; Bennani, Y. L.; Crispino, G. A.; Jeong, K.-S.; Kwong, H.-L.; Sharpless, K. B. *J. Am. Chem. Soc.* 1993, *115*, 12226.
261. Amberg, W.; Bennani, Y. L.; Chadha, R. K.; Crispino, G. A.; Davis, W. D.; Hartung, J.; Jeong, K.-S.; Y., O.; Shibata, T.; Sharpless, K. B. *J. Org. Chem.* 1993, *58*, 844.
262. Corey, E. J.; Jardine, D. P.; Virgil, S.; Yuen, P.-W.; Connell, R. D. *J. Am. Chem. Soc.* 1989, *111*, 9243.
263. Nakajima, M.; Tomioka, K.; Iitaka, Y.; Koga, K. *Tetrahedron* 1993, *49*, 10793.
264. Hanessian, S.; Meffre, P.; Girard, M.; Beaudoin, S.; Sancéau, J.-Y.; Bennani, Y. *J. Org. Chem.* 1993, *58*, 1991.
265. Crispino, G. A.; Jeong, K.-S.; Kolb, H. C.; Wang, Z.-M.; Xu, D.; Sharpless, K. B. *J. Org. Chem.* 1993, *58*, 3785.
266. Oishi, T.; Hirama, M. *Tetrahedron Lett.* 1992, *33*, 639.
267. Corey, E. J.; Noe, M. C. *J. Am. Chem. Soc.* 1993, *115*, 12579.
268. Sharpless, K. B.; Amberg, W.; Bennani, Y. L.; Crispino, G. A.; Hartung, J.; Jeong, K.-S.; Kwong, H.-L.; Morikawa, K.; Wang, Z.-M.; Xu, D.; Zhang, X.-L. *J. Org. Chem.* 1992, *57*, 2768.

269. Vidari, G.; Giori, A.; Dapiaggi, A.; Lanfranchi, G. *Tetrahedron Lett.* 1993, *34*, 6925.
270. Arrington, M. P.; Bennani, Y. L.; Göbel, T.; Walsh, P.; Zhao, S.-H.; Sharpless, K. B. *Tetrahedron Lett.* 1993, *34*, 7375.
271. Morikawa, K.; Park, J.; Andersson, P. G.; Hashiyama, T.; Sharpless, K. B. *J. Am. Chem. Soc.* 1993, *115*, 8463.
272. Wang, L.; Sharpless, K. B. *J. Am. Chem. Soc.* 1992, *114*, 7568.
273. VanNieuwenhze, M. S.; Sharpless, K. B. *J. Am. Chem. Soc.* 1993, *115*, 7864.
274. Pini, D.; Iuliano, A.; Rosini, C.; Salvadori, P. *Synthesis* 1990, 1023.
275. Xu, D.; Sharpless, K. B. *Tetrahedron Lett.* 1993, *34*, 951.
276. Rao, A. V. R.; Gurjar, M. K.; Joshi, S. V. *Tetrahedron: Asymmetry* 1990, *1*, 697.
277. Gao, Y.; Sharpless, K. B. *J. Am. Chem. Soc.* 1988, *110*, 7538.
278. Denmark, S. E. *J. Org. Chem.* 1981, *46*, 3144.
279. Lowe, G.; Salamone, S. J. *J. Chem. Soc., Chem. Commun.* 1983, 1392.
280. Oi, R.; Sharpless, K. B. *Tetrahedron Lett.* 1991, *32*, 999.
281. Gao, Y. PhD Thesis, Massachusetts Institute of Technology, 1988.
282. Lohray, B. B.; Gao, Y.; Sharpless, K. B. *Tetrahedron Lett.* 1989, *30*, 2623.
283. Berridge, M. S.; Franceschini, M. P.; Rosenfeld, E.; Tewson, T. J. *J. Org. Chem.* 1990, *55*, 1211.
284. Kim, B. M.; Sharpless, K. B. *Tetrahedron Lett.* 1989, *30*, 655.
285. Lohray, B. B.; Ahuja, J. R. *J. Chem. Soc., Chem. Commun.* 1991, 95.
286. Hirsenkorn, R. *Tetrahedron Lett.* 1990, *31*, 7591.
287. Deyrup, J. A. *Chem. Heterocycl. Comp.* 1983, *42 (pt.1)*, 1.
288. Duggan, M. E.; Karanewsky, D. S. *Tetrahedron Lett.* 1983, *24*, 2935.
289. Bates, G. S.; Varelas, M. A. *Can. J. Chem.* 1980, *58*, 2562.
290. Yahiro, N. *Chem. Lett.* 1982, 1479.
291. Pfister, J. R. *Synthesis* 1984, 969.
292. Kelly, J. W.; Eskew, N. L.; Evans, S. A. *J. Org. Chem.* 1986, *51*, 95.
293. Häner, R.; Olano, B.; Seebach, D. *Helv. Chim. Acta* 1987, *70*, 1676.
294. Berry, M. B.; Craig, D. *Synlett* 1992, 41.
295. Atkinson, R. S.; Fawcett, J.; Russell, D. R.; Tughan, G. *J. Chem. Soc., Chem. Commun.* 1986, 832.
296. Atkinson, R. S.; Tughan, G. *J. Chem. Soc., Chem. Commun.* 1986, 834.
297. Li, Z.; Conser, K. R.; Jacobsen, E. N. *J. Am. Chem. Soc.* 1993, *115*, 5326.
298. Evans, D. A.; Faul, M. M.; Bilodeau, M. T.; Anderson, B. A.; Barnes, D. M. *J. Am. Chem. Soc.* 1993, *115*, 5328.
299. Calot, S.; Urso, F.; Alper, H. *J. Am. Chem. Soc.* 1989, *111*, 931.
300. Bates, G. S.; Ramaswamy, S. *Can. J. Chem.* 1980, *58*, 716.
301. Ibuka, T.; Nakai, K.; Habashita, H.; Fujii, N. *Tetrahedron Lett.* 1993, *34*, 7421.
302. Ho, M.; Chung, J. K. K.; Tang, N. *Tetrahedron Lett.* 1993, *34*, 6513.
303. Schwan, A. L.; Refvik, M. D. *Tetrahedron Lett.* 1993, *34*, 4901.
304. Kawabata, T.; Kiryu, Y.; Sugiura, Y.; Fuji, K. *Tetrahedron Lett.* 1993, *34*, 5127.
305. Baldwin, J. E.; Spivey, A. C.; Schofield, C. J.; Sweeney, J. B. *Tetrahedron* 1993, *49*, 6309.
306. Baldwin, J. E.; Spivey, A. C.; Schofield, C. J. *Tetrahedron: Asymmetry* 1990, *1*, 881.
307. Wakamuja, T.; Mizuno, K.; Ukita, T.; Teshima, T. *Bull. Chem. Soc., Jpn.* 1978, *51*, 850.
308. Shiba, T.; Ukita, T.; Mizuno, K.; Teshima, T.; Wakamiya, T. *Tetrahedron Lett.* 1977, 2681.
309. Teshima, T.; Konishi, K.; Shiba, T. *Bull. Chem. Soc., Jpn.* 1980, *53*, 508.
310. Imae, K.; Kamachi, H.; Yamashita, H.; Okita, T.; Okuyama, S.; Tsuno, T.; Yamasaki, T.; Sawada, Y.; Ohbayashi, M.; Naito, T.; Oki, T. *J. Antiobiot.* 1991, *44*, 76.
311. Kogami, Y.; Okawa, K. *Bull. Chem. Soc., Jpn.* 1987, *60*, 2963.
312. Parry, R. J.; Naidu, M. V. *Tetrahedron Lett.* 1983, *24*, 1133.
313. Hata, Y.; Watanabe, M. *Tetrahedron* 1987, *43*, 3881.
314. Wakamiya, T.; Shimbo, K.; Shiba, T.; Nakajima, K.; Neya, M.; Okawa, K. *Bull. Chem. Soc., Jpn.* 1982, *55*, 3878.
315. Okawa, K.; Kinutano, T.; Sakai, K. *Bull. Chem. Soc., Jpn.* 1968, *41*, 1353.
316. Wade, T. N.; Gaynard, F.; Guedj, R. *Tetrahedron Lett.* 1979, 2681.
317. Wade, T. N.; Kheribet, R. *J. Chem. Res. (S)* 1980, 210.
318. Vedejs, E.; Moss, W. O. *J. Am. Chem. Soc.* 1993, *115*, 1607.
319. Matsubara, S.; Kodama, T.; Utimoto, K. *Tetrahedron Lett.* 1990, *31*, 6379.

Chapter 11

# OXIDATIONS OF FUNCTIONALIZED ALKENES

The incorporation of a functional group close to carbon–carbon unsaturation allows for high asymmetric induction. Unlike hydrogenations, oxidations can be performed on a wide variety of substrates, especially allyl alcohols. Oxidation systems have been developed that allow for high asymmetric induction and remove much of the uncertainty associated with the use of hydrogenation systems. In addition to the oxidation reactions, this chapter also covers the reactions of the product epoxides. The chapter is organized into four main sections: the first covers the formation of epoxides from allyl alcohols and derivatives. The second main topic is the reactions of 2,3-epoxy alcohols, omitting the methodology used to prepare the epoxide. The third portion covers epoxidation reactions of other functionalized alkenes and the reactions of the resultant epoxides. The last section deals with oxidation reactions that provide diols from alkenes. For these oxidation reactions, it will be seen that the appendant functionality on the alkene either plays a key role in the direction of the epoxidation reaction, or it modifies the reactions of the resultant epoxide. If neither of these criteria is fulfilled, the alkene is considered to be "isolated" and is thus discussed in Chapter 10.

## 11.1. EPOXIDATION OF ALLYL ALCOHOLS

The procedure for the asymmetric epoxidation of allyl alcohols has, in many respects, revolutionized organic synthetic methodology; very high asymmetric induction is observed with a wide range of substrates, allowing a chemical reaction to compete with an enzymatic process both in terms of chemical and optical yields. Several classes of reagents, mainly based on peroxy acids, vanadium, and molybdenum, have been utilized to perform stereoselective oxidations of allyl alcohols in addition to the Sharpless titanium methodology.[1-4] Many of the methods described for the stereoselective oxidation of isolated double bonds and allyl alcohols can often be used with highly functionalized substrates.[5,6]

### 11.1.1. METAL CATALYZED OXIDATIONS

The problems associated with peroxy acid oxidation of an allyl alcohol and its susceptibility to substituent effects [*11.1.3*], have led to the development and use of other oxidation systems, particularly ones that involve metals.[7,8] This work has culminated in the Sharpless method, but other metals, such as vanadium, tungsten, and molybdenum, do afford some selectivity.[9-21]

Allyl alcohols can be epoxidized with high stereoselectivity by the Sharpless method.[1,5,7] Transition metals and peracids form a chelate between the reagent and the substrate's allylic oxygen group; the oxidation is not normally affected by other remote stereogenic centers within the substrate, although exceptions have been observed.[22]

#### 11.1.1.1. With Vanadium

The key to vanadium catalyzed epoxidations of allyl alcohols lies in the ability of the hydroxy group to be a ligand for the metal.[23,24] This then allows for face selectivity. The preferred conformations for vanadium catalyzed oxidations are shown in Figure 11.1.[10,11,25]

As with peroxy acid oxidations, the use of the bulky trimethylsilyl group allows for high selectivity (Scheme 11.1).[11,26,27] The use of this bulky silyl group has a significant effect on the relative energies of the potential transition states. For Scheme 11.1 ($R^2 = SiMe_3$ in Figure 11.1), the silyl group — or other bulky substituent — can be incorporated into the substrate at $R^3$ and direct the reaction outcome.[11,26,28]

**FIGURE 11.1.** Conformations for vanadium catalyzed oxidations of allyl alcohols.

Whereas vanadium catalyzed epoxidations can, in some cases, offer higher selectivity than the corresponding peroxy acid reactions, for silyl-substituted allyl alcohols, the sense of induction for the two methods can be complimentary. Schemes 11.1 and 11.2 illustrate systems where high selectivity is observed.[11,12]

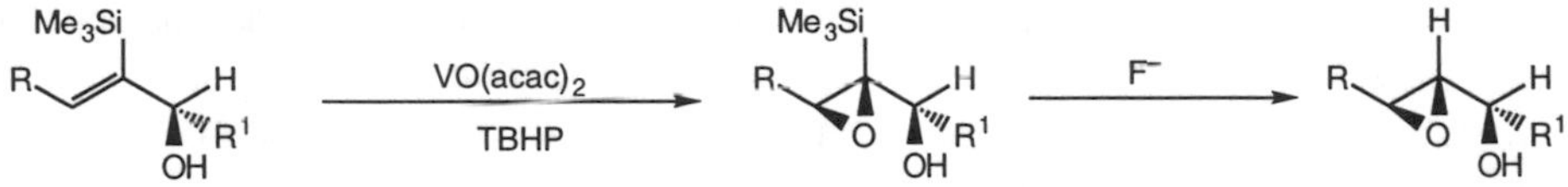

**SCHEME 11.1.**

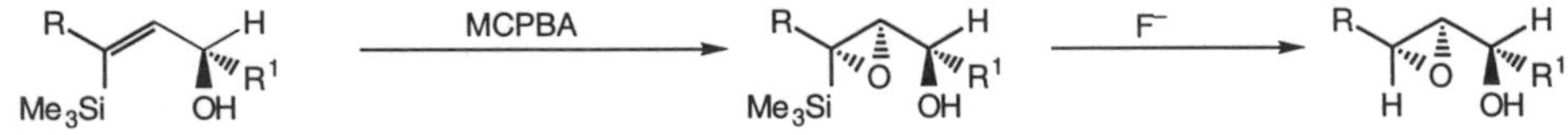

**SCHEME 11.2.**

Homoallyl alcohols can also be epoxidized with high stereoselective control when vanadium is used as the catalyst.[29-31] However, oxidation of the *cis*-vinylsilane (**11.1**, $R^2 = SiMe_3$) failed. When the silyl group was an allylsilane (**11.2**, $R^2 = CH_2SiMe_3$), the reaction gave good selection, and the silicon functionality could be utilized to provide functionalized 1,3-diols (Scheme 11.3).[32] When the silyl group is at the β-position of a homoallyl alcohol, stereoselection can be high.[33]

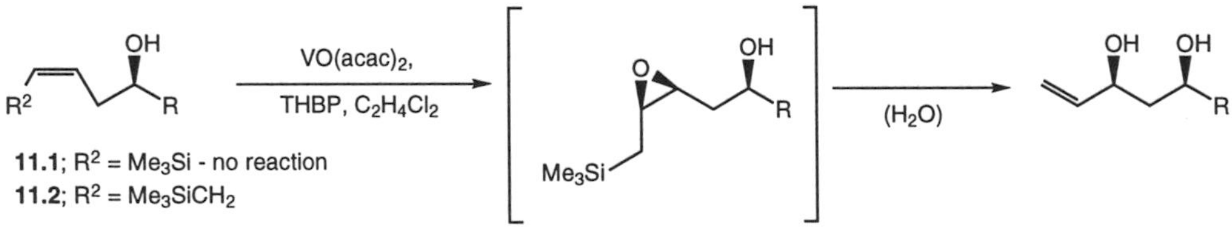

**11.1**; $R^2 = Me_3Si$ - no reaction
**11.2**; $R^2 = Me_3SiCH_2$

**SCHEME 11.3.**

The ability to control relative stereochemistry does, however, provide an excellent method for asymmetric synthesis if a chiral allyl alcohol is the substrate (Scheme 11.4).[23,34]

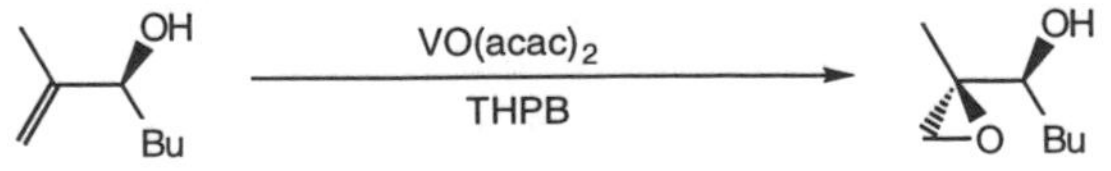

**SCHEME 11.4.**

Homoallylic alcohols can also provide high induction (Scheme 11.5). The analogous reaction with MCPBA gave a 1:1 mixture of diastereoisomers.[35-37]

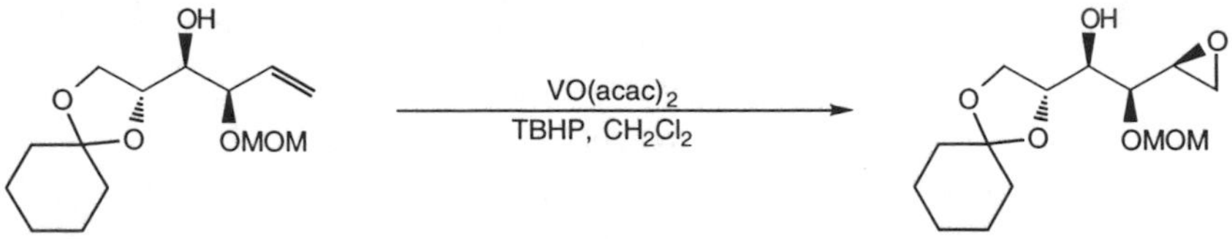

**SCHEME 11.5.**

The alkene can also bear additional functionality as illustrated by the oxidation of the homoallyl alcohol **11.3**, where the carbon–carbon unsaturation is part of an enol carbamate (Scheme 11.6). Again, diastereoselectivity was much higher than when MCPBA was used as an oxidant.[38]

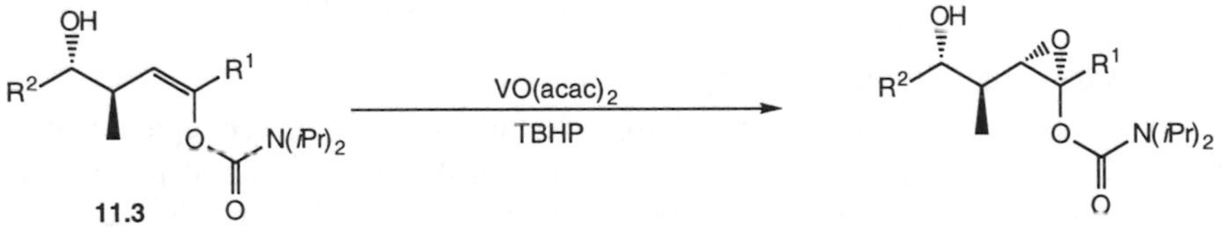

**SCHEME 11.6.**

The use of vanadium-pillared montmorillonite as the catalyst allows for the selective oxidation of di- or trisubstituted double bonds of allyl alcohols; terminal double bonds are not oxidized.[21]

The vanadium system has been modified to allow for the epoxidation of the trimethylsilyl ethers of allyl and homoallyl alcohols, but diastereoselectivity may suffer compared to the analogous reaction of the alcohol with vanadium or a peroxy acid.[27]

The use of chiral ligands with vanadium systems has shown only modest asymmetric induction.[39-41] For cyclic allyl alcohols, the conformation of the cyclic system can have significant implications for the reaction outcome.[42]

### 11.1.1.2. Molybdenum

As with vanadium, molybdenum provides methodology for the conversion of an allyl alcohol to a 2,3-epoxy alcohol. Again, the hydroxy group acts as a ligand, allowing for face differentiation of the carbon–carbon double bond. The sense of the selectivity follows that observed for peroxy acids.[12]

The stereoselection observed for the epoxidation of allylic amides was similar to that seen for the corresponding reaction with a peroxy acid.[43] The use of a chiral ligand does allow some asymmetric induction, but the optical yields are not high enough to be useful.[39,40]

### 11.1.1.3. Tungsten

The epoxidation of allyl alcohols catalyzed by tungsten proceeds with retention of configuration at the reaction centers. Reactivity towards the reagent is directly related to the nucleophilicity of the alkene; allyl alcohols are more reactive than homoallyl alcohols.[18,44] The major diastereoisomer is the same as that obtained when a vanadium system is employed.[19] However, if the hydroxy group in a homoallyl alcohol is protected with a bulky group, stereoselection can be reversed (Scheme 11.7).[29]

Oxidation

V(V)-THBP $R^1$ = H >30 : 1

$WO_5$.HMPA $R^1$ = TIS 1 : 7-16

where TIS = triisopropylsilyl

SCHEME 11.7.

#### 11.1.1.4. Other Metals

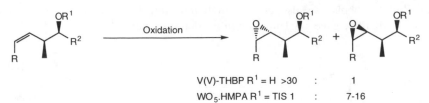

Secondary allyl alcohols can be epoxidized to the epoxy alcohols with an organoaluminum peroxide. The system also allows for subsequent oxidation to the epoxy ketone [*11.3.1*].[45]

Dibutyltin oxyperoxide epoxidizes allylic alcohols with high chemo- and stereoselectivity. The selectivity is usually higher than that observed for the analogous vanadium catalyzed reaction.[46]

This metal, as well as titanium, has been used to catalyze the conversion of allyl alcohols to epoxides by use of dioxygen. No asymmetric reactions were reported.[15]

### 11.1.2. SHARPLESS EPOXIDATION

One of the major advantages of the Sharpless[1,7] titanium asymmetric epoxidation is the simple method by which the stereochemical outcome of the reaction can be predicted.[47] The other powerful feature is the ability to change this selectivity to the other isomer by simple means (Figure 11.2).[48-50]

D-(–)-tartrate

"O"

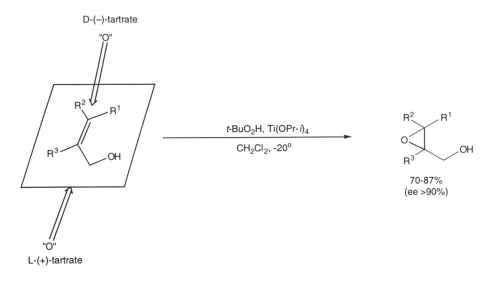

"O"

L-(+)-tartrate

*t*-$BuO_2H$, Ti(OPr-*i*)$_4$

$CH_2Cl_2$, -20°

70-87%
(ee >90%)

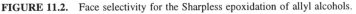

**FIGURE 11.2.** Face selectivity for the Sharpless epoxidation of allyl alcohols.

In addition to the reagent's ingredients being commercially available, the reaction is promiscuous and proceeds in good chemical yield with excellent enantiomeric excesses. The reaction, however, does suffer when bulky substituents are *cis* to the hydroxymethyl functionality ($R^1$ in Figure 11.2). For prochiral alcohols, the absolute stereochemistry of the transformation is predictable, while for a chiral alcohol, the diastereofacial selectivity of the reagent is often sufficient to override those preferences inherent in the substrate. When the chiral atom is in the *E*-β-position of the allyl alcohol ($R^2$ in Figure 11.2), then the epoxidation can be controlled to access either diastereoface of the alkene. In contrast, when the chirality is at either the α- or *Z*-β-positions ($R^1$ or $R^3$ in Figure 11.2), the process is likely to give selective access of the reagent from only one of the two diastereotopic faces.[48,51]

## TABLE 11.1
## Selectivity of the Sharpless Epoxidation Procedure

Bulky substituents *Z*-β to the $CH_2OH$ group are deleterious.
High optical purities usually result.
Absolute stereochemistry is predictable for prochiral alcohols (see Figure 11.2).
The procedure can be used for the kinetic resolution of a secondary alcohol.
When a chiral atom is attached to the alkene moiety in the *E*-β-position, the epoxidation process can access either diastereoface selectively.
When a chiral atom is attached to the alkene moiety in either the *Z*-β-position or the α-position, then it is likely that only a selective approach to one face will be observed.

The introduction of a *tert*-butyl group at each of the possible positions in the allyl alcohol resulted in no deviation from the principles outlined in Table 11.1.[52] Many examples of substrates for the epoxidation protocol are known (Table 11.2).[3,4] Even the prochiral divinylcarbinol undergoes epoxidation with high diastereo- and enantioselectivity.[53-56]

## TABLE 11.2
## Examples of Allyl Alcohol Substrates for the Sharpless Epoxidation[a]

| Substrate Type | Reference | Substrate Type | Reference |
|---|---|---|---|
| 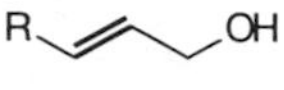 R OH | 47,48,52,57-63 | 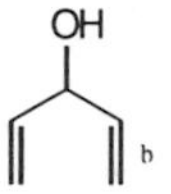 OH [b] | 53–55,64 |
| 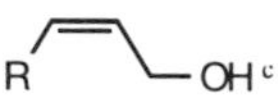 R OH[c] | 47,48,60,63, 65–70 | 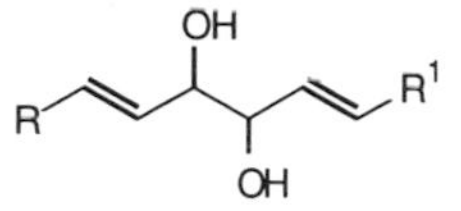 OH R R$^1$ OH | 71-74 |
| 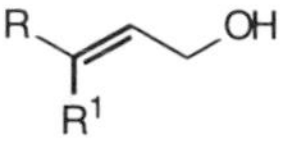 R OH R$^1$ | 75 | 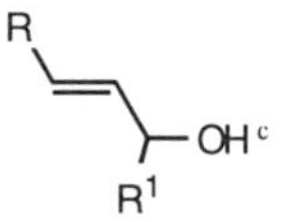 R OH[c] R$^1$ | 76–78 |
| 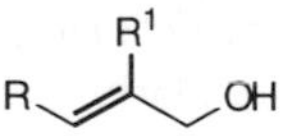 R$^1$ R OH | 47,48,60,79–81 | 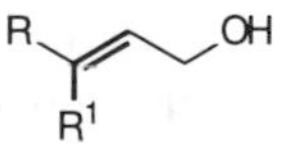 R OH R$^1$ | 47,82 |
| 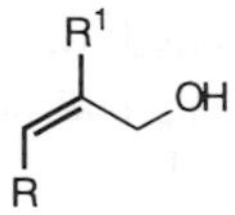 R$^1$ OH R | 48 | 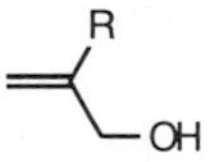 R OH | 47,52,82,83 |
| 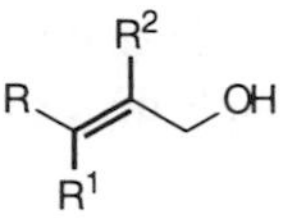 R$^2$ R OH R$^1$ | 48,84–86 | 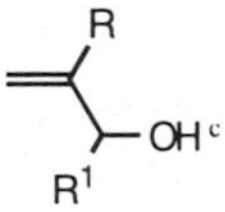 R OH[c] R$^1$ | 87 |
| 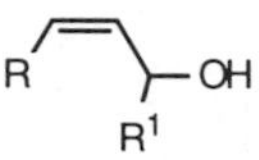 R OH R$^1$ | 76,88,89 |  R OH | 90 |
| 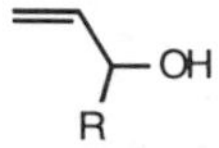 OH R | 91,92 | | |

[a] Unless a marked difference was observed for a different alkyl group, the reaction may be considered to be general. The selectivities of the two antipodes should be considered to be comparable. [b] The monoepoxide is formed. [c] Low ee's were obtained with R = *t*-Bu.[52]

An improved work-up procedure increases the yield for allyl alcohols containing a small number of carbon atoms.[60,93,94] Furthermore, less reactive substrates provide the epoxide readily.[95] The structure of the titanium–tartrate derivatives has been determined,[3,4,96-102] and is in accord with a frontier orbital interpretation.[103] Based on these observations, and the reaction selectivity, a mechanistic explanation has been proposed (Scheme 11.8).[104] The complex **11.4** contains a chiral titanium atom through the appendant tartrate ligands. The intramolecular hydrogen bond ensures that internal epoxidation is only favored at one face of the allyl alcohol. This explanation is in accord with the experimental observations that substrates with an α-substituent (b = alkyl; a = alkyl or hydrogen) react much slower than when this position is not substituted (b = hydrogen).

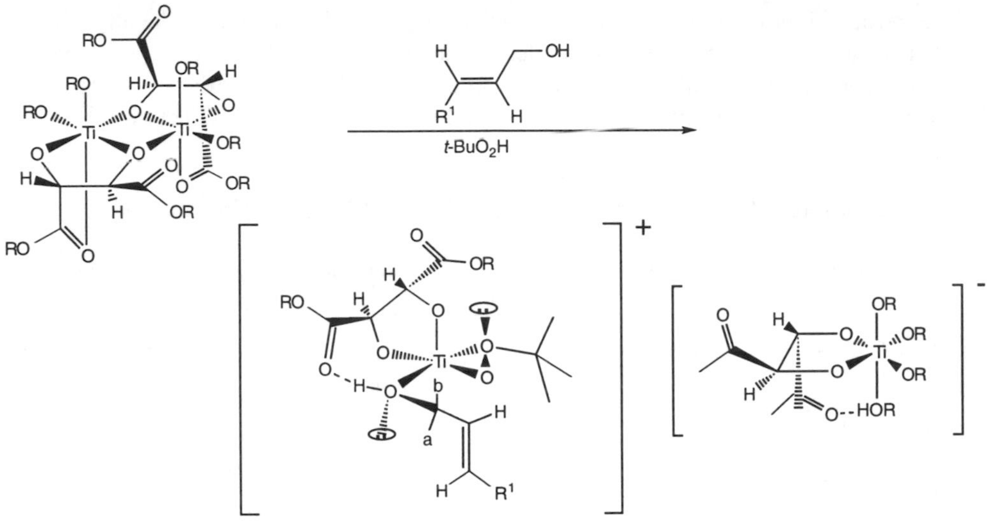

**SCHEME 11.8.**

Despite the widespread use of this asymmetric epoxidation procedure, some care must still be exercised in the choice of substrate, and not only from the standpoint of ensuring good stereochemical selection. Subsequent reactions of the resultant epoxide, particularly intramolecular ones, must be considered.[105]

The reaction time has been reduced dramatically by the addition of calcium hydride, silica gel, or montmorillonite catalysts, but these do not seem to have found widespread acceptance.[95,106-108] The use of polymer-bound tartrate esters has also been advocated.[109]

The protocol can also be used to epoxidize homoallyl alcohols, although the enantiofacial selection is opposite to that observed for allyl alcohols.[110,111] As an allyl alcohol usually reacts faster, the Sharpless epoxidation can be used to differentiate between two double bonds within a substrate molecule (Scheme 11.9).[112-114]

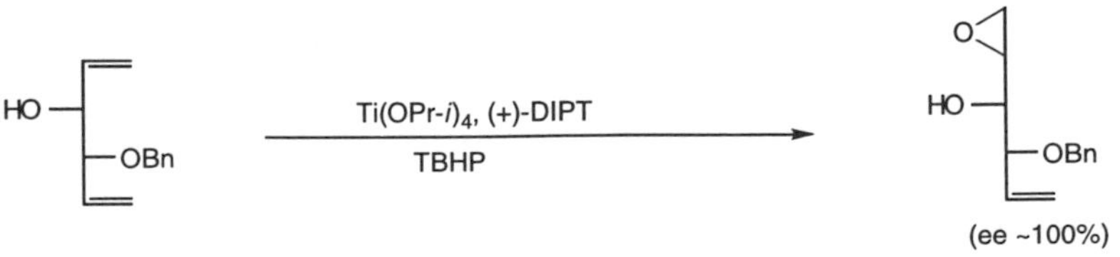

**SCHEME 11.9.**

Unsaturation that is not allylic or homoallylic is left untouched by the reagent system.[4,48] The procedure is also tolerated by a wide variety of functional groups.

Use of dichlorotitanium diisopropoxide in a 2:1 ratio in place of the tetraisopropoxide reverses the enantioselectivity (Scheme 11.10). The use of the more hindered tetrabutoxide gives higher yields in this series.[82]

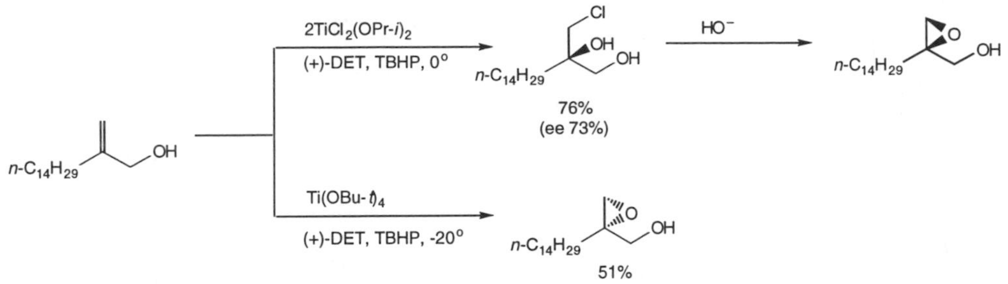

$2TiCl_2(OPr\text{-}i)_2$, (+)-DET, TBHP, 0° → 76% (ee 73%); $HO^-$ →

Ti(OBu-*t*)$_4$, (+)-DET, TBHP, -20° → 51%

**SCHEME 11.10.**

The use of tartamides can also reverse selectivity.[3] This has been rationalized in terms of the conformational disparity between the amides and esters.[115,116]

The power of the Sharpless epoxidation method is augmented by the versatility of the resultant 2,3-epoxy alcohols [*11.2.8*].

#### 11.1.2.1. Kinetic Resolution With the Sharpless Epoxidation Procedure

The ability of the Sharpless epoxidation catalyst to differentiate between the two enantiomers of an asymmetric allyl alcohol affords a powerful synthetic tool to obtain optically pure materials through kinetic resolution.[117] As the procedure relies on one enantiomer of a secondary allyl alcohol undergoing epoxidation at a much faster rate than its antipode, reactions are usually run to 50 to 55% completion.[118] In this way, resolution can often be impressive.[3,4,48,52,87,119] An increase in steric bulk at the olefin terminus increases the rate of reaction.[87,120] Examples of substrates that have been resolved by this methodology are given in Table 11.3.

This resolution method has been used to resolve furfuryl alcohols (the furan acts upon the alkene portion of the allyl alcohol, Scheme 11.11)[129,130,139] allyl propargyl alcohols,[136] and silylallyl alcohols,[133] and has also been used to determine the absolute configuration of cyclic alkenes.[140,141] Racemic β-hydroxyamines and furfuryl amines undergo kinetic resolution through *N*-oxide formation, by use of Sharpless' conditions.[121,123]

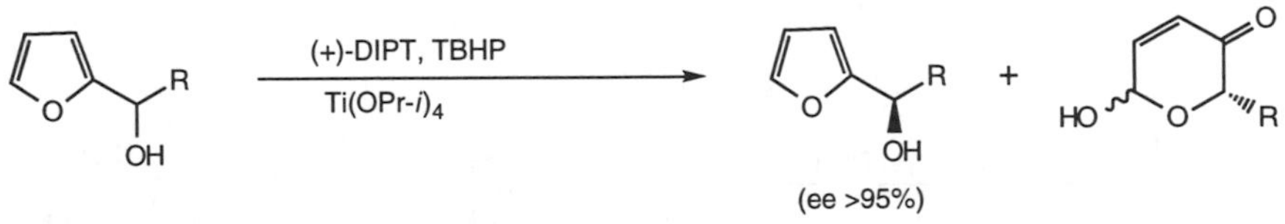

(+)-DIPT, TBHP, Ti(OPr-*i*)$_4$ → (ee >95%) +

**SCHEME 11.11.**

The resultant epoxy alcohols from this oxidation procedure can be used for a wide variety of transformations [*11.2*], including an alcohol transposition (Scheme 11.12).[134,142]

The kinetic resolution protocol can be reversed to provide methodology to resolved hydroperoxides from racemic mixtures.[143]

**TABLE 11.3**
**Examples of Substrates Used for the Kinetic Resolution with the Sharpless Epoxidation Procedure**

| Substrate | Reference | Substrate | Reference |
|---|---|---|---|
| 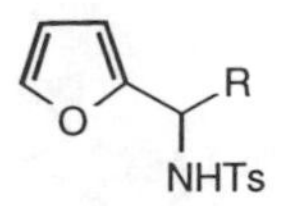 2-furyl–CH(NHTs)–R | 121,122 | 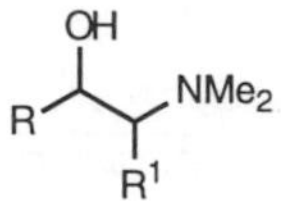 R–CH(OH)–CH(R¹)–NMe₂ | 123 |
| 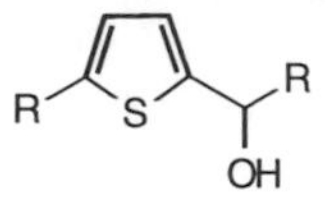 5-R-2-thienyl–CH(OH)–R | 124 | 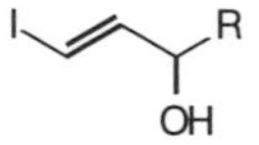 I–CH=CH–CH(OH)–R | 120,125–127 |
| 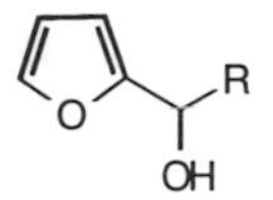 2-furyl–CH(OH)–R | 128–131 | 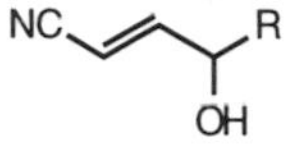 NC–CH=CH–CH(OH)–R | 132 |
| 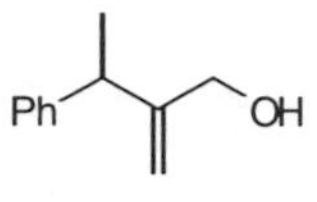 Ph–CH(CH₃)–C(=CH₂)–CH₂OH | 48 | 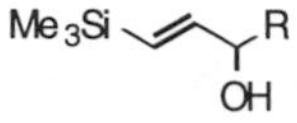 Me₃Si–CH=CH–CH(OH)–R | 125,126,133 |
| 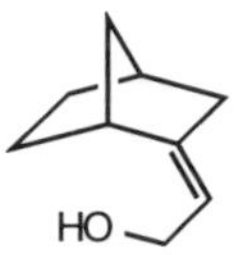 bicyclic alkylidene–CH₂OH (HO) | 48 | 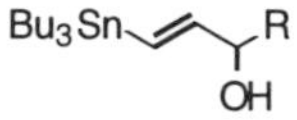 Bu₃Sn–CH=CH–CH(OH)–R | 126 |
| 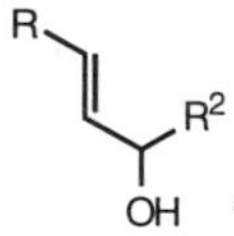 R–CH=CH–CH(OH)–R² [a] | 118,119,134 | 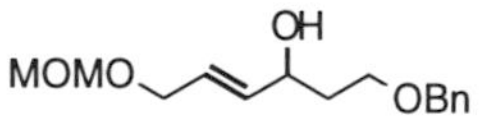 MOMO–CH₂–CH=CH–CH(OH)–CH₂CH₂–OBn | 135 |
| 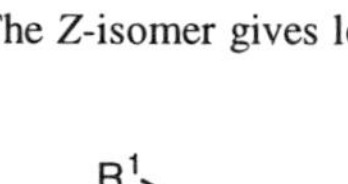 R¹–C≡C–CH(OH)–CH=CH–R² | 136 | 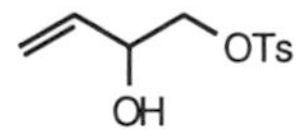 CH₂=CH–CH(OH)–CH₂–OTs | 137 |
| 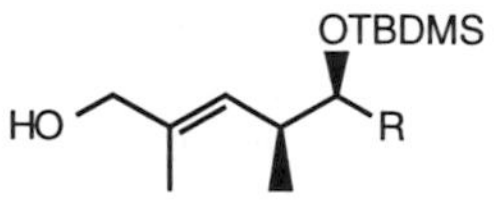 HO–CH₂–CH=C(CH₃)–CH(CH₃)–CH(OTBDMS)–R | 138 | | |

[a] The Z-isomer gives low ee's.[119,134]

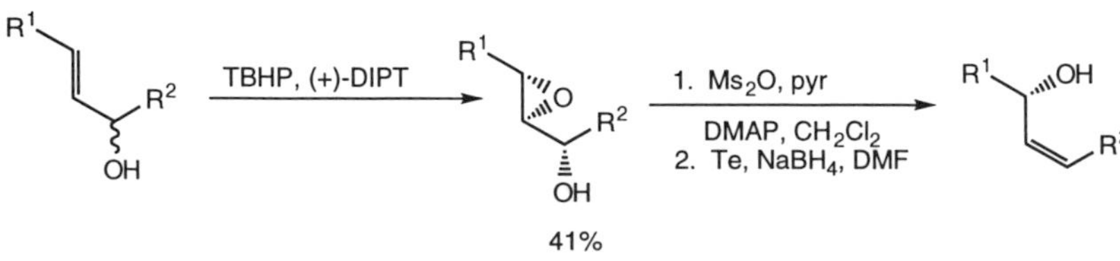

R¹ R² OH TBHP, (+)-DIPT R¹ O R² OH 41% 1. $Ms_2O$, pyr DMAP, $CH_2Cl_2$ 2. Te, $NaBH_4$, DMF R¹ OH R²

**SCHEME 11.12.**

## 11.1.3. PEROXY ACID OXIDATIONS OF ALLYL ALCOHOLS

Peroxy acids,[144] such as *m*-chloroperoxybenzoic acid (MCPBA), when reacted with an allyl alcohol, show a weak preference for formation of the *parf*-isomer of the product epoxide.[10] These observations have been interpreted in terms of a complex formation between the peroxy acid and the allyl alcohol, in which a dihedral angle of 120° is preferred. In consequence, the rotamer **11.5** is preferred over the alternative, **11.6** (Figure 11.3).

**11.5** **11.6**

**FIGURE 11.3.** Conformations for peroxy acid oxidations of allyl alcohols.

When the α-substituent is large (e.g., $R^2 = SiMe_3$), then epoxidation with MCPBA can be stereospecific for the formation of the *pref*-isomer, as the bulky substituent makes the rotamer **11.6** more favorable (Scheme 11.13).[11,33,138,145] As a silyl group can undergo protiodesilylation in the presence of a fluoride ion, this approach is an extremely powerful tool for the preparation of 2,3-epoxy alcohols of defined relative stereochemistry (*cf.* Scheme 11.2).

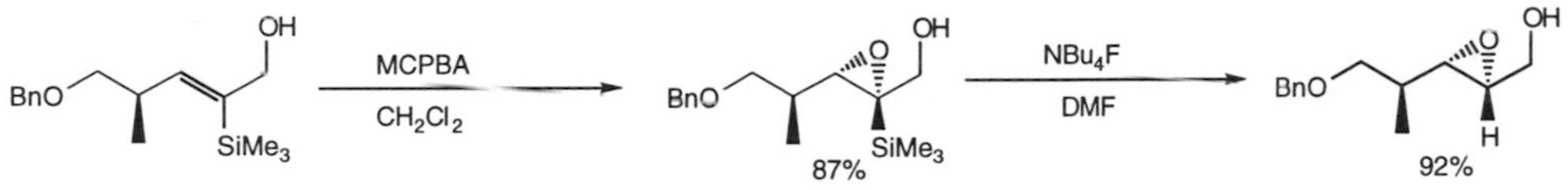

**SCHEME 11.13.**

A systematic study for *p*-nitroperoxybenzoic acid with various allyl alcohols also showed that the hydroxy group had a strong directing effect. This effect was lost when the allyl acetate was used in place of the alcohol. Stereoselectivity was compromised by the presence of a large group juxtaposed to the hydroxy group, or placed on the α-alkene carbon atom. β-Alkene substituents, however, allowed differentiation, particularly when that substituent was *cis* to the hydroxy moiety; these effects are very similar when MCPBA is the reagent.[10-12,24,146,147] The use of silyl ethers promotes formation of the *ancat*-isomer.[148-150]

When a large substituent is present in the β-Z-position, then a large preference is shown for the rotamer **11.5**, and hence formation of the *parf*-product.[26,151,152] Thus, high selection is seen for Z-allyl alcohols.[153]

In addition to the model just described, others have been put forward to predict the stereochemical outcome of an oxidation.[10,25,144,147,150,154-157] They are summarized as shown in Figure 11.4, which is consistent with *ab initio* calculations;[158-165] none of these models differs significantly in their product prediction.

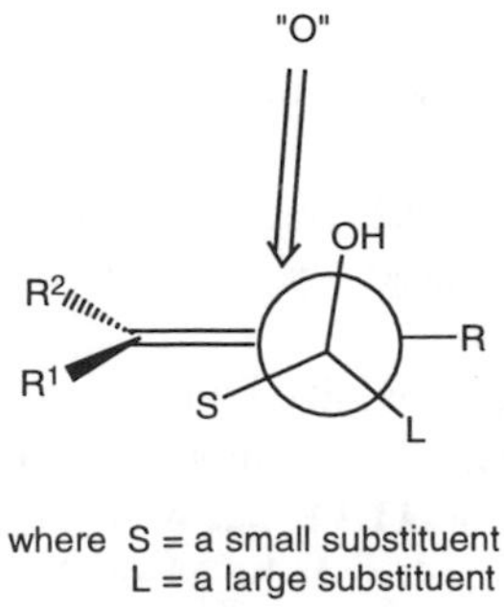

where S = a small substituent
L = a large substituent

**FIGURE 11.4.** Approach of a peroxy acid in the oxidation of an allyl alcohol.

The heavy dependence of the resultant stereochemistry upon the substituents of an allyl alcohol has made metal mediated oxidations preferable to the use of peroxy acids.[23] Despite this drawback, peroxy acid oxidation of allyl alcohols can sometimes provide a workable alternative to isomers of epoxy alcohols which are not readily available by a Sharpless epoxidation procedure (Scheme 11.14).[153,166-168] Thus, the methodology has been used in the preparation of higher carbon sugar derivatives,[13,169] while carbohydrates can also be used as chiral auxiliaries.[170] In cyclic cases, conformation considerations of the ring system usually result in high stereospecificity.[14] Cooperative effects can also be seen for 4-amino allylic alcohols (vide infra), so that good diastereoselection is obtained (Scheme 11.15).[168,171,172]

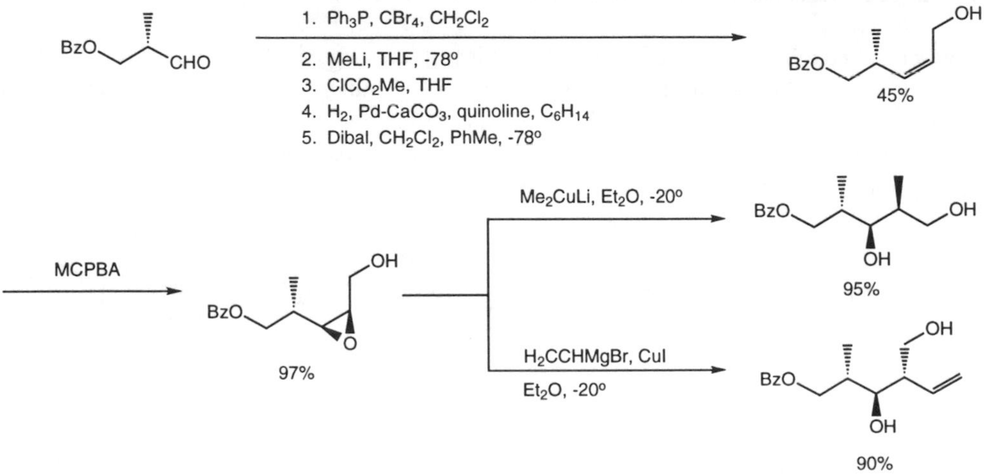

BzO CHO

1. $Ph_3P$, $CBr_4$, $CH_2Cl_2$
2. MeLi, THF, -78°
3. $ClCO_2Me$, THF
4. $H_2$, Pd-$CaCO_3$, quinoline, $C_6H_{14}$
5. Dibal, $CH_2Cl_2$, PhMe, -78°

BzO OH
45%

MCPBA

BzO OH O
97%

$Me_2CuLi$, $Et_2O$, -20°

BzO OH OH
95%

$H_2CCHMgBr$, CuI
$Et_2O$, -20°

BzO OH OH
90%

**SCHEME 11.14.**

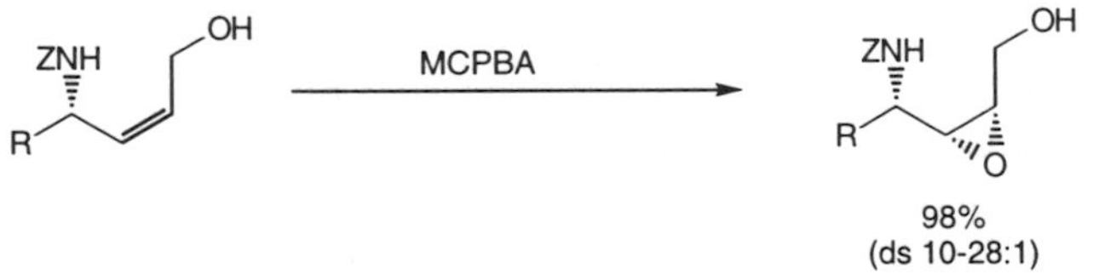

ZNH R OH

MCPBA

ZNH R OH O
98%
(ds 10-28:1)

**SCHEME 11.15.**

Protection of the hydroxy group can still lead to selective oxidation through the *cis*-addition of the oxygen, as observed with simple alkenes [*10.1.1*].[173,174]

Introduction of a second group which can also complex with the incoming peroxy acid can lead to high stereoselection (Scheme 11.14). The selectivity has been attributed to the preferred conformation **11.7**, as opposed to the conformers **11.8** and **11.9** (Figure 11.5); thus, this a variant of Houk's rule.[166,175]

The stereoselection is good for peroxy acid oxidation with functionalized allyl alcohols.[156]

Allylic amides have also been used successfully as substrates for peroxy acid oxidations.[43,172,176-178]

Homoallyl alcohols can also be epoxidized with peroxy acids. An example is provided in Kishi's synthesis of the polyether antibiotic, monensin. Treatment of the alcohol **11.10** with MCPBA led to the unstable epoxide **11.11** that was converted to the tetrahydrofuran **11.12** (Scheme 11.16) [*13.2.1*]. The selectivity can be rationalized by the conformation **11.13** of the alcohol being preferred over **11.14** (Figure 11.6).[179,180]

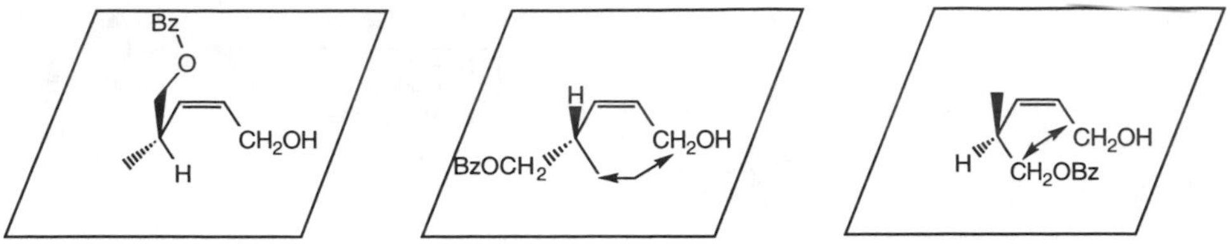

**11.7** **11.8** **11.9**

**FIGURE 11.5.** Possible conformations for 4-alkoxyallyl alcohols; **11.7** is preferred for peroxy acid oxidations.

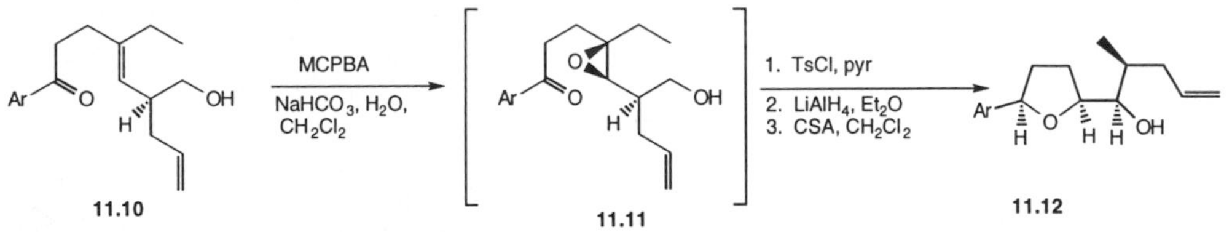

**SCHEME 11.16.**

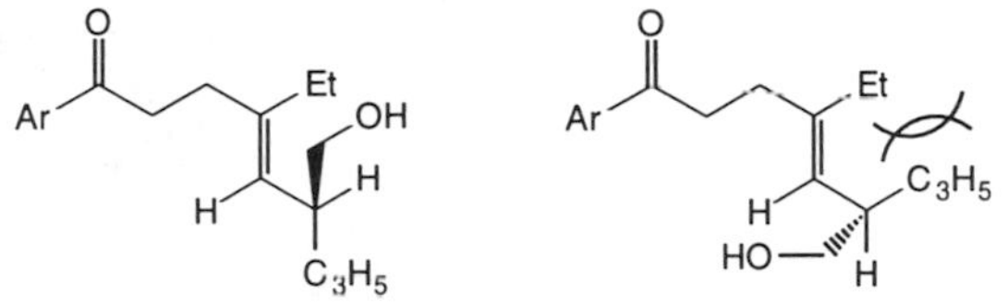

**11.13** **11.14**

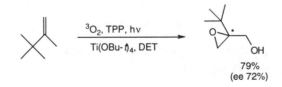

**FIGURE 11.6.** Selectivity for peroxy acid epoxidations.

## 11.1.4. OTHER METHODS TO 2,3-EPOXY ALCOHOLS

An ene reaction [*12.3*] allows for the conversion of an alkene to an epoxy alcohol by reaction with singlet oxygen. The reaction is catalyzed by titanium(IV) which, in turn, allows for the introduction of asymmetry (Scheme 11.17).[181-184]

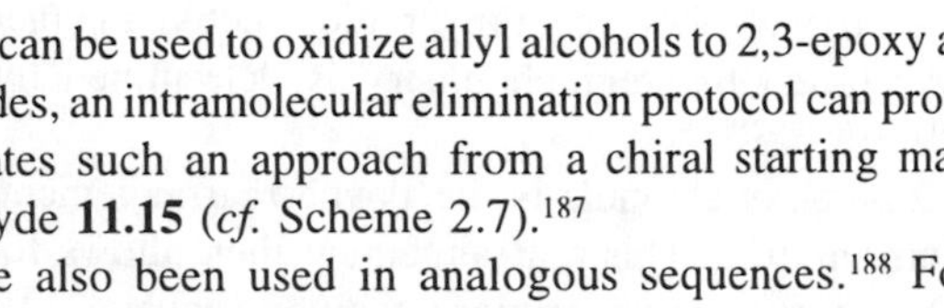

**SCHEME 11.17.**

Dimethyldioxirane can be used to oxidize allyl alcohols to 2,3-epoxy alcohols and enones.[185] As with isolated epoxides, an intramolecular elimination protocol can provide epoxy alcohols.[186] Scheme 11.18 illustrates such an approach from a chiral starting material, isopropylidene protected glyceraldehyde **11.15** (*cf.* Scheme 2.7).[187]

Other chirons have also been used in analogous sequences.[188] For example, a similar sequence is available from serine (Scheme 11.19).[189]

If the system allows the formation of a stable α-halo carbanion, a condensation approach can be used to obtain an epoxy ether (Scheme 11.20).[190]

Some of the methods used for the preparation of simple epoxides [*10.1.4*] can be extended to provide functionalized epoxides.[191]

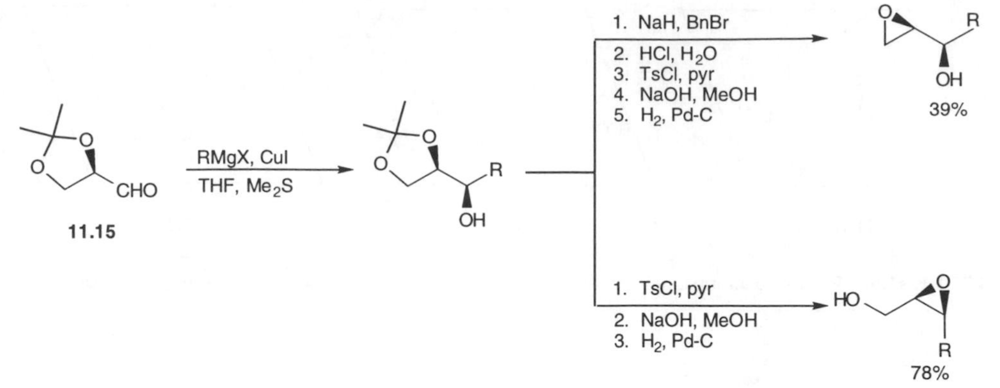

SCHEME 11.18.

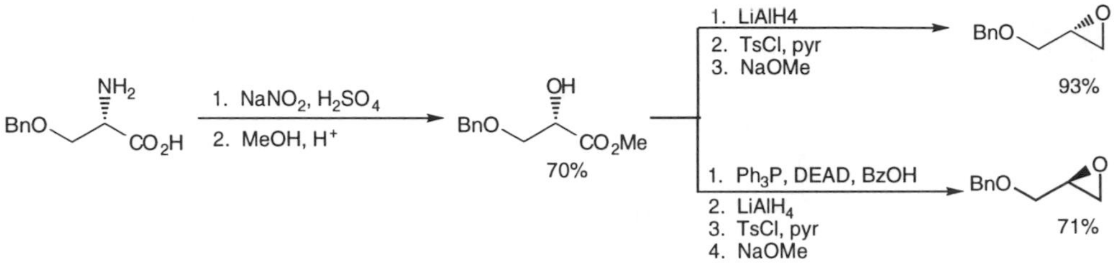

SCHEME 11.19.

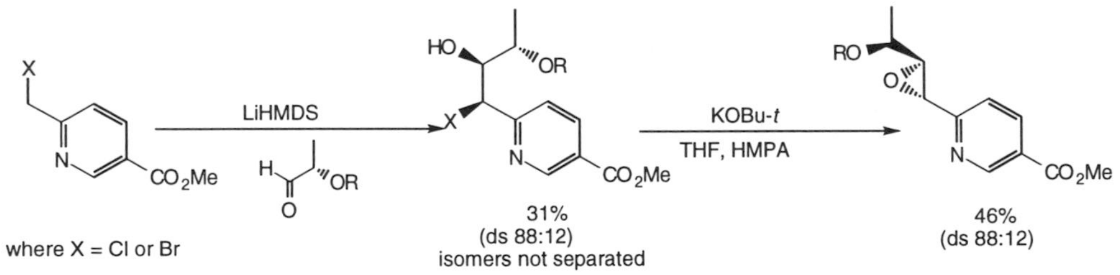

SCHEME 11.20.

## 11.2. REACTIONS OF 2,3-EPOXY ALCOHOLS AND RELATED SYSTEMS

As noted earlier, one reason for the powerful nature of the Sharpless epoxidation is the ability of the resultant epoxy alcohols to undergo regio- and stereoselective reaction with nucleophiles.[48-50,192,193] Epoxy alcohols react in a regioselective manner with a wide variety of nucleophiles.[5,49,169,192,194] Often the regiochemistry is determined by the functional group within the substrate.[90,138,176,195-219]

A key reaction of 2,3-epoxy alcohols is the Payne rearrangement, an isomerization that produces an equilibrium mixture. This rearrangement then allows for the selective reaction with a nucleophile at the most reactive, primary position (Scheme 11.21).[48,220] Reactions of a wide variety of nucleophiles with epoxy alcohols are summarized in Table 11.4.[§]

§ Table 11.4 describes general reactions. However, in many cases, the regioselectivity is dependent on the substrate structure. Thus, some entries may seem contradictory — no differentiation has been made between examples with primary or secondary alcohol groups, nor between terminal or substituted epoxides. Consultation of the original citations will define the scope and limitations of each reagent and reaction.

**TABLE 11.4**
**Examples of Reactions of Epoxyalcohols[a]**

| Reactant | Primary Product[b] | Reference | Reactant | Primary Product | Reference | Reactant | Primary Product | Reference |
|---|---|---|---|---|---|---|---|---|
| PhSNa | **1** | 63,222-224 | ROH, $BF_3{\cdot}OEt_2$ | **3**[c] | 235 | $R_2Mg$ | **3** | 251 |
| | **2** | 221,223 | ROH, CAN | **1**[c] | 236 | $LiAlH_4$ | **2** | 84,253 |
| | **3** | 221,225 | ROH, $Ti(OR)_4$ | **3** | 237 | | **3** | 16,54,64,254,255 |
| PhSH | **3** | 226,227 | $RCO_2H$ | **3** | 224,226,238–240 | Red-Al | **2** | 74,171,253,256–261 |
| KCN | **3** | 224 | $Ph_3P$, 3,5-$(O_2N)$ $C_6H_3CO_2H$, DEAD | **1** | 91 | Dibal | **3** | 244,257,258 |
| $NH_3$ | **3** | 57,222–224,228,229 | RNCO[d] | **1** | 65,171,241–243 | $LiBH_4$, Lewis acid | **3** | 262,263 |
| | **2** | 57 | $R_3Al$ | **3** | 203,244–246 | Li,$NH_3$,THF,*t*-BuOH[e] | **2** | 264 |
| $NaN_3$ | **1** | 92,223,230,231 | $ArAlR_3Li$ | **3** | 247 | $RuCl_3$, $NaIO_4$ | **1**[f] | 58,265,266 |
| | **3** | 224,232 | RLi | **1** | 77,88 | $R_3SiOTf$, $NR_3$ | **2**[g] | 267 |
| | **2**[h] | 233 | | **2** | 75 | 1. Swern, 2. $NaClO_2$ | **1**[f] | 61 |
| $HN_3$, $PPh_3$, DEAD | **1** | 234 | RMgBr,CuBr | **1** | 188 | $CrO_3$.2pyr | **1**[i] | 60,268,269 |
| HONa | **1** | 55,63,93,220 | $R_2CuLi$ | **2** | 23,153,175,203,248–250 | Swern | **1**[i] | 79,86,168,195,270,271 |
| | **3** | 55 | | **3** | 251,252 | PDC | **1**[i] | 272 |
| 1. $CrO_3$, pyr, 2. $NaClO_2$, $NaHPO_4$, $H_2O$ | **1**[f] | 273 | $TiCl_2(OiPr)_2$[j] | **3** | 82,282 | 1. $Ms_2O$, pyr, DMAP 2. Te, $NaBH_4$, DMF | **1**[k] | 134 |
| $Ph_3P$,$CCl_4$[l] | **1** | 274–277 | TsCl[m] | **1** | 283 | $CsCO_3$, $CO_2$ | **1**[n] | 285 |
| $Ph_3P$,$ZnI_2$[m] | **1** | 278 | 1. TsCl, $NEt_3$, 2. LiCl, DMF[m] | **1** | 277 | $CH_2O$, $K_2CO_3$ | **1**[o] | 286 |

**TABLE 11.4 (continued)**

| Reactant | Primary Product[b] | Reference | Reactant | Primary Product | Reference | Reactant | Primary Product | Reference |
|---|---|---|---|---|---|---|---|---|
| $R_3NHCl$, $Ti(OPr\text{-}i)_4$[p] | **3** | 279,280 | RX, base[m] | **1** | 53,284 | $NaBH_4$, NaOH | **1,2**[q] | 48 |
| $X_2$,$Ti(OPr\text{-}i)_4$[p] | **3** | 71,281 | $RCO_2H$, $TiX_4$ | **3** | 71,238 | PTS-LPTS | **3** | 229 |
| LiI | **3**[r] | 83 | BnOCOCl, pyr | **1**[n] | 171 | | | |
| TMSI | **3** | 201 | $Cp_2TiCl_2$ | **1**[k] | 132 | | | |

[a] The variations in selectivity are due to changes in reaction conditions. [b] This table indicates the major reaction pathway. The carbon atom (or oxygen atom) at which attack occurs is noted, the original hydroxyl group being labeled 1. [c] Epoxide opening also occurs to give the alkoxy triol. [d] The reagent is an electrophile, reaction occurs at oxygen, cyclization then occurs with inversion at C-2 to give the carbamate. [e] The oxygen must be protected as the triisopropylsilyl ether. [f] The primary alcohol is oxidized to the corresponding carboxylic acid. [g] The 3-siloxy aldehyde results from epoxide opening at this position. [h] A silyl group must be present at C-2. [i] The primary alcohol is oxidized to the corresponding aldehyde. [j] This reagent delivers chloride. [k] An allyl alcohol with the hydroxy at C-3 results. [l] The reagent is an electrophile, reaction occurs at oxygen, and the halide is the nucleophile. [m] The reagent is an electrophile and reaction occurs at oxygen. [n] The reagent is electrophilc and reaction occurs at oxygen. A cyclization reaction then follows with inversion at C-2 to give the cyclic carbonate. [o] The reagent is electrophilc and reaction occurs at oxygen. A cyclization reaction then follows with inversion at C-2 to give the 1,3-dioxolane. [p] The halogen is the nucleophile. [q] C-1 is converted to a methyl group; inversion occurs at C-2 to provide the diol. [r] In this example, C-3 was also primary.

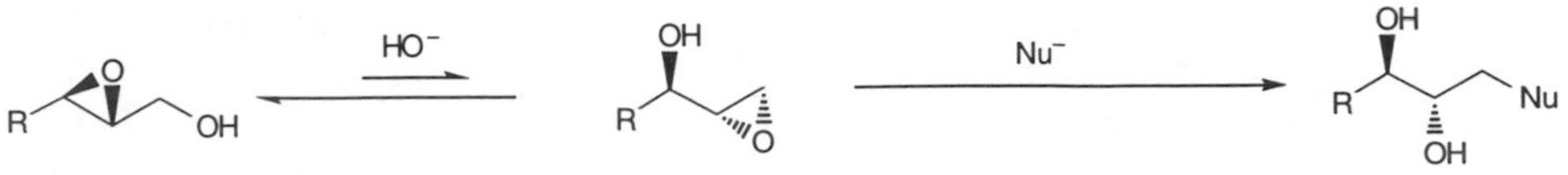

SCHEME 11.21.

Under Payne rearrangement conditions, sodium *t*-butylthiolate provides 1-*t*-butylthio-2,3-diols with very high regioselectivity [*11.2.3*]. The selectivity is, however, affected by many factors including reaction temperature, base concentration, and the rate of addition of the thiol. These sulfides can then be converted to the 1,2-epoxy-3-alcohols, which in turn react with a wide variety of nucleophiles specifically at the 1-position (Scheme 11.22). This methodology circumvents the problems associated with the instability of many nucleophiles under "Payne" conditions.[221]

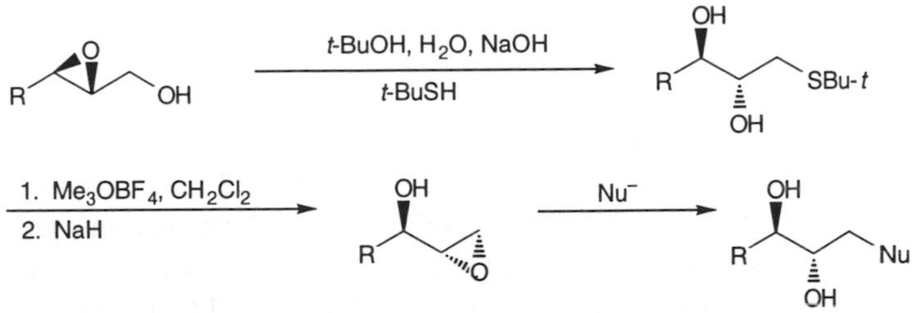

SCHEME 11.22.

With good nucleophiles, under relatively mild conditions, 2,3-epoxy alcohols will undergo epoxide ring opening at C-2 or C-3. In simple cases, nucleophilic attack at C-3 is the preferred mode of reaction. However, as the steric congestion at C-3 is increased, or if substituents play a significant electronic role, attack at C-2 can predominate.[223]

### 11.2.1. OXYGEN NUCLEOPHILES

The configuration of the hydroxyl group in a 1,2-epoxy-3-ol can be inverted by use of Mitsunobu conditions [*3.1.2*].[78,91] This approach has been exploited to provide the *parf*-1,2-epoxy alcohol, which is not readily available by a Sharpless protocol (Scheme 11.23).[238]

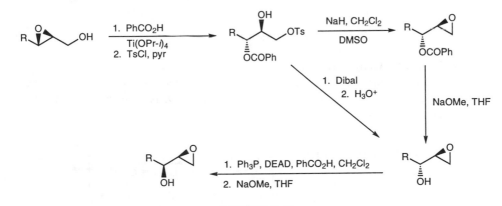

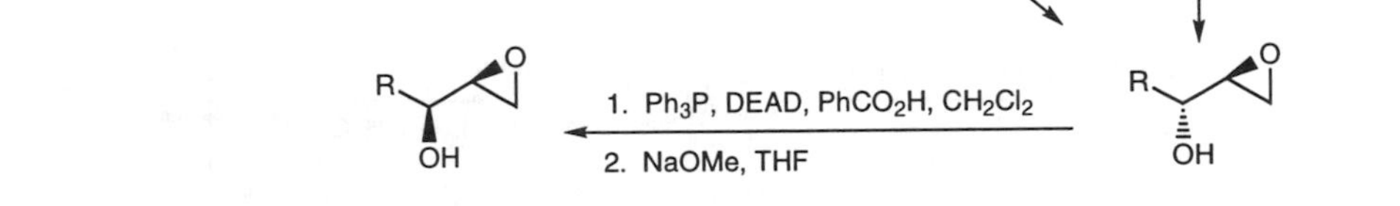

SCHEME 11.23.

Reaction of divinylcarbinol with the (+)-diethyl tartrate version of the Sharpless reagent afforded the epoxide **11.16**. Ring opening at pH 3 to 4 provided the triol **11.17** selectively. In contrast, treatment of the same epoxide with strong base provided the enantiomeric triol **11.18** as the major isomer by a double inversion (Scheme 11.24).[55]

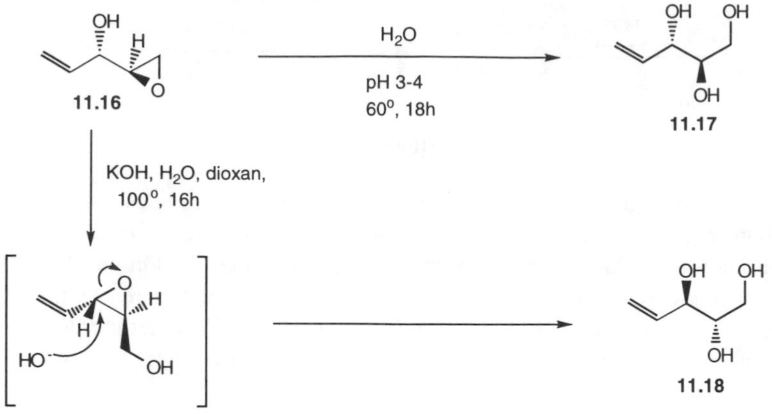

OH H 11.16 O — $H_2O$, pH 3-4, 60°, 18h → OH OH OH 11.17

KOH, $H_2O$, dioxan, 100°, 16h → [O H H HO⁻ OH] → OH OH OH 11.18

SCHEME 11.24.

## 11.2.2. NITROGEN-CONTAINING NUCLEOPHILES

An azide can be introduced at C-1 into a 2,3-epoxy alcohol under Payne rearrangement conditions.[231] Amines can be used as nucleophiles under Payne rearrangement conditions, provided excess is employed to overcome the regiochemical problems associated with the inherently slow reaction. However, attack at C-2 or C-3 can still be the preferred mode, and is dependent upon the substrate's structure.[62,232]

The use of cyanide as the nucleophile allows a variety of functional groups to be formed from the adduct (e.g., Scheme 11.25).[146]

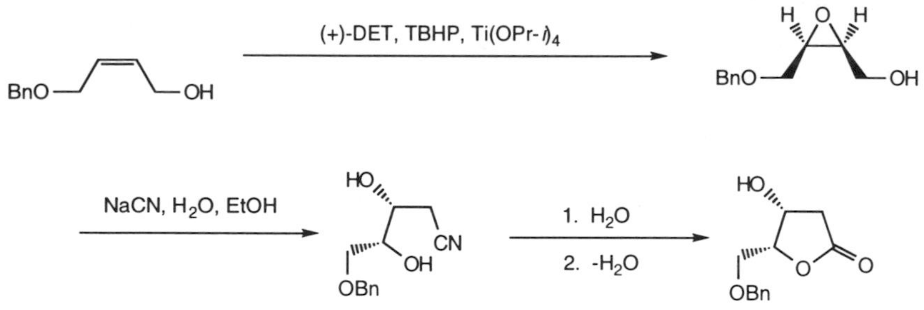

BnO—OH — (+)-DET, TBHP, Ti(OPr-$i$)$_4$ → H O H BnO OH

NaCN, $H_2O$, EtOH → HO CN OH OBn — 1. $H_2O$ 2. $-H_2O$ → HO O O OBn

SCHEME 11.25.

## 11.2.3. SULFUR NUCLEOPHILES

Use of a sulfur nucleophile has led to a general, systematic method for the synthesis of polyols (vide supra). Thus, the useful building block, **11.19**, is available by a Sharpless protocol (Scheme 11.26).[225]

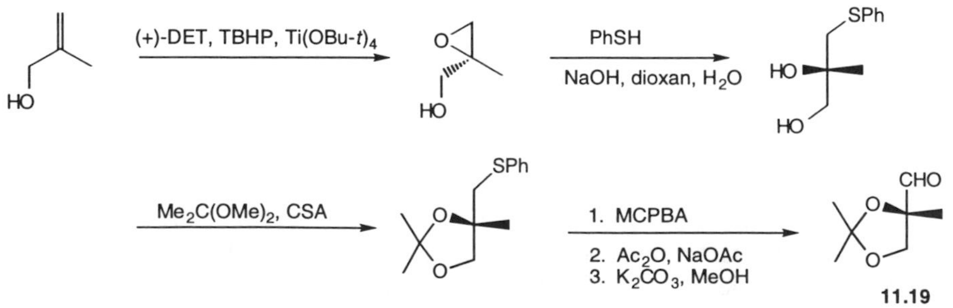

HO — (+)-DET, TBHP, Ti(OBu-$t$)$_4$ → O HO — PhSH, NaOH, dioxan, $H_2O$ → SPh HO HO

$Me_2C(OMe)_2$, CSA → SPh O O — 1. MCPBA 2. $Ac_2O$, NaOAc 3. $K_2CO_3$, MeOH → CHO O O 11.19

SCHEME 11.26.

This methodology has been employed for the preparation of L-threitol (**11.20**) (Scheme 11.27) and erythritol derivatives.[63] This simple iterative process has not only been used for the simple alditols,[63,287] deoxyalditols,[288] and aldoses,[63] but all of the L-hexoses.[222,289]

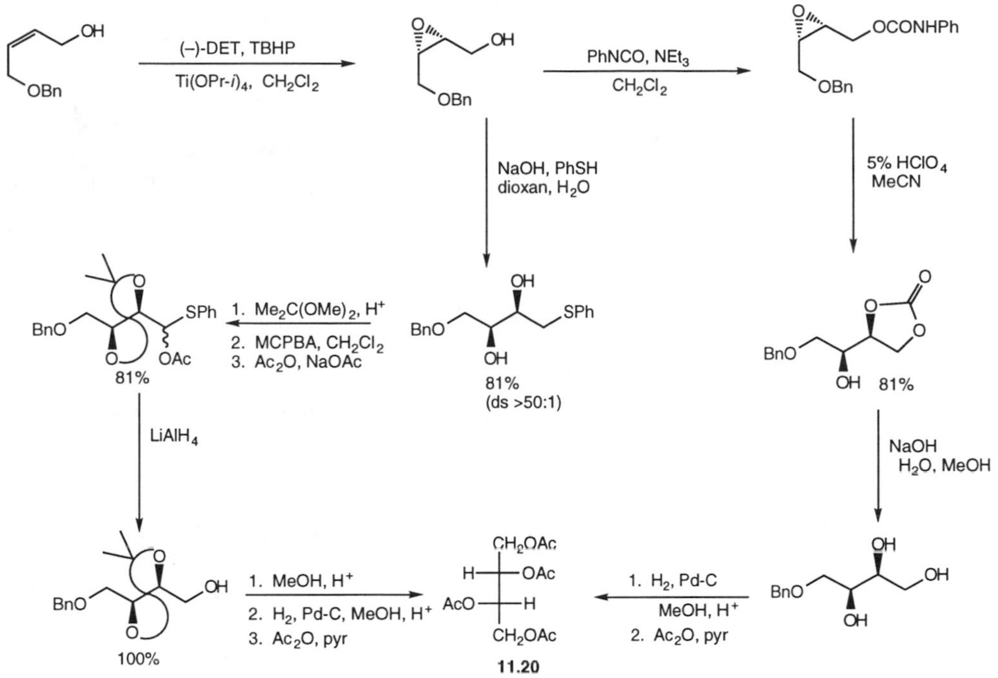

(−)-DET, TBHP
Ti(OPr-*i*)$_4$, $CH_2Cl_2$
PhNCO, $NEt_3$
$CH_2Cl_2$
NaOH, PhSH
dioxan, $H_2O$
5% $HClO_4$
MeCN
1. $Me_2C(OMe)_2$, $H^+$
2. MCPBA, $CH_2Cl_2$
3. $Ac_2O$, NaOAc
81%
81%
(ds >50:1)
81%
$LiAlH_4$
NaOH
$H_2O$, MeOH
1. MeOH, $H^+$
2. $H_2$, Pd-C, MeOH, $H^+$
3. $Ac_2O$, pyr
1. $H_2$, Pd-C
MeOH, $H^+$
2. $Ac_2O$, pyr
100%
**11.20**

SCHEME 11.27.

## 11.2.4. REACTIONS INVOLVING CYCLIC INTERMEDIATES

2,3-Epoxy alcohols react cleanly with isocyanates to provide the corresponding urethanes, which can then cyclize under acidic or basic conditions (Scheme 11.28).[243,290] The same transformation could also be achieved as a one pot reaction. The resultant isoxazolidinones **11.21** are cleanly opened by lithium hydroxide to afford 2-amino-1,3-diols (Scheme 11.29).[243,291]

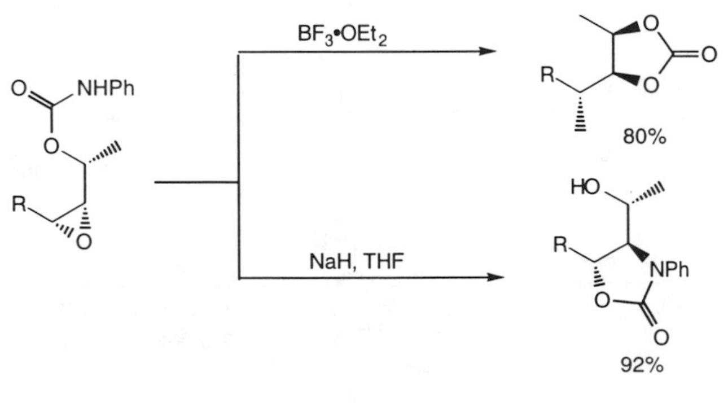

$BF_3$•$OEt_2$
80%
NaH, THF
92%

SCHEME 11.28.

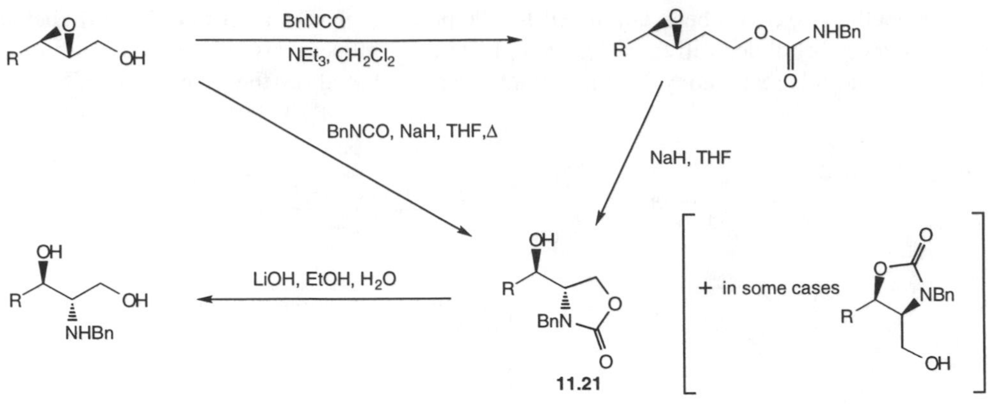

SCHEME 11.29.

This methodology has been used for the preparation of β-hydroxy-α-*N*-methylamino acids.[65,292] In contrast, acid treatment of the unsubstituted carbamate **11.22** afforded a mixture of cyclic carbonates (Scheme 11.30).[241]

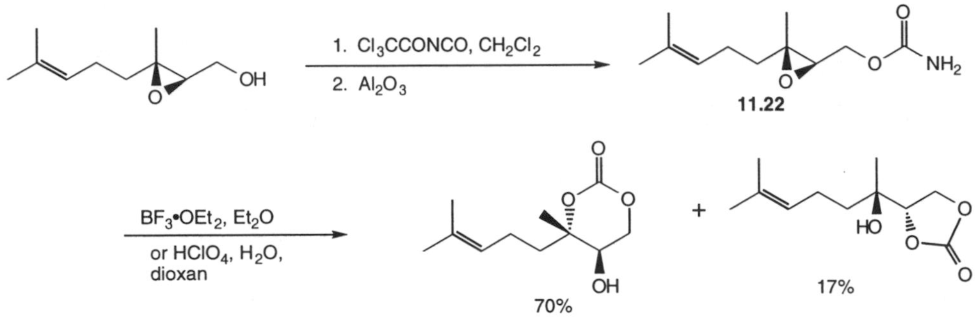

SCHEME 11.30.

As the Sharpless oxidation method can be also be used for kinetic resolutions (Schemes 11.31 and 11.32),[5] the approach has been employed for the synthesis of 2,6-dideoxyhexoses; a "chiral pool" methodology would have required many synthetic operations (Scheme 11.32).[¶] [89]

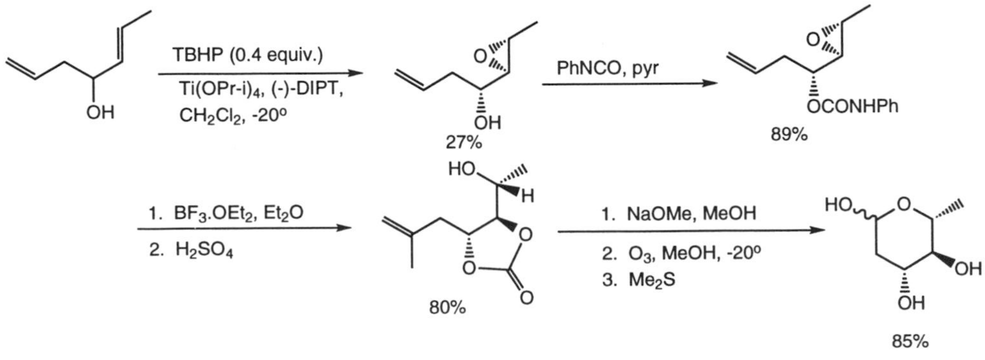

SCHEME 11.31.

¶ In some cases, the "chiral pool" does provide the requisite starting material in an expeditious manner.[293]

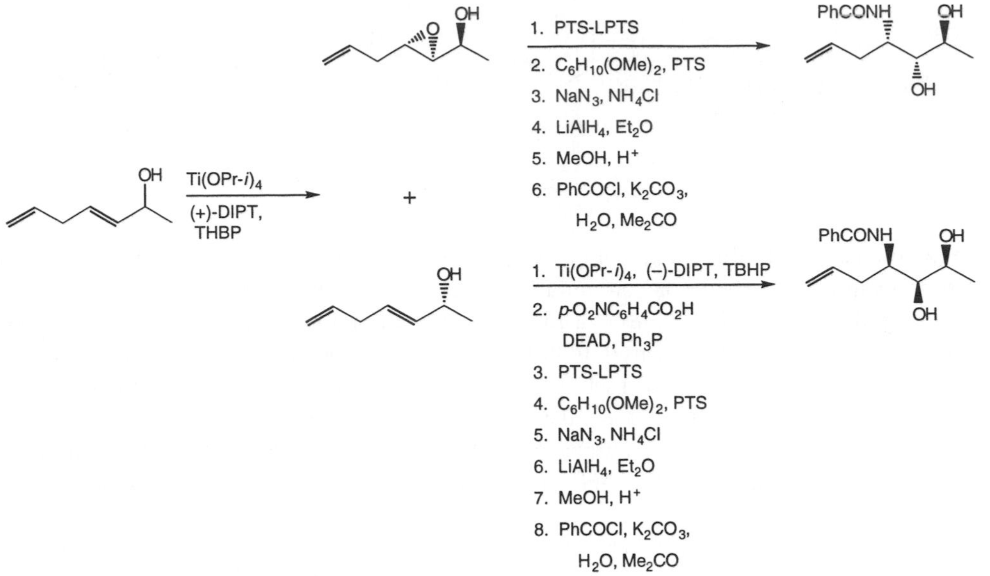

1. PTS-LPTS
2. $C_6H_{10}(OMe)_2$, PTS
3. $NaN_3$, $NH_4Cl$
4. $LiAlH_4$, $Et_2O$
5. MeOH, $H^+$
6. PhCOCl, $K_2CO_3$, $H_2O$, $Me_2CO$

$Ti(OPr\text{-}i)_4$, (+)-DIPT, THBP

1. $Ti(OPr\text{-}i)_4$, (−)-DIPT, TBHP
2. $p\text{-}O_2NC_6H_4CO_2H$, DEAD, $Ph_3P$
3. PTS-LPTS
4. $C_6H_{10}(OMe)_2$, PTS
5. $NaN_3$, $NH_4Cl$
6. $LiAlH_4$, $Et_2O$
7. MeOH, $H^+$
8. PhCOCl, $K_2CO_3$, $H_2O$, $Me_2CO$

**SCHEME 11.32.**

The intramolecular cyclization of the carbonate **11.23** does proceed as expected (Scheme 11.33).[89,171]

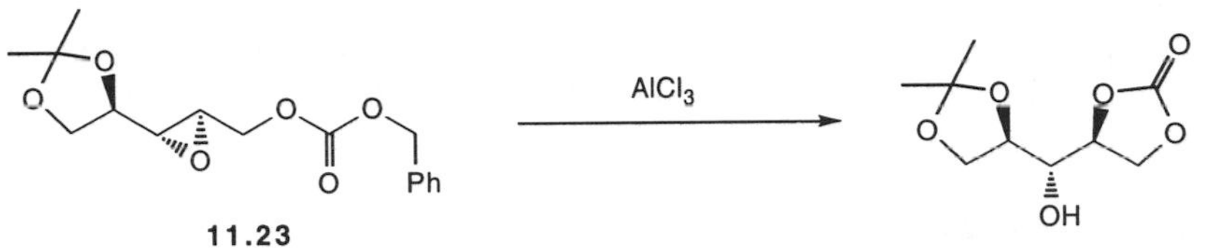

$AlCl_3$

**11.23**

**SCHEME 11.33.**

The presence of titanium tetraisopropoxide in the nucleophilic opening of 2,3-epoxy alcohols leads not only to a marked increase in the reaction rate, but also to an increase in the regioselectivity for attack at C-3 with a wide range of nucleophiles.[224,294] This is illustrated by the opening of **11.24** in a synthesis of D-ribose (**11.25**) (Scheme 11.34).[54,71,74]

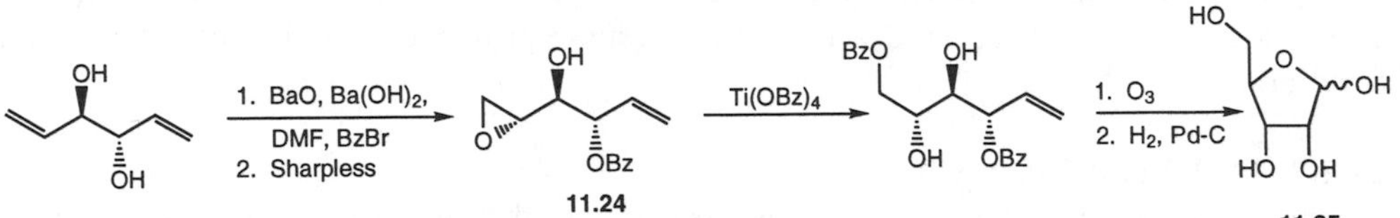

1. BaO, $Ba(OH)_2$, DMF, BzBr
2. Sharpless

$Ti(OBz)_4$

1. $O_3$
2. $H_2$, Pd-C

**11.24** **11.25**

**SCHEME 11.34.**

## 11.2.5. ORGANOMETALLIC REAGENTS

In addition to sulfur and nitrogen nucleophiles, regioselective reduction of an epoxy alcohol can be accomplished by sodium bis(methoxyethoxy)aluminum hydride (Red-Al).[257,258] Thus, this methodology can be used for a stereoselective synthesis of 1,3-diols, and has been extended to more complex systems (e.g., Scheme 11.35).[253,259]

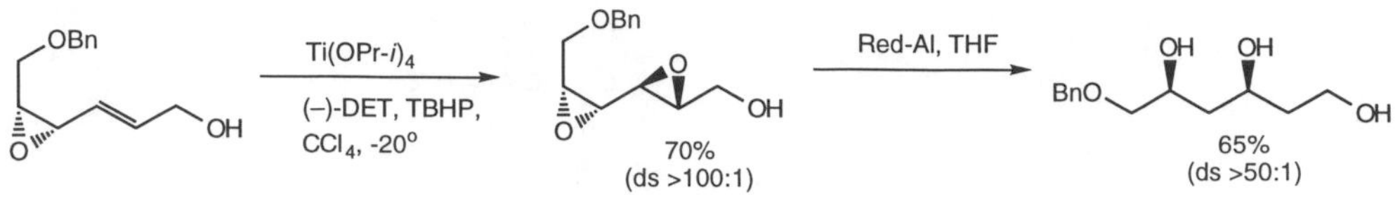

**SCHEME 11.35.**

Complementary to this methodology, epoxy alcohols undergo reaction at the 3-position with concomitant inversion of configuration when treated with organoaluminum reagents, including diisopropylaluminum hydride[257,258] and alkylaluminates.[244] The reaction of trialkylaluminates is catalyzed by butyllithium or lithium methoxide and proceeds best in hydrocarbon solvents when the hydroxyl group is protected as the benzyl ether.[245] This provides a method for the stereoselective synthesis of 1,2-diols.

Reduction of a 1,2-epoxy-3-ol, protected as the ethoxyethyl ether, with lithium aluminum hydride proceeds by attack at C-1 and formation of a 2,3-diol,[91] whereas attack at C-2 has been observed for a free alcohol.[84]

An alternative method for the reduction of 2,3-epoxy alcohols to 1,2-diols through regioselective delivery of hydride at C-3 can be achieved by use of lithium borohydride in the presence of titanium tetraisopropoxide with benzene as solvent.[262]

Epoxide opening with organocuprates is both regio- and stereoselective, and affords the substituted 1,3-diol (Scheme 11.14).[153,175,248,249] However, when the 2,3-epoxy alcohol is not branched at C-4, cuprate epoxide ring opening may not be regioselective. This problem has been overcome, to a certain degree, by the use of higher order cuprates; the preferred mode of attack is once again at C-2.[250]

The 1,2-epoxy-3-alcohols do react, as expected, with organometallic reagents at the primary position of the epoxide.[88] This aspect illustrates the importance of the isomerizations of epoxy alcohols (e.g., Scheme 11.22).

## 11.2.6. OXIDATIONS

2,3-Epoxy alcohols can be oxidized to the corresponding epoxy acids by treatment with ruthenium chloride in the presence of sodium periodate.[58] The analogous transformation to the aldehyde can be performed by use of Swern oxidation conditions.[195]

The coupling of two reactions, the regioselective opening of a 2,3-epoxy alcohol by benzoate followed by ruthenium oxidation, allows for the enantioselective preparation of α-hydroxy acids.[119] Treatment of a 2,3-epoxy alcohol with a silyl triflate, in the presence of a hindered nitrogen base, results in the formation of a β-siloxy aldehyde.[267]

A peroxy acid oxidation has provided a route to (+)-galactostatin (**11.26**) (Scheme 11.36).[295] The halogen is introduced at C-2 to allow for subsequent incorporation of the nitrogen functionality.

## 11.2.7. OTHER REACTIONS

Epoxy alcohols undergo rearrangement reactions in the presence of a Lewis acid. If an appropriate remote functional group is present, cyclization can be observed [Chapter 13], otherwise an enediol is formed (Scheme 11.37).[294]

A group can also migrate if it is attached to the 2-position (Scheme 11.38).[296,297]

## 11.2.8. REACTIONS OF MODIFIED 2,3-EPOXY ALCOHOLS

Reactions of 2,3-epoxy alcohols are often modified by conversion of the hydroxy group to a leaving group.

An alternative strategy to the 1,2-epoxy-3-ols is illustrated in Scheme 11.39. Under basic conditions, the 2,3-epoxy-1-sulfonate esters usually react through selective displacement of the sulfonate moiety, rather than through epoxide opening (*cf.* Scheme 10.9).[66,67,70,298] Under acidic conditions, reaction is regioselective for ring opening at C-3. Mild base treatment then provides the terminal epoxide.[62]

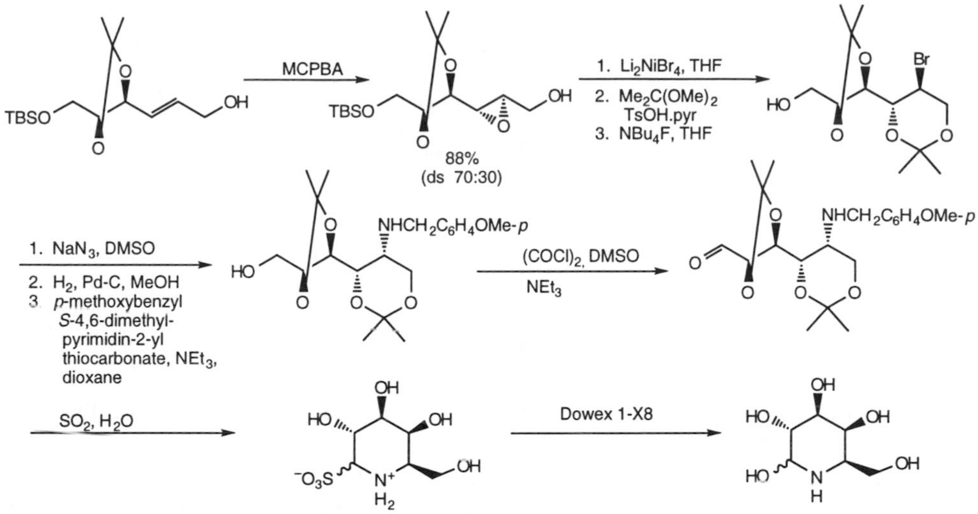

MCPBA; 88% (ds 70:30); 1. $Li_2NiBr_4$, THF 2. $Me_2C(OMe)_2$ TsOH.pyr 3. $NBu_4F$, THF; 1. $NaN_3$, DMSO 2. $H_2$, Pd-C, MeOH 3. *p*-methoxybenzyl *S*-4,6-dimethyl-pyrimidin-2-yl thiocarbonate, $NEt_3$, dioxane; $(COCl)_2$, DMSO, $NEt_3$; $SO_2$, $H_2O$; Dowex 1-X8; **11.26**

**SCHEME 11.36.**

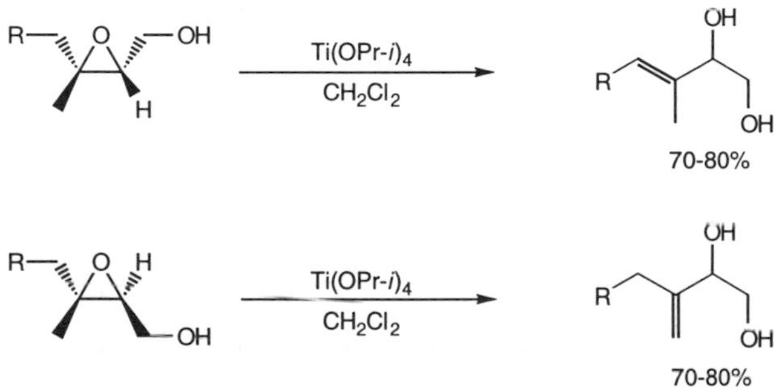

$Ti(OPr\text{-}i)_4$, $CH_2Cl_2$; 70-80%; $Ti(OPr\text{-}i)_4$, $CH_2Cl_2$; 70-80%

**SCHEME 11.37.**

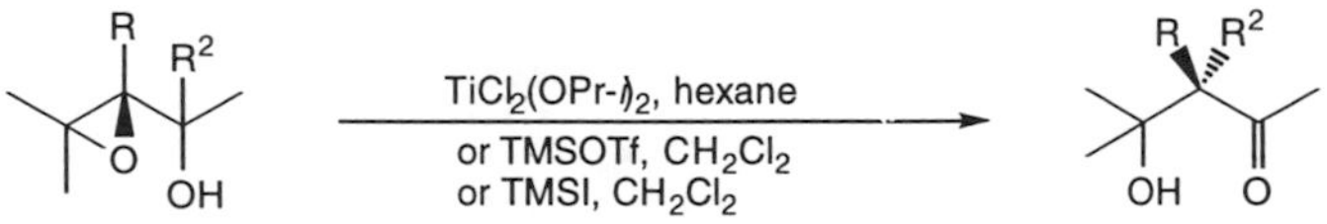

$TiCl_2(OPr\text{-}i)_2$, hexane or TMSOTf, $CH_2Cl_2$ or TMSI, $CH_2Cl_2$

**SCHEME 11.38.**

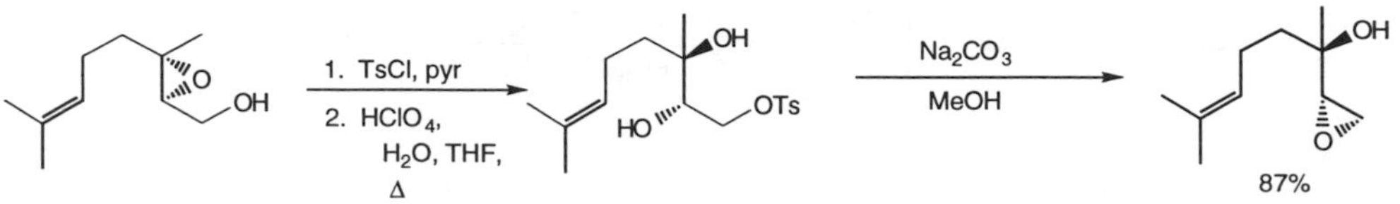

1. TsCl, pyr 2. $HClO_4$, $H_2O$, THF, Δ; $Na_2CO_3$, MeOH; 87%

**SCHEME 11.39.**

The tosylates derived from 2,3-epoxy alcohols can be displaced by halide with inversion of configuration at C-1. The allyl alcohol can then be generated by further treatment with zinc or trialkylstannate (Scheme 11.40; *cf.* Scheme 11.12).[283] The epoxy chloride can also be used to give chiral propargyl alcohols (Scheme 11.41).[275,299]

NaI, $Me_2CO$ ; Zn

**SCHEME 11.40.**

1. $Ph_3P$, $CCl_4$, $NaHCO_3$
2. $LiNH_2$, $NH_3$, or LDA, THF

**SCHEME 11.41.**

The Sharpless kinetic resolution provides a number of other routes to optically active propargyl alcohols through γ-iodoallyl alcohols.[126,132]

The epoxy alcohols can be reacted with a wide variety of reagents to modify the reactivity of the hydroxy group.[300] In addition to converting the hydroxy to a better leaving group, it can also be protected.†

Protected glycerol units are extremely useful for the preparation of many natural products. To achieve selective protection in glycerol, nucleophilic substitution of a sulfonate is a common approach; this is now augmented by the ring opening of an epoxysulfonate by an alcohol (Scheme 11.42).[62,301-305]

ROH, $BF_3 \cdot OEt_2$

**SCHEME 11.42.**

The mesylates or *m*-nitrobenzenesulfonates derived from 2,3-epoxy alcohols react with higher order cuprates at the least hindered epoxide center.[62,299,306,307] The resultant alkoxy mesylate does not ring-close to the epoxide at –78°, allowing stepwise reactions (Scheme 11.43).[306]

$Bu_2Cu(CN)Li_2$, $BF_3.OEt_2$, THF, -78°

BuLi, THF

$Bu_2Cu(CN)Li_2$, $BF_3.OEt_2$, THF, -78°

57% overall

**SCHEME 11.43.**

† The majority of these reactions are not included in Table 11.4.

The expected mode of attack, that is at the least substituted position of the epoxide, can be achieved with a wide range of nucleophiles if the hydroxy group of the epoxy alcohol is protected as an ether (Scheme 11.44).[304,308-310]

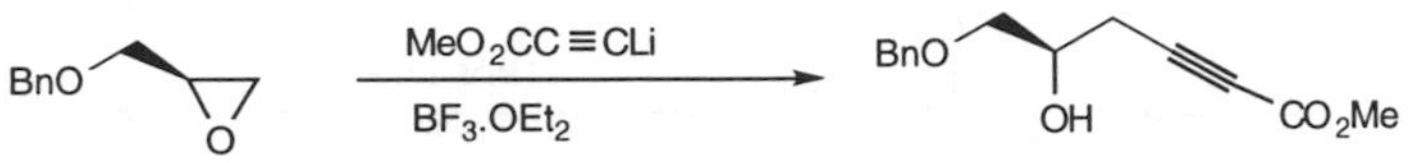

**SCHEME 11.44.**

The reaction has been extended to a bis-epoxide so that two ring openings can be performed concurrently.[311]

Protection of the hydroxyl function in a 2,3-epoxy alcohol can have a significant effect on the regiochemical outcome of a reaction. Acid catalysis during the ring opening of the α,β-epoxy ether **11.27** led to nucleophilic attack at C-2, as evidenced by the inversion of stereochemistry at this center (Scheme 11.45).[284,312]

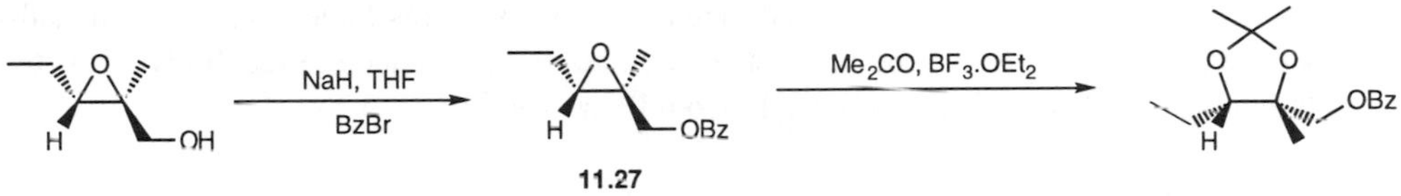

**SCHEME 11.45.**

Reaction of the silyl ether of an epoxy alcohol with the hindered aluminum reagent **11.28** leads to rearrangement and formation of the optically active β-siloxy aldehydes (Scheme 11.46).[313]

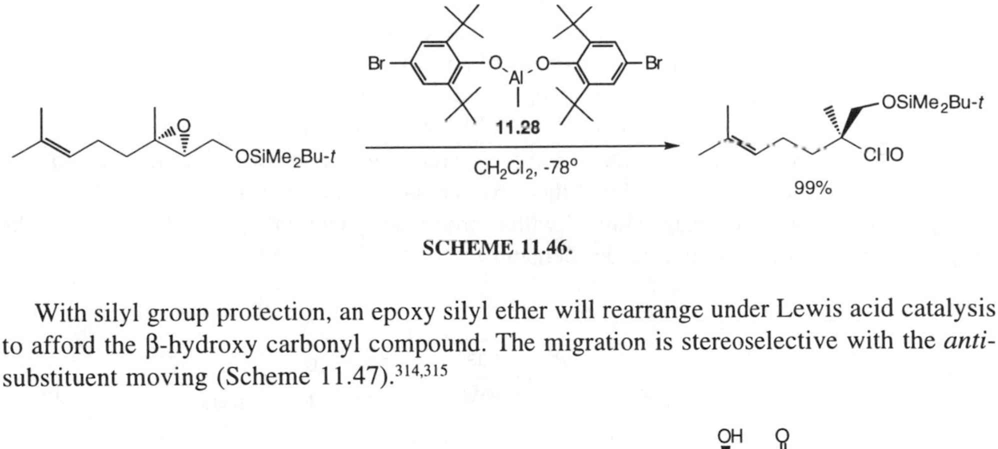

**SCHEME 11.46.**

With silyl group protection, an epoxy silyl ether will rearrange under Lewis acid catalysis to afford the β-hydroxy carbonyl compound. The migration is stereoselective with the *anti*-substituent moving (Scheme 11.47).[314,315]

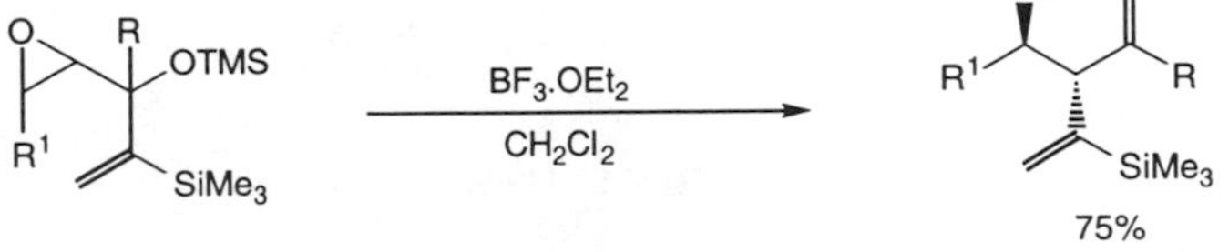

**SCHEME 11.47.**

For ethers derived from epoxy alcohols, a resolution procedure based on the use of dimethylsulfide addition can be used [*10.1.4*].[316] An epoxide can be "inverted" as shown in Scheme 11.48.[317]

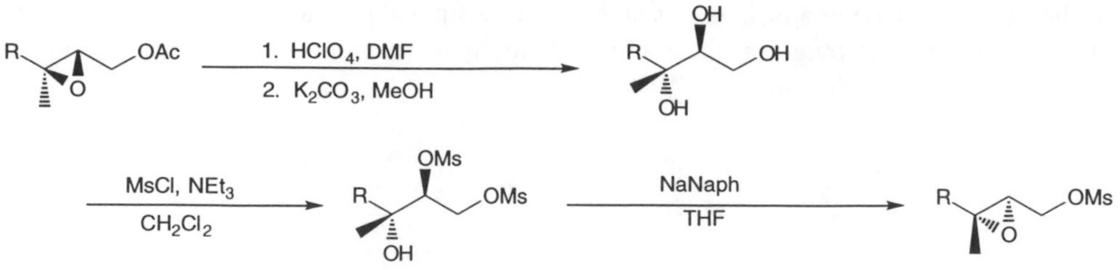

SCHEME 11.48.

Finally, notwithstanding their amenability to ring opening, epoxides are relatively stable entities and, with care, can survive a number of synthetic sequences.[318,319]

# 11.3. EPOXIDATIONS AND REACTIONS OF OTHER UNSATURATED SYSTEMS

In addition to allyl alcohols, α,β-unsaturated carbonyl systems have also served as substrates for oxidation reactions. The products can undergo stereoselective reactions that often complement those observed for their allyl alcohol analogs.[320]

## 11.3.1. CONJUGATED ENONES

### 11.3.1.1. Epoxidation

Enones have been converted to the corresponding α,β-epoxy ketones with a high degree of enantioselectivity by a triphasic system.[321-323]

Epoxidation of enones in the presence of an *N*-benzylquinine salt may give optically active epoxides.[324,325] The use of poly-L-leucine catalysis has been reported to provide high asymmetric induction for the oxidation of chalcones with hydrogen peroxide.[326]

α,β-Epoxy carbonyl compounds are also available by a dimethyldioxirane epoxidation,[327] a chromium induced oxidative rearrangement of tertiary allyl alcohols,[324] or by reaction of an arsenic ylide with an aldehyde.[328] An aldol-type condensation with α-halo ketones provides an alternative route to α,β-epoxy ketones.[329-331] The most useful approach to α,β-epoxy carbonyl compounds, when an asymmetric approach is required, is to use a Sharpless epoxidation of the corresponding allyl alcohol and then oxidize the resultant epoxy alcohol [*11.2.6*].

By use of a chiral auxiliary coupled with a bromolactonization, epoxy aldehydes can be prepared from α,β-unsaturated acids (Scheme 11.49).[332,333]

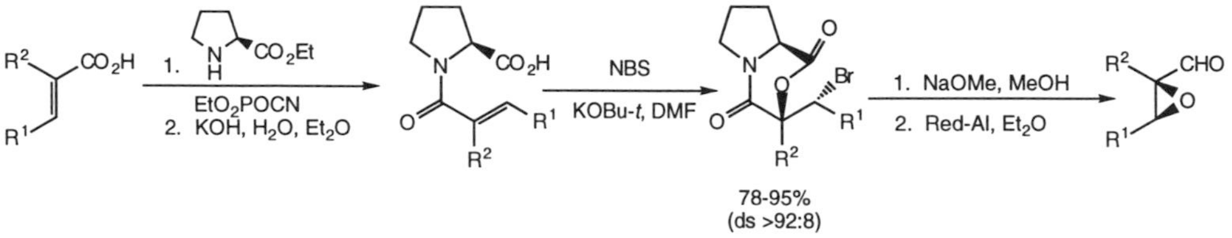

SCHEME 11.49.

### 11.3.1.2. Reactions

Treatment of an epoxy ketone with a selenide nucleophile provides an alternative route to an aldol-type product (Scheme 11.50).[334]

Although the stereochemistry for epoxidation may be difficult to control, relative stereochemistry can be controlled through reduction of the carbonyl group (Scheme 11.51).[335]

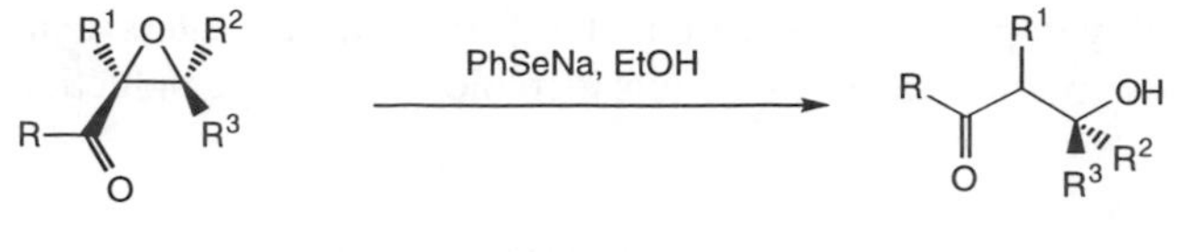

PhSeNa, EtOH

**SCHEME 11.50.**

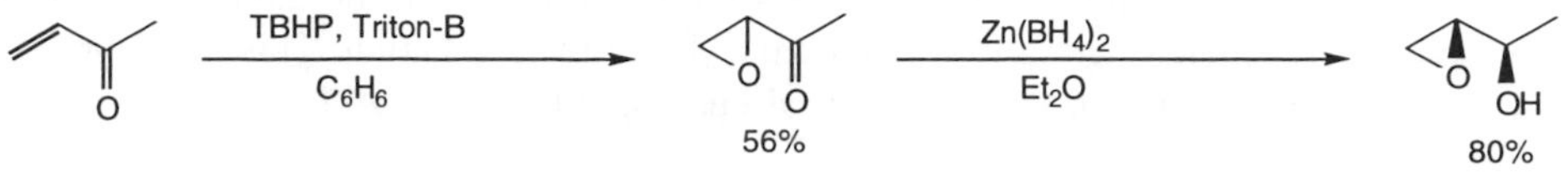

TBHP, Triton-B; $C_6H_6$; 56%; $Zn(BH_4)_2$; $Et_2O$; 80%

**SCHEME 11.51.**

The observed stereoselection can be rationalized in terms of a chelate model,[147] but the conformational model (as shown in Figure 11.7) seems more likely, in order to account for the effect of α-substituents.[30]

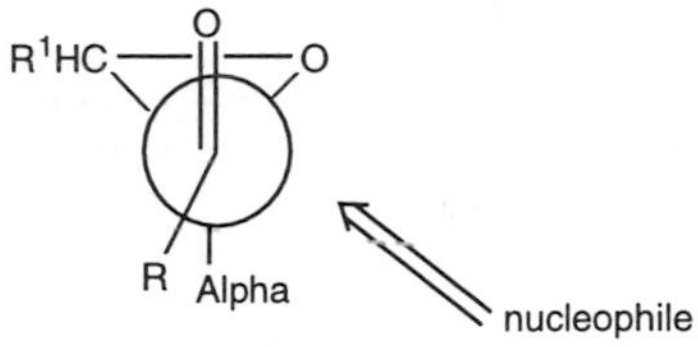

where alpha = alpha substituent

**FIGURE 11.7.** Approach of a nucleophile to an α,β-epoxy carbonyl compound.

The carbonyl group of α,β-epoxy ketones can also be reduced with high stereoselectivity by sodium borohydride, as long as no α-substituent is present (Scheme 11.52).[147,336] Use of zinc borohydride circumvents this limitation and results in the *pref*-isomer (*cf.* Scheme 11.51).[337]

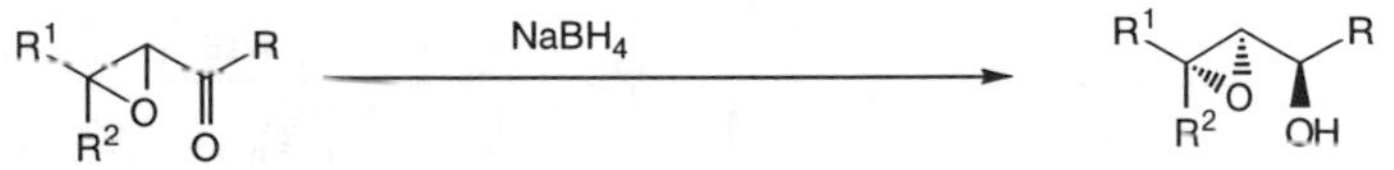

$NaBH_4$

**SCHEME 11.52.**

The α,β-epoxy carbonyl compounds can participate in the common reactions of the carbonyl group, such as a Wittig reaction, without disturbing the epoxide moiety.[338,339] The addition of nucleophiles, such as methyl lithium, takes place with reasonable stereoselectivity to give the product derived from no chelation control at the carbonyl group.[202,340]

## 11.3.2. α,β-UNSATURATED ESTERS

### 11.3.2.1. Epoxidation

Oxidation of α,β-unsaturated esters may be achieved with peroxy acid[341] or dimethyldioxirane.[342]

The Darzen condensation is a classical route to α,β-epoxy esters, but stereochemical control is difficult.[343-345] An alternative procedure is to react the dianion of a β-hydroxy ester with iodine, which provides the least hindered (*trans*) product, due to control in the condensation step [*cf.* *6.3*].[346,347]

Lithium *t*-butylhydroperoxide oxidation of α,β-unsaturated esters that contain a chiral auxiliary in the ester portion can result in considerable asymmetric induction (20 to 100%).[348]

#### 11.3.2.2. Dihydroxylation

α,β-Unsaturated esters are substrates for the asymmetric dihydroxylation methodology [*10.3.2*] (Scheme 11.53).[349] The resultant diol can then be used to form a cyclic sulfite, which can undergo subsequent ring opening reactions [*10.4*]. In addition, the 2,3-dihydroxyesters can be selectively reacted with arylsulfonyl chlorides at the 2-position. This then allows for epoxide formation or reaction with an external nucleophile (Scheme 11.53).[350]

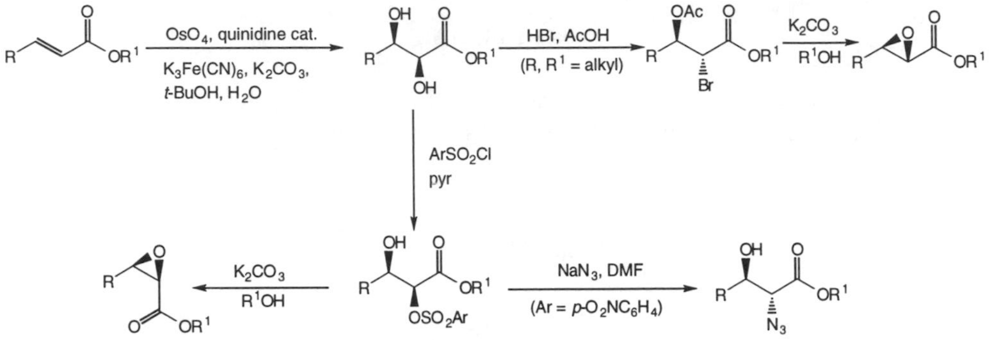

**SCHEME 11.53.**

#### 11.3.2.3. Reactions

Reaction of 2,3-epoxy esters with ammonia provides the 3-amino-2-hydroxy acid,[341,351] while reduction with lithium in ammonia provides the β-hydroxy ester.[352] In contrast, the epoxide can be opened with hydrazoic acid to provide β-hydroxy-α-amino acids (Scheme 11.54),[353,354] or aziridines,[355] or can be treated with organoaluminates to provide α-hydroxy esters.[356,357]

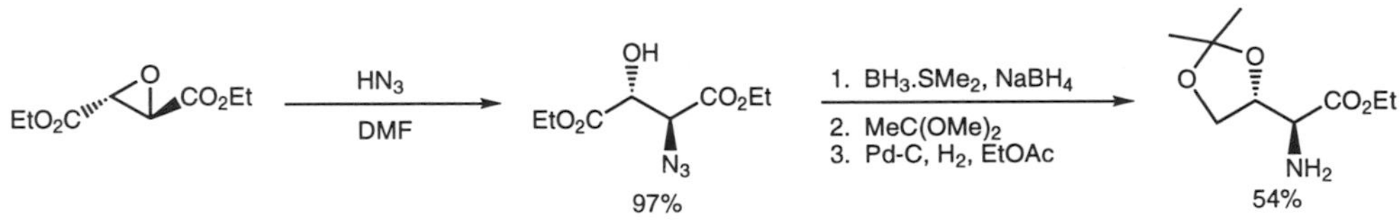

**SCHEME 11.54.**

α,β-Epoxy esters are attacked regioselectively by cuprates at C-2 to provide the β-hydroxy esters.[347,358,359] Many nucleophilic reactions are regioselective.[341,353]

α,β-Epoxy esters are reduced by samarium iodide to the β-hydroxy ester with retention of configuration at the β-position.[360] In contrast, use of magnesium iodide and tributyltin hydride affords the α-hydroxy ester with the α-configuration unaffected.[361]

### 11.3.3. α,β-UNSATURATED ACIDS

#### 11.3.3.1. Epoxidation

α,β-Epoxy acids are available by oxidation of the corresponding allyl alcohol with hydrogen peroxide in the presence of sodium tungstate; use of chiral amines then allows for resolution.[362-364] The epoxides of α,β-unsaturated acids are available from α-haloacids by a variant of the Darzen condensation.[190,365]

### 11.3.3.2. Reactions

The reduction of these epoxides can be regioselective and is influenced by the counterion (Scheme 11.55).[366]

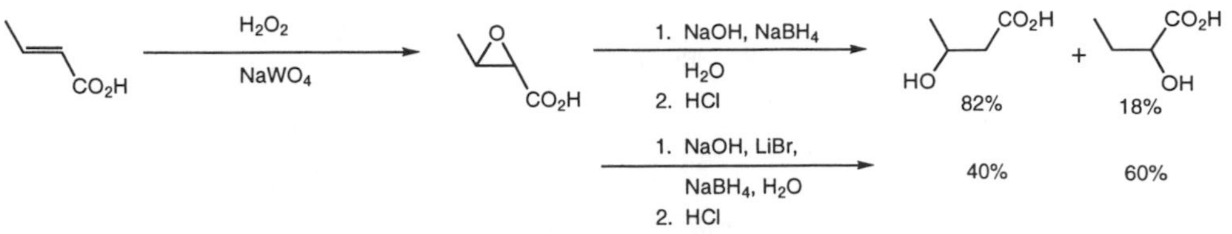

SCHEME 11.55.

2,3-Epoxy acids are opened regioselectively by cuprates (Scheme 11.56)[367] and other nucleophiles.[351]

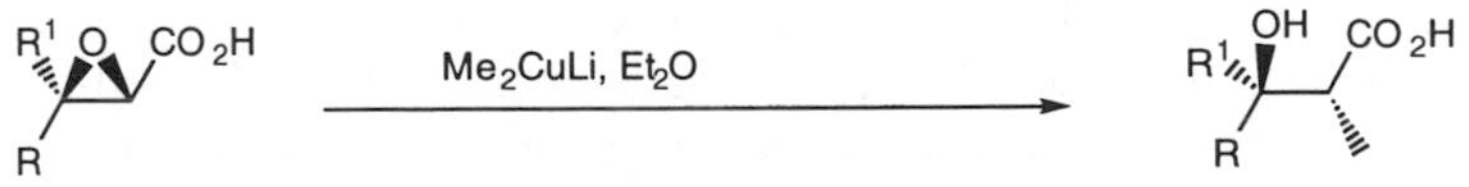

SCHEME 11.56.

## 11.3.4. EPOXIDES FROM HOMOALLYL ALCOHOLS

To a certain degree, the chemistry of these compounds has already been discussed [*11.2*], as has their synthesis by Sharpless methodology [*11.1.2*] or vandium oxidations [*11.1.1.1*]. To some extent, the regio- and stereochemical outcomes for the addition reactions of 3,4-epoxy alcohols are dependent upon substituents and the reagent; cuprates show a small preference for C-4 addition, while aluminates show a large preference for this mode of addition.[368]

## 11.3.5. CONJUGATED DIENES

### 11.3.5.1. Epoxidation

Conjugated dienes can be oxidized to monoepoxides in good chemical yields in the presence of molybdenum(VI), but the regioselectivity can be low.[9] An alternative strategy, the Wittig condensation with an epoxy aldehyde, does provide the monoepoxide of a conjugated diene stereoselectively;[369,370] telluride chemistry provides a further alternative.[371]

### 11.3.5.2. Reactions

Although reaction with a Grignard reagent can give rise to complex mixtures,[372-374] these monoepoxides undergo clean $S_N2'$ additions with organocopper reagents, (Scheme 11.57) among other nucleophilic reagents.[195,196,372-380]

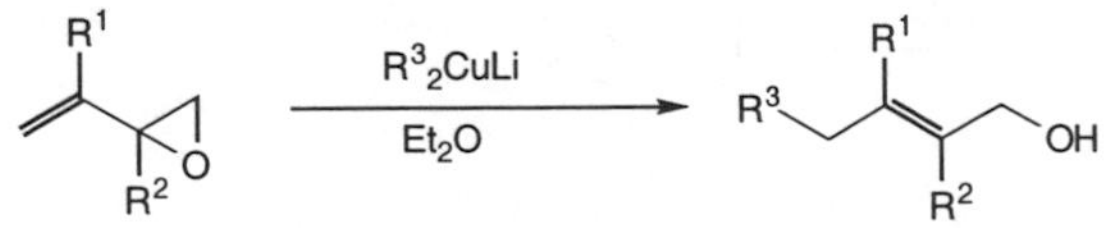

SCHEME 11.57.

Palladium(0) complexes have also proven to be a useful catalyst to bring about an $S_N2'$ reaction. A wide variety of nucleophilic and electrophilic species can be used in the reaction.[378,379,381-385]

By a careful choice of reaction conditions, regioselectivity problems can be overcome, as illustrated by the reaction of the monoepoxy diene **11.29** with hydrogen bromide (Scheme 11.58).[272]

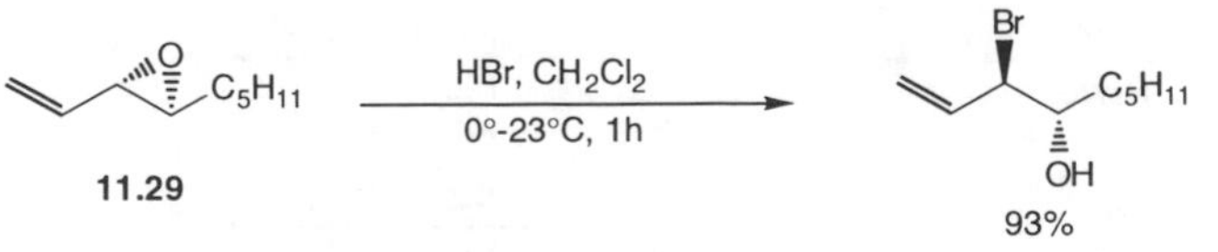

HBr, $CH_2Cl_2$, 0°-23°C, 1h

**11.29** → 93%

SCHEME 11.58.

## 11.3.6. OTHER SYSTEMS

α-Trimethylsilylepoxides, available from vinylsilanes,[386,387] can be used to prepare epoxides through fluoride ion displacement of the silyl group and subsequent reaction with an electrophile.[388,389] A tin group can be exchanged with an alkyl lithium and then the resultant α-lithioepoxide can be reacted with an electrophile in a similar manner.[390] In addition to exchange reactions, groups that stabilize an α-carbanion, such as silyl, sulfoxide, sulfone, cyano, carbon–carbon unsaturation, aryl, and esters, allow deprotonation to occur on the carbon atom of an epoxide that is bonded to one of these groups without disruption of the epoxide ring (Scheme 11.59).[388,389,391-395] The product **11.30** is also accessible by epoxidation of a silyl substituted allyl alcohol.[396,397]

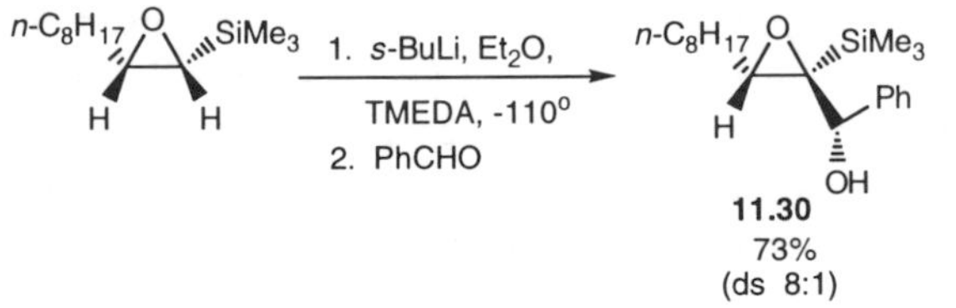

1. *s*-BuLi, $Et_2O$, TMEDA, -110°
2. PhCHO

**11.30** 73% (ds 8:1)

SCHEME 11.59.

The utility of an α,β-epoxy silane is extended by the simple acid catalyzed hydrolysis of these compounds to carbonyl compounds [*5.2.2*].[387,398] Nucleophilic attack at the silicon-bearing carbon gives rise to the opportunity for a Peterson-type elimination to afford a functionalized alkene;[232,399-402] methodology to *E*- or *Z*-allylsilanes is also available.[403] Epoxysilanes are precursors to silyl enol ethers [*5.1*].[404-406]

Oxidation of the chiral allylsilanes **11.31** provides a method to obtain allyl alcohols [*8.1.3.4*] (Scheme 11.60), but the ee was low for R = Ph (32%).[386]

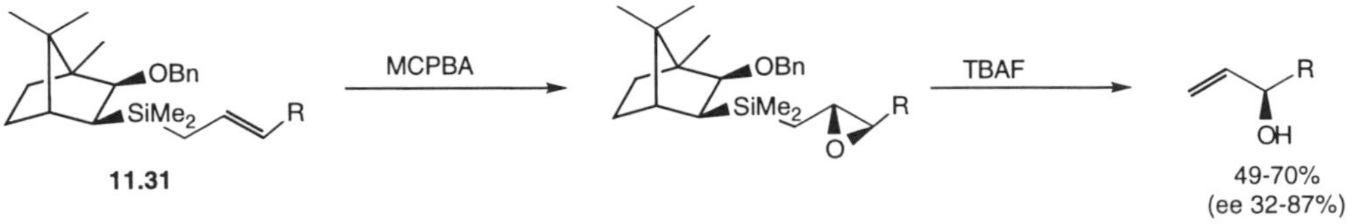

MCPBA → TBAF

**11.31** → 49-70% (ee 32-87%)

SCHEME 11.60.

β-Sulfinyl epoxides provide useful methodology for the preparation of homoallylic alcohols (Scheme 11.61) — note the use of chelation to control the reduction step[204] — while 2,3-epoxy-1-chloroalkanes provide an allyl alcohol synthesis.[407]

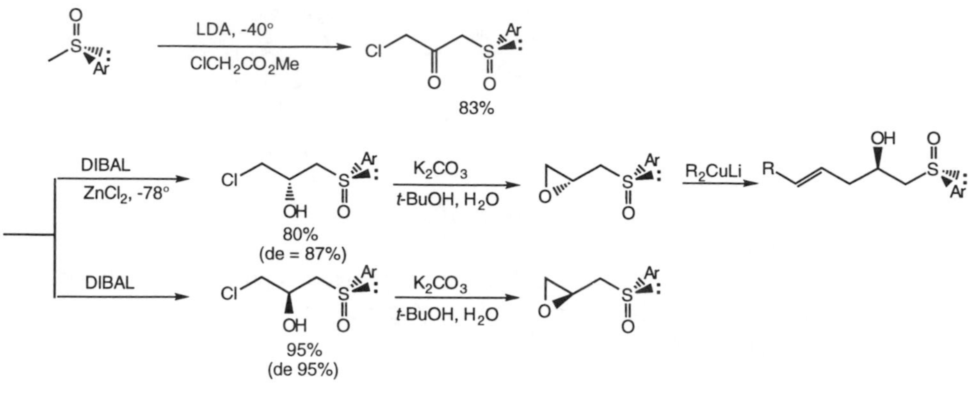

**SCHEME 11.61.**

## 11.4. OTHER OXIDATIONS

Other than osmium, the oxidation of functionalized alkenes has not found widespread use. The main problem lies in over oxidation that usually results in cleavage.[408-412]

### 11.4.1. OSMYLATION

The osmylation of alkenes possessing an allylic oxygen center is stereoselective with a predictable outcome.[413-429]

Osmium oxidizes allyl alcohols to afford a triol. An empirical rule has been advanced, where the reagent approaches from the face opposite to the pre-existing oxygen functionality (Figure 11.8).[413,415,416] Although the hydroxy group may be protected, the presence of an acyl group reduces stereoselectivity; *cis*-olefins provide better selection than do their *trans* counterparts.¶ [413-416] It should be noted that this is one empirical rule and other variants exist,[422] which will no doubt be modified as our understanding of the osmylation reaction expands. The dihydroxylation oxidation procedure for alkenes [*10.3.2*] has been extended to allyl alcohols[430-432] and allyl amines.[433,434]

High diastereoselectivity is also observed for the dihydroxylation of bis-allylic substrates,[435-437] even when no allylic oxygen group is present (Scheme 11.62).[438]

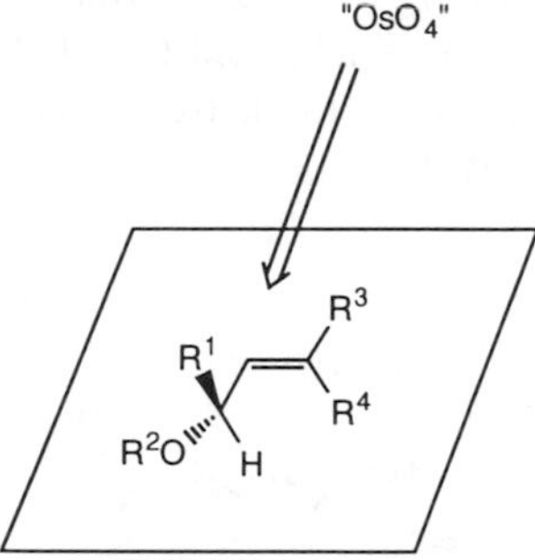


**FIGURE 11.8.** Approach of osmium tetroxide for the dihydroxylation of an allyl alcohol.

Cyclic substrates offer stereoselectivity through the inherent stereofacial requirements of the substrate,[5,169] or from chiral auxiliaries.[420]

¶ As the rule was proposed by Kishi, it is sometimes referred to as Kishi's rule [*2.2*].

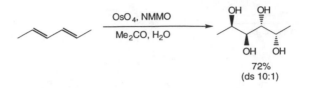

**SCHEME 11.62.**

The oxidation of allylsilanes[439] (e.g., Scheme 2.17) provides a powerful method to obtain allyl alcohols and triols.[440] The incorporation of an oxygen atom at C-1 allows for a rapid entry to *ancat* triols (Scheme 11.63).[419,423,435,441-443]



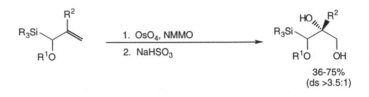

**SCHEME 11.63.**

α,β-Unsaturated esters can also be dihydroxylated.[435,444,445] The use of a chiral auxiliary, such as 8-phenylmenthyl, allows for some face selectivity (Scheme 11.64).[446]

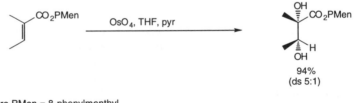



**SCHEME 11.64.**

The conjugated enone was oxidized by osmium tetroxide in the presence of a chiral ligand to afford the diol **11.32** (Scheme 11.65). In contrast, if the carbonyl group of **11.33** was protected as an ethylene glycol ketal prior to the oxidation step, the asymmetric induction of the resultant diol was low (36% ee).[447]



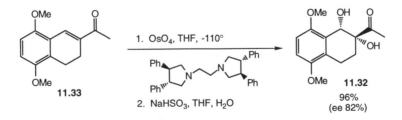

**SCHEME 11.65.**

An alternative approach that allows for the introduction of asymmetry, relies on the asymmetric dihydroxylation procedure developed for isolated alkenes [*10.3.2*]; α,β-unsaturated esters can also be used as reactant.[349,448,449] This asymmetric dihydroxylation approach can also be coupled with cyclic sulfate chemistry [*10.4*].[450] The diols can be selectively monotosylated or nosylated on the C-2 hydroxy group, which then allows for further reaction (Scheme 11.53).[350] Thus, the asymmetric dihydroxylation approach allows entry, through the use of functionalized substrates, to complex molecules, such as carbohydrate derivatives,[431] or kinetic resolutions in a similar manner to the Sharpless epoxidation.[451]

The hydroxylation of a γ-hydroxy-α,β-unsaturated ester by osmium tetroxide follows a complex formation model with the free hydroxy group to afford **11.34** (Scheme 11.66).[417]

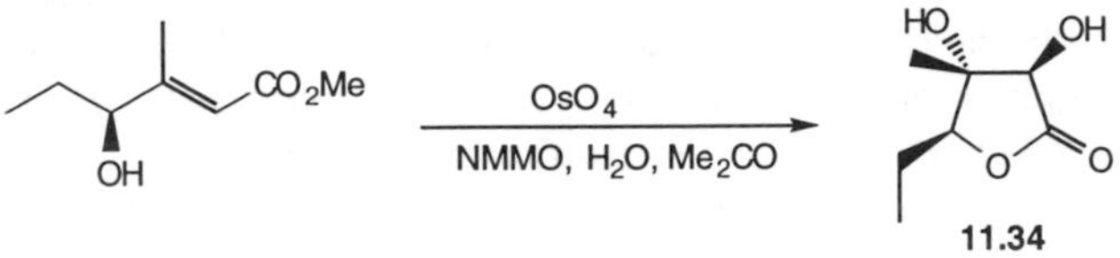

**SCHEME 11.66.**

The osmium catalyzed hydroxylation of a vinylsilane results in the formation of a diol that can be eliminated to provide silyl enol ethers [*5.1*].[452]

## 11.5. OTHER SYSTEMS

As with isolated alkenes, derivatives other than epoxides are available, but they have not found such widespread use in the synthetic repertoire.

A variant of the Darzen's reaction is available for the preparation of functionalized aziridines **11.35**; high *cis*-stereoselectivity is observed (Scheme 11.67).[453]

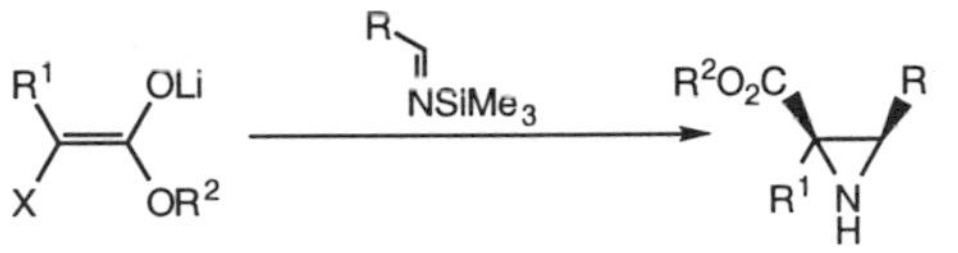

**SCHEME 11.67.**

The aziridines are also available from serine derivatives through a ring closure procedure [*10.5*].[454] The use of an *N*-protected thioester of an aziridine allowed for alkylation reactions to be undertaken — the lithiated species is configurationally stable (Scheme 11.68).[454]

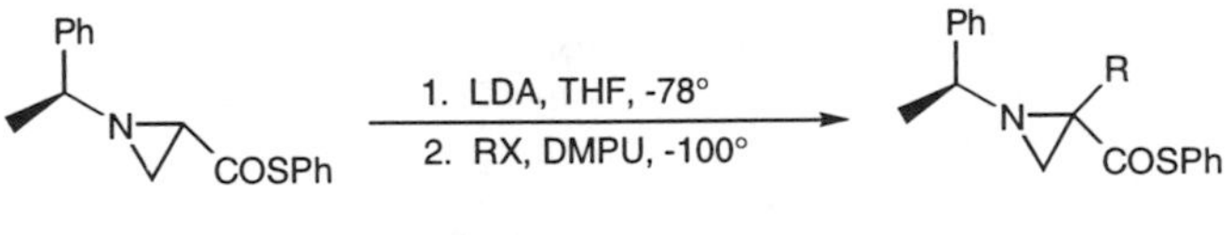

**SCHEME 11.68.**

A variation on iodolactonization that uses a phosphate cyclization process [*13.2.1*], provides a very useful method for the preparation of β,γ-epoxy alcohol derivatives that can, in turn, be used to obtain *syncat* 1,3-diols (Scheme 11.69).[31,455]

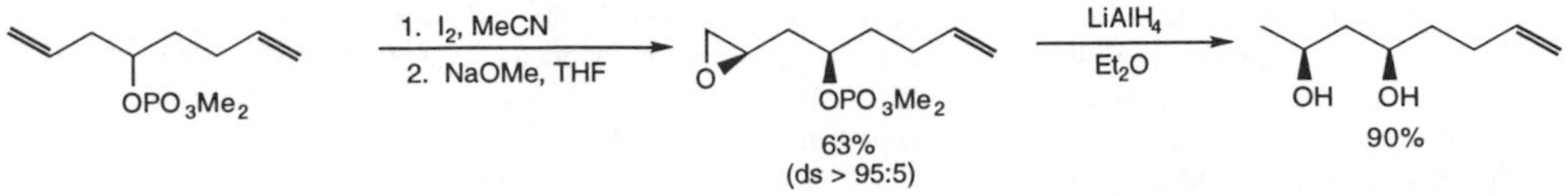

**SCHEME 11.69.**

## 11.6. SUMMARY

Asymmetric oxidations were the reactions that have given rise to much of today's success in the area of asymmetric synthesis. Unlike asymmetric hydrogenations, where a catalyst system may only be useful with a narrow class of substrates, the oxidation systems are, within limits, promiscuous. They showed chemists that reagent systems could be designed to rival stereoselectivities seen with biological reagents. The substrates in this chapter have a functional group that can form a complex to the oxidant and "guide" the reagent onto the face of an alkene. The Sharpless asymmetric epoxidation of allyl alcohols has now become firmly established, as it can be used with a wide range of substitution patterns on the allyl alcohol. In addition, the chemist has control of the resultant chirality through a choice of which tartrate isomer is employed in the reaction itself. Peroxy acid oxidations can complement the approach, as they allow some substitution patterns to be oxidized that provide only moderate asymmetric induction with the Sharpless methodology.

As the reaction is not temperature sensitive, the Sharpless method can be performed on a large scale.

A wide variety of reactions is available to modify 2,3-epoxy alcohols. This also makes the reactions outlined in this chapter extremely powerful.

Other functionalized alkenes can be epoxidized, but the selectivities are usually less than for allyl alcohols. Some of the more useful transformations, together with the scope and limitations of the chemistry of these functionalized epoxides, make this clear. In almost all cases, the oxidation of an allyl alcohol followed by conversion to the desired functionality is the approach of choice.

In addition to epoxides, functionalized alkenes can be transformed into diols with osmium reagents. An empirical rule has been formulated to predict the stereochemical outcome. The use of osmium is not as desirable as titanium on a large scale, but this is still feasible.

## 11.7. REFERENCES


1. Katsuki, T. *J. Syn. Org. Chem., Jpn.* 1987, *45*, 90.
2. Martens, J. *Chemiker-Zeitung* 1986, *110*, 169.
3. Finn, M. F.; Sharpless, K. B. In *Asymmetric Synthesis*; J. D. Morrison, Ed.; Academic Press: Orlando, FL, 1986; Vol. 5; pp 247.
4. Rossiter, B. E. In *Asymmetric Synthesis*; J. D. Morrison, Ed.; Academic Press: Orlando, FL, 1985; Vol. 5; pp 193.
5. Ager, D. J.; East, M. B. *Tetrahedron* 1992, *48*, 2803.
6. Backenstrass, F.; Streith, J.; Tschamber, T. *Tetrahedron Lett.* 1990, *31*, 2139.
7. Jorgensen, K. A. *Chem. Rev.* 1989, *89*, 431.
8. Sheldon, R. A. *J. Mol. Catal.* 1983, *20*, 1.
9. Sheng, M. N.; Zajacek, J. G. *J. Org. Chem.* 1970, *35*, 1839.
10. Rossiter, B. E.; Verhoeven, T. R.; Sharpless, K. B. *Tetrahedron Lett.* 1979, 4733.
11. Tomioka, H.; Suzuki, T.; Oshima, K.; Nozaki, H. *Tetrahedron Lett.* 1982, *23*, 3387.
12. Mihelich, E. D. *Tetrahedron Lett.* 1979, 4729.
13. Jarosz, S. *Carbohydr. Res.* 1988, *183*, 217.
14. Chou, D. T.-W.; Ganem, B. *J. Am. Chem. Soc.* 1980, *102*, 7987.
15. Lubben, T. V.; Wolczanski, P. T. *J. Am. Chem. Soc.* 1985, *107*, 701.
16. Mehendale, A. R.; Kulkarni, A.; Nagarajan, G. *Int. J. Chem.* 1988, *27B*, 305.
17. Arcoria, A.; Ballistreri, F. P.; Tomaselli, G. A.; Di Furia, F.; Modena, G. *J. Org. Chem.* 1986, *51*, 2374.
18. Prat, D.; Lett, R. *Tetrahedron Lett.* 1986, *27*, 707.
19. Prat, D.; Delpech, B.; Lett, R. *Tetrahedron Lett.* 1986, *27*, 711.
20. Sharpless, K. B.; Michaelson, R. C. *J. Am. Chem. Soc.* 1973, *95*, 6136.

21. Choudary, B. M.; Valli, V. L. K.; Prasad, A. D. *J. Chem. Soc., Chem. Commun.* 1990, 721.
22. Bessodes, M.; Egron, M.-J.; Antonakis, K. *J. Chem. Soc., Perkin Trans. 1* 1989, 2099.
23. Yasuda, A.; Tanaka, S.; Yamamoto, H.; Nozaki, H. *Bull. Chem. Soc., Jpn.* 1979, *52*, 1701.
24. Tanaka, S.; Yamamoto, H.; Nozaki, H.; Sharpless, K. B.; Michaelson, R. C.; Cutting, J. D. *J. Am. Chem. Soc.* 1974, *96*, 5254.
25. Itoh, T.; Jitsukawa, K.; Kaneda, K.; Teranishi, S. *J. Am. Chem. Soc.* 1979, *101*, 159.
26. Narula, A. S. *Tetrahedron Lett.* 1982, *23*, 5579.
27. Hiyama, T.; Obayashi, M. *Tetrahedron Lett.* 1983, *24*, 395.
28. Marson, C. M.; Walker, A. J.; Pickering, J.; Harper, S.; Wrigglesworth, R.; Edge, S. J. *Tetrahedron* 1993, *49*, 10317.
29. Hanamoto, T.; Katsuki, T.; Yamaguchi, M. *Tetrahedron Lett.* 1987, *28*, 6191.
30. Bartlett, P. A. *Tetrahedron* 1980, *36*, 3.
31. Bartlett, P. A.; Jernstedt, K. K. *J. Am. Chem. Soc.* 1977, *99*, 4829.
32. Mohr, P.; Tamm, C. *Tetrahedron Lett.* 1987, *28*, 391.
33. Kobayashi, Y.; Uchiyama, H.; Kanbara, H.; Sato, F. *J. Am. Chem. Soc.* 1985, *107*, 5541.
34. Roush, W. R.; Halterman, R. L. *J. Am. Chem. Soc.* 1986, *108*, 294.
35. Roush, W. R.; Michaelides, M. R. *Tetrahedron Lett.* 1986, *27*, 3353.
36. Depezay, J.-C.; LeMerrer, Y. *Tetrahedron Lett.* 1978, 2865.
37. Depezay, J.-C.; Dureault, A. *Tetrahedron Lett.* 1978, 2869.
38. Hoppe, D.; Lußmann, J.; Jones, P. G.; Schmidt, D.; Sheldrick, G. M. *Tetrahedron Lett.* 1986, *27*, 3591.
39. Yamada, S.-I.; Mashiko, T.; Terishima, S. *J. Am. Chem. Soc.* 1977, *99*, 1988.
40. Coleman-Kammula, S.; Dium-Koolstra, E. T. *J. Organomet. Chem.* 1983, *246*, 53.
41. Michaelson, R. C.; Palermo, R. E.; Sharpless, K. B. *J. Am. Chem. Soc.* 1977, *99*, 1990.
42. Dehnel, R. B.; Whitham, G. H. *J. Chem. Soc., Perkin Trans. 1* 1979, 953.
43. Roush, W. R.; Straub, J. A.; Brown, R. J. *J. Org. Chem.* 1987, *52*, 5127.
44. Mugdan, M.; Young, D. P. *J. Chem. Soc.* 1949, 2988.
45. Takai, K.; Oshima, K.; Nozaki, H. *Tetrahedron Lett.* 1980, *21*, 1657.
46. Kanemoto, S.; Nonaka, T.; Oshima, K.; Utimoto, K.; Nozaki, H. *Tetrahedron Lett.* 1986, *27*, 3387.
47. Katsuki, T.; Sharpless, K. B. *J. Am. Chem. Soc.* 1980, *103*, 5974.
48. Sharpless, K. B.; Behrens, C. H.; Katsuki, T.; Lee, A. W. M.; Martin, V. S.; Takatani, M.; Viti, S. M.; Walker, F. J.; Woodard, S. S. *Pure Appl. Chem.* 1983, *55*, 589.
49. Pfenninger, A. *Synthesis* 1986, 89.
50. Sato, F.; Kobayashi, Y. *Synlett* 1992, 849.
51. Kende, A. S.; Rizzi, J. P. *J. Am. Chem. Soc.* 1981, *103*, 4247.
52. Schweiter, M. J.; Sharpless, K. B. *Tetrahedron Lett.* 1985, *26*, 2543.
53. Hatakeyama, S.; Sakurai, K.; Takano, S. *J. Chem. Soc., Chem. Commun.* 1985, 1759.
54. Hatakeyama, S.; Sakurai, K.; Takano, S *Tetrahedron Lett.* 1986, *27*, 4485.
55. Hafele, B.; Schroter, D.; Jager, V. *Angew. Chem., Int. Ed. Engl.* 1986, *25*, 87.
56. Schreiber, S. L.; Schreiber, T. S.; Smith, D. B. *J. Am. Chem. Soc.* 1987, *109*, 1525.
57. Mori, K.; Umemura, T. *Tetrahedron Lett.* 1981, *22*, 4433.
58. Thijs, L.; Stokkingreef, E. H. M.; Lemmens, J. M.; Zwanenburg, B. *Tetrahedron* 1985, *41*, 2949.
59. Brimacombe, J. S.; Hanna, R.; Kabir, A. K. M. S. *J. Chem. Soc., Perkin Trans. 1* 1987, 2421.
60. Rossiter, B. E.; Katsuki, T.; Sharpless, K. B. *J. Am. Chem. Soc.* 1981, *103*, 464.
61. Thijs, L.; Dommerholt, F. J.; Leemhuis, F. M. C.; Zwanenburg, B. *Tetrahedron Lett.* 1990, *31*, 6589.
62. Klunder, J. M.; Onami, T.; Sharpless, K. B. *J. Org. Chem.* 1989, *54*, 1295.
63. Katsuki, T.; Lee, A. W. M.; Ma, P.; Martin, V. S.; Masamune, S.; Sharpless, K. B.; Tuddenham, D.; Walker, F. J. *J. Org. Chem.* 1982, *7*, 1373.
64. Babine, R. E. *Tetrahedron Lett.* 1986, *27*, 5791.
65. Sun, C.-Q.; Rich, D. H. *Tetrahedron Lett.* 1988, *29*, 5205.
66. Mori, K.; Kbata, T. *Tetrahedron* 1986, *42*, 3471.
67. Marczak, S.; Masnyk, M.; Wicha, J. *Tetrahedron Lett.* 1989, *30*, 2845.
68. Brimacombe, J. S.; Hanna, R.; Kabir, A. K. M. S. *Carbohydr. Res.* 1986, *153*, C7.
69. Brimacombe, J. S.; Kabir, A. K. M. S. *Carbohydr. Res.* 1986, *152*, 329.
70. Mori, K.; Ebata, T. *Tetrahedron Lett.* 1981, *22*, 4281.
71. Schmidt, R. R.; Frische, K. *Liebigs Ann. Chem.* 1988, 209.
72. Kobayashi, Y.; Kato, N.; Shimazaki, T.; Sato, F. *Tetrahedron Lett.* 1988, *29*, 6297.
73. Herunsalee, A.; Isobe, M.; Pikul, S.; Goto, T. *Synlett* 1991, 199.
74. Kufner, U.; Schmidt, R. R. *Angew. Chem., Int. Ed. Engl.* 1986, *25*, 89.
75. Yamada, S.; Shiraishi, M.; Ohmori, M.; Takayama, H. *Tetrahedron Lett.* 1984, *25*, 3347.
76. Roush, W. R.; Spada, A. P. *Tetrahedron Lett.* 1982, *23*, 3773.
77. White, J. D.; Kang, M.-C.; Sheldon, B. G. *Tetrahedron Lett.* 1983, *24*, 4539.

78. Page, P. C. B.; Rayner, C. M.; Sutherland, I. O. *J. Chem. Soc., Perkin Trans. 1* 1990, 2403.
79. Shimizu, I.; Hayashi, K.; Oshima, M. *Tetrahedron Lett.* 1990, *31*, 4757.
80. Mori, K.; Ueda, H. *Tetrahedron* 1981, *37*, 2581.
81. Meyers, A. I.; Hudspeth, J. P. *Tetrahedron Lett.* 1981, *22*, 3925.
82. Lu, L. D.-L.; Johnson, R. A.; Finn, M. G.; Sharpless, K. B. *J. Org. Chem.* 1984, *49*, 728.
83. Meister, C.; Scharf, H.-D. *Liebigs Ann. Chem.* 1983, 913.
84. Erickson, T. J. *J. Org. Chem.* 1986, *51*, 934.
85. Gosmini, C.; Dubuffet, T.; Sauvêtre, R.; Normant, J.-F. *Tetrahedron: Asymmetry* 1991, *2*, 223.
86. Oritani, T.; Yamashita, K. *Phytochem.* 1983, *22*, 1909.
87. Carlier, P. R.; Mungall, W. S.; Schroder, G.; Sharpless, K. B. *J. Am. Chem. Soc.* 1988, *110*, 2978.
88. White, J. D.; Avery, M. A.; Choudhry, S. C.; Dhingra, O. P.; Kang, M.-C.; Whittle, A. J. *J. Am. Chem. Soc.* 1983, *105*, 6517.
89. Roush, W. R.; Brown, R. J. *J. Org. Chem.* 1982, *47*, 1371.
90. Bernet, B.; Vasella, A. *Tetrahedron Lett.* 1983, *24*, 5491.
91. Mori, K.; Otsuka, T. *Tetrahedron* 1985, *41*, 553.
92. Bessodes, M.; Abushanab, E.; Antonakis, K. *Tetrahedron Lett.* 1984, *25*, 5899.
93. Mulzer, J.; Angermann, A.; Munch, W.; Schlichthorl, G.; Hentzschel, A. *Liebigs Ann. Chem.* 1987, 7.
94. Kang, J.; Park, M.; Shin, H. T.; Kim, J. K. *Bull. Korean Chem. Soc.* 1985, *6*, 376.
95. Wang, Z.-M.; Zhou, W.-S. *Tetrahedron* 1987, *43*, 2935.
96. Pedersen, S. F.; Dewan, J. C.; Eckman, R. R.; Sharpless, K. B. *J. Am. Chem. Soc.* 1987, *109*, 1279.
97. Burns, C. J.; Martin, C. A.; Sharpless, K. B. *J. Org. Chem.* 1989, *54*, 2826.
98. Carlier, P. R.; Sharpless, K. B. *J. Org. Chem.* 1989, *54*, 4016.
99. Woodard, S. S.; Finn, M. G.; Sharpless, K. B. *J. Am. Chem. Soc.* 1991, *113*, 106.
100. Finn, M. G.; Sharpless *J. Am. Chem. Soc.* 1991, *113*, 113.
101. Hawkins, J. M.; Sharpless, K. B. *Tetrahedron Lett.* 1987, *28*, 2825.
102. Potvin, P. G.; Bianchet, S. *J. Org. Chem.* 1992, *57*, 6629.
103. Jorgensen, K. A.; Wheeler, R. A.; Hoffmann, R. *J. Am. Chem. Soc.* 1987, *109*, 3240.
104. Corey, E. J. *J. Org. Chem.* 1990, *55*, 1693.
105. Corey, E. J.; Hashimoto, S.-I.; Barton, A. E. *J. Am. Chem. Soc.* 1981, *103*, 721.
106. Wang, Z.-M.; Zhou, W.-S.; Lin, C. Q. *Tetrahedron Lett.* 1985, *26*, 6221.
107. Choudary, B. M.; Valli, V. L. K.; Prasad, A. D. *J. Chem. Soc., Chem. Commun.* 1990, 1186.
108. Whang, Z.-M.; Zhou, W.-S. *Synth. Commun.* 1989, *19*, 2627.
109. Farrall, M. J.; Alexis, M.; Trecarten, M. *Nouv. J. Chim.* 1983, *7*, 449.
110. Marshall, J. A.; Jenson, T. M. *J. Org. Chem.* 1984, *49*, 1707.
111. Rossiter, B. E.; Sharpless, K. B. *J. Org. Chem.* 1984, *49*, 3707.
112. Ibuka, T.; Tanaka, M.; Yamamoto, Y. *J. Chem. Soc., Chem. Commun.* 1989, 967.
113. Smith, D. B.; Wang, Z.; Schreiber, S. L. *Tetrahedron* 1990, *46*, 4793.
114. Takano, S.; Iwabuchi, Y.; Ogasawara, K. *J. Am. Chem. Soc.* 1991, *113*, 2786.
115. Gawronski, J.; Gawronska, K.; Rychlewska, U. *Tetrahedron Lett.* 1990, *31*, 6071.
116. Gawronski, J.; Gawronski, K.; Rychlewska, G. U. *Tetrahedron Lett.* 1989, *30*, 6071.
117. Brown, J. M. *Chem. Int. (London)* 1988, 612.
118. Gao, Y.; Hanson, R. M.; Klunder, J. M.; Ko, S. Y.; Masamune, H.; Sharpless, K. B. *J. Am. Chem. Soc.* 1987, *109*, 5765.
119. Martin, V. S.; Woodard, S. S.; Katsuki, T.; Yamada, Y.; Ikeda, M.; Sharpless, K. B. *J. Am. Chem. Soc.* 1981, *103*, 6237.
120. Kitano, Y.; Matsumoto, T.; Wakasa, T.; Okamoto, S.; Shimazaki, T.; Kobayashi, Y.; Sato, F.; Miyaji, K.; Arai, K. *Tetrahedron Lett.* 1987, *28*, 6351.
121. Zhou, W.-S.; Lu, Z.-H.; Wang, Z.-M. *Tetrahedron Lett.* 1991, *32*, 1467.
122. Zhou, W.-S.; Lu, Z.-H.; Wang, Z.-M. *Tetrahedron* 1993, *49*, 2641.
123. Miyano, S.; Lu, L. D.-L.; Viti, S. M.; Sharpless, K. B. *J. Org. Chem.* 1985, *50*, 4350.
124. Kusakabe, M.; Kobayashi, Y.; Sato, F. *J. Org. Chem.* 1989, *54*, 994.
125. Matsumoto, T.; Kitanao, Y.; Sato, F. *Tetrahedron Lett.* 1988, *29*, 5685.
126. Ito, T.; Okamoto, S.; Sato, F. *Tetrahedron Lett.* 1989, *30*, 7083.
127. Ito, T.; Okamoto, S.; Sato, F. *Tetrahedron Lett.* 1990, *31*, 6399.
128. Kusakabe, M.; Sato, F. *J. Org. Chem.* 1989, *54*, 3486.
129. Kobayashi, Y.; Kusakabe, M.; Kitano, Y.; Sato, F. *J. Org. Chem.* 1988, *53*, 1586.
130. Kusakabe, M.; Kitano, Y.; Kobayashi, Y.; Sato, F. *J. Org. Chem.* 1989, *54*, 2085.
131. Honda, T.; Kametani, T.; Kanai, K.; Tatsuzaki, Y.; Tsubuki, M. *J. Chem. Soc., Perkin Trans. 1* 1990, 1733.
132. Yamakawa, I.; Urabe, H.; Kobayashi, Y.; Sato, F. *Tetrahedron Lett.* 1991, *32*, 2045.
133. Kitano, Y.; Matsumoto, T.; Sato, F. *J. Chem. Soc., Chem. Commun.* 1986, 1323.
134. Discordia, R. P.; Murphy, C. K.; Dittmer, D. C. *Tetrahedron Lett.* 1990, *31*, 5603.

135. Mulzer, J.; Scharp, M. *Synthesis* 1993, 615.
136. Yadav, J. S.; Radhakrishna, P. *Tetrahedron* 1990, *46*, 5825.
137. Neagu, C.; Hase, T. *Tetrahedron Lett.* 1993, *34*, 1629.
138. Isobe, M. I.; Kitamura, M.; Mio, S.; Goto, T. *Tetrahedron Lett.* 1982, *23*, 221.
139. Martin, S. F.; Zinke, P. W. *J. Am. Chem. Soc.* 1989, *111*, 2311.
140. Marshall, J. M.; Flynn, K. E. *J. Am. Chem. Soc.* 1982, *104*, 7430.
141. Marshall, J. A.; Audia, V. H. *J. Org. Chem.* 1987, *52*, 1106.
142. Dittmer, D. C.; Discordia, R. P.; Zhang, Y.; Murphy, C. K.; Kumar, A.; Pepito, A. S.; Wang, Y. *J. Org. Chem.* 1993, *58*, 718.
143. Höft, E.; Hamann, H.-J.; Kunath, A.; Rüffer, L. *Tetrahedron: Asymmetry* 1992, *3*, 507.
144. Kishi, Y. *Aldrichimica Acta* 1980, *13*, 23.
145. Hasan, I.; Kishi, Y. *Tetrahedron Lett.* 1980, *21*, 4229.
146. Takano, S.; Morimoto, M.; Ogasawara, K. *Synthesis* 1984, 834.
147. Chautemps, P.; Pierre, J.-L. *Tetrahedron* 1976, *32*, 549.
148. Chavdarian, C. G.; Heathcock, C. H. *Synth. Commun.* 1976, *6*, 277.
149. Schlessinger, R. H.; Lopes, A. *J. Org. Chem.* 1981, *46*, 5252.
150. McKittrick, B. A.; Ganem, B. *Tetrahedron Lett.* 1985, *26*, 4895.
151. Narula, A. S. *Tetrahedron Lett.* 1983, *24*, 5421.
152. Narula, A. S. *Tetrahedron Lett.* 1981, *22*, 2017.
153. Nagaoka, H.; Kishi, Y. *Tetrahedron* 1981, *37*, 3873.
154. Sharpless, K. B.; Verhoeven, T. R. *Aldrichimica Acta* 1979, *12*, 63.
155. Chamberlin, P.; Roberts, M. L.; Whitham, G. H. *J. Chem. Soc. (B)* 1970, 1374.
156. Johnson, M. R.; Kishi, Y. *Tetrahedron Lett.* 1979, 4347.
157. Boeckman, R. K.; Thomas, E. W. *J. Am. Chem. Soc.* 1979, *101*, 987.
158. Keck, G. E.; Abbott, D. E. *Tetrahedron Lett.* 1984, *25*, 1883.
159. Keck, G. E.; Boden, E. P. *Tetrahedron Lett.* 1984, *25*, 265.
160. Keck, G. E.; Boden, E. P. *Tetrahedron Lett.* 1984, *25*, 1879.
161. Yamamoto, Y.; Komatsu, T.; Maruyama, K. *J. Chem. Soc., Chem. Commun.* 1983, 191.
162. Yamamoto, Y.; Maruyama, K.; Matsumoto, K. *J. Chem. Soc., Chem. Commun.* 1983, 489.
163. Yamamoto, Y.; Yatagai, H.; Ishihara, Y.; Maeda, N.; Maruyama, K. *Tetrahedron* 1984, *40*, 2239.
164. Keck, G. E.; Abbott, D. E.; Boden, E. P.; Enholm, E. J. *Tetrahedron Lett.* 1984, *25*, 3927.
165. Yamamoto, Y.; Yatagai, H.; Naruta, Y.; Maruyama, K. *J. Am. Chem. Soc.* 1980, *102*, 7107.
166. Kishi, Y. *Pure Appl. Chem.* 1981, *53*, 1163.
167. Thomas, E. J.; Whitehead, J. W. F. *J. Chem. Soc., Perkin Trans. 1* 1989, 507.
168. Sakai, N.; Ohfune, Y. *Tetrahedron Lett.* 1990, *31*, 4151.
169. Ager, D. J.; East, M. B. *Tetrahedron* 1993, *49*, 5683.
170. Charette, A. B.; Côté, B. *Tetrahedron: Asymmetry* 1993, *4*, 2283.
171. Minami, N.; Ko, S. S.; Kishi, Y. *J. Am. Chem. Soc.* 1982, *104*, 1109.
172. Kogen, H.; Nishi, T. *J. Chem. Soc., Chem. Commun.* 1987, 311.
173. Shtacher, G.; Rubinstein, R.; Somane, P. *J. Med. Chem.* 1978, *21*, 678.
174. Mandville, G.; Ahmar, M.; Bloch, R. *Tetrahedron Lett.* 1993, *34*, 2119.
175. Johnson, M. R.; Nakata, T.; Kishi, Y. *Tetrahedron Lett.* 1979, 4343.
176. Ohfune, Y.; Kurokawa, N. *Tetrahedron Lett.* 1984, *25*, 1587.
177. Kocovsky, P.; Stary, I. *J. Org. Chem.* 1990, *55*, 3236.
178. Romeo, S.; Rich, D. H. *Tetrahedron Lett.* 1993, *34*, 7187.
179. Schmid, G. S.; Fukuyama, T.; Akasaka, K.; Kishi, Y. *J. Am. Chem. Soc.* 1979, *101*, 259.
180. Fukuyama, T.; Vranesic, B.; Negri, D. P.; Kishi, Y. *Tetrahedron Lett.* 1978, 2741.
181. Adam, W.; Griesbeck, A.; Staab, E. *Tetrahedron Lett.* 1986, *27*, 2839.
182. Adam, W.; Nestler, B. *Tetrahedron Lett.* 1993, *34*, 611.
183. Adam, W.; Brünker, H.-G. *J. Am. Chem. Soc.* 1993, *115*, 3008.
184. Adam, W.; Nestler, B. *J. Am. Chem. Soc.* 1993, *115*, 5041.
185. Adam, W.; Prechtl, F.; Richter, M. J.; Smerz, A. K. *Tetrahedron Lett.* 1993, *34*, 8427.
186. Overly, K. R.; Williams, J. M.; McGarvey, G. J. *Tetrahedron Lett.* 1990, *31*, 4573.
187. Sato, F.; Kobayashi, Y.; Takahashi, O.; Chiba, T.; Takeda, Y.; Kusakabe, M. *J. Chem. Soc., Chem. Commun.* 1985, 1636.
188. Kang, S.-K.; Kim, Y.-S.; Lim, J.-S.; Kim, K.-S.; Kim, S.-G. *Tetrahedron Lett.* 1991, *32*, 363.
189. De Witt, P.; Misiti, D.; Zappia, G. *Tetrahedron Lett.* 1989, *30*, 5505.
190. Andre-Barres, C.; Langlois, Y.; Gomez-Pacios, M. *Tetrahedron: Asymmetry* 1990, *1*, 571.
191. Reetz, M. T.; Binder, J. *Tetrahedron Lett.* 1989, *30*, 5425.
192. Behrens, C. H.; Sharpless, K. B. *Aldrichimica Acta* 1983, *16*, 67.
193. Hanson, R. M. *Chem. Rev.* 1991, *91*, 437.

194. Ward, R. S. *Chem. Soc. Rev.* 1990, *19*, 1.
195. Marshall, J. A.; Trometer, J. D. *Tetrahedron Lett.* 1987, *28*, 4985.
196. Marshall, J. A.; Trometer, J. D.; Blough, B. E.; Crute, T. D. *Tetrahedron Lett.* 1988, *29*, 913.
197. Chakraborty, T. K.; Joshi, S. P. *Tetrahedron Lett.* 1990, *31*, 2043.
198. Meyers, A. I.; Comins, D. L.; Roland, D. M.; Henning, R.; Shimizu, K. *J. Am. Chem. Soc.* 1979, *101*, 7104.
199. Trost, B. M.; Angle, S. R. *J. Am. Chem. Soc.* 1985, *107*, 6124.
200. Hungerbuhler, E.; Deebach, D.; Wasmuth, D. *Angew. Chem., Int. Ed. Engl.* 1979, *18*, 958.
201. Askin, D.; Volante, R. P.; Ryan, K. M.; Reamer, R. A.; Shinkai, I. *Tetrahedron Lett.* 1988, *29*, 4245.
202. Howe, G. P.; Wang, S.; Procter, G. *Tetrahedron Lett.* 1987, *28*, 2629.
203. Roush, W. R.; Adam, M. A.; Peseckis, S. M. *Tetrahedron Lett.* 1983, *24*, 1377.
204. Solladie, G.; Hamouchi, C.; Vicente, M. *Tetrahedron Lett.* 1988, *29*, 5929.
205. Page, P. C. B.; Rayner, C. M.; Sutherland, I. O. *J. Chem. Soc., Perkin Trans. 1* 1990, 1375.
206. Mulzer, J.; Schollhorn, B. *Angew. Chem., Int. Ed. Engl.* 1990, *29*, 1476.
207. Rao, A. V. R.; Bose, D. S.; Gurjar, M. K.; Ravindranathan, T. *Tetrahedron* 1989, *45*, 7031.
208. Marshall, J. A.; Blough, B. E. *J. Org. Chem.* 1991, *56*, 2225.
209. White, J. D.; Bolton, G. L. *J. Am. Chem. Soc.* 1990, *112*, 1626.
210. Rao, A. V. R.; Dhar, T. G. M.; Bose, D. S.; Chakraborty, T. K.; Gurjar, M. K. *Tetrahedron* 1989, *45*, 7361.
211. Jung, M. E.; Jung, Y. H. *Tetrahedron Lett.* 1989, *30*, 6637.
212. Wang, Z.; Schreiber, S. L. *Tetrahedron Lett.* 1990, *31*, 31.
213. Mori, Y.; Kuhara, M.; Takeuchi, A.; Suzuki, M. *Tetrahedron Lett.* 1988, *29*, 5419.
214. Murphy, P. J.; Procter, G. *Tetrahedron Lett.* 1990, *31*, 1059.
215. Hummer, W.; Gracza, T.; Jager, V. *Tetrahedron Lett.* 1989, *30*, 1517.
216. Mori, Y.; Takeuchi, A.; Kageyama, H.; Suzuki, M. *Tetrahedron Lett.* 1988, *29*, 5423.
217. Mori, Y.; Suzuki, M. *J. Chem. Soc., Perkin Trans. 1* 1990, 1809.
218. Mori, Y.; Kohchi, Y.; Ota, T.; Suzuki, M. *Tetrahedron Lett.* 1990, *31*, 2915.
219. Williams, N. R. *Adv. Carbohydr. Chem. Biochem.* 1970, *25*, 109.
220. Payne, G. B. *J. Org. Chem.* 1962, *27*, 3819.
221. Behrens, C. H.; Ko, S. Y.; Sharpless, K. B.; Walker, F. J. *J. Org. Chem.* 1985, *50*, 5687.
222. Ko, S. Y.; Lee, A. W. M.; Masamune, S.; Reed, L. A.; Sharpless, K. B.; Walker, F. J. *Science* 1983, *220*, 949.
223. Behrens, C. H.; Sharpless, K. B. *J. Org. Chem.* 1985, *50*, 5696.
224. Caron, M.; Sharpless, K. B. *J. Org. Chem.* 1985, *50*, 1557.
225. Dung, J.-S.; Armstrong, R. W.; Anderson, O. P.; Williams, R. M. *J. Org. Chem.* 1983, *48*, 3592.
226. Choudary, B. M.; Rani, S. S.; Kantam, M. L. *Synth. Commun.* 1990, *20*, 2313.
227. Lin, G.-Q.; Shi, Z.-C.; Zeng, C.-M. *Tetrahedron: Asymmetry* 1993, *4*, 1533.
228. Mori, K.; Umemura, T. *Tetrahedron Lett.* 1982, *23*, 3391.
229. Dai, L.-X.; Lou, B.-L.; Zhang, Y.-Z. *J. Am. Chem. Soc.* 1988, *110*, 5195.
230. Loibner, H.; Zbiral, E. *Helv. Chim. Acta* 1976, *59*, 2100.
231. Garner, P.; Park, J. M.; Rotello, V. *Tetrahedron Lett.* 1985, *26*, 3299.
232. Chakraborty, T. K.; Reddy, G. V. *Tetrahedron Lett.* 1990, *31*, 1335.
233. Chakraborty, T. K.; Reddy, G. V. *Tetrahedron Lett.* 1991, *32*, 679.
234. Saïah, M.; Bessodes, M.; Antonakis, K. *Tetrahedron: Asymmetry* 1991, *2*, 111.
235. Takano, S.; Yanase, M.; Ogasawara, K. *Synthesis* 1989, 39.
236. Iranpoor, N.; Baltork, I. M. *Synth. Commun.* 1990, *20*, 2789.
237. Raifeld, Y. E.; Nikitenko, A.; Arshava, B. M. *Tetrahedron* 1993, *49*, 2509.
238. Palazon, J. M.; Anorbe, B.; Martin, V. S. *Tetrahedron Lett.* 1986, *27*, 4987.
239. Martin, V. S.; Nuñez, M. T.; Tonn, C. E. *Tetrahedron Lett.* 1988, *29*, 2701.
240. Martín, V. S.; Ode, J. M.; Palazón, J. M.; Soler, M. A. *Tetrahedron: Asymmetry* 1992, *3*, 573.
241. Kocovsky, P. *Tetrahedron Lett.* 1986, *27*, 5521.
242. Wang, C.-L.; Gregory, W. A.; Wuonola, M. A. *Tetrahedron* 1989, *45*, 1323.
243. Roush, W. R.; Adam, M. A. *J. Org. Chem.* 1985, *50*, 3752.
244. Suzuki, T.; Saimoto, H.; Tomioka, H.; Oshima, K.; Nozaki, H. *Tetrahedron Lett.* 1982, *23*, 3597.
245. Pfaltz, A.; Mattenberger, A. *Angew. Chem., Int. Ed. Engl.* 1982, *21*, 71.
246. Yadav, J. S.; Kumar, T. K. P.; Maniyan, P. P. *Tetrahedron Lett.* 1993, *34*, 2969.
247. Ahn, K. H.; Kim, J. S.; Jin, C. S.; Kang, D. H.; Han, D. S.; Shin, Y. S.; Kim, D. H. *Synlett* 1992, 306.
248. Wood, R. D.; Ganem, B. *Tetrahedron Lett.* 1982, *23*, 707.
249. Ireland, R. E.; Wuts, P. G. M.; Ernst, B. *J. Am. Chem. Soc.* 1981, *103*, 3205.
250. Chong, J. M.; Cyr, D. R.; Mar, E. K. *Tetrahedron Lett.* 1987, *28*, 5009.
251. Paltz, A.; Mattenberger, A. *Angew. Chem., Suppl.* 1982, 161.
252. Mori, K.; Otsuka, T. *Tetrahedron* 1983, *39*, 3267.
253. Ma, P.; Martin, V. S.; Masamune, S.; Sharpless, K. B.; Viti, S. M. *J. Org. Chem.* 1982, *47*, 1380.
254. Johnston, B. D.; Oehlschlager, C. A. *Can. J. Chem.* 1984, *62*, 2148.

255. Lee, A. W. M. *J. Chem. Soc., Chem. Commun.* 1984, 578.
256. Roush, W. R.; Hoong, L. K.; Palmer, M. A. J.; Park, J. C. *J. Am. Chem. Soc.* 1990, *112*, 4109.
257. Finan, J. M.; Kishi, Y. *Tetrahedron Lett.* 1982, *23*, 2719.
258. Page, P. C. B.; Rayner, C. M.; Sutherland, I. O. *J. Chem. Soc., Perkin Trans. 1* 1990, 1615.
259. Viti, S. M. *Tetrahedron Lett.* 1982, *23*, 4541.
260. Kufner, U.; Schmidt, R. R. *Carbohydr. Res.* 1987, *161*, 211.
261. Hatakeyama, S.; Satoh, K.; Takano, S. *Tetrahedron Lett.* 1993, *34*, 7425.
262. Dai, L.-X.; Lou, B.-L.; Zhang, Y.-Z.; Guo, G.-Z. *Tetrahedron Lett.* 1986, *27*, 4343.
263. Sugita, K.; Onaka, M.; Izumi, Y. *Tetrahedron Lett.* 1990, *31*, 7467.
264. Evans, D. A.; Gauchet-Prunet, J. A.; Carreira, E. M.; Charette, A. B. *J. Org. Chem.* 1991, *56*, 741.
265. Still, W. C.; Ohmizu, H. *J. Org. Chem.* 1981, *46*, 5242.
266. Carlsen, P. H. J.; Katsuki, T.; Martin, V. S.; Sharpless, K. B. *J. Org. Chem.* 1981, *46*, 3936.
267. Jung, M. E.; D'Amico, D. C. *J. Am. Chem. Soc.* 1993, *115*, 12208.
268. Corey, E. J.; Marfat, A.; Munroe, J.; Kim, K. S.; Hopkins, P. B.; Brion, F. *Tetrahedron Lett.* 1981, *22*, 1077.
269. Baker, S. R.; Boot, J. R.; Morgan, S. E.; Osbourne, D. J.; Ross, W. J.; Shrubsall, P. R. *Tetrahedron Lett.* 1983, *24*, 4469.
270. Oshima, M.; Yamazaki, H.; Shimizu, I.; Nisar, M.; Tsuji, J. *J. Am. Chem. Soc.* 1989, *111*, 6280.
271. Nicolaou, K. C.; Uenishi, J. *J. Chem. Soc., Chem. Commun.* 1982, 1292.
272. Corey, E. J.; Mehrotra, M. M. *Tetrahedron Lett.* 1986, *27*, 5173.
273. Thijs, L.; Egenberger, D. M.; Zwanenburg, B. *Tetrahedron Lett.* 1989, *30*, 2153.
274. Takano, S.; Samizu, K.; Sugihara, T.; Ogasawara, K. *J. Chem. Soc., Chem. Commun.* 1989, 1344.
275. Yadav, J. S.; Deshpande, P. K.; Sharma, G. V. M. *Tetrahedron* 1990, *46*, 7033.
276. Mhaskar, S. Y.; Lakshminarayana, G. *Tetrahedron Lett.* 1990, *31*, 7227.
277. Takano, S.; Sugihara, T.; Ogasawara, K. *Synlett* 1990, 451.
278. Besson, T.; Rollin, P. *Synth. Commun.* 1990, *20*, 3039.
279. Jung, M. E.; Lew, W. *J. Org. Chem.* 1991, *56*, 1347.
280. Overman, L. E.; Thompson, A. S. *J. Am. Chem. Soc.* 1988, *110*, 2248.
281. Alvarez, E.; Nunez, M. T.; Martin, V. S. *J. Org. Chem.* 1990, *55*, 3429.
282. Ennis, M. D.; Baze, M. E. *Tetrahedron Lett.* 1986, *27*, 6031.
283. Corey, L. D.; Singh, S. M.; Oehlschlager, A. C. *Can. J. Chem.* 1987, *65*, 1821.
284. Colvin, E. W.; Robertson, A. D.; Wakharkar, S. *J. Chem. Soc., Chem. Commun.* 1983, 312.
285. Myers, A. G.; Widdowson, K. L. *Tetrahedron Lett.* 1988, *29*, 6389.
286. McCombie, S. W.; Metz, W. A. *Tetrahedron Lett.* 1987, *28*, 383.
287. Lee, A. W. M.; Martin, V. S.; Masamune, S.; Sharpless, K. B.; Walker, F. J. *J. Am. Chem. Soc.* 1982, *104*, 3515.
288. Ma, P.; Martin, V. S.; Masamune, S.; Sharpless, K. B.; Viti, S. M. *J. Org. Chem.* 1982, *47*, 1378.
289. Ko, S. Y.; Lee, A. W. M.; Masamune, S.; Reed, L. A.; Sharpless, K. B.; Walker, F. J. *Tetrahedron* 1990, *46*, 245.
290. Sunazuka, T.; Naganitsu, T.; Tanaka, H.; Omura, S.; Sprengler, P. A.; Smith, A. B. *Tetrahedron Lett.* 1993, *34*, 4447.
291. Wuts, P. G. M.; D'Costa, R.; Butler, W. *J. Org. Chem.* 1984, *49*, 2582.
292. Schmidt, U.; Respondek, M.; Lieberknecht, A.; Werner, J.; Fischer, P. *Synthesis* 1989, 256.
293. Schnurrenberger, P.; Hungerbuhler, E.; Seebach, D. *Liebigs Ann. Chem.* 1987, 733.
294. Morgans, D. J.; Sharpless, K. B.; Traynor, S. G. *J. Am. Chem. Soc.* 1981, *103*, 462.
295. Aoyagi, S.; Fujimaki, S.; Yamazaki, N.; Kibayashi, C. *J. Org. Chem.* 1991, *56*, 815.
296. Shimazaki, M.; Hara, H.; Suzuki, K.; Tsuchihashi, G.-I. *Tetrahedron Lett.* 1987, *28*, 5891.
297. Suzuki, K.; Miyazawa, M.; Tsuchihashi, G.-I. *Tetrahedron Lett.* 1987, *28*, 3515.
298. Bell, T. W.; Clacclo, J. A. *Tetrahedron Lett.* 1988, *29*, 865.
299. Burgos, C. E.; Nidy, E. G.; Johnson, R. A. *Tetrahedron Lett.* 1989, *30*, 5081.
300. Brunner, H.; Sicheneder, A. *Angew. Chem., Int. Ed. Engl.* 1988, *27*, 718.
301. Guivisdalsky, P. N.; Bittman, R. *J. Org. Chem.* 1989, *54*, 4643.
302. Guivisdalsky, P. N.; Bittman, R. *J. Org. Chem.* 1989, *54*, 4637.
303. Guivisdalsky, P. N.; Bittman, R. *J. Am. Chem. Soc.* 1989, *111*, 3077.
304. Hammadi, A.; Crouzel, C. *Tetrahedron: Asymmetry* 1990, *1*, 579.
305. Katsuki, T. *Tetrahedron Lett.* 1984, *25*, 2821.
306. Kurth, M. J.; Abreo, M. A. *Tetrahedron* 1990, *46*, 5085.
307. Kurth, M. J.; Abreo, M. A. *Tetrahedron Lett.* 1987, *28*, 5631.
308. Takano, S.; Shimazaki, Y.; Iwabuchi, Y.; Ogasawara, K. *Tetrahedron Lett.* 1990, *31*, 3622.
309. Takano, S.; Shimazaki, Y.; Ogasawara, K. *Tetrahedron Lett.* 1990, *31*, 3325.
310. Nicolaou, K. C.; Ramphal, J.; Abe, Y. *Synthesis* 1989, 898.
311. Schreiber, S. L.; Sammakia, T.; Uehling, D. E. *J. Org. Chem.* 1989, *54*, 15.

312. Lee, A. W. M. *Magn. Reson. Chem.* 1985, *23*, 468.
313. Maruoka, K.; Ooi, T.; Yamamoto, H. *J. Am. Chem. Soc.* 1989, *111*, 6431.
314. Maruoka, K.; Hasegawa, M.; Yamomoto, H.; Suzuki, K.; Shimazaki, M.; Tsuchihashi, G. *J. Am. Chem. Soc.* 1986, *108*, 3827.
315. Suzuki, K.; Miyazawa, M.; Shimazaki, M.; Tsuchihashi, G.-I. *Tetrahedron Lett.* 1986, *27*, 6237.
316. Cimetiere, B.; Jacob, L.; Julia, M. *Tetrahedron Lett.* 1986, *27*, 6329.
317. Hanson, R. M. *Tetrahedron Lett.* 1984, *25*, 3783.
318. Burke, S. D.; Cobb, J. E.; Takeuchi, K. *J. Org. Chem.* 1985, *50*, 3420.
319. Burke, S. D.; Cobb, J. E. *Tetrahedron Lett.* 1986, *27*, 4237.
320. Rao, A. S.; Paknikar, S. K.; Kirtane, J. G. *Tetrahedron* 1983, *39*, 2323.
321. Colonna, S.; Molinari, H.; Banfi, S.; Julia, S.; Masana, J.; Alvarez, A. *Tetrahedron* 1983, *39*, 1635.
322. Julia, S.; Guixer, J.; Masana, J.; Rocas, J.; Colonna, S.; Annuziata, R.; Molinari, H. *J. Chem. Soc., Perkin Trans. 1* 1982, 1317.
323. Julia, S.; Masana, J.; Vega, J. C. *Angew. Chem., Int. Ed. Engl.* 1980, *19*, 929.
324. Sundararaman, P.; Herz, W. *J. Org. Chem.* 1977, *42*, 813.
325. Hummelen, J. C.; Wynberg, H. *Tetrahedron Lett.* 1978, 1089.
326. Baures, P. W.; Eggleston, D. S.; Flisak, J. R.; Gombatz, K.; Lantos, I.; Mendelson, W.; Remich, J. J. *Tetrahedron Lett.* 1990, *31*, 6501.
327. Adam, W.; Hadjiarapoglou, L.; Nestler, B. *Tetrahedron Lett.* 1990, *31*, 331.
328. Chabert, P.; Oussert, J.; Mioskowski, C. *Tetrahedron Lett.* 1989, *30*, 179.
329. Mukaiyama, T.; Iwasawa, N.; Stevens, R. W.; Haga, T. *Tetrahedron* 1984, *40*, 1381.
330. Hünig, S.; Marschner, C. *Chem. Ber.* 1990, *123*, 107.
331. Shibata, I.; Yamasaki, H.; Baba, A.; Matsuda, H. *Synlett* 1990, 490.
332. Hayashi, M.; Terashima, S.; Koga, K. *Tetrahedron* 1981, *37*, 2797.
333. Terashima, S.; Hayashi, M.; Koga, K. *Tetrahedron Lett.* 1980, *21*, 2733.
334. Miyashita, M.; Suzuki, T.; Yoshikoshi, A. *Tetrahedron Lett.* 1987, *28*, 4293.
335. Amouroux, R.; Gerin, B.; Chastrette, M. *Tetrahedron Lett.* 1982, *23*, 4341.
336. Pierre, J.-L.; Chautemps, P. *Tetrahedron Lett.* 1972, 4371.
337. Nakata, T.; Tanaka, T.; Oishi, T. *Tetrahedron Lett.* 1981, *22*, 4723.
338. Bowden, M. C.; Patel, P.; Pattenden, G. *Tetrahedron Lett.* 1985, *26*, 4793.
339. Kitamura, M.; Isobe, M.; Ichikawa, Y.; Gota, T. *J. Am. Chem. Soc.* 1984, *106*, 3252.
340. Escudier, J.-M.; Baltas, M.; Gorrichon, L. *Tetrahedron* 1993, *49*, 5253.
341. Kato, K.; Saino, T.; Nishizawa, R.; Takita, T.; Umerzawa, H. *J. Chem. Soc., Perkin Trans. 1* 1980, 1618.
342. Bernardi, A.; Cardani, S.; Carugo, O.; Colombo, L.; Scolastico, C.; Villa, R. *Tetrahedron Lett.* 1990, *31*, 2779.
343. Ballester, M. *Chem. Rev.* 1955, *55*, 283.
344. Borch, R. *Tetrahedron Lett.* 1972, 3761.
345. Polniaszek, R. P.; Belmont, S. E. *Synth. Commun.* 1989, *19*, 221.
346. Kraus, G. A.; Taschner, M. J. *Tetrahedron Lett.* 1977, 4575.
347. Seebach, D.; Wasmuth, D. *Helv. Chim. Acta* 1980, *63*, 197.
348. Meth-Cohn, O.; Moore, C.; Taljaard, H. C. *J. Chem. Soc., Perkin Trans. 1* 1988, 2663.
349. Kim, B. M.; Sharpless, K. B. *Tetrahedron Lett.* 1990, *31*, 4317.
350. Fleming, P. R.; Sharpless, K. B. *J. Org. Chem.* 1991, *56*, 2869.
351. Luischitz, Y.; Rabinsohn, Y.; Perera, D. *J. Chem. Soc.* 1962, 1116.
352. van der Baan, J. L.; Barnick, J. W. F. K.; Bickelhaupt, B. F. *Synthesis* 1990, 897.
353. Saito, S.; Bunya, N.; Inaba, M.; Moriwake, T.; Torii, S. *Tetrahedron Lett.* 1985, *26*, 5309.
354. Saito, S.; Takahasi, N.; Ishikawa, T.; Moriwake, T. *Tetrahedron Lett.* 1991, *32*, 667.
355. Tanner, D.; Birgersson, C.; Dhaliwal, H. K. *Tetrahedron Lett.* 1990, *31*, 1903.
356. Abenhaim, D.; Namy, J.-L. *Tetrahedron Lett.* 1972, 1001.
357. Boireau, G.; Abenhaim, D.; Bernardon, C.; Henry-Basch, E.; Sabourault, B. *Tetrahedron Lett.* 1975, 2521.
358. Mulzer, J.; Lammer, O. *Chem. Ber.* 1986, *119*, 2178.
359. Hartman, B. C.; Livingstone, T.; Rickborn, B. *J. Org. Chem.* 1973, *38*, 4346.
360. Otsubo, K.; Inanaga, J.; Yamaguchi, M. *Tetrahedron Lett.* 1987, *28*, 4437.
361. Otsubo, K.; Inanaga, J.; Yamaguchi, M. *Tetrahedron Lett.* 1987, *28*, 4435.
362. Corey, E. J.; Trybulski, E. J.; Melvin, L. S.; Nicolaou, K. C.; Secrist, J. A.; Lett, R.; Sheldrake, P. W.; Falck, J. R.; Brunelle, D. J.; Haslanger, M. F.; Kim, S.; Yoo, S.-E. *J. Am. Chem. Soc.* 1978, *100*, 4618.
363. Harada, K. *J. Org. Chem.* 1966, *31*, 1407.
364. Harada, K.; Nakajima, Y. *Bull. Chem. Soc., Jpn.* 1974, *47*, 2911.
365. Johnson, C. R.; Bode, T. R. *J. Org. Chem.* 1982, *47*, 1205.
366. Mohrig, J. R.; Vreede, P. J.; Schultz, S. C.; Fierke, C. A. *J. Org. Chem.* 1981, *46*, 4655.
367. Chong, J. M.; Sharpless, K. B. *Tetrahedron Lett.* 1985, *26*, 4683.
368. Flippin, L. A.; Brown, P. A.; Jalali-Araghi, K. *J. Org. Chem.* 1989, *54*, 3588.

369. Masamune, S. *Aldrichimica Acta* 1978, *11*, 23.
370. Trost, B. M.; Sudhakar, A. R. *J. Am. Chem. Soc.* 1987, *109*, 3792.
371. Zhou, Z.-L.; Shi, L.-L.; Huang, Y.-Z. *Tetrahedron Lett.* 1990, *31*, 7657.
372. Rose, C. B.; Smith, C. W. *J. Chem. Soc., Chem. Commun.* 1969, 248.
373. Anderson, R. J. *J. Am. Chem. Soc.* 1970, *92*, 4978.
374. Herr, R. W.; Johnson, C. R. *J. Am. Chem. Soc.* 1970, *92*, 4979.
375. Lenox, R. S.; Katzenellenbogen, J. A. *J. Am. Chem. Soc.* 1973, *95*, 957.
376. Aithie, G. C. M.; Miller, J. A. *Tetrahedron Lett.* 1975, 4419.
377. Marshall, J. A. *Chem. Rev.* 1989, *89*, 1503.
378. Lentz, N. L.; Pett, N. P. *Tetrahedron Lett.* 1990, *31*, 811.
379. Larock, R. C.; Stolz-Dunn, S. K. *Tetrahedron Lett.* 1989, *30*, 3487.
380. Larock, R. C.; Ding, S. *J. Org. Chem.* 1993, *58*, 804.
381. Trost, B. M.; Hurnaus, R. *Tetrahedron Lett.* 1989, *30*, 3893.
382. Larock, R. C.; Ilkka, S. J. *Tetrahedron Lett.* 1986, *27*, 2211.
383. Trost, B. M.; Angle, S. R. *J. Am. Chem. Soc.* 1985, *107*, 6123.
384. Tsuji, J. *Pure Appl. Chem.* 1989, *61*, 1673.
385. Shimizu, I.; Oshima, M.; Nisar, M.; Tsuji, J. *Chem. Lett.* 1986, 1775.
386. Nativi, C.; Ravida, N.; Ricci, A.; Seconi, G.; Taddei, M. *J. Org. Chem.* 1991, *56*, 1951.
387. Ager, D. J. *Chem. Soc. Rev.* 1982, *11*, 493.
388. Molander, G. A.; Mautner, K. *Pure Appl. Chem.* 1990, *62*, 707.
389. Molander, G. A.; Mautner, K. *J. Org. Chem.* 1989, *54*, 4042.
390. Lohse, P.; Loner, H.; Acklin, P.; Sternfeld, F.; Pfaltz, A. *Tetrahedron Lett.* 1991, *32*, 615.
391. Eisch, J. J.; Galle, J. E. *J. Org. Chem.* 1990, *55*, 4835.
392. Ashwell, M.; Jackson, R. F. W. *J. Chem. Soc., Chem. Commun.* 1988, 645.
393. Hewkin, C. T.; Jackson, R. F. W. *Tetrahedron Lett.* 1988, *29*, 4889.
394. Ashwell, M.; Jackson, R. F. W. *J. Chem. Soc., Perkin Trans. 1* 1989, 835.
395. Hewkin, C. T.; Jackson, R. F. W. *Tetrahedron Lett.* 1990, *31*, 1877.
396. Yamamoto, K.; Kimura, T.; Tomo, Y. *Tetrahedron Lett.* 1985, *26*, 4505.
397. Yamamoto, K.; Kimura, T.; Tomo, Y. *Tetrahedron Lett.* 1984, *25*, 2155.
398. Gröbel, B.-T.; Seebach, D. *Angew. Chem., Int. Ed. Engl.* 1974, *13*, 83.
399. Ager, D. J. *Org. React.* 1990, *38*, 1.
400. Hudrlik, P. F.; Hudrlik, A. M.; Kulkarni, A. K. *Tetrahedron Lett.* 1985, *26*, 139.
401. Hudrlik, P. F.; Hudrlik, A. M.; Misra, R. N.; Peterson, D.; Withers, G. P.; Kulkarni, A. K. *J. Org. Chem.* 1980, *45*, 4444.
402. Chauret, D. C.; Chong, J. M. *Tetrahedron Lett.* 1993, *34*, 3695.
403. Soderquist, J. A.; Santiago, B. *Tetrahedron Lett.* 1989, *30*, 5693.
404. Hudrlik, P. F.; Schwartz, R. H.; Kulkarni, A. K. *Tetrahedron Lett.* 1979, 2233.
405. Hudrlik, P. F.; Wan, C.-N.; Withers, G. P. *Tetrahedron Lett.* 1976, 1449.
406. Hudrlik, P. F.; Misra, R. N.; Withers, G. F.; Hudrlik, A. M.; Rona, R. J.; Arcoleo, J. P. *Tetrahedron Lett.* 1976, 1453.
407. Barluenga, J.; Llavona, L.; Bernad, P. L.; Concellón, J. M. *Tetrahedron Lett.* 1993, *34*, 3173.
408. Wiberg, K. B.; Deutsch, C. J.; Rocek, J. *J. Am. Chem. Soc.* 1973, *95*, 3034.
409. Simandi, L. I.; Jaky, M. *J. Am. Chem. Soc.* 1976, *98*, 1995.
410. Lee, D. G.; Brown, K. C. *J. Am. Chem. Soc.* 1982, *104*, 5076.
411. Lee, D. G.; Brownridge, J. R. *J. Am. Chem. Soc.* 1974, *96*, 5517.
412. Stewart, R. *Oxidation Mechanisms*; Benjamin: New York, 1964.
413. Cha, J. K.; Christ, W. J.; Kishi, Y. *Tetrahedron Lett.* 1983, *24*, 3943.
414. Christ, W. J.; Cha, J. K.; Kishi, Y. *Tetrahedron Lett.* 1983, *24*, 3947.
415. Evans, D. A.; Kaldor, S. W. *J. Org. Chem.* 1990, *55*, 1698.
416. Cha, J. K.; Christ, W. J.; Kishi, Y. *Tetrahedron* 1984, *40*, 2247.
417. Stork, G.; Kahn, M. *Tetrahedron Lett.* 1983, *24*, 3951.
418. Brimacombe, J. S.; Hanna, R.; Kabir, M. S.; Bennett, F.; Taylor, I. D. *J. Chem. Soc., Perkin Trans. 1* 1986, 815.
419. Solladie, G.; Frechou, C.; Demailly, G. *Tetrahedron Lett.* 1986, *27*, 2867.
420. Johnson, C. R.; Barbachyn, M. R. *J. Am. Chem. Soc.* 1984, *106*, 2459.
421. Colombo, L.; Gennari, C.; Poli, G.; Scolastico, C.; Munari, S. D. *Tetrahedron Lett.* 1985, *26*, 5459.
422. Vedejs, E.; McClure, C. K. *J. Am. Chem. Soc.* 1986, *108*, 1094.
423. Solladie, G.; Hutt, J.; Frechou, C. *Tetrahedron Lett.* 1987, *28*, 61.
424. Annunziata, R.; Cinquini, M.; Cozzi, F. *Tetrahedron Lett.* 1987, *28*, 3139.
425. Fleming, I.; Sarkar, A. K.; Thomas, A. P. *J. Chem. Soc., Chem. Commun.* 1987, 157.
426. DeNinno, M. P.; Danishefsky, S. J.; Schulte, G. *J. Am. Chem. Soc.* 1988, *110*, 3925.

427. Vedejs, E.; Dent, W. H. *J. Am. Chem. Soc.* 1989, *111*, 6861.
428. Hauser, F. M.; Ellenberger, S. R.; Clardy, J. C.; Bass, L. S. *J. Am. Chem. Soc.* 1984, *106*, 2458.
429. Nakamura, M.; Tomioka, K.; Koga, K. *J. Synth. Org. Chem. Jpn.* 1989, *47*, 878.
430. Wang, Z.-M.; Sharpless, K. B. *Tetrahedron Lett.* 1993, *34*, 8225.
431. Ko, S. Y.; Malik, M. *Tetrahedron Lett.* 1993, *34*, 4675.
432. Wang, Z.-M.; Zhang, X.-L.; Sharpless, K. B. *Tetrahedron Lett.* 1993, *34*, 2267.
433. Walsh, P.; Bennani, Y. L.; Sharpless, K. B. *Tetrahedron Lett.* 1993, *34*, 5545.
434. Bennani, Y. L.; Sharpless, K. B. *Tetrahedron Lett.* 1993, *34*, 2079.
435. Saito, S.; Morikawa, Y.; Moriwake, T. *J. Org. Chem.* 1990, *55*, 5424.
436. Tani, K.; Sato, Y.; Okamoto, S.; Sato, F. *Tetrahedron Lett.* 1993, *34*, 4975.
437. Xu, D.; Crispino, G. A.; Sharpless, K. B. *J. Am. Chem. Soc.* 1992, *114*, 7570.
438. Park, C. Y.; Kim, B. M.; Sharpless, K. B. *Tetrahedron Lett.* 1991, *32*, 1003.
439. Fleming, I.; Sarker, A. K.; Thomas, A. P. *J. Chem. Soc., Chem. Commun.* 1987, 157.
440. Azzari, E.; Faggi, C.; Gelsomini, N.; Taddei, N. *J. Org. Chem.* 1990, *55*, 1106.
441. Panek, J. S.; Cirillo, P. F. *J. Am. Chem. Soc.* 1990, *112*, 4873.
442. Bravo, P.; Frigerio, M.; Fronza, G.; Ianni, A.; Resnati, G. *Tetrahedron* 1990, *46*, 997.
443. Panek, J. S.; Zhang, J. *J. Org. Chem.* 1993, *58*, 294.
444. Oishi, T.; Iida, K.-I.; Hirama, M. *Tetrahedron Lett.* 1993, *34*, 3573.
445. Ikemoto, N.; Schreiber, S. *J. Am. Chem. Soc.* 1990, *112*, 9657.
446. Hatakeyama, S.; Matsui, Y.; Suzuki, M.; Sakurai, K.; Takano, S. *Tetrahedron Lett.* 1985, *26*, 6485.
447. Tomioka, K.; Nakajima, M. *J. Chem. Soc., Chem. Commun.* 1989, 1921.
448. Sharpless, K. B.; Amberg, W.; Bennani, Y. L.; Crispino, G. A.; Hartung, J.; Jeong, K.-S.; Kwong, H.-L.; Morikawa, K.; Wang, Z.-M.; Xu, D.; Zhang, X.-L. *J. Org. Chem.* 1992, *57*, 2768.
449. Morikawa, K.; Sharpless, K. B. *Tetrahedron Lett.* 1993, *34*, 5575.
450. Gao, Y.; Sharpless, K. B. *J. Am. Chem. Soc.* 1988, *110*, 7538.
451. Lohray, B. B.; Bhushan, V. *Tetrahedron Lett.* 1993, *34*, 3911.
452. Hudrlik, P. F.; Hudrlik, A. M.; Kulkarni, A. K. *J. Am. Chem. Soc.* 1985, *107*, 4260.
453. Cainelli, G.; Panunzio, M.; Giacomini, D. *Tetrahedron Lett.* 1991, *32*, 121.
454. Häner, R.; Olano, B.; Seebach, D. *Helv. Chim. Acta* 1987, *70*, 1676.
455. Bartlett, P. A.; Jernstedt, K. K. *Tetrahedron Lett.* 1980, *21*, 1607.

Chapter 12

# PERICYCLIC REACTIONS

In this chapter we will discuss reactions that are perceived to proceed in a concerted manner. Some general rules have already been discussed [*2.5*]. Some transformations do not proceed in a concerted manner, but follow a stepwise mechanism.[1,2] Examples of the latter type of reactions have been included in this chapter when the stereochemical outcome is not altered by the reaction mechanism.

## 12.1. THE DIELS-ALDER REACTION

Diels-Alder reactions have found widespread use in organic synthesis because of their ability to control relative, and often absolute, stereochemistry while forming two carbon–carbon bonds.[3-6] In addition to the facial selectivity of the reaction,[7-12] considerable advances have been made to achieve asymmetric induction through the use of chiral dienophiles and catalysts.[6,13-49] The methodology also allows for a plethora of functionality to be introduced in a stereoselective and expeditious manner,[31] particularly if an intramolecular strategy is adopted.[50,51] The large number of examples in the literature attests to the methodology's utility. Our discussion concentrates on the factors that influence the stereochemical outcome of the cycloaddition reaction.

### 12.1.1. REGIO- AND STEREOCHEMISTRY

Any cycloaddition involving two unsymmetrical reactants can lead to regioisomers. For the Diels-Alder reaction, the *ortho* product[†] is, usually, the major isomer from 1-substituted dienes, while 2-substituted dienes provide the *para* product. Frontier molecular orbital theory has been used to predict and rationalize the regioselectivity.[52-54] The addition of Lewis acids can affect the ratio of these regioisomers. For example, thermal addition of 1,3-pentadiene and methyl acrylate results in a 9:1 *ortho*/*meta* ratio, but this is increased to 49:1 when aluminum chloride is used as a Lewis acid catalyst.[55]

Aside from the regiochemistry, Diels-Alder adducts may be formed by two types of approach: The most favored adduct is generally the *endo* isomer even though the *exo* addition product is thermodynamically preferred. This is referred to as the *endo*, or Alder, rule and can be aided by the addition of a Lewis acid catalyst [*2.5.2*]. This preference for the *endo* adduct has been attributed to additional stability gained by secondary molecular orbital overlap (Figure 12.1).[52,56,57]

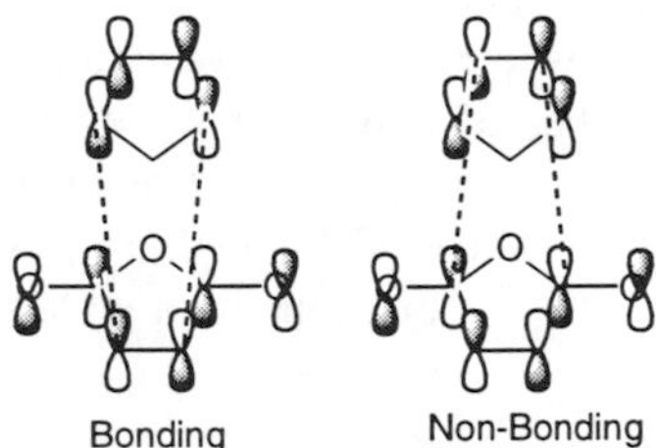


**FIGURE 12.1.** Secondary molecular orbital interactions for the Diels-Alder reaction of cyclopentadiene with maleic anhydride.

The Diels-Alder reaction is stereoselective; the relative stereochemistry of the dienophile is maintained in the product through *syn* addition. Hence, dimethyl maleate and dimethyl fumarate

[†] This nomenclature system follows the analogy of disubstituted aromatic systems.

react with butadiene to yield the *cis* and *trans* isomers of the cyclohexene adduct, respectively (Scheme 12.1). Selectivity is also observed when there is a preference for the *Si*- or *Re*-face of the diene or dienophile during the cycloaddition.[7-9,12] Where both the diene and the dienophile exhibit a preference, double asymmetric induction can be observed.[31,58,59]

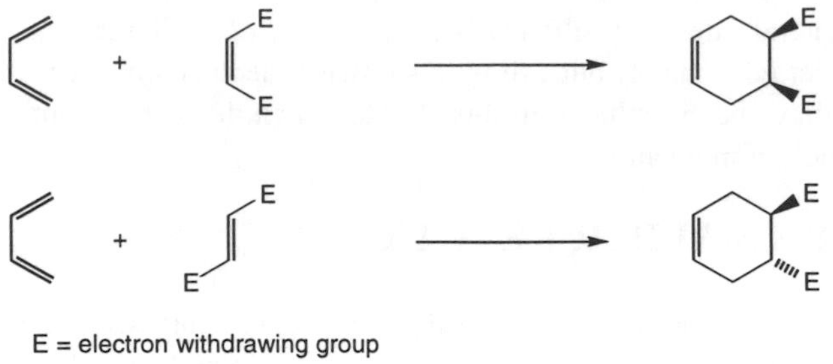

E = electron withdrawing group

SCHEME 12.1.

### 12.1.2. CHIRAL DIENOPHILES

Chiral dienophiles provide the vast majority of examples of asymmetric Diels-Alder reactions and, of these, acrylates are the most common (Table 12.1).[31] The use of a chiral auxiliary or group on the dienophile can provide for face selectivity. Chiral dienophiles have been divided into two categories: type I and type II reagents (Figure 12.2). Type I reagents, such as chiral acrylates, can incorporate a chiral group in a simple and straightforward manner. Type II reagents, where the chiral group is one atom closer to the double bond,[31] are more difficult to synthesize, and, often if the stereogenic center is not required in the final product, it is not trivial to recycle.

Type I　　Type II

**FIGURE 12.2.** Types of dienophiles for the Diels-Alder reaction.

High *endo–exo* selectivities have been achieved with bicyclic adducts, together with high asymmetric induction. Table 12.1 lists many of these dienophiles.

The success of chiral dienophiles to provide high asymmetric induction has, in part, led to the development of a large number of chiral auxiliaries but, to date, none is capable of asymmetric induction with a wide variety of dienes. However, there are some auxiliaries that do stand out, including menthol derivatives,[27,68] camphor derivatives (Scheme 12.2),[6,60-67] and oxazolidinones (Scheme 12.3)[71,107,118,119] that tolerate a wider variety of substrates.

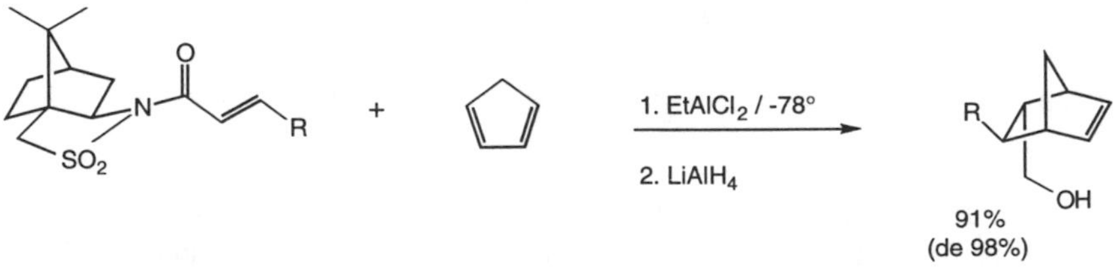

1. $EtAlCl_2$ / -78°
2. $LiAlH_4$

91%
(de 98%)

SCHEME 12.2.

**TABLE 12.1**
**Examples of Chiral Dienophiles that Have Been Used in the Diels-Alder Reaction[a]**

| Dienophile | Diene[b] | Catalyst | Ref. | Dienophile | Diene | Catalyst | Ref. |
|---|---|---|---|---|---|---|---|
| 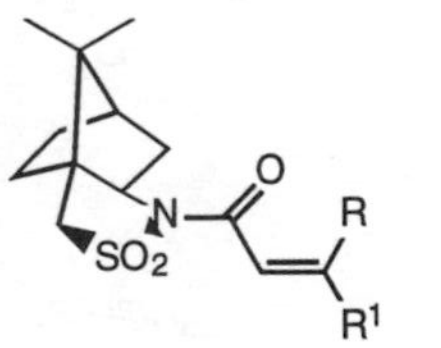 | A | $TiCl_4$<br>$EtAlCl_2$<br>$Me_2AlCl$ | 6,60 | 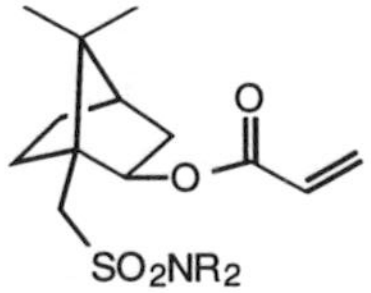 | A | $TiCl_2(OiPr)_2$ | 61 |
| 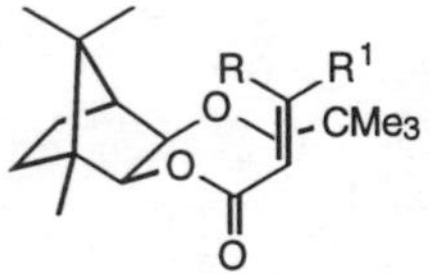 | A | $TiCl_2(OiPr)_2$ | 62,63 | 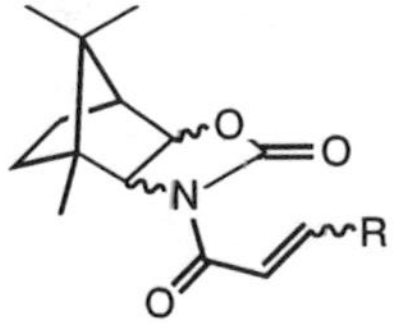 | A | $Et_2AlCl$<br>$BF_3 \cdot OEt_2$<br>$TiCl_4$<br>$TiCl_2(OiPr)_2$ | 64 |
| 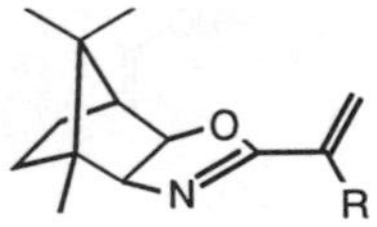 | A, D, E, O, P | — | 65–67 | 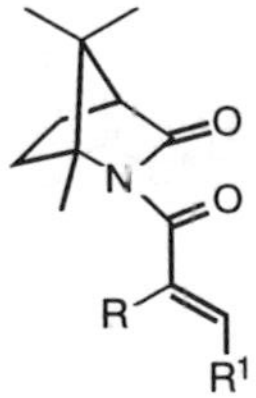 | A, B, C, D, E, G | $MeAlCl_2$ | 34 |
| 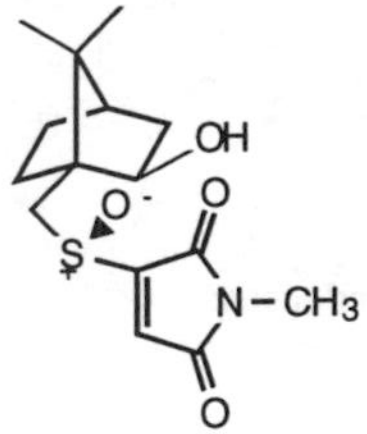 | A, H | $ZnCl_2$ | 18 | 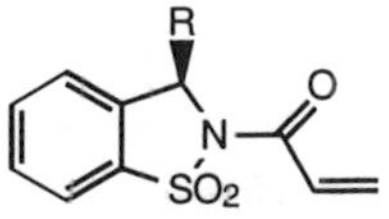 | A | $Me_2AlCl$<br>$EtAlCl_2$ | 19 |

**TABLE 12.1 (continued)**

| Dienophile | Diene[b] | Catalyst | Ref. | Dienophile | Diene | Catalyst | Ref. |
|---|---|---|---|---|---|---|---|
| 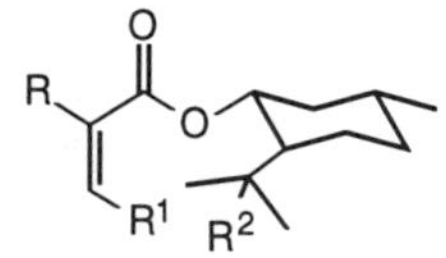 R, O, O, R[1], R[2] | A | $TiCl_4$<br>$Et_2AlCl$<br>Clay | 27,68<br>61,63,69–75 | 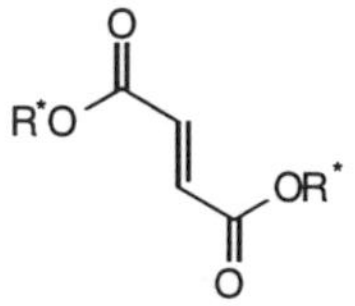 R*O, O, OR*, O | A, B, C,<br>D, E, F,<br>G | $Et_2AlCl$<br>$iBu_2AlCl$<br>$AlCl_3$ | 76–78 |
| 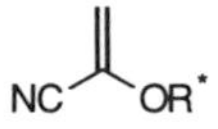 NC, OR* | H | $ZnX_2$ | 79,80 | 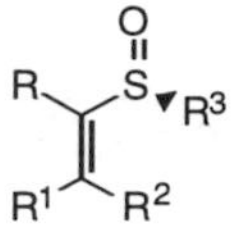 O, R, S, R[3], R[1], R[2] | A, B, C,<br>D, E, F,<br>H, L, M,<br>N, P | $ZnX_2$<br>$MgBr_2{\cdot}OEt_2$<br>$SiO_2$<br>$LiClO_4$<br>$Et_2AlCl$<br>$BF_3{\cdot}OEt_2$<br>Eu(fod)$_3$<br>$TiCl_4$ | 14–17,35<br>81–91 |
| 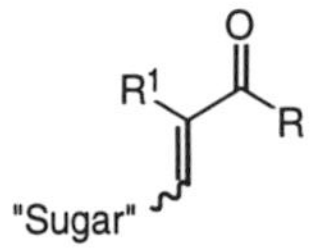 O, R[1], R, "Sugar" | A | —<br>$Et_3Al$<br>$EtAlCl_2$<br>$Et_2AlCl$<br>$Cp_2TiCl_2$ | 12,92–96 | 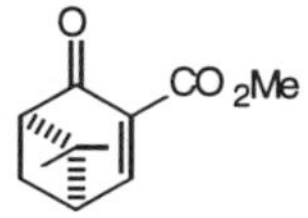 O, $CO_2Me$ | A, C, D,<br>E, I | $ZnCl_2$<br>$BF_3{\cdot}OEt_2$<br>$SnCl_4$<br>$Et_2AlCl$ | 97 |
| 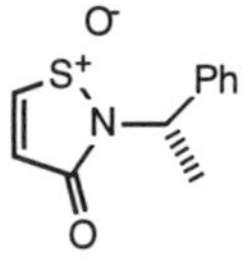 O⁻, S⁺, N, Ph, O | A, J | — | 98 | 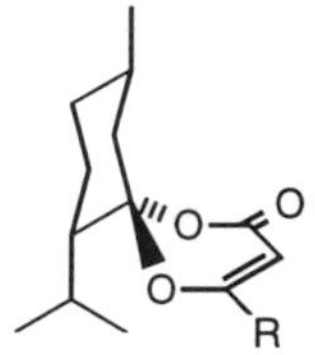 O, O, O, R | A | — | 99 |

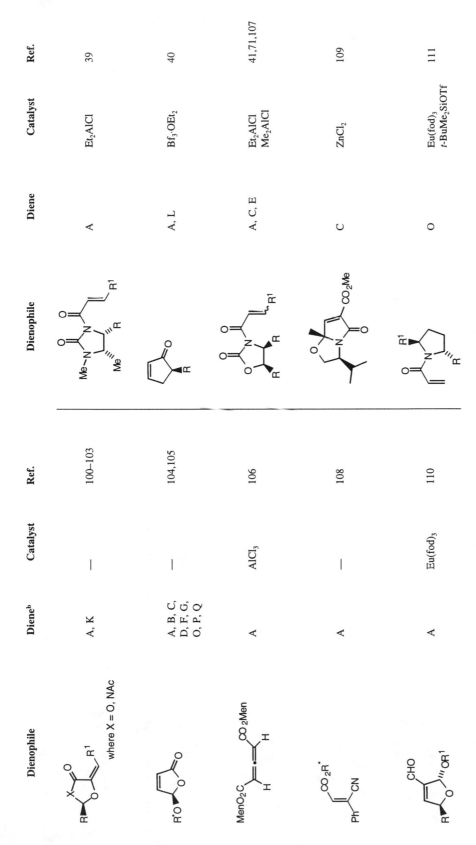

| Dienophile | Diene[b] | Catalyst | Ref. | Dienophile | Diene | Catalyst | Ref. |
|---|---|---|---|---|---|---|---|
| 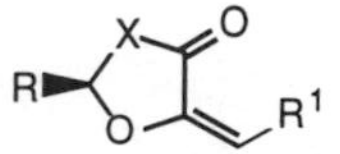 where X = O, NAc | A, K | — | 100–103 | 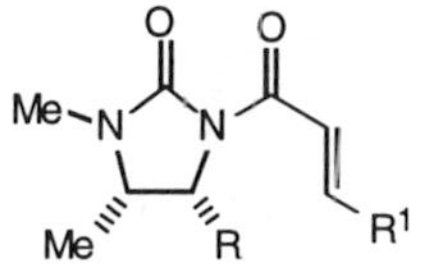 | A | $Et_2AlCl$ | 39 |
|  | A, B, C, D, F, G, O, P, Q | — | 104,105 | 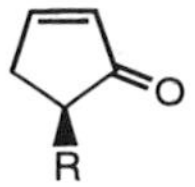 | A, L | $Bf_3 \cdot OEt_2$ | 40 |
| 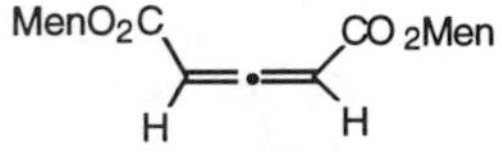 | A | $AlCl_3$ | 106 | 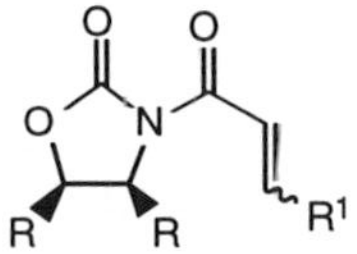 | A, C, E | $Et_2AlCl$<br>$Me_2AlCl$ | 41,71,107 |
| 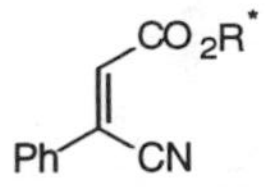 | A | — | 108 | 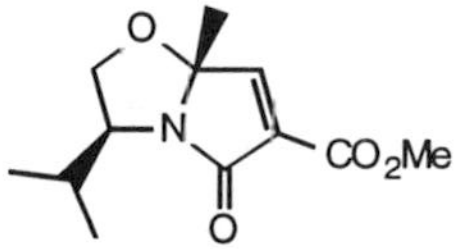 | C | $ZnCl_2$ | 109 |
| 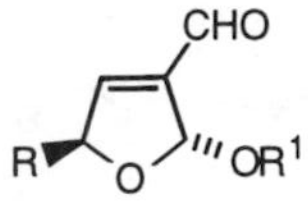 | A | $Eu(fod)_3$ | 110 | 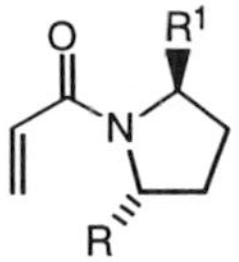 | O | $Eu(fod)_3$<br>*t*-$BuMe_2SiOTf$ | 111 |

**TABLE 12.1 (continued)**

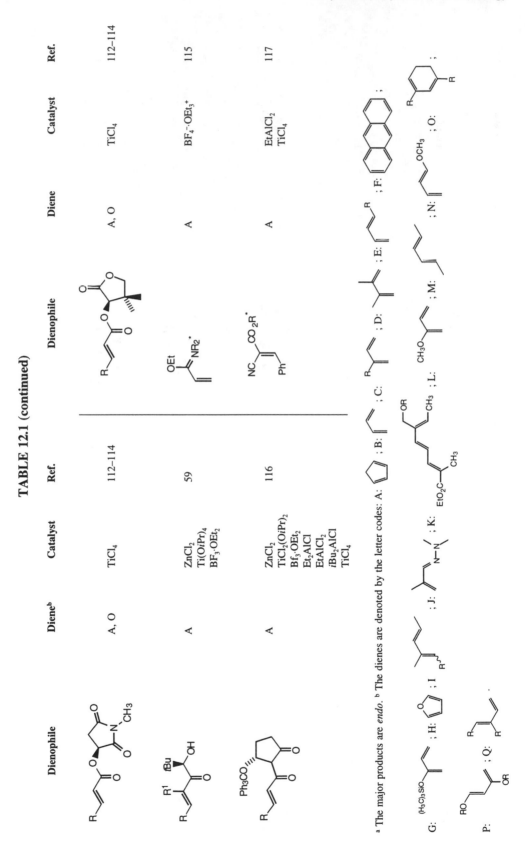

| Dienophile | Diene[b] | Catalyst | Ref. | Dienophile | Diene | Catalyst | Ref. |
|---|---|---|---|---|---|---|---|
| | A, O | $TiCl_4$ | 112–114 | | A, O | $TiCl_4$ | 112–114 |
| | A | $ZnCl_2$<br>$Ti(OiPr)_4$<br>$BF_3{\cdot}OEt_2$ | 59 | | A | $BF_4^-{\cdot}OEt_3^+$ | 115 |
| | A | $ZnCl_2$<br>$TiCl_2(OiPr)_2$<br>$Bf_3{\cdot}OEt_2$<br>$Et_2AlCl$<br>$EtAlCl_2$<br>$iBu_2AlCl$<br>$TiCl_4$ | 116 | | A | $EtAlCl_2$<br>$TiCl_4$ | 117 |

[a] The major products are *endo*. [b] The dienes are denoted by the letter codes: A: ; B: ; C: ; D: ; E: ; F: ; G: ; H: ; I ; J: ; K: ; L: ; M: ; N: ; O: ; P: ; Q: .

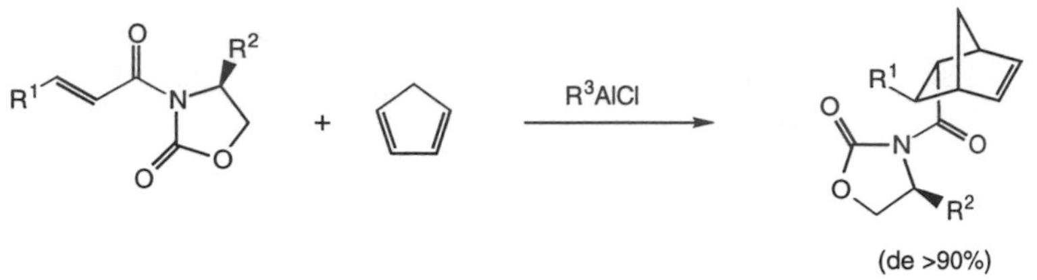

SCHEME 12.3.

In most asymmetric Diels-Alder reactions, significant induction is only observed in the presence of a Lewis acid (vide infra); this is especially true of chiral acrylates. However, when the dienophile contains a chiral center close to the reaction site, thermal Diels-Alder conditions can provide significant stereoselection (e.g., Scheme 12.4).[86]

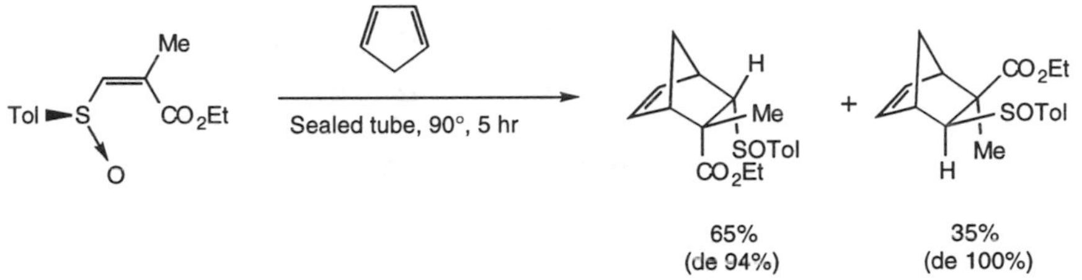

SCHEME 12.4.

### 12.1.3. LEWIS ACIDS

Frontier molecular orbital theory, in addition to the prediction of *endo-exo* selectivity, can also be used to explain the role of the Lewis acid as a catalyst in the Diels-Alder reaction.[52,120] Interaction of the Lewis acid with the dienophile lowers the energy of both the lowest and highest molecular orbital that results in a lower activation energy for the reaction. Indeed, the addition of a Lewis acid appears to be essential for the vast majority of Diels-Alder reactions to maximize selectivity.[32,121,122] This can be rationalized by the increase in reaction rate that allows the reaction to be run at lower temperatures and, thus, allows differentiation of the diastereomeric transition states. The increase in selectivity has also been explained by the type of complex formed between the Lewis acid and dienophile prior to the cycloaddition.[121] Complexation of the Lewis acid with the type II dienophile, as shown by X-ray crystallography, (**12.1**) (Figure 12.3) results in the *s-cis* conformation while complexation of dienophiles of type I (**12.2**) (Figure 12.4) gives *s-trans*. Further, it has been shown that selectivity can be reversed depending on the Lewis acid employed (Schemes 12.5 and 12.6).[76] This reversal has also been explained by the type of conformation the Lewis acid complex adopts.[121]

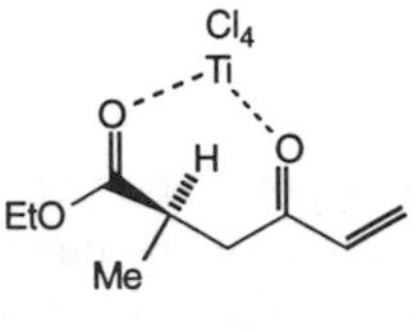

**12.1**

**FIGURE 12.3.** Type II dienophile-Lewis acid conformations.

**FIGURE 12.4.** Type I dienophile-Lewis acid conformations.

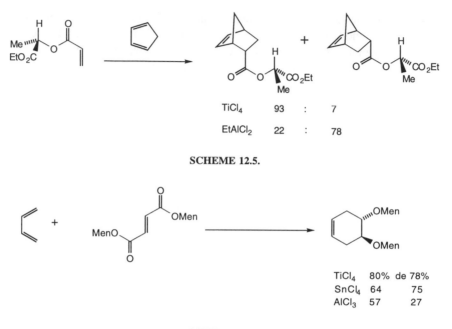

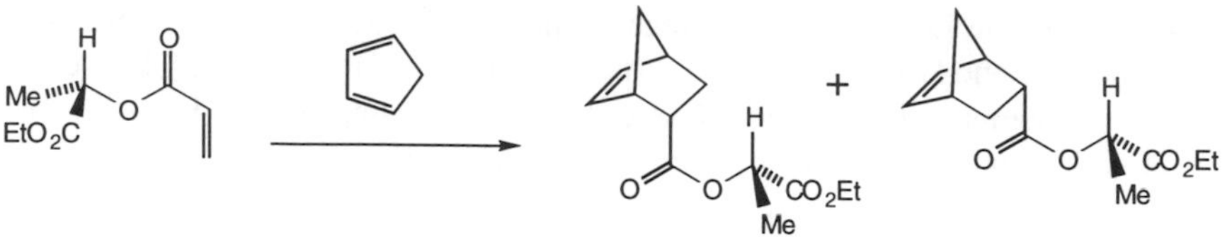

| | | | |
|---|---|---|---|
| $TiCl_4$ | 93 | : | 7 |
| $EtAlCl_2$ | 22 | : | 78 |

**SCHEME 12.5.**

| | | |
|---|---|---|
| $TiCl_4$ | 80% | de 78% |
| $SnCl_4$ | 64 | 75 |
| $AlCl_3$ | 57 | 27 |

**SCHEME 12.6.**

To minimize polymerization, preserve functional groups, and avoid substrate decomposition, Lewis acids have been used where strongly electron-withdrawing substituents, such as chloride, have been replaced by more moderate ligands, such as alkoxy, or weaker examples, such as lanthanide complexes, have been employed.[19,62,63,122-127] However, these changes also compromise the Lewis acidity, and the catalyst may be less effective. Recently, a Lewis acid–base combination has been advocated; a titanium-antimony complex in an asymmetric Diels-Alder reaction resulted in high diastereoselectivity with no evidence of polymerization.[122] The use of electrogenerated acids has also been advocated for labile dienophiles at low temperatures.[128] Clay catalysts suspended in organic solvents have provided comparable stereoselectivities but faster rates than the analogous reaction in water.[129]

### 12.1.3.1. Chiral Lewis Acids

The benefits of a chiral catalyst, as opposed to chiral auxiliaries, on an industrial scale are obvious. However, few examples are present in the literature and, their scope so far appears to be limited (Table 12.2). Research in this area is very active and, as our understanding of the role chiral catalysis plays in asymmetric induction increases, more general reagents should appear.

Chiral menthoxyaluminum catalyzes the reaction with methacrolein and cyclopentadiene to give adducts with ee's of up to 72% but, with other dienophiles, little if any, induction was noted.[143,144] A chiral cyclic amido aluminum complex (**12.3**) catalyzes the cycloaddition

**TABLE 12.2**
**Chiral Catalysts for the Diels-Alder Reaction**

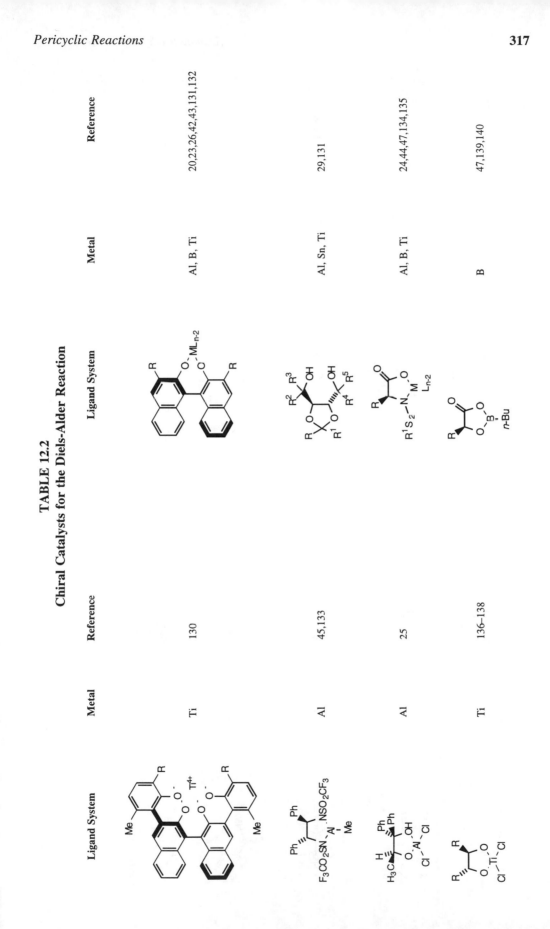

| Ligand System | Metal | Reference | Ligand System | Metal | Reference |
|---|---|---|---|---|---|
| 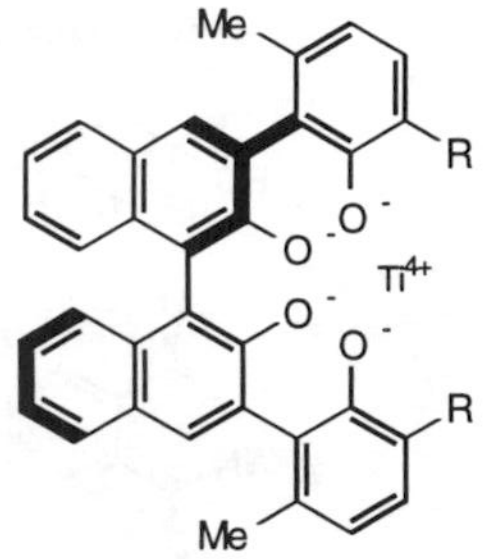 | Ti | 130 | 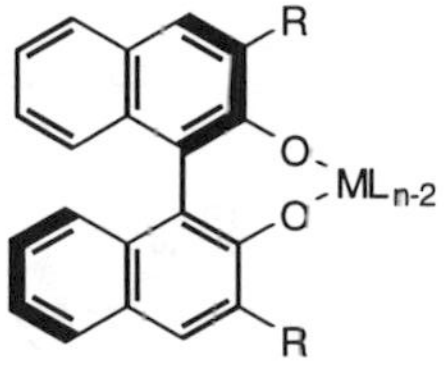 | Al, B, Ti | 20,23,26,42,43,131,132 |
| 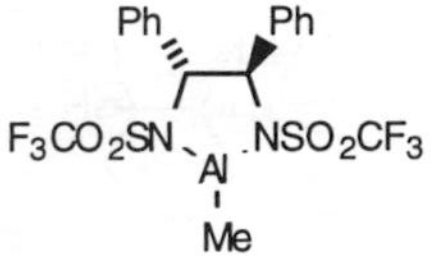 | Al | 45,133 | 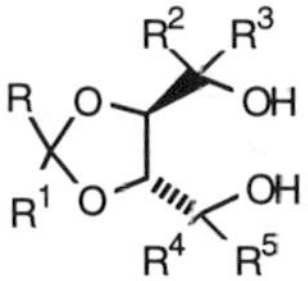 | Al, Sn, Ti | 29,131 |
| 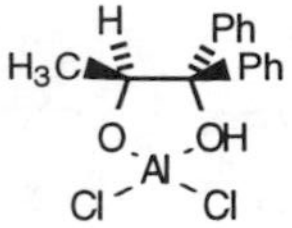 | Al | 25 | 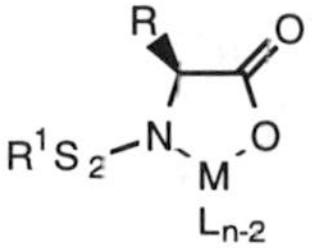 | Al, B, Ti | 24,44,47,134,135 |
| 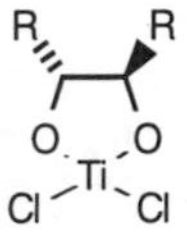 | Ti | 136–138 | 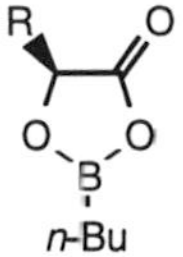 | B | 47,139,140 |

**TABLE 12.2 (continued)**

| Ligand System | Metal | Reference | Ligand System | Metal | Reference |
|---|---|---|---|---|---|
| 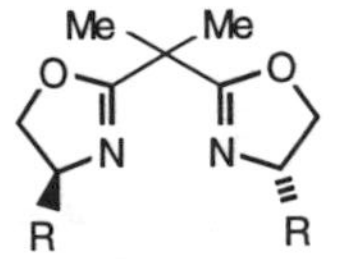 | Cu, Fe | 22,37,141 | 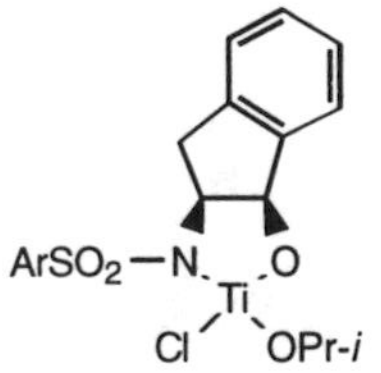 | Ti | 46 |
| 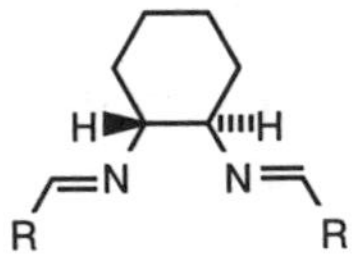 | Cu | 49 | 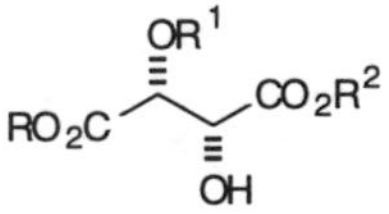 | Al, B, Sn, Ti | 20,21,131,142 |
| 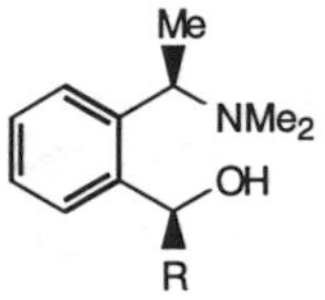 | Cr | 48 | 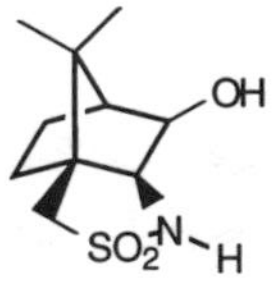 | Al, Sn, Ti | 131 |
| 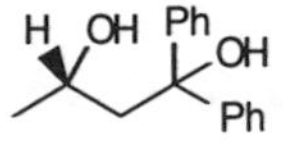 | Al | 131 | $R^*OAlCl_2$ | Al | 143,144 |
| CHIRAPHOS | Co | 28 | | | |

of cyclopentadiene with the *trans*-crotyl derivative (**12.4**) in good yield and selection (Scheme 12.7).[133] The chiral catalyst can easily be recovered.

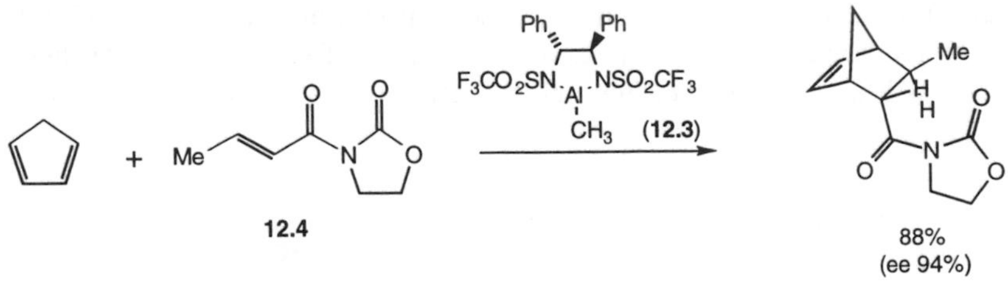

**12.4**

(**12.3**)

88%
(ee 94%)

SCHEME 12.7.

A chiral titanium catalyst, based on a diol derived from tartaric acid, has proven more successful (Scheme 12.8), although the reaction is very sensitive to water and solvent choice.[29] It appears that more sterically congested titanium complexes increase selectivity.[29,136] The use of a chiral tartrate ligand on boron also allows for enantioselectivity.[20,142]

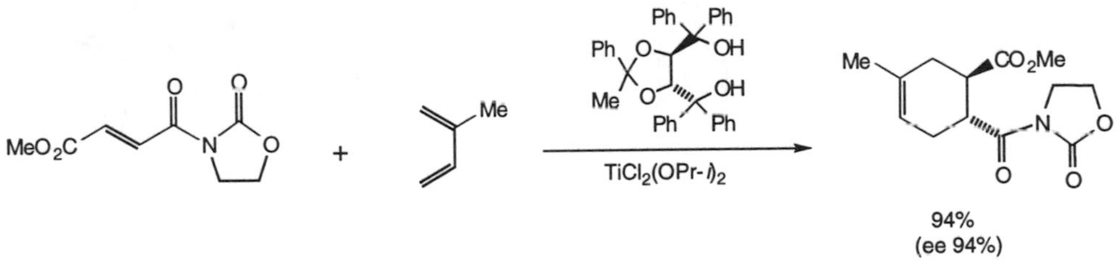

$TiCl_2(OPr\text{-}i)_2$

94%
(ee 94%)

SCHEME 12.8.

In addition, chiral Lewis acid catalysts based on iron (Scheme 12.9) and cobalt (Scheme 12.10) have shown encouraging asymmetric induction.[22,28]

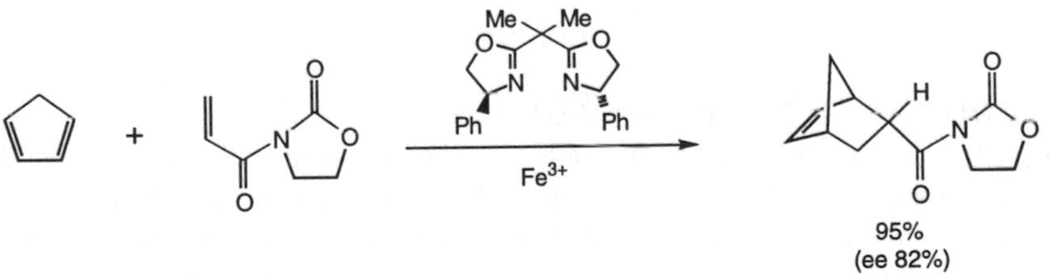

$Fe^{3+}$

95%
(ee 82%)

SCHEME 12.9.

$Co(acac)_3$, $Et_2AlCl$ (4 eq.)
Chiraphos

83%
(ee 91%)

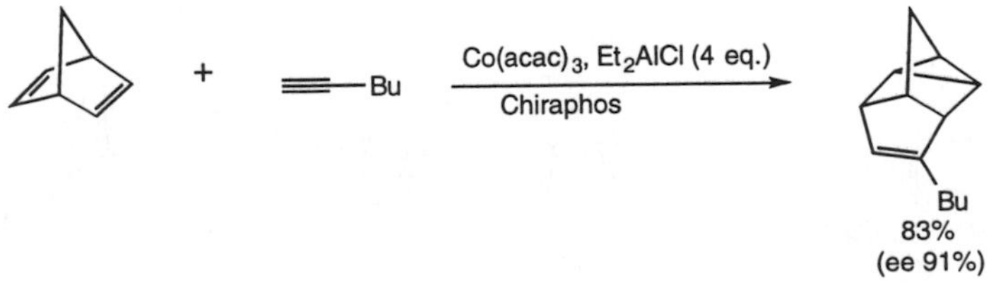

SCHEME 12.10.

## 12.1.4. CHIRAL DIENES

Dienes with a chiral substituent have proved to be less popular than their chiral dienophile counterpart. This is primarily due to the problem of designing a molecule that incorporates a chiral group or auxiliary, but also diastereoselectivities are often not as high,[11,145-165] as illustrated by the cycloaddition of the chiral butadiene (**12.5**) with acrolein (Scheme 12.11); greater success has been obtained through the use of double asymmetric induction (e.g., Scheme 12.12).[11,31,58,59]

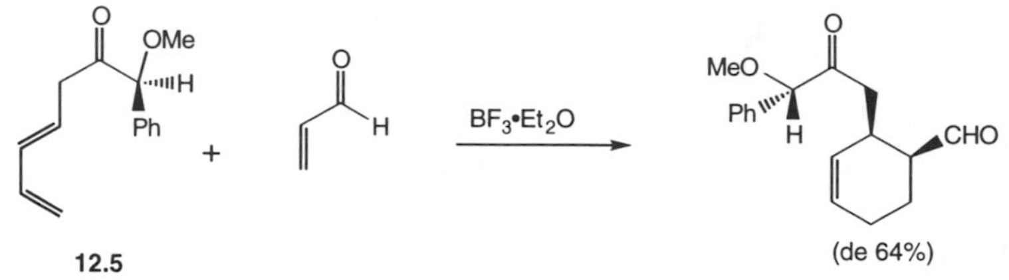

12.5 + acrolein —$BF_3\bullet Et_2O$→ (de 64%)

SCHEME 12.11.

$BF_3\bullet OEt_2$ (ds >130:1)

$BF_3\bullet OEt_2$ (ds 35:1)

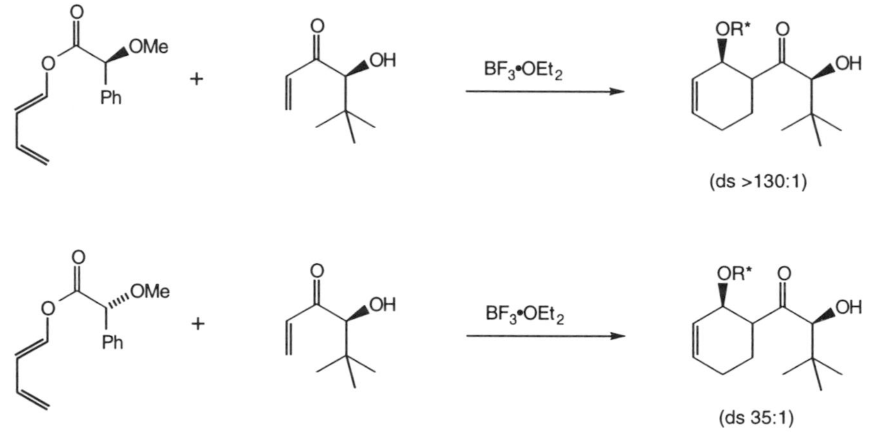

SCHEME 12.12.

## 12.1.5. COOPERATIVE BLOCKING GROUPS

Symmetry within substrates for cycloaddition reactions can be exploited; thus, the use of a chiral fumarate ester allows for the approach of the diene to be directed regardless of which face of the dienophile is attacked. In particular, dimenthyl fumarate (**12.6**) has been advocated for large scale due to its ready availability, good overall yields, and high diastereoselectivity.[32,76,77,166-171] Although other more exotic chiral auxiliaries may be used,[32] the use of **12.6** coupled with a homogeneous Lewis acid catalyst at low temperatures allows for remarkably high diastereoselectivity with a number of dienes (Schemes 12.6 and 12.13).[77]

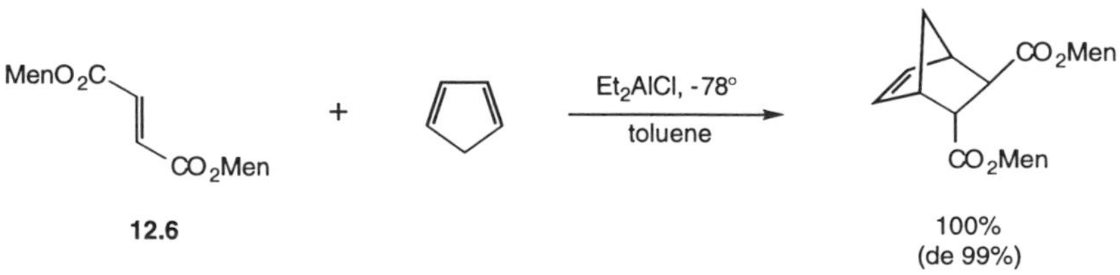

$MenO_2C$–CH=CH–$CO_2Men$ (12.6) + cyclopentadiene —$Et_2AlCl$, -78°, toluene→ $CO_2Men$ / $CO_2Men$ adduct, 100% (de 99%)

SCHEME 12.13.

### 12.1.6. BIOLOGICAL CATALYSTS

Although enzymes have not been particularly successful for the catalysis of the Diels-Alder reaction, they have been employed for the selective hydrolysis of a wide range of the resultant ester adducts [*14.4*].[172,173] A biocatalytic Diels-Alder reaction with baker's yeast has been reported where the *endo* rule was not always observed.[174] A wide variety of transformations has now been catalyzed by abzymes effecting regio- and stereoselective transformations [*14.4*],[175-178] including a Diels-Alder reaction.[175]

### 12.1.7. CHIRAL BASES

In addition to Lewis acid catalysts, bases have also been used to catalyze the Diels-Alder reaction. Although a number of alkaloids were investigated for the reaction of anthrone and *N*-methylmaleimide, only moderate asymmetric induction was observed (Scheme 12.14).[179]

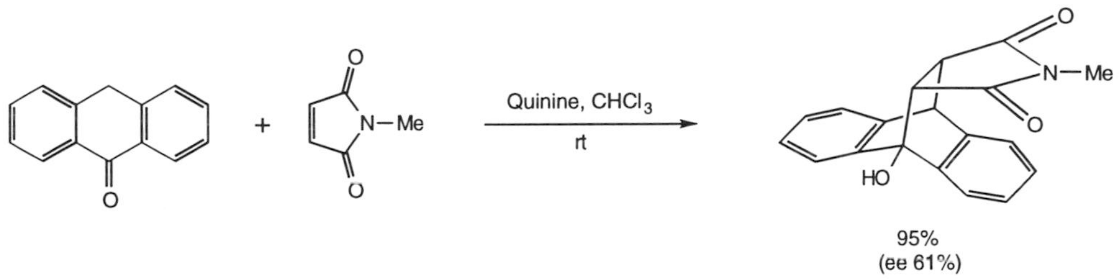

Quinine, $CHCl_3$
rt
N—Me
HO
95%
(ee 61%)

SCHEME 12.14.

### 12.1.8. SOLVENTS EFFECTS

Traditionally, the choice of solvent has had little, if any, effect on the Diels-Alder reaction.[180,181] However, there are some striking exceptions, such as the acceleration of the cycloaddition of *N*-ethylmaleimide with 9-hydroxymethyleneanthracene in water.[182] Although the addition of a Lewis acid might be expected to show solvent dependence, there appears to be little, if any, effect on asymmetric induction.[68,77] Recently it has been shown that *endo*, *exo* and diastereoselectivities correlate well with the polarity parameters of solvents in the cycloaddition of (–)-menthyl acrylate and cyclopentadiene.[183] Theoretical studies with this dienophile have also shown that with an increase in solvent polarity the *s-trans* transition state is stabilized more than *s-cis*.[184] Studies on the rates of the intramolecular Diels-Alder reaction with polar substrates show a strong dependence on the solvent.[185] It should be noted that the effect of solvents has not been rigorously investigated in the asymmetric Diels-Alder reaction.

### 12.1.9. PRESSURE EFFECTS

High pressure is often used in the Diels-Alder reaction to effect reluctant cycloadditions,[186-189] and the success has been attributed to the large negative activation volume of these reactions.[188,190-193] Although scale-up can prove to be problematic, especially with very high pressures, otherwise inaccessible cycloadducts can be synthesized.[188,194] Furan, an example of a diene that only reluctantly enters the Diels-Alder reaction,[195,196] reacts under high pressure with di-*l*-menthyl acetoxymethylenemalonate with moderate diastereoselectivity, whereas at ambient pressure no reaction was observed (Scheme 12.15).[197] Other systems have usually resulted in low asymmetric induction.[95,186,198,199]

### 12.1.10. INTRAMOLECULAR DIELS-ALDER REACTIONS

Intramolecular cycloadditions can provide good stereoselection, and have been valuable intermediates in the synthesis of a variety of natural products.[32,50,51,200,201] They have the advantage that the reactant can be constructed to permit either *endo* or *exo* attack.[202-207] Thus, the cyclization of the triene (**12.7**) in the presence of a magnesium salt allows only *exo* approach to give the strained *trans*-annellated product (**12.8**) (Scheme12.16).[208]

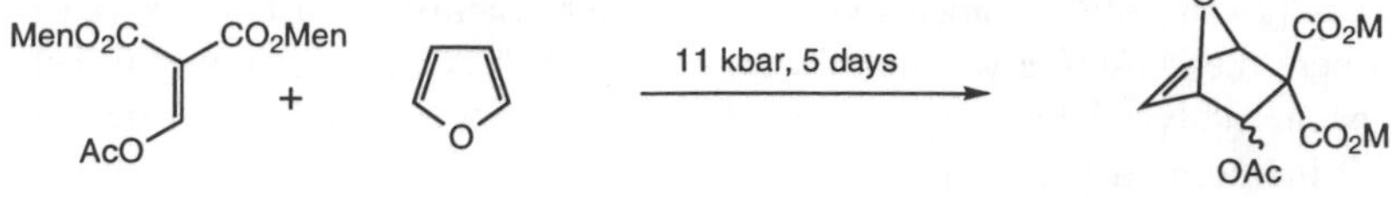

MenO2C CO2Men + AcO 11 kbar, 5 days CO2M CO2M OAc

| | | |
|---|---|---|
| *endo* | 40% | de 54% |
| *exo* | 20 | 61 |

**SCHEME 12.15.**

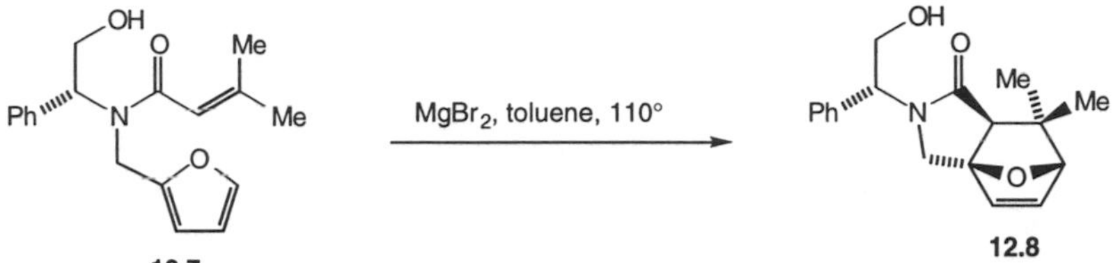

OH O Me Ph N Me MgBr2, toluene, 110° OH O Me Me Ph N O

**12.7** **12.8** (de 76%)

**SCHEME 12.16.**

However, where possible, the *endo* rule prevails, as illustrated by the reaction of camphor sultam derivative (**12.9**) (Scheme 12.17).[209,210] Chiral acyloxy boranes (CAB) have also proven to be efficient catalysts for the intramolecular Diels-Alder reaction (e.g., Scheme 12.18).[211] To avoid high temperatures, Lewis acid catalysis has been employed,[50,212,213] together with high pressure.[214,215] As with the parent reaction, solvent effects also play a role in the intramolecular varient.[216,217] There are a number of examples for hetero intramolecular Diels-Alder reactions.[50,218-221]

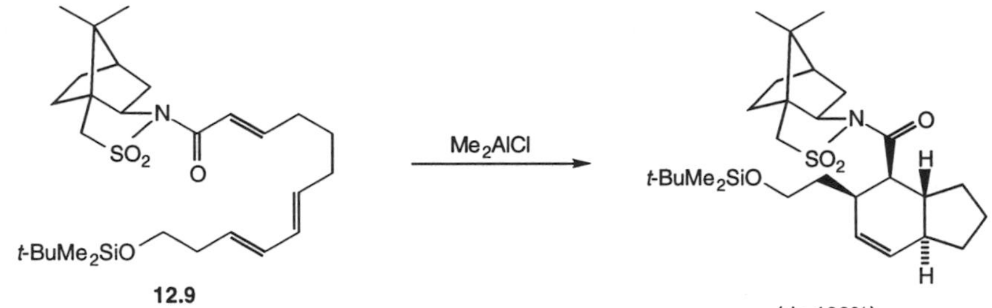

N SO2 O t-BuMe2SiO Me2AlCl N O SO2 H t-BuMe2SiO H

**12.9** (de 100%)

**SCHEME 12.17.**

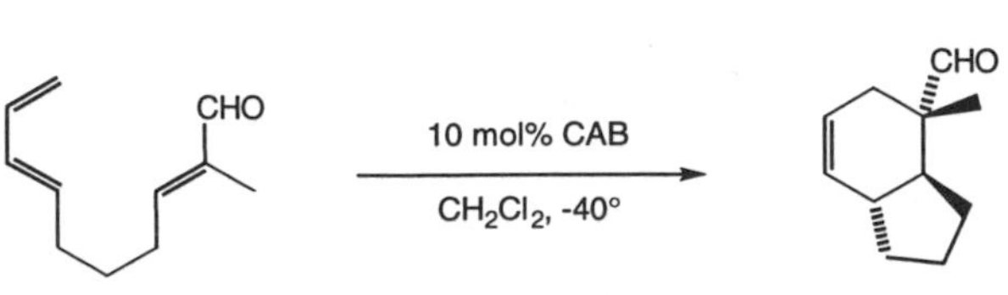

CHO 10 mol% CAB CH2Cl2, -40° CHO

(de 92%)

**SCHEME 12.18**

The power of this approach is illustrated by the spontaneous cyclization of the derivative (**12.10**), formed from the condensation of (*R*)-citronellal (**12.11**) and 5-*n*-pentyl-1,3-cyclohexanedione, with complete stereochemical control (Scheme 12.19).[222]

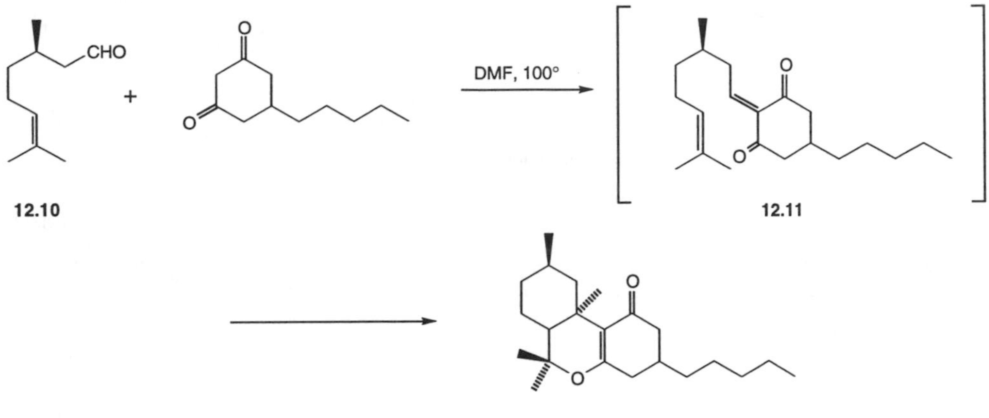

CHO + DMF, 100° 12.10 12.11

SCHEME 12.19.

## 12.1.11. RETRO DIELS-ALDER REACTIONS

Alkenes and dienes that are difficult to access by conventional methodologies [*8.1*] may be available from the retro-Diels-Alder reaction.[181,223-232] The reaction of the *meso* lactone (**12.12**) with a Grignard reagent results in a mixture of diastereomers that can be separated by simple recrystallization, to provide a stereoselective entry to butenolides after a retro-Diels-Alder reaction (Scheme 12.20).[233] As expected, the reaction can be catalyzed by Lewis acids (Scheme 12.21).[224]

1. 2 RMgX, $Et_2O$, THF 2. Separate diastereomers 12.12 130°

SCHEME 12.20.

$(CH_2)_6CO_2H$ $EtAlCl_2$ fumaronitrile $(CH_2)_6CO_2H$

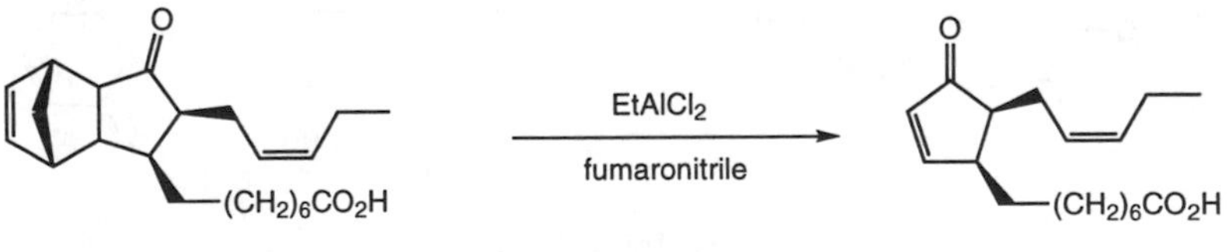

SCHEME 12.21.

## 12.1.12. INVERSE ELECTRON DEMAND DIELS-ALDER REACTIONS

The cycloaddition of electron-rich dienophiles is just beginning to be exploited.[234-242] Dienophiles of this type are often not trivial to synthesize nor set up for recycling, but impressive optical and chemical yields have been observed, such as the cycloaddition of an α-pyrone sulfone with a chiral vinyl ether (Scheme 12.22).[234,240] Isoquinolinium salts have also proved successful, especially when the chiral vinyl ether contained a phenyl substituent (Scheme 12.23).[235]

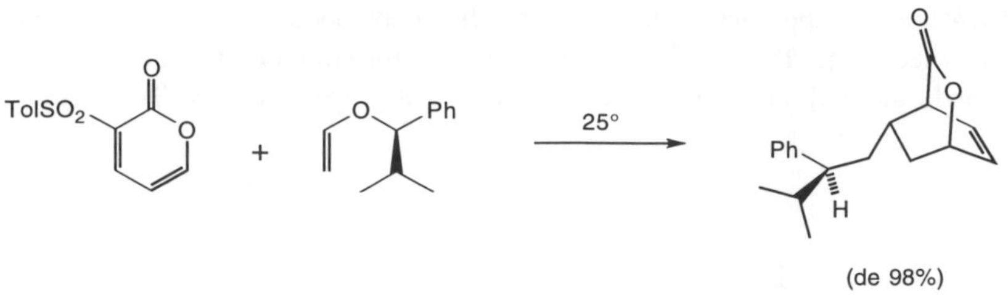

SCHEME 12.22.

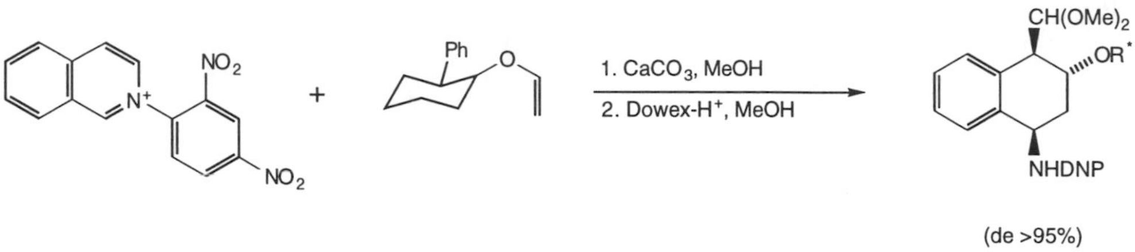

SCHEME 12.23.

### 12.1.13. REACTIONS OF DIENES WITH CARBONYL COMPOUNDS

The reactions of carbonyl compounds with dienes have been included in this section as they were originally believed to be Diels-Alder reactions.[3-5,243-245] The mechanism, however, has been established to proceed by a stepwise addition (vide infra), but the overall stereochemical outcome is the same whichever mechanistic pathway is followed. Early studies of the reaction of dienes with carbonyl compounds were thwarted by reactivity problems, although this was circumvented to a degree by the use of activated carbonyl components.[198,246-257]

Condensation of an aldehyde with the oxygenated diene **12.13**[258-261] in the presence of zinc chloride provides high Anh-Felkin selectivity (Scheme 12.24).[262] Enals also act as dienophiles to afford 2,3-dihydro-4-pyridinone products.[263] Similar reactions can also be achieved with formaldehyde,[264] methoxybutadienes,[265,266] and α-alkoxy aldehydes.[267,268]

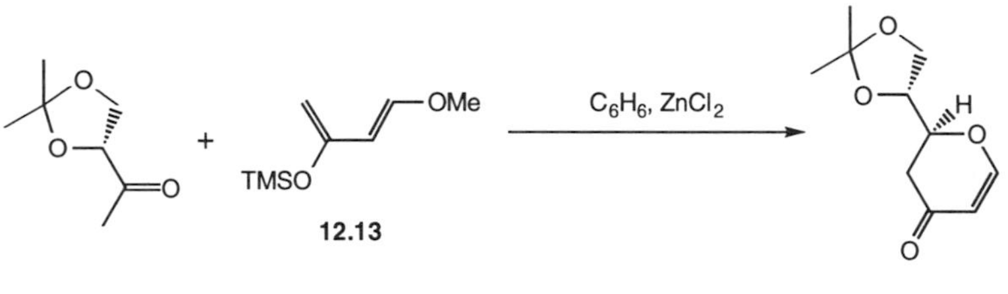

SCHEME 12.24.

The nature of the Lewis acid can determine the relative stereochemical outcome of the reaction (Scheme 12.25).[262,269]

In addition to zinc or boron Lewis acids,[270] the reaction is catalyzed by lanthanide metal complexes which allow fragile functionality to survive in the resultant adduct (Scheme 12.25).[123,124,271-273] The use of europium(III) complexes allows for high diastereoselectivity (>60:1),[267,268,274] and a reasonable amount of asymmetric induction (18 to 58% ee).[123,268,275] Homochiral pyranose derivatives have also been synthesized using a chiral auxiliary–chiral catalyst combination.[276,277] Chiral aluminum catalysts can also provide high levels of asymmetric induction,[36,278] while lower levels have been observed with chiral ruthenium complexes.[279] Modest diastereoselectivity has also been observed with the use of high pressure.[280]

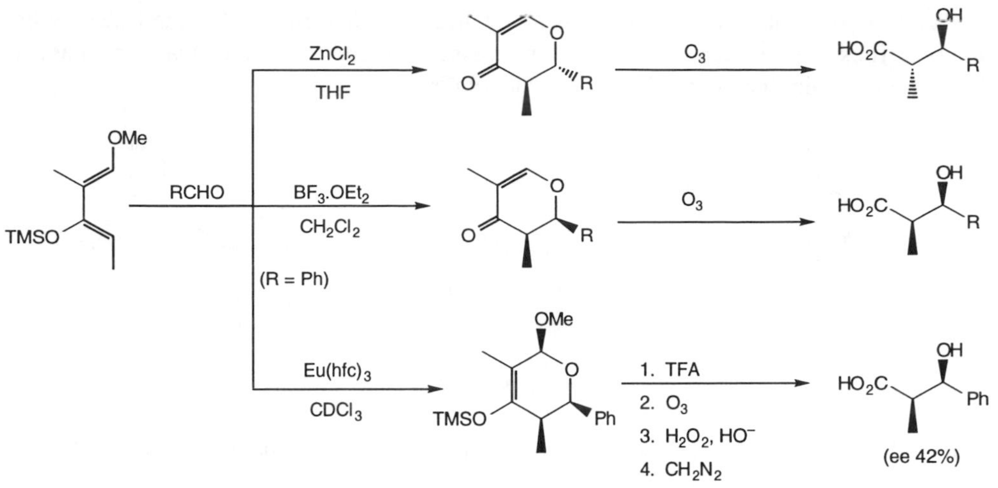

SCHEME 12.25.

Although the mechanism of the reaction can be described in terms of a pericyclic or aldol-cyclization [Chapter 7], the exact pathway followed can depend on the Lewis acid employed.[261,281] Magnesium bromide exhibits high diastereofacial control (Scheme 12.26),[282] and favors the *endo*-topology.[269] The geometry of the diene does control the relative geometry of substituents in the pyrone.[283]

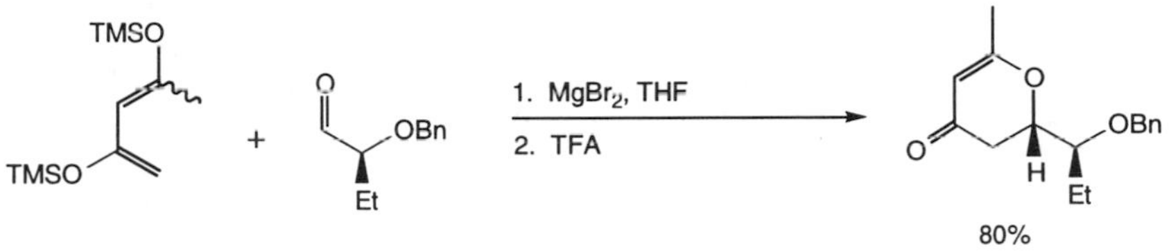

SCHEME 12.26.

The adducts from all of these reactions can be utilized as substrates for a wide variety of transformations,[284,285] as evident from the range of products which have been prepared.[243,266,286-296]

### 12.1.13.1. Related Cycloaddition Reactions

Other reactions that result in a pyranoside moiety exist, although as yet they have not been widely utilized in asymmetric synthesis. Vinyl ethers and allenic ethers condense with enals to give dihydropyrans (Scheme 12.27).[124,297-299] Similarly, electron deficient acyl ketenes react with vinyl ethers.[300] This approach also provides methodology to spiroketals[301,302] and spiroacetals.[292,303,304]

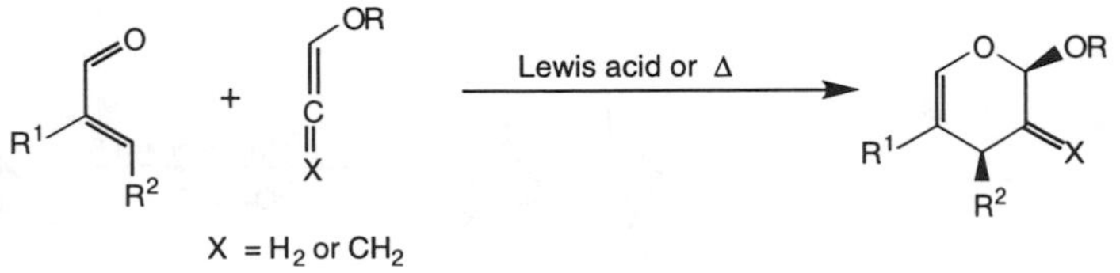

SCHEME 12.27.

Electron-deficient enamine carbaldehydes cyclize with vinyl ethers (Scheme 12.28). When the reactions are conducted under high pressure, there is a significant increase in diastereoselectivity for the resultant dihydropyran.[187]

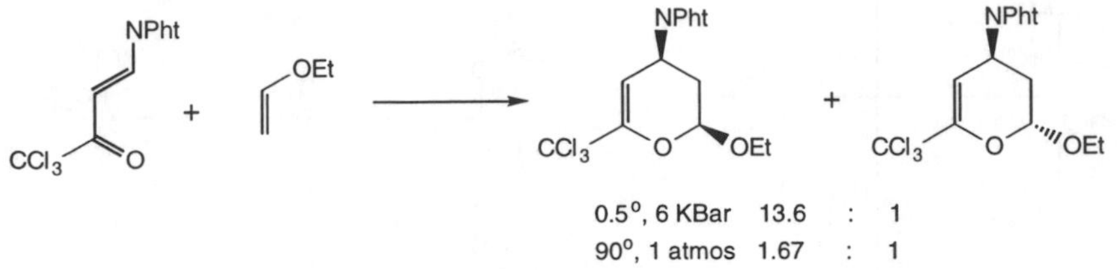
NPht
OEt
+
$CCl_3$
O
NPht
$CCl_3$
O
OEt
+
NPht
$CCl_3$
O
OEt
0.5°, 6 KBar 13.6 : 1
90°, 1 atmos 1.67 : 1

SCHEME 12.28.

In addition, there are a number of hetero dienophiles and dienes that have been utilized in a Diels-Alder type reaction that allow for both regio- and stereoselective control.[13,30,165,292,300,303-347]

## 12.2. CLAISEN REARRANGEMENTS

### 12.2.1. COPE REARRANGEMENT

Although the emphasis has been placed on the Claisen rearrangement, there are related reactions that are also synthetically useful. The Cope rearrangement is the carbon version of the Claisen rearrangement, and usually proceeds through a chair transition state (Scheme 12.29).[348,349]

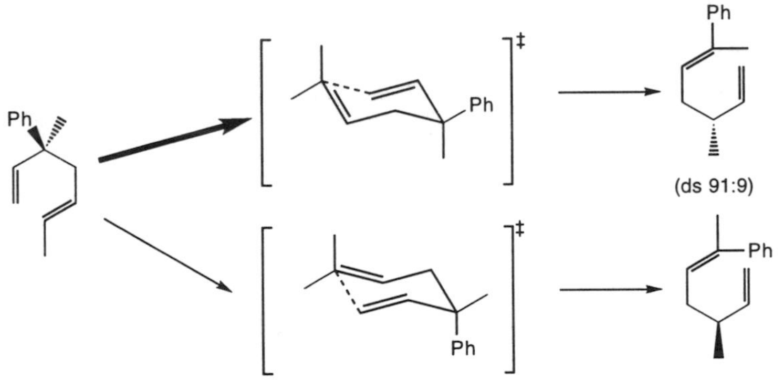
Ph
Ph
‡
Ph
(ds 91:9)
‡
Ph
Ph

SCHEME 12.29.

The aza-Cope reaction incorporates a nitrogen atom into the diene substrate.[350-352]

The reaction is reversible, but the introduction of an oxygen atom (the oxy-Cope rearrangement) results in the formation of an enolate whose stability drives the reaction to completion.[353,354] As the enolate is formed regioselectively, it can be used for further transformations *in situ* (Scheme 12.30).[355-389]

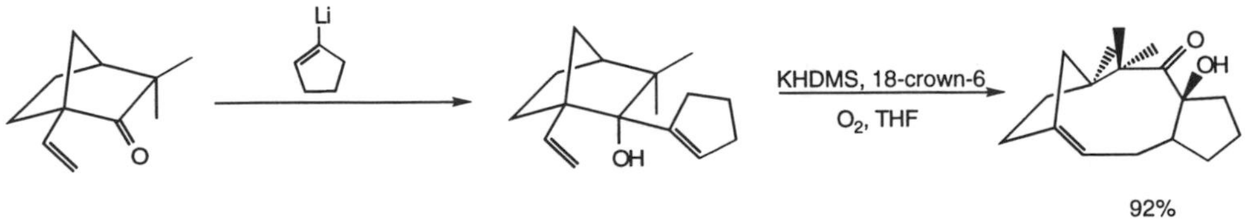
Li
O
OH
KHDMS, 18-crown-6
$O_2$, THF
O
OH
92%

SCHEME 12.30.

### 12.2.2. CLAISEN REARRANGEMENT

The Claisen rearrangement has proven extremely useful in the synthesis of natural products.[390-400] It has been adapted for a variety of derivatives including enol ethers,[401-406] amides,[407-415] esters and orthoesters,[405,416-424] acids,[425-427] oxazolines,[428] ketene acetals,[429,430] and thioesters.[431-433] Many of these variants rely on a cyclic primer to control stereochemistry, or to control only relative stereochemistry and alkene geometry. Thus, the use of a carbohydrate derivative allows exo-cyclic rearrangement to occur and afford geminal functionality at C-2 or C-3 of a hexapyranose system (e.g., Scheme 12.31).[434]

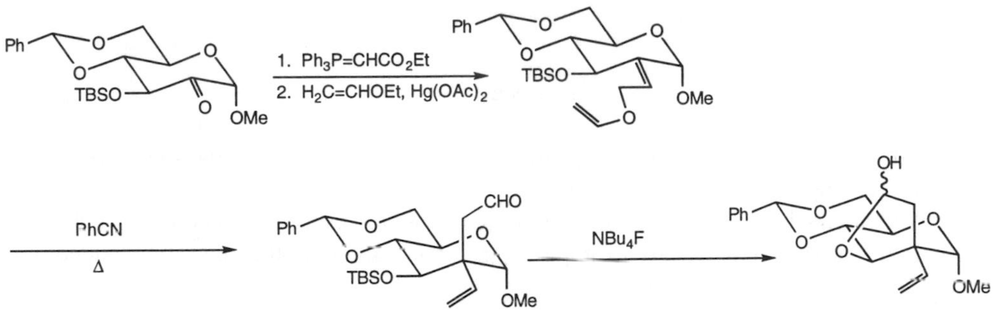

1. $Ph_3P{=}CHCO_2Et$
2. $H_2C{=}CHOEt$, $Hg(OAc)_2$

PhCN, Δ

$NBu_4F$

**SCHEME 12.31.**

Carbohydrates have also been employed as chiral auxiliaries; they also allow the rearrangement to be performed in an aqueous medium.[435] Indeed, water has been used to accelerate the Claisen rearrangement without the need to use an auxiliary to solubilize the substrate.[436,437] As an example of this "template" approach, consider the synthesis of thromboxane-$B_2$-glucose from D-glucose: the key reaction is the conversion of the allyl alcohol to the amide (Scheme 12.32).[407,408,411]

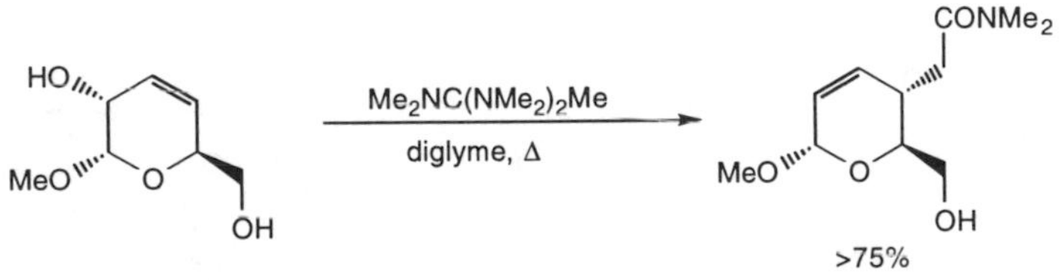

$Me_2NC(NMe_2)_2Me$, diglyme, Δ

>75%

**SCHEME 12.32.**

An example of an enol ether Claisen rearrangement is provided by the vinylsilane **12.14**, where the large silyl group adopts an equatorial position in the chair transition state (Scheme 12.33).[438]

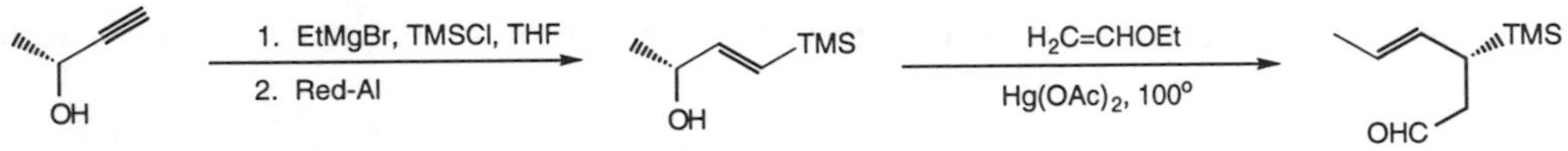

1. EtMgBr, TMSCl, THF
2. Red-Al

**12.14**

$H_2C{=}CHOEt$, $Hg(OAc)_2$, 100°

**SCHEME 12.33.**

The reaction provides a useful entry into chiral butyrolactones [*13.2.2*].[439] These reactions also provide examples of factors that can influence whether the reaction proceeds through a chair or boat transition state (Scheme 12.34).[398,439]

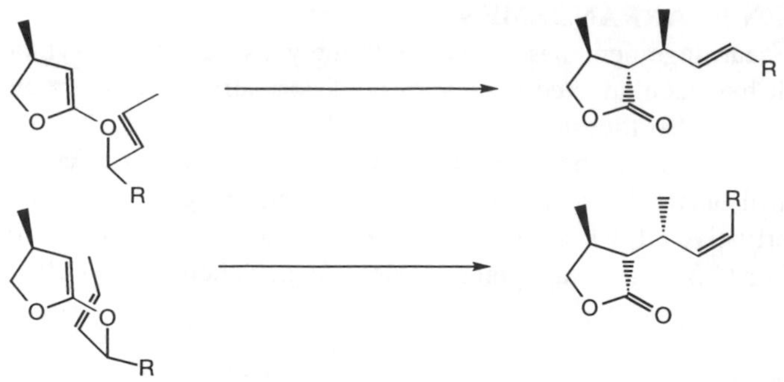

SCHEME 12.34.

The effect of allylic strain to differentiate between two potential pathways was used as an illustrative example for this phenomenon [*2.3*].[440]

Many elements have been incorporated into the reaction, especially heteroatoms close to carbon in the periodic table.[441] An example of an aza-Claisen rearrangement is provided in a synthesis of (+)-dihydropallescensin (**12.15**) (Scheme 12.35).[428,442]

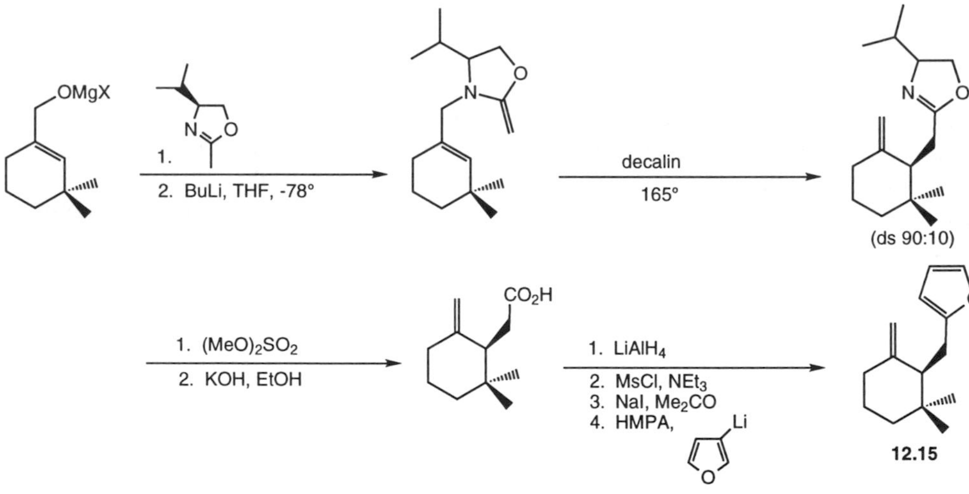

OMgX
1.
2. BuLi, THF, -78°
decalin
165°
(ds 90:10)
1. $(MeO)_2SO_2$
2. KOH, EtOH
$CO_2H$
1. $LiAlH_4$
2. MsCl, $NEt_3$
3. NaI, $Me_2CO$
4. HMPA,
**12.15**

SCHEME 12.35.

The Claisen rearrangement can be accelerated through π-electron donating groups at the 2-position of the vinyl portion of the ether.[443-447] The aliphatic Claisen rearrangement does proceed in the presence of organoaluminum compounds,[397,448,449] although other Lewis acids failed to bring about the same transformation.[397,401,450-452]

A palladium catalyzed [3,3]-sigmatropic rearrangement of an allylsilane proceeds with high selectivity (Scheme 12.36).[453-455]

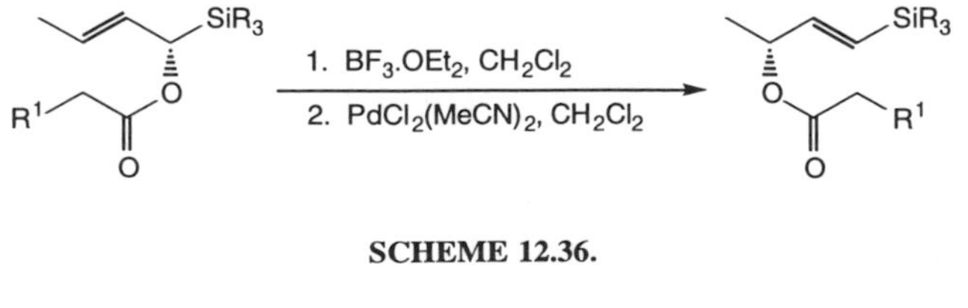

$SiR_3$
1. $BF_3.OEt_2$, $CH_2Cl_2$
2. $PdCl_2(MeCN)_2$, $CH_2Cl_2$

SCHEME 12.36.

The Lewis acid can also be incorporated into the substrate.[456] However, Claisen variants, such as the aza-Claisen, can be catalyzed by a wide variety of Lewis acids.[450] Another related reaction, the [3,3] thermal rearrangement of allylic imidates has been used in a synthesis of (+)-polyoxanic acid (**12.16**) (Scheme 12.37).[457-459]

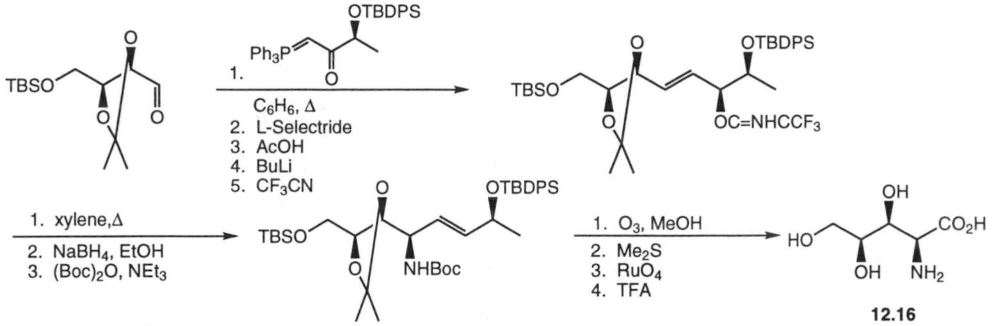

**SCHEME 12.37.**

### 12.2.3. ESTER ENOLATE CLAISEN REARRANGEMENT

The current, major application of the Claisen reaction utilizes allyl esters, as the overall reaction conditions are relatively mild.[427,460] The geometry of the initially-formed enol ether substrate is controlled by the solvent (Scheme 12.38) [*5.1*].[425,461-464] Thus, the methodology provides a useful alternative to an aldol approach.

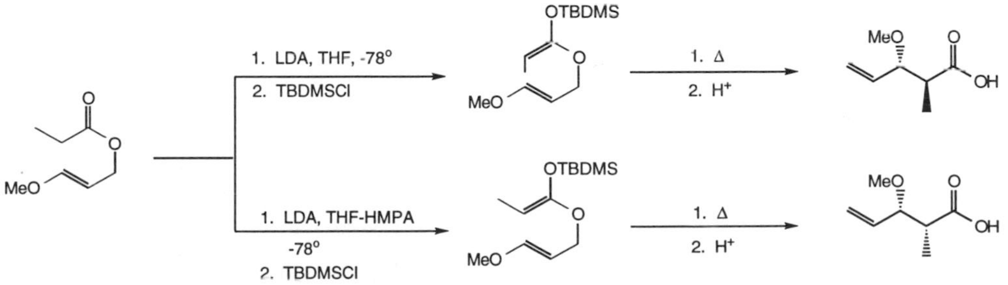

**SCHEME 12.38.**

The reaction even proceeds when a leaving group is β to the ester group.[465] As an anion is a key intermediate in this reaction, it is somewhat surprising that the alternative rearrangement, a [2,3]-Wittig rearrangement, does not compete.[466]

The stereochemistry of the allylic unsaturation does not have any effect on the stereochemical outcome of the rearrangement. Thus, the *E*-enolate provides the *ancat* product (Scheme 12.38, top line), while the *Z*-enolate gives the *syncat* product.[425] The silyl ketene acetals are usually employed with the ester enolate rearrangement, as this circumvents the problems of competing aldol reactions; the use of *t*-butyldimethylsilyl ethers also avoids *C*-silylation.[427]

The geometry of the enolate is, by far, the most important aspect for the stereochemical control of the reaction. However, other substituents can influence the reaction pathway. As with the Claisen rearrangement itself, substituents may cause the transition state to be a boat rather than a chair. However, in unhindered acyclic cases, the chair transition state is preferred. In cyclic systems a boat transition state can be the predominant transition state. For the rearrangement of cyclohexenal propanoate, the (*E*)-silyl ketene acetal rearranges through a chair transition state, while the (*Z*)-acetal proceeds by a boat.[467,468] The rationale for this difference is shown in Figure 12.5. In the case of the *E*-isomer, there is an unfavorable interaction between the $R^1$ group (for this example $R^1$ = Me) and the cyclohexyl ring in a possible boat transition state. Conversely, there is a large, unfavorable interaction between the ring system and the *O*-silyl group in a potential chair transition state with the *Z*-enolate.

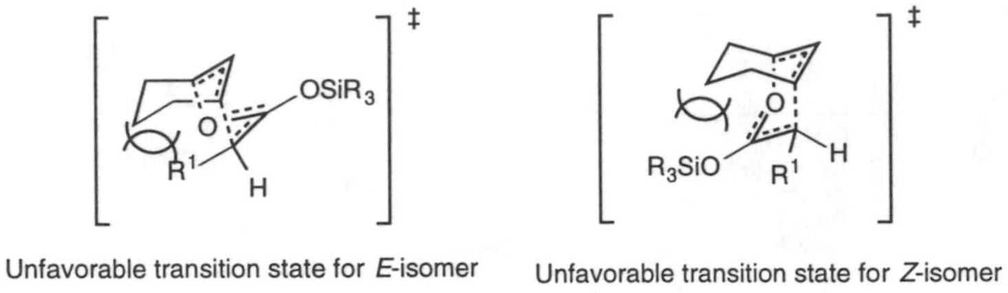

Unfavorable transition state for *E*-isomer

Unfavorable transition state for *Z*-isomer

**FIGURE 12.5.** Transition states for the Claisen rearrangement of silyl ketene acetals.

The examples where the rearrangement is performed with a pyranoid compound go through a boat transition state for both the *E*- and *Z*-enolates.[468] The transition states are shown in Figure 12.6. In a similar manner, cyclopentenyl compounds rearrange by a chair transition state, while the furanoid analogs proceed through a boat geometry (Figure 12.7).

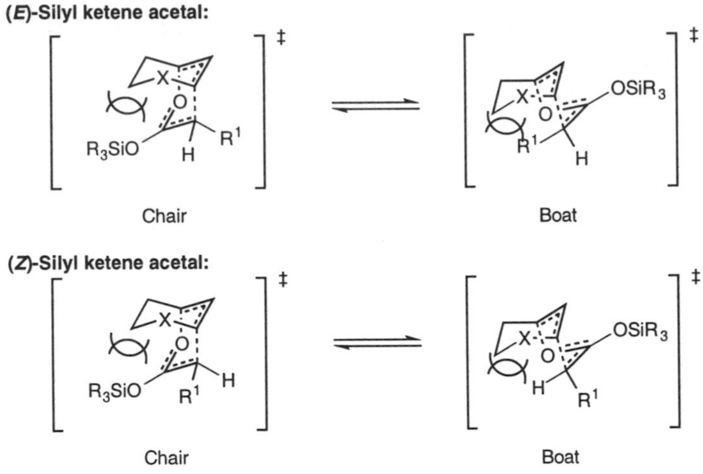

**(*E*)-Silyl ketene acetal:**

Chair

Boat

**(*Z*)-Silyl ketene acetal:**

Chair

Boat

**FIGURE 12.6.** Transition states for pyranoid compounds in the Claisen silyl ketene acetal rearrangement.

The subtle differences between the carbocycle and oxygen heterocycles suggest that stereoelectronic effects, rather than steric factors, are the key difference.[460]

For the isomeric ketene acetals generated from methoxy allyl propionate, rearrangement results in a mixture of carboxylic acids, although the isomer that results from a chair transition state predominates. Use of isotope effects has led to the proposal that the aliphatic Claisen rearrangement resembles an oxyallyl-allyl radical pair (Scheme 12.39).[469] This results in a transition state with more advanced O(3)–C(4) bond breaking than in the parent system, and so an electron donor group at C-6 will tend to favor a chair transition state.[460]

Examples of the application of this reaction[470] are provided by syntheses of nonactic acid,[471,472] tetronates,[473] the Prelog-Djerassi lactone,[474] prostanoids,[475] tirandamycic acid (Scheme 12.40),[476] sesqui- and diterpenes,[477-482] macrolides,[207,483-489] ionophores,[465,490-495] *C*-glycosides,[496] alkaloids,[497] and bicyclic and polycyclic systems,[477,498,499] including steroids,[500,501] iridoids,[502,503] and medium to large carbocycles.[502,504] Methodology to other functionality is also available (vide infra).

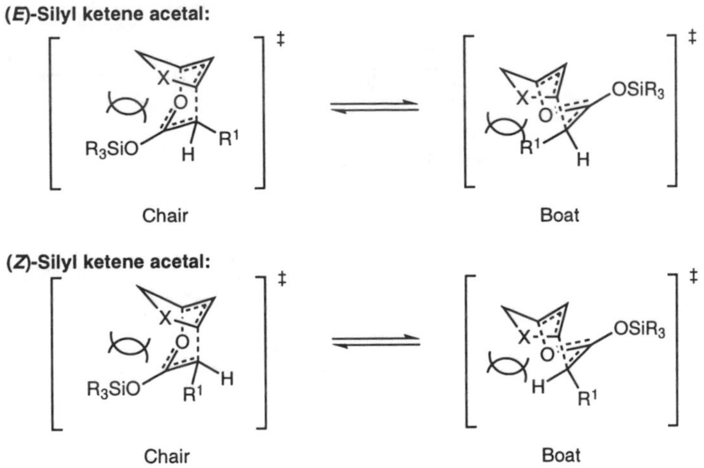
(E)-Silyl ketene acetal:
X
O
R3SiO
H
R1
Chair
OSiR3
R1
H
Boat
(Z)-Silyl ketene acetal:
R3SiO
R1
H
Chair
OSiR3
H
R1
Boat

**FIGURE 12.7.** Transition states for furanoid compounds in the Claisen silyl ketene acetal rearrangement.

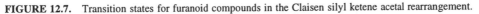

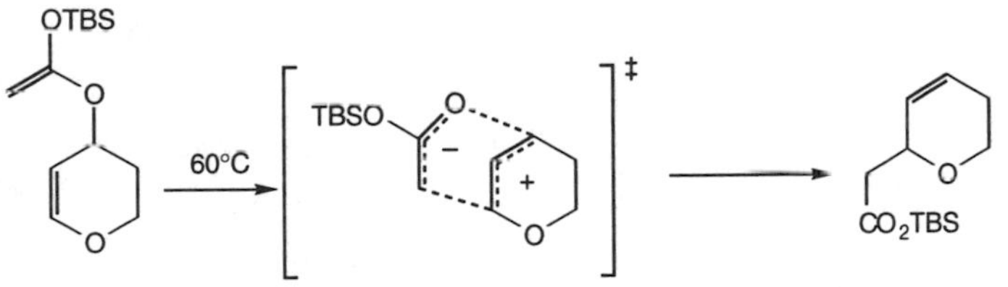
OTBS
O
60°C
TBSO
O
CO2TBS

**SCHEME 12.39.**

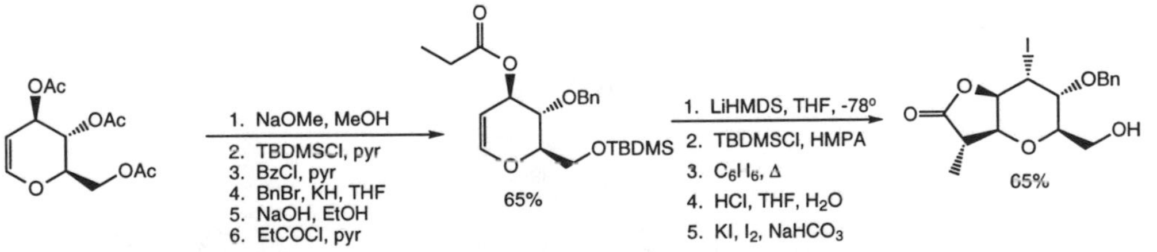
OAc
OAc
OAc
1. NaOMe, MeOH
2. TBDMSCl, pyr
3. BzCl, pyr
4. BnBr, KH, THF
5. NaOH, EtOH
6. EtCOCl, pyr
OBn
OTBDMS
65%
1. LiHMDS, THF, -78°
2. TBDMSCl, HMPA
3. C6H6, Δ
4. HCl, THF, H2O
5. KI, I2, NaHCO3
OBn
OH
65%

**SCHEME 12.40.**

The Ireland Claisen ester enolate rearrangement utilizes the silyl ketene acetal as described above. Other groups have also been incorporated into the system as variations and to allow chelation.[505] Boron can be used to control enolate geometry; this has been exploited in an allylic ester rearrangement (Scheme 12.41),[506] where high ee's can be observed.

The rearrangement methodology has been extended to allow the preparation of α-alkoxy esters, with diastereoselection,[138,453,454,507-514] α-phenylthio esters,[510,515,516] α-amino acids,[495,514,517-519] α-fluoro esters,[520] cycloalkenes,[479,521] tetronic acids,[522] and dihydropyrans (Scheme 12.42).[476,523,524] The dioxanone to dihydropyran transformation allows a pyran nucleus to be built up rapidly with impressive stereochemical control.[486,512,525]

The use of a heteroatom α to the ester carbonyl group allows for the formation of a chelate with the metal counterion; hence, the geometry of the ester enolate can be assured.[507,509] This approach has been used effectively in a synthesis of (+)-blastmycinone (**12.17**) (Scheme 12.43).[508]

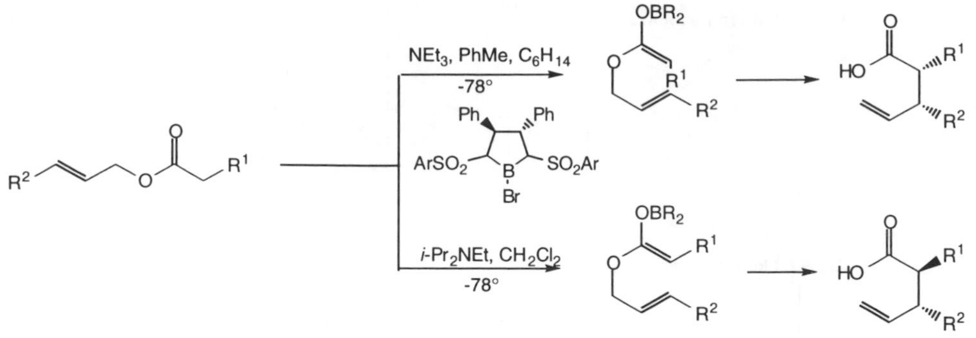

SCHEME 12.41.

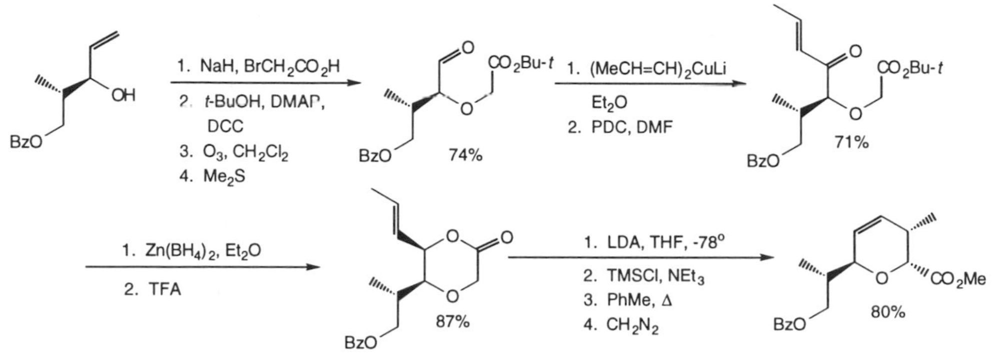

SCHEME 12.42.

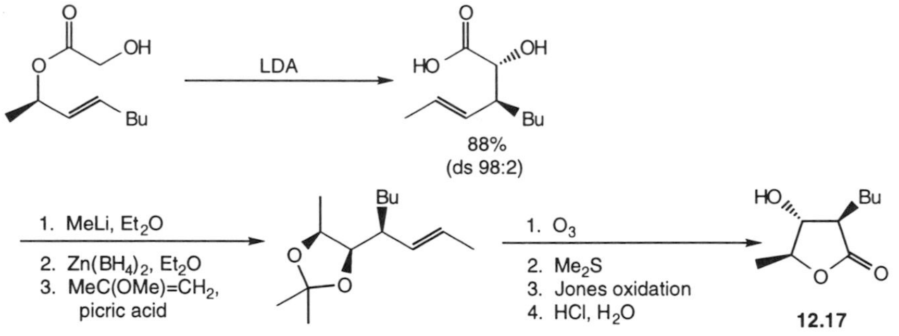

SCHEME 12.43.

The reaction also provides methods to chiral stannanes.[412] The power of the rearrangement lies in its ability to control a number of chiral centers. This is impressively illustrated by a "double" rearrangement (Scheme 12.44).[526] The ketone is not involved in the reaction sequence and no epimerization occurs.

In a few cases an alternative, competing pathway which involves a retro-Diels-Alder reaction followed by a Diels-Alder addition has been observed rather than the Claisen rearrangement; the major product is the same, whichever mechanism is followed.[487] Of course, the problems associated with a chair vs. boat transition state are then alleviated.[354]

The thio-Claisen rearrangement can be used to provide substituted thioamides and β-hydroxy dithioesters.[527,528] The ketal version of the rearrangement provides a regioselective method to γ,δ-enones from allyl alcohols.[529] The aza-Claisen rearrangement with amide enolates is analogous to the ester reactions.[414]

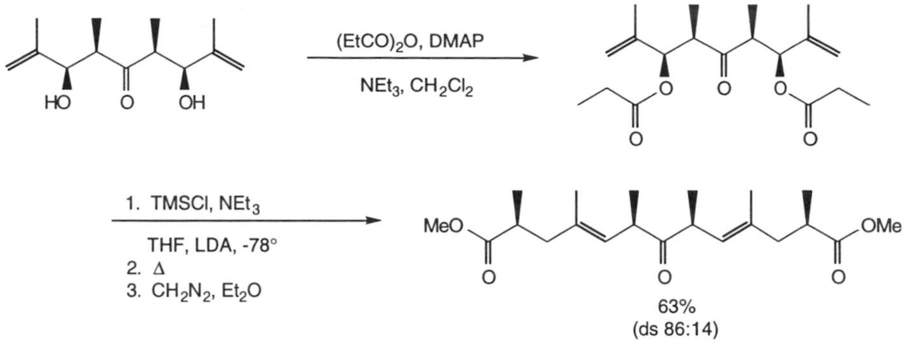

(EtCO)$_2$O, DMAP

NEt$_3$, CH$_2$Cl$_2$

1. TMSCl, NEt$_3$

THF, LDA, -78°

2. Δ

3. CH$_2$N$_2$, Et$_2$O

63%
(ds 86:14)

**SCHEME 12.44.**

The closely related ester enolate Carroll rearrangement affords entry into β-keto acids or γ,δ-enones.[530,531]

## 12.3. ENE REACTION

The ene reaction,[6,532-536] like the Wittig rearrangement, allows for chirality transfer.[537-540] The reaction is closely related to the Diels-Alder reaction as a σ-bond is formed at the expense of a π-bond. In addition, relative stereochemistry can also be controlled through the use of Lewis acid catalysis (Scheme 12.45).[541-543]

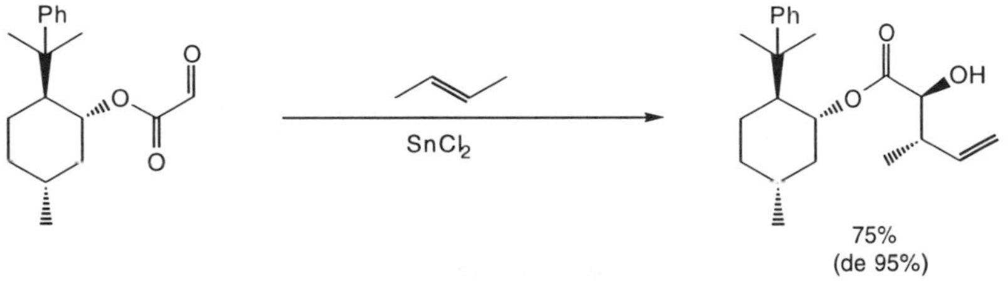

SnCl$_2$

75%
(de 95%)

**SCHEME 12.45.**

The use of a Lewis acid catalyst not only alleviates the need for high temperatures, but allows for the simple introduction of a chiral agent (vide infra).[532,535] Indeed, the Lewis acid can be incorporated into one of the reagents to result in a "metallo-ene" reaction. Such an example is provided by the addition of an allylborane to a carbonyl compound [*4.1.5.1*].[6,535] The Lewis acid methodology has been used in an approach to brassinosteroids, plant growth promotion regulators (Scheme 12.46).[544-546]

Very impressive stereoselection can be observed with intramolecular ene reactions. With Lewis acid catalysis, this selectivity can be carried over to intermolecular reactions (Schemes 12.45 and 12.47).[547-549]

The majority of applications to date have centered around carbocyclic preparations.[534,550] Not only can alkenes be used in ene reactions, but masked functionality, such as latent carbonyl derivatives, can also be employed (Scheme 12.48).[535,551-554]

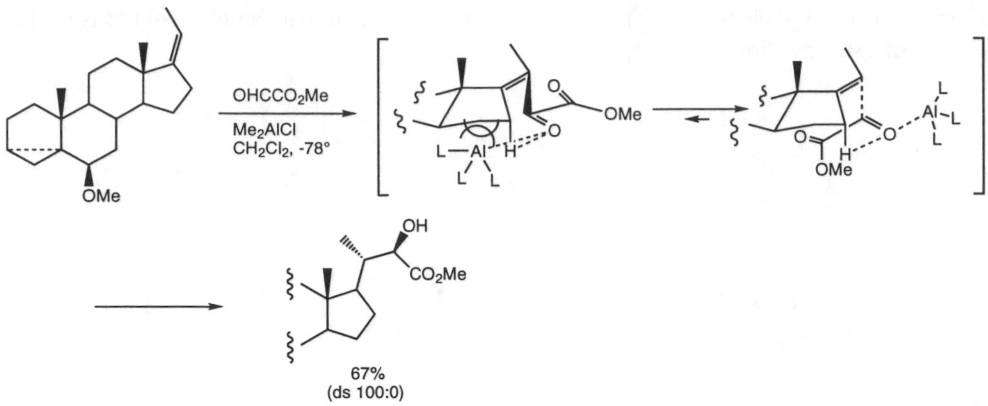

SCHEME 12.46.

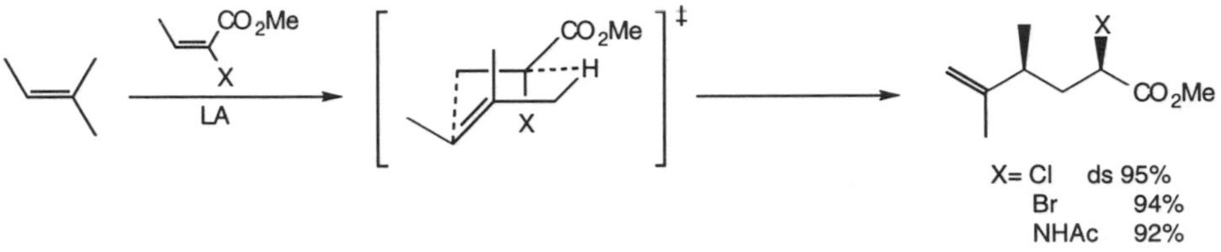

SCHEME 12.47.

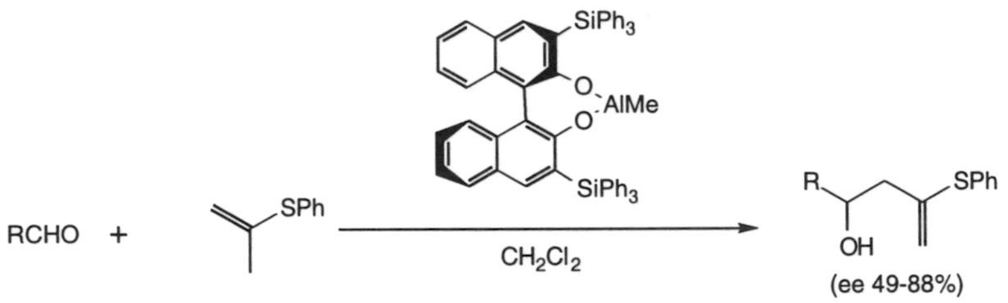

SCHEME 12.48.

The methodology allows for a selective preparation of cyclic compounds, including tetrahydrofurans and pyrans (Scheme 12.49).[555-558]

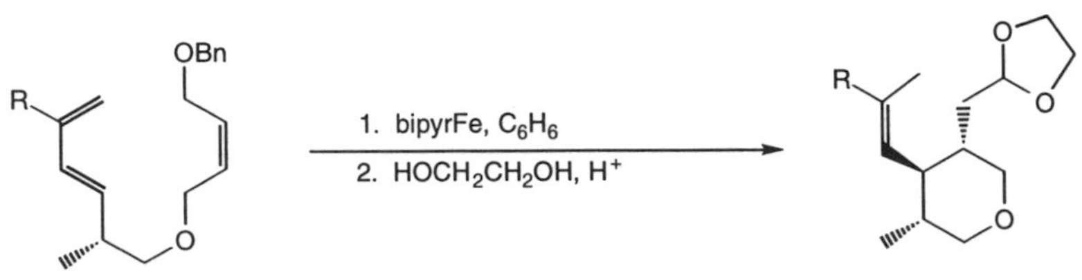

SCHEME 12.49.

Some acyclic applications provide useful stereochemical control when a cyclic primer is employed.[537,541,544,559-567] An example is provided by a synthesis of (±)-khusimone (**12.18**) (Scheme 12.50).[568] This intramolecular "magnesium-ene" reaction gave no detectable amounts of the isomer of **12.19**. This observation has been rationalized in terms of kinetic stereoselection with the transition state **12.20** being favored over **12.21** with its severe steric interactions (Figure 12.8).

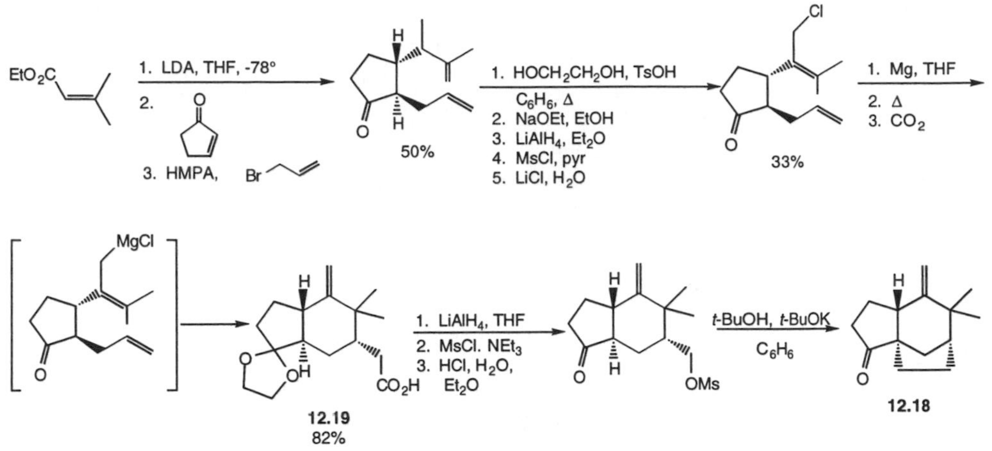

1. LDA, THF, -78°
2. cyclopentenone
3. HMPA, allyl bromide
50%
1. $HOCH_2CH_2OH$, TsOH, $C_6H_6$, Δ
2. NaOEt, EtOH
3. $LiAlH_4$, $Et_2O$
4. MsCl, pyr
5. LiCl, $H_2O$
33%
1. Mg, THF
2. Δ
3. $CO_2$
**12.19** 82%
1. $LiAlH_4$, THF
2. MsCl. $NEt_3$
3. HCl, $H_2O$, $Et_2O$
*t*-BuOH, *t*-BuOK, $C_6H_6$
**12.18**

**SCHEME 12.50.**

**12.20** **12.21**

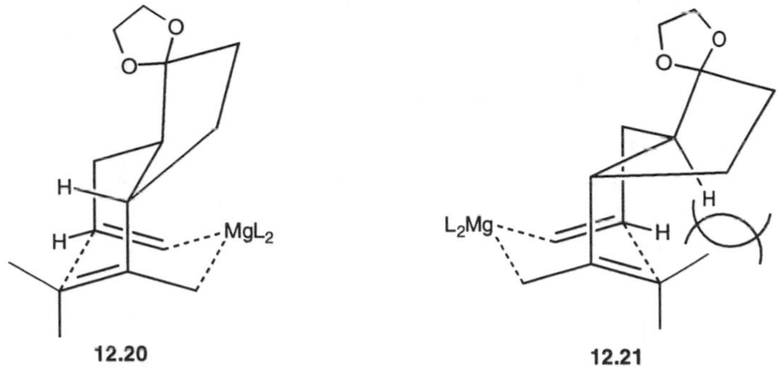

**FIGURE 12.8.** Possible transition states for a "magnesium-ene" reaction.

Chelation control can be used to obtain high stereoselectivity (Scheme 12.51).[569-571] Chirality can also be induced from the Lewis acid catalyst (*cf.* Scheme 12.48).[543,572,573] A silicon group can be used to promote a "magnesium-ene" reaction.[574-576]

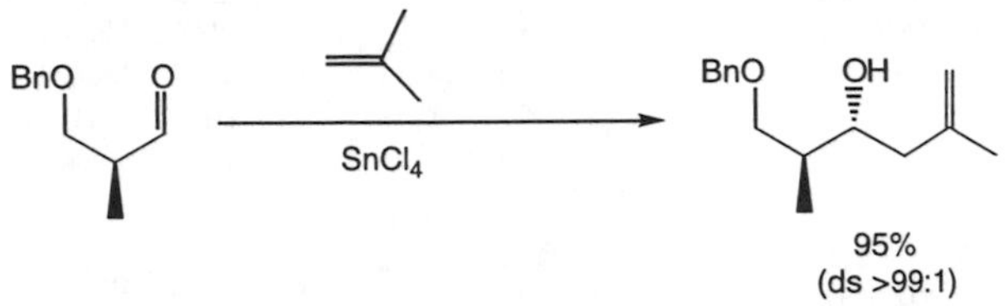

$SnCl_4$
95% (ds >99:1)

**SCHEME 12.51.**

Singlet oxygen can react with alkenes by an ene reaction to provide allyl alcohols.[577] In a synthetic example, the allyl alcohols formed by an ene reaction could be cyclized selectively to give the (–)-rose oxides (**12.22**) by an $S_N2'$ reaction (Scheme 12.52).[578]

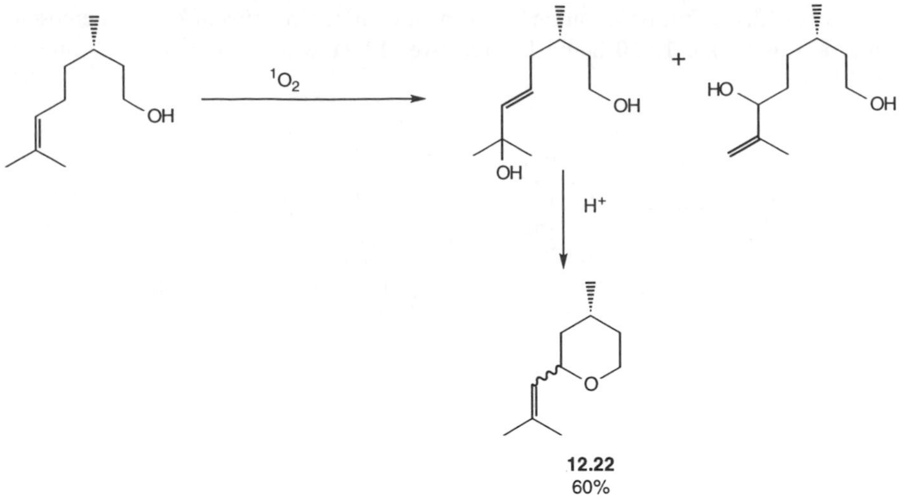

$^1O_2$

OH OH HO OH OH

$H^+$

O

**12.22**
60%

SCHEME 12.52.

## 12.4. DIPOLAR CYCLOADDITIONS

This class of reactions[3,4] is closely related to the Diels-Alder reaction, but allows for the preparation of a five-membered adduct including cyclopentane derivatives.[579-588]

Substituted furans readily undergo condensation with oxyallyl cations to form a bicyclo[3.2.1]oct-6-en-3-one system.[589-592] A variety of transformations can be performed on these adducts including alkylation adjacent to the carbonyl moiety.[593-595] An example of the use of an adduct for the preparation of the Prelog-Djerassi lactone (**12.23**) is given in Scheme 12.53.[596] Note the excision of the acetyl appendage, ring cleavage and the relatively large number of functional group manipulations.

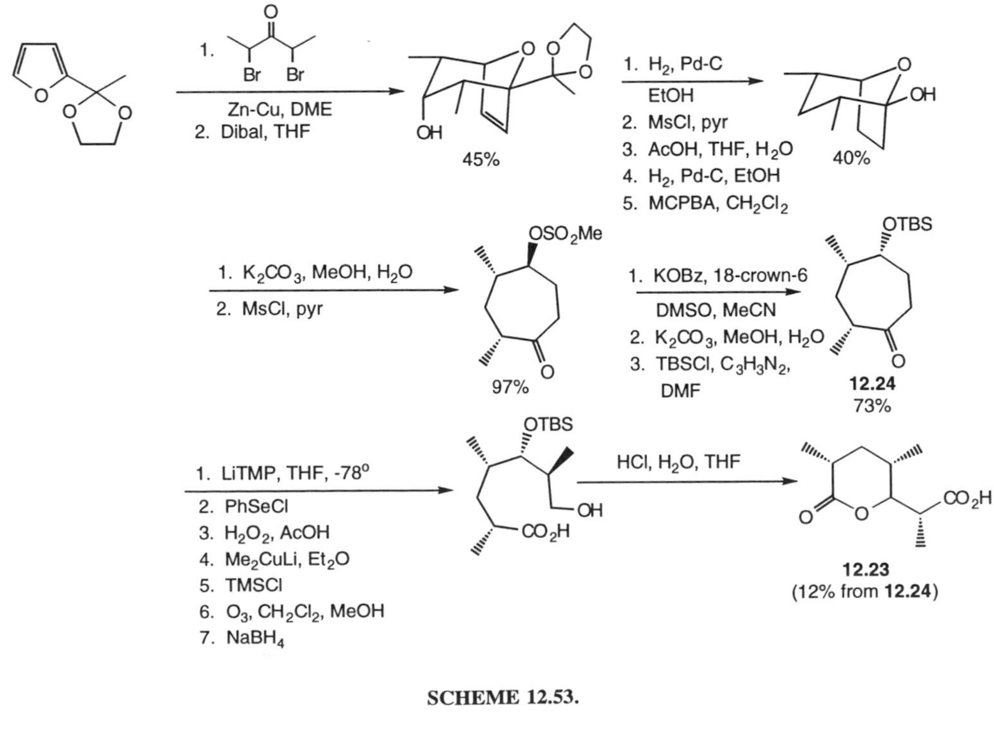

1. Br Br (dibromoketone)
Zn-Cu, DME
2. Dibal, THF

45%

1. $H_2$, Pd-C, EtOH
2. MsCl, pyr
3. AcOH, THF, $H_2O$
4. $H_2$, Pd-C, EtOH
5. MCPBA, $CH_2Cl_2$

40%

1. $K_2CO_3$, MeOH, $H_2O$
2. MsCl, pyr

$OSO_2Me$

97%

1. KOBz, 18-crown-6, DMSO, MeCN
2. $K_2CO_3$, MeOH, $H_2O$
3. TBSCl, $C_3H_3N_2$, DMF

OTBS

**12.24**
73%

1. LiTMP, THF, -78°
2. PhSeCl
3. $H_2O_2$, AcOH
4. $Me_2CuLi$, $Et_2O$
5. TMSCl
6. $O_3$, $CH_2Cl_2$, MeOH
7. $NaBH_4$

OTBS OH $CO_2H$

HCl, $H_2O$, THF

$CO_2H$

**12.23**
(12% from **12.24**)

SCHEME 12.53.

As furan can be regarded as a 1,4-dicarbonyl compound, the use of the latter compounds can be considered an extension of this approach. The use of a carbon dipole then provides functionalized seven-[597-602] and eight-membered heterocycles,[603-605] as exemplified by the reaction of the bissilyl enol ether **12.25** (Scheme 12.54).[586]

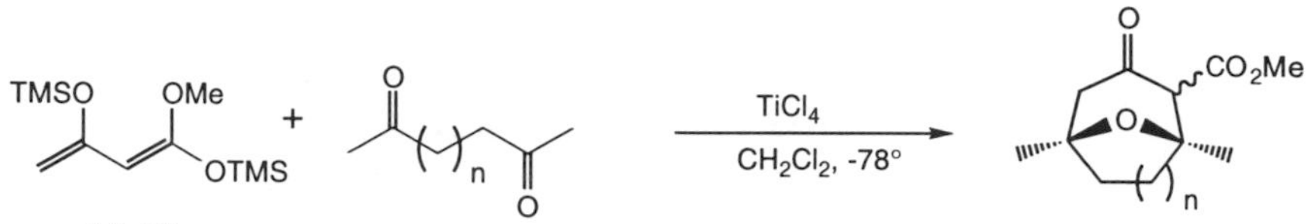

**SCHEME 12.54.**

Use of a homoenolate equivalent with an acetylenic ester provides an entry to cyclopentenones by a formal [3 + 2] cycloaddition, but there is no stereochemistry to control.[606]

Condensation of a nitrile oxide with a chiral allylic ether usually affords an isoxazoline with selectivity for the *pref*-isomer. This selectivity increases with the size of the alkyl substituent and is insensitive to the size of the allyl oxygen substituent. Allyl alcohols tend to form the *parf*-isomer preferentially, but the selectivity is low. These observations have been interpreted, through calculation, that allyl ethers prefer a transition state conformation with the alkoxy substituent inside and the alkyl substituent occupying the *anti* ($R^1$) position (Figure 12.9).[607] In contrast, allyl alcohols react preferentially with the hydroxyl group outside (i.e., $R^2$ = OH). An alternative model with a Felkin-type transition state has also been proposed,[608-612] while intramolecular reactions can be rationalized in terms of allylic strain.[613]

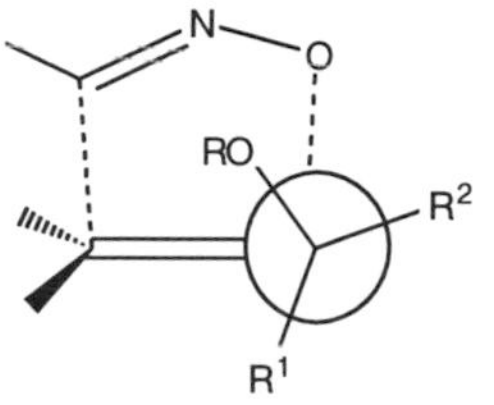

**FIGURE 12.9.** Transition state for 1,3-dipolar cycloadditions.

Use of an allylsilane, to improve selectivity in the cycloaddition through the introduction of a large group, still gave a mixture of adducts (Scheme 12.55).[614,615] Unmasking the isoxazole to the β-hydroxy ketone allows for a Peterson olefination reaction.

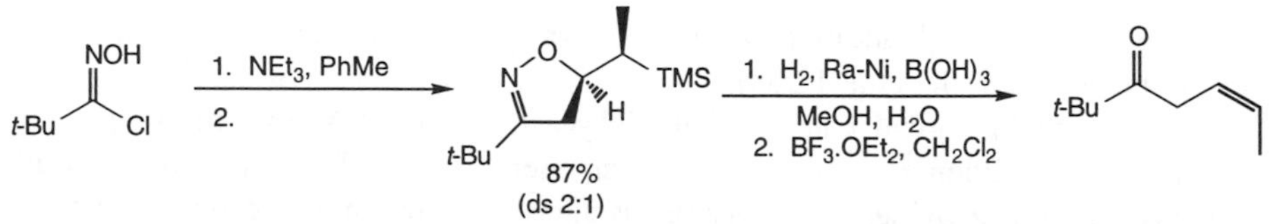

**SCHEME 12.55.**

The condensation of chiral acrylate esters with nitrile oxides shows modest diastereoselectivity.[32,616] This problem has been circumvented by the use of a chiral sultam **12.26** (Scheme 12.56),[567,617-619] or by an intramolecular approach.[620] Vinyl sulfur compounds can also act as dipolarophiles,[621,622] even with unactivated alkenes,[623] but selectivity may be low.[624] Unactivated alkenes can be functionalized by such a dipolar addition; *Z*-alkenes give good regioselectivity, whereas the *E*-isomers give mixtures. Only control of relative stereochemistry is available by this approach,[623] although use of a chromium complex derived from a styrene substrate does provide some selectivity.[625]

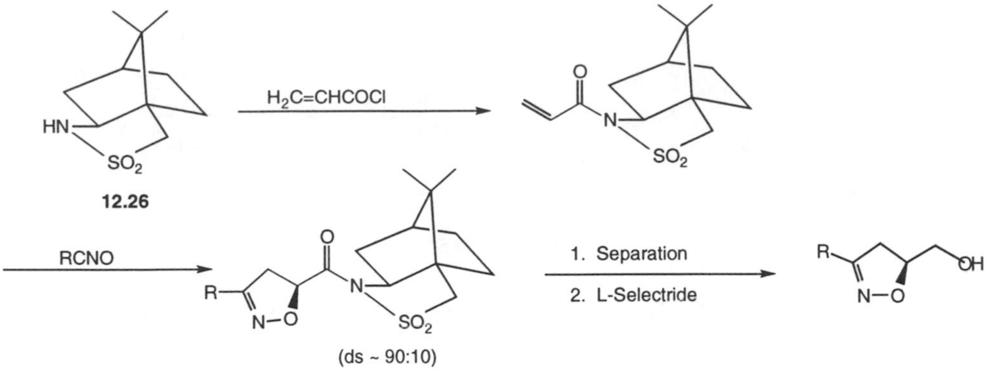

HN SO2 12.26 H2C=CHCOCl RCNO (ds ~ 90:10) 1. Separation 2. L-Selectride

**SCHEME 12.56.**

In some cases, high diastereoselectivities can be observed when the nitrone contains two asymmetric centers (vide infra).[626,627] This should be noted in the context that achiral nitrones react with, for example, α,β-unsaturated esters with high selection of relative stereochemistry,[628,629] but the use of a chiral nitrone usually results in almost no control of absolute configuration,[244,630,631] even for an intramolecular reaction.[632,633] This selectivity problem is also observed for carbon dipoles. Thus, palladium catalyzed cycloaddition of the acetoxysilane **12.27**[587,634] does not see useful induction when chiral phosphine ligands are employed on the metal catalyst,[635,636] or a chiral auxiliary is part of the dipolarophile.[60,107] However, addition of the derivative to an acceptor that has a relatively rigid conformation does provide some selectivity (Scheme 12.57).[637]

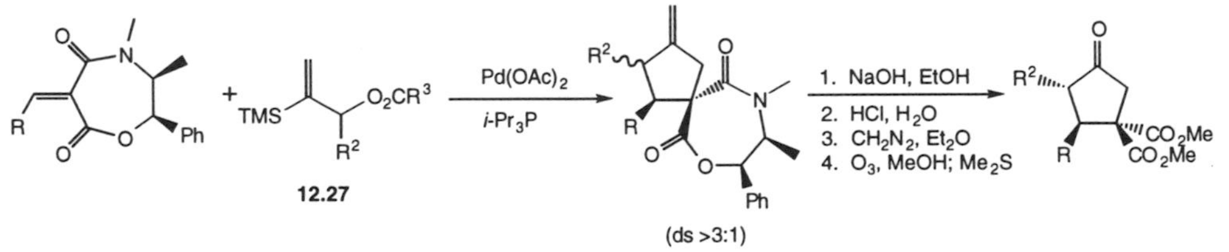

TMS O2CR3 R2 12.27 Pd(OAc)2 *i*-Pr3P (ds >3:1) 1. NaOH, EtOH 2. HCl, H2O 3. CH2N2, Et2O 4. O3, MeOH; Me2S CO2Me CO2Me

**SCHEME 12.57.**

This type of [3 + 2] cycloaddition has been shown, through kinetic observations, to follow a stepwise mechanism (Scheme 12.58).[583]

The use of the biological catalyst, baker's yeast, has been shown to bring about an asymmetric cycloaddition between achiral components [*14.3*].[638,639] This dipolar addition reaction based on nitrile oxides is very useful, as the isoxazoline moiety can be converted into a wide variety of functional groups under relatively mild conditions.[640-661] As an example, the addition can be used to prepare β-hydroxy ketones (Scheme 12.59).[640]

The isoxazoline group can also be used as a method of protection, and can even influence relative stereochemistry through chelation control (Scheme 12.60).[662,663] The high diastereoselection with a magnesium counterion is attributed to formation of a chelate (**12.28**) in the transition state.[662]

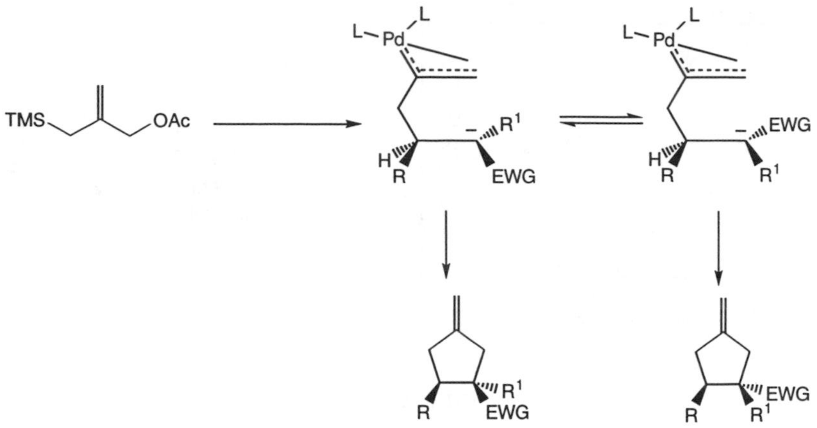

**SCHEME 12.58.**

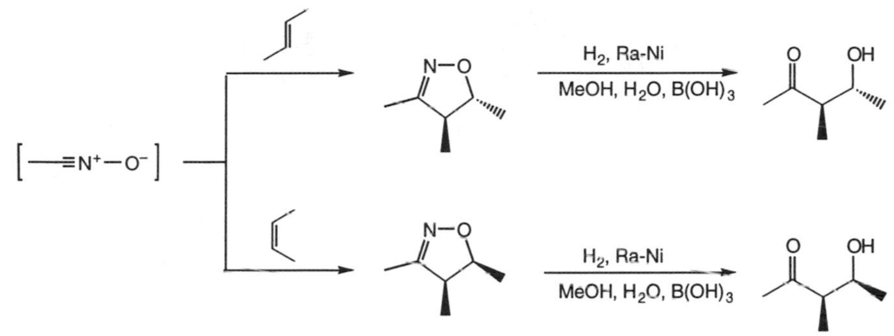

**SCHEME 12.59.**

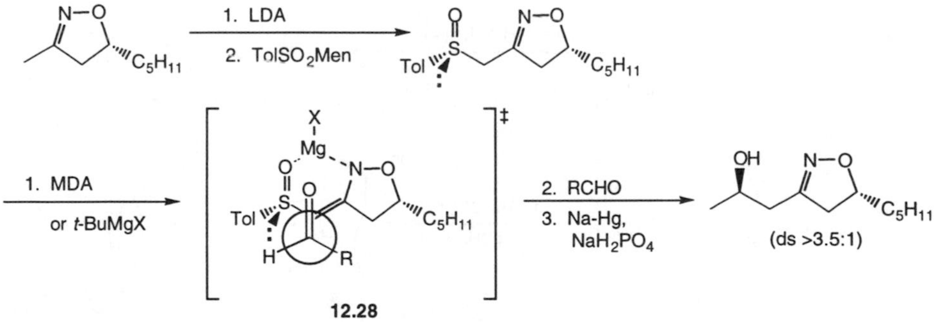

**SCHEME 12.60.**

2-Deoxy-D-ribose (**12.29**) has been prepared by condensation of an allyl ether **12.30** with a nitrile oxide (Scheme 12.61).[608,610,655,664] The nature of the nitrile oxide can be varied significantly; Scheme 12.61 shows three separate methods.

Other examples of carbohydrate synthesis are provided by derivatives of lividosamine (**12.31**) (Scheme 12.62),[665,666] D-allosamine hydrochloride,[666] and daunosamine (*cf.* **12.32**) (Scheme 12.63), where the dipole contains two asymmetric centers;[667] high diastereoselection is observed, and other dipolarophiles give similar selectivity.[653,668] The latter example illustrates that nitrones also participate in 1,3-dipolar cycloadditions,[669] which can be used to prepare alcohols with moderate asymmetric induction.[624,670-674]

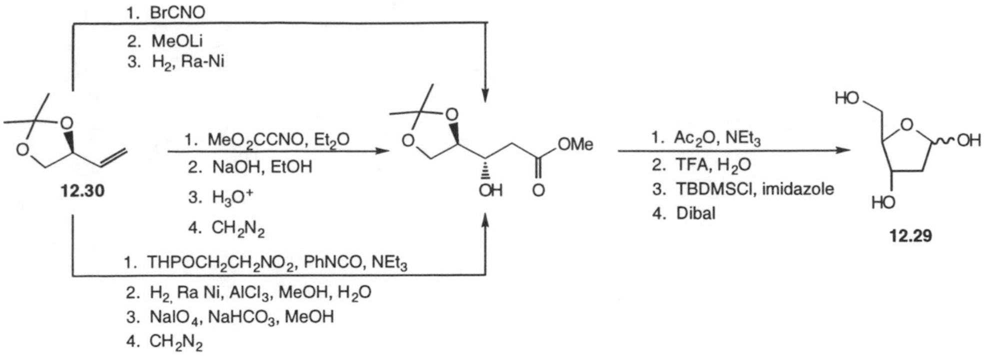

1. BrCNO
2. MeOLi
3. $H_2$, Ra-Ni

12.30

1. $MeO_2CCNO$, $Et_2O$
2. NaOH, EtOH
3. $H_3O^+$
4. $CH_2N_2$

OMe OH O

1. THPOCH$_2$CH$_2$NO$_2$, PhNCO, NEt$_3$
2. $H_2$, Ra Ni, $AlCl_3$, MeOH, $H_2O$
3. $NaIO_4$, $NaHCO_3$, MeOH
4. $CH_2N_2$

1. $Ac_2O$, $NEt_3$
2. TFA, $H_2O$
3. TBDMSCl, imidazole
4. Dibal

HO OH HO

12.29

**SCHEME 12.61.**

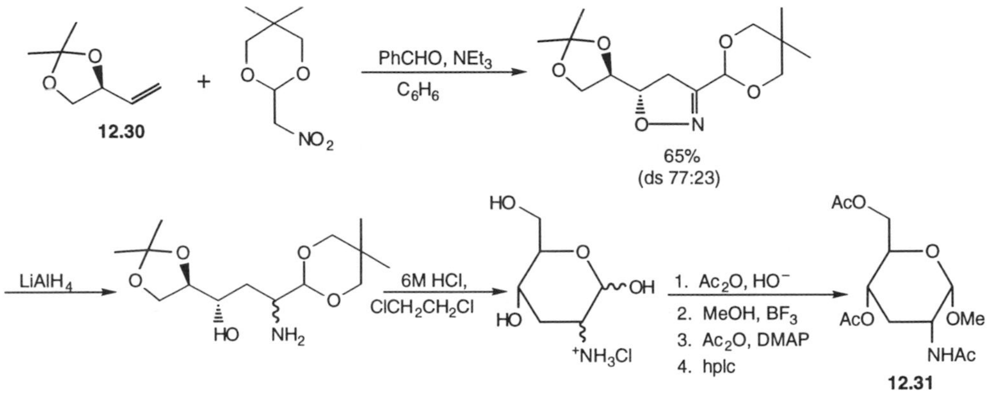

12.30 + $NO_2$

PhCHO, $NEt_3$
$C_6H_6$

65%
(ds 77:23)

$LiAlH_4$

HO $NH_2$

6M HCl,
$ClCH_2CH_2Cl$

HO OH HO $^+NH_3Cl$

1. $Ac_2O$, $HO^-$
2. MeOH, $BF_3$
3. $Ac_2O$, DMAP
4. hplc

AcO AcO OMe NHAc

12.31

**SCHEME 12.62.**

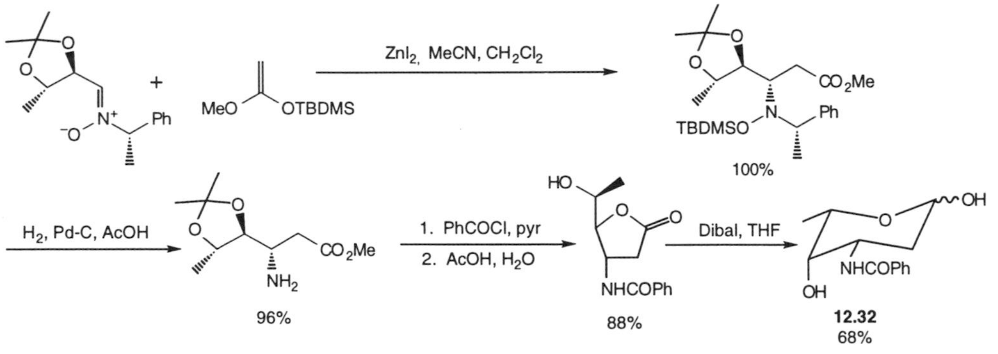

+ MeO OTBDMS

$ZnI_2$, MeCN, $CH_2Cl_2$

$CO_2Me$ TBDMSO N Ph

100%

$H_2$, Pd-C, AcOH

$CO_2Me$ $NH_2$

96%

1. PhCOCl, pyr
2. AcOH, $H_2O$

HO NHCOPh

88%

Dibal, THF

OH NHCOPh OH

12.32
68%

**SCHEME 12.63.**

Carbohydrate derivatives are available through the use of furan as a dipolarophile. This methodology has been used to prepare an analogue of 5-epi-norjirimycin (**12.33**) (Scheme 12.64).[647]

Furanones are also useful dipolarophiles in this type of approach,[675-681] as are other unsaturated lactones.[682,683]

However, an intramolecular version of this reaction with furanoid compounds can provide low diastereoselectivity even though complete regiochemical control is observed.[221] This is in contrast to other intramolecular examples which give very high selectivity in favor of the *syn*-fused five-membered products (Scheme 12.65).[620,684,685]

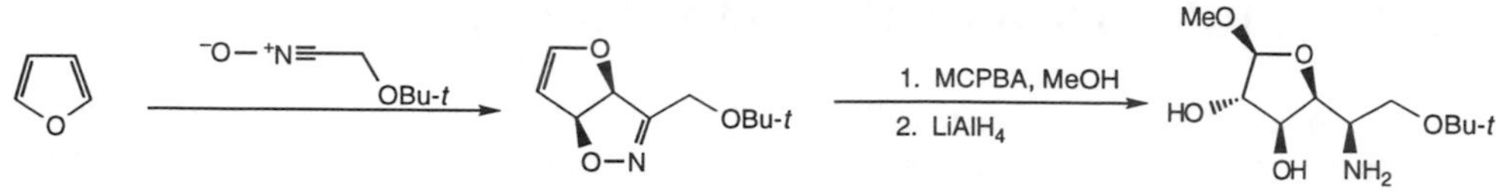

SCHEME 12.64.

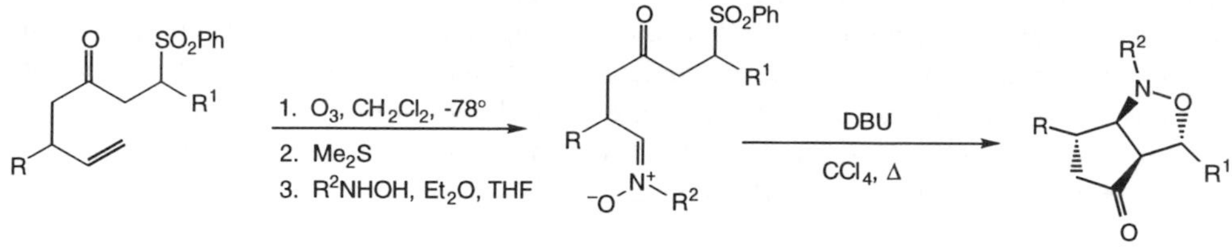

SCHEME 12.65.

With a bicyclic alkene, such as 2-substituted 7-oxabicyclo[2.2.1]hept-5-enes (**12.34**) or norbornadienes, addition of nitrogen dipoles results in formation of the *cis*-product, but the regioselectivity depends upon the electronic and steric nature of the 2-substituents;[195,314,686,687] this effect is also seen with chiral dipolarophiles.[688]

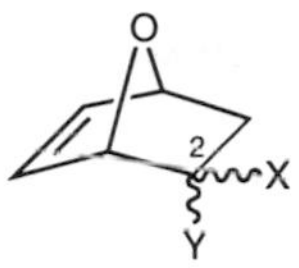

**12.34**

In addition to nitrile oxides, imines derived from amino esters will undergo cycloadditions with α,β-unsaturated esters. The dipoles can be generated thermally,[689] through use of a metal salt,[690-692] such as silver or lithium, and triethylamine in THF. The regiochemistry can be reversed through the use of a titanium(IV) complex (Scheme 12.66).[693]

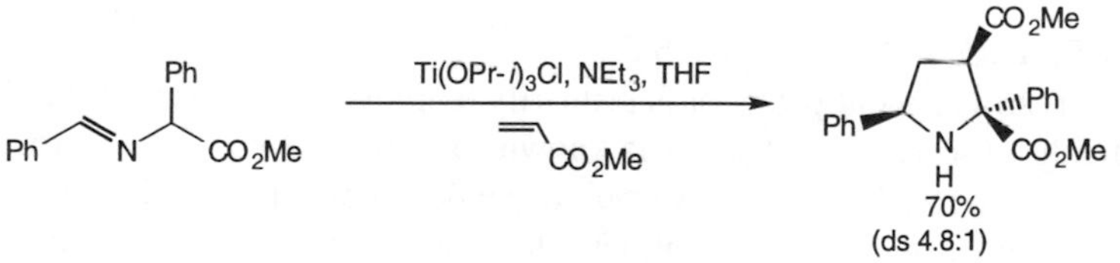

SCHEME 12.66.

The high diastereoselectivity of the cycloaddition, when a chiral template is derived from a carbohydrate, allows for a simple entry to higher sugars, such as the pyranopyran **12.35** (Scheme 12.67).[609,611,694-699]

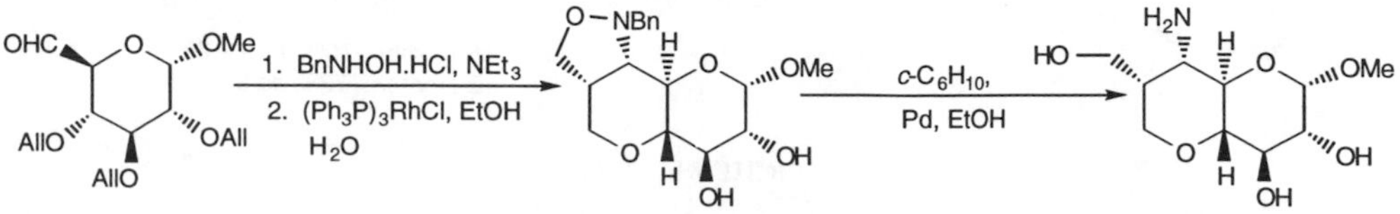

SCHEME 12.67.

## 12.5. [2,3] SIGMATROPIC REARRANGEMENTS

[2,3]-Sigmatropic rearrangements,[700,701] of which the Wittig rearrangement is an example, allow for the transfer of stereochemistry.[702,703] These rearrangements proceed through a five-membered transition state, which means that heteroatoms have to be incorporated into the substituents to accommodate lone pairs or charges (Scheme 12.68).[700] In addition, there is a greater flexibility in the transition state when compared to a [3,3] sigmatropic rearrangement.[704] The most useful examples of this class of reactions are illustrated by the Wittig and Evans rearrangement reactions. Numerous other examples for a wide variety of systems are known, but the electronic and stereochemical arguments employed for these two major reactions allow similarities and extrapolations to be drawn with confidence.[700] The [2,3]-rearrangement can often be accomplished at low temperature, minimizing competition from a [3,3]-rearrangement or nonconcerted process.[705-708]

Type I

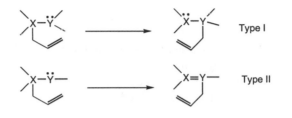

Type II

SCHEME 12.68.

As one example, a [2,3]-rearrangement of an allyl alcohol to an amide has been accomplished to obtain optically active isoprenoid synthons (Scheme 12.69).[703]

OH

$Me_2NCH(OMe)_2$

xylene, Δ

$Me_2N$

O

SCHEME 12.69.

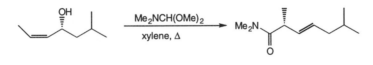

### 12.5.1. WITTIG REARRANGEMENT

The Wittig rearrangement protocol is primarily employed to convert an allyl alcohol to a homoallyl alcohol (Scheme 12.70), but is known for other systems.[709] The resultant stereochemistry of the double bond is, however, dependent upon the particular system, as the transition state geometry plays an important role to determine the reaction outcome (Scheme 12.71).[702,705,708-717]

$SnBu_3$

O

*i*-Pr

O

Bn

BuLi, THF, HMPA

*i*-Pr

O

Bn

OH

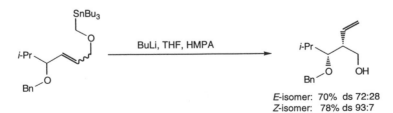

*E*-isomer: 70% ds 72:28
*Z*-isomer: 78% ds 93:7

SCHEME 12.70.

**SCHEME 12.71.**

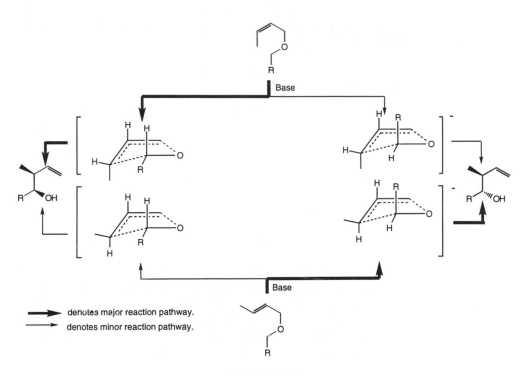

The diastereoselectivity arises from the relative stabilities of possible transition state geometries. In general terms, use of a *Z*-alkene will result in the *parf*-isomer; *E*-alkenes show *pref*-selectivity but this does depend on the nature of the substituent. In addition, the substituent can play a role in stereoselection, as it will influence the structure of the carbanion. The presence of a chiral substituent can lead to high asymmetric induction.[718-724]

Extension of the Wittig rearrangement to allyl ethers derived from secondary allyl alcohols, rather than primary, introduces a new stereochemical problem — the configuration of the resultant alkene.[725,726] The presence of an allylic stereocenter usually results in the formation of the *pref* product isomer (Scheme 12.72). The use of the *Z*-isomer as substrate usually results in higher selectivity for the *pref*-isomer than when the *E*-alkene is the reactant; these results have been rationalized by the transition state models **12.36** and **12.37** for the *Z*- and *E*-alkenes, respectively (Figure 12.10).[727-730]



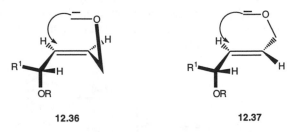

**FIGURE 12.10.** Potential transition states for the Wittig rearrangement.

The use of a substituted alkene provides the opportunity for the rearrangement to yield the *E*-alkene as the major product; presumably the substituent on the alkene prefers to adopt a pseudoequatorial position.[700,731]

When a chiral substituent is present in the allyl moiety, 1,3-asymmetric induction is observed (Scheme 12.72).[718,725,732]

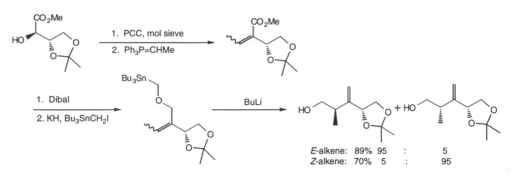

1. PCC, mol sieve
2. $Ph_3P{=}CHMe$

1. Dibal
2. KH, $Bu_3SnCH_2I$

BuLi

| | | | |
|---|---|---|---|
| *E*-alkene: | 89% | 95 : | 5 |
| *Z*-alkene: | 70% | 5 : | 95 |

**SCHEME 12.72.**

Although the rearrangement does not seem to be dependent on the metal counterion,[733-735] the stereochemical outcome is dependent upon the electronic and steric demands within the substrate itself,[727,728,732,736,737] and the stereochemistry of the incipient anionic center.[738] However, the use of a metal, such as zirconium that can complex to a number of oxygen atoms within the substrate, and hence hold it in a specific conformation, will influence the stereochemical outcome of a rearrangement.[721,722,739] An example is provided by the synthesis of a family of insect pheromones (Scheme 12.73).[740]

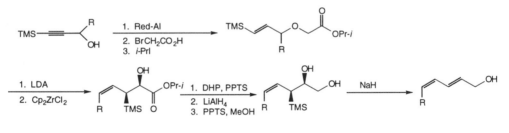

1. Red-Al
2. $BrCH_2CO_2H$
3. *i*-PrI

1. LDA
2. $Cp_2ZrCl_2$

1. DHP, PPTS
2. $LiAlH_4$
3. PPTS, MeOH

NaH

**SCHEME 12.73.**

Titanium has also been used successfully to coordinate a substrate to a specific conformation.[737] Another example of the effect of the metal counterion is provided by the rearrangement of the oxazoline derivative **12.38**. With the methyl ether (**12.38**, R = Me) and a lithium counterion, the ester **12.39** was formed as the *R*-isomer (**12.39R**) with an ee of 38 to 78%. With the alcohol (**12.38**, R = OH) and a potassium counterion, the stereochemical outcome was reversed to provide the *S*-product. The addition of 18-crown-6, however, reverted the product to the *R*-configuration (Scheme 12.74).[712,720,724]

**12.38** —Base→ (R = Me) **12.39R**; (R = H) **12.39S**

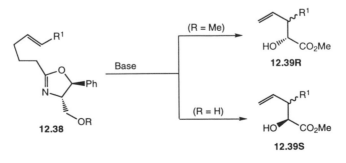

**SCHEME 12.74.**

The use of a chiral base with acyclic substrates failed to give optically active [2,3]-sigmatropic rearrangement products,[741,742] although enantioselectivity has been observed with cyclic substrates.[743,744]

The use of a functionalized propargyl ether as substrate allows for the stereoselective preparation of allenyl carbinols through the rearrangement (Scheme 12.75).[741,745]

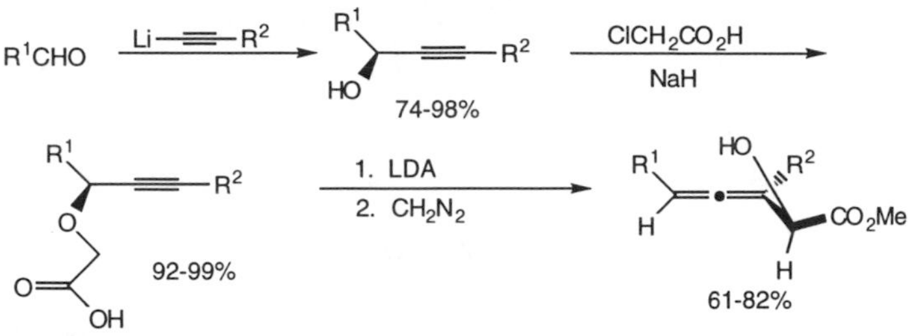

SCHEME 12.75.

The rearrangement has also been extended to bis-allylic ethers, although a [1,2]-shift can compete.[715,746,747] This latter problem does not arise for allylic propargyl ethers.[701,748-751] The relative stereochemistry at the saturated centers is unaffected (Scheme 12.76).[714,726,752]

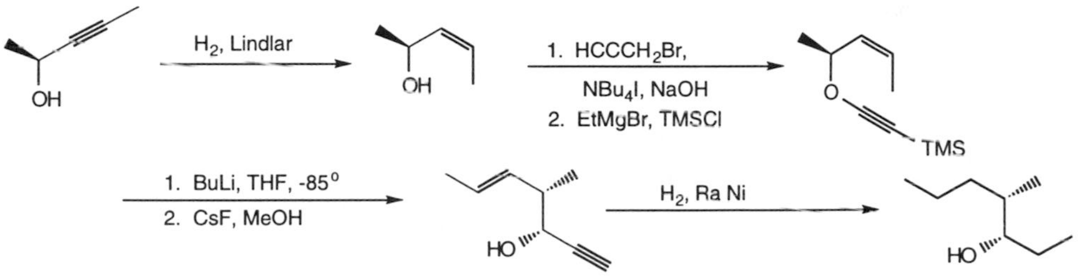

SCHEME 12.76.

A further variant involves the propargylic allylic ethers **12.40**, where the stereochemical outcome of the reaction can be controlled by the terminal alkyne group (Scheme 12.77).[701,753]

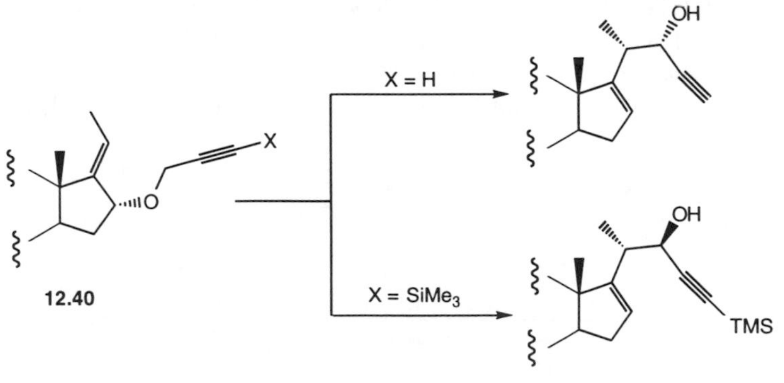

SCHEME 12.77.

A further extension of the protocol provides a method to selectively protected 1,2-diols, but is limited by the necessity for an anion stabilizing group to be present.[754] 1,3-Polyols are also available from a strategy that employs a Wittig rearrangement.[755-757]

Wittig rearrangement of allylic glycolic esters **12.41**, either in the presence of base or silyl triflate, allows for the stereoselective synthesis of γ,δ-unsaturated-α-hydroxy esters **12.42** (Scheme 12.78).[713,714,739,748]

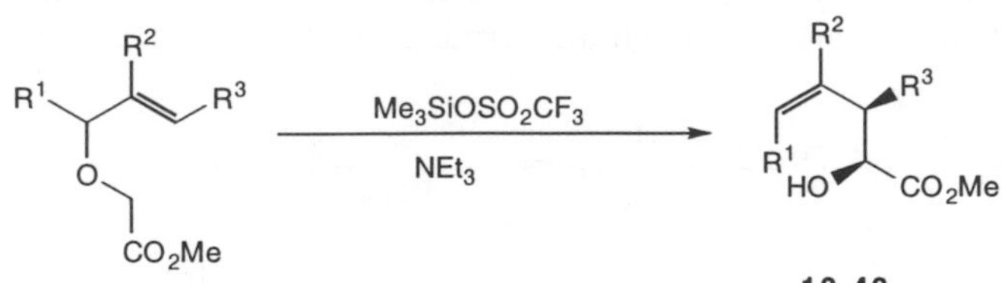

SCHEME 12.78.

The utility of the Wittig rearrangement is readily demonstrated by the diversity of synthesis which relies on the methodology,[740,743,744,749,750,753,758-770] ranging from large antibiotics,[759,765] to dihydrofurans,[741,771] to steroid derivatives,[753,762,764] to the Prelog-Djerassi lactone ester (**12.23**) (Scheme 12.79).[758,759]

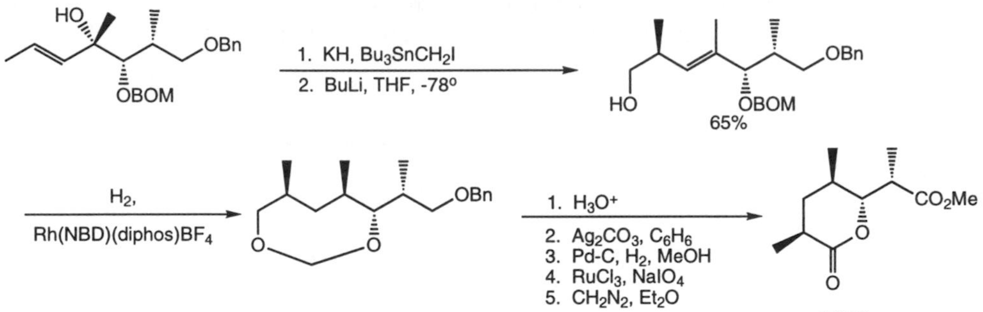

SCHEME 12.79.

## 12.5.2. EVANS REARRANGEMENT

In addition to the Wittig rearrangement, another [2,3]-reaction, the Evans rearrangement,[772] can be extended to prepare functionalized allyl alcohols.[773,774] The Evans rearrangement is the conversion of an allylic sulfoxide to an allyl alcohol.[772] The reverse reaction is also available to convert an allyl alcohol to the analogous sulfoxide with allylic rearrangement. The "push-pull" nature of the two reactions (Scheme 12.80), coupled with the chirality transfer, provides a method to invert an allyl alcohol with isomerization of the alkene moiety (Scheme 12.81).[700,775]

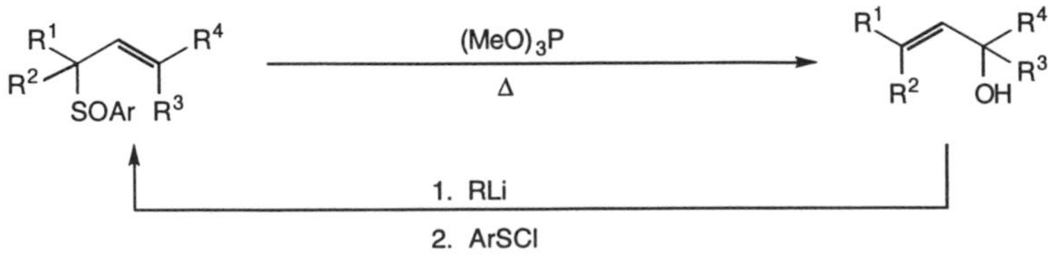

SCHEME 12.80.

The overall reaction involves a [2,3]-sigmatropic rearrangement that can be driven to completion by the addition of a thiophile, such as trimethylphosphite (Scheme 12.80).[776-780] This strategy allows the chemistry of the allyl phenyl sulfoxide, or other sulfur precursor, to be exploited before the allyl alcohol is unmasked.[772,781-785] The use of the addition of phenylsulfenyl chloride to an alkene, followed by the elimination of hydrogen chloride and rearrangement, provides a useful oxidative method to allyl alcohols.[783,786] As the sigmatropic rearrangement is concerted, stereochemical transfer can be achieved.[787,788] The reverse reaction involves the treatment of an allyl alcohol with a base followed by arylsulfenyl chloride to produce an allyl sulfoxide.[704,789-791] A variation of the approach allows for the allylic amination of *Z*-alkenes.[792,793]

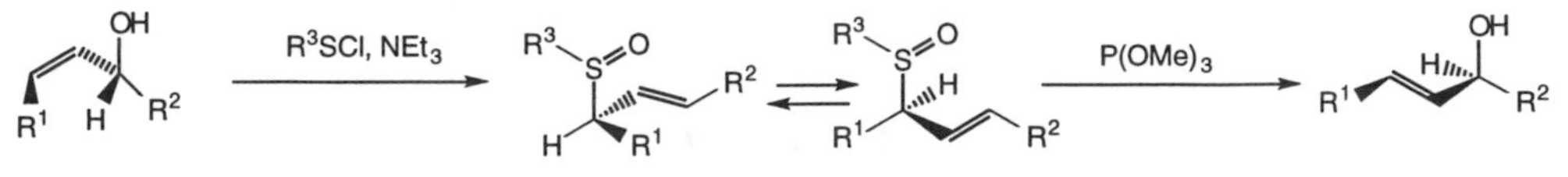

$R^3SCl$, $NEt_3$; $P(OMe)_3$

**SCHEME 12.81.**

There is also a potential for neighboring group participation in the Evans rearrangement, although a two-step process — rearrangement followed by acyl group transfer — cannot be ruled out (Scheme 12.82).[794]

MCPBA

**SCHEME 12.82.**

An example of the rearrangement in synthesis is provided by the stereoselective introduction of the 15α-hydroxy group in an approach to carbocyclin analogs (Scheme 12.83).[787,795-797]

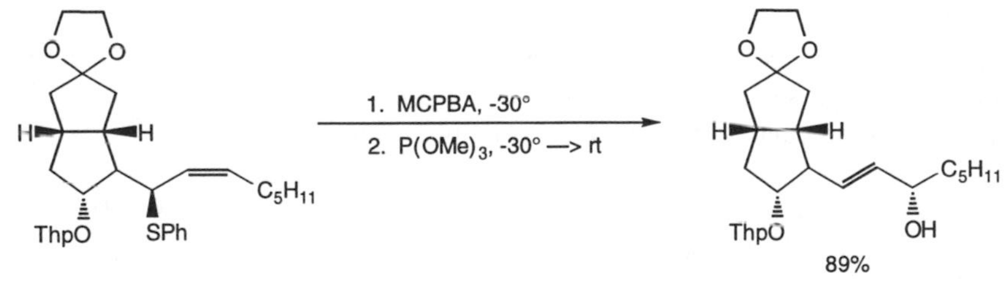

1. MCPBA, -30°
2. $P(OMe)_3$, -30° —> rt

89%

**SCHEME 12.83.**

The rearrangement has a high preference to afford the *E*-product when a substituent is branched at the β-position from the sulfinyl group [$R^1$ is large in Scheme 12.80].[774,777,781,798,799] This is due to $R^1$ preferring to occupy the pseudoequatorial position in the transition state.[700,772]

An analogous [2,3]-rearrangement of allyl sulfites to allyl sulfonates has also been observed with a palladium(II) catalyst.[800] There are also many examples with selenium in place of the sulfur atom.[700,801]

## 12.6. PHOTOCHEMICAL REACTIONS

Photochemical reactions are often dismissed by synthetic chemists, as they do not have the required equipment, or consider them difficult to scale-up. The plethora of reactions available by this approach often provides products not directly accessible by other methods.[303,802] The first in-roads to asymmetric synthesis by photochemical means were made in the solid phase or organized assemblies;[803] few examples were known in solution.[804]

Circular polarized light is a chiral agent; it has the potential to induce asymmetric synthesis or degradation, but to date, useful ee's have not been obtained.[802]

In contrast, a [2 + 2] photocycloaddition can provide useful selectivity, as illustrated by the reaction of *trans*-stilbene with chiral dialkyl fumarates (Scheme 12.84).[170,805]

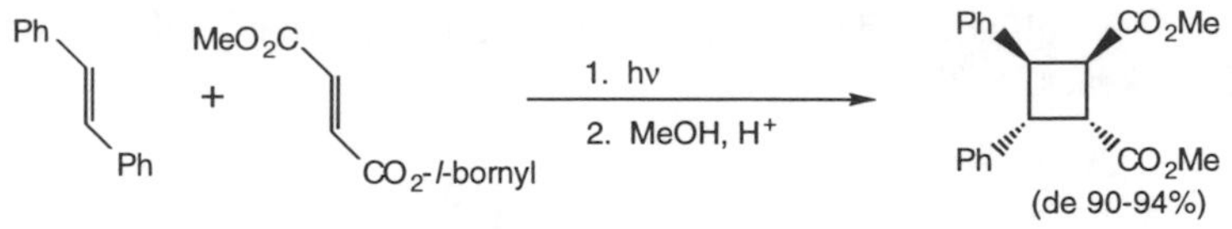

**SCHEME 12.84.**

The two alkenes can also be tethered (Scheme 12.85), but the selectivity does depend on the nature of the chiral backbone. With D-mannitol, 1,6-dibenzoate 2,3,4,5-tetracinnamate and (–)-di-*O*-methyl-L-erythritol 1,2-dicinnamate give the highest de's (~85%).[806-808]

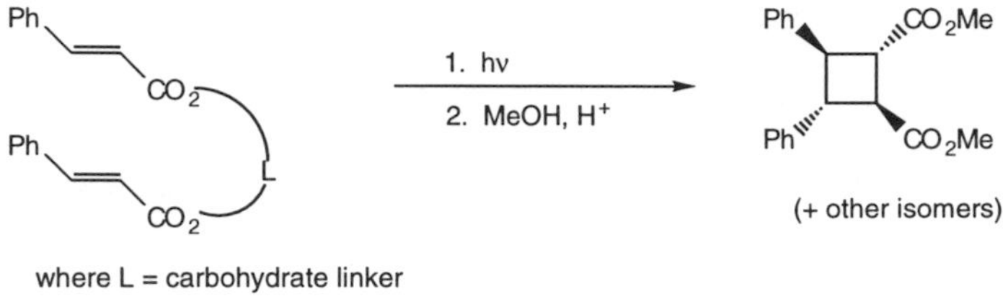

**SCHEME 12.85.**

The highest degrees of induction for photochemical reactions, are seen with chiral substrates (vide infra). The use of a chiral solvent for a pinacol reaction, for example, resulted in ee's of <25%.[809-811]

Photochemically induced annelation of an alkene with methyl diformylacetate (Scheme 12.86) allows for high stereoselectivity but poor regioselection.[812] This limitation has been overcome to a certain degree in a cyclic case, by use of an allylsilane.[813]

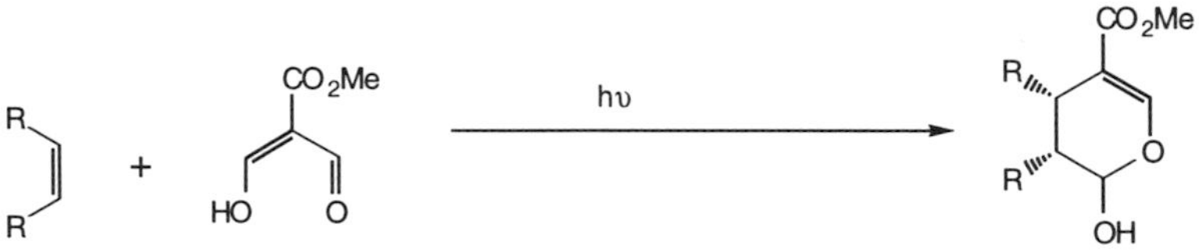

**SCHEME 12.86.**

Photochemical addition of diethylketene acetal to a phenylglyoxylate ester affords good induction, although chemical yields are moderate at best.[814] Malic acid is available from an asymmetric [2 + 2] cycloaddition involving ketene and catalyzed by quinidine.[815] Photocycloadditions of aldehydes to furans provide adducts with a plethora of functionality.[816-821]

One common reaction for photochemistry is to induce a [2 + 2] cycloaddition. An example is provided by a synthesis of a subunit of cytochalasin C (Scheme 12.87). The differentiation between the two products arises from facial selectivity.[822]

Enantiomeric excesses of up to 70% have been observed for the photochemical deconjugation of α,β-unsaturated esters in the presence of a chiral proton donor (Scheme 12.88) [*5.4.2*].[823]

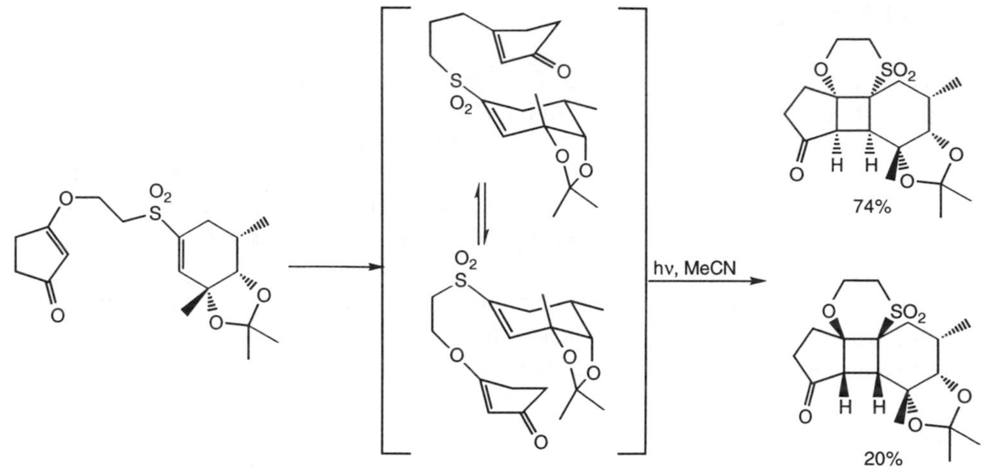

SCHEME 12.87.

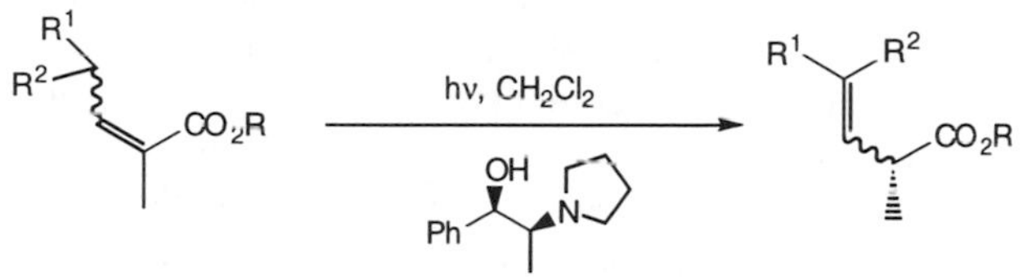

SCHEME 12.88.

## 12.7. OTHER REACTIONS

Many types of pericyclic and cycloaddition reactions have been documented. It is not feasible to list examples of all of the types within this chapter. One other class of reactions, however, does bear mention — the [2 + 2] cycloaddition, especially of a ketene, with an electron-rich alkene, although other systems can also be used.[824,825] The resultant stereochemistry is *cis*, and functionalized ketenes have been employed. Usually, the ketene is generated *in situ* as illustrated in Scheme 12.89.[826]

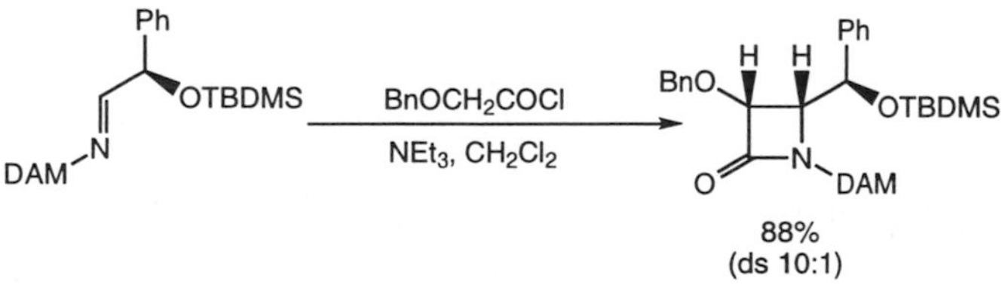

SCHEME 12.89.

The condensation of an imine with a Reformatsky type reagent can be considered a [2 + 2] cycloadditon [*7.2.7*], while the photochemical [2 + 2] has already been covered [*12.6*].[170,807,814,815,818,819,827-851]

In addition, tandem examples of cycloaddition reactions have been omitted; in these, the product of one pericyclic reaction is the starting material for the next. As the two steps can be separated to predict the overall reaction outcome, there is little need for further discussion.[747]

## 12.8. SUMMARY

This chapter describes the most useful types of the pericyclic reaction; it was not intended to be comprehensive nor illustrative of the power of "tandem reactions." With the potential to form two carbon–carbon bonds in a single step, and control enantioselectivity by external means, such as through the use of a Lewis acid catalyst, the Diels-Alder reaction is a very powerful weapon in the synthetic arsenal. Even when existing chirality within a substrate is used to control stereochemistry, the approach still has awesome power. The ability to change the product relative stereochemistry through a simple change in the substrate, such as the *cis*-product is available from a *cis*-alkene while the *trans*-product is available from a *trans*-dienophile, enhances the scope of the reaction.

Although a large number of asymmetric Diels-Alder reactions are known, great variations in asymmetric induction can occur with minor changes in substituents or Lewis acids. Despite the large amount of work in this area, much of this cannot be used for comparison as more than one parameter was changed, or too few examples were used to show the versatility of a given system. The recent focus on chiral catalysts holds great promise although, again, they often seem system specific. As our understanding of the many factors that influence and control these cycloadditions increases, we may be able to predict conditions and reagents as well as the amount of variation a given catalyst, dienophile, or diene will tolerate. The scale-up of many of these asymmetric cycloadditions will also present problems due to the use of "exotic" reagents as well as the use of low temperatures. However, the potential of chiral catalysts that can be recovered does alleviate some of these problems.

The Claisen rearrangement also is useful for the formation of carbon–carbon bonds. One variant, the Ireland Claisen ester enolate, has shown the power of a pericyclic reaction for the control of product stereochemistry through the simple procedure of enolate geometry.

The ene reaction and [2,3]-dipolar additions have the advantages outlined for the Claisen rearrangement, but also provide access to different types of functionality and classes of compounds. In this regard, many of the reactions described in this chapter are complimentary. All of these reactions can be carried out on large scales. Although those reactions that employ carbanions, such as the Wittig rearrangement, or rely on complex formation, may also carry the burden of needing low temperature.

Photochemical reactions can provide the opportunity for unique reactions. Some are efficient, but scale-up can present some unique problems even within the laboratory scale. For this reason alone, synthetic chemists tend to shy away from this class of reactions, and may, in some instances, be missing a useful and efficient transformation.

## 12.9. REFERENCES

1. Houk, K. N.; Li, Y.; Evanseck, J. D. *Angew. Chem., Int. Ed. Engl.* 1992, *31*, 682.
2. Berson, J. A. *Acc. Chem. Res.* 1991, *24*, 215.
3. Desimoni, G.; Tacconi, G.; Barco, A.; Polinni, G. P. In *Natural Product Synthesis Through Pericyclic Reactions;* American Chemical Society: Washington, D. C., 1983; pp 119.
4. Kocovsky, P.; Turecek, F.; Hajicek, J. In *Synthesis of Natural Products: Problems of Stereoselection;* CRC Press: Boca Raton, FL, 1986; Vol. 1.
5. Danishefsky, S. *Acc. Chem. Res.* 1981, *14*, 400.
6. Oppolzer, W. *Angew. Chem., Int. Ed. Engl.* 1984, *23*, 876.
7. Gleiter, R.; Paquette, L. A. *Acc. Chem. Res.* 1983, *16*, 328.
8. Gree, R.; Kessabi, J.; Mosset, P.; Martelli, J.; Carrie, R. *Tetrahedron Lett.* 1984, *25*, 3697.
9. Franck, R. W.; Argade, S.; Subramaniam, C. S.; Frechet, D. M. *Tetrahedron Lett.* 1985, *26*, 3187.

10. Kaila, N.; Franck, R. W.; Dannenberg, J. J. *J. Org. Chem.* 1989, *54*, 4206.
11. Trost, B. M.; O'Krongly, D.; Belletire, J. L. *J. Am. Chem. Soc.* 1980, *102*, 7595.
12. Franck, R. W.; John, T. V.; Olejniczak, K.; Blount, J. F. *J. Am. Chem. Soc.* 1982, *104*, 1106.
13. Hamada, T.; Zenkoh, T.; Sato, H.; Yonemitsu, O. *Tetrahedron Lett.* 1991, *32*, 1649.
14. Ronan, B.; Kagan, H. B. *Tetrahedron: Asymmetry* 1991, *2*, 75.
15. Lopez, R.; Carretero, J. C. *Tetrahedron: Asymmetry* 1991, *2*, 93.
16. Fuji, K.; Tanaka, K.; Abe, H.; Itoh, A.; Node, M.; Taga, T.; Miwa, Y.; Shiro, M. *Tetrahedron: Asymmetry* 1991, *2*, 179.
17. Alonso, I.; Carretero, J. C.; Garcia Ruano, J. L. *Tetrahedron Lett.* 1991, *32*, 947.
18. Arai, Y.; Matsui, M.; Koizumi, T.; Shiro, M. *J. Org. Chem.* 1991, *56*, 1983.
19. Oppolzer, W.; Wills, M.; Kelly, M. J.; Signer, M.; Blagg, J. *Tetrahedron Lett.* 1990, *31*, 5015.
20. Yamamoto, H.; Maruoka, K.; Furuta, K.; Naruse, Y. *Pure Appl. Chem.* 1989, *61*, 419.
21. Furuta, K.; Shimizu, S.; Miwa, Y.; Yamamoto, H. *J. Org. Chem.* 1989, *54*, 1481.
22. Corey, E. J.; Imai, N.; Zhang, H.-Y. *J. Am. Chem. Soc.* 1991, *113*, 728.
23. Terada, M.; Mikami, K.; Nakai, T. *Tetrahedron Lett.* 1991, *32*, 935.
24. Takasu, M.; Yamamoto, H. *Synlett* 1990, 194.
25. Rebiere, F.; Riant, O.; Kagan, H. B. *Tetrahedron: Asymmetry* 1990, *1*, 199.
26. Kaufmann, D.; Boese, R. *Angew. Chem., Int. Ed. Engl.* 1990, *29*, 545.
27. Oppolzer, W.; Kurth, M.; Reichlin, D.; Chapuis, C.; Mohnhaupt, M.; Moffatt, F. *Helv. Chim. Acta* 1981, *64*, 2802.
28. Lautens, M.; Lautens, J. C.; Smith, A. C. *J. Am. Chem. Soc.* 1990, *112*, 5627.
29. Narasaka, K.; Iwasawa, N.; Inoue, M.; Yamada, T.; Nakashima, M.; Sugimori, J. *J. Am. Chem. Soc.* 1989, *111*, 5340.
30. Boger, D. L.; Curran, T. T. *J. Org. Chem.* 1990, *55*, 5439.
31. Masamune, S.; Choy, W.; Petersen, J. S.; Sita, L. R. *Angew. Chem., Int. Ed. Engl.* 1985, *24*, 1.
32. Paquette, L. A. In *Asymmetric Synthesis*; J. D. Morrison, Ed.; Academic Press: Orlando, FL, 1984; Vol. 3; pp 455.
33. Nógrádi, M. *Stereoselective Synthesis*; VCH: New York, 1986.
34. Boeckman, R. K.; Nelson, S. G.; Gaul, M. D. *J. Am. Chem. Soc.* 1992, *114*, 2258.
35. Alonso, I.; Carretero, J. C.; Ruano, J. L. G. *J. Org. Chem.* 1993, 3231.
36. Matuoka, K.; Yamamoto, H. *J. Am. Chem. Soc.* 1989, *111*, 789.
37. Evans, D. A.; Miller, S. J.; Lectka, T. *J. Am. Chem. Soc.* 1993, *115*, 6460
38. Carretero, C. J. C.; Ruano, G. J. L.; Lorente, A.; Yuste, F. *Tetrahedron: Asymmetry* 1993, *4*, 177.
39. Jensen, K. N.; Roos, G. H. P. *Tetrahedron: Asymmetry* 1992, *3*, 1553.
40. Hoffmann, H.; Bolte, M.; Berger, B.; Hoppe, D. *Tetrahedron Lett.* 1993, *34*, 6537.
41. Kimura, K.; Murata, K.; Otsuka, K.; Ishizuka, T.; Haratake, M.; Kunieda, T. *Tetrahedron Lett.* 1992, *33*, 4461.
42. Bao, J.; Wulff, W. D.; Rheongold, A. L. *J. Am. Chem. Soc.* 1993, *115*, 3814.
43. Kobayashi, S.; Hachiya, I.; Ishitani, H.; Araki, M. *Tetrahedron Lett.* 1993, *34*, 2581.
44. Corey, E. J.; Loh, T.-P. *Tetrahedron Lett.* 1993, *34*, 3979.
45. Corey, E. J.; Sarshar, S. *J. Am. Chem. Soc.* 1992, *114*, 7938.
46. Corey, E. J.; Roper, T. D.; Ishihara, K.; Sarakinos, G. *Tetrahedron Lett.* 1993, *34*, 8399.
47. Corey, E. J.; Loh, T.-P.; Roper, T. D.; Azimioara, M. D.; Noe, M. C. *J. Am. Chem. Soc.* 1992, *114*, 8290.
48. Uemura, M.; Hayashi, Y. *Tetrahedron: Asymmetry* 1993, *4*, 2291.
49. Evans, D. A.; Lectka, T.; Miller, S. J. *Tetrahedron Lett.* 1993, *34*, 7027.
50. Fallis, A. G. *Can. J. Chem.* 1984, *62*, 183.
51. Takano, S. *Pure Appl. Chem.* 1987, *59*, 353.
52. Fleming, I. *Frontier Orbitals and Organic Chemical Reactions*; John Wiley & Sons: New York, 1976.
53. Houk, K. N. *Acc. Chem. Res.* 1975, *8*, 361.
54. Fringuelli, F.; Taticchi, A. *Dienes in the Diels-Alder Reaction*; John Wiley & Sons: New York, 1990.
55. Inukai, T.; Kojima, T. *J. Org. Chem.* 1971, *36*, 924.
56. Houk, K. N.; Strozier, R. W. *J. Am. Chem. Soc.* 1973, *95*, 4094.
57. Chao, T.-M.; Baker, J.; Hehre, W. J.; Kahn, S. D. *Pure Appl. Chem.* 1991, *63*, 283.
58. Masamune, S.; Reed, L. A. I.; Choy, W. *J. Org. Chem.* 1983, *48*, 1137.
59. Masamune, S.; Reed, L. A.; Davis, J. T.; Choy, W. *J. Org. Chem.* 1983, *48*, 4441.
60. Oppolzer, W.; Rodriguez, I.; Blagg, J.; Bernardinelli, G. *Helv. Chim. Acta* 1989, *72*, 123.
61. Oppolzer, W.; Chapuis, C.; Bernardinelli, G. *Tetrahedron Lett.* 1984, *25*, 5885.
62. Oppolzer, W.; Chopius, C. *Tetrahedron Lett.* 1983, *24*, 4665.
63. Oppolzer, W.; Chapius, C.; Dao, G. M.; Reichlin, D.; Godel, T. *Tetrahedron Lett.* 1982, *23*, 4781.
64. Tanaka, K.; Uno, H.; Osuga, H.; Suzuki, H. *Tetrahedron: Asymmetry* 1993, *4*, 629.
65. Langlois, Y.; Pouilhes, A. *Tetrahedron: Asymmetry* 1991, *2*, 1223.

66. Pouilhes, A.; Uriarte, E.; Kouklovsky, C.; Langlois, N.; Langlois, Y.; Chiaroni, A.; Riche, C. *Tetrahedron Lett.* 1989, *30*, 1395.
67. Kouklovsky, C.; Pouilhes, A.; Langlois, Y. *J. Am. Chem. Soc.* 1990, *112*, 6672.
68. Oppolzer, W.; Kurth, M.; Reichlin, D.; Moffatt, F. *Tetrahedron Lett.* 1981, 22, 2545.
69. Cativiela, C.; Figueras, F.; Fraile, J. M.; Garcia, J. I.; Mayoral, J. A. *Tetrahedron: Asymmetry* 1991, *2*, 953.
70. Oppolzer, W.; Chapuis, C.; Kelly, M. J. *Helv. Chim. Acta* 1983, *66*, 2358.
71. Evans, D. A.; Chapman, K. T.; Bisaha, J. *J. Am. Chem. Soc.* 1984, *106*, 4261.
72. Corey, E. J.; Ensley, H. E. *J. Am. Chem. Soc.* 1975, *97*, 6908.
73. Cativiela, C.; Lopez, P.; Mayoral, J. A. *Tetrahedron: Asymmetry* 1990, *1*, 61.
74. Takayama, H.; Iyobe, A.; Koizumi, T. *J. Chem. Soc., Chem. Commun.* 1986, 771.
75. Gras, J. L.; Poncet, A.; Nouguier, R. *Tetrahedron Lett.* 1992, *33*, 3323.
76. Walborsky, H. M.; Barush, L.; Davis, T. C. *Tetrahedron* 1963, *19*, 2333.
77. Furuta, K.; Iwanaga, K.; Yamamoto, H. *Tetrahedron Lett.* 1986, *27*, 4507.
78. Waldemann, H.; Drager, M. *Tetrahedron Lett.* 1989, *30*, 4227.
79. Reymond, J.-L.; Vogel, P. *J. Chem. Soc., Chem. Commun.* 1990, 1070.
80. Chen, Y.; Vogel, P. *Tetrahedron Lett.* 1992, *33*, 4917.
81. Fuji, K.; Tanaka, K.; Abe, H. *Tetrahedron: Asymmetry* 1992, *3*, 609.
82. Alonso, I.; Cid, M. B.; Carretero, J. C.; Ruano, J. L.; Hoyos, M. A. *Tetrahedron: Asymmetry* 1991, *2*, 1193.
83. Carreno, M. C.; Ruano, J. L. G.; Urbano, A. *Tetrahedron Lett.* 1989, *30*, 4003.
84. Arai, Y.; Kuwayama, S.-I.; Takeuchi, Y.; Koizumi, T. *Tetrahedron Lett.* 1985, *26*, 6205.
85. Alonso, I.; Carretero, J. C.; Ruano, J. L. G. *Tetrahedron Lett.* 1989, *30*, 3853.
86. Koizumi, T.; Hakamada, I.; Yoshii, E. *Tetrahedron Lett.* 1984, *25*, 87.
87. Takahashi, T.; Iyobe, A.; Arai, Y.; Koizumi, T. *Synthesis* 1989, 189.
88. Takahashi, T.; Kotsubo, H.; Iyobe, A.; Namiki, T.; Koizumi, T. *J. Chem. Soc., Perkin Trans. 1* 1990, 3065.
89. Arai, Y.; Matsui, M.; Koizumi, T. *J. Chem. Soc., Perkin Trans. 1* 1990, 1233.
90. Takayama, H.; Iyobe, A.; Koizumi, T. *Chem. Pharm. Bull.* 1987, *35*, 433.
91. Takayama, H.; Hayashi, K.; Koizumi, T. *Tetrahedron Lett.* 1986, *27*, 5509.
92. Horton, D.; Machinami, T. *J. Chem. Soc., Chem. Commun.* 1981, 88.
93. Horton, D.; Machinami, T.; Takagi, Y. *Carbohydrate Res.* 1983, *121*, 135.
94. Chen, Z.; Ortuno, R. N. *Tetrahedron: Asymmetry* 1992, *3*, 621.
95. Jurczak, J.; Tkacz, M. *Synthesis* 1979, 42.
96. Kim, K. S.; Cho, I. H.; Joo, Y. H.; Yoo, I. J.; Song, J. H.; H., K. J. *Tetrahedron Lett.* 1992, *33*, 4029.
97. Liu, H.-J.; Chew, S. Y.; Browne, E. N. C. *Tetrahedron Lett.* 1991, *32*, 2005.
98. Waldner, A. *Tetrahedron Lett.* 1989, *30*, 3061.
99. Sato, M.; Orii, C.; Sakaki, J.-I.; Kaneko, C. *J. Chem. Soc., Chem. Commun.* 1989, 1435.
100. Roush, W. R.; Brown, B. B. *Tetrahedron Lett.* 1989, *30*, 7309.
101. Kneer, G.; Mattay, J.; Raabe, G.; Kruger, C.; Lauterwein, J. *Synthesis* 1990, 599.
102. Mattay, J.; Mertes, J.; Maas, G. *Chem. Ber.* 1989, *122*, 327.
103. Roush, W. R.; Essenfeld, A. P.; Warmus, J. S.; Brown, B. B. *Tetrahedron Lett.* 1989, *30*, 7305.
104. De Jong, J. C.; van Bolhuis, F.; Feringa, B. L. *Tetrahedron: Asymmetry* 1991, *2*, 1247.
105. De Jong, J. C.; Jansen, J. F. G. A.; Feringa, B. L. *Tetrahedron Lett.* 1990, *31*, 3047.
106. Aso, M.; Ikeda, I.; Kawabe, T.; Shiro, M.; Kanematsu, K. *Tetrahedron Lett.* 1992, *33*, 5789.
107. Evans, D. A.; Chapman, K. T.; Bisaha, J. *J. Am. Chem. Soc.* 1988, *110*, 1238.
108. Cativiela, C.; Mayoral, J. A.; Avenoza, A.; Peregrina, J. M.; Lahoz, F. J.; Gimeno, S. *J. Org. Chem.* 1992, *57*, 4664.
109. Meyers, A. I.; Busacca, C. A. *Tetrahedron Lett.* 1989, *30*, 6977.
110. Rehnberg, N.; Sundin, A.; Magnusson, G. *J. Org. Chem.* 1990, *55*, 5477.
111. Lamy-Schelkens, H.; Ghosez, L. *Tetrahedron Lett.* 1989, *30*, 5891.
112. Linz, G.; Weetmen, J.; Hady, A. F. A.; Helmchan, G. *Tetrahedron Lett.* 1989, *30*, 5599.
113. Helmchen, G.; Hady, A. F. A.; Hartmann, H.; Karge, R.; Krotz, A.; Sartor, K.; Urmann, M. *Pure Appl. Chem.* 1989, *61*, 409.
114. Poll, T.; Hady, A. F. A.; Karge, R.; Linz, G.; Weetman, J.; Helmchen, G. *Tetrahedron Lett.* 1989, *30*, 5595.
115. Jung, M. E.; Vaccaro, W. D.; Buszek, K. R. *Tetrahedron Lett.* 1989, *30*, 1893.
116. Tomioka, K.; Hamada, N.; Suenaga, T.; Koga, K. *J. Chem. Soc., Perkin Trans. 1* 1990, 426.
117. Avenoza, A.; Cativiela, C.; Mayoral, J. A.; Peregrina, J. M.; Sinou, D. *Tetrahedron: Asymmetry* 1990, *1*, 765.
118. Evans, D. A.; Chapman, K. T.; Bisaha, J. *Tetrahedron Lett.* 1984, *25*, 4071.
119. Evans, D. A.; Chapman, K. T.; Hung, D. T.; Hawaguchi, A. T. *Angew. Chem., Int. Ed. Engl.* 1987, *26*, 1184.
120. Nelson, D. J. *J. Org. Chem.* 1986, *51*, 3185.
121. Poll, T.; Metter, J. O.; Helmchen, G. *Angew. Chem., Int. Ed. Engl.* 1985, *24*, 112.
122. Yamamoto, Y.; Suzuki, I. *J. Org. Chem.* 1993, *58*, 4783.
123. Bednarski, M.; Danishefsky, S. *J. Am. Chem. Soc.* 1983, *105*, 3716.

124. Danishefsky, S.; Bednarski, M. *Tetrahedron Lett.* 1984, *25*, 721.
125. Smith, D. A.; Houk, K. N. *Tetrahedron Lett.* 1991, *32*, 1549.
126. Grieco, P. A.; Nunes, J. J.; Gaul, M. D. *J. Am. Chem. Soc.* 1990, *112*, 4595.
127. Bonnesen, P. V.; Puckett, C. L.; Honeychuck, R. V.; Hersh, W. H. *J. Am. Chem. Soc.* 1989, *111*, 6070.
128. Inokuchi, T.; Tanigawa, S.-I.; Torii, S. *J. Org. Chem.* 1990, *55*, 3958.
129. Laszlo, P.; Luchetti, J. *Tetrahedron Lett.* 1984, *25*, 2147.
130. Maruoka, K.; Murase, N.; Yamamoto, H. *J. Org. Chem.* 1993, *58*, 2938.
131. Ketter, A.; Glahsl, G.; Herrmann, R. *J. Chem. Res. (S)* 1990, 278.
132. Hattori, K.; Yamamoto, H. *J. Org. Chem.* 1992, *57*, 3264.
133. Corey, E. J.; Imwinkelried, R.; Pikul, S.; Xiang, Y. B. *J. Am. Chem. Soc.* 1989, *111*, 5493.
134. Sartor, D.; Saffrich, J.; Helmchen, G. *Synlett* 1990, 197.
135. Seerden, J.-P. G.; Scheeren, H. W. *Tetrahedron Lett.* 1993, *34*, 2669.
136. Devine, P. N.; Oh, T. *Tetrahedron Lett.* 1991, *32*, 883.
137. Devine, P. N.; Oh, T. *J. Org. Chem.* 1992, *57*, 396.
138. Oh, T.; Wrobel, Z.; Devine, P. N. *Synlett* 1992, 81.
139. Ishihara, K.; Gao, Q.; Yamamoto, H. *J. Org. Chem.* 1993, *58*, 6917.
140. Gao, Q.; Maruyama, T.; Mouri, M.; Yamamoto, H. *J. Org. Chem.* 1992, *57*, 1951.
141. Corey, E. J.; Ishihara, K. *Tetrahedron Lett.* 1992, *33*, 6807.
142. Furuta, K.; Miwa, Y.; Iwanaga, K.; Yamamoto, H. *J. Am. Chem. Soc.* 1988, *110*, 6254.
143. Takemura, H.; Komeshima, N.; Takashito, I.; Hashimoto, S.-I.; Ikota, N.; Tomioka, K.; Koga, K. *Tetrahedron Lett.* 1987, *28*, 5687.
144. Hashimoto, S.-i.; Komeshima, N.; Koga, K. *J. Chem. Soc., Chem. Commun.* 1979, 437.
145. McDougal, P. G.; Jump, J. M.; Rojas, C.; Rico, J. G. *Tetrahedron Lett.* 1989, *30*, 3897.
146. Rieger, R.; Btreitmaier, E. *Synthesis* 1990, 697.
147. Beagley, B.; Larsen, D. S.; Pritchard, R. G.; Stoodley, R. J. *J. Chem. Soc., Perkin Trans. 1* 1990, 3113.
148. Giuliano, R. M.; Jordan, A. D. J.; Gauthier, A. D.; Hoogstein, K. *J. Org. Chem.* 1993, *58*, 4979.
149. Marazano, C.; Yannic, S.; Genisson, Y.; Mehmandoust, M.; Das, B. C. *Tetrahedron Lett.* 1990, *31*, 1995.
150. Gupta, R. C.; Raynor, C. M.; Stoodley, R. J.; Slawin, A. M. Z.; Williams, D. J. *J. Chem. Soc., Perkin Trans. 1* 1988, 1773.
151. Lubineau, A.; Auge, J.; Lubin, N. *J. Chem. Soc., Perkin Trans. 1* 1990, 3011.
152. Lyssikatos, J. P.; Bednarski, M. D. *Synlett* 1990, 230.
153. Burnouf, C.; Lopez, J. C.; Calvo-Flores, F. G.; Laborde, M. d. l. A.; Olesker, A.; Lukacs, G. *J. Chem. Soc., Chem. Commun.* 1990, 823.
154. Tripathy, R.; Carroll, P. J.; Thornton, E. R. *J. Am. Chem. Soc.* 1990, *112*, 6743.
155. Hamada, T.; Sato, H.; Hikota, M.; Yonemitsu, O. *Tetrahedron Lett.* 1989, *30*, 6405.
156. Thiem, R.; Rotscheidt, K.; Breitmaier, E. *Synthesis* 1989, 836.
157. Menezes, R. F.; Zezza, C. A.; Sheu, J.; Smith, M. B. *Tetrahedron Lett.* 1989, *30*, 3295.
158. Larsen, D. S.; Stoodley, R. J. *J. Chem. Soc., Perkin Trans. 1* 1990, 1339.
159. Larsen, D. S.; Stoodley, R. J. *J. Chem. Soc., Perkin Trans. 1* 1989, 1841.
160. Gupta, R. C.; Larsen, D. S.; Stoodley, R. J.; Slawin, A. M. Z.; Williams, D. J. *J. Chem. Soc., Perkin Trans. 1* 1989, 739.
161. Trost, B. M.; Godleski, S. A.; Ippen, J. *J. Org. Chem.* 1978, *43*, 4559.
162. Bird, C. W.; Lewis, A. *Tetrahedron Lett.* 1989, *30*, 6227.
163. Bloch, R.; Chaptal-Gradoz, N. *Tetrahedron Lett.* 1992, *33*, 6147.
164. Kozikowski, A. P.; Nieduzak, T. R. *Tetrahedron Lett.* 1986, *27*, 819.
165. Defoin, A.; Pires, J.; Tissot, I.; Tschamber, T.; Bur, D.; Zehnder, M.; Streith, J. *Tetrahedron: Asymmetry* 1991, 2, 1209.
166. Corey, E. J. *Pure Appl. Chem.* 1990, *62*, 1209.
167. Tolbert, L. M.; Ali, M. B. *J. Am. Chem. Soc.* 1981, *103*, 2104.
168. Walborsky, H. M.; Barash, L.; Davis, T. C. *J. Org. Chem.* 1961, *26*, 4778.
169. Misumi, A.; Iwanaga, K.; Furuta, K.; Yamamoto, H. *J. Am. Chem. Soc.* 1985, *107*, 3343.
170. Tolbert, L. M.; Ali, M. B. *J. Am. Chem. Soc.* 1982, *104*, 1742.
171. Tolbert, L. M.; Ali, M. B. *J. Am. Chem. Soc.* 1984, *106*, 3806.
172. Van der Eycken, J.; Vandewalle, M.; Heinemann, G.; Laumen, K.; Schneider, M. P.; Kredel, J.; Sauer, J. *J. Chem. Soc., Chem. Commun.* 1989, 306.
173. Janssen, A. J. M.; Klunder, A. J. H.; Zwanenburg, B. *Tetrahedron Lett.* 1990, *31*, 7219.
174. Rao, K. R.; Srinivasan, T. N.; Bhanumathi, N. *Tetrahedron Lett.* 1990, *31*, 5959.
175. Braisted, A. C.; Schultz, P. G. *J. Am. Chem. Soc.* 1990, *112*, 7430.
176. Hilvert, D.; Hill, K. W.; Nared, K. D.; Auditor, M.-T. M. *J. Am. Chem. Soc.* 1989, *111*, 9261.
177. Iverson, B. L.; Iverson, S. A.; Roberts, V. A.; Getzoff, E. D.; Tainer, J. A.; Benkovic, S. J.; Lerner, R. A. *Science* 1990, *249*, 659.

178. Blackburn, G. M.; Kang, A. S.; Kingsbury, G. A.; Burton, D. R. *Biochem. J.* 1989, *262*, 381.
179. Riant, O.; Kagan, H. B. *Tetrahedron Lett.* 1989, *30*, 7403.
180. Reichardt, C. *Solvents and Solvent Effects in Organic Chemistry*; VCH: Weinheim, 1988.
181. Sauer, J.; Sustmann, R. *Angew. Chem., Int. Ed. Engl.* 1980, *19*, 779.
182. Breslow, R.; Rideout, D. *J. Am. Chem. Soc.* 1980, *102*, 7816.
183. Cativiela, C.; Garcia, J. I.; Mayoral, J. A.; Royo, A. J.; Salvatella, L.; Assfeld, X.; Ruiz-Lopez, M. F. *J. Phys. Org. Chem.* 1992, *5*, 230.
184. Ruiz-Lopez, M. F.; Assfeld, X.; Garcia, J. I.; Mayoral, J. A.; Salvatella, L. *J. Am. Chem. Soc.* 1994, *115*, 8780.
185. Jung, M. E. *Synlett* 1990, 186.
186. Jurczak, J. *Bull. Chem. Soc., Jpn.* 1979, *52*, 3438.
187. Buback, M.; Tost, W.; Hubsch, T.; Voss, E.; Tietze, L. F. *Chem. Ber.* 1989, *122*, 1179.
188. Le Noble, W. J.; Asano, T. *Chem. Rev.* 1978, *78*, 407.
189. Isaacs, N. S.; George, A. V. *Chem. Brit.* 1987, 47.
190. McCabe, J. R.; Eckert, C. A. *Acc. Chem. Res.* 1974, *7*, 251.
191. Firestone, R. A.; Smith, G. M. *Chem. Ber.* 1989, *122*, 1089.
192. Dauben, W. G.; Kessel, C. R.; Takemura, K. H. *J. Am. Chem. Soc.* 1980, *102*, 6893.
193. Chung, W.-S.; Turro, N. J.; Mertes, J.; Mattay, J. *J. Org. Chem.* 1989, *54*, 4881.
194. Le Noble, W. J.; Kelm, H. *Angew. Chem., Int. Ed. Engl.* 1980, *19*, 841.
195. Ager, D. J.; East, M. B. *Tetrahedron* 1993, *49*, 5683.
196. Kotsuki, H.; Nishizawa, H.; Ochi, M.; Matsuoka, K. *Bull. Chem. Soc., Jpn.* 1982, *55*, 496.
197. Katagiri, N.; Akatsuka, H.; Kaneko, C.; Sera, A. *Tetrahedron Lett.* 1988, *29*, 5397.
198. Jurczak, J.; Bauer, T. *Tetrahedron* 1986, *42*, 5045.
199. Jurczak, J.; Tkacz, M. *J. Org. Chem.* 1979, *44*, 3347.
200. Brieger, G.; Bennett, J. N. *Chem. Rev.* 1980, *80*, 63.
201. Jung, M. E.; Zimmerman, C. N.; Lowen, G. T.; Khan, S. I. *Tetrahedron Lett.* 1993, *34*, 4453.
202. Davidson, A. H.; Moloney, B. A. *J. Chem. Soc., Chem. Commun.* 1989, 445.
203. Gibbs, R. A.; Bartels, K.; Lee, R. W. K.; Okamura, W. H. *J. Am. Chem. Soc.* 1989, *111*, 3717.
204. Pyne, S. G.; Hensel, M. J.; Fuchs, P. L. *J. Am. Chem. Soc.* 1982, *104*, 5719.
205. Pyne, S. P.; Spellmeyer, D. C.; Chen, S.; Fuchs, P. L. *J. Am. Chem. Soc.* 1982, *104*, 5728.
206. Roush, W. R.; Kageyama, M.; Riva, R.; Brown, B. B.; Warmus, J. S.; Moriarty, K. J. *J. Org. Chem.* 1991, *56*, 1192.
207. Coe, J. W.; Roush, W. R. *J. Org. Chem.* 1989, *54*, 915.
208. Mukaiyama, T.; Iwasawa, N. *Chem. Lett.* 1981, 29.
209. Oppolzer, W.; Dupuis, D.; Poli, G.; Raynham, T. M.; Bernardinelli, G. *Tetrahedron Lett.* 1988, *29*, 5885.
210. Oppolzer, W.; Dupuis, D. *Tetrahedron Lett.* 1985, *26*, 5437.
211. Furuta, K.; Kanematsu, A.; Yamamoto, H.; Takaoka, S. *Tetrahedron Lett.* 1989, *30*, 7231.
212. Bailey, M. S.; Brisdon, B. J.; Brown, D. W.; Stark, K. M. *Tetrahedron Lett.* 1983, *24*, 3037.
213. Roush, W. R.; Gillis, H. R.; Ko, A. I. *J. Am. Chem. Soc.* 1982, *104*, 2269.
214. Isaacs, N. S.; Van der Beeke, P. *Tetrahedron Lett.* 1982, *23*, 2147.
215. Keay, B. A.; Dibble, P. W. *Tetrahedron Lett.* 1989, *30*, 1045.
216. Jung, M. E.; Gervay, J. *J. Am. Chem. Soc.* 1991, *113*, 224.
217. Jung, M. E.; Gervay, J. *J. Am. Chem. Soc.* 1989, *111*, 5469.
218. Sisko, J.; Weinreb, S. M. *Tetrahedron Lett.* 1989, *30*, 3037.
219. Levin, J. I. *Tetrahedron Lett.* 1989, *30*, 2355.
220. Uyehara, T.; Suzuki, I.; Yamamoto, Y. *Tetrahedron Lett.* 1990, *31*, 3753.
221. Annunziata, R.; Cinquini, M.; Cozzi, F.; Raimondi, L. *Tetrahedron Lett.* 1989, *30*, 5013.
222. Tietze, L.-F.; von Kiedrowski, G.; Harms, K.; Clegg, W.; Sheldrick, G. *Angew. Chem., Int. Ed. Engl.* 1980, *19*, 134.
223. Lehd, M.; Jensen, F. *J. Org. Chem.* 1990, *55*, 1034.
224. Grieco, P. A.; Abood, N. *J. Org. Chem.* 1989, *54*, 6008.
225. Beckmann, M.; Hildebrandt, H.; Winterfeldt, E. *Tetrahedron: Asymmetry* 1990, *1*, 335.
226. Ripoll, J. L.; Rouessac, A.; Rouessac, F. *Tetrahedron* 1978, *34*, 19.
227. Perrier, M.; Rouessac, F. *Nouv. J. Chim.* 1977, *1*, 367.
228. Lasne, M.-C.; Ripoll, J.-L. *Synthesis* 1985, 121.
229. Kwart, H.; King, K. *Chem. Rev.* 1968, *68*, 415.
230. Knapp, S.; Ornaf, R. M.; Rodriques, K. E. *J. Am. Chem. Soc.* 1983, *105*, 5494.
231. Magnus, P.; Cairns, P. M. *J. Am. Chem. Soc.* 1986, *108*, 217.
232. Anderson, W. K.; Milowsky, A. S. *J. Org. Chem.* 1985, *50*, 5423.
233. Canonne, P.; Akssira, M.; Lemay, G. *Tetrahedron Lett.* 1981, *22*, 2611.
234. Posner, G. H.; Wettlaufer, D. G. *Tetrahedron Lett.* 1986, *27*, 667.
235. Choudhury, A.; Franck, R. W.; Gupta, R. B. *Tetrahedron Lett.* 1989, *30*, 4921.

236. Arnold, T.; Reissig, H.-U. *Synlett* 1990, 514.
237. Boger, D. L.; Corbett, W. L.; Curran, T. T.; Kasper, A. M. *J. Am. Chem. Soc.* 1991, *113*, 1713.
238. Backvall, J.-E.; Rise, F. *Tetrahedron Lett.* 1989, *30*, 5347.
239. Mattay, J.; Kneer, G.; Mertes, J. *Synlett* 1990, 145.
240. Posner, G. H.; Wettlaufer, D. G. *J. Am. Chem. Soc.* 1986, *108*, 7373.
241. Greene, A. E.; Charbonnier, F.; Luche, M. J.; Moyano, A. *J. Am. Chem. Soc.* 1987, *109*, 4752.
242. Boger, D. L.; Mullican, M. D. *Tetrahedron Lett.* 1982, *23*, 4551.
243. Danishefsky, S. J. *Aldrichimica Acta* 1986, *19*, 59.
244. Kametani, T.; Chu, S.-D.; Honda, T. *Heterocycles* 1987, *25*, 241.
245. Petrzilka, M.; Grayson, J. I. *Synthesis* 1981, 753.
246. Zamojski, A.; Banaszek, A.; Grynkiewicz, G. *Adv. Carbohydr. Chem. Biochem.* 1982, *40*, 1.
247. Weinreb, S. M.; Staib, R. R. *Tetrahedron* 1982, *38*, 3087.
248. Needleman, S. B.; Kuo, M. C. C. *Chem. Rev.* 1962, *62*, 405.
249. Angerbauer, R.; Schmidt, R. *Carbohydr. Res.* 1981, *89*, 193.
250. Mieczkiwski, J.; Zamojski, A. *Carbohydr. Res.* 1977, *55*, 177.
251. Schmidt, R. R.; Angerbauer, R. *Carbohydr. Res.* 1979, *72*, 272.
252. Schmidt, R. R.; Angerbauer, R. *Angew. Chem., Int. Ed. Engl.* 1977, *16*, 782.
253. Konowa, A.; Jurczak, J.; Zamojski, A. *Tetrahedron* 1976, *32*, 2957.
254. Stambouli, A.; Chastrette, M.; Soufiaoui, M. *Tetrahedron Lett.* 1991, *32*, 1723.
255. Danishefsky, S.; Kitahara, T. *J. Am. Chem. Soc.* 1974, *96*, 7807.
256. Danishefsky, S.; Kato, N.; Askin, D.; Kerwin, J. F. J. *J. Am. Chem. Soc.* 1982, *104*, 360.
257. Daniewski, A. R.; Uskokovic, M. R. *Tetrahedron Lett.* 1990, *31*, 5599.
258. Belanger, J.; Landry, N. L.; Pare, J. R. J.; Jankowski, K. *J. Org. Chem.* 1982, *47*, 3649.
259. Danishefsky, S.; Bednarski, M.; Izawa, T.; Maring, C. *J. Org. Chem.* 1984, *49*, 2290.
260. Danishefsky, S.; Kerwin, J. F. J.; Kobayashi, S. *J. Am. Chem. Soc.* 1982, *104*, 358.
261. Larson, E. R.; Danishefsky, S. *J. Am. Chem. Soc.* 1982, *104*, 6458.
262. Danishefsky, S.; Larson, E. R.; Askin, D. *J. Am. Chem. Soc.* 1982, *104*, 6457.
263. Danishefsky, S.; Kerwin, J. F. *J. Org. Chem.* 1982, *47*, 3183.
264. Danishefsky, S.; Webb, R. R. *J. Org. Chem.* 1984, *49*, 1955.
265. Chmieleuski, M.; Jurczak, J. *J. Org. Chem.* 1981, *46*, 2230.
266. Wender, P. A. *J. Am. Chem. Soc.* 1987, *109*, 4390.
267. Midland, M. M.; Koops, R. W. *J. Org. Chem.* 1990, *55*, 5058.
268. Midland, M. M.; Koops, R. W. *J. Org. Chem.* 1990, *55*, 4647.
269. Danishefsky, S. J.; Pearson, W. H.; Harvey, D. F. *J. Am. Chem. Soc.* 1984, *106*, 2456.
270. Danishefsky, S.; Kobayashi, S.; Kerwin, J. F. *J. Org. Chem.* 1982, *47*, 1981.
271. Danishefsky, S.; Bednarski, M. *Tetrahedron Lett.* 1985, *26*, 3411.
272. Danishefsky, S.; Harvey, D. F.; Quallich, G.; Uang, B. J. *J. Org. Chem.* 1984, *49*, 392.
273. Castellino, S.; Sims, J. J. *Tetrahedron Lett.* 1984, *25*, 2307.
274. Midland, M. M.; Graham, R. S. *J. Am. Chem. Soc.* 1984, *106*, 4294.
275. Bednarski, M.; Maring, C.; Danishefsky, S. *Tetrahedron Lett.* 1983, *24*, 3451.
276. Bednarski, M.; Danishefsky, S. *J. Am. Chem. Soc.* 1983, *105*, 6968.
277. Page, P. C. B.; Prodger, J. C. *Synlett* 1991, 84.
278. Maruoka, K.; Itoh, T.; Shirasaka, T.; Yamamoto, H. *J. Am. Chem. Soc.* 1988, *110*, 310.
279. Faller, J. W.; Smart, C. J. *Tetrahedron Lett.* 1989, *30*, 1189.
280. Golebiowski, A.; Jurczak, J. *Tetrahedron* 1991, *47*, 1037.
281. Larson, E. R.; Danishefsky, S. *Tetrahedron Lett.* 1982, *23*, 1975.
282. Danishefsky, S. J.; Uang, B. J.; Quallich, G. *J. Am. Chem. Soc.* 1984, *106*, 2453.
283. Danishefsky, S. J.; Selnick, H. G.; DeNinno, M. P.; Zelle, R. E.; *J. Am. Chem. Soc.* 1987, *109*, 1572.
284. Danishefsky, S.; DeNinno, M. P.; Phillips, G. B.; Zelle, R. E.; Lartey, P. A. *Tetrahedron* 1986, *42*, 2809.
285. Fraser-Reid, B.; Rahman, M. A.; Kelly, D. R.; Srivastava, R. M. *J. Org. Chem.* 1984, *49*, 1835.
286. Danishefsky, S. J.; Pearson, W. H.; Harvey, D. F.; Maring, C. J.; Springer, J. P. *J. Am. Chem. Soc.* 1985, *107*, 1256.
287. Danishefsky, S. J.; Myles, D. C.; Harvey, D. F. *J. Am. Chem. Soc.* 1987, *109*, 862.
288. Danishefsky, S. J.; Armistead, D. M.; Wincott, F. E.; Selnick, H. G.; Hungate, R. *J. Am. Chem. Soc.* 1987, *109*, 8117.
289. Egbertson, M.; Danishefsky, S. J. *J. Org. Chem.* 1989, *54*, 11.
290. Danishefsky, S. J.; Selnick, H. G.; Armistead, D. M.; Wincott, F. E. *J. Am. Chem. Soc.* 1987, *109*, 8119.
291. Danishefsky, S. J.; Armistead, D. M.; Wincott, F. E.; Selnick, H. G.; Hungate, R. *J. Am. Chem. Soc.* 1989, *111*, 2967.
292. Tietze, L. F.; Schnieder, C. *J. Org. Chem.* 1991, *56*, 2476.
293. Danishefsky, S. J.; DeNinno, M. P. In *Trends in Synthetic Carbohydrate Chemistry*; D. Horton and L. D. Hawkins, Eds.; American Chemical Society: Washington, D. C., 1989; pp 160.

294. Danishefsky, S. J.; DeNinno, M. P.; Audia, J. E.; Schulte, G. In *Trends in Synthetic Carbohydrate Chemistry*; D. Horton and L. D. Hawkins, Eds.; American Chemical Society: Washington, D. C., 1989; pp 176.
295. Danishefsky, S. J.; DeNinno, M. P. *Angew. Chem., Int. Ed. Engl.* 1987, *26*, 15.
296. Golebiowski, A.; Jurczak, J. *J. Chem. Soc., Chem. Commun.* 1989, 263.
297. Tietze, L.-F.; Glusenkamp, K.-H. *Angew. Chem., Int. Ed. Engl.* 1983, *22*, 887.
298. Yamauchi, M.; Katayama, S.; Baba, O.; Watanabe, T. *J. Chem. Soc., Chem. Commun.* 1983, 281.
299. Conrads, M.; Mattay, J.; Runsink, J. *Chem. Ber.* 1989, *122*, 2207.
300. Coleman, R. S.; Grant, E. B. *Tetrahedron Lett.* 1990, *31*, 3677.
301. Ireland, R. E.; Habich, D. *Chem. Ber.* 1981, *114*, 1418.
302. Ireland, R. E.; Habich, D. *Tetrahedron Lett.* 1980, *21*, 1389.
303. Leblanc, Y.; Fitzsimmons, B. J. *Tetrahedron Lett.* 1989, *30*, 2889.
304. Leblanc, Y.; Fitzsimmons, B. J.; Springer, J. P.; Rokach, J. *J. Am. Chem. Soc.* 1989, *111*, 2995.
305. Weinreb, S. M.; Scola, P. M. *Chem. Rev.* 1989, *89*, 1525.
306. Golebiowski, A.; Raczko, J.; Jacobsson, U.; Jurczak, J. *Tetrahedron* 1991, *47*, 1053.
307. Golebiowski, A.; Jurczak, J. *Tetrahedron* 1991, *47*, 1045.
308. Jurczak, J.; Golebiowski, A.; Racko, J. *J. Org. Chem.* 1989, *54*, 2495.
309. Taylor, E. C.; Macor, J. E. *J. Org. Chem.* 1989, *54*, 4984.
310. Le Coz, L.; Veyrat-Martin, C.; Wartski, L.; Scyden-Penne, J.; Bois, C.; Philoche-Levisalles, M. *J. Org. Chem* 1990, *55*, 4870.
311. Giuliano, R. M.; Buzby, J. H.; Marcopulos, N.; Springer, J. P. *J. Org. Chem.* 1990, *55*, 3555.
312. Backenstrass, F.; Streith, J.; Tschamber, T. *Tetrahedron Lett.* 1990, *31*, 2139.
313. Teng, M.; Fowler, F. W. *J. Org. Chem.* 1990, *55*, 5646.
314. Arjona, O.; Dominguez, C.; Pradilla, R. F. d. l.; Mallo, A.; Manzano, C.; Plumet, J. *J. Org. Chem.* 1989, *54*, 5883.
315. Padwa, A.; Harrison, B.; Norman, B. H. *Tetrahedron Lett.* 1989, *30*, 3259.
316. Barbaro, G.; Battaglia, A.; Giorgianni, P.; Bonini, B. F.; Maccagnani, G.; Zani, P. *J. Org. Chem.* 1991, *56*, 2512.
317. Barluenga, J.; Aznar, F.; Fustero, S.; Tomas, M. *Pure Appl. Chem.* 1990, *62*, 1957.
318. Cabral, J.; Laszlo, P. *Tetrahedron Lett.* 1989, *30*, 7237.
319. Maggini, M.; Prato, M.; Scorrano, G. *Tetrahedron Lett.* 1990, *31*, 6243.
320. Reissig, H.-U.; Hippeli, C.; Arnold, T. *Chem. Ber.* 1990, *123*, 2403.
321. Bell, S. I.; Parvez, M.; Weinreb, S. M. *J. Org. Chem.* 1991, *56*, 373.
322. Boger, D. L.; Nakahara, S. *J. Org. Chem.* 1991, *56*, 880.
323. Tietze, L. F.; Hartfiel, U. *Tetrahedron Lett.* 1990, *31*, 1697.
324. Bailey, P. D.; Wilson, R. D.; Brown, G. R. *Tetrahedron Lett.* 1989, *30*, 6781.
325. Allcock, S. J.; Gilchrist, T. L.; King, F. D. *Tetrahedron Lett.* 1991, *32*, 125.
326. Le Coz, L.; Wartski, L.; Seyden-Penne, J.; Charpin, P.; Nierlich, M. *Tetrahedron Lett.* 1989, *30*, 2795.
327. Barluenga, J.; Gondalez, F. J.; Fustero, S. *Tetrahedron Lett.* 1989, *30*, 2685.
328. Barluenga, J.; Gonzalez, F. J.; Fustero, S. *Tetrahedron Lett.* 1990, *31*, 397.
329. Corey, E. J.; Yuen, P.-W. *Tetrahedron Lett.* 1989, *30*, 5825.
330. Lamy-Schelkens, H.; Giomi, D.; Ghosez, L. *Tetrahedron Lett.* 1989, *30*, 5887.
331. Baruah, P. D.; Mukherjee, S.; Mahajan, M. P. *Tetrahedron* 1990, *46*, 1951.
332. Hassner, A.; Fischer, B. *Tetrahedron* 1989, *45*, 3535.
333. Gouverneur, V.; Ghosez, L. *Tetrahedron: Asymmetry* 1990, *1*, 363.
334. Defoin, A.; Brouillard-Poichet, A.; Streith, J. *Helv. Chim. Acta* 1991, *74*, 103.
335. Miller, A.; Procter, G. *Tetrahedron Lett.* 1990, *31*, 1041.
336. Stella, L.; Abraham, H.; Feneau-Dupont, J.; Tinant, B.; Declercq, J. P. *Tetrahedron Lett.* 1990, *31*, 2603.
337. Brouillard-Poichet, A.; Defoin, A.; Streith, J. *Tetrahedron Lett.* 1989, *30*, 7061.
338. Midland, M. M.; Afonso, M. M. *J. Am. Chem. Soc.* 1989, *111*, 4368.
339. Morley, A. D.; Hollinshead, D. M.; Procter, G. *Tetrahedron Lett.* 1990, *31*, 1047.
340. Schmidt, R. R. *Acc. Chem. Res.* 1986, *19*, 250.
341. Boger, D.; Weinreb, S. M. *Hetero Diels-Alder Methodology in Organic Synthesis*; Academic Press: San Diego, 1987.
342. Kametani, T.; Hibino, S. *The Synthesis of Heterocyclic Natural Products by Hetero Diels-Alder Cycloaddition Reactions*; Academic Press: Orlando, FL, 1987.
343. Miller, A.; Paterson, T. M.; Procter, G. *Synlett* 1989, 32.
344. Miller, A.; Procter, G. *Tetrahedron Lett.* 1990, *31*, 1043.
345. Bauer, T.; Chapuis, C.; Kozak, J.; Jurczak, J. *Helv. Chim. Acta* 1989, *72*, 482.
346. Kirby, G. W.; Nazeer, M. *Tetrahedron Lett.* 1988, *29*, 6173.
347. Waldmann, H.; Braun, M.; Dräger, M. *Tetrahedron: Asymmetry* 1991, *2*, 1231.
348. Hill, R. K.; Gilman, N. W. *J. Chem. Soc., Chem. Commun.* 1967, 617.

349. Dewar, M. J. S.; Jie, C. *Acc. Chem. Res.* 1992, *25*, 537.
350. Gilchrist, T. L.; Gonsalves, A. M. d. R.; Pinho e Melo, T. M. V. *Tetrahedron Lett.* 1993, *34*, 6945.
351. Agami, C.; Couty, F.; Poursoulis, M. *Synlett* 1992, 847.
352. Agami, C.; Couty, F.; Lin, J.; Mikaeloff, A.; Poursoulis, M. *Tetrahedron* 1993, *49*, 7239.
353. Paquette, L. A.; Maynard, G. D. *Angew. Chem., Int. Ed. Engl.* 1991, *30*, 1368.
354. Paquette, L. A. *Angew. Chem., Int. Ed. Engl.* 1990, *29*, 609.
355. Paquette, L. A.; DeRussy, D. T.; Pegg, N. A.; Taylor, R. T.; Zydowsky, T. M. *J. Org. Chem.* 1989, *54*, 4576.
356. Evans, D. A.; Nelson, J. V. *J. Am. Chem. Soc.* 1980, *102*, 774.
357. Evans, D. A.; Golob, A. M.; Mandel, N. S.; Mandel, G. S. *J. Am. Chem. Soc.* 1978, *100*, 8170.
358. Still, W. C.; Murata, S.; Revial, G.; Yoshihara, K. *J. Am. Chem. Soc.* 1983, *105*, 625.
359. Still, W. C. *J. Am. Chem. Soc.* 1979, *101*, 2493.
360. Still, W. C. *J. Am. Chem. Soc.* 1977, *99*, 4186.
361. Jung, M. E.; Hudspeth, J. P. *J. Am. Chem. Soc.* 1978, *100*, 4309.
362. Jung, M. E.; Hudspeth, J. P. *J. Am. Chem. Soc.* 1980, *102*, 2463.
363. Koreeda, M.; Tanaka, Y.; Schwartz, A. *J. Org. Chem.* 1980, *45*, 1172.
364. Fleming, I.; Terrett, N. K. *Tetrahedron Lett.* 1984, *24*, 5103.
365. Kozikowski, A.; Shmiesing, R. *J. Chem. Soc., Chem. Commun.* 1979, 106.
366. Schreiber, S. L.; Santini, C. *Tetrahedron Lett.* 1981, *22*, 4651.
367. Martin, S. F.; White, J. B.; Wagner, R. *J. Org. Chem.* 1982, *47*, 3190.
368. Ticew, C. M.; Heathcock, C. H. *J. Org. Chem.* 1981, *46*, 9.
369. Kuroda, C.; Nakamura, T.; Hirota, H.; Enomoto, K.; Takahashi, T. *Bull. Chem. Soc., Jpn.* 1985, *58*, 146.
370. Paquette, L. A.; Crouse, G. D.; Sharma, A. K. *J. Am. Chem. Soc.* 1980, *102*, 3972.
371. Utagawa, A.; Hirota, H.; Ohno, S.; Takahashi, T. *Bull. Chem. Soc., Jpn.* 1988, *61*, 1207.
372. Rigby, J. H.; Denis, J.-P. *Synth. Commun.* 1986, *16*, 1789.
373. Ohmuna, T.; Hata, N.; Miyachi, N.; Wakamatsu, T.; Ban, Y. *Tetrahedron Lett.* 1986, *27*, 819.
374. Annunziata, R.; Cinquini, M.; Cozzi, F. *Tetrahedron Lett.* 1987, *28*, 3139.
375. Paquette, L. A.; DeRussy, D. T.; Cottrell, C. E. *J. Am. Chem. Soc.* 1988, *110*, 890.
376. Paquette, L. A.; Romine, J. L.; Lin, H.-S. *Tetrahedron Lett.* 1987, *28*, 31.
377. Paquette, L. A.; Poupart, M.-A. *Tetrahedron Lett.* 1988, *29*, 273.
378. Paquette, L. A.; Oplinger, J. A. *Tetrahedron* 1989, *45*, 107.
379. Oplinger, J. A.; Paquette, L. A. *Tetrahedron Lett.* 1987, *28*, 5441.
380. Paquette, L. A.; Crouse, G. D.; Sharma, A. K. *J. Am. Chem. Soc.* 1982, *104*, 4411.
381. Crouse, G. D.; Paquette, L. A. *Tetrahedron Lett.* 1981, *22*, 3167.
382. Wender, P. A.; Sieburth, S. M.; Petraitis, J. J.; Singh, S. K. *Tetrahedron* 1981, *37*, 3967.
383. Paquette, L. A.; DeRussy, D. T.; Gallucci, J. C. *J. Org. Chem.* 1989, *54*, 2278.
384. Paquette, L. A.; Learn, K. S.; Romine, J. L.; Lin, H.-S. *J. Am. Chem. Soc.* 1988, *110*, 879.
385. Uma, R.; Rajogopalan, K.; Swaminathan, S. *Tetrahedron* 1986, *42*, 2757.
386. Schrieber, S. L.; Santini, C. *J. Am. Chem. Soc.* 1984, *106*, 4038.
387. Paquette, L. A.; He, W.; Rogers, R. D. *J. Org. Chem.* 1989, *54*, 2291.
388. Janardhanam, S.; Devan, B.; Rajagopalan, K. *Tetrahedron Lett.* 1993, *34*, 6761.
389. Sprules, T. J.; Galpin, J. D.; Macdonald, D. *Tetrahedron Lett.* 1993, *34*, 247.
390. Ziegler, F. E. *Acc. Chem. Res.* 1977, *10*, 227.
391. Rhoads, S. J.; Rawlins, N. R. *Org. React.* 1975, *22*, 1.
392. Ziegler, F. E. *Chem. Rev.* 1988, *88*, 1423.
393. Blechert, S. *Synthesis* 1989, 71.
394. Bennett, G. B. *Synthesis* 1977, *589.*,
395. Tarbell, D. S. *Chem. Rev.* 1940, *27*, 495.
396. Tarball, D. S. *Org. Reactions* 1944, 2, 1.
397. Takai, K.; Mori, I.; Oshima, K.; Nozaki, H. *Tetrahedron Lett.* 1981, 22, 3985.
398. Hansen, H.-J.; Schmid, H. *Tetrahedron* 1974, *30*, 1959.
399. Ferrier, R. J.; Vethavijasar, N. *J. Chem. Soc., Perkin Trans. 1* 1973, 1791.
400. Mandai, T.; Ueda, M.; Hasegawa, S.-I.; Kawada, M.; Tsuji, J. *Tetrahedron Lett.* 1990, *31*, 4041.
401. van der Baan, J. L.; Bickelhaupt, F. *Tetrahedron Lett.* 1986, *27*, 6267.
402. Daub, G. W.; Griffith, D. A. *Tetrahedron Lett.* 1986, *27*, 6311.
403. Hill, R. K.; Edwards, A. G. *Tetrahedron Lett.* 1964, 3239.
404. Faulkner, D. J.; Petersen, M. R. *Tetrahedron Lett.* 1969, 3243.
405. Faulkner, D. J.; Peterson, M. R. *J. Am. Chem. Soc.* 1973, *95*, 553.
406. Reetz, M. T.; Gansäuer, A. *Tetrahedron* 1993, *49*, 6025.
407. Wick, A. E.; Felix, D.; Steen, K.; Eschenmoser, A. *Helv. Chim. Acta* 1964, *47*, 2425.
408. Felix, D.; Gschwend-Steen, D.; Wick, A. E.; Eschenmoser, A. *Helv. Chim. Acta* 1969, *52*, 1030.
409. Ficini, J.; Barbara, C. *Tetrahedron Lett.* 1966, 6425.

410. Bartlett, P. A.; Hahne, W. F. *J. Org. Chem.* 1979, *44*, 882.
411. Corey, E. J.; Shibasaki, M.; Knolle, J. *Tetrahedron Lett.* 1977, 1625.
412. Ritter, K. *Tetrahedron Lett.* 1990, *31*, 869.
413. Tsunoda, T.; Sasaki, O.; Ito, S. *Tetrahedron Lett.* 1990, *31*, 727.
414. Ito, S.; Tsunoda, T. *Pure Appl. Chem.* 1990, *62*, 1405.
415. Nubbemeyer, U. *Synthesis* 1993, 1120.
416. Johnson, W. S.; Wetherman, L.; Bartlett, W. R.; Brocksom, T. J.; Li, T.; Faulkner, D. J.; Peterson, M. R. *J. Am. Chem. Soc.* 1970, *92*, 941.
417. Ziegler, F. E.; Stirchak, E. P.; Wester, R. T. *Tetrahedron Lett.* 1986, *27*, 1229.
418. Lythgoe, B.; Metcalfe, D. A. *Tetrahedron Lett.* 1975, 2447.
419. Cave, R. J.; Lythgoe, B.; Metcalfe, D. A.; Waterhouse, I. *J. Chem. Soc., Perkin Trans. 1* 1977, 1218.
420. Chapleo, C. B.; Hallett, P.; Lythgoe, B.; Waterhouse, I.; Wright, P. W. *J. Chem. Soc., Perkin Trans. 1* 1977, 1211.
421. Johnson, W. S.; Werthemann, L.; Bartlatt, W. R.; Brocksom, T. J.; Li, T.-T.; Faulkner, D. J.; Petersen, M. R. *J. Am. Chem. Soc.* 1970, *92*, 741.
422. Jones, G. B.; Huber, R. S.; Chau, S. *Tetrahedron* 1993, *49*, 369.
423. Mulzer, J.; Scharp, M. *Synthesis* 1993, 615.
424. Bao, R.; Valverde, S.; Herradón, B. *Synlett* 1992, 217.
425. Ireland, R. E.; Willard, A. K. *Tetrahedron Lett.* 1975, 3975.
426. Arnold, R. T.; Searles, S. *J. Am. Chem. Soc.* 1949, *71*, 1150.
427. Ireland, R. E.; Mueller, R. H. *J. Am. Chem. Soc.* 1972, *94*, 5897.
428. Kurth, M.; Soares, C. J. *Tetrahedron Lett.* 1987, *28*, 1031.
429. Funk, R. L.; Abelman, M. M.; Munger, J. D. *Tetrahedron* 1986, *42*, 2831.
430. Nubbemeyer, U.; Öhrlein, R.; Gonda, J.; Ernst, B.; Bellus, D. *Angew. Chem., Int. Ed. Engl.* 1991, *30*, 1465.
431. Malherbe, R.; Bellus, D. *Helv. Chim. Acta* 1978, *61*, 295.
432. Rosini, G.; Spineti, G. G.; Foresti, E.; Pradella, G. *J. Org. Chem.* 1981, *46*, 2228.
433. Beslin, P.; Perrio, S. *Tetrahedron* 1993, *49*, 3131.
434. Tulshian, D. B.; Tsang, R.; Fraser-Reid, B. *J. Org. Chem.* 1984, *49*, 2347.
435. Lubineau, A.; Auge, J.; Bellanger, N.; Caillebourdin, S. *Tetrahedron Lett.* 1990, *31*, 4147.
436. Grieco, P. A.; Brandes, E. B.; McCann, S.; Clark, J. D. *J. Org. Chem.* 1989, *54*, 5849.
437. Brandes, E.; Grieco, P. A.; Gajewski, J. J. *J. Org. Chem.* 1989, *54*, 515.
438. Mikami, K.; Maeda, T.; Kishi, N.; Nakai, T. *Tetrahedron Lett.* 1984, *25*, 5151.
439. Ziegler, F. E.; Thottathil, J. K. *Tetrahedron Lett.* 1982, *23*, 3531.
440. Takahashi, T.; Yamada, H.; Tsuji, J. *Tetrahedron Lett.* 1982, *23*, 233.
441. Hill, R. K.; Gilman, N. W. *Tetrahedron Lett.* 1967, 1421.
442. Gonda, J.; Helland, A.-C.; Ernst, B.; Bellus, D. *Synthesis* 1993, 729.
443. Koreeda, M.; Luengo, J. I. *J. Am. Chem. Soc.* 1985, *107*, 5572.
444. Denmark, S. E.; Harmata, M. A. *J. Am. Chem. Soc.* 1982, *104*, 4972.
445. Denmark, S. E.; Rajendra, G.; Marlin, J. E. *Tetrahedron Lett.* 1989, *30*, 2469.
446. Denmark, S. E.; Harmata, M. A.; White, K. S. *J. Am. Chem. Soc.* 1989, *111*, 8878.
447. Denmark, S. E.; Marlin, J. E. *J. Org. Chem.* 1991, *56*, 1003.
448. Maruoka, K.; Banno, H.; Nonoshita, K.; Yamamoto, H. *Tetrahedron Lett.* 1989, *30*, 1265.
449. Nonoshita, K.; Banno, H.; Maruoka, K.; Yamamoto, H. *J. Am. Chem. Soc.* 1990, *112*, 316.
450. Hill, R. K.; Khatri, H. N. *Tetrahedron Lett.* 1978, 4337.
451. Maruoka, K.; Banno, H.; Yamamoto, H. *J. Am. Chem. Soc.* 1990, *112*, 7791.
452. Yamamoto, H.; Maruoka, K. *Pure Appl. Chem.* 1990, *62*, 2063.
453. Panek, J. S.; Sparks, M. A. *J. Org. Chem.* 1990, *55*, 5564.
454. Sparks, M. A.; Panek, J. S. *J. Org. Chem.* 1991, *56*, 3431.
455. Panek, J. S.; Clark, T. D. *J. Org. Chem.* 1992, *57*, 4323.
456. Normant, J. F.; Quirion, J. C.; Alexakis, A.; Masuda, Y. *Tetrahedron Lett.* 1989, *30*, 3955.
457. Savage, I.; Thomas, E. J. *J. Chem. Soc., Chem. Commun.* 1989, 717.
458. Overman, L. E. *J. Am. Chem. Soc.* 1976, *98*, 2901.
459. Metz, P.; Mues, C. *Synlett* 1990, 97.
460. Pereira, S.; Srebnik, M. *Aldrichimica Acta* 1993, *26*, 17.
461. Narula, A. S. *Tetrahedron Lett.* 1981, *22*, 2017.
462. Ireland, R. E.; Wilcox, C. S. *Tetrahedron Lett.* 1977, 2839.
463. Ireland, R. E.; Wipf, P.; Armstrong, J. D. *J. Org. Chem.* 1991, *56*, 650.
464. Seebach, D.; Mukhopadhyay *Helv. Chim. Acta* 1982, *65*, 385.
465. Ireland, R. E.; Norbeck, D. W. *J. Am. Chem. Soc.* 1985, *107*, 3279.
466. Takahashi, O.; Maeda, T.; Mikami, K.; Nakai, T. *Chem. Lett.* 1986, 1355.
467. Bartlett, P. A.; Pizzo, C. F. *J. Org. Chem.* 1981, *46*, 3896.

468. Ireland, R. E.; Wipf, P.; Xiang, J. *J. Org. Chem.* 1991, *56*, 3572.
469. Gajewski, J. J.; Conrad, N. D. *J. Am. Chem. Soc.* 1979, *101*, 6693.
470. Paterson, I.; N., H. A. *Tetrahedron Lett.* 1990, *31*, 7513.
471. Ireland, R. E.; Vevert, J.-P. *J. Org. Chem.* 1980, *45*, 4259.
472. Ireland, R. E.; Vevert, J.-P. *Can. J. Chem.* 1981, *59*, 572.
473. Ireland, R. E.; Smith, M. G. *J. Am. Chem. Soc.* 1988, *110*, 854.
474. Ireland, R. E.; Daub, J. P. *J. Org. Chem.* 1981, *46*, 479.
475. Ireland, R. E.; Mueller, R. H.; Willard, A. K. *J. Org. Chem.* 1976, *41*, 986.
476. Ireland, R. E.; Wuts, P. G. M.; Ernst, B. *J. Am. Chem. Soc.* 1981, *103*, 3205.
477. Ireland, R. E.; Dow, W. C.; Godfrey, J. D.; Thaisrivongs, S. *J. Org. Chem.* 1984, *49*, 1001.
478. Avery, M. A.; Jennings-White, C.; Chong, W. K. M. *J. Org. Chem.* 1989, *54*, 1789.
479. Danishefsky, S.; Tsuzuki, K. *J. Am. Chem. Soc.* 1980, *102*, 6891.
480. Ireland, R. E.; Godfrey, J. D.; Thaisrivongs, S. *J. Am. Chem. Soc.* 1981, *103*, 2446.
481. Begley, M. J.; Cameron, A. G.; Knight, D. W. *J. Chem. Soc., Chem. Commun.* 1984, 827.
482. Piers, E.; Fleming, F. F. *J. Chem. Soc., Chem. Commun.* 1989, 1665.
483. Cameron, A. G.; Knight, D. W. *Tetrahedron Lett.* 1982, *23*, 5455.
484. Burke, S. D.; Schoenen, F. J.; Nair, M. S. *Tetrahedron Lett.* 1987, *28*, 4143.
485. Burke, S. D.; Chandler, A. C.; Nair, M. S.; Campopiano, O. *Tetrahedron Lett.* 1987, *28*, 4147.
486. Burke, S. D.; Schoenen, F. J.; Murtiashaw, C. W. *Tetrahedron Lett.* 1986, *27*, 449.
487. Burke, S. D.; Armistead, D. M.; Shankaran, K. *Tetrahedron Lett.* 1986, *27*, 6295.
488. McDougal, P. G.; Oh, Y.-I.; VanDerveer, D. *J. Org. Chem.* 1989, *54*, 91.
489. Holmes, C. P.; Bartlett, P. A. *J. Org. Chem.* 1989, *54*, 98.
490. Ireland, R. E.; Anderson, R. C.; Badoud, R.; Fitzsimmons, B. J.; McGarvey, G. J.; Thaisrivongs, S.; Wilcox, C. S. *J. Am. Chem. Soc.* 1983, *105*, 1988.
491. Edwards, M. P.; Ley, S. V.; Lister, S. G.; Palmer, B. D. *J. Chem. Soc., Chem. Commun.* 1983, 630.
492. Ireland, R. E.; Habich, D.; Norbeck, D. W. *J. Am. Chem. Soc.* 1985, *107*, 3271.
493. Ireland, R. E.; Norbeck, D. W.; Mandel, G. S.; Mandel, N. S. *J. Am. Chem. Soc.* 1985, *107*, 3285.
494. Ireland, R. E.; Thaisrivongs, S.; Wilcox, C. S. *J. Am. Chem. Soc.* 1980, *102*, 1155.
495. Bartlett, P. A.; Holm, K. H.; Morimoto, A. *J. Org. Chem.* 1985, *50*, 5179.
496. Ireland, R. E.; Wilcox, C. S.; Thaisrivong, S.; Vanier, N. R. *Can. J. Chem.* 1979, *57*, 1743.
497. Mulzer, J.; Shanyoor, M. *Tetrahedron Lett.* 1993, *34*, 6545.
498. Danishefsky, S. J.; Audia, J. E. *Tetrahedron Lett.* 1988, *29*, 1371.
499. Rowley, M.; Tsukamoto, M.; Kishi, Y. *J. Am. Chem. Soc.* 1989, *111*, 2735.
500. Shu, A. Y. L.; Djerassi, C. *J. Chem. Soc., Perkin Trans. 1* 1987, 1291.
501. Nemoto, H.; Satoh, A.; Ando, M.; Fukumoto, K. *J. Chem. Soc., Chem. Commun.* 1990, 1001.
502. Abelman, M. M.; Funk, R. L.; Munger, J. D. *J. Am. Chem. Soc.* 1982, *104*, 4030.
503. Kim, B. H.; Jacobs, P. B.; Elliott, R. L.; Curran, D. P. *Tetrahedron* 1988, *44*, 3079.
504. Cameron, A. G.; Knight, D. W. *J. Chem. Soc., Perkin Trans. 1* 1986, 161.
505. Krafft, M. E.; Jarrett, S.; Dasse, O. *Tetrahedron Lett.* 1993, *34*, 8209.
506. Corey, E. J.; Lee, D. H. *J. Am. Chem. Soc.* 1991, *113*, 4026.
507. Sato, T.; Tajima, K.; Fujia, T. *Tetrahedron Lett.* 1983, *24*, 729.
508. Fujisawa, T.; Kohama, H.; Tajima, K.; Sato, T. *Tetrahedron Lett.* 1984, *25*, 5155.
509. Kallmerten, J.; Gould, T. J. *Tetrahedron Lett.* 1983, *24*, 5177.
510. Ager, D. J.; Cookson, R. C. *Tetrahedron Lett.* 1982, *23*, 3419.
511. Bartlett, P. A.; Tanzella, D. J.; Barstow, J. F. *J. Org. Chem.* 1982, *47*, 3941.
512. Burke, S. D.; Pacofsky, G. J. *Tetrahedron Lett.* 1986, *27*, 445.
513. Tadano, K.-I.; Minami, M.; Ogawa, S. *J. Org. Chem.* 1990, *55*, 2108.
514. Tsunoda, T.; Tatsuki, S.; Shiraishi, Y.; Akasaka, M.; Itô, S. *Tetrahedron Lett.* 1993, *34*, 3297.
515. Lythgoe, B.; Milner, J. R.; Tideswell, J. *Tetrahedron Lett.* 1975, 2593.
516. Richardson, S. K.; Sabol, M. R.; Watt, D. S. *Synth. Commun.* 1989, *19*, 359.
517. Bartlett, P. A.; Barstow, J. F. *Tetrahedron Lett.* 1982, *23*, 623.
518. Bartlett, P. A.; Barstow, J. F. *J. Org. Chem.* 1982, *47*, 3933.
519. Baumann, H.; Duthaler, R. O. *Helv. Chim. Acta* 1988, *71*, 1025.
520. Welch, J. T.; Plummer, J. S.; Chou, T.-S. *J. Org. Chem.* 1991, *56*, 353.
521. Danishefsky, S.; Funk, R. L.; Kerwin, J. F. *J. Am. Chem. Soc.* 1980, *102*, 6889.
522. Brandage, S.; Flodman, L.; Norberg, A. *J. Org. Chem.* 1984, *49*, 927.
523. Burke, S. D.; Cobb, J. E. *Tetrahedron Lett.* 1986, *27*, 4237.
524. Ireland, R. E.; Thaisrivongs, S.; Vanier, N.; Wilcox, C. S. *J. Org. Chem.* 1980, *45*, 48.
525. Burke, S. D.; Porter, W. J.; Rancourt, J.; Kaltenbach, R. F. *Tetrahedron Lett.* 1990, *31*, 5285.
526. Paterson, I.; Hulme, A. N.; Wallace, D. J. *Tetrahedron Lett.* 1991, *32*, 7601.
527. Tamuru, Y.; Amino, Y.; Furukawa, Y.; Kagotani, M.; Yoshida, Z. *J. Am. Chem. Soc.* 1982, *104*, 4018.

528. Beslin, P.; Perrio, S. *J. Chem. Soc., Chem. Commun.* 1989, *414.*,
529. Daub, G. W.; Sanchez, M. G.; Crommer, R. A.; Gibson, L. L. *J. Org. Chem.* 1982, *47*, 743.
530. Wilson, S. R.; Price, M. F. *J. Org. Chem.* 1984, *49*, 722.
531. Ouvrard, N.; Rodriguez, J.; Santelli, M. *Tetrahedron Lett.* 1993, *34*, 1149.
532. Snider, B. B. *Acc. Chem. Res.* 1980, *13*, 426.
533. Oppolzer, W. *Angew. Chem., Int. Ed. Engl.* 1989, *28*, 38.
534. Oppolzer, W. *Pure Appl. Chem.* 1981, *53*, 1181.
535. Mikami, K.; Shimizu, M. *Chem. Rev.* 1992, *92*, 1021.
536. Thomas, B. E.; Houk, K. N. *J. Am. Chem. Soc.* 1993, *115*, 790.
537. Oppolzer, W.; Begley, T.; Ashcroft, A. *Tetrahedron Lett.* 1984, *25*, 825.
538. Oppolzer, W.; Robbiani, C.; Battig, K. *Tetrahedron* 1984, *40*, 1391.
539. Oppolzer, W.; Jacobsen, E. J. *Tetrahedron Lett.* 1986, *27*, 1141.
540. Lehmkuhl, H. *Bull. Soc. Chim., Fr.* 1981, 87.
541. Whitesell, J. K.; Bhattacharya, A.; Aguilar, D. A.; Henke, K. *J. Chem. Soc., Chem. Commun.* 1982, 989.
542. Whitesell, J. K.; Deyo, D.; Bhattacharya, A. *J. Chem. Soc., Chem. Commun.* 1983, 802.
543. Mikami, K.; Shimizu, M.; Nakai, T. *J. Org. Chem.* 1991, *56*, 2952.
544. Mikami, K.; Loh, T.-P.; Nakai, T. *Tetrahedron Lett.* 1988, *29*, 6305.
545. Houston, T. A.; Tanaka, Y.; Koreeda, M. *J. Org. Chem.* 1993, *58*, 4287.
546. Shimizu, M.; Mikami, K. *J. Org. Chem.* 1992, *57*, 6105.
547. Duncia, J. V.; Lansbury, P. T.; Miller, T.; Snider, B. B. *J. Am. Chem. Soc.* 1982, *104*, 1930.
548. Snider, B. B.; Duncia, J. V. *J. Am. Chem. Soc.* 1980, *102*, 5926.
549. Snider, B. B.; Duncia, J. V. *J. Org. Chem.* 1981, *46*, 3223.
550. Sarko, T. K.; Gosh, S. H.; Subba Rao, P. S. V.; Satapathi, T. K. *Tetrahedron Lett.* 1990, *31*, 3461.
551. Maruoka, K.; Hoshino, Y.; Shirasaka, T.; Yamamoto, H. *Tetrahedron Lett.* 1988, *29*, 3967.
552. Snider, B. B.; Cartaya-Mari, C. P. *J. Org. Chem.* 1984, *49*, 1688.
553. Tanino, K.; Nakamura, T.; Kuwajima, I. *Tetrahedron Lett.* 1990, *31*, 2165.
554. Mikami, K.; Terada, M.; Narisawa, S.; Nakai, T. *Synlett* 1992, 255.
555. Mikami, K.; Takahashi, K.; Nakai, T. *Tetrahedron Lett.* 1989, *30*, 357.
556. Takacs, J. M.; Anderson, L. G.; Greswell, M. W.; Takacs, B. E. *Tetrahedron Lett.* 1987, *28*, 5627.
557. Oppolzer, W.; Keller, T. H.; Kuo, D. L.; Pachinger, W. *Tetrahedron Lett.* 1990, *31*, 1265.
558. Oppolzer, W. *Pure Appl. Chem.* 1990, *62*, 1941.
559. Snider, B. B.; van Straten, J. W. *J. Org. Chem.* 1979, *44*, 3567.
560. Snider, B. B.; Rodini, D. J.; Cann, R. S. E.; Sealfon, S. *J. Am. Chem. Soc.* 1979, *101*, 5283.
561. Snider, B. B.; Rodini, D. J. *Tetrahedron Lett.* 1980, *21*, 1815.
562. Whitesell, J. K.; Bhattacharya, A.; Buchanan, C. M.; Chen, H. H.; Deyo, D.; James, D.; Liu, C.-L.; Minton, M. A. *Tetrahedron* 1986, *40*, 2993.
563. Whitesell, J. K.; Nabona, K.; Deyo, D. *J. Org. Chem.* 1989, *54*, 2258.
564. Whitesell, J. J.; Carpenter, J. F.; Yaser, H. K.; Machajewski, T. *J. Am. Chem. Soc.* 1990, *112*, 7653.
565. Benner, J. P.; Gill, G. B.; Parrott, S. J.; Wallace, B.; Belay, M. J. *J. Chem. Soc., Perkin Trans. 1* 1984, 314.
566. Mikami, K.; Kaneko, M.; Loh, T.-P.; Terada, M.; Nakai, T. *Tetrahedron Lett.* 1990, *31*, 3909.
567. Kim, B. H.; Lee, J. Y.; Kim, K.; Whang, D. *Tetrahedron: Asymmetry* 1991, *2*, 27.
568. Oppolzer, W.; Pitteloud, R. *J. Am. Chem. Soc.* 1982, *104*, 6478.
569. Mikami, K.; Loh, T.-P.; Nakai, T. *Tetrahedron: Asymmetry* 1990, *1*, 13.
570. Mikami, K.; Loh, T.-P.; Nakai, T. *J. Chem. Soc., Chem. Commun.* 1991, 77.
571. Nakamura, T.; Tanino, K.; Kuwajima, I. *Tetrahedron Lett.* 1993, *34*, 477.
572. Mikami, K.; Terada, M.; Nakai, T. *J. Am. Chem. Soc.* 1989, *111*, 1940.
573. Mikami, K.; Terada, M.; Nakai, T. *J. Am. Chem. Soc.* 1990, *112*, 3949.
574. Kocienski, P.; Love, C.; Roberts, D. A. *Tetrahedron Lett.* 1989, *30*, 6753.
575. Utimoto, K.; Imi, K.; Shiragami, H.; Fujikara, S.; Nozaki, H. *Tetrahedron Lett.* 1985, *26*, 2101.
576. Lehmkuhl, H.; Hausthid, H.; Bellenbaum, M. *Chem. Ber.* 1984, *117*, 383.
577. Wasserman, H. H.; Ives, J. L. *Tetrahedron* 1981, *37*, 1825.
578. Ohloff, G.; Klein, E.; Schenck, G. O. *Angew. Chem.* 1961, *73*, 578.
579. Trost, B. M.; King, S.; Nanninga, T. N. *Chem. Lett.* 1987, 15.
580. Trost, B. M.; Mignani, S. M. *Tetrahedron Lett.* 1985, *26*, 6313.
581. Trost, B. M.; Lynch, J.; Renaut, P.; Steinman, D. H. *J. Am. Chem. Soc.* 1986, *108*, 284.
582. Trost, B. M.; Bonk, P. J. *J. Am. Chem. Soc.* 1985, *107*, 1778.
583. Trost, B. M.; Miller, M. L. *J. Am. Chem. Soc.* 1988, *110*, 3687.
584. Trost, B. M.; Seoane, P.; Mignani, S.; Acemoglu, M. *J. Org. Chem.* 1989, *54*, 7487.
585. Trost, B. M.; Acemoglu, M. *Tetrahedron Lett.* 1989, *30*, 1495.
586. Molander, G. A.; Andrews, S. W. *Tetrahedron Lett.* 1989, *30*, 2351.
587. Trost, B. M. *Angew. Chem., Int. Ed. Engl.* 1986, *25*, 1.

588. Horiguchi, Y.; Suchiro, I.; Sasaki, A.; Kuwajima, I. *Tetrahedron Lett.* 1993, *34*, 6077.
589. Mann, J.; Holland, H. J.; Lewis, T. *Tetrahedron* 1987, *43*, 2533.
590. Barbosa, L.-C. A.; Mann, J.; Wilde, P. D. *Tetrahedron* 1989, *45*, 4619.
591. Ohno, M.; Mori, K.; Hattori, T.; Eguchi, S. *J. Org. Chem.* 1990, *55*, 6086.
592. Murray, D. H.; Albizati, K. F. *Tetrahedron Lett.* 1990, *31*, 4109.
593. Bowers, K. G.; Mann, J.; Markson, A. J. *J. Chem. Res. (S)* 1986, 424.
594. Lautens, M.; Abd-El-Aziz, A. S.; Lough, A. *J. Org. Chem.* 1990, *55*, 5305.
595. Lautens, M.; Chiu, P. *Tetrahedron Lett.* 1993, *34*, 773.
596. White, J. D.; Fukuyama, Y. *J. Am. Chem. Soc.* 1979, *101*, 226.
597. Wender, P. A.; Fisher, K. *Tetrahedron Lett.* 1986, *27*, 1857.
598. Lee, T. V.; Boucher, R. J.; Ellis, K. L.; Richardson, K. A. *Tetrahedron Lett.* 1988, *29*, 685.
599. Boger, D. L.; Brotherton, C. E. *J. Am. Chem. Soc.* 1986, *108*, 6695.
600. Heilmann, W.; Koshinsky, R.; Mayr, H. *J. Org. Chem.* 1987, *52*, 1989.
601. Böhshar, M.; Heydt, H.; Maas, G.; Gümbel, H.; Regitz, M. *Angew. Chem., Int. Ed. Engl.* 1985, *24*, 597.
602. Trost, B. M.; MacPherson, D. T. *J. Am. Chem. Soc.* 1987, *109*, 3483.
603. Wender, P. A.; Snapper, M. L. *Tetrahedron Lett.* 1987, *28*, 2221.
604. Wender, P. A.; Ihle, N. C. *Tetrahedron Lett.* 1987, *28*, 2451.
605. Schreiber, S. L.; Sammakia, T.; Crowe, W. E. *J. Am. Chem. Soc.* 1986, *108*, 3128.
606. Crimmins, M. T.; Nanternet, P. G. *J. Org. Chem.* 1990, *55*, 4235.
607. Houk, K. N.; Moses, S. R.; Wu, Y.-D.; Rondan, N. G.; Jager, V.; Schohe, R.; Fronczek, F. R. *J. Am. Chem. Soc.* 1984, *106*, 3880.
608. Kozikowski, A. P.; Ghosh, A. K. *J. Org. Chem.* 1984, *49*, 2762.
609. Vasella, A. *Helv. Chim. Acta* 1977, *60*, 426.
610. Kozikowski, A. P.; Ghosh, A. K. *J. Am. Chem. Soc.* 1982, *104*, 5788.
611. Huber, R.; Vasella, A. *Tetrahedron* 1990, *46*, 33.
612. Kanemasa, S.; Tsuruoka, T.; Wada, E. *Tetrahedron Lett.* 1993, *34*, 87.
613. Hoffmann, R. W. *Chem. Rev.* 1989, *89*, 1841.
614. Curran, D. P.; Kim, B. H. *Synthesis* 1990, 312.
615. Panek, J. S.; Beresis, R. T. *J. Am. Chem. Soc.* 1993, *115*, 7898.
616. Olsson, T.; Stern, K.; Westman, G.; Sundell, S. *Tetrahedron* 1990, *46*, 2473.
617. Curran, D. P.; Kim, B. H.; Daugherty, J.; Heffner, T. A. *Tetrahedron Lett.* 1988, *29*, 3555.
618. Curran, D. P.; Heffner, T. A. *J. Org. Chem.* 1990, *55*, 4585.
619. Kim, B. H.; Curran, D. P. *Tetrahedron* 1993, *49*, 293.
620. Annunziata, R.; Cinquini, M.; Cozzi, F.; Raimondi, L. *Tetrahedron Lett.* 1988, *29*, 2881.
621. Caramella, P.; Albini, E.; Bandiera, T.; Coda, A. C.; Grananger, P.; Albani, F. M. *Tetrahedron* 1983, *39*, 689.
622. Annunziata, R.; Cinquini, M.; Cozzi, F.; Giaroni, P.; Raimondi, L. *Tetrahedron: Asymmetry* 1990, *1*, 251.
623. Martin, S. F.; Dupre, B. *Tetrahedron Lett.* 1983, *24*, 1337.
624. Belzecki, C.; Panfil, I. *J. Chem. Soc., Chem. Commun.* 1977, 303.
625. Baldoni, C.; Del Butero, P.; Maiorana, S.; Zecchi, G.; Moret, M. *Tetrahedron Lett.* 1993, *34*, 2529.
626. Bérranger, T.; André-Barrès, C.; Kobayakawa, M.; Langlois, Y. *Tetrahedron Lett.* 1993, *34*, 5079.
627. McGraig, A. E.; Wightman, R. H. *Tetrahedron Lett.* 1993, *34*, 3939.
628. Stevens, R. V.; Albizati, K. *J. Chem. Soc., Chem. Commun.* 1982, 104.
629. Kametani, T.; Huang, S.-P.; Nakayama, A.; Honda, T. *J. Org. Chem.* 1982, *47*, 2328.
630. Kametani, T.; Nagahara, T.; Honda, T. *J. Org. Chem.* 1985, *50*, 2327.
631. Ito, Y.; Kimura, Y.; Terashima, S. *Bull. Chem. Soc., Jpn.* 1987, *60*, 3337.
632. Ihara, M.; Takashima, M.; Kukumoto, K.; Kametani, T. *J. Chem. Soc., Chem. Commun.* 1988, 9.
633. Ihara, M.; Takahashi, M.; Fukumoto, K.; Kametani, T. *Heterocycles* 1988, *27*, 327.
634. Binger, P.; Buch, M. *Top. Curr. Chem.* 1987, *135*, 77.
635. Trost, B. M.; Nanninga, T. N. *J. Am. Chem. Soc.* 1985, *107*, 1075.
636. Yamamoto, A.; Ito, Y.; Hayashi, T. *Tetrahedron Lett.* 1989, *30*, 375.
637. Trost, B. M.; Yang, B.; Miller, M. L. *J. Am. Chem. Soc.* 1989, *111*, 6482.
638. Rao, K. R.; Bhanumathi, N.; Srinivasan, T. N.; Sattur, P. B. *Tetrahedron Lett.* 1990, *31*, 899.
639. Rao, K. R.; Bhanumathi, N.; Sattur, P. B. *Tetrahedron Lett.* 1990, *31*, 3201.
640. Curran, D. P. *J. Am. Chem. Soc.* 1983, *105*, 5826.
641. Curran, D. P.; Kim, B. H. *Synthesis* 1986, 312.
642. Jager, V.; Grund, H. *Angew. Chem., Int. Ed. Engl.* 1976, *15*, 50.
643. Jager, V.; Grund, H.; Schwab, W. *Angew. Chem., Int. Ed. Engl.* 1979, *18*, 78.
644. Jager, V.; Buss, V.; Schwab, W. *Tetrahedron Lett.* 1978, 3133.
645. Kaluoda, J.; Kaufmann, H. *J. Chem. Soc., Chem. Commun.* 1976, 209.
646. Kaluoda, J.; Kaufmann, H. *J. Chem. Soc., Chem. Commun.* 1976, 210.
647. Muller, I.; Jager, V. *Tetrahedron Lett.* 1982, *23*, 4777.

648. Wade, P. A.; Hinney, H. R. *J. Am. Chem. Soc.* 1979, *101*, 1319.
649. Brandi, A.; De Sarlo, F.; Guarna, A.; Speroni, G. *Synthesis* 1982, 719.
650. Kunig, W. A.; Hass, W. *Liebigs Ann. Chem.* 1982, 1615.
651. Kozikowski, A. P.; Ishida, H. *J. Am. Chem. Soc.* 1980, *102*, 4265.
652. Kozikowski, A. P.; Chen, Y. Y. *J. Org. Chem.* 1981, *46*, 5248.
653. Kozikowski, A. P.; Scripko, J. G. *J. Am. Chem. Soc.* 1984, *106*, 353.
654. Confalone, P. N.; Lollar, E. D.; Pizzalato, G.; Uskokovic, M. R. *J. Am. Chem. Soc.* 1978, *100*, 6291.
655. Caldirola, P.; Ciancaglione, M.; De Amici, M.; De Micheli, C. *Tetrahedron Lett.* 1986, *27*, 4647.
656. Smith, A. B.; Schow, S. R.; Bloom, J. P.; Thompson, A. S.; Wizenberg, K. N. *J. Am. Chem. Soc.* 1982, *104*, 4015.
657. Kurth, M. J.; Rodriguez, M. J.; Olmstead, M. M. *J. Org. Chem.* 1990, *55*, 283.
658. Huisgen, R.; Gristl, M. *Chem. Ber.* 1973, *106*, 3291.
659. Saksena, A. K.; Lovey, R. G.; Girijavallabhan, V. M.; Guzik, H.; Ganguly, A. K. *Tetrahedron Lett.* 1993, *34*, 3267.
660. Kamimura, A.; Kakehi, A.; Hori, K. *Tetrahedron* 1993, *49*, 7637.
661. Sibi, M.; Gaboury, J. A. *Synlett* 1992, 83.
662. Annunziata, R.; Cinquini, M.; Cozzi, F.; Restelli, A. *J. Chem. Soc., Chem. Commun.* 1984, 1253.
663. Kamimura, A.; Marumo, S. *Tetrahedron Lett.* 1990, *31*, 5053.
664. Boyd, E. C.; Paton, R. M. *Tetrahedron Lett.* 1993, *34*, 3169.
665. Jager, V.; Schohe *Tetrahedron* 1984, *40*, 2199.
666. Jager, V.; Schroter, D. *Synthesis* 1990, *556.*,
667. Kita, Y.; Itoh, F.; Tamura, O.; Ke, Y. Y.; Tamura, Y. *Tetrahedron Lett.* 1987, *28*, 1431.
668. Freer, A.; Overton, K.; Tomanek, R. *Tetrahedron Lett.* 1990, *31*, 1471.
669. Burdisso, M.; Gandolfi, R.; Grunanger, P.; Rastelli, A. *J. Org. Chem.* 1990, *55*, 3427.
670. Grigg, R.; Markandu, J.; Surendrakumar, S. *Tetrahedron Lett.* 1990, *31*, 1191.
671. Annunziata, R.; Cinquini, M.; Cozzi, F.; Raimondi, L. *J. Org. Chem.* 1990, *55*, 1901.
672. Annunziata, R.; Cinquini, M.; Cozzi, F.; Giaroni, P.; Raimondi, L. *Tetrahedron Lett.* 1991, *32*, 1659.
673. Kasahara, K.; Iida, H.; Kibayashi, C. *J. Org. Chem.* 1989, *54*, 2225.
674. Ihara, M.; Takahashi, M.; Fukumoto, K.; Kametani, T. *J. Chem. Soc., Perkin Trans. 1* 1989, 2215.
675. Keller, E.; de Lange, B.; Rispens, M. T.; Feringa, B. L. *Tetrahedron* 1993, *49*, 8899.
676. De Lange, B.; Feringa, B. L. *Tetrahedron Lett.* 1988, *29*, 5317.
677. Tufanello, J. J.; Tette, J. P. *J. Org. Chem.* 1975, *40*, 3866.
678. Figueredo, M.; Font, J.; de March, P. *Chem. Ber.* 1989, *122*, 1701.
679. Figueredo, M.; Font, J.; de March, P. *Chem. Ber.* 1990, *123*, 1595.
680. Panfil, I.; Chmielewski, M. *Tetrahedron* 1985, *41*, 4713.
681. Banerji, A.; Basu, S. *Tetrahedron* 1992, *48*, 3335.
682. Alonso-Perarnau, D.; de March, P.; Figueredo, M.; Font, J.; Soria, A. *Tetrahedron* 1993, *49*, 4267.
683. Cid, P.; de March, P.; Figueredo, M.; Font, J.; Milán, S.; Soria, A.; Virgili, A. *Tetrahedron* 1993, *49*, 3857.
684. Le Bel, N. A.; Banucci, E. G. *J. Org. Chem.* 1971, *36*, 2440.
685. Annunziato, R.; Cinquini, M.; Cozzi, F.; Raimondi, L. *Tetrahedron* 1987, *43*, 4051.
686. Arjona, O.; de Dios, A.; Fernandez de la Pradilla, R.; Mallo, A.; Plumet, J. *Tetrahedron* 1990, *46*, 8179.
687. Cristina, D.; Amici, M. D.; Micheli, C. D.; Gandolfi, R. *Tetrahedron* 1981, *37*, 1349.
688. Feringa, B. L.; de Lange, B.; de Jong, J. C. *J. Org. Chem.* 1989, *54*, 2471.
689. Grigg, R. *Chem. Soc. Rev.* 1987, *16*, 89.
690. Barr, D. A.; Grigg, R.; Gunaratne, H. Q. N.; Kemp, J.; McMeekin, P.; Sridharan, V. *Tetrahedron* 1988, *44*, 557.
691. Pätzel, M.; Galley, G.; Jones, P. G.; Chrapkowsky, A. *Tetrahedron Lett.* 1993, *34*, 5707.
692. Pandey, G.; Lakshmaiah, C. *Tetrahedron Lett.* 1993, *34*, 4861.
693. Barr, D. A.; Grigg, R.; Sridharan, V. *Tetrahedron Lett.* 1989, *30*, 4727.
694. Collins, P. M.; Ashwood, M. S.; Eder, H.; Wright, S. H. B.; Kennedy, D. J. *Tetrahedron Lett.* 1990, *31*, 2055.
695. Bhattacharja, A.; Chattopadhyay, P.; McPhail, A. T.; McPhail, D. R. *J. Chem. Soc., Chem. Commun.* 1990, 1508.
696. De Amici, M.; De Micheli, C.; Ortisi, A.; Gatti, G.; Gandolfi, R.; Toma, L. *J. Org. Chem.* 1989, *54*, 793.
697. Tatsuta, K.; Niwata, Y.; Umezawa, K.; Toshima, K.; Nakata, M. *Tetrahedron Lett.* 1990, *31*, 1171.
698. Takahashi, T.; Nakazawa, M.; Sakamoto, Y.; Houk, K. N. *Tetrahedron Lett.* 1993, *34*, 4075.
699. Stack, J. A.; Heffner, T. A.; Geib, S. J.; Curran, D. P. *Tetrahedron* 1993, *49*, 995.
700. Hoffmann, R. W. *Angew. Chem., Int. Ed. Engl.* 1979, *18*, 563.
701. Tomooka, K.; Watanabe, M.; Nakai, T. *Tetrahedron Lett.* 1990, *31*, 7353.
702. Marshall, J. A.; Jenson, T. M. *J. Org. Chem.* 1984, *49*, 1707.
703. Chan, K.-K.; Saucy, G. *J. Org. Chem.* 1977, *42*, 3828.
704. Bickart, P.; Carson, F. W.; Jacobus, J.; Miller, E. G.; Mislow, K. *J. Am. Chem. Soc.* 1968, *90*, 4869.

705. Rautenstrauch, V. *J. Chem. Soc., Chem. Commun.* 1970, 4.
706. Schollkopf, U.; Fellenberger, V.; Rizk, M. *Liebigs Ann. Chem.* 1970, *734*, 106.
707. Baldwin, J. E.; DeBernardis, J.; Patrick, J. E. *Tetrahedron Lett.* 1970, 353.
708. Baldwin, J. E.; Patrick, J. E. *J. Am. Chem. Soc.* 1971, *93*, 3556.
709. Felkin, H.; Frajerman, C. *Tetrahedron Lett.* 1977, 3485.
710. Still, W. C.; Mitra, A. *J. Am. Chem. Soc.* 1978, *100*, 1927.
711. Mikami, K.; Kimura, Y.; Kishi, N.; Nakai, T. *J. Org. Chem.* 1983, *48*, 279.
712. Mikami, K.; Fujimoto, K.; Kasuga, T.; Nakai, T. *Tetrahedron Lett.* 1984, *25*, 6011.
713. Mikami, K.; Takahashi, O.; Tabei, T.; Nakai, T. *Tetrahedron Lett.* 1986, *27*, 4511.
714. Mikami, K.; Maeda, T.; Nakai, T. *Tetrahedron Lett.* 1986, *27*, 4189.
715. Rautenstrauch, V.; Buchi, G.; Wuest, H. *J. Am. Chem. Soc.* 1974, *96*, 2576.
716. Hoffmann, R.; Brückner, R. *Angew. Chem., Int. Ed. Engl.* 1992, *31*, 647.
717. Antoniotti, P.; Tonachini, G. *J. Org. Chem.* 1993, *58*, 3622.
718. Nakai, E.-i.; Nakai, T. *Tetrahedron Lett.* 1988, *29*, 4587.
719. Kakinuma, K.; Li, H.-Y. *Tetrahedron Lett.* 1989, *30*, 4157.
720. Mikami, K.; Kasuga, T.; Fujimoto, K.; Nakai, T. *Tetrahedron Lett.* 1986, *27*, 4185.
721. Uchikawa, M.; Hanamoto, T.; Katsuki, T.; Yamaguchi, M. *Tetrahedron Lett.* 1986, *27*, 4577.
722. Uchikawa, M.; Katsuki, T.; Yamaguchi, M. *Tetrahedron Lett.* 1986, *27*, 4581.
723. Keegan, D. S.; Midland, M. M.; Werley, R. T.; McLoughlin, J. I. *J. Org. Chem.* 1991, *56*, 1185.
724. Mikami, K.; Fujimoto, K.; Nakai, T. *Tetrahedron Lett.* 1983, *24*, 513.
725. Tsai, D. J.-S.; Midland, M. M. *J. Org. Chem.* 1984, *49*, 1842.
726. Sayo, N.; Azuma, K.-I.; Mikami, K.; Nakai, T. *Tetrahedron Lett.* 1984, *25*, 565.
727. Priepke, H.; Bruckner, R. *Chem. Ber.* 1990, *123*, 153.
728. Bruckner, R. *Chem. Ber.* 1989, *122*, 193.
729. Priepke, H.; Bruckner, R.; Harms, K. *Chem. Ber.* 1990, *123*, 555.
730. Bruckner, R. *Chem. Ber.* 1989, *122*, 703.
731. Hoffmann, R. W.; Goldman, S.; Gerlach, R.; Maak, N. *Chem. Ber.* 1980, *113*, 845.
732. Scheuplein, S. W. K., A.; Bruckner, R.; Harms, K. *Chem. Ber.* 1990, *123*, 917.
733. Kruse, B.; Bruckner, R. *Tetrahedron Lett.* 1990, *31*, 4425.
734. Kruse, B.; Bruckner, R. *Chem. Ber.* 1989, *122*, 2023.
735. Broka, C. A.; Shen, T. *J. Am. Chem. Soc.* 1989, *111*, 2981.
736. Ishikkawa, A.; Uchiyama, H.; Katsuki, T.; Yamaguchi, M. *Tetrahedron Lett.* 1990, *31*, 2415.
737. Yamaguchi, M. *Pure Appl. Chem.* 1989, *61*, 413.
738. Tomooka, K.; Igarashi, T.; Nakai, T. *Tetrahedron Lett.* 1993, *34*, 8139.
739. Kuroda, S.; Sakaguchi, S.-I.; Ikegami, S.; Hanamoto, T.; Katsuki, T.; Yamaguchi, M. *Tetrahedron Lett.* 1988, *29*, 4763.
740. Kuroda, S.; Katsuki, T.; Yamaguchi, M. *Tetrahedron Lett.* 1987, *28*, 803.
741. Marshall, J. A.; Wang, X.-J. *J. Org. Chem.* 1990, *55*, 2995.
742. Cox, P. J.; Simpkins, N. S. *Tetrahedron: Asymmetry* 1991, *2*, 1.
743. Marshall, J. A.; Lebreton, J. *Tetrahedron Lett.* 1987, *28*, 3323.
744. Marshall, J. A.; Lebreton, J. *J. Org. Chem.* 1988, *53*, 4108.
745. Marshall, J. A.; Wang, X.-J. *J. Org. Chem.* 1991, *56*, 960.
746. Nakai, T.; Mikami, K.; Taya, S.; Fujita, Y. *J. Am. Chem. Soc.* 1981, *103*, 6492.
747. Sayo, N.; Kimara, Y.; Nakai, T. *Tetrahedron Lett.* 1982, *23*, 3931.
748. Nakai, T.; Mikami, K.; Taya, S.; Kimura, Y.; Mimura, T. *Tetrahedron Lett.* 1981, *22*, 69.
749. Marshall, J. A.; Nelson, D. J. *Tetrahedron Lett.* 1988, *29*, 741.
750. Marshall, J. A.; Robinson, E. D.; Lebreton, J. *Tetrahedron Lett.* 1988, *29*, 3547.
751. Nakai, T.; Mikami, K. *Chem. Rev.* 1986, *86*, 885.
752. Mikami, K.; Azuma, K.-I.; Nakai, T. *Tetrahedron* 1984, *40*, 2303.
753. Mikami, K.; Kawamoto, K.; Nakai, T. *Tetrahedron Lett.* 1985, *26*, 5799.
754. Nakai, E.-I.; Nakai, N. *Tetrahedron Lett.* 1988, *29*, 5409.
755. Schreiber, S. L.; Goulet, M. T.; Schulte, G. *J. Am. Chem. Soc.* 1987, *109*, 4718.
756. Nicolaou, K. C.; Uenishi, J. *J. Chem. Soc., Chem. Commun.* 1982, 1292.
757. Nicolaou, K. C.; Ahn, K. H. *Tetrahedron Lett.* 1989, *30*, 1217.
758. Balestra, M.; Kallmerten, J. *Tetrahedron Lett.* 1988, *29*, 6901.
759. Balestra, M.; Wittman, M. D.; Kallmerten, J. *Tetrahedron Lett.* 1988, *29*, 6905.
760. Ikegami, S.; Katsuki, T.; Yamaguchi, M. *Tetrahedron Lett.* 1988, *29*, 5285.
761. Marshall, J. A.; Lebreton, J.; DeHoff, B. S.; Jenson, T. M. *Tetrahedron Lett.* 1987, *28*, 723.
762. Midland, M. M.; Kwon, Y. C. *Tetrahedron Lett.* 1985, *26*, 5021.
763. Tsai, D. J.-S.; Midland, M. M. *J. Am. Chem. Soc.* 1985, *107*, 3915.
764. Fujimoto, Y.; Ohhana, M.; Terasawa, T.; Ikekawa, N. *Tetrahedron Lett.* 1985, *26*, 3239.

765. Bruckner, R. *Tetrahedron Lett.* 1988, *29*, 5747.
766. Kaye, A. D.; Pattenden, G.; Roberts, S. M. *Tetrahedron Lett.* 1986, *27*, 2033.
767. Kido, F.; Sinha, S. C.; Abiko, T.; Watanabe, M.; Yoshikoshi, A. *Tetrahedron* 1990, *46*, 4887.
768. Frauenrath, H.; Arenz, T.; Raabe, G.; Zorn, M. *Angew. Chem., Int. Ed. Engl.* 1993, *32*, 83.
769. Kress, M. H.; Kaller, B. F.; Kishi, Y. *Tetrahedron Lett.* 1993, *34*, 8047.
770. You, Z.; Koreeda, M. *Tetrahedron Lett.* 1993, *34*, 2597.
771. Marshall, J. A.; Robinson, E. D.; Zapata, A. *J. Org. Chem.* 1989, *54*, 5854.
772. Evans, D. A.; Andrews, G. C. *Acc. Chem. Res.* 1974, *7*, 147.
773. Burgess, K.; Henderson, I. *Tetrahedron Lett.* 1989, *30*, 4325.
774. Sato, T.; Otera, J.; Nozaki, H. *J. Org. Chem.* 1989, *54*, 2779.
775. Goldmann, S.; Hoffmann, R. W.; Maak, N.; Geueke, K. J. *Chem. Ber.* 1980, *113*, 831.
776. Evans, D. A.; Andrews, G. C.; Sims, C. L. *J. Am. Chem. Soc.* 1971, *93*, 4956.
777. Grieco, P. A. *J. Chem. Soc., Chem. Commun.* 1972, 702.
778. Evans, D. A.; Andrews, G. C. *J. Am. Chem. Soc.* 1972, *94*, 3672.
779. Babler, J. H.; Haack, R. A. *J. Org. Chem.* 1982, *47*, 4801.
780. Hoffmann, R. W.; Goldmann, S.; Maak, N.; Gerlach, R.; Frickel, F.; Steinbach, G. *Chem. Ber.* 1980, *113*, 819.
781. Ogiso, A.; Kitazawa, E.; Kurabayashi, M.; Sato, A.; Takahashi, S.; Noguchi, H.; Kuwano, H.; Kobayashi, S.; Mishima, H. *Chem. Pharm. Bull.* 1978, *26*, 3117.
782. Evans, D. A.; Andrews, G. C.; Fujimoto, T. T.; Wells, D. *Tetrahedron Lett.* 1973, 1389.
783. Masaki, Y.; Hashimoto, K.; Sakuma, K.; Kaji, K. *J. Chem. Soc., Chem. Commun.* 1979, 855.
784. Evans, D. A.; Andrews, G. C.; Fujimoto, T. T.; Wells, D. *Tetrahedron Lett.* 1973, 1385.
785. Trost, B. M.; Rigby, J. H. *J. Org. Chem.* 1978, *43*, 2938.
786. Masaki, Y.; Hashimoto, K.; Kaji, K. *Tetrahedron Lett.* 1978, 4539.
787. Kojima, K.; Koyama, K.; Ameniya, S. *Tetrahedron* 1985, *41*, 4449.
788. Miller, J. G.; Kurz, W.; Untch, K. G.; Stork, G. *J. Am. Chem. Soc.* 1974, *96*, 6774.
789. Tang, R.; Mislow, K. *J. Am. Chem. Soc.* 1970, *92*, 2100.
790. Hoffmann, R. W.; Maak, N. *Tetrahedron Lett.* 1976, 2237.
791. Baudin, J.-B.; Julai, S. A. *Tetrahedron Lett.* 1989, *30*, 1963.
792. Whitesell, J. K.; Yaser, H. K. *J. Am. Chem. Soc.* 1991, *113*, 3526.
793. Delenis, G.; Dunogues, J.; Gadras, A. *Tetrahedron* 1988, *44*, 4243.
794. Wittman, M. D.; Halcomb, R. L.; Danishefsky, S. J. *J. Org. Chem.* 1990, *55*, 1979.
795. Taber, D. F. *J. Am. Chem. Soc.* 1977, *99*, 3513.
796. Kondo, K.; Umemoto, T.; Takahatake, Y.; Tunemoto, D. *Tetrahedron Lett.* 1977, 113.
797. Kondo, K.; Umemoto, T.; Takahatake, Y.; Tunemoto, D. *Tetrahedron Lett.* 1978, 3927.
798. Grieco, P. A.; Finkelhur, R. S. *J. Org. Chem.* 1973, *38*, 2245.
799. Baudin, J.-B.; Julia, S. A. *Tetrahedron Lett.* 1988, *29*, 3251.
800. Tamaru, Y.; Nagao, K.; Bando, T.; Yoshida, Z.-I. *J. Org. Chem.* 1990, *55*, 1823.
801. Bock, T.; Lai, E. K. Y.; Muralidhara, K. R. *J. Org. Chem.* 1990, *55*, 4595.
802. Inoue, Y. *Chem. Rev.* 1992, *92*, 741.
803. Rau, H. *Chem. Rev.* 1983, *83*, 535.
804. Kagan, H. B.; Fiaud, J. C. *TSC* 1978, *10*, 175.
805. Kaupp, G.; Haak, M. *Angew. Chem., Int. Ed. Engl.* 1993, *32*, 694.
806. Green, B. S.; Rabinsohn, Y.; Rejtö, M. *CarbRes* 1975, *45*, 115.
807. Green, B. S.; Rabinsohn, Y.; Rejtö, M. *J. Chem. Soc., Chem. Commun.* 1975, 313.
808. Green, B. S.; Hagler, A. T.; Robinsohn, Y.; Rejtö, M. *IsJC* 1977, *15*, 124.
809. Seebach, D.; Daum, H. *J. Am. Chem. Soc.* 1971, *93*, 2795.
810. Seebach, D.; Oei, H.-A. *Chem. Ber.* 1977, *110*, 2316.
811. Seebach, D.; Oei, H. A. *Angew. Chem., Int. Ed. Engl.* 1975, *14*, 634.
812. Partridge, J. J.; Chadha, N. K.; Uskokovic, M. R. *J. Am. Chem. Soc.* 1973, *95*, 532.
813. Chaudhuri, R. K.; Ikeda, T.; Hutchinson, R. C. *J. Am. Chem. Soc.* 1984, *106*, 6004.
814. Koch, H.; Runsink, J.; Scharf, H.-D. *Tetrahedron Lett.* 1983, *24*, 3217.
815. Wynberg, H.; Staring, E. G. J. *J. Am. Chem. Soc.* 1982, *104*, 166.
816. Schreiber, S. L.; Desmaele, D.; Porco, J. A. *Tetrahedron Lett.* 1988, *29*, 6689.
817. Kozluk, T.; Zamojski, A. *Tetrahedron* 1983, *39*, 805.
818. Zamojski, A.; Jarosz, S. *Tetrahedron* 1982, *38*, 1447.
819. Zamojski, A.; Jarosz, S. *Tetrahedron* 1982, *38*, 1453.
820. Schreiber, S. L. *Science* 1985, *225*, 857.
821. Schreiber, S. L.; Satake, K. *Tetrahedron Lett.* 1986, *23*, 2575.
822. Musser, A. K.; Fuchs, P. L. *J. Org. Chem.* 1982, *47*, 3121.
823. Pira, O.; Henin, F.; Muzart, J.; Pete, J.-P. *Tetrahedron Lett.* 1987, *28*, 4825.
824. Engler, T. A.; Ali, M. H.; Velde, D. V. *Tetrahedron Lett.* 1989, *30*, 1761.

825. Marouka, K.; Concepcion, A. B.; Yamamoto, H. *Synlett* 1992, 31.
826. Kobayashi, Y.; Takemoto, Y.; Ito, Y.; Terashima, S. *Tetrahedron Lett.* 1990, *31*, 3031.
827. Ojima, I.; Inaba, S.-I. *Tetrahedron Lett.* 1980, *21*, 2081.
828. Ojima, I.; Inaba, S. *Tetrahedron Lett.* 1980, *21*, 2077.
829. Houge, C.; Frisque-Hesbain, A. M.; Mockel, A.; Ghosez, L.; Declercq, J. P.; Germain, G.; Van Meerssche, M. *J. Am. Chem. Soc.* 1982, *104*, 2920.
830. Gluchouski, C.; Cooper, L.; Bergbreiter, D. E.; Newcomb, M. *J. Org. Chem.* 1980, *45*, 3413.
831. Gotthardt, H.; Lenz, W. *Angew. Chem., Int. Ed. Engl.* 1979, *18*, 868.
832. Gotthardt, H.; Lenz, W. *Tetrahedron Lett.* 1979, 2879.
833. Bernstein, J.; Green, B. S.; Rejto, M. *J. Am. Chem. Soc.* 1980, *102*, 323.
834. Elgavi, A.; Green, B. S.; Schmidt, G. M. *J. Am. Chem. Soc.* 1973, *95*, 2058.
835. Addadi, L.; Lahav, M. *J. Am. Chem. Soc.* 1978, *100*, 2838.
836. Green, B. S.; Hagler, A. T.; Rabinsohn, Y.; Rejto, M. *Isr. J. Chem.* 1977, *15*, 24.
837. Baldwin, J. E.; Roy, U. V. *J. Chem. Soc., Chem. Commun.* 1969, 1225.
838. Bertrand, M.; Le Gras, J.; Goré, J. *Tetrahedron Lett.* 1972, 2499.
839. Duncan, W. G.; Weyler, W.; Moore, H. W. *Tetrahedron Lett.* 1973, 4391.
840. Bampfield, H. A.; Brook, P. R. *J. Chem. Soc., Chem. Commun.* 1974, 171.
841. Bampfield, H. A.; Brook, P. R. *J. Chem. Soc., Chem. Commun.* 1975, 132.
842. Moore, W. R.; Bach, R.; Ozretich, T. M. *J. Am. Chem. Soc.* 1969, *91*, 5918.
843. Elgavi, A.; Green, B. S.; Schmidt, G. M. J. *J. Am. Chem. Soc.* 1973, *95*, 2058.
844. Green, B. S.; Hagler, A. T.; Rabinsohn, Y.; Rejto, M. *Isr. J. Chem.* 1976-77, *15*, 124.
845. Tietze, L. F.; Glüsenkamp, K.-H.; Nakame, M.; Hutchinson, C. R. *Angew. Chem., Int. Ed. Engl.* 1982, *21*, 70.
846. Paquette, L. A.; Freeman, J. P. *J. Am. Chem. Soc.* 1969, *91*, 7548.
847. Paquette, L. A.; Freeman, J. P.; Mawrana, S. *Tetrahedron* 1971, *27*, 2599.
848. Belzecki, C.; Krawczyk, Z. *J. Chem. Soc., Chem. Commun.* 1977, 302.
849. Fukukawa, M.; Okawara, T.; Noguchi, Y.; Terawaki, Y. *Chem. Pharm. Bull.* 1979, *27*, 2795.
850. Kamiya, T.; Oku, T.; Nakaguchi, O.; Takeno, H.; Hashimoto, M. *Tetrahedron Lett.* 1978, 5119.
851. Hatanaka, N.; Ojima, I. *J. Chem. Soc., Chem. Commun.* 1981, 344.

Chapter 13

# CYCLIZATION REACTIONS AND REACTIONS OF CYCLIC SYSTEMS

The purpose of this chapter is not to discuss the multitude of possible cyclic systems nor how stereochemistry can be controlled and manipulated within specific systems.[1-4] This book has discussed asymmetric methodology with a strong emphasis on the control of acyclic stereochemistry.[§] This theme is continued in this chapter; reactions that can be employed through a cyclic compound to ultimately control acyclic stereochemistry have provided the majority of the material. This firmly places heterocycles at the forefront of the subject matter. It should, however, be remembered that carbocycles can be used to provide acyclic compounds by ring cleavage reactions, such as ozonolysis or the Baeyer-Villiger oxidation,[6-10] and to provide a transition between cyclic and acyclic frameworks.[11,12]

Some carbocyclic chemistry will still be found within this chapter, to illustrate some of the concepts of the use of cyclic systems. In addition, subjects such as cyclopropanes have been included in this chapter, as this was the obvious place for this material.

Reactions of cyclic systems have been discussed as appropriate in other chapters;[†] this includes pericyclic reactions [Chapter 12],[13] chiral cyclic acetals acting as carbonyl equivalents,[14] and examples of bicyclic systems.[15-17]

With regard to the heterocyclic systems discussed in this chapter, the emphasis has obviously been placed on oxygen heterocycles, especially lactones, because of the simple methods — such as hydrolysis and reduction — that are available to convert them to acyclic systems.

The use of Baldwin's rules for the formation of cyclic systems can rationalize many of the ring-forming reactions [*2.4*]. Cyclization reactions have been omitted when the product's steric requirements are the overriding factors, such as a *cis*-cyclopropane in a polycyclic system, and allow for no flexibility in the approach.

## 13.1. CARBOCYCLES

### 13.1.1. CYCLOPROPANATIONS

The cyclopropane ring is strained and, thus must be *cis* if attached to another cyclic system. Formation of the three-membered ring system is controlled by face selectivity,[18] usually by reaction of a carbenoid with an alkene, although ring closure reactions are available.[19-22] A chiral auxiliary that can guide a Simmons-Smith reagent to one face can provide high face selectivity (Scheme 13.1).[23-27]

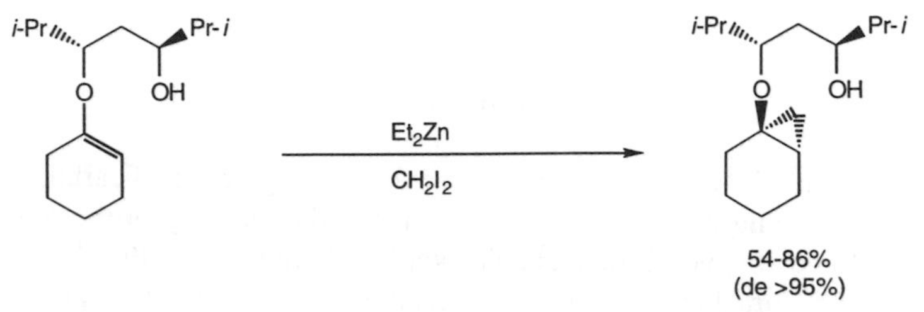

i-Pr Pr-i O OH → $Et_2Zn$ / $CH_2I_2$ → i-Pr Pr-i O OH

54-86%
(de >95%)

SCHEME 13.1.

§ Another book in this series covers cyclization reactions.[5]

† Specific examples can be found through the index.

An "external" chiral auxiliary or catalyst can also be employed (e.g., Scheme 13.2).[28-41]

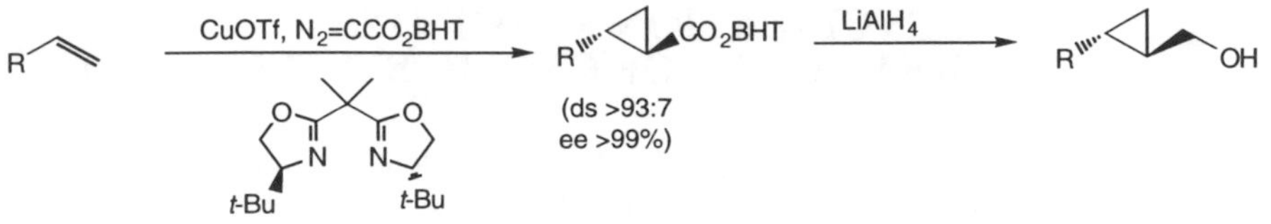

SCHEME 13.2.

A ring contraction of the pyrazoline, derived from a 1,3-dipolar addition of diazomethane with an alkene to the cyclopropane, can derive its induction from the initial ring formation reaction.[42]

A method that leads to a wide variety of cyclopropanone derivatives involves reaction of an organocuprate with a cyclopropenone analog (Scheme 13.3).[43]

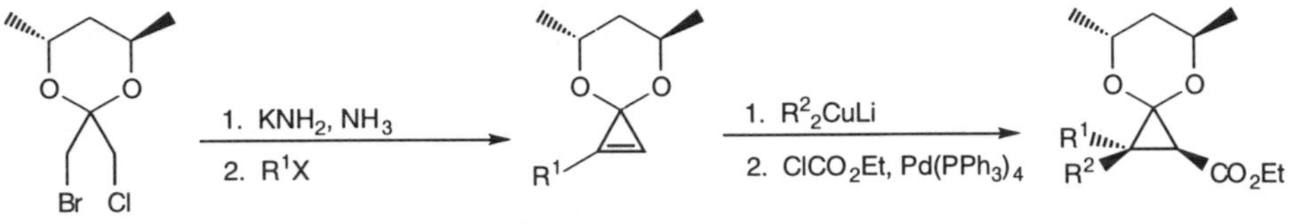

SCHEME 13.3.

## 13.1.2. CYCLIZATIONS

Many ring closure reactions rely on an $S_N2$-type reaction, and many examples have already been seen with regard to Baldwin's rules [*2.4*]. In nature, an enzyme catalyzes the cyclization of squalene to lanosterol.[44,45] A similar polycyclization reaction is available based on an electrophilic cyclization from an epoxide (vide infra) (Scheme 13.4),[46-48] or closely related chemistry.[49-51]

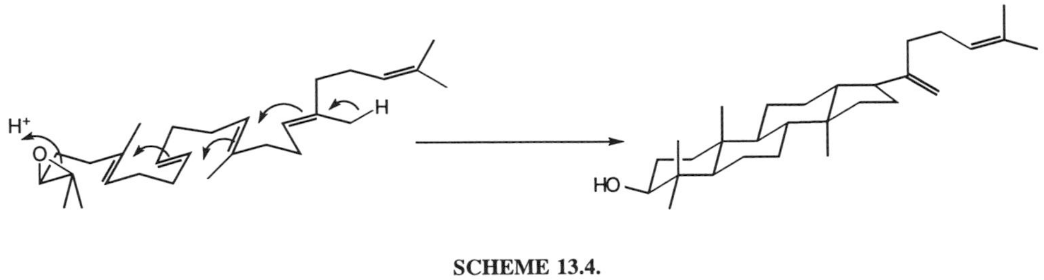

SCHEME 13.4.

Many synthetic methods have been developed for the preparation of carbocycles.[52-58] As we are concerned with control of acyclic stereochemistry, the majority of these methods are not relevant here. However, one of the reactions worthy of comment is the Robinson annelation in the presence of a chiral ligand or base. This reaction is a variation of an intra-molecular aldol reaction [Chapter 7] where selectivity has been observed (Scheme 7.3).[59,60] Unfortunately, many of the methods for the Robinson annelation are system specific or provide less than useful levels of induction.[61-63] However, many chiral auxiliaries have been incorporated into annelation methodologies, especially for the first step, 1,4-addition of the enolate [*9.5.1*].[64-73] Aldol reactions can be influenced by proline (*cf.* Scheme 7.3), and this methodology has been extended for the preparation of the bicyclic enone **13.1** (Scheme 13.5).[74]

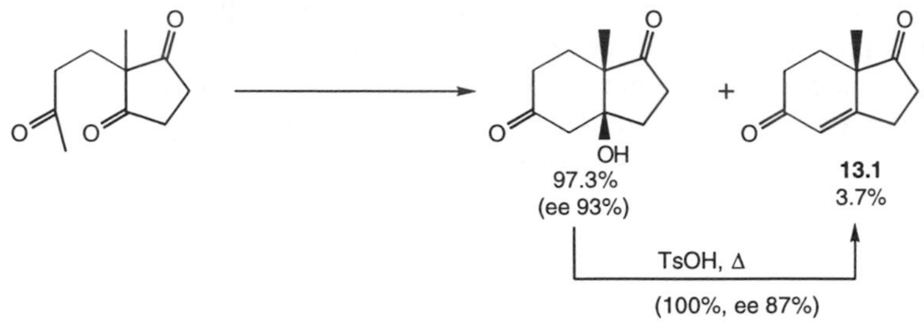

OH
97.3%
(ee 93%)

**13.1**
3.7%

TsOH, Δ
(100%, ee 87%)

SCHEME 13.5.

### 13.1.3. REACTIONS OF CARBOCYCLES

As a ring system can impart significant facial selectivity to a reaction, a discussion on this subject could be voluminous. We have, therefore, limited our coverage to reactions already outlined in other chapters. Of course, face selectivity may be inherent in a cyclic system, especially if it is polycyclic; the preponderance for axial reactions may result in selectivity with cyclic substrates not observed in acyclic counterparts.† Reactions where the selectivity of the product is dependent upon the reaction itself, rather than the position of a substituent, are particularly useful; and an example is provided by the aldol reaction of a silyl enol ether with a substituted cyclohexanone acetal where equatorial attack is the preferred reaction pathway.[75]

One area that has attracted particular attention is the reduction of cyclic ketones and nitrogen derivatives [*4.5.1.2*]; larger reducing agents tend to provide axial alcohols and amines.[76-86]

Oxidation of cyclic allyl alcohols can be very selective, as the hydroxy group will guide the oxidant to just one face as long as complex is formed (e.g., Scheme 13.6).[87-92]

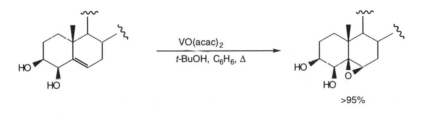

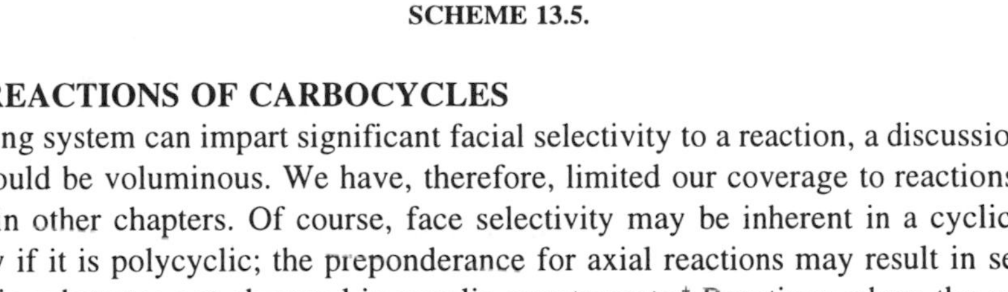

HO
HO
$VO(acac)_2$
$t$-BuOH, $C_6H_6$, Δ
HO
HO
O
>95%

SCHEME 13.6.

Epoxides attached to another ring system can often provide reactions with high stereoselection, as axial attack by the nucleophile is the normal reaction pathway.[93-97]

Reductions can also be directed by a hydroxy group on a cyclic substrate.[98-100]

## 13.2. PREPARATIONS OF LACTONES AND CYCLIC ETHERS

### 13.2.1. ELECTROPHILIC CYCLIZATIONS

Intramolecular cyclization is a useful method for the preparation of lactones and cyclic ethers.[101] The most common examples are iodolactonization and iodoetherification, the difference being that the former uses a carboxylic acid derivative as the nucleophile, while the latter relies on a hydroxy group. Thus, butyrolactones are available from γ,δ-unsaturated carboxylic acid derivatives,[102-112] while unsaturated alcohols lead to cyclic ethers.[113-125] Lactones are also available from a wide variety of nucleophiles such as carbonates,[126]

† Examples of this type can be found for asymmetric deprotonations [*5.4.1*].

orthoesters,[127] or carbamates,[128,129] which can all be used in place of a carboxylate anion.[129,130] The approach has been used to prepare 1,3-diols [*9.7*].

The intermediate iodonium ion controls the relative stereochemistry of the cyclization. An asymmetric center present within the substrate, therefore, allows for enantioselectivity (Scheme 13.7).[131] The reaction is very susceptible to the steric interactions within the transition state.[128,131-133]

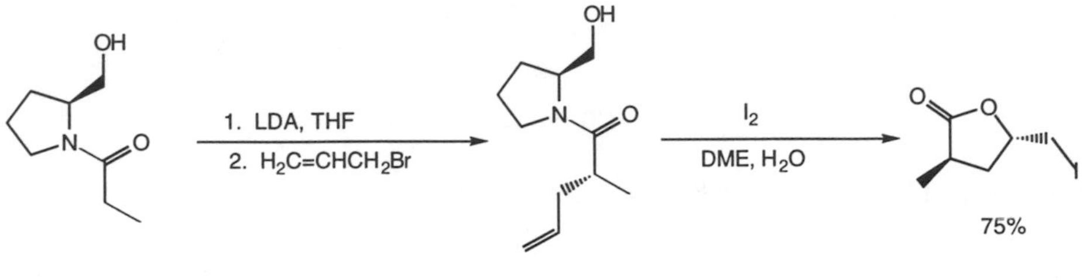

OH
1. LDA, THF
2. H2C=CHCH2Br
OH
$I_2$
DME, H2O
75%

**SCHEME 13.7.**

The resultant iodo compound can be used for further elaboration, as illustrated by the *in situ* reaction of an epoxide (Scheme 13.8).[111,134]

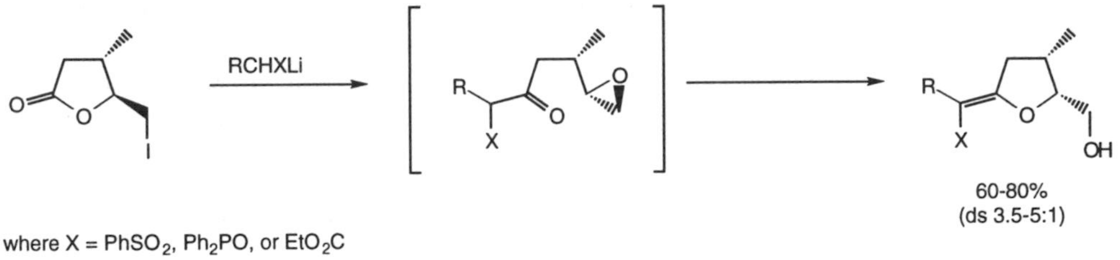

RCHXLi
R
X
R
X
OH
60-80%
(ds 3.5-5:1)
where X = PhSO2, Ph2PO, or EtO2C

**SCHEME 13.8.**

Variations on this cyclization methodology include the use of bromine, rather than iodine; again, both lactones and ethers can be prepared.[118,126,127,129,135-140]

An asymmetric bromolactonization procedure affords α-hydroxy acids with good asymmetric induction.[141-146] The procedure can be modified to prepare α,β-epoxy aldehydes.[147,148]

The selective oxidation of the diene **13.2**, followed by bromoetherification leads to the *cis*-tetrahydrofuran **13.3** (Scheme 13.9).[149]

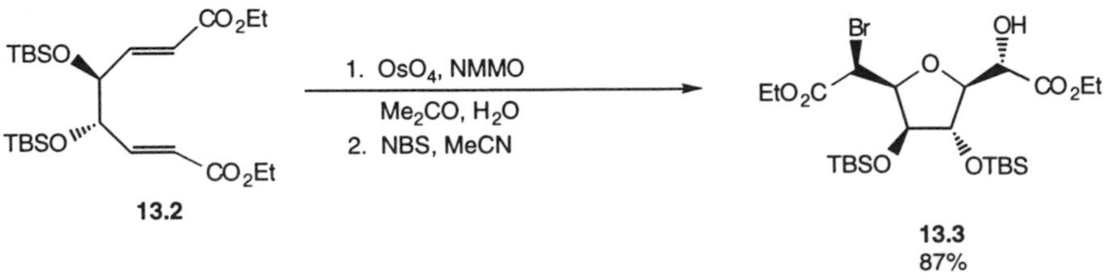

CO2Et
TBSO
TBSO
CO2Et
13.2
1. OsO4, NMMO
Me2CO, H2O
2. NBS, MeCN
Br
OH
EtO2C
CO2Et
TBSO
OTBS
13.3
87%

**SCHEME 13.9.**

A further variation utilizes selenium or sulfur as the electrophilic species to induce cyclization. The diversity of organoselenium or organosulfur chemistry is then available for further modification of the product.[150-152] Both lactonizations with selenium[153-157] and sulfur[158] are known. However, the majority of examples form cyclic ethers (Scheme 13.10).[152,159-167]

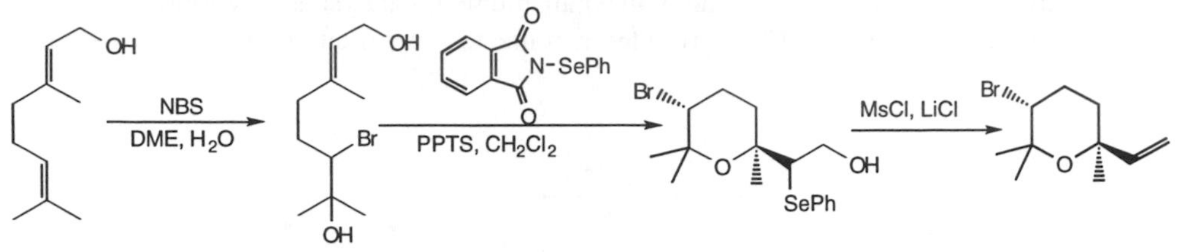

SCHEME 13.10.

In addition to halogens and selenium compounds, metal salts such as those of titanium,[168-170] lead,[171] thallium,[172] and mercury,[173-177] can be used to activate an alkene.

Oxidative methods can be used to induce cyclization. Use of hypervalent iodine compounds provides an electrophilic center adjacent to a carbonyl group (Scheme 13.11).[178] These reagents can also be used with carbon–carbon unsaturated compounds.[179,180] A *p*-methoxystyrene derivative, upon oxidation with DDQ, provides an electrophile for intramolecular attack by a predisposed hydroxy group.[181,182] Oxidative cyclizations of 5,6-dihydroxyalkenes with Cr(VI) species afford *cis*-tetrahydrofuran diols.[183]

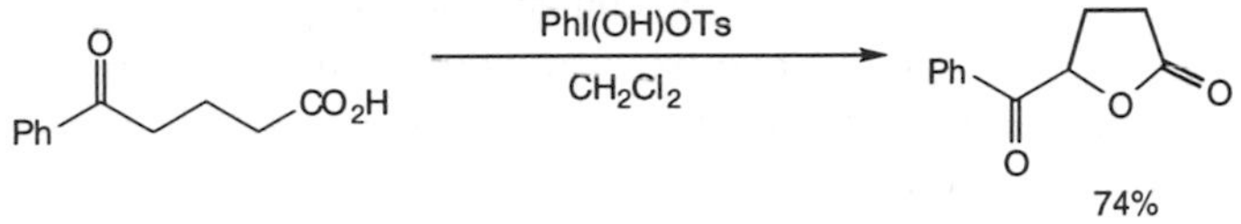

SCHEME 13.11.

An alkene can also be converted to an epoxide; this then allows cyclization to an ether or lactone.[184-193] Such an approach was used to establish four stereogenic centers from a single one, as in the allyl alcohol **11.10** (Scheme 11.16).[135,194-198] These reactions usually derive their stereoselectivity from the oxidation step, as the epoxide opening occurs by an $S_N2$ mechanism. In addition to five-membered ring compounds, six-membered oxygen heterocycles[199] and carbocycles are also available by this approach,[200-202] as are nitrogen heterocycles.[203] Cascade reactions are also possible.[204-206]

As intramolecular nucleophilic attack on an epoxide will provide the smaller ring compound when Baldwin's rules are followed [*2.4*], an alternative strategy has to be employed if the tetrahydropyran derivative is the desired product. One of the alternative methodologies has been to use a 6-*exo-trig* cyclization based on a Michael reaction; the stereochemical control is increased by the use of low temperature (Scheme 13.12).[186] Many variations on this theme exist.[207]

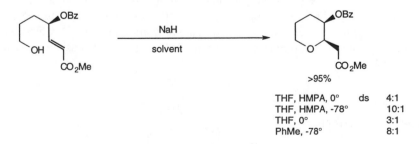

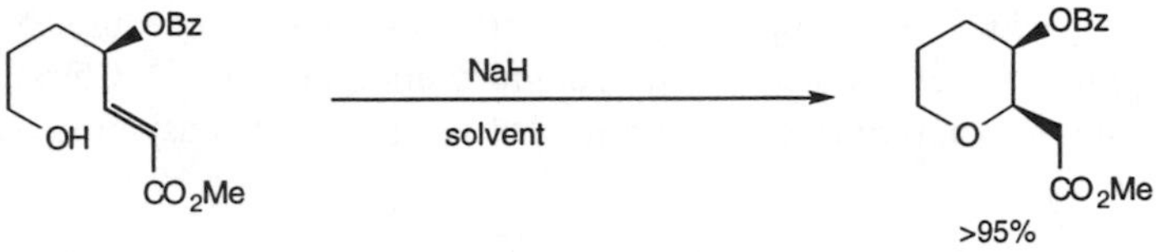

| | ds |
|---|---|
| THF, HMPA, 0° | 4:1 |
| THF, HMPA, -78° | 10:1 |
| THF, 0° | 3:1 |
| PhMe, -78° | 8:1 |

SCHEME 13.12.

Cyclization of a 1-hydroxy-5-alkene with a palladium(II) catalyst and hydride elimination controlled by the solvent, DMSO, provides a route to *cis*-2,6-tetrahydropyrans (Scheme 13.13).[208-224]

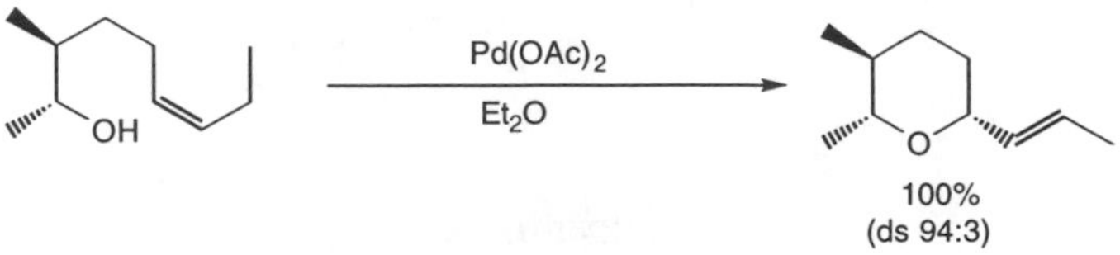

$Pd(OAc)_2$, $Et_2O$; 100% (ds 94:3)

**SCHEME 13.13.**

δ,ε-Unsaturated aldehydes and ketones, where the unsaturation also bears an electron-withdrawing group, can be reductively cyclized to the cyclopentanol derivative with samarium iodide, but control of stereochemistry is mediocre unless a bicyclic system results.[225]

A lactone ring expansion, with a 2 + 2 addition providing the lactone, provides *syncat* tetrahydrofurans.[226]

## 13.2.2. PREPARATIONS OF BUTYROLACTONES

γ-Butyrolactones can provide some useful template chemistry for the preparation of acyclic compounds.[2,3,15,227] Cyclization of a γ-hydroxy acid often occurs spontaneously to afford the butyrolactone.[228-235] This allows for resolution techniques to be applied to lactones by a ring opening–resolution–ring closure sequence.[236]

One of the most direct entries into an asymmetric example of this class of compounds is from L-glutamic acid (Scheme 1.1).[237-241] (*S*)-3-Hydroxy-γ-butyrolactone (**13.4**) is available from (*S*)-(−)-malic acid (**13.5**) (Scheme 13.14).[242] Other natural products that afford butyrolactones are carbohydrates,[243,244] such as mannose,[245-250] mannitol,[251-253] ribose,[248] citramalic acid,[254] and tartaric acid.[255] Chiral α-hydroxy esters can be converted to 4-alkyl-2-hydroxytetronic acids by a Claisen rearrangement [*12.2*].[256]

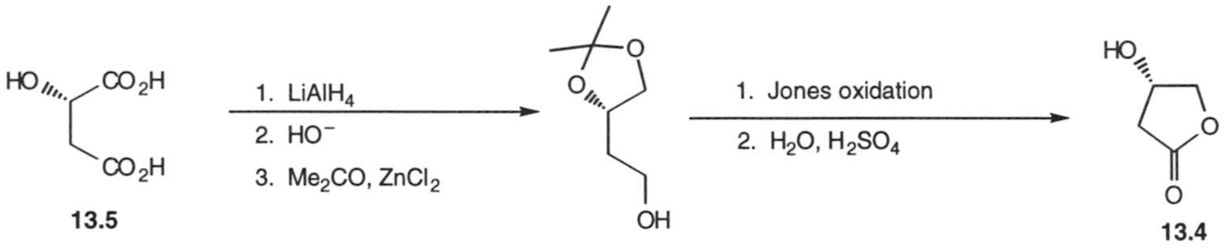

**13.5** → 1. $LiAlH_4$; 2. $HO^-$; 3. $Me_2CO$, $ZnCl_2$ → 1. Jones oxidation; 2. $H_2O$, $H_2SO_4$ → **13.4**

**SCHEME 13.14.**

The above methods all rely on readily available natural products as starting materials. Some lactones can be resolved by an enantioselective kinetic protonation procedure.[236] The cyclic system is available by a variety of means, including alkylation methods.[257] Chiral oxazolines provide methodology to γ-butyrolactones with moderate (64 to 73%) asymmetric induction (Scheme 13.15).[258]

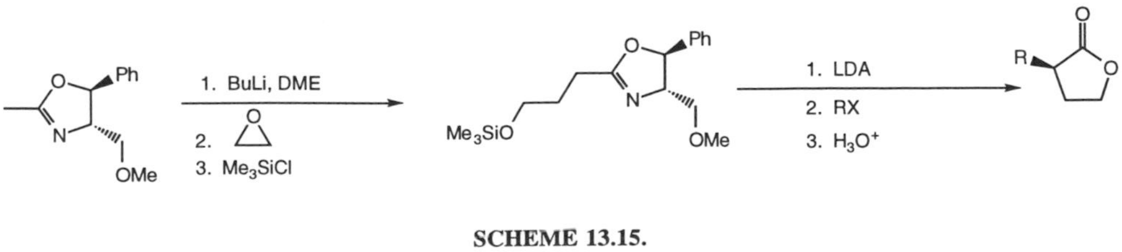

1. BuLi, DME; 2. (ethylene oxide); 3. $Me_3SiCl$ → 1. LDA; 2. RX; 3. $H_3O^+$

**SCHEME 13.15.**

Although ester enolates do not undergo a facile reaction with epoxides, the usc of an aluminum enolate not only provides reasonable chemical yields, but high diastereoselection with the *cis*-isomer predominant.[259]

Reaction of a γ-aldehyde ester (**13.6**) with a cuprate results in chelation controlled addition (Scheme 13.16). The reaction is, however, very solvent dependent.[260]

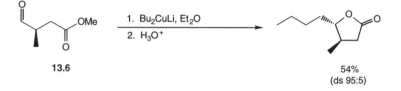

1. $Bu_2CuLi$, $Et_2O$
2. $H_3O^+$

**13.6**

54%
(ds 95:5)

**SCHEME 13.16.**

Reductions of γ-oxo esters under asymmetric hydrogenation conditions provide a powerful entry to 4-alkylbutyrolactones, as either enantiomer becomes available.[261]

An alternative condensation approach uses a chiral sulfoxide **13.7** to control stereochemistry (Scheme 13.17).[262-264]

1. LDA, THF, -75°
2. RCHO

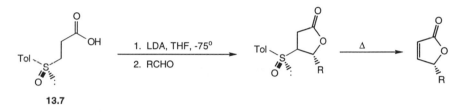

Δ

**13.7**

**SCHEME 13.17.**

A further variant utilizes a chiral enolate derived from an organoiron compound.[265,266] A related methodology, that employs zirconium, can convert optically active propargyl alcohols to butenolides.[267] Indeed, propargyl alcohols, which can be obtained by asymmetric reduction, provide a number of useful synthetic methods to butyrolactones.[268-271]

A chiral sulfoxide can be incorporated into a butyrolactone (Scheme 13.18).[272,273] The butenolide **13.8** can then be used as a Michael acceptor for further modification [*9.5.1*].[272-279] The sulfoxide group can also be used as masked unsaturation.[263]

1. TBDMSCl
2. *t*-BuLi
3. $TolSO_2Men$
4. NBuF

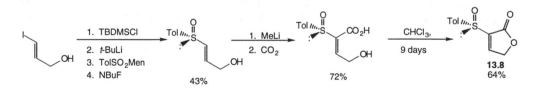

43%

1. MeLi
2. $CO_2$

72%

$CHCl_3$,
9 days

**13.8**
64%

**SCHEME 13.18.**

An approach that allows for the preparation of either isomer of a 4-substituted butyrolactone relies on the stereospecific addition of dichloroketene to an allyl sulfoxide (Scheme 13.19).[264,280-285]

Furans can also provide a facile entry to butenolides.[286-289] 2-Trimethylsiloxyfuran (**13.9**) is a four-carbon donor and a masked butenolide; it enters into stereoselective additions with carbonyl compounds (Scheme 13.20).[290-296]

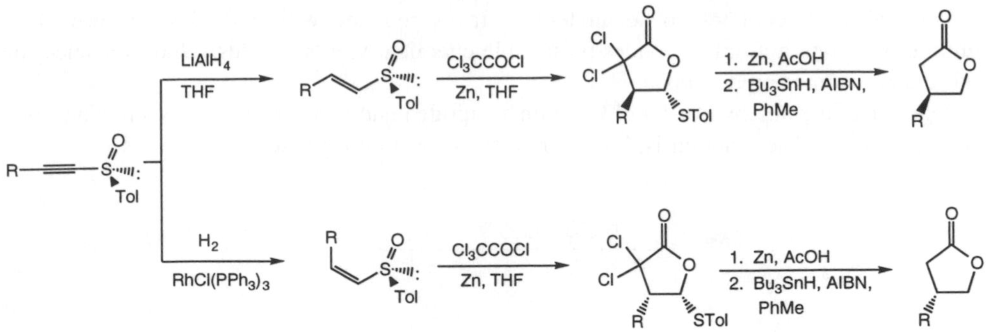

SCHEME 13.19.

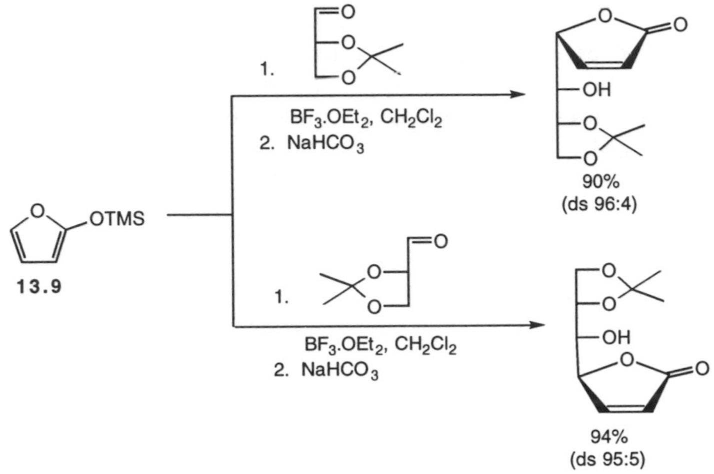

SCHEME 13.20.

In a closely related approach, β-acylvinyl anions add to carbonyl compounds to yield 4-alkylbutyrolactones.[297,298] Substituted γ-butyrolactones are available from condensation of an allyl anion with a carbonyl compound [*4.1.5*].[299,300] A similar approach is based on a Michael addition, but optical yields are low.[301] The use of α-alkoxyorganocuprates does allow the preparation of *cis*-substituted butyrolactones, but the approach is compromised by the relatively high amount of competing 1,2-addition.[233] α-Alkoxyorganolithium reagents react with α,β-unsaturated carbonyl derivatives to provide γ-hydroxy carboxylic acids and, hence, butyrolactone derivatives enantioselectively.[302]

Anh-Felkin addition of a silyl ketene acetal **13.10** to an α-amino aldehyde **13.11** provides the lactone **13.12** (Scheme 13.21).[303]

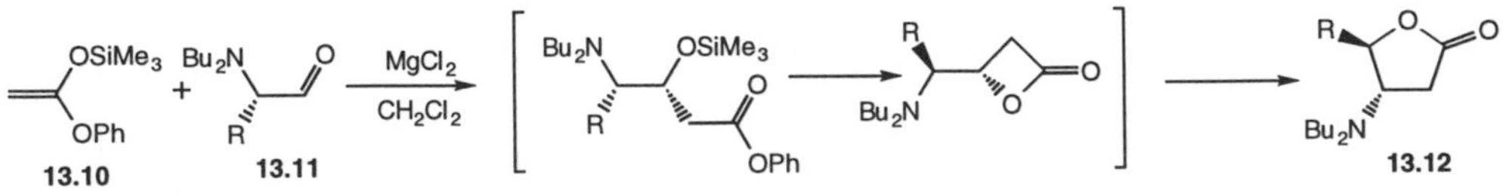

SCHEME 13.21.

The larger number of asymmetric routes to chiral butyrolactones, however, relies on a 2,3-epoxy alcohol, either derived by conversion of a carbohydrate,[304] or by a Sharpless protocol.[305,306] Opening of an epoxide by a carboxylate anion provides the basis for a general method to butyrolactones (Scheme 13.22).[190,195,204,243,304,307-313]

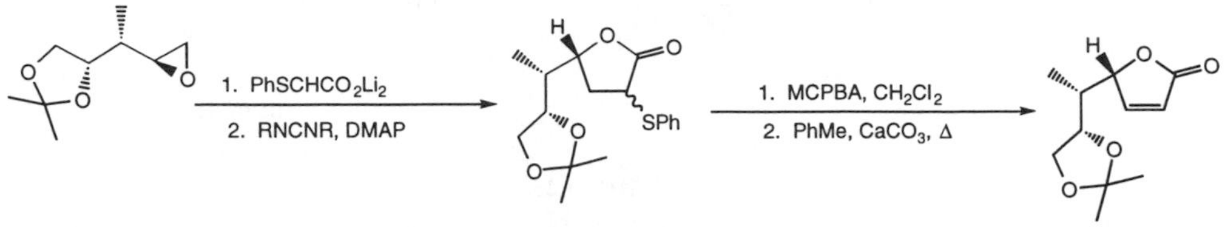

1. $PhSCHCO_2Li_2$
2. RNCNR, DMAP

1. MCPBA, $CH_2Cl_2$
2. PhMe, $CaCO_3$, Δ

**SCHEME 13.22.**

The use of an allylsilane substrate **13.13** for a peroxy acid oxidation allows access to butyrolactones with the stereochemistry of the cyclization being controlled by the carboxylic acid derivative (Scheme 13.23).[314,315]

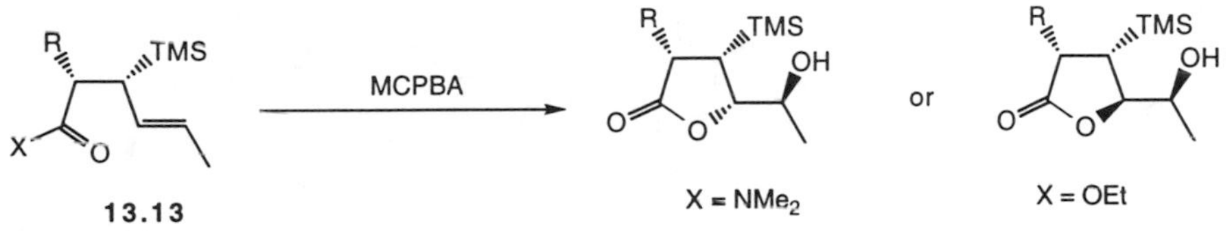

MCPBA

**13.13**

X = $NMe_2$ or X = OEt

**SCHEME 13.23.**

The stereochemistry of the lactone prepared by an acyclic method invariably results from stereochemical control during the synthesis of the precursor, as illustrated by the selective hydroxylation procedure (Scheme 13.24).[267,316]

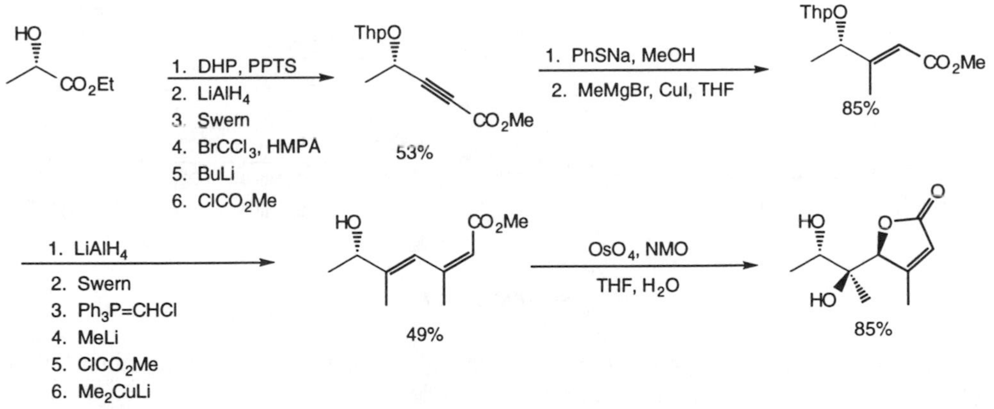

1. DHP, PPTS
2. $LiAlH_4$
3. Swern
4. $BrCCl_3$, HMPA
5. BuLi
6. $ClCO_2Me$

53%

1. PhSNa, MeOH
2. MeMgBr, CuI, THF

85%

1. $LiAlH_4$
2. Swern
3. $Ph_3P$=CHCl
4. MeLi
5. $ClCO_2Me$
6. $Me_2CuLi$
7. PPTS, MeOH

49%

$OsO_4$, NMO
THF, $H_2O$

85%

**SCHEME 13.24.**

A bicyclic system framework that can then be removed provides a stereoselective route to butenolides with secondary Grignard reagents. The bicyclic adduct can be separated from its diastereoisomer by simple recrystallization (Scheme 12.20).[317]

Lactones can be prepared by radical cyclization.[318-330] The use of α-halo esters and a chiral substrate allows for asymmetric induction (Scheme 13.25).[331]

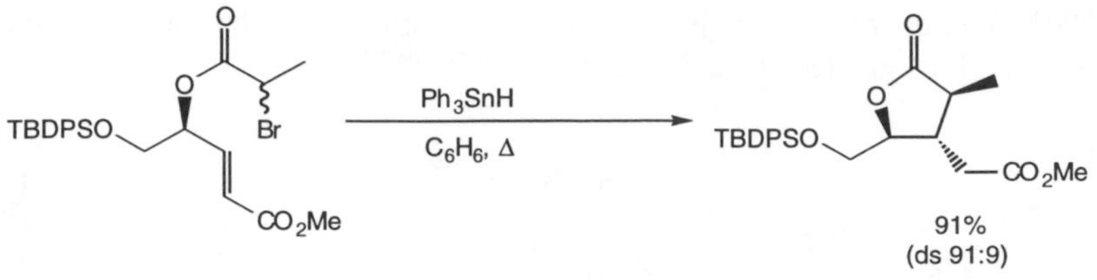

**SCHEME 13.25.**

### 13.2.3. PREPARATION OF VALEROLACTONES AND RELATED SYSTEMS

Many of the methods described for the preparation of butyrolactones can be modified to prepare valerolactones.[2,227,332] There are many routes to valerolactone derivatives,[333-339] as evident by the various syntheses of the well-known Prelog-Djerassi lactone (**13.14**) (e.g., Scheme 13.26).[340-343]

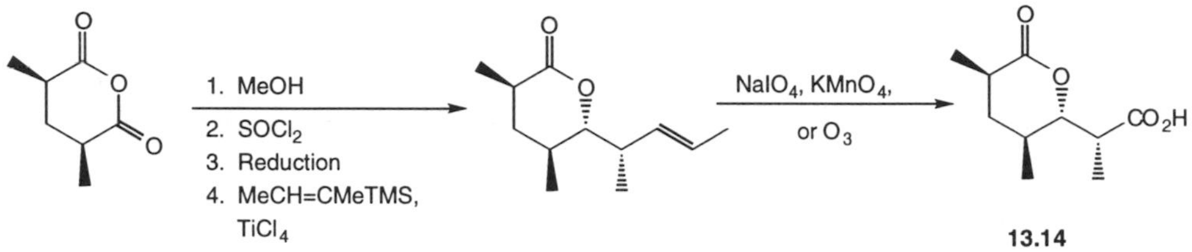

**SCHEME 13.26.**

The use of chiral auxiliaries, such as oxazolines (Scheme 13.15),[344,345] oxathianes (Scheme 13.27),[346,347] and sulfoxides[262,348] allows the preparation of butyro- or valerolactones.

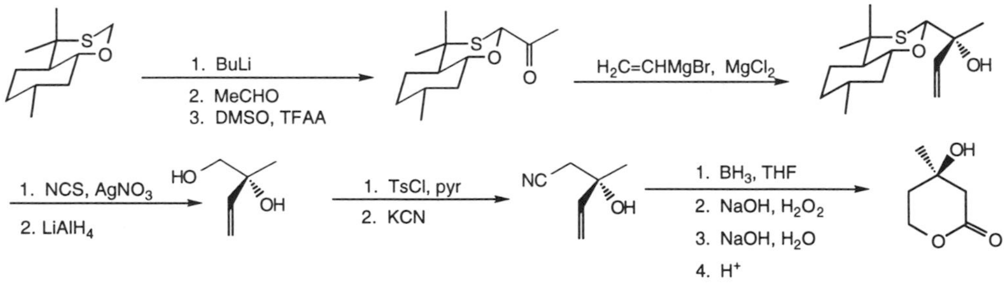

**SCHEME 13.27.**

The chiral center for the preparation of a chiral valerolactone can be an integral part of the substrate molecule (Schemes 13.26 and 13.27).[337,349-352]

Oxidation of a furan derivative followed by rearrangement gives the functionalized valerolactone (**13.15**) (Scheme 13.28).[353-355]

Another access to valerolactones relies on a Sharpless protocol to introduce the desired chirality.[356] A Sharpless kinetic resolution can be performed on the 2-furylcarbinol; one isomer is converted to the functionalized valerolactone, while the other remains as the furan compound [*11.1.2.1*].[357]

Good stereochemical control for the preparation of valerolactones can also be obtained from an acetal opening approach (Scheme 13.29) (*cf.* Scheme 7.24),[358-360] or an allyl addition [*4.1.5*].[361]

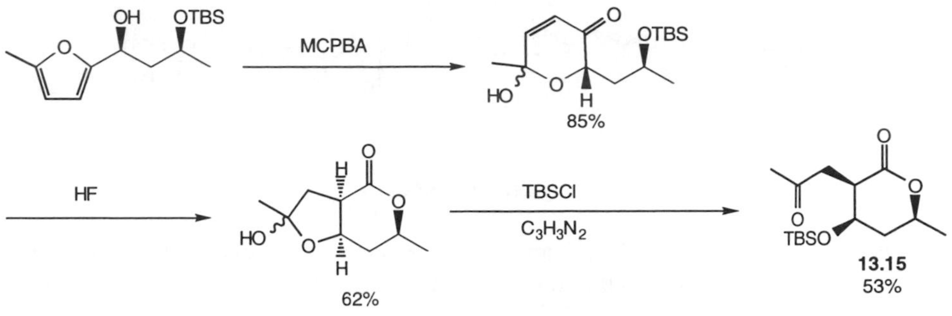

OH OTBS MCPBA O OTBS HO O H 85%

HF H O O HO O H 62% TBSCl $C_3H_3N_2$ O O O TBSO 13.15 53%

SCHEME 13.28.

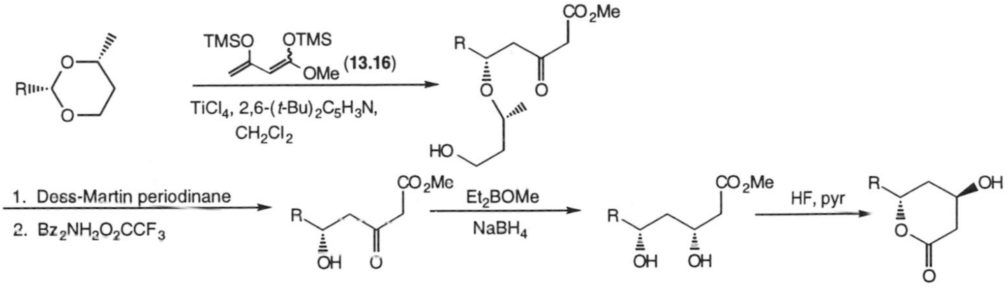

TMSO OTMS OMe (13.16) $TiCl_4$, 2,6-(t-Bu)$_2C_5H_3N$, $CH_2Cl_2$ R O O $CO_2Me$ HO

1. Dess-Martin periodinane 2. $Bz_2NH_2O_2CCF_3$ R OH O $CO_2Me$ $Et_2BOMe$ $NaBH_4$ R OH OH $CO_2Me$ HF, pyr R O O OH

SCHEME 13.29.

The stereochemistry from a homoallyl alcohol can be retained in the product γ-lactone by use of a hydroformylation reaction. This is illustrated by another example of the synthesis of the Prelog-Djerassi lactone (**13.14**) (Scheme 13.30).[362]

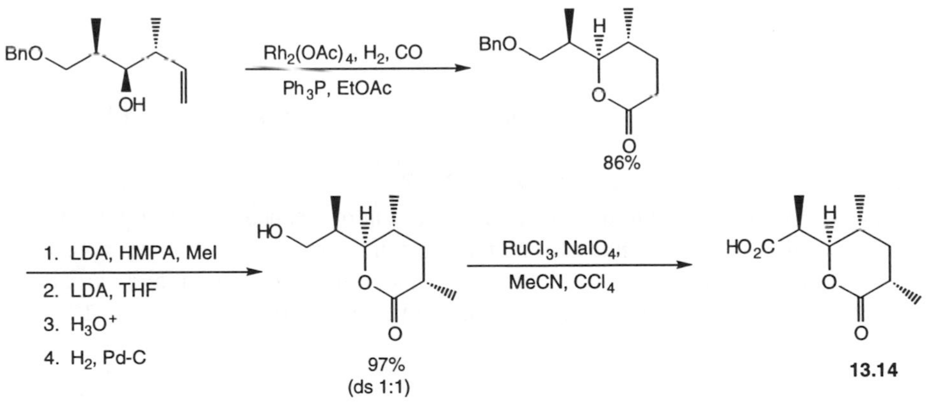

BnO OH $Rh_2(OAc)_4$, $H_2$, CO $Ph_3P$, EtOAc BnO H O O 86%

1. LDA, HMPA, MeI 2. LDA, THF 3. $H_3O^+$ 4. $H_2$, Pd-C HO H O O 97% (ds 1:1) $RuCl_3$, $NaIO_4$, MeCN, $CCl_4$ $HO_2C$ H O O 13.14

SCHEME 13.30.

2,3-Dihydro-4*H*-pyran-4-ones are available by a Mukaiyama reaction of 2,4-bissilyl-1,3-pentadienes (**13.16**) with aldehydes or ketones. The reaction is under chelation control and, therefore, does provide some diastereoselection (*cf.* Scheme 13.28).[363,364]

*Cis*-Dihydropyrones (**13.17**) are available by an aldol approach, which also sets up the requisite stereochemistry (Scheme 13.31).[365,366]

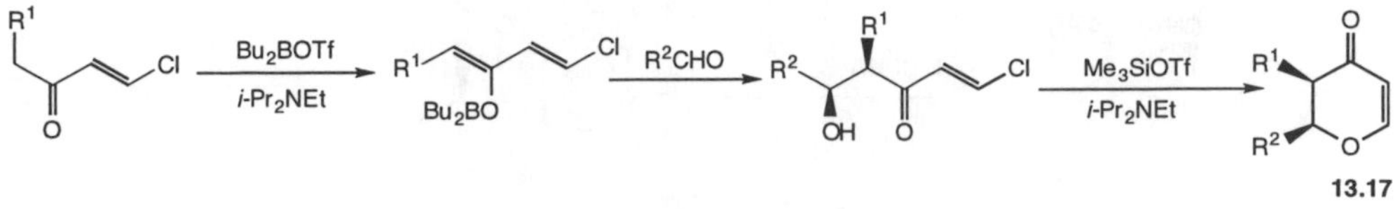

R1
Cl
O
Bu2BOTf
i-Pr2NEt
R1
Cl
Bu2BO
R2CHO
R2
R1
OH
O
Cl
Me3SiOTf
i-Pr2NEt
O
R1
R2
O
13.17

SCHEME 13.31.

Of course, tetrahydropyranyl derivatives can be prepared by cyclization reactions that utilize stereogenic centers within the substrate.[367] Thus, the pyrone (**13.18**) was prepared from the β-keto ester **13.19** (Scheme 13.32).[368,369]

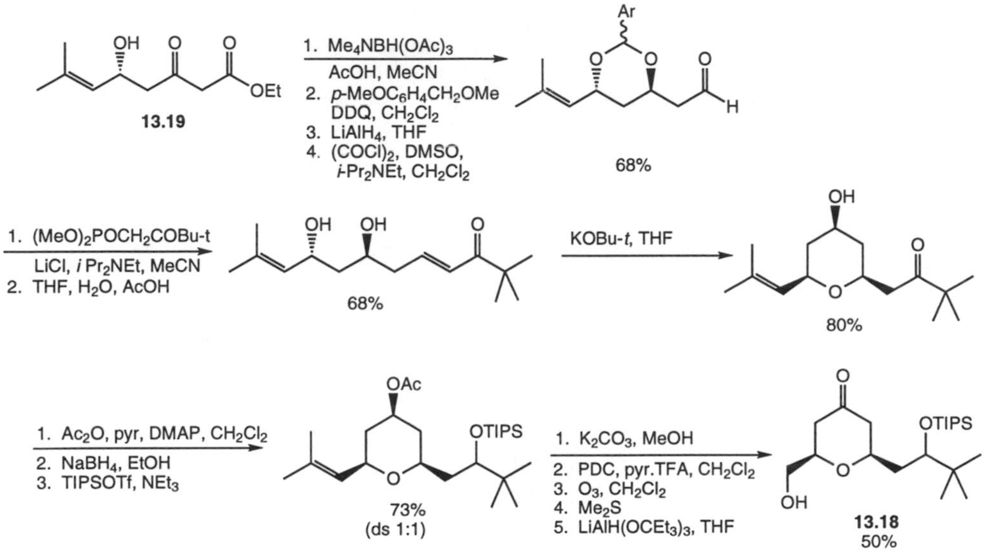

OH O O
OEt
13.19
1. Me4NBH(OAc)3
AcOH, MeCN
2. p-MeOC6H4CH2OMe
DDQ, CH2Cl2
3. LiAlH4, THF
4. (COCl)2, DMSO,
i-Pr2NEt, CH2Cl2
Ar
O O O
H
68%
1. (MeO)2POCH2COBu-t
LiCl, i Pr2NEt, MeCN
2. THF, H2O, AcOH
OH OH O
68%
KOBu-t, THF
OH
O
O
80%
1. Ac2O, pyr, DMAP, CH2Cl2
2. NaBH4, EtOH
3. TIPSOTf, NEt3
OAc
OTIPS
O
73%
(ds 1:1)
1. K2CO3, MeOH
2. PDC, pyr.TFA, CH2Cl2
3. O3, CH2Cl2
4. Me2S
5. LiAlH(OCEt3)3, THF
O
OTIPS
O
OH
13.18
50%

SCHEME 13.32.

# 13.3. REACTIONS OF LACTONES AND RELATED SYSTEMS

## 13.3.1. REACTIONS OF BUTYROLACTONES

The steric constraints within the five-membered ring of a substituted butyrolactone allow for the stereoselective introduction of functional groups[2] either by reaction juxtaposed to the carbonyl group, or at the β-position.[370,371] Functional group manipulation and extrusion of a carbon allow for the conversion of an α-substituent to a β-substituent (Scheme 13.33).[372]

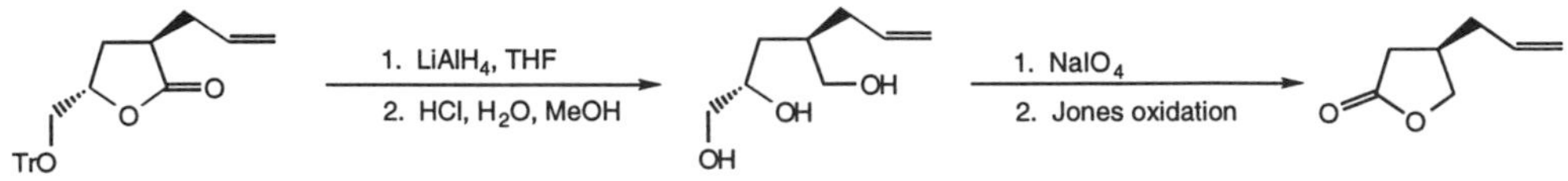

O
O
TrO
1. LiAlH4, THF
2. HCl, H2O, MeOH
OH
OH
OH
1. NaIO4
2. Jones oxidation
O
O

SCHEME 13.33.

Alkylation or aldol condensation of the dianion derived from lactone **13.20** results in stereoselective introduction of the electrophilic moiety (Scheme 13.34).[242,373-375]

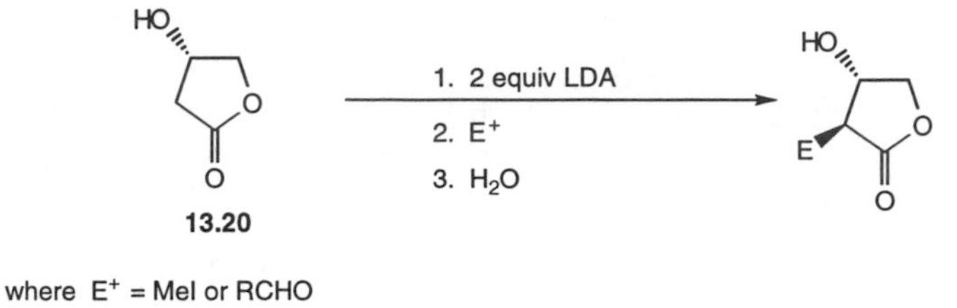

**SCHEME 13.34.**

Conversion of the hydroxy group of a 2-hydroxy butyrolactone to an *O*-trifluoromethanesulfonate ester followed by treatment with lithium iodide trihydrate in tetrahydrofuran provides the 2-deoxylactone.[376]

The sidechain of a chiral butyrolactone can control the facial selectivity next to the carbonyl moiety (vide infra).[313,370] The use of a chiral 5-alkoxy substituent on a butenolide also allows for asymmetric induction (Scheme 13.35).[377-383]

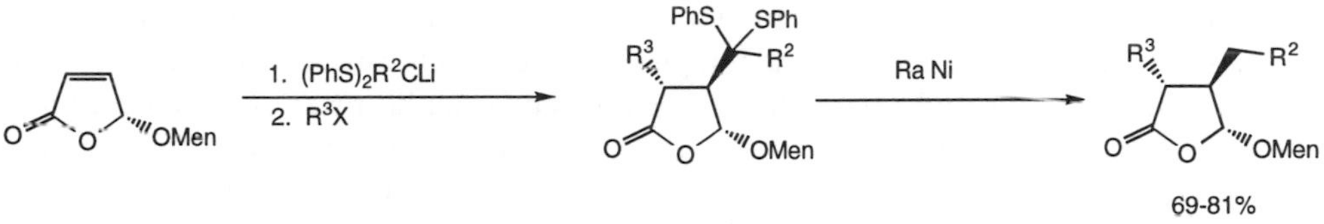

**SCHEME 13.35.**

The facial selectivity imparted by a 5-substituent also allows for the consecutive, stereoselective introduction of two alkyl groups at C-2 in a butyrolactone.[240] Indeed, protonation of the lithium enolate of butyrolactone-bearing substituents at the $\alpha$- and $\gamma$-positions provides the *syn*-isomer selectively.[371] A $\beta$-substituent allows for the introduction of an electrophile at the $\alpha$-position with the two resultant substituents being *trans*.[384]

Similar methodology can be used to introduce a hydroxy group, although the facial selectivity is slightly reduced (Scheme 13.36).[312,385]

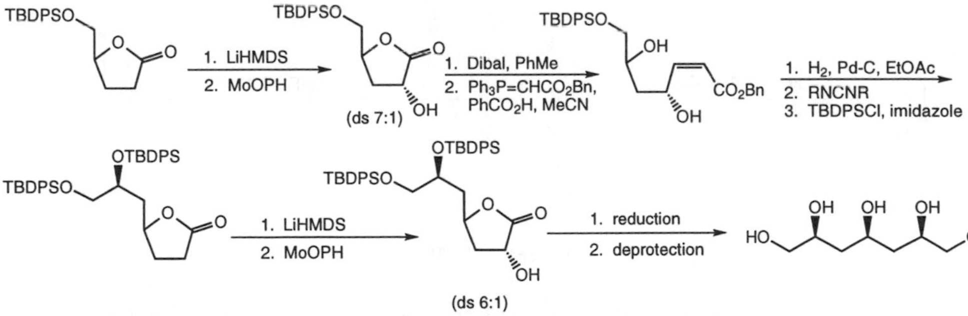

**SCHEME 13.36.**

An enolate protocol can be useful for the conversion of a butyrolactone to a butenolide.[238,377,386,387]

Conjugate additions can also be controlled by the stereochemistry of a $\gamma$-substituent (Scheme 13.37).[388,389]

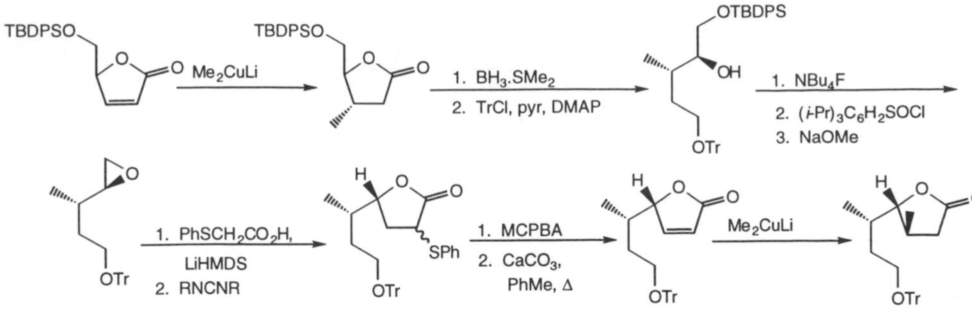

SCHEME 13.37.

These two methods, conjugate additions and alkylations, can be utilized to introduce two groups in one operation (Scheme 13.35).[245,385,390]

### 13.3.2. REACTIONS OF SIX-MEMBERED COMPOUNDS

A number of the methods developed for five-membered ring compounds can be used with six-membered rings. These methodologies include the chemistry of lactol ethers and vinyl ethers. Six-membered ring lactones undergo nucleophilic reaction at the carbonyl center as expected;[7,391] the use of cerium reagents has been advocated to prevent ring opening.[392] As with butyrolactones, alkylations adjacent to the carbonyl group can be controlled by a β-substituent.[384,393] An alkyl group can be introduced at C-4 of an enolone by a radical reaction.[394]

### 13.3.3. REACTIONS OF OTHER OXYGEN SYSTEMS

Several methods have been developed for the preparation of cyclic vinyl ether anions, which allow for the introduction of a wide range of functionality,[191,205,395-413] and are available in chiral form.[414,415,416] The approach has been extended to other dihydropyran derivatives.[417-421] The use of lactol analogs also provides for direct substitution adjacent to the ring oxygen with the cyclic system being the nucleophilic species (e.g., Scheme 13.38).[422] The *trans* product is presumably formed as the axial lithium is more stable due to stereoelectronic interactions with the ring oxygen atom.

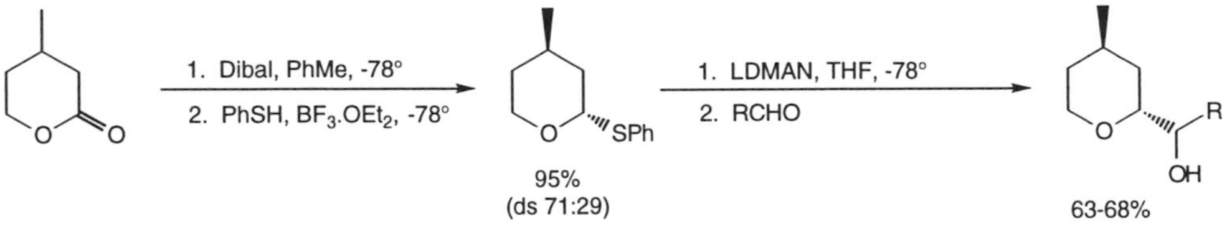

SCHEME 13.38.

2-(Benzenesulfonyl)tetrahydro-2*H*-pyrans (**13.21**), available by reaction of benzenesulfinic acid with dihydropyrans, lactols, or lactol ethers,[423,424] can be used with a wide variety of carbon nucleophiles to provide tetrahydropyrans, dihydropyrans, and even lactols (Scheme 13.39).[424-432]

Another approach involving an anion at the 2-position of a cyclic ether involves phosphorane formation.[396,397]

The reaction of a dihydrofuran or dihydropyran and an α-hydroxyester allows for the resolution of the resultant acetal.[433-435]

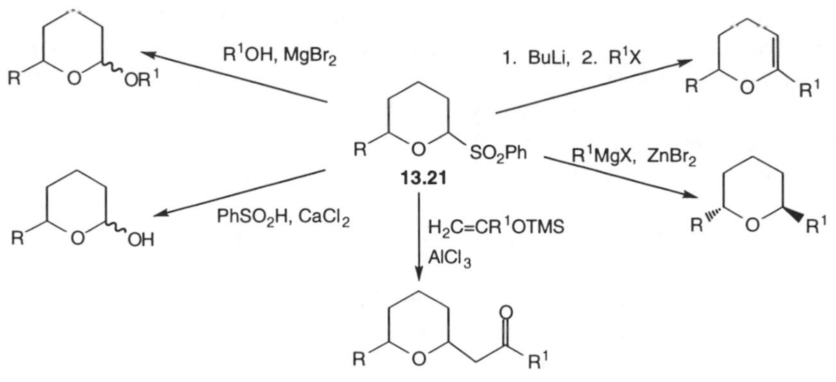

SCHEME 13.39.

Some reactions of cyclic compounds produce surprising results. An example is provided by electrophilic additions to chiral 3-acyl-2-oxazolones where two reagents provide the opposite diastereofacial selectivity (Scheme 13.40).[436,437]

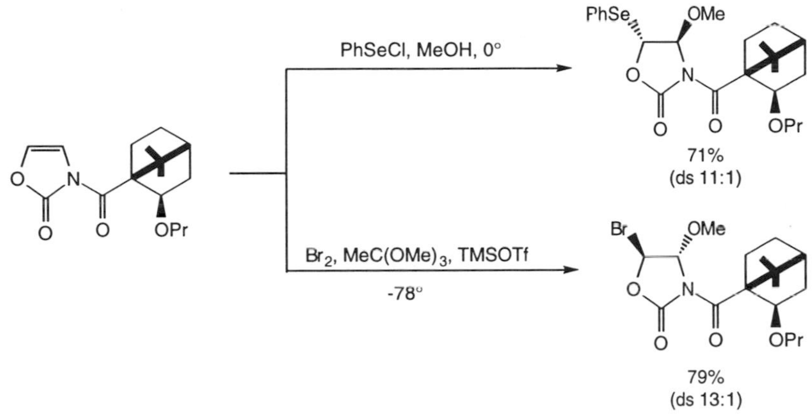

SCHEME 13.40.

Nucleophilic addition to α-chiral lactols provides 1,4-diols with a high control of diastereoselection.[438] Lactols and their esters provide 2-substituted tetrahydrofurans with Reformatsky and other nucleophilic agents.[439,440]

Dihydrofurans and pyrans can be hydroborated by chiral boranes to provide β-hydroxy ethers with high asymmetric induction.[441,442]

α-Lithiation can be performed on Boc-protected piperidines, providing methodology to 2-substituted piperidines.[443]

#### 13.3.3.1. Larger Ring Systems

The cyclic system itself often provides considerable control over relative stereochemistry.[444] However, as the ring size increases, some of this control can be lost due to the increased flexibility of the system.

A large amount of work has been performed with regard to the preparation and reactions of medium and large ring lactones,[445] including conformation effects.[446] One of the reasons behind these investigations is the chemistry of macrolide antibiotics.

Cyclic ethers larger than six-membered rings are also found in nature, and there are many preparations of these ring systems.[447] The chemistry of cyclic ethers, and, in particular, bicyclic systems, allows for the oxygen to be used as masked functionality [*12.4*].[15,442,448-451] The larger cyclic ethers can be formed by ring closure reactions analogous to five- and six-membered systems, or by pericyclic reactions.[121,452-456] Some of the reactions of these medium ring ethers arise from trans-annular interactions that can be specific to that ring system.

## 13.4. REACTIONS OF CYCLIC SULFUR COMPOUNDS

As sulfur can be reductively removed, it provides a unique opportunity to use a cyclic system to control stereochemistry before conversion to an acyclic derivative.[457-459] This philosophy was, perhaps, brought to its pinnacle in Woodward's synthesis of erythromycin where cyclic sulfides were used as synthons for 1,3-dimethyl functionality (Schemes 7.3 and 13.41).[59,60]

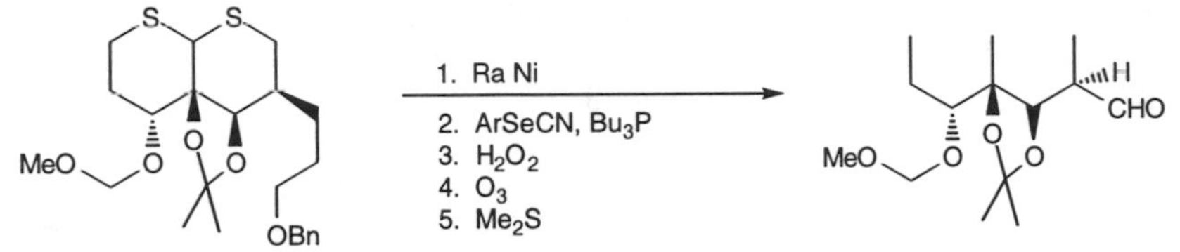

1. Ra Ni
2. ArSeCN, $Bu_3P$
3. $H_2O_2$
4. $O_3$
5. $Me_2S$

**SCHEME 13.41.**

Another useful example of cyclic sulfide chemistry is illustrated by the preparation of 1,2-diols (Scheme 13.42).[460]

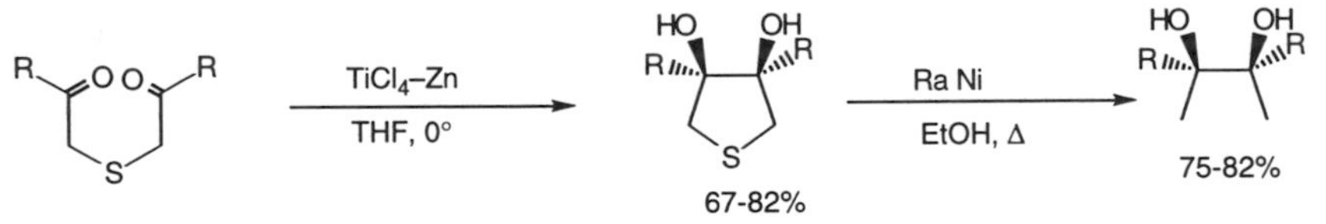

$TiCl_4$–Zn, THF, 0° → 67-82%; Ra Ni, EtOH, Δ → 75-82%

**SCHEME 13.42.**

The use of cyclic sulfur compounds as substrates for asymmetric deprotonations, as well as their incorporation into chiral auxiliaries and functional group masking, continues to spur work in this area.[461]

## 13.5. SUMMARY

Cyclic systems, through their inherent face selectivity, can provide extremely efficient methodology for the introduction of a new stereogenic center. The cyclic system itself often defines this facial preference. Many chiral auxiliaries are based on cyclic systems to provide their facial selectivity. This chapter discusses the reactions of heterocyclic systems that can be used as alternatives to acyclic polyfunctional systems. Cyclic ethers and lactones have provided the majority of the discussion. However, many examples of the reactions of cyclic systems can be found in other chapters, such as the self-regeneration of stereogenic centers.

Two ring-based methodologies that have proven useful in synthetic methodology have also been discussed: the use of cyclic sulfides to control anion chemistry, and the sulfur then being removed by reduction to release the acyclic system.

Cyclic systems provide a very reliable method to control face selectivity, as the degrees of rotational freedom for the system are significantly reduced. The stereochemical outcome of a reaction can be predicted with a fair amount of certainty, especially if the system is very rigid, as in a polycycle, or the reagents have a strong preference for axial or equatorial reactions. A very large amount of literature exists on these aspects of stereochemical control.

## 13.7. REFERENCES


1. Frenking, G.; Köhler, K. F.; Reetz, M. T. *Angew. Chem., Int. Ed. Engl.* 1991, *30*, 1146.
2. Boivin, T. L. B. *Tetrahedron* 1987, *43*, 3309.
3. Thebtaranonth, C.; Thebtaranonth, Y. *Tetrahedron* 1990, *46*, 1385.
4. Harmange, J.-C.; Figadère, B. *Tetrahedron: Asymmetry* 1993, *4*, 1711.
5. Thebtaranonth, C.; Thebtaranonth, Y. *Cyclization Reactions*; CRC Press: London, 1994.
6. Hassell, C. H. *Org. Reactions* 1957, *9*, 73.
7. Martinez, G. R.; Grieco, P. A.; Williams, E.; Kanai, K.-I.; Srinivasan, C. V. *J. Am. Chem. Soc.* 1982, *104*, 1436.
8. Page, P. C. B.; Carefull, J. F.; Powell, L. H.; Sutherland, I. O. *J. Chem. Soc., Chem. Commun.* 1985, 822.
9. Paterson, I. *Tetrahedron Lett.* 1983, *24*, 1311.
10. Asaoka, M.; Hayaslube, S.; Sonada, S.; Takei, H. *Tetrahedron Lett.* 1990, *31*, 4761.
11. Trost, B. M.; Hiroi, K. *J. Am. Chem. Soc.* 1975, *97*, 6911.
12. Grieco, P. A.; Yokoyama, Y.; Gilman, S.; Ohfune, Y. *J. Chem. Soc., Chem. Commun.* 1977, 870.
13. Mauvais, A.; Winterfeldt, E. *Tetrahedron* 1993, *49*, 5817.
14. Alexakis, A.; Mangeney, P. *Tetrahedron: Asymmetry* 1990, *1*, 477.
15. Ager, D. J.; East, M. B. *Tetrahedron* 1993, *49*, 5683.
16. Meyers, A. I.; Harre, M.; Garland, R. *J. Am. Chem. Soc.* 1984, *106*, 1146.
17. Bloch, R.; Brillet, C. *Tetrahedron: Asymmetry* 1992, *3*, 333.
18. Davies, H. M. L.; Clark, T. J.; Church, L. A. *Tetrahedron Lett.* 1989, *30*, 5057.
19. Schaumann, E.; Friese, C. *Tetrahedron Lett.* 1989, *30*, 7033.
20. Tanaka, K.; Minami, K.; Funaki, I.; Suzuki, H. *Tetrahedron Lett.* 1990, *31*, 2727.
21. Ojima, I.; Clos, N.; Bastos, C. *Tetrahedron* 1989, *45*, 6901.
22. Brunner, H. *Angew. Chem., Int. Ed. Engl.* 1992, *31*, 1183.
23. Charette, A. B.; Marcoux, J.-F. *Tetrahedron Lett.* 1993, *34*, 7157.
24. Charette, A. B.; Côte, B. *J. Org. Chem.* 1993, *58*, 933.
25. Sugimura, T.; Futagawa, T.; Yoshikawa, M.; Tai, A. *Tetrahedron Lett.* 1989, *30*, 3807.
26. Mash, E. A.; Torok, D. S. *J. Org. Chem.* 1989, *54*, 250.
27. Mash, E. A.; Math, S. K.; Arterburn, J. B. *J. Org. Chem.* 1989, *54*, 4951.
28. Evans, D. A.; Woerpel, K. A.; Hinman, M. M.; Faul, M. M. *J. Am. Chem. Soc.* 1991, *113*, 726.
29. Evans, D. A.; Woerpel, K. A.; Scott, M. J. *Angew. Chem., Int. Ed. Engl.* 1992, *31*, 430.
30. Bolm, C. *Angew. Chem., Int. Ed. Engl.* 1991, *30*, 542.
31. Hall, J.; Lehn, J.-M.; De Cian, A.; Fischer, J. *Helv. Chim. Acta* 1991, *74*, 1.
32. Lowenthal, R. E.; Abiko, A.; Masamune, S. *Tetrahedron Lett.* 1990, *31*, 6005.
33. Bolm, C.; Weickhardt, K.; Zehnder, M.; Ranff, T. *Chem. Ber.* 1991, *124*, 1173.
34. Muller, D.; Umbricht, G.; Weber, B.; Pfaltz, A. *Helv. Chim. Acta* 1991, *74*, 232.
35. Corey, E. J.; Imai, N.; Zhang, H.-Y. *J. Am. Chem. Soc.* 1991, *113*, 728.
36. Helmchen, G.; Krotz, A.; Ganz, K.-T.; Hansen, D. *Synlett* 1991, 257.
37. Davies, H. M. L.; Hutcheson, D. K. *Tetrahedron Lett.* 1993, *34*, 7243.
38. Ito, K.; Katsuki, T. *Tetrahedron Lett.* 1993, *34*, 2661.
39. Davies, H. M. L.; Huby, N. J. S.; Cantrell, W. R.; Olive, J. L. *J. Am. Chem. Soc.* 1993, *115*, 9468.
40. Denmark, S. E.; Edwards, J. P. *Synlett* 1992, 229.
41. Ito, K.; Tabuchi, S.; Katsuki, T. *Synlett* 1992, 575.
42. Fernández, M. D.; de Frutos, M. P.; Marco, J. L.; Férnandez-Alvarez, E.; Bernabé, M. *Tetrahedron Lett.* 1989, *30*, 3101.
43. Isaka, M.; Nakamura, E. *J. Am. Chem. Soc.* 1990, *112*, 7428.
44. Xiao, X.-Y.; Sen, S. E.; Prestwich, G. D. *Tetrahedron Lett.* 1990, *31*, 2097.
45. Abe, I.; Rohmer, M.; Prestwich, G. D. *Chem. Rev.* 1993, *93*, 2189.

46. Johnson, W. S. *Acc. Chem. Res.* 1968, *1*, 1.
47. Johnson, W. S. *Chimia* 1975, *29*, 310.
48. Johnson, W. S. *Angew. Chem., Int. Ed. Engl.* 1976, *15*, 9.
49. Brinkmeyer, R. S. *Tetrahedron Lett.* 1979, 207.
50. Yamada, S.-I.; Shibasaki, M.; Terashima, S. *Tetrahedron Lett.* 1973, 381.
51. Cohen, N. *Acc. Chem. Res.* 1976, *9*, 412.
52. Pearson, A. J. *Synlett* 1990, 10.
53. Curran, D. P. *Synlett* 1991, 63.
54. Schultz, A. G. *Acc. Chem. Res.* 1990, *23*, 207.
55. Porter, N., A.; Giese, B.; Curran, D. P. *Acc. Chem. Res.* 1991, *24*, 296.
56. Padwa, A. *Acc. Chem. Res.* 1991, *24*, 22.
57. RajanBabu, T. V. *Acc. Chem. Res.* 1991, *24*, 139.
58. Jasperse, C. P.; Curran, D. P.; Fevig, T. L. *Chem. Rev.* 1991, *91*, 1237.
59. Woodward, R. B.; Logusch, E.; Nambiar, K. P.; Sakan, K.; Ward, D. E.; Au-Yeung, B.-W.; Balaram, P.; Browne, L. J.; Card, P. J.; Chen, C. H.; Chenevert, R. B.; Fliri, A.; Frobel, K.; Gais, H.-J.; Garratt, D. G.; Hayakawa, K.; Heggie, W.; Hesson, D. P.; Hoppe, D.; Hoppe, I.; et al. *J. Am. Chem. Soc.* 1981, *103*, 3210.
60. Woodward, R. B.; Logusch, E.; Nambiar, K. P.; Sakan, K.; Ward, D. E.; Au-Yeung, B.-W.; Balaram, P.; Browne, L. J.; Card, P. J.; Chen, C. H.; Chenevert, R. B.; Fliri, A.; Frobel, K.; Gais, H.-J.; Garratt, D. G.; Hayakawa, K.; Heggie, W.; Hesson, D. P.; Hoppe, D.; Hoppe, I.; et al. *J. Am. Chem. Soc.* 1981, *103*, 3213.
61. Coisne, J.-M.; Pecher, J.; Declercq, J.-P.; Germain, G.; Van Meerssche, M. *Bull. Chim. Soc. Belg.* 1981, *90*, 481.
62. Eder, U.; Sauer, G.; Wiechert, R. *Angew. Chem., Int. Ed. Engl.* 1971, *10*, 496.
63. Hiroi, K.; Yamada, S.-I. *Chem. Pharm. Bull.* 1975, *23*, 1103.
64. Stork, G.; Logusch, E. W. *Tetrahedron Lett.* 1979, 3361.
65. d'Angelo, J.; Guingant, A. *Tetrahedron Lett.* 1988, *29*, 2667.
66. Otani, G.; Yamada, S.-I. *Chem. Pharm. Bull.* 1973, *21*, 2130.
67. Otani, G.; Yamada, S.-I. *Chem. Pharm. Bull.* 1973, *21*, 2112.
68. Dumas, F.; d'Angelo, J. *Tetrahedron: Asymmetry* 1990, *1*, 167.
69. Takahashi, T.; Okumoto, H.; Tsuji, J. *Tetrahedron Lett.* 1984, *25*, 1925.
70. Revial, G. *Tetrahedron Lett.* 1989, *30*, 4121.
71. d'Angelo, J.; Revial, G.; Volpe, T.; Pfau, M. *Tetrahedron Lett.* 1988, *29*, 4427.
72. Yamada, S.-I.; Otani, G. *Tetrahedron Lett.* 1969, 4237.
73. d'Angelo, J.; Revial, G.; Guingant, A.; Riche, C.; Chiaroni, A. *Tetrahedron Lett.* 1989, *30*, 2645.
74. Hajos, Z. G.; Parrish, D. R. *J. Org. Chem.* 1974, *39*, 1615.
75. Nakamura, E.; Horiguchi, Y.; Shimada, J.-I.; Kuwajima, I. *J. Chem. Soc., Chem. Commun.* 1983, 796.
76. Wigfield, D. C. *Tetrahedron* 1979, *35*, 449.
77. Brown, H. C.; Krishnamurthy, S. *J. Am. Chem. Soc.* 1972, *94*, 7159.
78. Brown, H. C.; Krishnamurthy, S.; Kim, S. C. *J. Chem. Soc., Chem. Commun.* 1973, 391.
79. Boone, J. R.; Ashby, E. C. *Top. Stereochem.* 1979, *11*, 53.
80. Cieplak, A. S. *J. Am. Chem. Soc.* 1981, *103*, 4540.
81. Wrobel, J. E.; Ganem, B. *Tetrahedron Lett.* 1981, *22*, 3447.
82. Hutchins, R. O.; Su, W.-Y.; Sivakumar, R.; Cistone, F.; Stercho, Y. P. *J. Org. Chem.* 1983, *48*, 3412.
83. Rausser, R.; Weber, L.; Hershberg, E. B.; Oliveto, E. P. *J. Org. Chem.* 1966, *31*, 1342.
84. Rausser, R.; Weber, L.; Hershberg, E. B.; Oliveto, E. P. *J. Org. Chem.* 1966, *31*, 1346.
85. Ashby, E. C.; Laemmle, J. T. *Chem. Rev.* 1975, *75*, 521.
86. Hutchins, R. O.; Su, W.-Y. *Tetrahedron Lett.* 1984, *25*, 695.
87. Sharpless, K. B.; Michaelson, R. C. *J. Am. Chem. Soc.* 1973, *95*, 6136.
88. Dehnel, R. B.; Whitham, G. H. *J. Chem. Soc., Perkin Trans. 1* 1979, 953.
89. Kocovsky, P.; Stary, I. *J. Org. Chem.* 1990, *55*, 3236.
90. Kocovsky, P. *Tetrahedron Lett.* 1988, *29*, 2475.
91. Henbest, H. B. *Chem. Soc., Spec. Publ.* 1965, *19*, 83.
92. Henbest, H. B.; Wilson, R. A. L. *J. Chem. Soc.* 1957, 1958.
93. Eliel, E. L. *Stereochemistry of Carbon Compounds*; McGraw-Hill Kogakusha: Tokyo, 1962.
94. Danishefsky, S.; Tsai, M.-Y.; Kitahara, T. *J. Org. Chem.* 1977, *42*, 394.
95. Nagendrappa, G. *Tetrahedron* 1982, *38*, 2429.
96. Berti, G.; Canedoli, S.; Crotti, P.; Macchia, F. *JCSPI1183* 1984, 1183.
97. Prior, M. J.; Whitham, G. H. *J. Chem. Soc., Perkin Trans. 1* 1986, 683.
98. Kogure, T.; Ojima, I. *J. Organometal. Chem.* 1982, *234*, 249.
99. Crabtree, R. H.; Davis, M. W. *Organometallics* 1983, *2*, 681.
100. Machado, A. S.; Oleseker, A.; Castillion, S.; Lukacs, G. *J. Chem. Soc., Chem. Commun.* 1985, 330.
101. Cardillo, G.; Orena, M. *Tetrahedron* 1990, *46*, 3321.
102. Bartlett, P. A. *Tetrahedron* 1980, *36*, 3.

103. Tamaru, Y.; Mizutani, M.; Furukawa, Y.; Kawamura, S.-I.; Yoshida, Z.-I.; Yanagi, K.; Minobe, M. *J. Am. Chem. Soc.* 1984, *106*, 1079.
104. Bartlett, P. A.; Richardson, D. P.; Myerson, J. *Tetrahedron* 1984, *40*, 2317.
105. Collum, D. C.; McDonald, J. H.; Still, W. C. *J. Am. Chem. Soc.* 1980, *102*, 2118.
106. Davidson, A. H.; Moloney, B. A. *J. Chem. Soc., Chem. Commun.* 1989, 445.
107. Chamberlin, A. R.; Dezube, M.; Dussault, P. *Tetrahedron Lett.* 1981, *22*, 4611.
108. Bartlett, P. A.; Myerson, J. *J. Am. Chem. Soc.* 1978, *100*, 3950.
109. Holmes, C. P.; Bartlett, P. A. *J. Org. Chem.* 1989, *54*, 98.
110. Kurth, M. J.; Bloom, S. H. *J. Org. Chem.* 1989, *54*, 411.
111. Batmangheerlich, S.; Davidson, A. H.; Procter, G. *Tetrahedron Lett.* 1983, *24*, 2889.
112. Simonot, B.; Rousseau, G. *Tetrahedron Lett.* 1993, *34*, 4527.
113. Amouroux, R.; Gerin, B.; Chastrette, M. *Tetrahedron Lett.* 1982, *23*, 4341.
114. Williams, D. R.; White, F. H. *Tetrahedron Lett.* 1986, *27*, 2195.
115. Berrada, S.; Metzner, P. *Tetrahedron Lett.* 1987, *28*, 409.
116. Baldwin, S. W.; McIver, J. M. *J. Org. Chem.* 1987, *52*, 322.
117. Kim, Y. G.; Cha, J. K. *Tetrahedron Lett.* 1988, *29*, 2011.
118. Al-Dulayymi, J.; Baird, M. S. *Tetrahedron Lett.* 1989, *30*, 253.
119. Labelle, M.; Guindon, Y. *J. Am. Chem. Soc.* 1989, *111*, 2204.
120. Iqbal, J.; Pandey, A. *Synth. Commun.* 1990, *20*, 665.
121. Haufe, G. *Tetrahedron Lett.* 1984, *25*, 4365.
122. Johnson, R. A.; Lincoln, F. H.; Thompson, J. L.; Nidy, E. G.; Mizsak, S. A.; Axen, U. *J. Am. Chem. Soc.* 1977, *99*, 4162.
123. Rychnovsky, S. D.; Bartlett, P. A. *J. Am. Chem. Soc.* 1981, *103*, 3963.
124. Labelle, M.; Morton, H. E.; Guidon, Y.; Springer, J. P. *J. Am. Chem. Soc.* 1988, *110*, 4533.
125. Kang, S. H.; Lee, S. B. *Tetrahedron Lett.* 1993, *34*, 7579.
126. Bongini, A.; Cardillo, G.; Orena, M.; Porzi, G.; Sandri, S. *J. Org. Chem.* 1982, *47*, 4626.
127. Williams, D. R.; Harigaya, Y.; Moore, J. L.; D'asa, A. *J. Am. Chem. Soc.* 1984, *106*, 2641.
128. Hirama, M.; Uei, M. *Tetrahedron Lett.* 1982, *23*, 5307.
129. Kobayashi, S.; Isobe, T.; Ohno, M. *Tetrahedron Lett.* 1984, *25*, 5079.
130. Friesen, R. W. *Tetrahedron Lett.* 1990, *31*, 4249.
131. Najdi, S.; Reichlin, D.; Kurth, M. J. *J. Org. Chem.* 1990, *55*, 6241.
132. Kurth, M. J.; Beard, R. L.; Olmstead, M.; Macmillan, J. G. *J. Am. Chem. Soc.* 1989, *111*, 3712.
133. Fuji, K.; Node, M.; Naniwa, Y.; Kawabata, T. *Tetrahedron Lett.* 1990, *31*, 3175.
134. Baldwin, S. W.; McIver, J. M. *J. Org. Chem.* 1987, *52*, 320.
135. Fukuyama, T.; Wang, C.-L. J.; Kishi, Y. *J. Am. Chem. Soc.* 1979, *101*, 260.
136. Tonn, C. E.; Palazon, J. M.; Ruiz-Perez, C.; Rodriguez, M. L.; Martin, V. S. *Tetrahedron Lett.* 1988, *29*, 3149.
137. Jung, M. E.; Lew, W. *J. Org. Chem.* 1991, *56*, 1347.
138. Balko, T. W.; Brinkmeyer, R. S.; Terando, N. H. *Tetrahedron Lett.* 1989, *30*, 2045.
139. Ting, P. C.; Bartlett, P. A. *J. Am. Chem. Soc.* 1984, *106*, 2668.
140. Bradbury, R. H.; Revill, J. M.; Rivett, J. E.; Waterson, D. *Tetrahedron Lett.* 1989, *30*, 3845.
141. Jew, S.-s.; Terashima, S.; Koga, K. *Tetrahedron* 1979, *35*, 2337.
142. Terashima, S.; Jew, S.-S. *Tetrahedron Lett.* 1977, 1005.
143. Terashima, S.; Jew, S.-S.; Koga, K. *Chem. Lett.* 1977, 1109.
144. Terashima, S.; Jew, S.-S.; Koga, K. *Tetrahedron Lett.* 1977, 4507.
145. Terashima, S.; Jew, S.-S.; Koga, K. *Tetrahedron Lett.* 1978, 4937.
146. Jew, S.-S.; Terashima, S.; Koga, K. *Chem. Pharm. Bull.* 1979, *27*, 2351.
147. Jew, S.-S.; Terashima, S.; Koga, K. *Tetrahedron* 1979, *35*, 2345.
148. Hayashi, M.; Terashima, S.; Koga, K. *Tetrahedron* 1981, *37*, 2797.
149. Saito, S.; Morikawa, Y.; Moriwake, T. *Synlett* 1990, 523.
150. Nicolaou, K. C. *Tetrahedron* 1981, *37*, 4097.
151. Nicolaou, K. C.; Barnette, W. E. *J. Chem. Soc., Chem. Commun.* 1977, 331.
152. Murata, S.; Suzuki, T. *Tetrahedron Lett.* 1987, *28*, 4415.
153. Clive, D. L. J.; Chittattu, G. *J. Chem. Soc., Chem. Commun.* 1977, 484.
154. Tiecco, M.; Testaferri, L.; Tingoli, M.; Bartoli, D. *Tetrahedron* 1989, *45*, 6819.
155. Nicolaou, K. C.; Sipio, W. J.; Magolda, R. J.; Claremon, D. A. *J. Chem. Soc., Chem. Commun.* 1979, 83.
156. Nicolaou, K. C.; Seitz, S. P.; Sipio, W. J.; Blount, J. F. *J. Am. Chem. Soc.* 1979, *101*, 3884.
157. Bartlett, P. A.; Holm, K. H.; Morimoto, A. *J. Org. Chem.* 1985, *50*, 5179.
158. Nicolaou, K. C.; Lysenko, Z. *J. Chem. Soc., Chem. Commun.* 1977, 293.
159. Nicolaou, K. C.; Barnette, W. E.; Magolda, R. L. *J. Am. Chem. Soc.* 1981, *103*, 3480.
160. Wrensford, G.; Grab, L. A.; Salvino, J. M.; Williard, P. G. *Tetrahedron Lett.* 1990, *31*, 4257.
161. Nicolaou, K. C.; Barnette, W. E.; Magolda, R. L. *J. Am. Chem. Soc.* 1981, *103*, 3486.

162. Nicolaou, K. C.; Barnette, W. E.; Magolda, R. L. *J. Am. Chem. Soc.* 1978, *100*, 2567.
163. Uemura, S.; Toshimitsu, A.; Aoai, T.; Okano, M. *Tetrahedron Lett.* 1980, *21*, 1533.
164. Kang, S. H.; Hwang, T. S.; Kim, W. J.; Lim, J. K. *Tetrahedron Lett.* 1990, *31*, 5917.
165. Mihelich, E. *J. Am. Chem. Soc.* 1990, *112*, 8995.
166. Nicolaou, K. C.; Claremon, D. A.; Barnette, W. E.; Seitz, S. P. *J. Am. Chem. Soc.* 1979, *101*, 3704.
167. Current, S.; Sharpless, K. B. *Tetrahedron Lett.* 1978, 5075.
168. Morgans, D. J.; Sharpless, K. B.; Traynor, S. G. *J. Am. Chem. Soc.* 1981, *103*, 462.
169. Factor, A.; Traylor, T. G. *J. Org. Chem.* 1968, *33*, 2607.
170. Nishitani, K.; Yamakawa, K. *Tetrahedron Lett.* 1991, *32*, 387.
171. Corey, E. J.; Gross, A. W. *Tetrahedron Lett.* 1980, *21*, 1819.
172. Michael, J. P.; Ting, P. C.; Bartlett, P. A. *J. Org. Chem.* 1985, *50*, 2416.
173. McKillop, A.; Ford, M. E.; Taylor, E. C. *J. Org. Chem.* 1974, *39*, 2434.
174. Larock, R. C.; Stinn, D. E.; Kuo, M.-Y. *Tetrahedron Lett.* 1990, *31*, 17.
175. Walkup, R. D.; Park, G. *J. Am. Chem. Soc.* 1990, *112*, 1597.
176. Aranda, V. G.; Barluenga, J.; Yus, M.; Asensio, G. *Synthesis* 1974, 806.
177. Amoroso, R.; Cardillo, G.; Tomasini, C. *Tetrahedron Lett.* 1990, *31*, 6413.
178. Moriarty, R. M.; Vaid, R. K.; Hopkins, T. E.; Vaid, B. K.; Prakash, O. *Tetrahedron Lett.* 1990, *31*, 201.
179. Tiecco, M.; Testaferri, L.; Tingoli, M.; Bartoli, M.; Bartoli, D. *Tetrahedron* 1990, *46*, 7139.
180. Nicolaou, K. C.; Prasad, C. V. C.; Somers, P. K.; Hwang, C.-K. *J. Am. Chem. Soc.* 1989, *111*, 5330.
181. Noda, I.; Horita, K.; Oikawa, Y.; Yonemitsu, O. *Tetrahedron Lett.* 1986, *27*, 1917.
182. Oikawa, Y.; Horita, K.; Yonemitsu, O. *Heterocycles* 1985, *23*, 553.
183. Walba, D. M.; Stoudt, G. S. *Tetrahedron Lett.* 1982, *23*, 727.
184. Ochiai, M.; Takaoka, Y.; Masaki, Y. *Tetrahedron Lett.* 1989, *30*, 6701.
185. Paterson, I.; Craw, P. A. *Tetrahedron Lett.* 1989, *30*, 5799.
186. Martin, V. S.; Nunez, M. T.; Ramirez, M. A.; Soler, M. A. *Tetrahedron Lett.* 1990, *31*, 763.
187. Cory, R. M.; Ritchie, B. M.; Shrier, A. M. *Tetrahedron Lett.* 1990, *31*, 6789.
188. Murphy, P. J.; Russell, A. T.; Procter, G. *Tetrahedron Lett.* 1990, *31*, 1055.
189. Suzuki, T.; Sato, O.; Hirama, M. *Tetrahedron Lett.* 1990, *31*, 4747.
190. McWhorter, W. W.; Kang, S. H.; Kishi, Y. *Tetrahedron Lett.* 1983, *24*, 2243.
191. Kocienski, P.; Wadman, S.; Cooper, K. *Tetrahedron Lett.* 1988, *29*, 2357.
192. Doherty, A. M.; Ley, S. V. *Tetrahedron Lett.* 1986, *27*, 105.
193. de Laszlo, S. E.; Ford, M. J.; Ley, S. V.; Maw, G. N. *Tetrahedron Lett.* 1990, *31*, 5525.
194. Fukuyama, T.; Vranesic, B.; Negri, D. P.; Kishi, Y. *Tetrahedron Lett.* 1978, 2741.
195. Lußmann, J.; Hoppe, D.; Jones, P. G.; Fittschen, C.; Sheldrick, G. M. *Tetrahedron Lett.* 1986, *27*, 3595.
196. Schmid, G. S.; Fukuyama, T.; Akasaka, K.; Kishi, Y. *J. Am. Chem. Soc.* 1979, *101*, 259.
197. Hoppe, D.; Lußmann, J.; Jones, P. G.; Schmidt, D.; Sheldrick, G. M. *Tetrahedron Lett.* 1986, *27*, 3591.
198. Fukuyama, T.; Akasaka, K.; Karanewsky, D. S.; Wang, C.-L. J.; Schmid, G.; Kishi, Y. *J. Am. Chem. Soc.* 1979, *101*, 262.
199. Ho, P.-T. *Can. J. Chem.* 1982, *60*, 90.
200. Hatekeyama, S.; Sugawara, K.; Kawamura, M.; Takano, S. *Synlett* 1990, 691.
201. Hatekeyama, S.; Numata, H.; Osanai, K.; Takano, S. *J. Chem. Soc., Chem. Commun.* 1989, 1893.
202. Hatekeyama, S.; Osanai, K.; Numata, H.; Takano, S. *Tetrahedron Lett.* 1989, *30*, 4845.
203. Wasserman, H. H.; Rodriques, K.; Kucharczyk, R. *Tetrahedron Lett.* 1989, *30*, 6077.
204. Paterson, I.; Boddy, I.; Mason, I. *Tetrahedron Lett.* 1987, *28*, 5205.
205. Amouroux, R.; Folefoc, G.; Chastrette, F.; Chastrette, M. *Tetrahedron Lett.* 1981, *22*, 2259.
206. Dolle, R. E.; Nicolaou, K. C. *J. Am. Chem. Soc.* 1985, *107*, 1691.
207. Lee, E.; Tae, J. S.; Lee, C.; Park, C. M. *Tetrahedron Lett.* 1993, *34*, 4831.
208. Semmelhack, M. F.; Zhang, N. *J. Org. Chem.* 1989, *54*, 4483.
209. Stork, G.; Poirier, J. M. *J. Am. Chem. Soc.* 1983, *105*, 1073.
210. Semmelhack, M. F.; Kim, C. R.; Dobler, W.; Meier, M. *Tetrahedron Lett.* 1989, *30*, 4925.
211. Semmelhack, M. F.; Kim, C.; Zhang, N.; Bodurow, C.; Sanner, M.; Dobler, W.; Meier, M. *Pure Appl. Chem.* 1990, *62*, 2035.
212. Takahashi, T.; Kataoka, H.; Tsuji, J. *J. Am. Chem. Soc.* 1983, *105*, 147.
213. Takahashi, T.; Ootake, A.; Tsuji, J. *Tetrahedron Lett.* 1984, *25*, 1921.
214. Tsuji, J.; Kataoka, H.; Kobayashi, Y. *Tetrahedron Lett.* 1981, *22*, 2575.
215. Trost, B. M.; Molander, G. A. *J. Am. Chem. Soc.* 1981, *103*, 5969.
216. Trost, B. M.; Warner, R. W. *J. Am. Chem. Soc.* 1983, *105*, 5940.
217. Wicha, J.; Kabat, M. M. *J. Chem. Soc., Chem. Commun.* 1983, 985.
218. Stork, G.; Schoofs, A. R. *J. Am. Chem. Soc.* 1979, *101*, 5081.
219. Semmelhack, M. F.; Bodurow, C. *J. Am. Chem. Soc.* 1984, *106*, 1496.
220. Semmelhack, M. F.; Zask, A. *J. Am. Chem. Soc.* 1983, *105*, 2034.

221. Semmelhack, M. F.; Keller, L.; Sato, T.; Spiess, E. *J. Org. Chem.* 1982, *47*, 4382.
222. Semmelhack, M. F.; Bodurow, C.; Baum, M. *Tetrahedron Lett.* 1984, *25*, 3171.
223. Semmelhack, M. F.; Epa, W. R. *Tetrahedron Lett.* 1993, *34*, 7205.
224. van der Heide, T. A. J.; van der Baan, J. L.; Bijpost, E. A. *Tetrahedron Lett.* 1993, *34*, 4655.
225. Enholm, E. J.; Trivellas, A. *Tetrahedron Lett.* 1989, *30*, 1063.
226. Mead, K.-T.; Yang, H.-L. *Tetrahedron Lett.* 1989, *30*, 6829.
227. Petragnani, N.; Yonashiro, M. *Synthesis* 1982, 521.
228. Baudouy, R.; Crabbe, P.; Greene, A. E.; Le Drian, C.; Orr, A. F. *Tetrahedron Lett.* 1977, 2973.
229. Brandage, S.; Flodman, L.; Norberg, A. *J. Org. Chem.* 1984, *49*, 927.
230. Kurihara, T.; Nakajima, Y.; Mitsunobu, O. *Tetrahedron Lett.* 1976, 2455.
231. Partridge, J. J.; Chadha, N. K.; Uskokovic, M. R. *J. Am. Chem. Soc.* 1973, *95*, 7171.
232. Mori, K.; Takigawa, T.; Matsui, M. *Tetrahedron Lett.* 1976, 3953.
233. Linderman, R. J.; McKenzie, J. R. *Tetrahedron Lett.* 1988, *29*, 3911.
234. Lygo, B.; O'Connor, N. *Synlett* 1992, 529.
235. Saïah, M.; Bessodes, M.; Antonakis, K. *Tetrahedron Lett.* 1993, *34*, 1597.
236. Fuji, K.; Node, M.; Murata, M.; Terada, S.; Hashimoto, K. *Tetrahedron Lett.* 1986, *27*, 5381.
237. Gringore, O. H.; Rouessac, F. P. *Org. Synth.* 1984, *63*, 121.
238. Taniguchi, M.; Koga, K.; Yamada, S. *Tetrahedron* 1974, *30*, 3547.
239. Yoda, H.; Naito, S.; Takabe, K.; Tanaka, N.; Hosoya, K. *Tetrahedron Lett.* 1990, *31*, 7623.
240. Takano, S.; Kasahara, C.; Ogasawara, K. *J. Chem. Soc., Chem. Commun.* 1981, 637.
241. Attwood, M. R.; Carr, M. G.; Jordan, S. *Tetrahedron Lett.* 1990, *31*, 283.
242. Shieh, H.-M.; Prestwich, G. D. *J. Org. Chem.* 1981, *46*, 4319.
243. Takano, S. *Pure Appl. Chem.* 1987, *59*, 353.
244. Alker, D.; Jones, N.; Taylor, G. M.; Wood, W. W. *Tetrahedron Lett.* 1991, *32*, 1667.
245. Hanessian, S.; Murray, P. J. *Tetrahedron* 1987, *43*, 5055.
246. Mann, J.; Partlett, N. K.; Thomas, A. *J. Chem. Res. (S)* 1987, 369.
247. Hamada, Y.; Kawai, A.; Shioiri, T. *Tetrahedron Lett.* 1984, *25*, 5413.
248. Attwood, S. V.; Barrett, A. G. M. *J. Chem. Soc., Perkin Trans. 1* 1984, 1315.
249. Barrett, A. G. M.; Sheth, H. G. *J. Chem. Soc., Chem. Commun.* 1982, 170.
250. Corey, E. J.; Su, W.-G. *Tetrahedron Lett.* 1990, *31*, 2089.
251. Takano, S.; Goto, E.; Hirama, M.; Ogasawara, K. *Heterocycles* 1981, *16*, 381.
252. Jager, V.; Wehner, V. *Angew. Chem., Int. Ed. Engl.* 1989, *28*, 469.
253. Chattopadhyay, S.; Mamdapur, V. R.; Chadha, M. S. *Tetrahedron* 1990, *46*, 3667.
254. Gill, M.; Smrdel, A. F. *Tetrahedron: Asymmetry* 1990, *1*, 453.
255. Ortuno, R. M.; Alonso, D.; Font, J. *Tetrahedron Lett.* 1986, *27*, 1079.
256. Witiak, D. T.; Tehim, A. K. *J. Org. Chem.* 1990, *55*, 1112.
257. Seebach, D.; Wasmuth, D. *Helv. Chim. Acta* 1980, *63*, 197.
258. Meyers, A. I.; Mihelich, E. D. *J. Org. Chem.* 1975, *40*, 1186.
259. Sturm, T.-J.; Marelewski, A. E.; Rezenka, D. S.; Taylor, S. K. *J. Org. Chem.* 1989, *54*, 2039.
260. Janowitz, A.; Kunz, T.; Handke, G.; Reissig, H.-U. *Synlett* 1989, 24.
261. Ohkuma, T.; Kitamura, M.; Noyori, R. *Tetrahedron Lett.* 1990, *31*, 5509.
262. Solladie, G.; Matloubi-Maghadam, F. *J. Org. Chem.* 1982, *47*, 91.
263. Bravo, P.; Carrera, P.; Resnati, G.; Ticozzi, C. *J. Chem. Soc., Chem. Commun.* 1984, 19.
264. Marino, J. P.; Pradilla, R. F. D. L.; Laborde, E. *Synthesis* 1987, 1088.
265. Davies, S. G.; Middlemiss, D.; Naylor, A.; Wills, M. *Tetrahedron Lett.* 1989, *30*, 587.
266. Davies, S. G.; Polywka, R.; Warner, P. *Tetrahedron* 1990, *46*, 4847.
267. Buchwald, S. L.; Fang, Q.; King, S. M. *Tetrahedron Lett.* 1988, *29*, 3445.
268. Stork, G.; Kahn, M. *Tetrahedron Lett.* 1983, *24*, 3951.
269. Nishizawa, M.; Yamada, M.; Noyori, R. *Tetrahedron Lett.* 1981, *22*, 247.
270. Midland, M. M.; Nguyen, N. H. *J. Org. Chem.* 1981, *46*, 4107.
271. Sugai, T.; Ohsawa, S.; Yamada, H.; Ohta, H. *Synthesis* 1990, 1112.
272. Posner, G. H.; Kogan, T. P.; Haines, S. R.; Frye, L.-L. *Tetrahedron Lett.* 1984, *25*, 2627.
273. Holton, R. A.; Kim, H.-B. *Tetrahedron Lett.* 1986, *27*, 2191.
274. Posner, G. H.; Kogan, T. P.; Hulce, M. *Tetrahedron Lett.* 1984, *25*, 383.
275. Posner, G. H.; Mallamo, J. P.; Muira, K. *J. Am. Chem. Soc.* 1981, *103*, 2886.
276. Posner, G. H.; Mallamo, J. P.; Muira, K.; Hulce, M. *Pure Appl. Chem.* 1981, *53*, 2307.
277. Abbot, D. J.; Colonna, S.; Stirling, C. J. M. *J. Chem. Soc., Perkin Trans. 1* 1976, 49.
278. Posner, G. H.; Hulce, M. *Tetrahedron Lett.* 1984, *25*, 379.
279. Posner, G. H.; Mallamo, J. P.; Hulce, M.; Frye, L. L. *J. Am. Chem. Soc.* 1982, *104*, 4180.
280. Kosugi, H.; Tagami, K.; Takahashi, A.; Kanna, H.; Uda, H. *J. Chem. Soc., Perkin Trans. 1* 1989, 935.
281. Marino, J. P.; Neisser, M. *J. Am. Chem. Soc.* 1981, *103*, 7687.

282. Marino, J. P.; Perez, A. D. *J. Am. Chem. Soc.* 1984, *106*, 7643.
283. Marino, J. P.; Pradilla, R. F. D. L. *Tetrahedron Lett.* 1985, *26*, 5381.
284. Posner, G. H.; Asirvatham, E.; Ali, S. F. *J. Chem. Soc., Chem. Commun.* 1985, 542.
285. Kosugi, H.; Kitaoka, M.; Tagami, K.; Takahashi, A.; Uda, H. *J. Org. Chem.* 1987, *52*, 1078.
286. Kuwajima, I.; Urabe, H. *Tetrahedron Lett.* 1981, *22*, 5191.
287. Shono, T.; Matsumura, Y.; Yamane, S.-I. *Tetrahedron Lett.* 1981, *22*, 3269.
288. Tanis, S. P.; Head, D. B. *Tetrahedron Lett.* 1984, *25*, 4451.
289. Marco, J. L. *Tetrahedron* 1989, *45*, 1475.
290. Casiraghi, G.; Colombo, L.; Rassu, G.; Spanu, P.; Fava, G. G.; Belicchi, M. F. *Tetrahedron* 1990, *46*, 5807.
291. Casiraghi, G.; Colombo, L.; Rassu, G.; Spanu, P. *J. Org. Chem.* 1990, *55*, 2565.
292. Jefford, C. W.; Sledeski, A. W.; Boukouvalas, J. *Helv. Chim. Acta* 1989, *72*, 1362.
293. Casiraghi, G.; Colombo, L.; Rassu, G.; Spanu, P. *J. Org. Chem.* 1991, *56*, 2135.
294. Casiraghi, G.; Colombo, L.; Rassu, G.; Spanu, P. *Tetrahedron Lett.* 1989, *30*, 5325.
295. Rassu, G.; Casiraghi, G.; Spanu, P.; Pinna, L.; Fava, G. G.; Ferrari, M. B.; Pelosi, G. *Tetrahedron: Asymmetry* 1992, *3*, 1035.
296. Rassu, G.; Pinna, L.; Spanu, P.; Ulgheri, F.; Cornia, M.; Zanardi, F.; Casiraghi, G. *Tetrahedron* 1993, *49*, 6489.
297. Najera, C.; Yus, M. *J. Chem. Soc., Perkin Trans. 1* 1989, 1387.
298. Jatke, H.; Evertz, U.; Schmidt, R. R. *Synlett* 1990, 191.
299. Hoppe, D.; Bronneke, A. *Tetrahedron Lett.* 1983, *24*, 1687.
300. Moret, E.; Schlosser, M. *Tetrahedron Lett.* 1984, *25*, 4491.
301. Wakabayashi, T.; Kato, Y.; Watanabe, K. *Chem. Lett.* 1976, 1283.
302. Chong, J. M.; Mar, E. K. *Tetrahedron Lett.* 1990, *31*, 1981.
303. Reetz, M.; Schmidt, A.; Holdgrun, X. *Tetrahedron Lett.* 1989, *30*, 5421.
304. Takano, S.; Morimoto, M.; Ogasawara, K. *Synthesis* 1984, 834.
305. Hiyama, T.; Mishima, T.; Sawada, H.; Nozaki, H. *J. Am. Chem. Soc.* 1975, *97*, 1626.
306. Page, P. C. B.; Carefull, J. F.; Powell, L. H.; Sutherland, I. O. *J. Chem. Soc., Chem. Commun.* 1985, 827.
307. Baker, R.; Gibson, C. L.; Swain, C. J.; Tapolczay, D. J. *J. Chem. Soc., Chem. Commun.* 1984, 619.
308. Marschall, H.; Penninger, J.; Weyerstahe, P. *Liebigs Ann. Chem.* 1982, 49.
309. Mettemich, R.; Ludi, W. *Tetrahedron Lett.* 1988, *29*, 3923.
310. Scheeren, J. W.; Lange, J. *Tetrahedron Lett.* 1984, *25*, 1609.
311. Takano, S.; Sekiguchi, Y.; Ogasawara, K. *J. Chem. Soc., Chem. Commun.* 1987, 555.
312. Hanessian, S.; Sahoo, S. P.; Murray, P. J. *Tetrahedron Lett.* 1985, *26*, 5631.
313. Hanessian, S.; Murray, P. J.; Sahoo, S. P. *Tetrahedron Lett.* 1985, *26*, 5623.
314. Russell, A. T.; Procter, G. *Tetrahedron Lett.* 1987, *28*, 2041.
315. Russell, A. T.; Procter, G. *Tetrahedron Lett.* 1987, *28*, 2045.
316. Cha, J. K.; Cooke, R. J. *Tetrahedron Lett.* 1987, *28*, 5473.
317. Canonne, P.; Akssira, M.; Lemay, G. *Tetrahedron Lett.* 1981, *22*, 2611.
318. RajanBabu, T. V.; Kukunaga, T.; Reddy, G. S. *J. Am. Chem. Soc.* 1989, *111*, 1759.
319. Curran, D. P.; Chang, C.-T. *J. Org. Chem.* 1989, *54*, 3140.
320. Stork, G.; Mah, R. *Heterocycles* 1989, *28*, 723.
321. Belletire, J. L.; Mahmoodi, N. O. *Tetrahedron Lett.* 1989, *30*, 4363.
322. Clough, J. M.; Pattenden, G.; Wright, P. G. *Tetrahedron Lett.* 1989, *30*, 7469.
323. Ueno, Y.; Chino, K.; Watanabe, M.; Moriya, O.; Okawara, M. *J. Am. Chem. Soc.* 1982, *104*, 5564.
324. Ueno, Y.; Chino, K.; Watanabe, M.; Moriya, O.; Okawara, M. *J. Chem. Soc., Perkin Trans. 1* 1986, 1351.
325. Stork, G.; Mook, R.; Biller, S. A.; Rynovsky, S. D. *J. Am. Chem. Soc.* 1983, *105*, 3741.
326. Stork, G.; Sher, P. M. *J. Am. Chem. Soc.* 1983, *105*, 6765.
327. Stork, G.; Sher, P. M.; Chen, H.-L. *J. Am. Chem. Soc.* 1986, *108*, 6384.
328. Stork, G.; Sher, P. M. *J. Am. Chem. Soc.* 1986, *108*, 303.
329. Stork, G. *Bull. Chem. Soc., Jpn.* 1988, *61*, 149.
330. Koreeda, M.; George, I. A. *J. Am. Chem. Soc.* 1986, *108*, 8098.
331. Hanessian, S.; Fabio, R. D.; Marcoux, J.-F.; Prud'homme, M. *J. Org. Chem.* 1990, *55*, 3436.
332. Masamune, S. *Aldrichimica Acta* 1978, *11*, 23.
333. Masamune, S.; Lu, L. D.-L.; Jackson, W. P.; Kaiko, T.; Toyoda, T. *J. Am. Chem. Soc.* 1982, *104*, 5523.
334. Fiaud, J.-C.; Kagan, H. B. *Tetrahedron Lett.* 1971, 1019.
335. Izawa, T.; Mukaiyama, T. *Chem. Lett.* 1978, 409.
336. Short, R. P.; Masamune, S. *Tetrahedron Lett.* 1987, *28*, 2841.
337. Roush, W. R.; Spada, A. P. *Tetrahedron Lett.* 1983, *24*, 3693.
338. Carretero, J. C.; Ghosez, L. *Tetrahedron Lett.* 1988, *29*, 2059.
339. Trost, B. M.; Edstrom, E. D. *Angew. Chem., Int. Ed. Engl.* 1990, *29*, 520.
340. Santelli-Rouvier, C. *Tetrahedron Lett.* 1984, *25*, 4371.
341. Still, W. C.; Shaw, K. R. *Tetrahedron Lett.* 1981, *22*, 3725.

342. Honda, M.; Katsuki, T.; Yamaguchi, M. *Tetrahedron Lett.* 1984, *25*, 3857.
343. Ziegler, F. E.; Wester, R. T. *Tetrahedron Lett.* 1986, *27*, 1225.
344. Meyers, A. I.; Whitten, C. E. *Tetrahedron Lett.* 1976, 1947.
345. Meyers, A. I.; Yamamoto, Y.; Mihelich, E. D.; Bell, R. A. *J. Org. Chem.* 1980, *45*, 2792.
346. Eliel, E. L.; Soai, K. *Tetrahedron Lett.* 1981, *22*, 2859.
347. Kogure, T.; Eliel, E. L. *J. Org. Chem.* 1984, *49*, 576.
348. Abushanab, E.; Reed, D.; Suzuki, F.; Sih, C. J. *Tetrahedron Lett.* 1978, 3415.
349. Bernardi, R.; Ghiringhelli, D. *Synthesis* 1989, 938.
350. Ghiringhelli, D. *Tetrahedron Lett.* 1983, *24*, 287.
351. Bernardi, R.; Cardillo, R.; Ghiringhelli, D.; Pava, V. D. P. *J. Chem., Soc., Perkin Trans. I* 1987, 1607.
352. Ford, M. J.; Ley, S. V. *Synlett* 1990, 771.
353. Honda, T.; Kametani, T.; Kanai, K.; Tatsuzaki, Y.; Tsubuki, M. *J. Chem. Soc., Perkin Trans. 1* 1990, 1733.
354. DeShong, P.; Simpson, D. M.; Lin, M.-T. *Tetrahedron Lett.* 1989, *30*, 2885.
355. Martin, S. F.; Zinke, P. W. *J. Am. Chem. Soc.* 1989, *111*, 2311.
356. Still, W. C.; Ohmizu, H. *J. Org. Chem.* 1981, *46*, 5242.
357. Kusakabe, M.; Kitano, Y.; Kobayashi, Y.; Sato, F. *J. Org. Chem.* 1989, *54*, 2085.
358. Johnson, W. S.; Kelson, A. B.; Elliott, J. D. *Tetrahedron Lett.* 1988, *29*, 3757.
359. Frauenrath, H.; Sawicki, M. *Tetrahedron Lett.* 1990, *31*, 649.
360. Mash, E. A.; Arterburn, J. B. *J. Org. Chem.* 1991, *56*, 885.
361. Yamamoto, Y.; Taniguchi, K.; Maruyama, K. *J. Chem. Soc., Chem. Commun.* 1985, 1429.
362. Wuts, P. G. M.; Obrzut, M. L.; Thompson, P. A. *Tetrahedron Lett.* 1984, *25*, 4051.
363. Peterson, J. R.; Kirchhoff, E. W. *Synlett* 1990, 394.
364. Chan, T.-H.; Arya, P. *Tetrahedron Lett.* 1989, *30*, 4065.
365. Martin, V. A.; Perron, F.; Albizati, K. F. *Tetrahedron Lett.* 1990, *31*, 301.
366. Paterson, I.; Osborne, S. *Tetrahedron Lett.* 1990, *31*, 2213.
367. Wei, Z. Y.; Wang, D.; Li, J. S.; Chan, T. H. *J. Org. Chem.* 1989, *54*, 5768.
368. Shao, L.; Seki, T.; Kawano, H.; Saburi, M. *Tetrahedron Lett.* 1991, *32*, 7699.
369. Evans, D. A.; Carreira, E. M. *Tetrahedron Lett.* 1990, *31*, 4703.
370. Poss, A. J.; Smyth, M. S. *Tetrahedron Lett.* 1988, *29*, 5723.
371. Takano, S.; Kudo, J.; Takahashi, M.; Ogasawara, K. *Tetrahedron Lett.* 1986, *27*, 2405.
372. Takano, S.; Tamura, N.; Ogasawara, K. *J. Chem. Soc., Chem. Commun.* 1981, 1155.
373. Chamberlin, A. R.; Dezube, M. *Tetrahedron Lett.* 1982, *23*, 3055.
374. Wengel, A. S.; Reffstrup, T.; Boll, P. M. *Tetrahedron* 1979, *35*, 2181.
375. Takano, S.; Yonaga, M.; Ogasawara, K. *J. Chem. Soc., Chem. Commun.* 1981, 1153.
376. Elliott, R. P.; Fleet, G. W. J.; Gyoung, Y. S.; Ramsden, N. G.; Smith, C. *Tetrahedron Lett.* 1990, *31*, 3785.
377. Wilson, L. J.; Liotta, D. *Tetrahedron Lett.* 1990, *31*, 1815.
378. Jansen, J. F. G. A.; Feringa, B. L. *Tetrahedron Lett.* 1989, *30*, 5481.
379. Jansen, J. F. G. A.; Jansen, C.; Feringa, B. L. *Tetrahedron: Asymmetry* 1991, *2*, 109.
380. Ong, B. S. *Tetrahedron Lett.* 1980, *21*, 4225.
381. Pelter, A.; Ward, R. S.; Jones, D. M.; Maddocks, P. *Tetrahedron: Asymmetry* 1990, *1*, 857.
382. Ito, T.; Yamakawa, I.; Okamoto, S.; Kobayashi, Y.; Sato, F. *Tetrahedron Lett* 1991, *32*, 371.
383. Ferringa, B. L.; de Lange, B. *Tetrahedron Lett.* 1988, *29*, 1303.
384. Tomioka, K.; Yasuda, K.; Kawasaki, H.; Koga, K. *Tetrahedron Lett.* 1986, *27*, 3247.
385. Hanessian, S.; Murray, P. J. *Can. J. Chem.* 1986, *64*, 2231.
386. Martin, S. F.; Moore, D. R. *Tetrahedron Lett.* 1976, 4459.
387. Koga, K.; Taniguchi, M.; Yamada, S.-I. *Tetrahedron Lett.* 1971, 263.
388. Hanessian, S.; Murray, P. J.; Sahoo, S. P. *Tetrahedron Lett.* 1985, *26*, 5627.
389. Lange, B. D.; Bolhuis, F. V.; Feringa, B. L. *Tetrahedron* 1989, *45*, 6799.
390. Hanessian, S.; Murray, P. J. *J. Org. Chem.* 1987, *52*, 1170.
391. Baker, R.; Boyes, R. H. O.; Broom, D. M. P.; Devlin, J. A.; Swain, C. J. *J. Chem. Soc., Chem. Commun.* 1983, 829.
392. Mudryk, B.; Shook, C. A.; Cohen, T. *J. Am. Chem. Soc.* 1990, *112*, 6389.
393. Tomioka, K.; Kawasaki, H.; Yasuda, K.; Koga, K. *J. Am. Chem. Soc.* 1988, *110*, 3597.
394. Giese, B.; Witzel, T. *Tetrahedron Lett.* 1987, *28*, 2571.
395. Cohen, T.; Bhupathy, M. *Acc. Chem. Res.* 1989, 22, 152.
396. Ley, S. V.; Lygo, B.; Organ, H. M.; Wonnacott, A. *Tetrahedron* 1985, *41*, 3825.
397. Ousset, J. B.; Mioskowski, C.; Yang, Y.-L.; Falck, J. R. *Tetrahedron Lett.* 1984, *25*, 5903.
398. Godoy, J.; Ley, S. V.; Lygo, B. *J. Chem. Soc., Chem. Commun.* 1984, 1381.
399. Kocienski, P.; Yeates, C. *J. Chem. Soc., Perkin Trans. 1* 1985, 1879.
400. Rychnovsky, S. D.; Mickus, D. E. *Tetrahedron Lett.* 1989, *30*, 3011.
401. Sawyer, J. S.; Kucerovy, A.; Macdonald, T. L.; McGarvey, G. J. *J. Am. Chem. Soc.* 1988, *110*, 842.

402. Cohen, T.; Matz, J. R. *J. Am. Chem. Soc.* 1980, *102*, 6900.
403. Wadman, S.; Whitby, R.; Yeates, C.; Kocienski, P.; Cooper, K. *J. Chem. Soc., Chem. Commun.* 1987, 241.
404. Kocienski, P.; Dixon, N. J.; Wadman, S. *Tetrahedron Lett.* 1988, *29*, 2353.
405. Kocienski, P.; Wadman, S.; Cooper, K. *J. Am. Chem. Soc.* 1989, *111*, 2363.
406. Kocienski, P.; Dixon, N. J. *Synlett* 1989, 52.
407. Kocienski, P.; Love, C.; Whitby, R.; Roberts, D. A. *Tetrahedron Lett.* 1988, *29*, 2867.
408. Kocienski, P.; Barber, C. *Pure Appl. Chem.* 1990, *62*, 1933.
409. Boeckman, R. K.; Bruza, K. J. *Tetrahedron* 1981, *37*, 3997.
410. Boeckman, R. K.; Charette, A. B.; Asberom, T.; Johnston, B. H. *J. Am. Chem. Soc.* 1987, *109*, 7553.
411. Boeckman, R. K.; Bruza, K. J. *Tetrahedron Lett.* 1977, 4187.
412. Stocks, M.; Kocienski, P. *Tetrahedron Lett.* 1990, *31*, 1637.
413. Amouroux, R. *Heterocycles* 1984, *22*, 1489.
414. Ireland, R. E.; Wilcox, C. S.; Thaisrivongs, S. *J. Org. Chem.* 1978, *43*, 786.
415. Barber, C.; Jarowicki, K.; Kocienski, P. *Synlett* 1991, 197.
416. Ozawa, F.; Kobatake, Y.; Hayashi, T. *Tetrahedron Lett.* 1993, *34*, 2505.
417. Jamart-Gregoire, B.; Grand, V.; Ianelli, S.; Nardelli, M.; Caubere, P. *Tetrahedron Lett.* 1990, *31*, 7603.
418. Alexakis, A.; Hanaizi, J.; Jachiet, D.; Normant, J.-F. *Tetrahedron Lett.* 1990, *31*, 1271.
419. Jarowicki, K.; Kocienski, P.; Marczak, S.; Willson, T. *Tetrahedron Lett.* 1990, *31*, 3433.
420. Qiu, D.; Schmidt, R. R. *Synthesis* 1990, 875.
421. Schmidt, R. R.; Kast, J. *Tetrahedron Lett.* 1986, *27*, 4007.
422. Cohen, T.; Lin, M.-J. *J. Am. Chem. Soc.* 1984, *106*, 1130.
423. Ley, S. V.; Lygo, B.; Sternfeld, F.; Wonnacott, A. *Tetrahedron* 1986, *42*, 4333.
424. Brown, D. S.; Bruno, M.; Davenport, R. J.; Ley, S. V. *Tetrahedron* 1989, *45*, 4293.
425. Brown, D. S.; Bruno, M.; Ley, S. V. *Heterocycles* 1989, *28, Special Issue no. 2*, 773.
426. Ley, S. V.; Lygo, B.; Wonnacott, A. *Tetrahedron Lett.* 1985, *26*, 535.
427. Prandi, J.; Audin, C.; Beau, J.-M. *Tetrahedron Lett.* 1991, *32*, 769.
428. Greck, C.; Grice, P.; Ley, S. V.; Wonnacott, A. *Tetrahedron Lett.* 1986, *27*, 5277.
429. Brown, D. S.; Ley, S. V. *Tetrahedron Lett.* 1988, *29*, 4873.
430. Diez-Martin, D.; Grice, P.; Kolb, H. C.; Ley, S. V.; Madin, A. *Tetrahedron Lett.* 1990, *31*, 3445.
431. Brown, D. S.; Charreau, P.; Ley, S. V. *Synlett* 1990, 749.
432. Crich, D.; Lim, L. B. L. *Tetrahedron Lett.* 1990, *31*, 1897.
433. Mash, E. A.; Fryling, J. A. *J. Org. Chem.* 1991, *56*, 1094.
434. Mash, E. A.; Arterburn, J. B.; Fryling, J. A.; Mitchell, S. H. *J. Org. Chem.* 1991, *56*, 1088.
435. Mash, E. A.; Arterburn, J. B.; Fryling, J. A. *Tetrahedron Lett.* 1989, *30*, 7145.
436. Ishizuka, T.; Ishibuchi, S.; Kunieda, T. *Tetrahedron Lett.* 1989, *30*, 3449.
437. Ishizuka, T.; Ishibuchi, S.; Kunieda, T. *Tetrahedron* 1993, *49*, 1841.
438. Tomooka, K.; Okinaga, T.; Suzuki, K.; Tsuchihashi, G.-I. *Tetrahedron Lett.* 1989, *30*, 1563.
439. Hayashi, M.; Sugiyama, M.; Toba, T.; Oguni, N. *J. Chem. Soc., Chem. Commun.* 1990, 767.
440. Schmitt, A.; Reissig, H.-U. *Synlett* 1990, 40.
441. Brown, H. C.; Prasad, J. V. N. V. *J. Am. Chem. Soc.* 1986, *108*, 2049.
442. Ager, D. J.; East, M. B. *Tetrahedron* 1992, *48*, 2803.
443. Beak, P.; Lee, W. K. *J. Org. Chem.* 1990, *55*, 2578.
444. Roxburgh, C. J. *Tetrahedron* 1993, *49*, 10749.
445. Nicolaou, K. C. *Tetrahedron* 1977, *33*, 683.
446. Still, W. C.; Galynker, I. *Tetrahedron* 1981, *37*, 3981.
447. Overman, L. E.; Thompson, A. S. *J. Am. Chem. Soc.* 1988, *110*, 2248.
448. Ager, D. J.; East, M. B. *J. Chem. Soc., Chem. Commun.* 1989, 178.
449. Vogel, P.; Fattori, D.; Gasparini, F.; Le Drian, C. *Synlett* 1990, 173.
450. Lautens, M.; Smith, A. C.; Abd-El-Aziz, A. S.; Huboux, A. H. *Tetrahedron Lett.* 1990, *31*, 3253.
451. Molander, G. A.; Andrews, S. W. *Tetrahedron Lett.* 1989, *30*, 2351.
452. Hajos, Z. G.; Wachter, M. P.; Werblood, H. M.; Adams, R. E. *J. Org. Chem.* 1984, *49*, 2600.
453. Kim, H.-J.; Schlecht, M. F. *Tetrahedron Lett.* 1988, *29*, 1771.
454. Zárraga, M.; Alveraz, E.; Ravelo, J. L.; Rodriguez, V.; Rodriguez, M. L.; Martín, J. D. *Tetrahedron Lett.* 1990, *31*, 1633.
455. Alvarez, E.; Díaz, M. T.; Rodríguez, M. L.; Martín, J. D. *Tetrahedron Lett.* 1990, *31*, 1629.
456. Evans, P. A.; Holmes, A. B.; Russell, K. *Tetrahedron: Asymmetry* 1990, *1*, 593.
457. Vedejs, E.; Krafft, G. A. *Tetrahedron* 1982, *38*, 2857.
458. Kondo, K.; Negishi, A.; Matsui, K.; Tunemoto, D. *J. Chem. Soc., Chem. Commun.* 1972, 1311.
459. Stork, G.; Stotter, P. L. *J. Am. Chem. Soc.* 1969, *91*, 7780.
460. Nakayama, J.; Yamaoka, S.; Hoshino, M. *Tetrahedron Lett.* 1987, *28*, 1799.
461. Armer, R.; Simpkins, N. S. *Tetrahedron Lett.* 1993, *34*, 363.

Chapter 14

# BIOCHEMICAL METHODS

The use of enzymes in organic synthesis has become one of the fastest growing areas for the development of new methodology, and has been comprehensively reviewed.[1-14]

The aim of this chapter is to highlight some enzyme reactions that could be useful in a chemical approach to chiral compounds. It must be remembered that some enzymes are promiscuous, while others are very specific to a particular functional group, or even compound. Much of the data is presented in tabular form; the more generic the structure, the more promiscuous the system is. However, not all substrates may provide a clean reaction at a useful rate; the reader is referred to the original citations.

As many enzymatic transformations are known, only reactions that are of interest for synthetic methodology have been included. The reader is again referred to suitable texts for enzymes that perform only one specific transformation with a very specific substrate. Thus, neither the cleavage of a polypeptide next to a specific amino acid residue, nor the enzyme that performs a specific methylation in the synthesis of vitamin $B_{12}$ will be discussed. In addition, transformations catalyzed by modified enzymes have been omitted, as initial attempts at a reaction of interest would surely be performed with the native enzyme. However, as catalytic antibodies do provide the potential for the development of "designer enzymes," a discussion on this topic has been included in this chapter.

Table 14.1 lists the chemical reaction types that can be mediated by enzymes or microorganisms.[15,16]

**TABLE 14.1**
**Transformations Mediated by Enzymes and Microorganisms**

| Class | Reaction Type |
|---|---|
| Oxidation | Dehydrogenation of C–C |
| | Epoxidation |
| | Hydroxylation |
| | Oxidation of hetero-functions |
| | Oxidations of alcohols |
| | Oxidations of aldehydes |
| | Oxidative deamination |
| | Oxidative degradation of alkyl chains |
| | Oxidative degradation of carboxyalkyl chains |
| | Oxidative degradation of oxoalkyl chains |
| | Oxidative removal of substituents |
| | Oxidative ring fission |
| Reductions | Dehydroxylations |
| | Hydrogenations of C–C unsaturation |
| | Reductions of aldehydes |
| | Reductions of hetero-functions |
| | Reductions of ketones |
| | Reductions of organic acids |
| | Reductive aminations of substituents |
| | Reductions of nitrogen compounds |
| Formation of | C–C bonds |
| | Heteroatom bonds |
| | Ring systems |
| Hydrolysis | Hydrations of C–C unsaturation |
| | Hydrations of epoxides |
| | Hydrolysis of amides |
| | Hydrolysis of amines |
| | Hydrolysis of esters |
| | Hydrolysis of ethers |
| | Hydrolysis of lactams |
| | Hydrolysis of lactones |
| | Aminations |
| | Hydrolysis of alkyl halides |
| | Hydrolysis of nitriles |
| Condensations | Dehydrations |
| | Esterifications |
| | Glycosidations |
| | Lactonizations |
| | *N*-Acylations |
| | *O*-Acylations |
| Isomerizations | Migrations of double bonds |
| | Migrations of oxygen functionality |
| | Racemization |
| | Rearrangements |
| Transfers | Transesterification |
| | Transaminations |
| | Hydroxymethyl group transfers |

Many enzymatic processes are reversible — oxidations and reductions usually being the exceptions as the reaction is driven to completion by a secondary transformation — and many uses involve kinetic resolution. A number of equations have been derived from common reaction types to allow correlations between the degree of conversion of the racemic substrate and the enantiomeric excess.[17,18]

Biological systems can be very specific, and they can also be used on a large scale. In some instances, chemical and biological approaches may compete head-to-head in a scale-up program. If an enzyme and chemical catalyst are known, the deciding factor is invariably the economics of the alternatives — it may be possible to determine this after only a few experiments.

The development of a process "from scratch," however, is a different matter, and parallel chemical and biological development pathways may be the short-term solution. The factors that can affect the choice of the final approach are listed in Table 14.2.

**TABLE 14.2**
**Comparison of Chemical and Biological Approaches**

| | **Chemical** | **Biological** |
|---|---|---|
| Throughput | High | Low |
| Catalyst cost | Variable | Variable |
| Selectivity | Variable | High |
| Yield | Medium–High | High[a] |
| Isolation/Processing | Simple | Can be complex |
| Operating conditions | Usual chemical | May need careful control |
| Catalysts design | Often empirical | Through screening/engineering |

[a] As the process has to be close to optimal for commercial success. The high in this instance refers to "simple" transformations and not multi-step processes.

It is the experience of the authors that a chemical approach will often provide the "first" process that can be used to prepare a compound, but a biological approach will be the most economical for large scale production. However, the research investment for the biological approach can be large. The use of a biological resolution, however, with the potential for high selectivity can often be the method of choice, as the problems associated with removal and recycling of the "wrong" isomer have to be dealt with in both the chemical and biological methodologies.[19] The use of immobilized enzymes and cell systems can often simplify isolation procedures — this approach is not simple to implement for a chemical transformation.

## 14.1. ESTERS AND RELATED COMPOUNDS

The use of lipases has provided means to resolve alcohols, acids, and esters. *Meso*-substrates provide a powerful entry to a single enantiomer, as all of the substrate is converted to the desired product. If this approach is not available, racemization of the undesired isomer and recycling become a necessity as the scale of the reaction increases.[20] As lipase enzymes catalyze equilibria, esters can be prepared from alcohols, particularly if an activated ester, such as vinyl acetate, is used as the donor ester in a transesterification.[21]

Despite the relatively large number of substrates used with lipases, in particular pig liver esterase (PLE), it is not always possible to predict the stereospecificity of a particular lipase with a substrate.[5,22] This is one of the reasons that enzymes have not found complete acceptance in the repertoire of chiral reagents for chemical reactions. This problem is being addressed through the development of "easy-to-use" active site models. Some of these models predict selectivity, as for PLE[23] and *Candida rugosa* lipase (CRL),[24,25] while others give an estimation of the active site size so that they can be used for more general, predictable applications.[22,26] Many of the models are empirical and, hence, cannot be relied on for all potential substrates. From a rule developed by Prelog,[27] that predicts a yeast reduction occurs with hydrogen delivery from the front face (Figure 14.1), the concept has been extended to predict which enantiomer of a secondary alcohol will react faster in the presence of *Pseudomonas cepacia* or *Candida rugosa* lipases (Figure 14.2).[28]

from top face

O

$R^S$ $R^L$

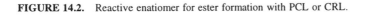

where $R^S$ = small group
$R^L$ = large group

**FIGURE 14.1.** Direction of hydride delivery in a yeast reduction.

H OH

$R^M$ $R^L$

$R^M$ = medium group

**FIGURE 14.2.** Reactive enatiomer for ester formation with PCL or CRL.

This also applies to esterification reactions.[29] The ability to predict the isomer that is most easily hydrolyzed by a lipase has led to a number of similar models.[28,30-36] In addition to the prediction of the stereochemical outcome, correlation of the model's stereochemical requirements to those of the substrate can be used to increase efficiency through substrate modification prior to the enzymatic step.

As evident from Table 14.3, PLE has been used for a wide variety of substrates. An active site model has been proposed that defines the volume of an active site (Figure 14.3).[37,38] The two important binding regions for specificity determination are the hydrophobic pockets $H_{L(arge)}$ and $H_{S(mall)}$ with the two more polar pockets, $P_{F(ront)}$ and $P_{B(ack)}$. A few simple rules can then be used to make a stereochemical prediction:

1. The carbon atom of the (methyl) ester carbonyl group to be hydrolyzed must be within the carbon–oxygen single bond length of the serine sphere.
2. The remaining groups are then matched to the most appropriate areas with:
   a. the smaller nonpolar (aliphatic) groups binding in $H_S$ most efficiently unless they are too big;
   b. if the substrate is a 1,2- or similar diester, the unhydrolyzed group will occupy $P_F$;
   c. groups capable of forming hydrogen bonds with the $P_B$ region will do so.[26]

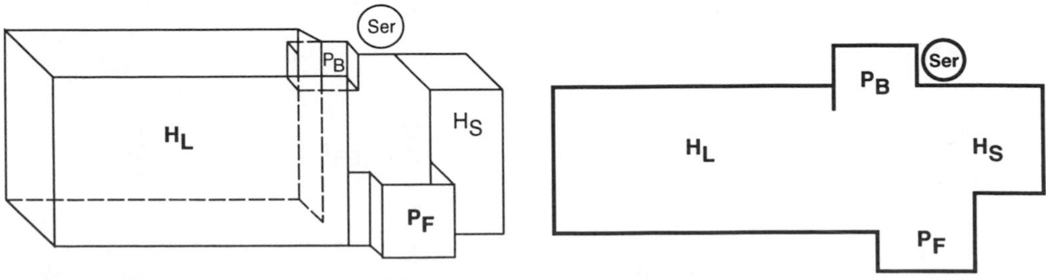


**FIGURE 14.3.** Active site model for PLE.

**TABLE 14.3**
**Ester Hydrolysis and Related Reactions Catalyzed by Enzymes and Organisms**

| Substrate | Enzyme or Organism | Product(s)[a] | Ref. |
|---|---|---|---|
| $RO_2CR^1$ | *Candida cylindracea* lipase, protease, *subtilisin Carlsberg* | ROH[b] | 41–43 |
| 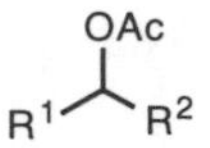 OAc; $R^1$, $R^2$ | *Rhizopus nigricans*, pig liver acetone powder, Lipase amano P., yeast lipase | 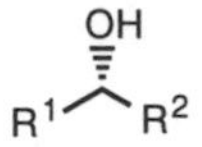 OH; $R^1$, $R^2$ | 33,44–46 |
| 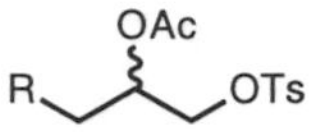 OAc; R, OTs | LP-80-lipase | 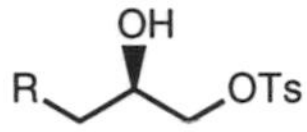 OH; R, OTs | 47 |
| 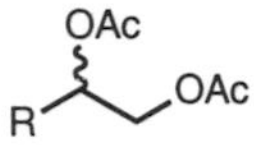 OAc; R, OAc | PPL | 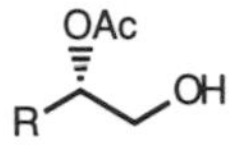 OAc; R, OH | 48 |
| 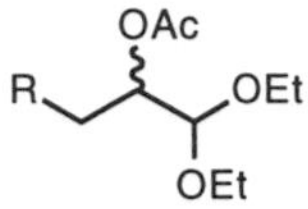 OAc; R, OEt, OEt | LP-80-lipase | 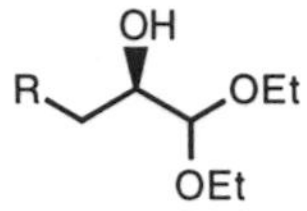 OH; R, OEt, OEt | 47,49 |
| 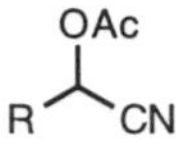 OAc; R, CN | PFL, *Candida tropicalis* | 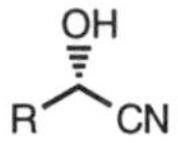 OH; R, CN | 50–53 |
| 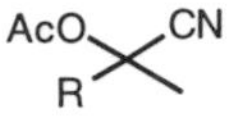 AcO, CN; R | *Pichia miso*[b] | 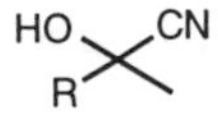 HO, CN; R | 54,55 |
| 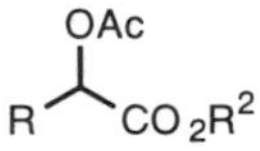 OAc; R, $CO_2R^2$ | CCL | 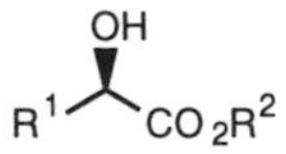 OH; $R^1$, $CO_2R^2$ | 46,56 |
| 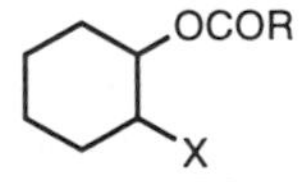 OCOR; X<br>X = Ar, R, RO, $N_3$, Cl, Br, I | *Pseudomonas* sp. | 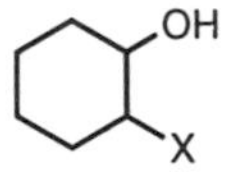 OH; X | 57–60 |
| 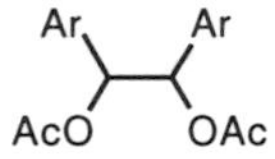 Ar, Ar; AcO, OAc | *Trichoderma viride* (IFO 9065) | 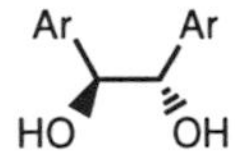 Ar, Ar; HO, OH | 61 |
| 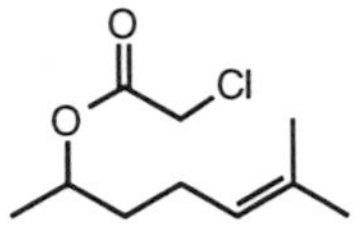 O; Cl; O | PPL, CCL | 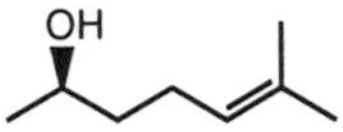 OH | 62,63 |
| 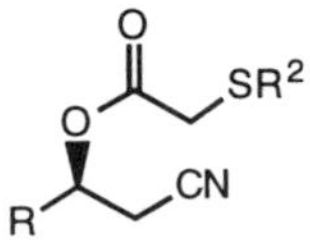 O; $SR^2$; O; R, CN | Lipase P (*Pseudomonas* sp.) | 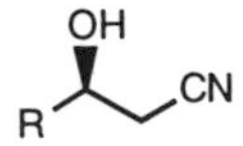 OH; R, CN | 64 |
| 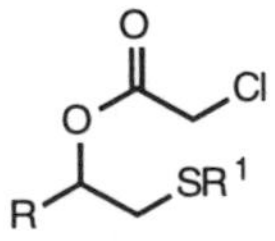 O; Cl; O; R, $SR^1$ | Lipase SAM-1 | 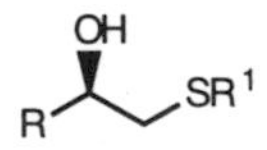 OH; R, $SR^1$ | 65 |
| 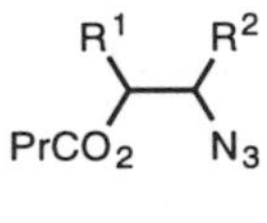 $R^1$, $R^2$; $PrCO_2$, $N_3$ | CCL, PFL | 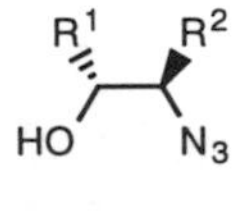 $R^1$, $R^2$; HO, $N_3$ | 66 |

**TABLE 14.3 (continued)**

| Substrate | Enzyme or Organism | Product(s)[a] | Ref. |
|---|---|---|---|
| 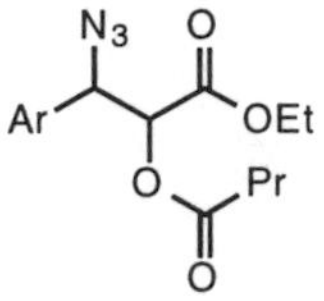 $N_3$, O, Ar, OEt, O, Pr, O | Lipase P (amano) | 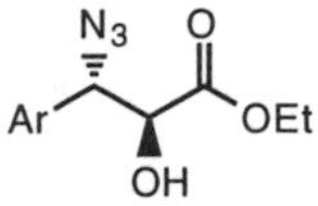 $N_3$, O, Ar, OEt, OH | 67 |
| 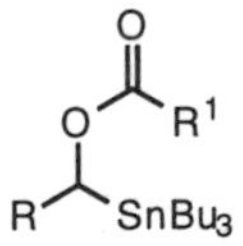 O, O, $R^1$, R, $SnBu_3$ | Lipase P | 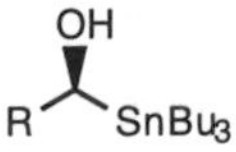 OH, R, $SnBu_3$ | 68 |
| 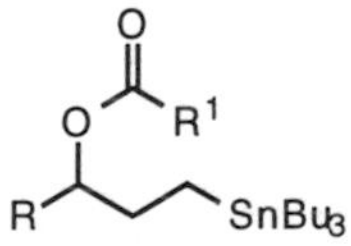 O, O, $R^1$, R, $SnBu_3$ | Lipase P | 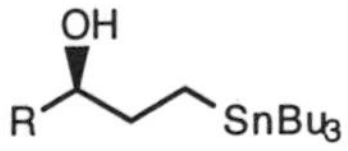 OH, R, $SnBu_3$ | 69 |
| 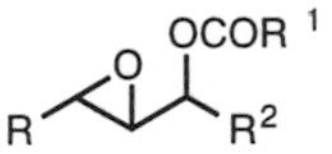 O, $OCOR^1$, R, $R^2$ | PPL | 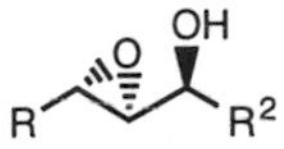 O, OH, R, $R^2$ | 70 |
| 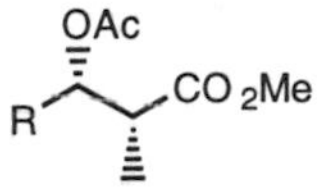 OAc, R, $CO_2Me$ (both isomers) | Amano A lipase | 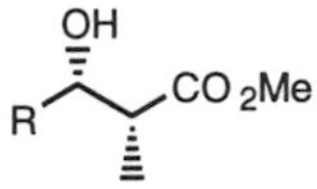 OH, R, $CO_2Me$ | 71 |
| 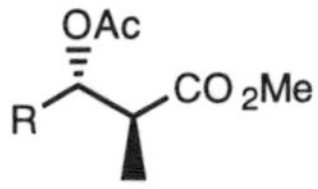 OAc, R, $CO_2Me$ (both isomers) | Amano A lipase | 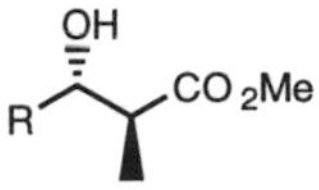 OH, R, $CO_2Me$ | 71 |
| 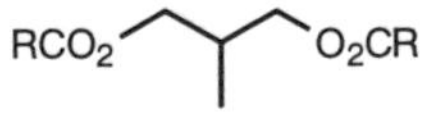 $RCO_2$, $O_2CR$ | PFL |  $RCO_2$, OH | 72–74 |
| 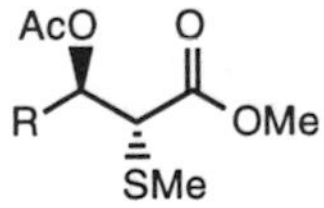 AcO, O, R, OMe, SMe | Lipase A6 *aspergillus* sp.[c] | 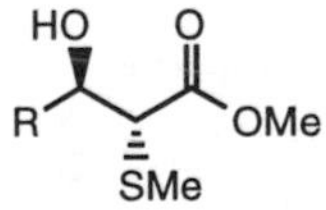 HO, O, R, OMe, SMe | 75 |
| 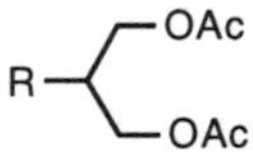 R, OAc, OAc | PPL, PFL | 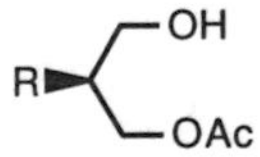 R, OH, OAc | 47,76–83 |
| 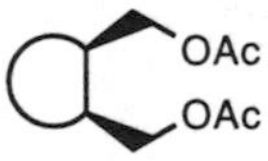 OAc, OAc | PPL, PFL | 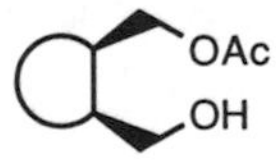 OAc, OH | 84–86 |
| 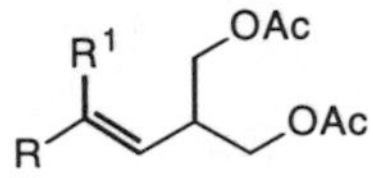 $R^1$, R, OAc, OAc | PPL | 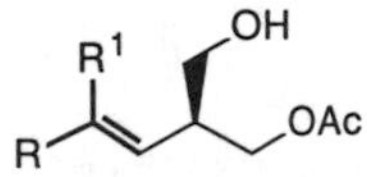 $R^1$, R, OH, OAc | 47,87–90 |
| 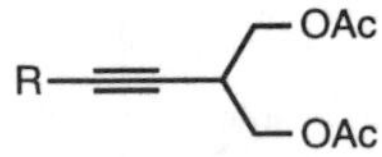 R, OAc, OAc | PPL | 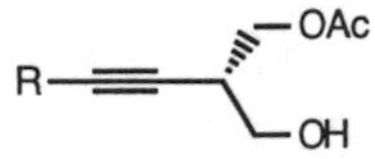 R, OAc, OH | 89,90 |
| 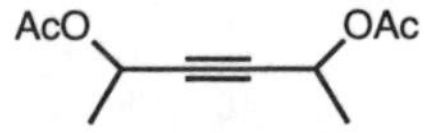 AcO, OAc | PFL |  HO, OAc | 91 |

## TABLE 14.3 (continued)

| Substrate | Enzyme or Organism | Product(s)[a] | Ref. |
|---|---|---|---|
| 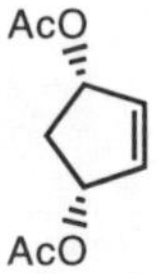 AcO AcO | *Bacillus subtillis* | 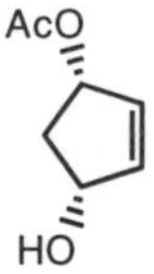 AcO HO | 92 |
| 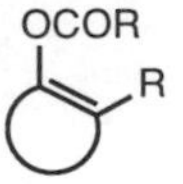 OCOR R | *Pichia miso* | 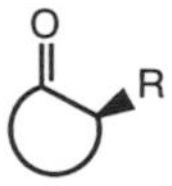 O R | 93 |
| 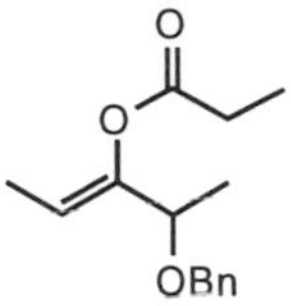 O O OBn | Lipase OF | 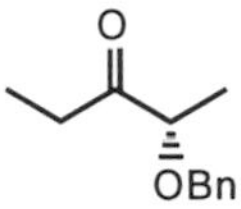 O OBn | 94,95 |
| 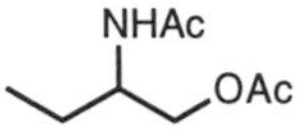 NHAc OAc | PPL | 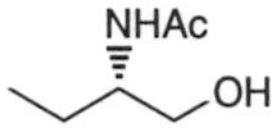 NHAc OH | 96 |
| 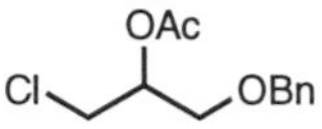 OAc Cl OBn | PPL | 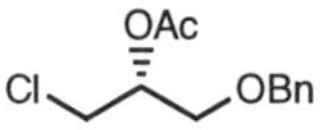 OAc Cl OBn | 47,97 |
| 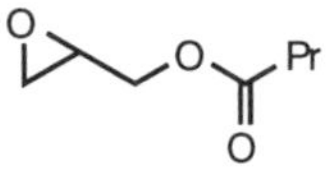 O O Pr O | PPL | 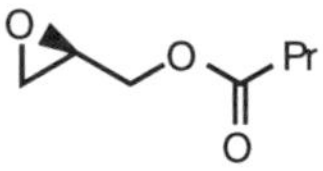 O O Pr O | 98 |
| 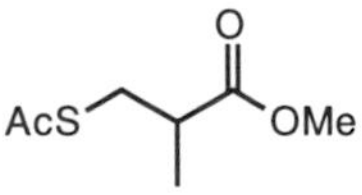 O AcS OMe | Lipase P | 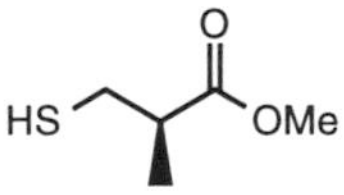 O HS OMe | 99 |
| 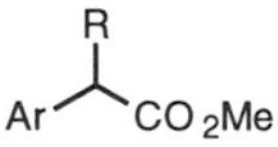 R Ar $CO_2Me$ | Horse liver esterase, CCL, A form, penicillin acylase | 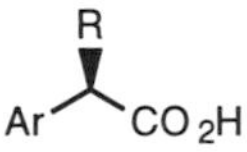 R Ar $CO_2H$ | 100–103 |
| 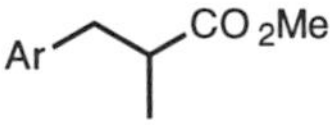 Ar $CO_2Me$ | Lipase P | 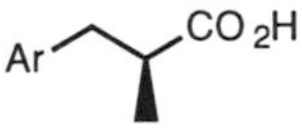 Ar $CO_2H$ | 104 |
| 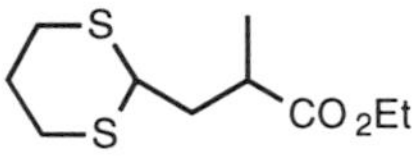 S S $CO_2Et$ | PPL | 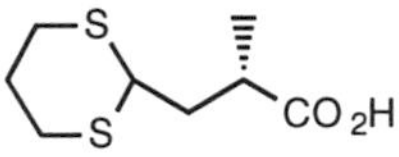 S S $CO_2H$ | 105 |
| 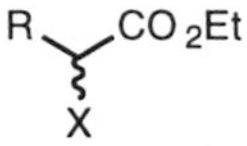 R $CO_2Et$ X<br>X = F,Br, or OH | Lipase P-30 amano | 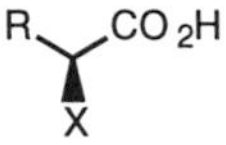 R $CO_2H$ X | 106 |
| 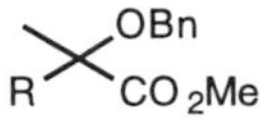 OBn R $CO_2Me$ | Lipase OF<br>*Candida cylindacea* | 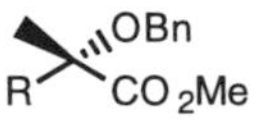 OBn R $CO_2Me$ | 107 |
| 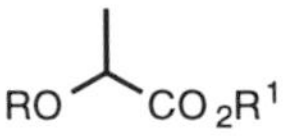 RO $CO_2R^1$ | BSA, β-cyclodextrin | 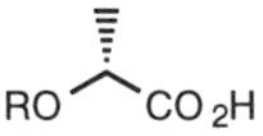 RO $CO_2H$ | 108 |
| 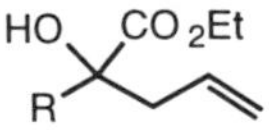 HO $CO_2Et$ R | PLE | 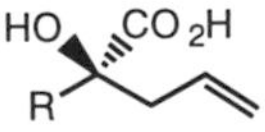 HO $CO_2H$ R | 109 |
| 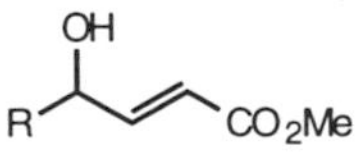 OH R $CO_2Me$ | PPL, PLE | 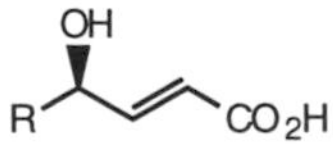 OH R $CO_2H$ | 110 |

**TABLE 14.3 (continued)**

| Substrate | Enzyme or Organism | Product(s)[a] | Ref. |
|---|---|---|---|
| cis-cycloalkane with $CO_2Me$ and X; n = 2 or 3; X = Me, $CO_2Me$ or $CH_2Br$ | PLE | cycloalkane with $CO_2H$ and X | 111 |
| lactone $(CH_2)_{n-3}$, R | PPL, PLE, HLE | $(CH_2)_{n-3}$ with $CO_2H$, OH, R, H | 112 |
| 5-R-hydantoin (R, H, N, O, O, N, H) | *Bacillus brevis* or *Agrobacterium radiobacter* D-hydantoinase | $H_2NCONH$–CH(R)–$CO_2H$ | 113–120 |
| γ-lactone with $NHCOR^1$, R, O, O | PPL[d] | γ-lactone with NHCOX, R, O, O (* *) | 121 |
| TMS–$CH_2$–C(R)($CO_2R^1$)($CO_2R^1$) | PLE | TMS–$CH_2$–C(R)($CO_2H$)($CO_2R^1$) | 122 |
| cyclic cis-diester ($CO_2Me$, $CO_2Me$) | PLE | cyclic $CO_2H$, $CO_2Me$ | 123–130 |
| RO, RO, OMe, OMe, O, O (tartrate diester) | PLE, CCL | RO, RO, OMe, OH, O, O | 131 |
| $R^2$–CH($CH_2CO_2R$)($CH_2CO_2R^1$) | PLE | $R^2$–CH($CH_2CO_2R$)($CH_2CO_2H$) | 128,129,132,133 |
| $MeO_2C$–$CH_2$–C($R^1$)($R^2$)–$CH_2$–$CO_2Me$ | PLE | $R^3O_2C$–$CH_2$–C($R^1$)($R^2$)–$CH_2$–$CO_2R^4$ | 23,40,134–137 |
| $EtO_2C$–$CH_2$–CH(OH)–$CH_2$–$CO_2Et$ | Esterase 30000 | $EtO_2C$–$CH_2$–CH(OH)–$CH_2$–$CO_2H$ | 138 |
| 2-oxocyclohexane with $CO_2Et$, R | PLE[e] | 2-oxocyclohexane with $CO_2Et$, R | 139 |
| $R^1$–CH(OH)–$CH_2$–CH(R)–$CO_2Me$ | PPL | γ-lactone with R, $R^1$, O, O | 140 |

## TABLE 14.3 (continued)

| Substrate | Enzyme or Organism | Product(s)[a] | Ref. |
|---|---|---|---|
| 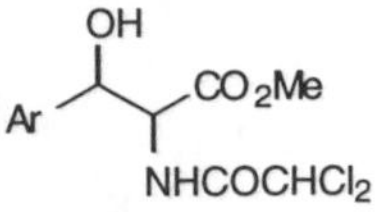 OH, Ar, $CO_2Me$, $NHCOCHCl_2$ | Subtilisin | 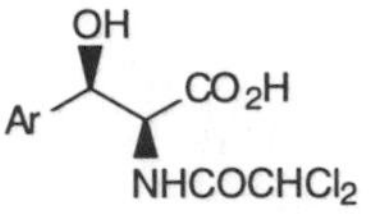 OH, Ar, $CO_2H$, $NHCOCHCl_2$ | 141 |
| 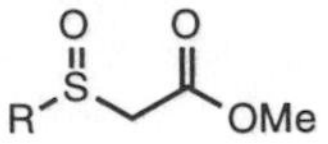 O, O, R, S, OMe | *Pseudomonas K-10* | 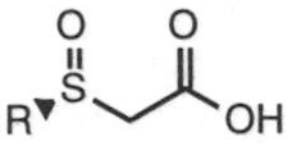 O, O, R, S, OH | 142 |
| 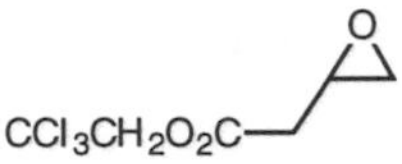 O, $CCl_3CH_2O_2C$ | PPL[f] | 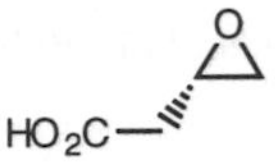 O, $HO_2C$ | 143 |
| 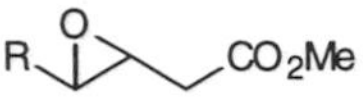 R, O, $CO_2Me$ | PLE |  R, O, $CO_2H$ | 144 |
| 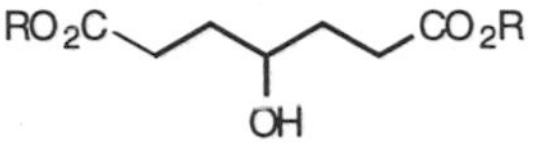 $RO_2C$, $CO_2R$, OH | PPL | 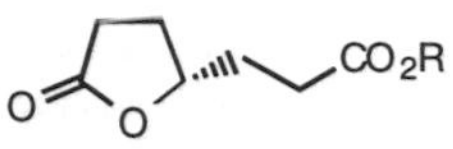 O, O, $CO_2R$ | 145 |
| 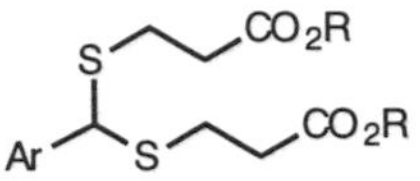 S, $CO_2R$, Ar, S, $CO_2R$ | CCL, Lipase (*Pseudomonas* sp.) | 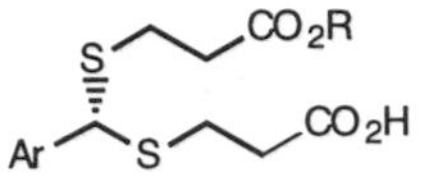 S, $CO_2R$, Ar, S, $CO_2H$ | 146,147 |
| 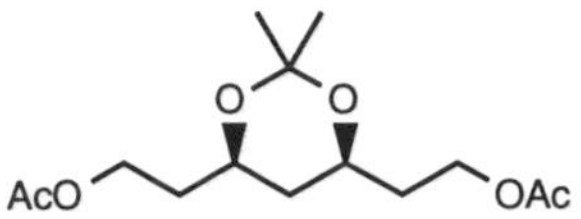 O, O, AcO, OAc | PFL | 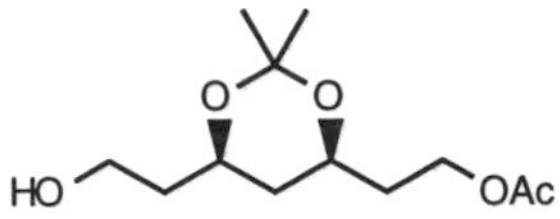 O, O, HO, OAc | 148 |

[a] All products are optically active unless noted otherwise; omission of stereochemistry implies that the original citation either did not report the absolute configuration or that the stereochemical outcome of the reaction is substrate dependent. When only one isomer is given as the product this implies that the enzyme can be used for a kinetic resolution. Consult the original reference for details about enantioselectivity. [b] Resolution occurs by ester hydrolysis. [c] The diastereoisomer is not affected by this enzyme. [d] Proceeds by lactone hydrolysis. [e] The other isomer is hydrolyzed and this results in decarboxylation of the β-keto acid. [f] In the presence of polyethylene glycol; a subsequent saponification is required.

As an example, consider the hydrolysis of dimethyl cyclohexane-1,2-dicarboxylate (**14.1**) (Scheme 14.1).[39,40]

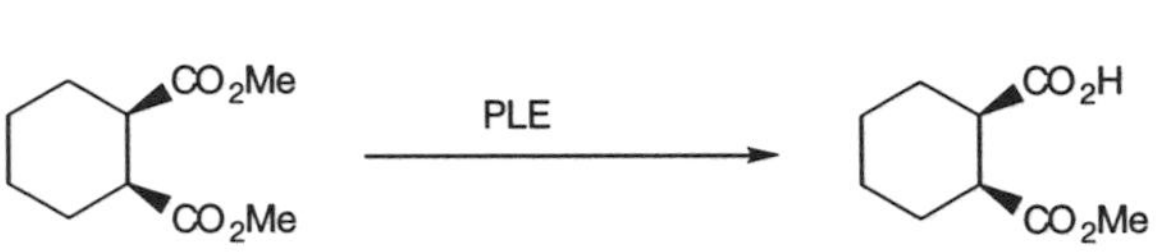

$CO_2Me$, $CO_2Me$ —PLE→ $CO_2H$, $CO_2Me$

**14.1** **14.2**

**SCHEME 14.1.**

The two possible modes that the diester **14.1** can fit into the active site model show why the product formed is **14.2**, as the complex (Figure 14.4) does not have the size exclusion problems of Figure 14.5.[37]

This model also rationalizes why apparent reversal is observed for the diester of the cyclobutane analog, while the cyclopentane dialog gives essentially racemic product.[39,40]

As well as enantioselective reactions, regio- and stereoselective reactions are also possibilities for enzyme catalysis, but often the limitations imposed by the structural requirements of the substrate are greatly increased. Examples of transformations performed with lipases are given in Tables 14.3 and 14.4.

**TABLE 14.4**
**Ester Formation and Related Reactions Catalyzed by Enzymes and Organisms**

| Substrate | Enzyme or Organism | Product(s)[a] | Ref. |
|---|---|---|---|
| 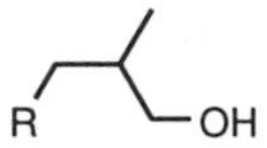 R, OH | PFL[b] | 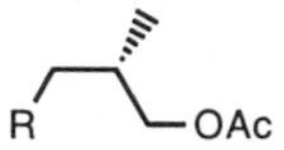 R, OAc | 104,149–151 |
| 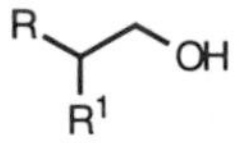 R, OH, $R^1$ | Lipase[b] | 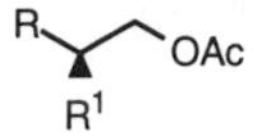 R, OAc, $R^1$ | 152 |
| 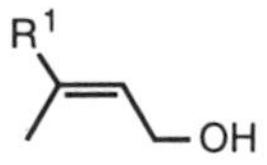 $R^1$, OH | Pancreatic powder lipase[c] | 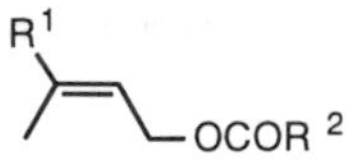 $R^1$, $OCOR^2$ | 153,154 |
| 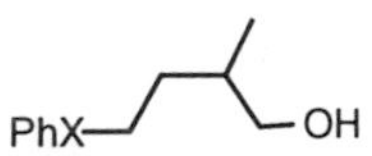 PhX, OH<br>where X = S, Se, or $SO_2$ | PFL[b] | 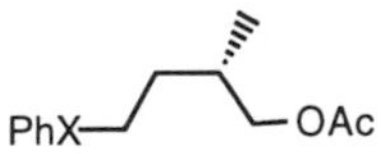 PhX, OAc | 149 |
| 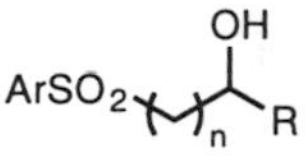 OH, $ArSO_2$, n, R | PPL[c] | 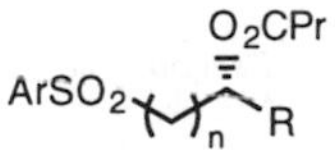 $O_2CPr$, $ArSO_2$, n, R | 155 |
| 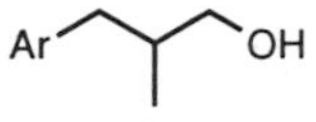 Ar, OH | Lipase P[b] |  Ar, OAc | 104 |
| 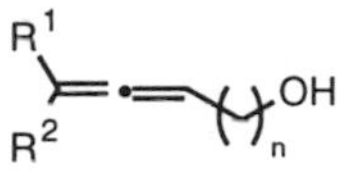 $R^1$, $R^2$, OH, n | CCL[d] | 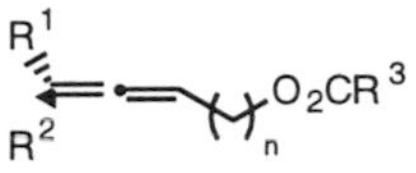 $R^1$, $R^2$, $O_2CR^3$, n | 156 |
| 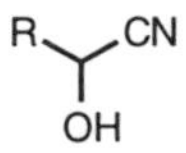 R, CN, OH | PSL (*Pseudomonas* lipoprotein lipase)[b], PPL[b] | 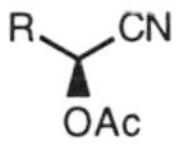 R, CN, OAc | 157,158 |
| 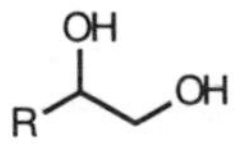 OH, R, OH | Lipase PS[b] | 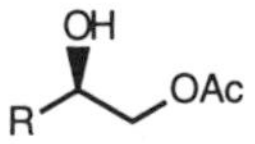 OH, R, OAc | 159 |
| 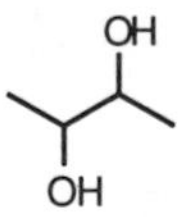 OH, OH | Amano lipase PS[b] | 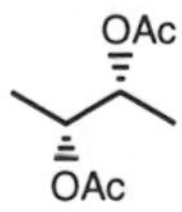 OAc, OAc | 160 |
| 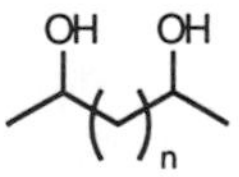 OH, OH, n<br>n = 0–2 | Lipase[e] | 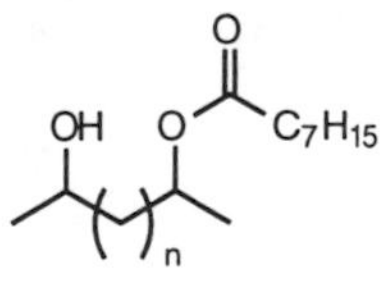 OH, O, O, $C_7H_{15}$, n | 161 |
| 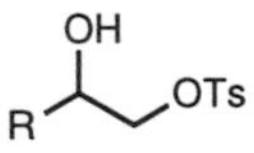 OH, R, OTs | PFL[b] | 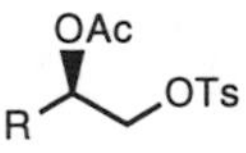 OAc, R, OTs | 162 |
| 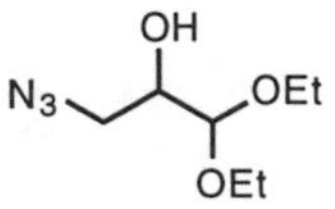 OH, $N_3$, OEt, OEt | XAD-8 lipoprotein lipase[b] | 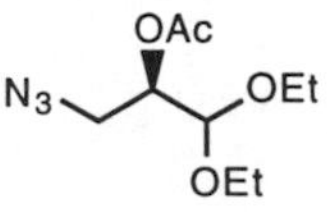 OAc, $N_3$, OEt, OEt | 163 |
| 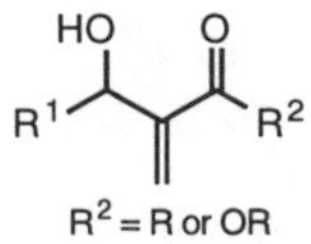 HO, O, $R^1$, $R^2$<br>$R^2$ = R or OR | *Pseudomonas AK* | 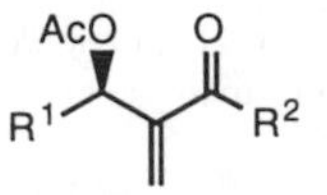 AcO, O, $R^1$, $R^2$ | 164,165 |

**TABLE 14.4 (continued)**

| Substrate | Enzyme or Organism | Product(s)[a] | Ref. |
|---|---|---|---|
| 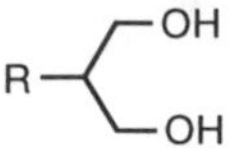 | PPL, Lipase P (*Pseudomonas fluorescens*)[b] | 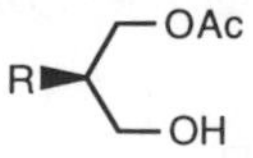 | 72,76,79,150, 166–169 |
| 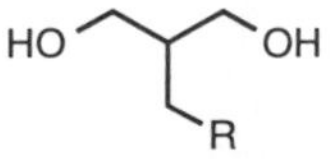 | Lipase P[b] | 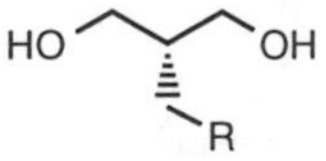 | 170 |
| 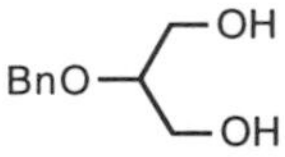 | *Pseudomonas* lipase | 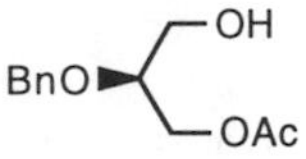 | 166 |
| 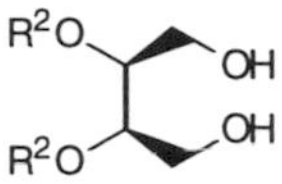 | PPL, PAN[f] | 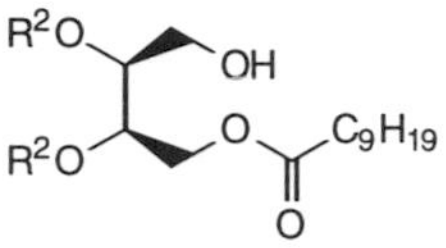 | 131 |
| 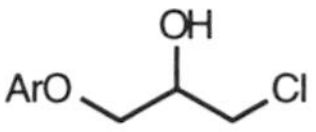 | *Pseudomonas* sp. lipase[b] | 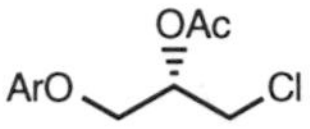 | 97 |
| 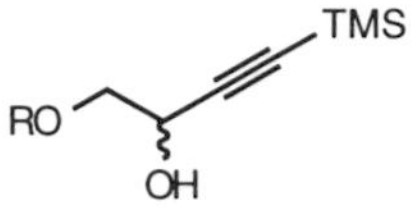 | Lipase PS[b] | 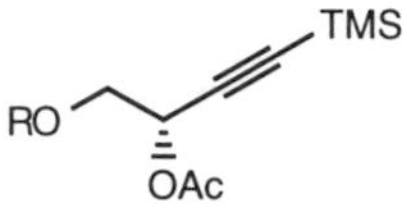 | 171 |
| 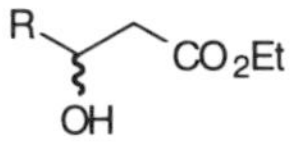 | CAL[g] | 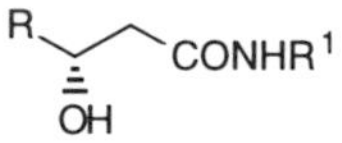 | 172,173 |
| 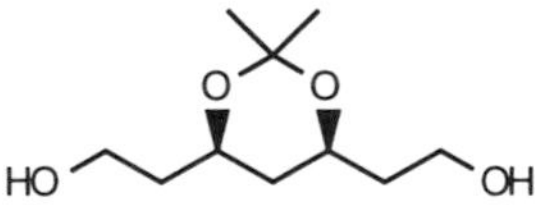 | PPL[h] | 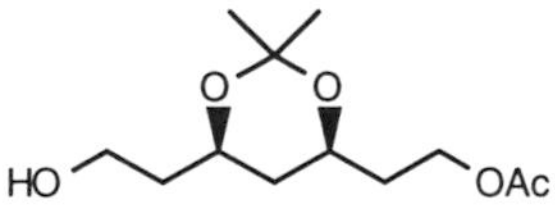 | 148 |
| 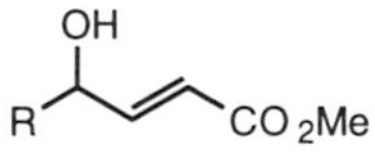 | CCL[h], PPL[h] | 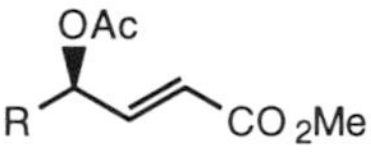 | 174 |
| 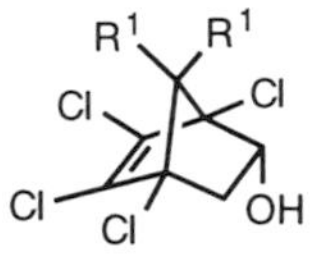 | Lipase AY-30 (*Candida cylindracea*)[c] | 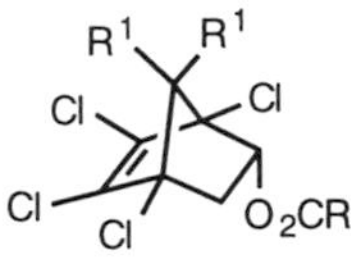 | 175 |
| 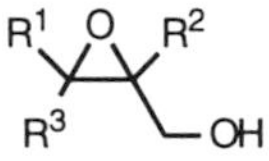 | PPL[h] | 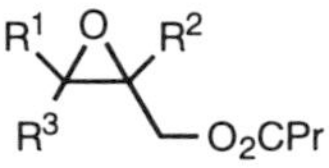 | 176,177 |
| 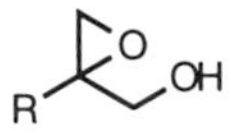 | PFL[h] | 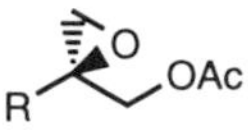 | 178 |
| 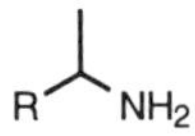 | CAL[i] | 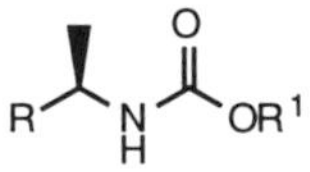 | 179,180 |
| 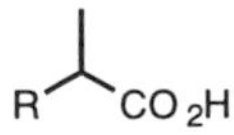 | CCL[j], *Candida rugosa* lipase[j] | 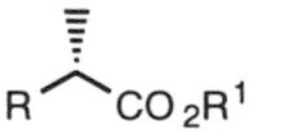 | 181,182 |
| 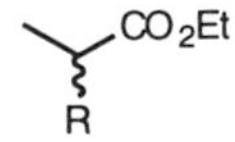 | CAL[f] | 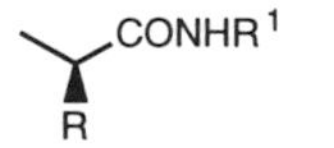 | 183 |

**TABLE 14.4 (continued)**

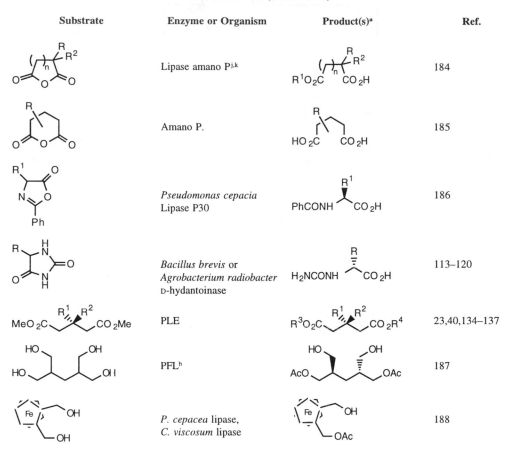

| Substrate | Enzyme or Organism | Product(s)[a] | Ref. |
|---|---|---|---|
| | Lipase amano P[j,k] | | 184 |
| | Amano P. | | 185 |
| | *Pseudomonas cepacia* Lipase P30 | | 186 |
| | *Bacillus brevis* or *Agrobacterium radiobacter* D-hydantoinase | | 113–120 |
| | PLE | | 23,40,134–137 |
| | PFL[b] | | 187 |
| | *P. cepacea* lipase, *C. viscosum* lipase | | 188 |

[a] All products are optically active unless noted otherwise; omission of stereochemistry implies that the original citation either did not report the absolute configuration or that the stereochemical outcome of the reaction is substrate dependent. When only one isomer is given as the product this implies that the enzyme can be used for a kinetic resolution. Consult the original reference for details about enantioselectivity. [b] In the presence of vinyl acetate. [c] In the presence of an acylating agent. [d] In the presence of lauric acid. [e] In the presence of ethyl thiooctanoate. [f] In the presence of $Cl_3CCH_2OCOC_9H_{19}$. [g] In the presence of an amine. See also Table 14.6. [h] In the presence of $PrO_2CCH_2CF_3$. [i] In the presence of octyl or butyl vinyl carbonate. [j] In the presence of an alcohol, $R^1OH$. [k] The degree of selectivity is variable.

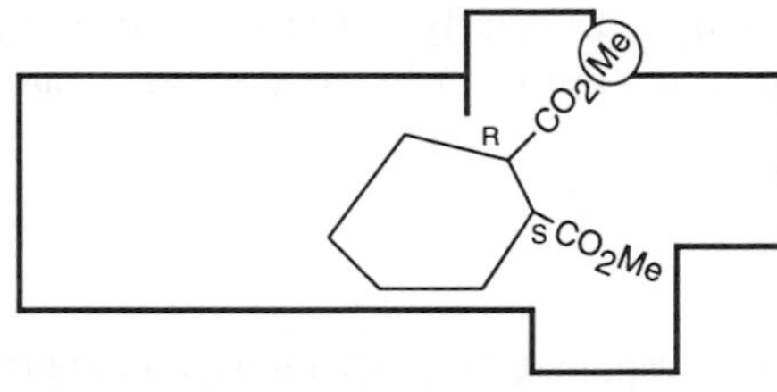


**FIGURE 14.4.** **14.1** as substrate in PLE active site model.

The enzymatic approach has been expanded by the use of enzymes in nonaqueous media; this allows for the use of water insoluble substrates,[29,189-193] and control of the stereochemical outcome of the reaction through manipulation of these variables. NMR solid state studies of an enzyme in an organic solvent have shown that the integrity is preserved.[194]

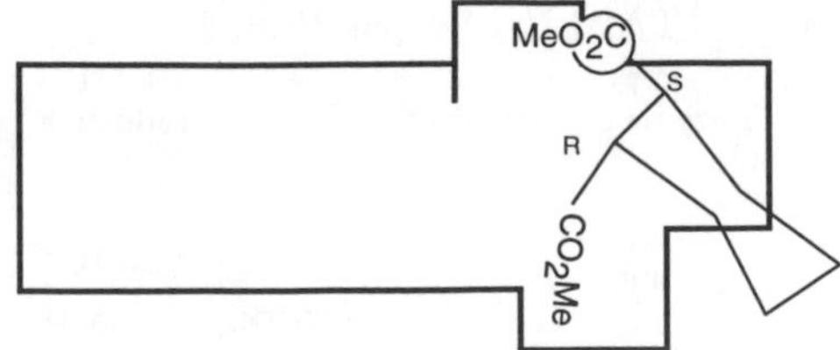


**FIGURE 14.5.** Complex needed for **14.1** to give antipode from PLE hydrolysis.

Enzymes used for selective hydrolyses can provide kinetic resolutions (see Tables 14.3 and 14.4).[112] Such selectivity allows differentiation within *meso*-esters and the preparation of selectively protected glycerol (Scheme 14.2).[77,84]

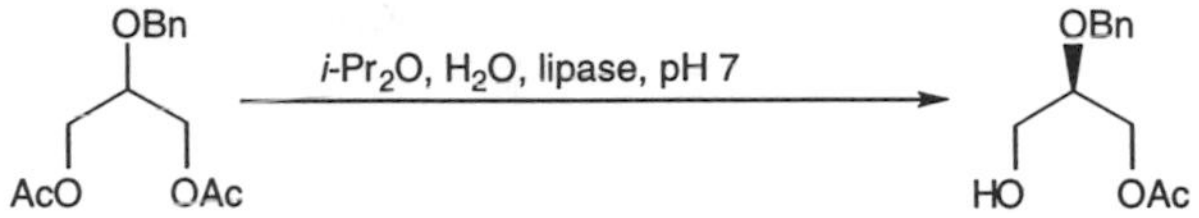

**SCHEME 14.2.**

The methodology can be taken one step further through the use of an enzyme for ester hydrolysis or formation. This opens up the potential to either isomer (Scheme 14.3).[76]

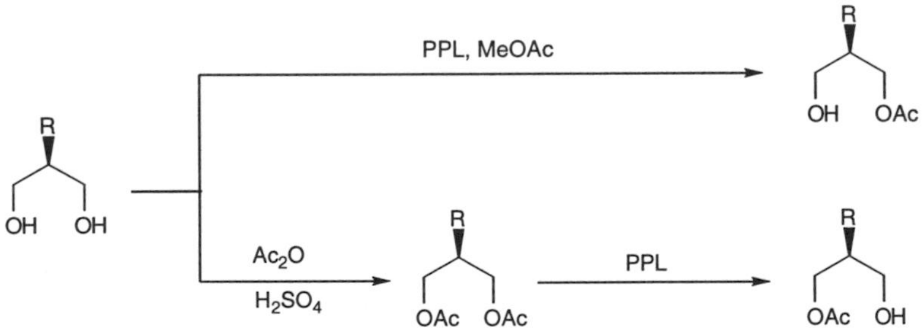

**SCHEME 14.3.**

The use of catalytic antibodies now allows for the design and production of an enzyme-like system that can catalyze specific ester hydrolyses [*14.4*]. It is also possible to mutate an enzyme to increase stability, particularly in organic solvent systems, without affecting stereospecificity.[195,196] Indeed, the use of an organic solvent can effect the outcome of an enzymatic reaction.[197-199]

## 14.2. REDUCTIONS

The reduction of carbonyl compounds by yeast allows for the direct introduction of a chiral center.[8,200-202] A number of classes of compounds have been reduced by this tolerant system as illustrated in Table 14.5. Related reductions have also been included in this Table. Baker's yeast is, by far, the most common agent for these reductions due to the wide range of substrates that can be employed, and the low cost.[6,8]

A rule has been developed by Prelog[27] that predicts that hydrogen addition by the yeast *Culvaria lunata* will occur from the front face (Figure 14.1). This model also demonstrates that if a system does not conform to the requirements, stereoselection could be low if the

## TABLE 14.5
## Reductions Catalyzed by Biological Systems

| Substrate | Enzyme or Organism | Product(s)[a] | Ref. |
|---|---|---|---|
| $R^1C(O)R^2$ | *C. lunata*, alcohol dehydrogenase, *Sporotrichium exile* (QM-1250), baker's yeast, *Thermoanaerobium brockii* alcohol dehydrogenase, HLADH | $R^1CH(OH)R^2$ | 203–209 |
| $RC(=NOH)R^2$ | Baker's yeast | $RCH(NH_2)R^2$ | 210 |
| $RCH_2C(O)C(OH)R^1R^2$ | Baker's yeast | $RCH_2CH(OH)C(OH)R^1R^2$ | 211–224 |
| $R^1C(O)CH_2OR^2$ | Baker's yeast | $R^1CH(OH)CH_2OR^2$ | 207,211, 221,225 |
| $(RX)_2CHCH_2C(O)R$ where X = O or S | Baker's yeast[b] | $(RX)_2CHCH_2CH(OH)R$ | 214,226–230 |
| $PhC(O)CH_2X$ where X = F, Cl, Br | Baker's yeast | $PhCH(OH)CH_2X$ | 231 |
| $RC(O)CO_2H$ | L-Lactate dehydrogenase, *Geotricium* glycerol dehydrogenase, *Bacillus stearothermophilus* L-lactate dehydrogenase (BSLDH) | $RCH(OH)CO_2H$ | 232–237 |
| $R^1C(O)CO_2R^2$ | Baker's yeast | $R^1CH(OH)CO_2R^2$ | 238–241 |
| $PhC(O)C(O)CH_3$ | Baker's yeast | $PhCH(OH)CH(OH)CH_3$ | 242 |
| $PhC(O)C(O)CH_3$ | Baker's yeast[c] | $PhC(O)CH(OH)CH_3$ | 243 |
| Ph-epoxide-C(O)R | Baker's yeast | $PhCH(OH)CH(OH)CH(OH)R$ | 244 |
| $HC\equiv CC(O)CH_3$ | *Thermoanaerobium brockii* alcohol dehydrogenase | $HC\equiv CCH(OH)CH_3$ | 245 |
| $RCH_2CH(Cl)C(O)CH_3$ | Baker's yeast | $RCH_2CH(Cl)CH(OH)CH_3$ | 220 |

## TABLE 14.5 (continued)

| Substrate | Enzyme or Organism | Product(s)[a] | Ref. |
|---|---|---|---|
| 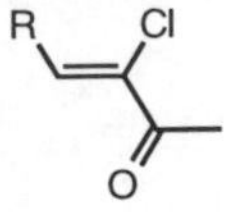 | Baker's yeast | 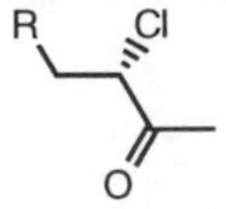 | 220 |
| 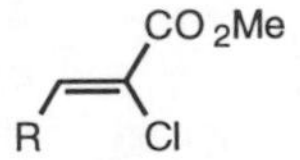 | Baker's yeast | 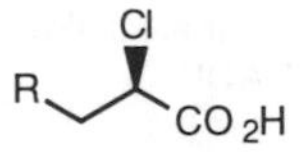 | 215,246 |
| 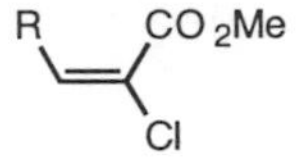 | Baker's yeast | 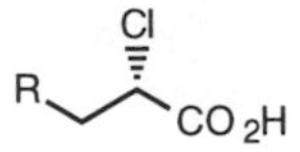 | 215,246 |
| 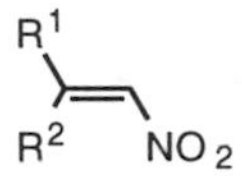 | Baker's yeast | 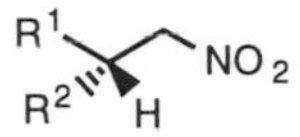 | 247 |
| 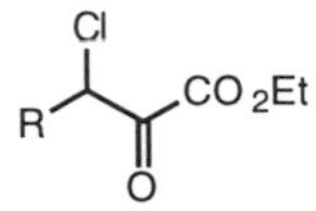 | Baker's yeast | 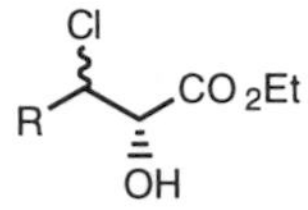 | 217,248 |
| 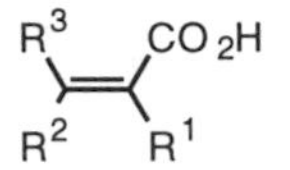 | *Clostrydium kluyeri*[d] | 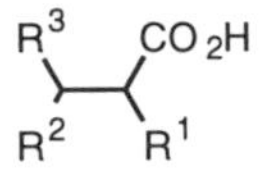 | 249 |
| 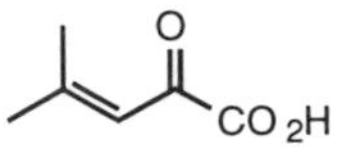 | *Bacillus stearothermophilis* lactate dehydrogenase | 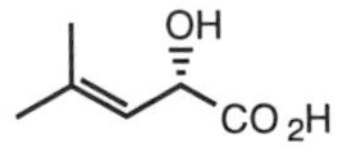 | 235,237 |
| 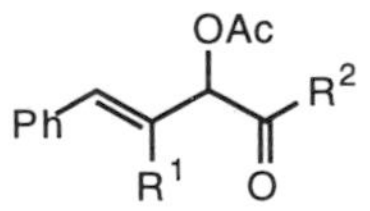 | Baker's yeast[e] | 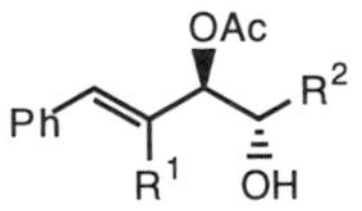 | 219,250 |
| 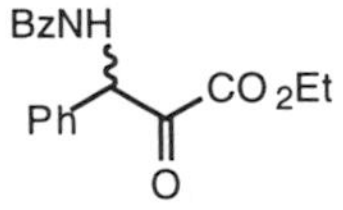 | *Hansenula polymorph* | 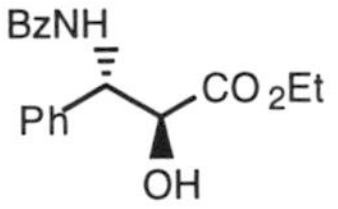 | 251 |
| 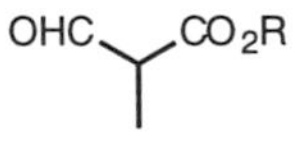 | Baker's yeast, *Thermoanaerobium brockii, Candida humiola* | 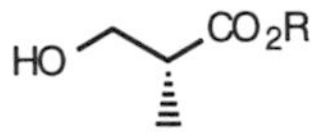 | 72,252–256 |
| 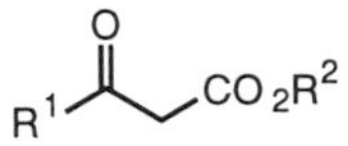 | Baker's yeast, immobilized *Nicotiana tabacum* | 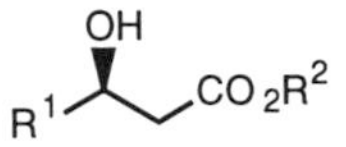 | 238,257–276 |
| 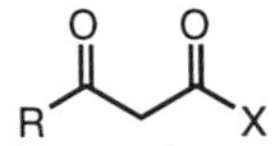 where X = OEt or NHPh | *Saccharomyces cerevisial, Geotrichium candidum, Candida parapsilosis, Rhodococcus erythropolis* | 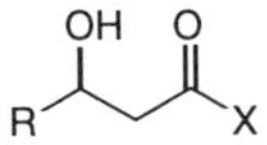 | 238,277,278 |
| 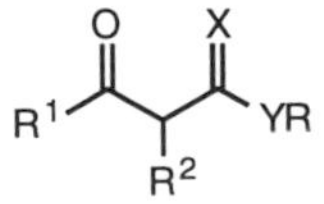 X=Y=S<br>X=O, Y=S<br>Y=NH<br>X=Y=O | Baker's yeast,[f,g] *Aspergillus niger* lipase, *G. candidum*[h] | 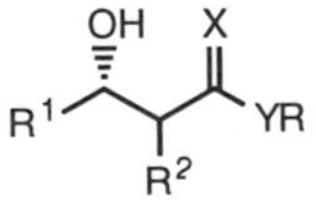 | 71,238,252, 257,258,266, 268,279–303 |

**TABLE 14.5 (continued)**

| Substrate | Enzyme or Organism | Product(s)[a] | Ref. |
|---|---|---|---|
| 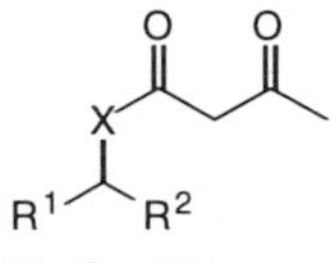 X = O or NH | Baker's yeast | 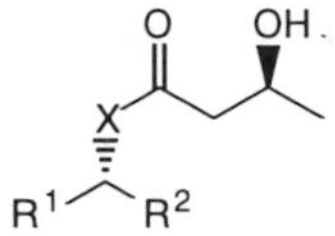 | 252,304,305 |
| 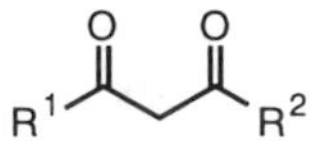 | *Geotrichum candida, A. niger, Mortierella isabellina, Candida rugosa, Saccharomyces cerevisiae* | 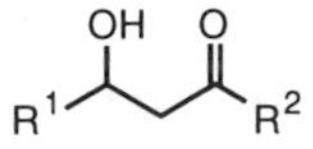 | 306–308 |
| 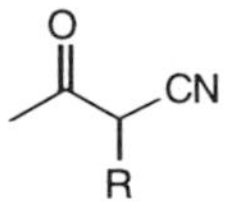 | Baker's yeast | 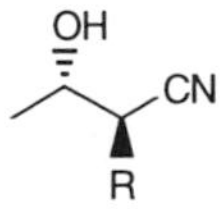 | 309 |
| 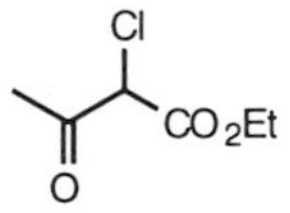 | Baker's yeast | 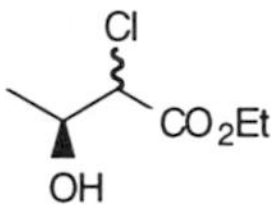 | 310 |
| 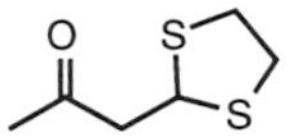 | Yeast | 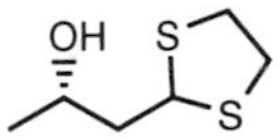 | 226 |
| 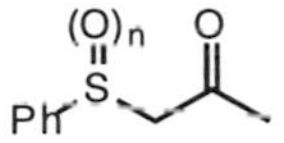 where n = 1 or 2 | Baker's yeast | 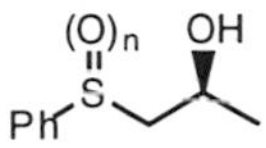 | 311–313 |
| 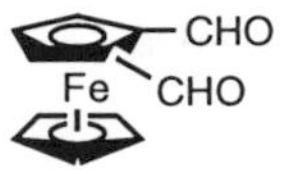 | HLADH | 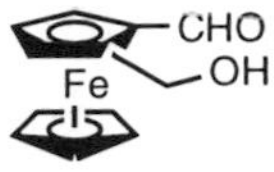 | 314 |
| 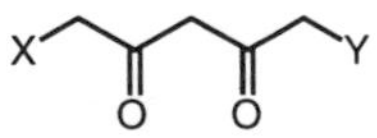 where x = Cl, $N_3$, or NHAc; Y = NHR or $R^1$ | Baker's yeast | 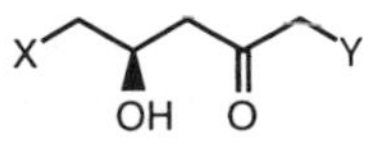 | 315 |
| 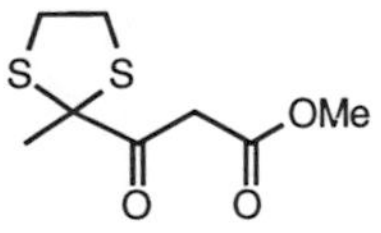 | Baker's yeast | 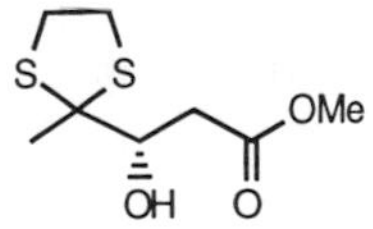 | 316 |
| 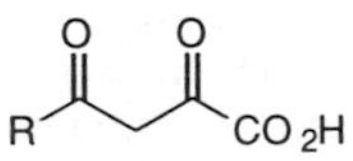 | Lactate dehydrogenase, *P. vulgaris*[i] | 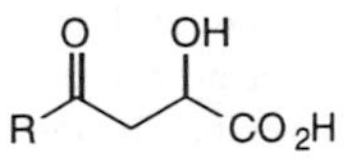 | 317,318 |
| 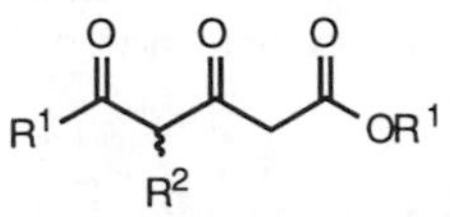 | Baker's yeast[j] | 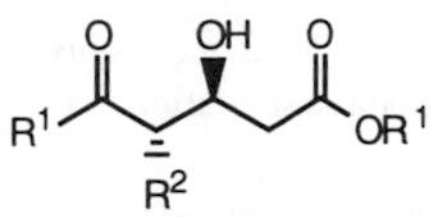 | 319 |
| 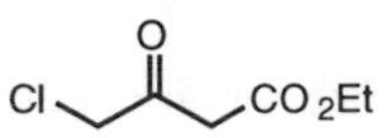 | Baker's yeast, *lactobacillus kefir, Sulfolobus sulfataricu* | 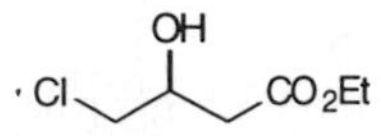 | 320,321 |
| 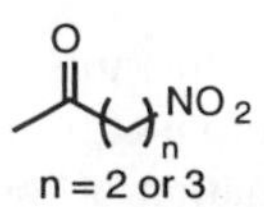 n = 2 or 3 | Baker's yeast, *Sacc. cerevisiae* | 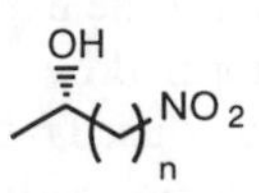 | 233,322–324 |

**TABLE 14.5 (continued)**

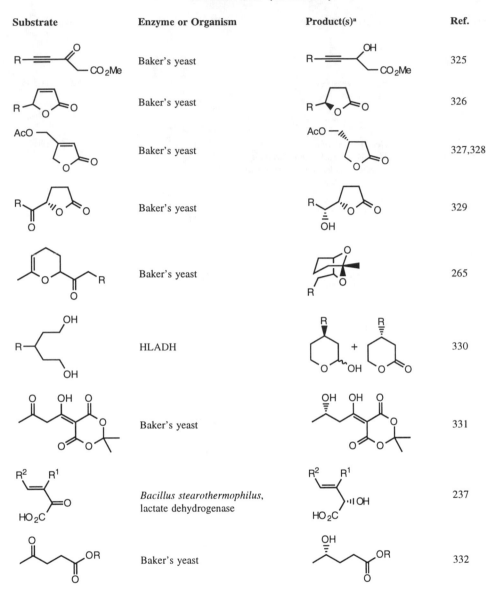

| Substrate | Enzyme or Organism | Product(s)[a] | Ref. |
|---|---|---|---|
| 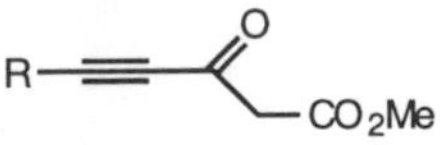 | Baker's yeast | 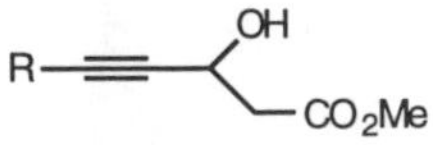 | 325 |
| 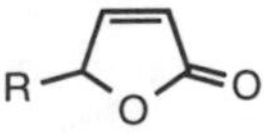 | Baker's yeast | 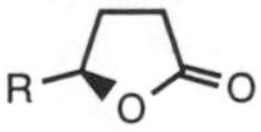 | 326 |
| 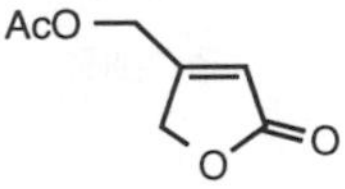 | Baker's yeast | 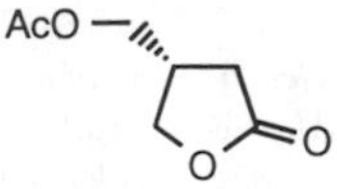 | 327,328 |
| 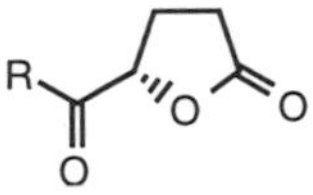 | Baker's yeast | 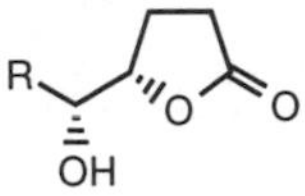 | 329 |
| 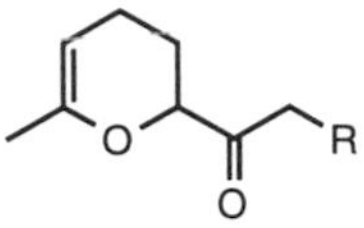 | Baker's yeast |  | 265 |
| 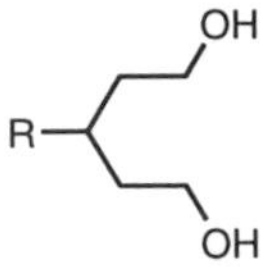 | HLADH | 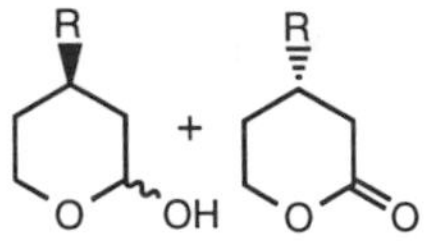 | 330 |
| 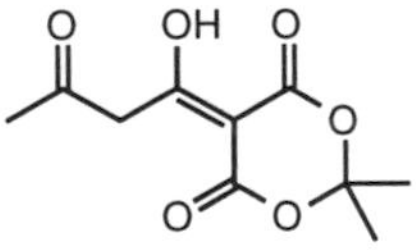 | Baker's yeast | 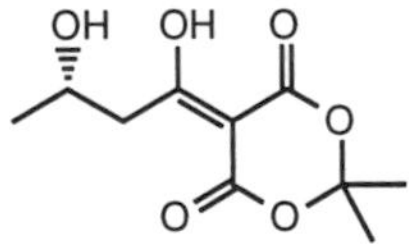 | 331 |
| 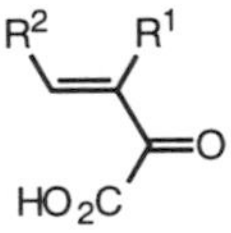 | *Bacillus stearothermophilus*, lactate dehydrogenase | 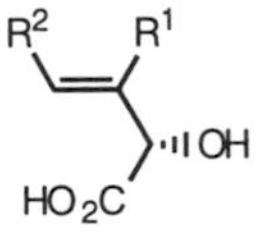 | 237 |
| 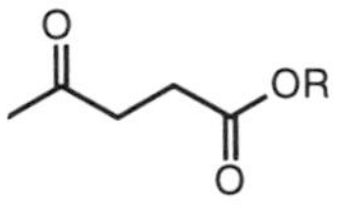 | Baker's yeast | 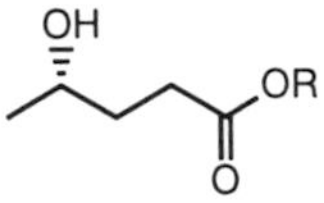 | 332 |

[a] When no stereochemistry is indicated, the reaction outcome is dependent upon the organism and conditions employed. [b] The yeast can be free or immobilized on calcium alginate. [c] At pH 5. [d] In the presence of hydrogen gas. [e] The product is the *ancat* isomer, but the absolute stereochemistry is dependent on $R^2$. [f] The stereochemistry of the product depends upon $R^2$; also the yeast yeast can be free or immobilized on magnesium alginate. [g] Can be performed in the presence of methyl vinyl ketone. [h] This can be immobilized. [i] In the presence of hydrogen gas or formate, with R = Ph. [j] $R^2$ has to contain unsaturation.

compound is a substrate. Yeast, however, is fairly tolerant, as illustrated by the number of examples in Table 14.5.

Exceptions do exist to Prelog's rule, but slight modifications to the substrate can direct the course of the reaction toward the expected direction (Scheme 14.4).[267]

Reductions, when performed by an enzymatic system, usually require a cofactor; this is why organisms are invariably used rather than purified enzymes. Isolation of the product from the reaction media may become a significant cost and convenience factor; formation of the analogous methyl ester from a carboxylic acid through use of diazomethane is a common technique.[246,257] For many of the β-keto ester examples, the highest selectivity is observed when the alkyl ester group is relatively long, such as with octyl.[258]

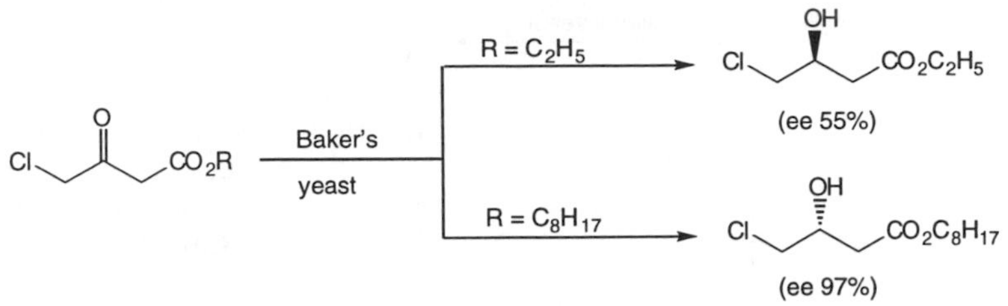

**SCHEME 14.4.**

In some cases, enzymatic systems are available that allow for the complimentary, selective production of either of the two possible isomers (Scheme 14.5).[291,307]

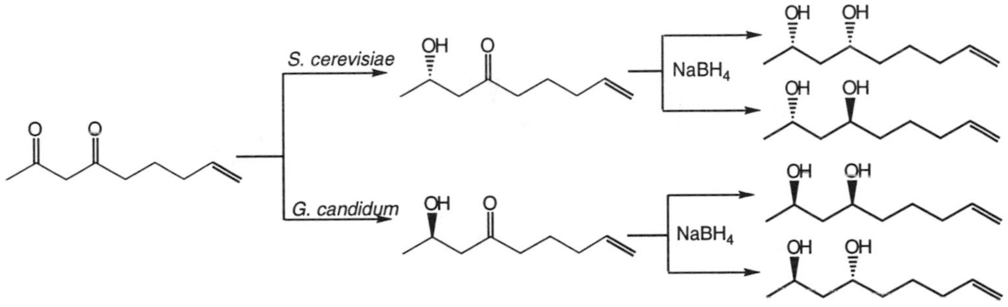

**SCHEME 14.5.**

The entry for the reduction of β-keto esters in Table 14.5 may appear to be somewhat confusing — both isomers of the resultant β-hydroxy ester are available. Reduction with baker's yeast generally affords a mixture of the *syncat* and *ancat* isomers, with a large preference for *syncat*.[296] However, by the use of an additive (e.g., Scheme 14.6),[293] or a mold, such as *Geotrichum candidum*, the *ancat* isomer can be obtained as the major product (Scheme 14.7).[282,291]

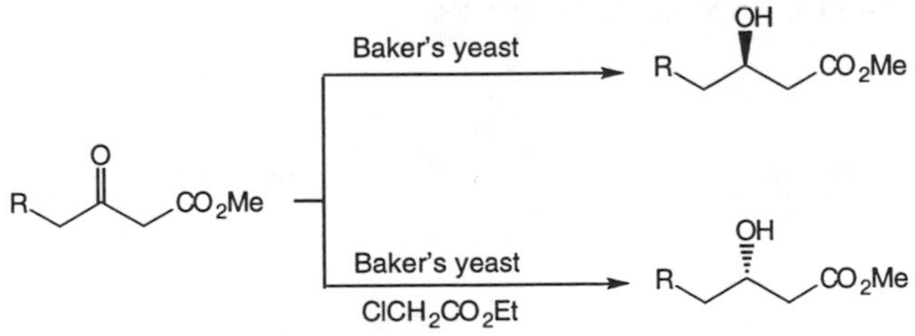

**SCHEME 14.6.**

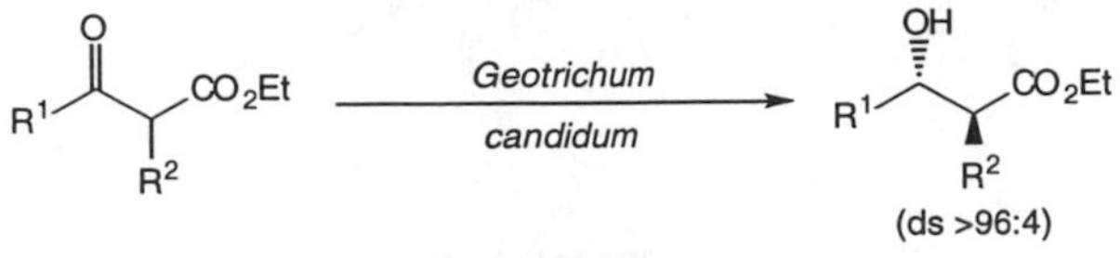

**SCHEME 14.7.**

For the use of baker's yeast, the selectivity is also dependent upon the nature of the alkyl group and the ester group (Scheme 14.8).[283]

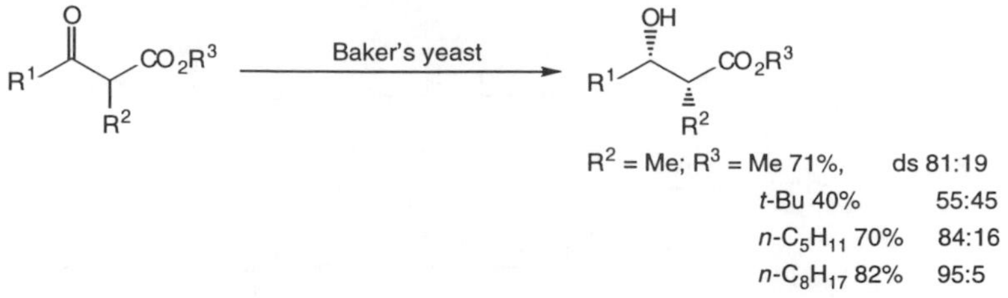

**SCHEME 14.8.**

Even the growth medium can have a pronounced effect on the stereochemical outcome of the yeast reduction of a β-keto ester.[281]

A yeast reduction methodology has been used to prepare L-deoxy sugars (e.g., **14.3**) (Scheme 14.9).[218,219]

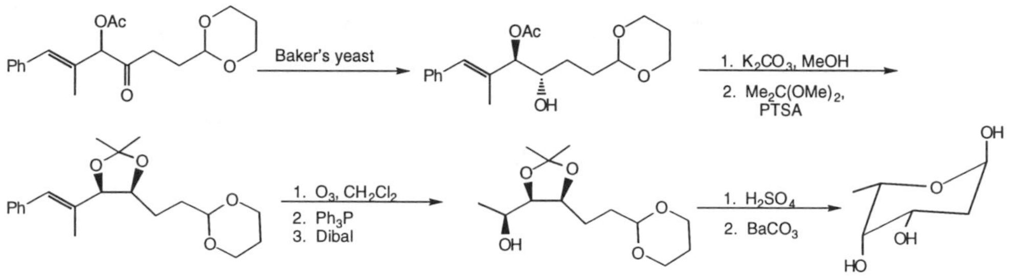

**SCHEME 14.9.**

## 14.3. OTHER REACTIONS

Enzymatic systems have been used to catalyze a wide variety of transformations; some of the relevant and useful ones are summarized in Table 14.6.

The oxidation of an aromatic substrate has provided a synthesis of 2,3-isopropylidene L-ribonic γ-lactone (**14.4**) (Scheme 14.10).[393]

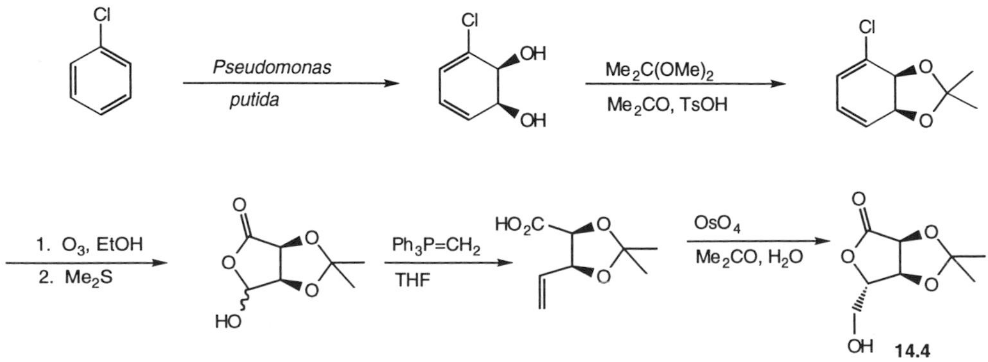

**SCHEME 14.10.**

**TABLE 14.6**
**Other Transformations Catalyzed by Biological Systems**

| Substrate | Enzyme or Organism | Product(s)[a] | Ref. |
|---|---|---|---|
| Aldol type reactions | Baker's yeast, fructose diphosphate aldolase, acylneumaminate pyruvate lyase, RAMA | See text | 333–343 |
| 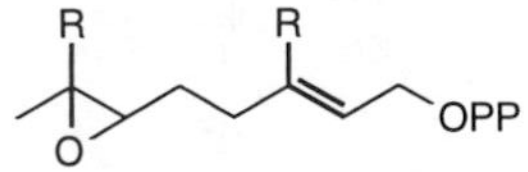 R R O OPP | Farnesyl diphosphate synthetase[a] | 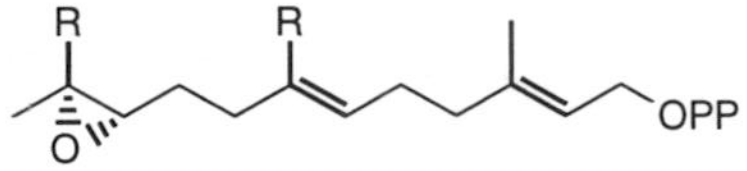 R R O OPP | 344 |
| RCHO | Flavoprotein D-oxynitrilase, powdered almond meal | 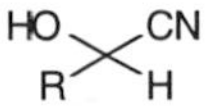 HO CN R H | 345–349 |
| RCHO + HCN | Oxynitrilase, cyclo-S-Phe-S-His[b] | 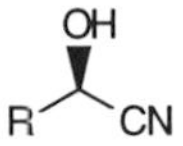 OH R CN | 346,348,350–352 |
| RCHO + TMSCN, then $Ac_2O$ | Ester hydrolyase, *Pseudomonas* sp. | 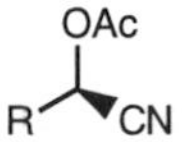 OAc R CN | 50 |
| 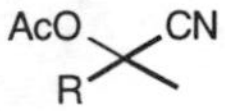 AcO CN R | *Pichia miso* |  HO CN R | 54,353 |
| RCHO | D-Hydroxynitrile lyase | 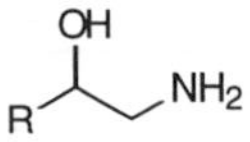 OH R $NH_2$ | 354 |
| 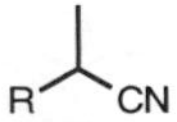 R CN | *Rhodococcus butanica*, *Rhodococcus rhodochrous* K22 | 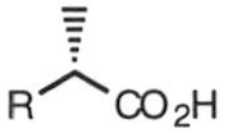 R $CO_2H$ | 355–357 |
| $RNH_2$ | Lipase SP382 (*Candida* sp.)[c] | RNHCOPr | 358 |
| 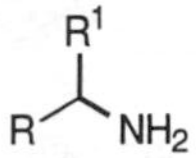 $R^1$ R $NH_2$ | Subtilisin[d] | 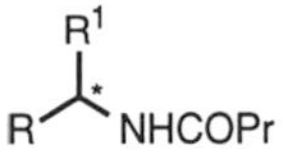 $R^1$ R * NHCOPr | 359 |
| 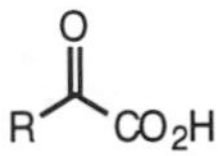 O R $CO_2H$ | *Pseudomonas fluorescens,* E. coli, *Bacillus spaeccus* aminotransferase | 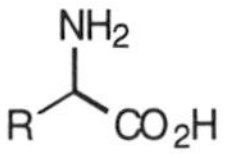 $NH_2$ R $CO_2H$ | 360–364 |

**TABLE 14.6 (continued)**

| Substrate | Enzyme or Organism | Product(s)[a] | Ref. |
|---|---|---|---|
| 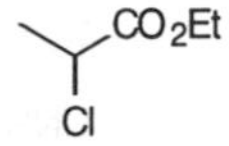 | CCL, subtilisin protease[e] | 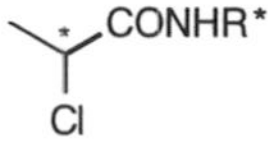 | 365,366 |
| 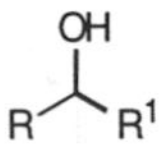 | Baker's yeast, *Bacillus stearothermophilus*, fatty alcohol oxidase[f] | 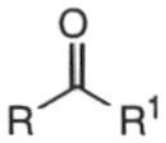 | 367–369 |
| 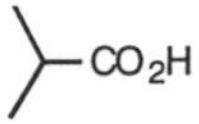 | *Pseudomonas putida, Candida rugosa* | 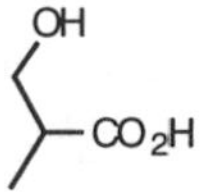 | 72,370–373 |
| 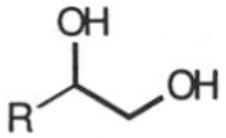 | Galactose oxidase | 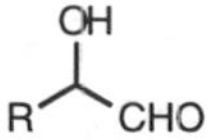 | 374 |
| 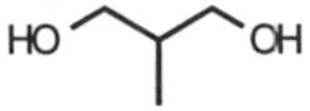 | *Gluconobacter roseus* | 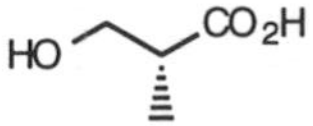 | 72,375 |
| 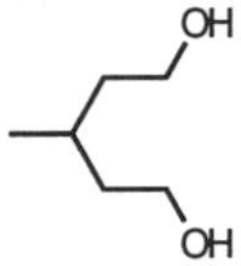 | Horse liver alcohol dehydrogenase | 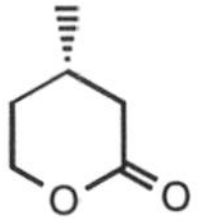 | 314,330 |
| 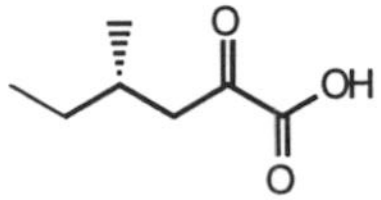 | α-Ketoisocaproate dioxygenase[g] | 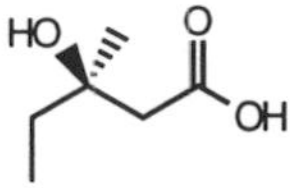 | 376 |
| 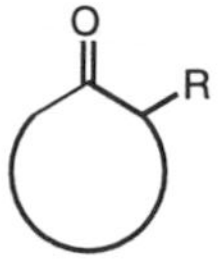 | *Acinetobacter calcoaceticus* NC1B981B | 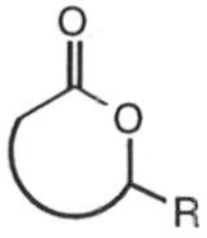 | 377 |

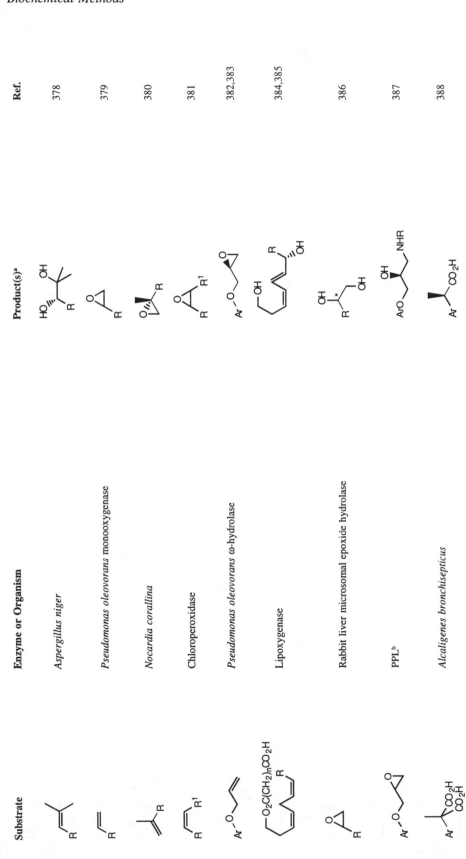

| Substrate | Enzyme or Organism | Product(s)[a] | Ref. |
|---|---|---|---|
| R | *Aspergillus niger* | HO, OH, R | 378 |
| R | *Pseudomonas oleovorans* monooxygenase | O, R | 379 |
| R | *Nocardia corallina* | O, R | 380 |
| R, $R^1$ | Chloroperoxidase | O, R, $R^1$ | 381 |
| Ar, O | *Pseudomonas oleovorans* ω-hydrolase | Ar, O, O | 382,383 |
| $O_2C(CH_2)_nCO_2H$, R | Lipoxygenase | OH, R, OH | 384,385 |
| O, R | Rabbit liver microsomal epoxide hydrolase | OH, R, *, OH | 386 |
| Ar, O, O | PPL[h] | OH, ArO, NHR | 387 |
| Ar, $CO_2H$, $CO_2H$ | *Alcaligenes bronchisepticus* | Ar, $CO_2H$ | 388 |

**TABLE 14.6 (continued)**

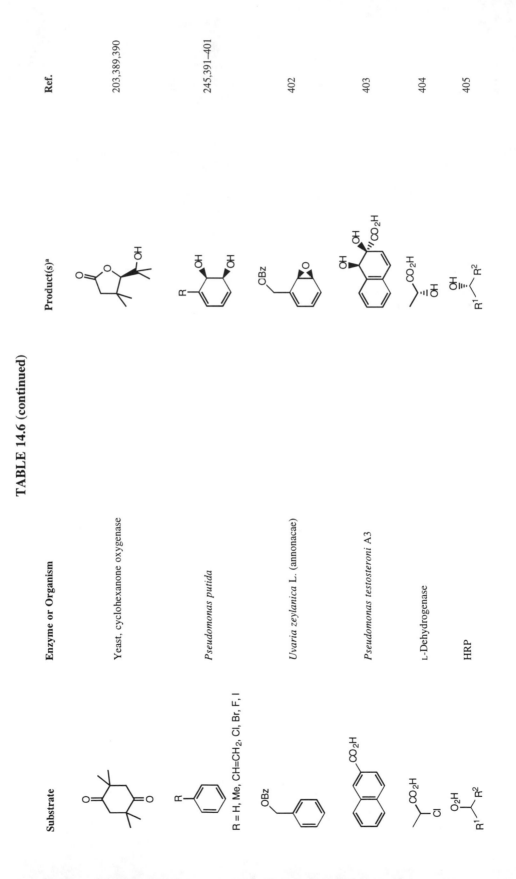

| Substrate | Enzyme or Organism | Product(s)[a] | Ref. |
|---|---|---|---|
| 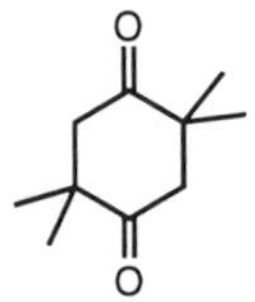 | Yeast, cyclohexanone oxygenase | 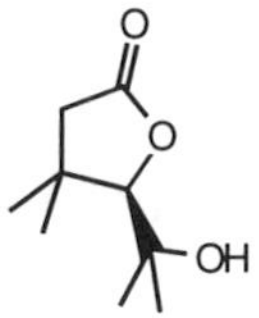 OH | 203,389,390 |
| 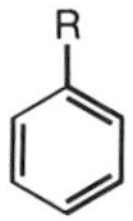 R = H, Me, CH=$CH_2$, Cl, Br, F, I | *Pseudomonas putida* | 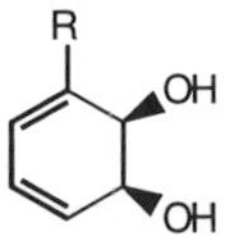 R, OH, OH | 245,391–401 |
| 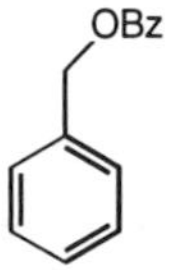 OBz | *Uvaria zeylanica* L. (annonacae) | 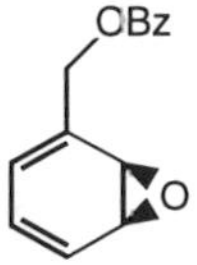 OBz, O | 402 |
| 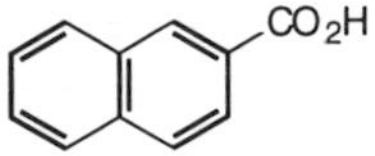 $CO_2H$ | *Pseudomonas testosteroni* A3 | 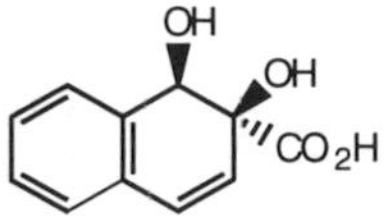 OH, OH, $CO_2H$ | 403 |
| 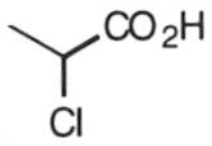 $CO_2H$, Cl | L-Dehydrogenase | 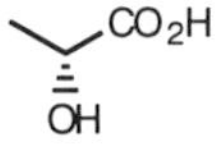 $CO_2H$, OH | 404 |
| 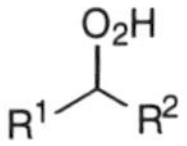 $O_2H$, $R^1$, $R^2$ | HRP | 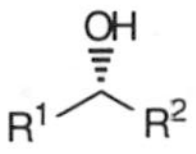 OH, $R^1$, $R^2$ | 405 |

| Substrate | Enzyme or Organism | Product(s)[a] | Ref. |
|---|---|---|---|
| 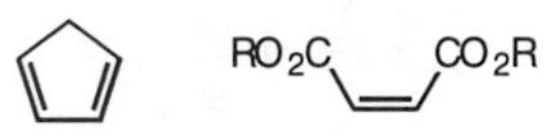 cyclopentadiene + $RO_2C$–CH=CH–$CO_2R$ | Baker's yeast | 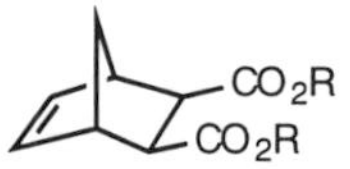 bicyclic adduct bearing $CO_2R$, $CO_2R$ | 406,407 |
| 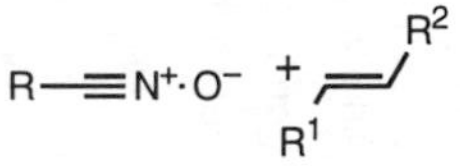 R—≡$N^+$·$O^-$ + $R^1$–CH=CH–$R^2$ where $R^2$ = H or $CO_2R^3$ | Baker's yeast[i] | 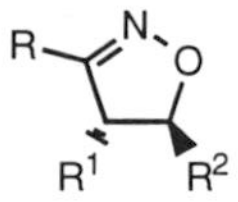 isoxazoline (R, N, O, $R^1$, $R^2$) | 408,409 |

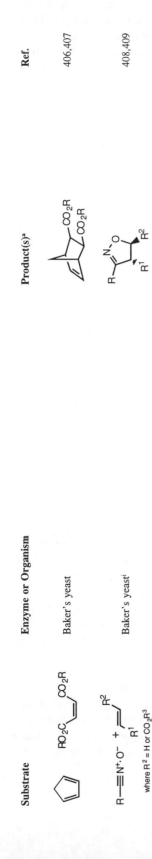

[a] In the presence of isopentenyl diphosphate. [b] This is a catalyst rather than an enzyme, but its stereoselectivity gives the appearance of an enzyme. [c] In the presence of ethyl butyrate. [d] In the presence of trifuoroethyl butyrate. [e] In the presence of an amine that can also be a racemate. [f] This reaction is specific for one alcohol isomer, and can, therefore, be used to obtain the other alcohol isomer by a resolution protocol. [g] In the presence of oxygen. [h] In the presence of an amine. [i] This transformation can also be performed in the presence of cyclodextrin.

Mention must also be made of the enzyme asymmetric hydroxylation reactions catalyzed by an iron porphyrin that allows for a 40% yield (41% ee) in the oxidation of ethylbenzene to 1-phenylethanol.[410] As our understanding of these types of oxidation reactions increases, the obvious potential for synthetic chemistry should come to fruition. Other members of this class of oxidation are known.[10,411,412] Perhaps one of the more impressive examples of an oxidation is in the kinetic resolution for the oxidative degradation of the sidechain for the preparation of ibuprofen (**14.5**) (Scheme 14.11).[413]

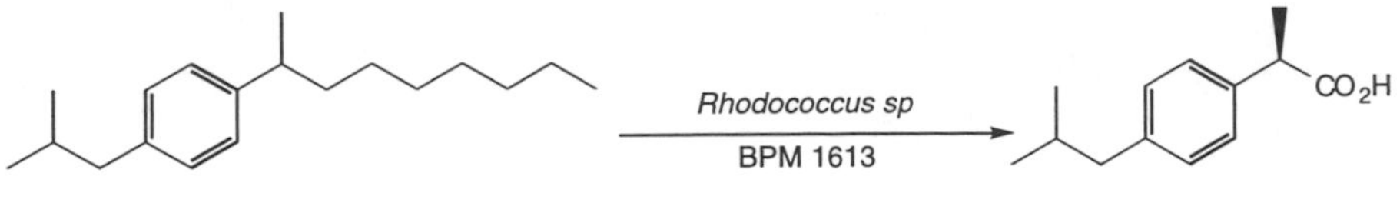

*Rhodococcus sp*

BPM 1613

$CO_2H$

**14.5**
17%

**SCHEME 14.11.**

Biological systems also catalyze pericyclic reactions; examples are cited in Table 14.6.

### 14.3.1. ALDOL REACTIONS

The enzymes that catalyze carbon–carbon bond formation reactions, such as the aldol, are of particular interest.[333,340] The power of these aldol enzymatic procedures for organic synthesis, that include specific enolate formation,[414] is illustrated by a route to 1-deoxymannojirimycin (**14.6**) (Scheme 14.12).[333,415,416]

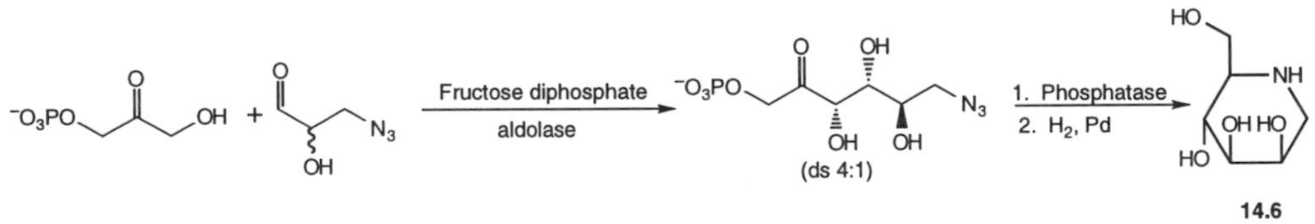

$^-O_3PO$ OH + OH $N_3$ Fructose diphosphate aldolase $^-O_3PO$ O OH $N_3$ OH OH (ds 4:1) 1. Phosphatase 2. $H_2$, Pd HO NH OH HO HO

**14.6**

**SCHEME 14.12.**

Rabbit muscle aldolase (RAMA) [D-fructose-1,6-bisphosphate aldolase] is proving to be a very powerful catalyst for an aldol approach to monosaccharides.[1,45,334,415-423] The preparation of 1-deoxymannojirimycin (**14.6**) and 1-deoxynojirimycin (**14.7**) are illustrated in Scheme 14.13; anion exchange chromatography was used to separate the azides **14.8** and **14.9**.[418]

The use of RAMA as a catalyst has these advantages: the hydroxy groups need no protection; the stereochemistry generated is D-threo (*parf*); and a wide variety of aldehydes can be employed as substrate.[1,417,419,424,425] In contrast, dihydroxyacetone phosphate is required as substrate, and ketoses are the product (*cf.* Scheme 14.13). The use of a half-protected dialdehyde substrate allows for an aldose synthesis, as illustrated by the preparation of L-xylose (**14.10**) (Scheme 14.14).[419]

The limitations of RAMA have led to the search for alternative enzymes. A bacterial fuculose-1-phosphate aldolase provides *pref*-isomer product stereochemistry,[426] as does a 2-deoxyribose-5-phosphate aldolase (DERA) (Scheme 14.15).[427] For aminosugars, acylneuraminate pyruvate lyase catalyzes an aldol condensation (Scheme 14.16) and has a broad specificity. This allows the enzyme to be used with non-amino monosaccharides.[336,428,429]

1. RAMA
2. $BaCl_2$

70%

1. Dowex 50Wx8 $H^+$
2. acid phosphatase (EC3.1.3.2)
3. Dowex 1x8, $HCO_2^-$

**14.8** ($\alpha,\beta$ = 28:72)

**14.9** ($\alpha,\beta$ = 87:13)

$H_2$, Pt-C
$H_2O$

**14.6**

**14.7**

SCHEME 14.13.

1. RAMA
2. acid phosphatase

L-iditol dehydrogenase
NADH, formate dehydrogenase
$NaO_2CH$

HCl
$H_2O$, THF

**14.10**

SCHEME 14.14.

DERA

SCHEME 14.15.

Acylneuraminate pyruvate lyase

SCHEME 14.16.

## 14.4. CATALYTIC ANTIBODIES

A large number of transformations can now be catalyzed by antibodies ("abzymes"). In many regards, this approach allows for the development of designer catalysts ("enzymes"). The selectivity can be exquisite and the reactions catalyzed range from selective deprotection[430] to an ester hydrolysis.[431] Indeed, a large number of reactions have now been reported, and the scope of the approach continues to grow. For example, antibodies can also be modified semi-synthetically to improve reactivity.[432,433] However, as there are still some issues surrounding their use, we will refrain from a large catalog of examples, as the area has been reviewed.[434-439] There is still considerable work to be done around the scale-up of this approach. Unlike most chemical reactions, a hapten (a compound that resembles the transition state of the desired reaction) has to be synthesized, and conjugated to a protein carrier, before antibodies can be raised. As a large number of antibodies are produced, a screening has to be undertaken to find ones with the desired catalytic activity. Molecular biology techniques can then be used to obtain more of the catalytic antibody. For the chemist, therefore, this approach is not viable on a small scale, as considerable effort has to be extended in the preparation of the hapten, before any biological work can be performed. The potential use, however, must lie in process development. At the time of this writing, impressive rate accelerations have been observed with abzymes, including pericyclic reactions [Chapter 12].[440] A metal co-ordination site can be engineered into the light chain, opening up the potential for redox and other reactions that require co-factors.[441] The unique selectivity provided by this approach provides large potential, but only reduction to practice will show if the methodology is commercially viable.

Although not usually a major consideration, the hapten has to resemble the transition state of the desired reaction — in some instances, this may not be known.

## 14.5. SUMMARY

Our use and knowledge base of whole cell and isolated enzyme systems to produce chiral materials continues to expand. With a large number of systems available, kinetic resolution of carboxylic acid derivatives and alcohols has become a commercial methodology. Although some enzymes are very specific, others are not, and a wide range of substrates may be converted by one enzyme. For the chemist, this means that it may be possible to take a reagent from the shelf and know that, with a reasonable degree of certainty, it will perform the desired transformation; some testing or screening may be necessary to find a suitable enzyme system.

Reductions can also be performed by biological systems, in particular baker's yeast. As empirical models have been developed to explain stereoselectivity, this approach is becoming more commonplace. Certainly, the yeast is inexpensive, and can be used on a large scale.

Catalytic antibodies hold promise for the future, but the amount of work involved at present excludes them from short-term solutions to synthesis problems.

## 14.6. REFERENCES

1. Toone, E. J.; Simon, E. S.; Bednarski, M. D.; Whitesides, G. M. *Tetrahedron* 1989, *45*, 5365.
2. Roberts, S. M. *Chem. Int. (London)* 1988, 384.
3. Jones, J. B. *Tetrahedron* 1986, *42*, 3351.
4. Wong, C.-H. *Science* 1989, *244*, 1145.
5. Ohno, M.; Otsuka, M. *Org. Reactions* 1989, *37*, 1.
6. Servi, S. *Synthesis* 1990, 1.

7. Drueckhammer, D. G.; Hennen, W. J.; Pederson, R. L.; Barbas, C. F.; Gautheron, C. M.; Krach, T.; Wong, C.-H. *Synthesis* 1991, 499.
8. Csuk, R.; Glanzer, B. I. *Chem. Rev.* 1991, *91*, 49.
9. Zhu, L.-M.; Tedford, M. C. *Tetrahedron* 1990, *46*, 6587.
10. Santaniello, E.; Ferraboschi, P.; Grisenti, P.; Manzocchi, A. *Chem. Rev.* 1992, *92*, 1071.
11. Wong, C.-H.; Drueckhammer, D. G.; Durrwachter, J.; Sweers, H. M.; Smith, G. L.; Yang, J.-S.; Henner, W. J. In *Trends in Synthetic Carbohydrate Chemistry*; D. Horton; L. D. Hawkins and G. J. McGarvey, Eds.; American Chemical Society: Washington, D. C., 1989; pp 317.
12. Fuganti, C. In *Trends in Synthetic Carbohydrate Chemistry*; D. Horton and L. D. Hawkins, Eds.; American Chemical Society: Washington, D. C., 1989; pp 305.
13. *ACS Symposium Series, No. 466: Enzymes in Carbohydrate Synthesis*; Bednarski, M. D.; Simon, E. S., Eds.; American Chemical Society: Washington, D.C., 1991.
14. Menger, F. M. *Acc. Chem. Res.* 1993, *26*, 206.
15. Leuenberger, H. G. W. *Pure Appl. Chem.* 1990, *62*, 753.
16. Leuenberger, H. G. W. In *Biotechnology*; H. J. Rehm and G. Reed, Ed.; Verlag Chemie: Weinheim, 1984; Vol. 6a; pp 5.
17. Chen, C.-S.; Fujimoto, Y.; Girdaukas, G.; Sih, C. J. *J. Am. Chem. Soc.* 1982, *104*, 7294.
18. Williams, B. A.; Toone, E. J. *J. Org. Chem.* 1993, *58*, 3507.
19. Roberts, S. M. and Wiggins, K.; *Preparative Biotransformations, Whole Cell and Isolated Enzymes in Organic Synthesis*; John Wiley & Sons: Chichester, 1992.
20. Straathof, A. J. J.; Rakels, J. L. L.; Heijen, J. J. *Biocatalysis* 1990, *4*, 89.
21. Degueil-Castaing, M.; de Jeso, B.; Drouillard, S.; Maillard, B. *Tetrahedron Lett.* 1987, *28*, 953.
22. Jones, J. B. *Pure Appl. Chem.* 1990, *62*, 1445.
23. Mohr, P.; Waespe-Sarcevic, N.; Tamm, C.; Gawronska, K.; Gawronski, J. K. *Helv. Chim. Acta* 1983, *66*, 2501.
24. Oberhauser, T.; Faber, K.; Griengl, H. *Tetrahedron* 1989, *45*, 1679.
25. Exl, C.; Hönig, H.; Renner, G.; Rogi-Kohlenprath, R.; Seebauer, V.; Seufer-Wasserthal, P. *Tetrahedron: Asymmetry* 1992, *3*, 1391.
26. Toone, E. J.; Werth, M. J.; Jones, J. B. *J. Am. Chem. Soc.* 1990, *112*, 4946.
27. Prelog, V. *Pure Appl. Chem.* 1964, *9*, 119.
28. Kazlauskas, R. J.; Weissfloch, A. N. E.; Rappaport, A. T.; Cuccia, L. A. *J. Org. Chem.* 1991, *56*, 2656.
29. Lutz, D.; Guldner, A.; Thums, R.; Schreier, P. *Tetrahedron: Asymmetry* 1990, *1*, 783.
30. Ziffer, H.; Kawai, K.; Kasai, M.; Froussios, C. *J. Org. Chem.* 1983, *48*, 3017.
31. Kasai, M.; Kawai, K.; Imuta, M.; Ziffer, H. *J. Org. Chem.* 1984, *49*, 675.
32. Charton, M.; Ziffer, H. *J. Org. Chem.* 1987, *52*, 2400.
33. Ito, S.; Kasai, M.; Ziffer, H.; Silverton, J. V. *Can. J. Chem.* 1987, *65*, 574.
34. Roberts, S. M. *Phil. Trans. R. Soc. London B* 1989, *324*, 557.
35. Umemura, T.; Hirohara, H. In *Biocatalysis in Agricultural Biotechnology*; J. R. Whitaker and P. E. Sonnet, Eds.; American Chemical Society: Washington, D. C., 1989.
36. Xie, Z. F.; Suemune, H.; Sakai, K. *Tetrahedron: Asymmetry* 1990, *1*, 395.
37. Jones, J. B. *Aldrichimica Acta* 1994, *26*, 105.
38. Provencher, L.; Wynn, H.; Jones, J. B. *Tetrahedron: Asymmetry* 1993, *4*, 2025.
39. Sabbioni, G.; Jones, J. B. *J. Org. Chem.* 1987, *52*, 4565.
40. Lam, L. K. P.; Lui, R. A. H. F.; Jones, J. B. *J. Org. Chem.* 1986, *51*, 2047.
41. Langrand, G.; Baratti, J.; Buono, G.; Triantaphylides, C. *Tetrahedron Lett.* 1986, *27*, 29.
42. Gil, G.; Ferre, E.; Meou, A.; Le Petit, J.; Triantaphylides, C. *Tetrahedron Lett.* 1987, *28*, 1647.
43. Sime, J. T.; Pool, C. R.; Tyler, J. W. *Tetrahedron Lett.* 1987, *28*, 5169.
44. Basavaiah, D.; Rao, P. D. *Synth. Commun.* 1990, *20*, 2945.
45. Whitesides, G. M.; Wong, C.-H. *Angew. Chem., Int. Ed. Engl.* 1985, *24*, 617.
46. Bevinakatti, H. S.; Banerji, A. A.; Newadkar, R. V. *J. Org. Chem.* 1989, *54*, 2453.
47. Pederson, R. L.; Liu, K. K.-C.; Rutan, J. F.; Chen, L.; Wong, C.-H. *J. Org. Chem.* 1990, *55*, 4897.
48. Poppe, L.; Novák, L.; Kajtár-Peredy, M.; Szántay, C. *Tetrahedron: Asymmetry* 1993, *4*, 2211.
49. Therisod, M.; Klibanov, A. M. *J. Am. Chem. Soc.* 1987, *109*, 3977.
50. Almsick, A. v.; Buddrus, J.; Honicke-Schmidt, P.; Laumen, K.; Schneider, M. P. *J. Chem. Soc., Chem. Commun.* 1989, 1391.
51. Ohta, H.; Hiraga, S.; Miyamoto, K.; Tsuchihashi, G.-I. *Agric. Biol. Chem.* 1988, *52*, 3023.
52. Effenberger, F.; Gutterer, B.; Ziegler, T.; Eckardt, E.; Aichholz, R. *Liebigs Ann. Chem.* 1991, 47.
53. Ohta, H.; Miyamae, Y.; Tsuchihashi, G.-I. *Agric. Biol. Chem.* 1989, *53*, 281.
54. Ohta, H.; Kimura, Y.; Sugano, Y. *Tetrahedron Lett.* 1988, *29*, 6957.
55. Matsuo, N.; Ohno, N. *Tetrahedron Lett.* 1985, *26*, 5533.
56. Lu, Y.; Miet, C.; Kunesch, N.; Poisson, J. E. *Tetrahedron: Asymmetry* 1991, 2, 871.
57. Laumen, K.; Breitgoff, D.; Seemayer, R.; Schneider, M. P. *J. Chem. Soc., Chem. Commun.* 1989, 148.

58. Faber, K.; Hönig, H.; Seufer-Wasserthal, P. *Tetrahedron Lett.* 1988, *29*, 1903.
59. Hönig, H.; Seufer-Wasserthal, P. *Synthesis* 1990, 1137.
60. Hönig, H.; Seufer-Wasserthal, P.; Fülöp, F. *J. Chem. Soc., Perkin Trans. 1* 1989, 2341.
61. Yamamoto, K.; Ando, H.; Shuetake, T.; Chikamatsu, H. *J. Chem. Soc., Chem. Commun.* 1989, 754.
62. Rabiller, C. G.; Konigsberger, K.; Faber, K.; Griengl, H. *Tetrahedron* 1990, *46*, 4231.
63. Liang, S.; Paquette, L. A. *Tetrahedron: Asymmetry* 1990, *1*, 445.
64. Itoh, T.; Takagi, Y.; Nishiyama, S. *J. Org. Chem.* 1991, *56*, 1521.
65. Goergens, U.; Schneider, M. P. *Tetrahedron: Asymmetry* 1992, *3*, 1149.
66. Foelsche, E.; Hickel, A.; Honig, H.; Seufer-Wasserthal, P. *J. Org. Chem.* 1990, *55*, 1749.
67. Honig, H.; Seufer-Wasserthal, P. S.; Weber, H. *Tetrahedron Lett.* 1990, *31*, 3011.
68. Itoh, T.; Ohta, T. *Tetrahedron Lett.* 1990, *31*, 6407.
69. Itoh, T.; Ohta, T.; Sano, M. *Tetrahedron Lett.* 1990, *31*, 6387.
70. Marples, B. A.; Rogers-Evans, M. *Tetrahedron Lett.* 1989, *30*, 261.
71. Akita, H.; Matsukura, H.; Oishi, T. *Tetrahedron Lett.* 1986, *27*, 5241.
72. Banfi, L.; Guanti, G. *Synthesis* 1993, 1029.
73. Xie, Z.; Suemune, H.; Sakai, K. *J. Chem. Soc., Chem. Commun.* 1988, 1638.
74. Wirz, B.; Schmid, R.; Walther, W. *Biocatal.* 1990, *3*, 159.
75. Itoh, T.; Kuroda, K.; Tomosada, M.; Takagi, Y. *J. Org. Chem.* 1991, *56*, 797.
76. Tombo, G. M. R.; Schar, H.-P.; Busquets, X. F. I.; Ghisalba, O. *Tetrahedron Lett.* 1986, *27*, 5707.
77. Kerscher, V.; W., K. *Tetrahedron Lett.* 1987, *28*, 531.
78. Guanti, G.; E., N.; Podgorski, T.; Thea, S.; Williams, A. *Tetrahedron* 1990, *46*, 7081.
79. Wang, Y.-F.; Wong, C.-H. *J. Org. Chem.* 1988, *53*, 3127.
80. Breitgoff, D.; Laumen, K.; Schneider, M. P. *J. Chem. Soc., Chem. Commun.* 1986, 1523.
81. Ciuffreda, P.; Colombo, D.; Ronchetti, F.; Toma, L. *J. Org. Chem.* 1990, *55*, 4187.
82. Guanti, G.; Banfi, L.; Narisano, E. *Tetrahedron: Asymmetry* 1990, *1*, 721.
83. Xie, Z.-F.; Suemune, H.; Sakai, K. *Tetrahedron: Asymmetry* 1993, *4*, 973.
84. Guanti, G.; Banfi, L.; Narisano, E.; Riva, R.; Thea, S. *Tetrahedron Lett.* 1986, *27*, 4639.
85. Ader, U.; Breitgoff, D.; Klein, P.; Laumen, K. E.; Schneider, M. P. *Tetrahedron Lett.* 1989, *30*, 1793.
86. Holla, E. W. *Angew. Chem., Int. Ed. Engl.* 1989, *28*, 220.
87. Guanti, G.; Banfi, L.; Narisano, E. *Tetrahedron Lett.* 1990, *31*, 6421.
88. Guanti, G.; Banfi, L.; Narisano, E. *Tetrahedron Lett.* 1989, *30*, 2697.
89. Guanti, G.; Banfi, L.; Merlo, V.; Narisano, E.; Thea, S. *Tetrahedron* 1993, *49*, 9501.
90. Guanti, G.; Banfi, L.; Narisano, E. *J. Org. Chem.* 1992, *57*, 1540.
91. Adjé, N.; Breuilles, P.; Uguen, D. *Tetrahedron Lett.* 1993, *34*, 4631.
92. Takano, S.; Tanigawa, K.; Ogasawara, K. *J. Chem. Soc., Chem. Commun.* 1976, 189.
93. Ohta, H.; Matsumoto, K.; Tsutsumi, S.; Ihori, T. *J. Chem. Soc., Chem. Commun.* 1989, 485.
94. Matsumoto, K.; Suzuki, N.; Ohta, H. *Tetrahedron Lett.* 1990, *31*, 7159.
95. Matsumoto, K.; Suzuki, N.; Ohta, H. *Tetrahedron Lett.* 1990, *31*, 7163.
96. Bevinakatti, H. S.; Newadkar, R. V. *Tetrahedron: Asymmetry* 1990, *1*, 583.
97. Ader, U.; Schneider, M. P. *Tetrahedron: Asymmetry* 1992, *3*, 521.
98. Ladner, W. E.; Whitesides, G. M. *J. Am. Chem. Soc.* 1984, *106*, 7250.
99. Bianchi, D.; Cesti, P. *J. Org. Chem.* 1990, *55*, 5657.
100. Wu, S.-H.; Guo, Z.-W.; Sih, C. J. *J. Am. Chem. Soc.* 1990, *112*, 1990.
101. Guo, Z.-W.; Sih, C. J. *J. Am. Chem. Soc.* 1989, *111*, 6836.
102. Ahmar, M.; Girard, C.; Bloch, R. *Tetrahedron Lett.* 1989, *30*, 7053.
103. Waldmann, H. *Tetrahedron Lett.* 1989, *30*, 3057.
104. Delinck, D. L.; Margolin, A. L. *Tetrahedron Lett.* 1990, *31*, 6797.
105. Gu, R.-L.; Sih, C. J. *Tetrahedron Lett.* 1990, *31*, 3283.
106. Kalaritis, P.; Regenye, R. W.; Partridge, J. J.; Coffen, D. L. *J. Org. Chem.* 1990, *55*, 812.
107. Sugai, T.; Kakeya, H.; Ohta, H. *Tetrahedron* 1990, *46*, 3463.
108. Kamal, A.; Ramalingam, T.; Venugopal, N. *Tetrahedron: Asymmetry* 1991, *2*, 39.
109. Mooriag, H.; Kellogg, R. M.; Kloosterman, M.; Kaptein, B.; Kamphuis, J.; Schoemaker, H. E. *J. Org. Chem.* 1990, *55*, 5878.
110. Allevi, P.; Anastasia, M.; Ciuffreda, P.; Sanvito, A. M. *Tetrahedron: Asymmetry* 1993, *4*, 1397.
111. Toone, E. J.; Jones, J. B. *Tetrahedron: Asymmetry* 1991, *2*, 207.
112. Blanco, L.; Guibe-Jampel, E.; Rousseau, G. *Tetrahedron Lett.* 1988, *29*, 1915.
113. Yamada, H.; Takahashi, S.; Yoshiaki, K.; Kumagai, H. *J. Ferment. Technol.* 1978, *56*, 484.
114. Cecere, F.; Galli, G.; Morisi, F. *FEBS Lett.* 1975, *57*, 192.
115. Guivarch, M.; Gillonnier, C.; Brunie, J.-C. *Bull. Soc. Chim. Fr.* 1980, 91.
116. Bommarius, A. S.; Drauz, K.; Groeger, U.; Wandrey, C. In *Chirality in Industry*; A. N. Collins; G. N. Sheldrake, and J. Crosby, Eds.; John Wiley & Sons: Chichester, 1992; pp 372.

117. Drauz, K.; Kottenhahn, M.; Makryaleas, K.; Klenk, H.; Bernd, M. *Angew. Chem., Int. Ed. Engl.* 1991, *30*, 712.
118. Yokozeki, K.; Nakamori, S.; Eguchi, C.; Yamada, K.; Mitsugi, K. *Agric. Biol. Chem.* 1987, *51*, 355.
119. Olivieri, R.; Fascetti, E.; Angelini, L.; Degen, L. *Biotechnol. Bioeng.* 1981, *23*, 2173.
120. Gaebler, O. H.; Keltch, A. K. *J. Biol. Chem.* 1926, *70*, 763.
121. Gutman, A. L.; Zuobi, K.; Guibe-Jampel, E. *Tetrahedron Lett.* 1990, *31*, 2037.
122. De Jeso, B.; Belair, N.; Deleuze, H.; Rascle, M.-C.; Maillard, B. *Tetrahedron Lett.* 1990, *31*, 653.
123. Schneider, M.; Engel, N.; Honicke, P.; Heinemann, G.; Gorisch, H. *Angew. Chem., Int. Ed. Engl.* 1984, *23*, 67.
124. Van der Eycken, J.; Vandewalle, M.; Heinemann, G.; Laumen, K.; Schneider, M. P.; Kredel, J.; Sauer, J. *J. Chem. Soc., Chem. Commun.* 1989, 306.
125. Kobayashi, S.; Kamiyama, O., M. *J. Org. Chem.* 1990, *55*, 1169.
126. Metz, P. *Tetrahedron* 1989, *45*, 7311.
127. Kobayashi, S.; Kamiyama, K.; Ohno, M. *J. Org. Chem.* 1990, *55*, 1169.
128. Lam, L. K. P.; Brown, C. M.; De Jeso, B.; Lym, L.; Toone, E. J.; Jones, J. B. *J. Am. Chem. Soc.* 1988, *110*, 4409.
129. Klunder, A. J. H.; van Gastel, F. J. C.; Zwanenburg, B. *Tetrahedron Lett.* 1988, *29*, 2697.
130. Kobayashi, S.; Sato, M.; Eguchi, Y.; Ohno, M. *Tetrahedron Lett.* 1992, *33*, 1081.
131. Bestmann, H. J.; Philipp, U. C. *Angew. Chem., Int. Ed. Engl.* 1991, *30*, 86.
132. Nakada, M.; Kobayashi, S.; Ohno, M. *Tetrahedron Lett.* 1988, 29, 3951.
133. Andruszkiewicz, R.; Barrett, A. G. M.; Silverman, R. B. *Synth. Commun.* 1990, *20*, 159.
134. Huang, F.-C.; Lee, L. F. H.; Mittal, R. S. D.; Ravikumar, P. R.; Chan, J. A.; Sih, C. J.; Capsi, E.; Eck, C. R. *J. Am. Chem. Soc.* 1975, *97*, 4144.
135. Herold, P.; Mohr, P.; Tamm, C. *Helv. Chim. Acta* 1983, *66*, 744.
136. Ohno, M.; Kobayashi, S.; Iimori, T.; Wang, Y.-F.; Izawa, T. *J. Am. Chem. Soc.* 1981, *103*, 2405.
137. Adachi, K.; Kobayashi, S.; Ohno, M. *Chimia* 1986, *40*, 311.
138. Monteiro, J.; Braun, J.; Goffic, F. L. *Synth. Commun.* 1990, *20*, 315.
139. Westerman, B.; Scharmann, H. G.; Kortmann, I. *Tetrahedron: Asymmetry* 1993, *4*, 2119.
140. Gutman, A. L.; Zuobi, K.; Bravdo, T. *J. Org. Chem.* 1990, *55*, 3546.
141. Chenevert, R.; Thiboutot *Synthesis* 1989, 444.
142. Burgess, K.; Henderson, I. *Tetrahedron Lett.* 1989, *30*, 3633.
143. Wallace, J. S.; Reda, K. B.; Williams, M. E.; Morrow, C. J. *J. Org. Chem.* 1990, *55*, 3544.
144. Mohr, P.; Rosslein, L.; Tamm, C. *Tetrahedron Lett.* 1989, *30*, 2513.
145. Gutman, A. L.; Bravdo, T. *J. Org. Chem.* 1989, *54*, 4263.
146. Hughes, D. L.; Bergan, J. J.; Amato, J. S.; Bhupathy, M.; Leazer, J. L.; McNamara, J. M.; Sidler, D. R.; Reider, P. J.; Grabowski, E. J. J. *J. Org. Chem.* 1990, *55*, 6252.
147. Hughes, D. L.; Bergan, J. J.; Amato, J. S.; Reider, P. J.; Grabowski, E. J. J. *J. Org. Chem.* 1989, *54*, 1787.
148. Bonini, C.; Racioppi, R.; Righi, G.; Viggiani, L. *J. Org. Chem.* 1993, *58*, 802.
149. Ferraboschi, P.; Grisenti, P.; Manzocchi, A.; Santaniello, E. *J. Org. Chem.* 1990, *55*, 6214.
150. Ferraboschi, P.; Grisenti, P.; Santaniello, E. *Synlett* 1990, 545.
151. Ferraboschi, P.; Brembilla, D.; Grisenti, P.; Santaniello, E. *Synlett* 1991, 310.
152. Barth, S.; Effenberger, F. *Tetrahedron: Asymmetry* 1993, *4*, 823.
153. Fourneron, J.-D.; Chiche, M.; Pieroni, G. *Tetrahedron Lett.* 1990, *31*, 4875.
154. Morgan, B.; Oehlschlager, A. C. *Tetrahedron: Asymmetry* 1993, *4*, 907.
155. Chinchilla, R.; Najera, C.; Pardo, J.; Yus, M. *Tetrahedron: Asymmetry* 1990, *1*, 575.
156. Gil, G.; Ferre, E.; Meou, A.; le Petit, J.; Triantaphylides, C. *Tetrahedron Lett.* 1987, *28*, 1648.
157. Wang, Y.-F.; Chen, S.-T.; Liu, K. K.-C.; Wong, C.-H. *Tetrahedron Lett.* 1989, *30*, 1917.
158. Kanerva, L. T.; Kiljunen, E.; Huuhtanen, T. T. *Tetrahedron: Asymmetry* 1993, *4*, 2355.
159. Theil, F.; Weidner, J.; Ballschuh, S.; Kunath, A.; Schick, H. *Tetrahedron Lett.* 1993, *34*, 305.
160. Caron, G.; Kazlauskas, R. J. *Tetrahedron: Asymmetry* 1993, *4*, 1995.
161. Mattson, A.; Öhrner, N.; Hult, K.; Norin, T. *Tetrahedron: Asymmetry* 1993, *4*, 925.
162. Chen, C.-S.; Liu, Y.-C. *Tetrahedron Lett.* 1989, *30*, 7165.
163. Hsu, S.-H.; Wu, S.-S.; Wang, Y.-F.; Wong, C.-H. *Tetrahedron Lett.* 1990, *31*, 6403.
164. Burgess, K.; Jennings, L. D. *J. Org. Chem.* 1990, *55*, 1138.
165. Burgess, K.; Jennings, L. D. *J. Am. Chem. Soc.* 1990, *112*, 7434.
166. Wang, Y.-F.; Lalonde, J. J.; Momongan, M.; Bergbreiter, D. E.; Wong, C.-H. *J. Am. Chem. Soc.* 1988, *110*, 7200.
167. Tsuji, K.; Terao, Y.; Achiwa, K. *Tetrahedron Lett.* 1989, *30*, 6189.
168. Santaniello, E.; Ferraboschi, P.; Grisenti, P. *Tetrahedron Lett.* 1990, *31*, 5657.
169. Barnett, C. J.; Wilson, T. M. *Tetrahedron Lett.* 1989, *30*, 6291.
170. Atsuumi, S.; Nakano, M.; Koike, Y.; Tanaka, S.; Ohkubo, M.; Yonezawa, T.; Funabashi, H.; Hashimoto, J.; Morishima, H. *Tetrahedron Lett.* 1990, *31*, 1601.

171. Takano, S.; Setoh, M.; Yamada, O.; Ogasawara, K. *Synthesis* 1993, 1253.
172. García, M. J.; Rebolledo, F.; Gotor, V. *Tetrahedron: Asymmetry* 1992, *3*, 1519.
173. García, M. J.; Rebolledo, F.; Gotor, V. *Tetrahedron: Asymmetry* 1993, *4*, 2199.
174. Burgess, K.; Henderson, I. *Tetrahedron: Asymmetry* 1990, *1*, 57.
175. Berger, B.; Rabiller, C. G.; Konigsberger, K.; Faber, K.; Griengl, H. *Tetrahedron: Asymmetry* 1990, *1*, 541.
176. Kanerva, L. T.; Vänttinen, E. *Tetrahedron: Asymmetry* 1993, *4*, 85.
177. Vänttinen, E.; Kanerva, L. T. *Tetrahedron: Asymmetry* 1992, *3*, 1529.
178. Ferraboschi, P.; Casati, S.; Grisenti, P.; Santaniello, E. *Tetrahedron: Asymmetry* 1993, *4*, 9.
179. Pozo, M.; Gotor, V. *Tetrahedron* 1993, *49*, 4321.
180. Pozo, M.; Gotor, V. *Tetrahedron* 1993, *49*, 10725.
181. Engel, K.-H. *Tetrahedron: Asymmetry* 1991, *2*, 165.
182. Berglund, P.; Holmquist, M.; Hedenström, E.; Hult, K.; Högberg, H.-E. *Tetrahedron: Asymmetry* 1993, *4*, 1869.
183. Quirós, M.; Sánchez, V. M.; Brieva, R.; Rebolledo, F.; Gotor, V. *Tetrahedron: Asymmetry* 1993, *4*, 1105.
184. Hiratake, J.; Yamamoto, K.; Yamamoto, Y.; Oda, J. *Tetrahedron Lett.* 1989, *30*, 1555.
185. Yamamoto, K.; Nishioka, T.; Oda, J.; Yamamoto, Y. *Tetrahedron Lett.* 1988, *29*, 1717.
186. Crich, J. Z.; Brieva, R.; Marquart, P.; Gu, R.-L.; Flemming, S.; Sih, C. J. *J. Org. Chem.* 1993, *58*, 3252.
187. Breuilles, P.; Schmittnerger, T.; Uguen, D. *Tetrahedron Lett.* 1993, *34*, 4205.
188. Nicolosi, G.; Morrone, R.; Patti, A.; Piattelli, M. *Tetrahedron: Asymmetry* 1992, *3*, 753.
189. Guo, Z.-W.; Wu, S.-H.; Chen, C.-S.; Girdaukas, G.; Sih, C. J. *J. Am. Chem. Soc.* 1990, *112*, 4942.
190. Parmar, V. S.; Prasad, A. K.; Singh, P. K.; Gupta, S. *Tetrahedron: Asymmetry* 1992, *3*, 1395.
191. Secundo, F.; Riva, S.; Carrea, G. *Tetrahedron: Asymmetry* 1992, *3*, 267.
192. Terradas, F.; Teston-Henry, M.; Fitzpatrick, P. A.; Klibanov, A. M. *J. Am. Chem. Soc.* 1993, *115*, 390.
193. Parida, S.; Dordick, J. S. *J. Org. Chem.* 1993, *58*, 3238.
194. Burke, P. A.; Smith, S. O.; Bachovchin, W. M.; Klibanov, A. M. *J. Am. Chem. Soc.* 1989, *111*, 8290.
195. Wong, C.-H.; Chen, S.-T.; Hennen, W. J.; Bibbs, J. A.; Wang, Y.-F.; Liu, J. L.-C.; Pantoliano, M. W.; Whitlow, M.; Bryan, P. N. *J. Am. Chem. Soc.* 1990, *112*, 945.
196. Zhong, Z.; Liu, J. L.-C.; Dinterman, L. M.; Finkelman, M. A.; Mueller, W. T.; Rollence, M. L.; Whitlow, M.; Wong, C.-H. *J. Am. Chem. Soc.* 1991, *113*, 683.
197. Rubio, E.; Fernandez-Mayorales, A.; Klibanov, A. M. *J. Am. Chem. Soc.* 1991, *113*, 695.
198. Fitzpatrick, P. A.; Klibanov, A. M. *J. Am. Chem. Soc.* 1991, *113*, 3166.
199. Klibanov, A. M. *Acc. Chem. Res.* 1990, *23*, 114.
200. Yamada, H.; Shimizu, S. *Angew. Chem., Int. Ed. Engl.* 1988, *27*, 622.
201. Sih, C. J.; Chen, C.-S. *Angew. Chem., Int. Ed. Engl.* 1984, *23*, 570.
202. Simon, H.; Bader, J.; Gunther, H.; Neumann, S.; Thanos, J. *Angew. Chem., Int. Ed. Engl.* 1985, *24*, 539.
203. Ouazzani-Chahdi, J.; Buisson, D.; Azerod, R. *Tetrahedron Lett.* 1987, *28*, 1109.
204. Keinan, E.; Hafeli, E. K.; Seth, K. K.; Lamed, R. *J. Am. Chem. Soc.* 1986, *108*, 162.
205. Uskokovic, M. R.; Pruess, D. L.; Despreaux, C. W.; Shiuey, S.-J.; Pizzolato, G.; Gutzwiller, J. *Helv. Chim. Acta* 1973, *56*, 2834.
206. Nakamura, K.; Yoneda, T.; Miyai, T.; Ushido, K.; Oka, S.; Ohno, A. *Tetrahedron Lett.* 1988, *29*, 2453.
207. Ferraboschi, P.; Grisenti, P.; Manzocchi, A.; Santaniello, E. *J. Chem. Soc., Perkin Trans. 1* 1990, 2469.
208. Bucciarelli, M.; Forni, A.; Moretti, I.; Torre, G. *J. Chem. Soc., Chem. Commun.* 1978, 456.
209. Gopalan, A. S.; Jacobs, H. K. *Tetrahedron Lett.* 1990, *31*, 5575.
210. Gibbs, D. E.; Barnes, D. *Tetrahedron Lett.* 1990, *31*, 5555.
211. Guette, J.-P.; Spassky, N.; Boucherot, D. *Bull. Soc. Chim., Fr.* 1972, 4217.
212. Aragozzini, F.; Maconi, E.; Potenza, D.; Scolastico, C. *Synthesis* 1989, 225.
213. Kodama, M.; Minami, H.; Mima, Y.; Fukayama, Y. *Tetrahedron Lett.* 1990, *31*, 4025.
214. Guanti, G.; Banfi, L.; Narisano, E. *Tetrahedron Lett.* 1986, *27*, 3547.
215. Utaka, M.; Konishi, S.; Okubo, T.; Tsuboi, S.; Takeda, A. *Tetrahedron Lett.* 1987, *28*, 1447.
216. Sato, T.; Mizutani, T.; Okumura, Y.; Fujisawa, T. *Tetrahedron Lett.* 1989, *30*, 3701.
217. Tsuboi, S.; Furutani, H.; Utaka, M.; Takeda, A. *Tetrahedron Lett.* 1987, *28*, 2709.
218. Fronza, G.; Fuganti, C.; Grasselli, P.; Servi, S. *Tetrahedron Lett.* 1985, *26*, 4961.
219. Fronza, G.; Fuganti, C.; Grasselli, P.; Servi, S. *Tetrahedron Lett.* 1986, *27*, 4363.
220. Utaka, M.; Konishi, S.; Takeda, A. *Tetrahedron Lett.* 1986, *27*, 4737.
221. Levene, P. A.; Walti, A. *Org. Synth.* 1943, *CV II*, 545.
222. Barry, J.; Kagan, H. B. *Synthesis* 1981, 453.
223. Ramaswamy, S.; Oehlschlager, A. C. *Tetrahedron* 1991, *47*, 1145.
224. Schrötter, E.; Weidner, J.; Schick, H. *Liebigs Ann. Chem.* 1991, 397.
225. Manzocchi, A.; Fiecchi, A.; Santaniello, E. *J. Org. Chem.* 1988, *53*, 4405.
226. Deschenaux, P.-F.; Kallimopoulos, T.; Jacot-Guillarmod, A. *Helv. Chim. Acta* 1989, *72*, 1259.
227. Dasaradhi, L.; Fadnavis, N. W.; Bhalerao, U. T. *J. Chem. Soc., Chem. Commun.* 1990, 729.
228. Fujisawa, T.; Kojima, E.; Itoh, T.; Sato, T. *Chem. Lett.* 1985, 1751.

229. Han, C. O.; Di Tullio, D.; Wang, Y. F.; Sih, C. J. *J. Org. Chem.* 1986, *51*, 1253.
230. Guanti, G.; Banfi, L.; Guaragna, A.; Narisano, E. *J. Chem. Soc., Chem. Commun.* 1986, 138.
231. de Carvalho, M.; Okamoto, M. T.; Samenho Moran, P. J.; Rodrigues, J. A. R. *Tetrahedron* 1991, *47*, 2073.
232. Kim, M.-J.; Whitesides, G. M. *J. Am. Chem. Soc.* 1988, *110*, 2959.
233. Nakamura, K.; Inoue, Y.; Shibahara, J.; Oka, S.; Ohno, A. *Tetrahedron Lett.* 1988, *29*, 4769.
234. Bur, D.; Luyton, M. A.; Wynn, H.; Provencher, L. R.; Jones, J. B.; Gold, M.; Friesen, J. D.; Clarke, A. R.; Holbrook, J. J. *Can. J. Chem.* 1989, *67*, 1065.
235. Casy, G.; Lee, T. V.; Lovell, H.; Nichols, B. J.; Sessions, R. B.; Holbrook, J. J. *J. Chem. Soc., Chem. Commun.* 1992, 924.
236. Kallwass, H. K. W.; Luyten, M. A.; Parris, W.; Gold, M.; Kay, C. M.; Jones, J. B. *J. Am. Chem. Soc.* 1992, *114*, 4551.
237. Casy, G.; Lee, T. V.; Lovell, H. *Tetrahedron Lett.* 1992, *33*, 817.
238. Deol, B. S.; Ridley, D. D.; Simpson, G. W. *Aust. J. Chem.* 1976, *29*, 2459.
239. Barton, D. H. R.; Brown, B. D.; Ridley, D. D.; Widdowson, D. A.; Keys, A. J.; Leaver, C. J. *J. Chem. Soc., Perkin Trans. 1* 1975, 2069.
240. Nakamura, K.; Inoue, K.; Ushio, K.; Oka, S.; Ohno, A. *J. Org. Chem.* 1988, *53*, 2598.
241. Ridley, D. D.; Stralow, M. *J. Chem. Soc., Chem. Commun.* 1975, 400.
242. Brenelli, E. C. S.; Moran, P. J. S.; Rodrigues, A. R. *Synth. Commun.* 1990, *20*, 261.
243. Chenevert, R.; Thiboutot, S. *Chem. Lett.* 1988, 1191.
244. Fouche, G.; Horak, R. M.; Meth-Cohn, O. In *Molecular Mechanisms in Bioorganic Processes*; C. Bleasdale and B. T. Golding, Eds.; Royal Society of Chemistry: London, 1990; pp 350.
245. Amici, M. D.; Micheli, C. D.; Carrea, G.; Spezia, S. *J. Org. Chem.* 1989, *54*, 2646.
246. Utaka, M.; Konishi, S.; Mizuoka, A.; Ohkubo, T.; Sakai, T.; Tsuboi, S.; Takeda, A. *J. Org. Chem.* 1989, *54*, 4989.
247. Ohta, H.; Kobayashi, N.; Ozaki, K. *J. Org. Chem.* 1989, *54*, 1802.
248. Tsuboi, S.; Furutani, H.; Ansari, M. H.; Sakai, T.; Utaka, M.; Takeda, A. *J. Org. Chem.* 1993, *58*, 486.
249. Simon, H.; Rambeck, B.; Hachimoto, H.; Günther, H.; Nohymek, G.; Neumann, H. *Angew. Chem., Int. Ed. Engl.* 1974, *13*, 608.
250. Fronza, G.; Fuganti, C.; Grasselli, P.; Servi, S. *Tetrahedron Lett.* 1986, *26*, 4961.
251. Patel, R. N.; Banerjee, A.; Howell, J. M.; McNomee, C. G.; Brozozowski, D.; Mirfakhrae, D.; Narduri, V.; Thottathil, J. K.; Szarka, L. J. *Tetrahedron: Asymmetry* 1993, *4*, 2069.
252. Nakamura, K.; Miyai, T.; Ushio, K.; Oka, S.; Ohno, A. *Bull. Chem. Soc., Jpn.* 1988, *61*, 2089.
253. Züger, M. F.; Giovannini, F.; Seebach, D. *Angew. Chem., Int. Ed. Engl.* 1983, *22*, 1012.
254. Seebach, D.; Züger, M. F.; Giovannini, F.; Sonnleitner, B.; Fiechter, A. *Angew. Chem., Int. Ed. Engl.* 1984, *23*, 151.
255. Sonnleitner, B.; Giovannini, F.; Fiechter, A. *J. Biotech.* 1985, *3*, 33.
256. Matzinger, P. K.; Leuenberger, H. G. W. *Appl. Microbiol. Technol.* 1985, *22*, 208.
257. Utaka, M.; Watabu, H.; Higashi, H.; Sakai, T.; Tsuboi, S.; Torii, S. *J. Org. Chem.* 1990, *55*, 3917.
258. Afonso, C. M.; Barros, M. T.; Godinho, L.; Maycock, C. D. *Tetrahedron Lett.* 1989, *30*, 2707.
259. Ghosh, S. K.; Chattopadhyay, S.; Mamdapur, V. R. *Tetrahedron* 1991, *47*, 3089.
260. Bennett, F.; Knight, D. M. *Tetrahedron Lett.* 1988, *29*, 4625.
261. Bennett, F.; Knight, D. M.; Fenton, G. *J. Chem. Soc., Perkin Trans. 1* 1991, 133.
262. Bennett, F.; Knight, D. M.; Fenton, G. *J. Chem. Soc., Perkin Trans. 1* 1991, 1543.
263. Gopolan, A. S.; Jacobs, H. K. *J. Chem. Soc., Perkin Trans. 1* 1990, 1897.
264. Watabu, H.; Ohkubo, M.; Matsubara, H.; Sakai, T.; Tsuboi, S.; Utaka, M. *Chem. Lett.* 1989, 2183.
265. Ramaswamy, S.; Oehlachlager, A. C. *J. Org. Chem.* 1989, *54*, 255.
266. Nakamura, K.; Kawai, Y.; Oka, S.; Ohno, A. *Tetrahedron Lett.* 1989, *30*, 2245.
267. Zhou, B.-N.; Gopalan, A. S.; VanMiddleworth, F.; Shieh, W.-R.; Sih, C. J. *J. Am. Chem. Soc.* 1983, *105*, 5925.
268. Frater, G. *Helv. Chim. Acta* 1979, *62*, 2829.
269. Seebach, D.; Eberle, M. *Synthesis* 1986, 37.
270. Heidlas, J.; Engel, K. H.; R., T. *Eur. J. Biochem.* 1988, *172*, 633.
271. Seebach, D.; Renaud, P.; Schweizer, W. B.; Zueger, M. F. *Helv. Chim. Acta* 1984, *67*, 1843.
272. Hirama, M.; Shimizu, M.; Iwashita, M. *J. Chem. Soc., Chem. Commun.* 1983, 599.
273. Lemieux, R. U.; Giguere, J. *Can. J. Chem.* 1951, *29*, 678.
274. Nakamura, K.; Higaki, M.; Ushio, M.; Oka, S.; Ohno, A. *Tetrahedron Lett.* 1985, *26*, 4213.
275. Naoshima, Y.; Akakabe, Y. *J. Org. Chem.* 1989, *54*, 4237.
276. Jayasinghe, L. Y.; Smallridge, A. J.; Trewhella, M. A. *Tetrahedron Lett.* 1993, *34*, 3949.
277. Buisson, D.; Azerad, R.; Sanner, C.; Larchevêque, M. *Tetrahedron: Asymmetry* 1991, *2*, 987.
278. Peters, J.; Zelinski, T.; Minuth, T.; Kula, M.-R. *Tetrahedron: Asymmetry* 1993, *4*, 1683.
279. Seebach, D.; Zuger, M. X. *Helv. Chim. Acta* 1982, *65*, 495.
280. Mori, K. *Tetrahedron* 1981, *37*, 1341.
281. Ushio, K.; Inouye, K.; Nakamura, K.; Oka, S.; Ohno, A. *Tetrahedron Lett.* 1986, *27*, 2657.

282. Buisson, D.; Sanner, C.; Larcheveque, M.; Azerad, R. *Tetrahedron Lett.* 1987, *28*, 3939.
283. Namamura, K.; Miyai, T.; Nozaki, K.; Ushido, K.; Oka, S.; Ohno, A. *Tetrahedron Lett.* 1986, *27*, 3155.
284. Gopalan, A. S.; Jacobs, H. K. *Tetrahedron Lett.* 1989, *30*, 5705.
285. Frater, G. W. *Helv. Chim. Acta* 1980, *63*, 1383.
286. Kumar, A.; Ner, D. H.; Dike, S. Y. *Tetrahedron Lett.* 1991, *32*, 1901.
287. Deschenaux, P.-F.; Kallimopoulos, T.; Stoeckli-Evans, H.; Jacot-Guillarmod, A. *Helv. Chim. Acta* 1989, *72*, 731.
288. Gopalan, A. S.; Jacobs, H. K. *J. Chem. Soc., Chem. Commun.* 1990, 1897.
289. Chikashita, H.; Motozawa, T.; Itoh, K. *Synth. Commun.* 1989, *19*, 1119.
290. Frater, G.; Muller, U.; Gunther, W. *Tetrahedron* 1984, *40*, 1269.
291. Buisson, D.; Henrot, S.; Larcheveque, M.; Azerad, R. *Tetrahedron Lett.* 1987, *28*, 5033.
292. Itoh, T. I.; Yonekawa, Y.; Sato, T.; Fujisawa, T. *Tetrahedron Lett.* 1986, *27*, 5405.
293. Nakamura, K.; Kawai, Y.; Ohno, A. *Tetrahedron Lett.* 1990, *31*, 267.
294. Spiliotis, V.; Papahatjis, D.; Ragoussis, N. *Tetrahedron Lett.* 1990, *31*, 1615.
295. Ferreira, J. T. B.; Simonelli, F. *Tetrahedron* 1990, *46*, 6311.
296. Akita, H.; Furuichi, A.; Koshiji, H.; Oishi, T. *Chem. Pharm. Bull.* 1983, *31*, 4376.
297. Sato, T.; Tsurumaki, M.; Fujisawa, T. *Chem. Lett.* 1986, 1367.
298. Hoffmann, R. W.; Ladner, W.; Steinback, K.; Massa, W.; Schmidt, R.; Snatzke, G. *Chem. Ber.* 1981, *114*, 2786.
299. Akita, H.; Furuichi, A.; Koshiji, H.; Horikoshi, K.; Oishi, T. *Tetrahedron Lett.* 1983, *24*, 2009.
300. Fujisawa, T.; Itoh, T.; Sato, T. *Tetrahedron Lett.* 1984, *25*, 5083.
301. Ehrler, J.; Giovannini, F.; Lamatsch, B.; Seebach, D. *Chimia* 1986, *40*, 172.
302. Shieh, W.-R.; Sih, C. J. *Tetrahedron: Asymmetry* 1993, *4*, 1259.
303. Nakamura, K.; Takano, S.; Ohno, A. *Tetrahedron Lett.* 1993, *34*, 6087.
304. Hudlicky, T.; Tsunoda, T.; Gadamasetti, K. G.; Murry, J. A.; Keck, G. E. *J. Org. Chem.* 1991, *56*, 3619.
305. Hudlicky, T.; Gillman, G.; Andersen, C. *Tetrahedron: Asymmetry* 1992, *3*, 281.
306. Fauve, A.; Veschambre, H. *J. Org. Chem.* 1988, *53*, 5215.
307. Dauphin, G.; Fauve, A.; Veschambre, H. *J. Org. Chem.* 1989, *54*, 2238.
308. Jian-Xin, G.; Zu-Yi, L.; Guo-Qiang, L. *Tetrahedron* 1993, *49*, 5805.
309. Itoh, T.; Takagi, Y.; Fujisawa, T. *Tetrahedron Lett.* 1989, *30*, 3811.
310. Akita, H.; Todoroki, R.; Endo, H.; Ikari, Y.; Oishi, T. *Synthesis* 1993, 513.
311. Crumbie, R. L.; Ridley, D. D.; Simpson, G. W. *J. Chem. Soc., Chem. Commun.* 1977, 315.
312. Kozikowski, A. P.; Mugrage, B. B.; Li, C. S.; Felder, L. *Tetrahedron Lett.* 1986, *27*, 4817.
313. Robin, S.; Huet, F.; Fauve, A.; Veschambre, H. *Tetrahedron: Asymmetry* 1993, *4*, 239.
314. Yamazaki, Y.; Hosoni, K. *Tetrahedron Lett.* 1988, *29*, 5769.
315. Fuganti, C.; Grasselli, P.; Seneci, P. F.; Casati, P. *Tetrahedron Lett.* 1986, *27*, 5275.
316. Afonso, C. A. M.; Barros, M. T.; Godinho, L. S.; Maycock, C. D. *Tetrahedron* 1993, *49*, 4283.
317. Casy, G. *Tetrahedron Lett.* 1992, *33*, 8159.
318. Schummer, A.; Yu, H.; Simon, H. *Tetrahedron* 1991, *47*, 9019.
319. Arslan, T.; Benner, S. A. *J. Org. Chem.* 1993, *58*, 2260.
320. Trincone, A.; Nicolaus, B.; Lama, L.; Marsiglia, F.; Gambacorta, A. *Biotechnol. Lett.* 1991, *13*, 31.
321. Aragozzini, F.; Valenti, R.; Santaniello, E.; Ferraboschi, P.; Grisenti, P. *Biocatalysis* 1992, *5*, 325.
322. Nakamura, K.; Kitayama, T.; Inoue, Y.; Ohno, A. *Tetrahedron* 1990, *46*, 7471.
323. Nakamura, K.; Kitayama, T.; Inoue, Y.; Ohno, A. *Bull. Chem. Soc., Jpn.* 1990, *63*, 91.
324. Fantin, G.; Fogagnolo, M.; Guerzoni, M. E.; Marotta, E.; Medici, E.; Medici, A.; Pedrini, P. *Tetrahedron: Asymmetry* 1992, *3*, 947.
325. Ansari, M. H.; Kusumoto, T.; Hiyama, T. *Tetrahedron Lett.* 1993, *34*, 8271.
326. Fronza, G.; Fuganti, C.; Grasselli, P.; Mele, A. *Tetrahedron Lett.* 1993, *34*, 6467.
327. Takabe, K.; Hiyoshi, H.; Sawada, H.; Tanaka, M.; Miyazaki, A.; Yamada, T.; Katagiri, T.; Yoda, H. *Tetrahedron: Asymmetry* 1992, *3*, 1399.
328. Takabe, K.; Tanaka, M.; Sugimoto, M.; Yamada, T.; Yoda, H. *Tetrahedron: Asymmetry* 1992, *3*, 1385.
329. Fronza, G.; Fuganti, C.; Grasselli, P.; Pulido-Fernandez, R.; Servi, S.; Tagliani, A.; Terreni, M. *Tetrahedron* 1991, *47*, 9247.
330. Irwin, A. J.; Jones, J. B. *J. Am. Chem. Soc.* 1977, *99*, 556.
331. Sato, M.; Sakaki, J.-I.; Sugita, Y.; Nakano, T.; Kaneko, C. *Tetrahedron Lett.* 1990, *31*, 7463.
332. Jacobs, H.; Berrymen, K.; Jones, J.; Gopalan, A. *Synth. Commun.* 1990, *20*, 999.
333. Durrwachter, J. R.; Sweers, H. M.; Nozaki, K.; Wong, C.-H. *Tetrahedron Lett.* 1986, *27*, 1261.
334. Bednarski, M. D.; Waldmann, H. J.; Whitesides, G. M. *Tetrahedron Lett.* 1986, *27*, 5807.
335. Kim, M.-J.; Hennen, W. J.; Sweers, H. M.; Wong, C.-H. *J. Am. Chem. Soc.* 1988, *110*, 6481.
336. Auge, C.; Gautheron, C. *J. Chem. Soc., Chem. Commun.* 1987, 859.
337. Fuganti, C.; Grasselli, P.; Marinoni, G. *Tetrahedron Lett.* 1979, 1161.
338. Bernardi, R.; Fuganti, C.; Grasselli, P.; Mariononi, G. *Synthesis* 1980, 50.
339. Fuganti, C.; Grasselli, P.; Servi, S.; Zirotti, C. *Tetrahedron Lett.* 1982, *23*, 4269.

340. Fessner, W.-D.; Sincrius, G.; Schneider, A.; Dreyer, M.; Schulz, G. E.; Badia, J.; Aguilar, J. *Angew. Chem., Int. Ed. Engl.* 1991, *30*, 555.
341. Valentin, M.-L.; Bolte, J. *Tetrahedron Lett.* 1993, *34*, 8103.
342. Augé, C.; Delest, V. *Tetrahedron: Asymmetry* 1993, *4*, 1165.
343. Dalmas, V.; Demuynck, C. *Tetrahedron: Asymmetry* 1993, *4*, 1169.
344. Koyama, T.; Inoue, H.; Ohnuma, S.-I.; Ogura, K. *Tetrahedron Lett.* 1990, *31*, 4189.
345. Becker, W.; Pfeil, E. *J. Am. Chem. Soc.* 1966, *88*, 4299.
346. Niedermeyer, U.; Kula, M.-R. *Angew. Chem., Int. Ed. Engl.* 1990, *29*, 386.
347. Effenberger, F.; Ziegler, T.; Forster, S. *Angew. Chem., Int. Ed. Engl.* 1987, *26*, 458.
348. Effenberger, F.; Horsch, B.; Forster, S.; Ziegler, T. *Tetrahedron Lett.* 1990, *31*, 1249.
349. Huuhtanen, T. T.; Kanerva, L. T. *Tetrahedron: Asymmetry* 1992, *3*, 1223.
350. Ziegler, T.; Horsch, B.; Effenberger, F. *Synthesis* 1990, 575.
351. Tanaka, K.; Mori, A.; Inoue, S. *J. Org. Chem.* 1990, *55*, 181.
352. Brusse, J.; Loos, W. T.; Kruse, C. G.; Van der Gen, A. *Tetrahedron* 1990, *46*, 979.
353. Ohta, H.; Kimura, Y.; Sugano, Y.; Sugai, T. *Tetrahedron* 1989, *45*, 5469.
354. Becker, W.; Freund, H.; Pfeil, E. *Angew. Chem., Int. Ed. Engl.* 1965, *4*, 1079.
355. Kakeya, H.; Sakai, N.; Sugai, T.; Ohta, H. *Tetrahedron Lett.* 1991, *32*, 1343.
356. Cohen, M. A.; Sawden, J.; Turner, N. J. *Tetrahedron Lett.* 1990, *31*, 7223.
357. Kobayashi, M.; Yanaka, N.; Nagasawa, T.; Yamada, H. *Tetrahedron* 1990, *46*, 5587.
358. Djeghaba, Z.; Deleuze, H.; Jeso, B. d.; Messadi, D.; Maillard, B. *Tetrahedron Lett.* 1991, *32*, 761.
359. Kitaguchi, H.; Fitzpatrick, P. A.; Huber, J. E.; Klibanov, A. M. *J. Am. Chem. Soc.* 1989, *111*, 3094.
360. Stirling, D. I. In *Chirality in Industry*; A. N. Collins; G. N. Sheldrake, and J. Crosby, Eds.; John Wiley & Sons: New York, 1992; pp 209.
361. Evans, C. T.; Choma, C.; Peterson, W.; Misawa, M. *Biotech, Bioeng.* 1987, *30*, 1067.
362. Evans, C. T.; Peterson, W.; Choma, C.; Misawa, M. *Appl. Microbiol. Biotech.* 1987, *26*, 305.
363. Calton, G. J.; Wood, L.; Updike, M. H.; Lantz, L.; Hammann, J. P. *Biotech.* 1986, *4*, 317.
364. Wichmann, R.; Wanang, C.; Bückmann, A. F.; Kula, M. R. *Biotech. Bioeng.* 1981, *23*, 2789.
365. Brieva, R.; Rebolledo, F.; Gotor, V. *J. Chem. Soc., Chem. Commun.* 1990, 1386.
366. Gotor, V.; Brieva, R.; Rebolledo, F. *Tetrahedron Lett.* 1988, *29*, 6973.
367. Müller, H.-G.; Güntherberg, H.; Drechsler, H.; Mauersberger, S.; Kortus, K.; Oehme, G. *Tetrahedron Lett.* 1991, *32*, 2009.
368. Fantin, G.; Fogagnolo, M.; Medici, A.; Pedrini, P.; Poli, S.; Sinigaglia, M. *Tetrahedron Lett.* 1993, *34*, 883.
369. Fantin, G.; Fogagnolo, M.; Medici, A.; Pedrini, P.; Poli, S.; Gardini, F. *Tetrahedron: Asymmetry* 1993, *4*, 1607.
370. Hasegawa, J.; Ogura, M.; Hamaguchi, S.; Shimazaki, M.; Kawaharada, H.; Watanabe, K. *J. Ferment. Technol.* 1981, *59*, 203.
371. Mori, K.; Senda, S. *Tetrahedron* 1985, *41*, 541.
372. Goodhue, C. T.; Schaeffer, J. R. *Biotech. Bioeng.* 1971, *13*, 203.
373. Cohen, N.; Eichel, W. F.; Lopresti, R. J.; Neukom, C.; Sauay, G. *J. Org. Chem.* 1976, *41*, 3505.
374. Root, R. L.; Durrwachter, J. R.; Wong, C.-H. *J. Am. Chem. Soc.* 1985, *107*, 2997.
375. Ohta, H.; Tetsukawa, H.; Noto, N. *J. Org. Chem.* 1982, *47*, 2400.
376. Han, H.; Pascal, R. A. *J. Org. Chem.* 1990, *55*, 5173.
377. Alphand, V.; Archelas, A.; Furstoss, R. *J. Org. Chem.* 1990, *55*, 347.
378. Fourneron, J. D.; Archelas, A.; Furstoss, R. *J. Org. Chem.* 1989, *54*, 4686.
379. Colbert, J. E.; Katopodis, A. G.; May, S. W. *J. Am. Chem. Soc.* 1990, *112*, 3993.
380. Takahashi, O.; Umezawa, J.; Furuhashi, K.; Takagi, M. *Tetrahedron Lett.* 1989, *30*, 1583.
381. Allain, E. J.; Hager, L. P.; Deng, L.; Jacobsen, E. N. *J. Am. Chem. Soc.* 1993, *115*, 4415.
382. Fu, H.; Shen, G.-J.; Wong, C.-H. *REC* 1991, *110*, 167.
383. Fu, H.; Newcomb, M.; Wong, C.-H. *J. Am. Chem. Soc.* 1991, *113*, 5878.
384. Datcheva, V. K.; Kiss, K.; Solomon, L.; Kyler, K. S. *J. Am. Chem. Soc.* 1991, *113*, 270.
385. Zhang, P.; Kyler, K. S. *J. Am. Chem. Soc.* 1989, *111*, 9241.
386. Bellucci, G.; Chiappe, C.; Conti, L.; Marioni, F.; Pierini, G. *J. Org. Chem.* 1989, *54*, 5978.
387. Kamal, A.; Damayanthi, Y.; Rao, M. V. *Tetrahedron: Asymmetry* 1992, *3*, 1361.
388. Miyamoto, K.; Ohta, H. *J. Am. Chem. Soc.* 1990, *112*, 4077.
389. Taschner, M.; Black, D. J. *J. Am. Chem. Soc.* 1988, *110*, 6892.
390. Taschner, M. J.; Black, D. J.; Chen, Q.-Z. *Tetrahedron: Asymmetry* 1993, *4*, 1387.
391. Ley, S. V.; Sternfeld, F.; Taylor, S. *Tetrahedron Lett.* 1987, *28*, 225.
392. Hudlicky, T.; Luna, H.; Price, J. D.; Rulin, F. *Tetrahedron Lett.* 1989, *30*, 4053.
393. Hudlicky, T.; Price, J. D. *Synlett* 1990, 159.
394. Ley, S. V.; Sternfeld, F. *Tetrahedron* 1989, *45*, 3463.
395. Ley, S. V.; Sternfeld, F. *Tetrahedron Lett.* 1988, *29*, 5305.
396. Carless, H. A. J. *Tetrahedron: Asymmetry* 1992, *3*, 795.

397. Ley, S. V.; Yeung, L. L. *Synlett* 1992, 997.
398. Dumortier, L.; Liu, P.; Dobbelaere, S.; Van der Eycken, J.; Vandewalle, M. *Synlett* 1992, 243.
399. Ley, S. V.; Yeung, L. L. *Synlett* 1992, 291.
400. Königsberger, K.; Hudlicky, T. *Tetrahedron: Asymmetry* 1993, *4*, 2469.
401. Hudlicky, T.; Boros, E. E.; Boros, C. H. *Tetrahedron: Asymmetry* 1993, *4*, 1365.
402. Hudlicky, T.; Seoane, G.; Pettus, T. *J. Org. Chem.* 1989, *54*, 4239.
403. Knackmuss, H.-J.; Beckmann, W.; Otting, W. *Angew. Chem., Int. Ed. Engl.* 1976, *15*, 549.
404. Motosugi, K.; Esaki, N.; Soda, K. *Biotech. Bioeng.* 1984, *26*, 805.
405. Adam, W.; Hoch, U.; Saha-Möller, C. R.; Schreier, P. *Angew. Chem., Int. Ed. Engl.* 1993, *32*, 1737.
406. Corey, E. J.; Imwinkelried, R.; Pikul, S.; Xiang, Y. B. *J. Am. Chem. Soc.* 1989, *111*, 5493.
407. Rao, K. R.; Srinivasan, T. N.; Bhanumathi, N. *Tetrahedron Lett.* 1990, *31*, 5959.
408. Rao, K. R.; Bhanumathi, N.; Srinivasan, T. N.; Sattur, P. B. *Tetrahedron Lett.* 1990, *31*, 899.
409. Rao, K. R.; Bhanumathi, N.; Sattur, P. B. *Tetrahedron Lett.* 1990, *31*, 3201.
410. Groves, J. T.; Viski, P. *J. Am. Chem. Soc.* 1989, *111*, 8537.
411. Sirimanne, S. R.; May, S. W. *J. Am. Chem. Soc.* 1988, *110*, 7560.
412. Yamamoto, H.; Oritani, T.; Koga, H.; Horiuchi, T.; Yamashita, K. *Agric. Biol. Chem.* 1990, *54*, 1915.
413. Sugai, T.; Mori, K. *Agric. Biol. Chem.* 1984, *48*, 2501.
414. Pollack, R. M. *Tetrahedron* 1989, *45*, 4913.
415. Pederson, R. L.; Kim, M.-J.; Wong, C.-H. *Tetrahedron Lett.* 1988, *29*, 4645.
416. von der Osten, C. H.; Sinskey, A. J.; Barbas, C. F.; Pederson, R. L.; Wang, Y.-F.; Wong, C.-H. *J. Am. Chem. Soc.* 1989, *111*, 3924.
417. Bednarski, M. D.; Simon, E. S.; Bischofberger, N.; Fessner, W.-D.; Kim, M.-J.; Lees, W.; Saito, T.; Waldmann, H.; Whitesides, G. M. *J. Am. Chem. Soc.* 1989, *111*, 627.
418. Straub, A.; Effenberger, F.; Fischer, P. *J. Org. Chem.* 1990, *55*, 3926.
419. Borysenko, C. W.; Spaltenstein, A.; Straub, J. A.; Whitesides, G. M. *J. Am. Chem. Soc.* 1989, *111*, 9275.
420. Durrwachter, J. R.; Wong, C.-H. *J. Org. Chem.* 1988, *53*, 4175.
421. Wong, C.-H. In *Enzymes as Catalysts in Organic Synthesis*; C. M. P. Schneider, Ed.; Reidel: Dortrecht, 1986; pp 199.
422. Matsumoto, K.; Shimagaki, M.; Nakata, T.; Oishi, T. *Tetrahedron Lett.* 1993, *34*, 4935.
423. Lees, W. J.; Whitesides, G. M. *J. Org. Chem.* 1993, *58*, 1887.
424. Schmid, W.; Whitesides, G. M. *J. Am. Chem. Soc.* 1990, *112*, 9670.
425. Schultz, M.; Waldmann, H.; Vogt, W.; Kunz, H. *Tetrahedron Lett.* 1990, *31*, 867.
426. Ozaki, A.; Toone, E. J.; Osten, C. H. v. d.; Sinskey, A. J.; Whitesides, G. M. *J. Am. Chem. Soc.* 1990, *112*, 4970.
427. Barbas, C. F.; Wang, Y.-F.; Wong, C.-H. *J. Am. Chem. Soc.* 1990, *112*, 2013.
428. Auge, C.; Bouxom, B.; Cavaye, B.; Gautheron, C. *Tetrahedron Lett.* 1989, *30*, 2217.
429. Auge, C.; David, S.; Gautheron, C.; Malleron, A.; Cavaye, B. *Nouv. J. Chim.* 1988, *12*, 733.
430. Iverson, B. L.; Cameron, K. E.; Jahangiri, G. K.; Pasternak, D. S. *J. Am. Chem. Soc.* 1990, *112*, 5320.
431. Jacobs, J. W.; Schultz, P. G.; Sugasawara, R.; Powell, M. *J. Am. Chem. Soc.* 1987, *109*, 2174.
432. Pollack, S. J.; Schultz, P. G. *J. Am. Chem. Soc.* 1989, *111*, 1929.
433. Iverson, B. L.; Iverson, S. A.; Roberts, V. A.; Getzoff, E. D.; Tainer, J. A.; Benkovic, S. J.; Lerner, R. A. *Science* 1990, *249*, 659.
434. Lerner, R. A.; Benkovic, S. J.; Schultz, P. G. *Science* 1991, *252*, 659.
435. Blackburn, G. M.; Kang, A. S.; Kingsbury, G. A.; Burton, D. R. *Biochem. J.* 1989, *262*, 381.
436. Schultz, P. G.; Lerner, R. A. *Acc. Chem. Res.* 1993, *26*, 391.
437. Hilvert, D. *Acc. Chem. Res.* 1993, *26*, 552.
438. Stewart, J. D.; Liotta, L. J.; Benkovic, S. J. *Acc. Chem. Res.* 1993, *26*, 396.
439. Stewart, J. D.; Benkovic, S. J. *Chem. Soc. Rev.* 1993, *22*, 213.
440. Jackson, D. Y.; Liang, M. N.; Bartlett, P. A.; Schultz, P. G. *Angew. Chem., Int. Ed. Engl.* 1992, *31*, 182.
441. Wade, W. S.; Koh, J. S.; Han, N.; Hoekstra, D. M.; Lerner, R. A. *J. Am. Chem. Soc.* 1993, *115*, 4449.

# INDEX

## A

## E

## H

## I

## J



## K

## L

## N

## P

## Q



## R

## S

## T

## U

## V

## W



## X



## Y



## Z